Student's Solutions Manual

Calculus with Applications
Ninth Edition
and
Calculus with Applications
Brief Version
Ninth Edition

Margaret L. Lial

American River College

Raymond N. Greenwell

Hofstra University

Nathan P. Ritchey

Youngstown State University

Boston San Francisco New York
London Toronto Sydney Tokyo Singapore Madrid
Mexico City Munich Paris Cape Town Hong Kong Montreal

Reproduced by Pearson Addison-Wesley from electronic files supplied by the author.

Publishing as Pearson Addison-Wesley, 75 Arlington Street, Boston, MA 02116.

ISBN-13: 978-0-321-45569-7
ISBN-10: 0-321-45569-X

2 3 4 5 6 BB 10 09 08

CONTENTS

CHAPTER 3 THE DERIVATIVE

CHAPTER 4 CALCULATING THE DERIVATIVE

CHAPTER 5 GRAPHS AND THE DERIVATIVE

CHAPTER 6 APPLICATIONS OF THE DERIVATIVE

CHAPTER 7 INTEGRATION

CHAPTER 8 FURTHER TECHNIQUES AND APPLICATIONS OF INTEGRATION

CHAPTER 9 MULTIVARIABLE CALCULUS

CHAPTER 10 DIFFERENTIAL EQUATIONS

CHAPTER 11 PROBABILITY AND CALCULUS

CHAPTER 12 SEQUENCES AND SERIES

PREFACE

This book provides solutions for many of the exercises in *Calculus with Applications,* Seventh Edition, by Margaret L. Lial, Raymond N. Greenwell, and Nathan P. Ritchey. Solutions are included for odd–numbered exercises. Solutions are not provided for exercises with open–response answers. Sample tests are provided at the end of each chapter to help you determine if you have mastered the concepts in a given chapter.

This book should be used as an aid as you work to master your coursework. Try to solve the exercises that your instructor assigns before you refer to the solutions in this book. Then, if you have difficulty, read these solutions to guide you in solving the exercises. The solutions have been written so that they are consistent with the methods used in the textbook.

You may find that some of the solutions are presented in greater detail than others. Thus, if you cannot find an explanation for a difficulty that you encountered in one exercise, you may find the explanation in the solution for a similar exercise elsewhere in the exercise set.

In addition to solutions, you will find a list of suggestions on how to be successful in mathematics. A careful reading will be helpful for many students.

The following people have made valuable contributions to the production of this *Student's Solutions Manual:* LaurelTech Integrated Publishing Services, editors; Judy Martinez and Sheri Minkner, typists.

We also want to thank Tommy Thompson of Cedar Valley Community College for the essay "To the Student: Success in Mathematics."

TO THE STUDENT

TO THE STUDENT: SUCCESS IN MATHEMATICS

The main reason students have difficulty with mathematics is that they don't know how to study it. Studying mathematics *is* different from studying subjects like English or history. The key to success is regular practice.

This should not be surprising. After all, can you learn to play the piano or to ski well without a lot of regular practice? The same thing is true for learning mathematics. Working problems nearly every day is the key to becoming successful. Here is a list of things you can do to help you succeed in studying mathematics.

1. *Attend class regularly.* Pay attention in class to what your instructor says and does, and make careful notes. In particular, note the problems the instructor works on the board and copy the complete solutions. Keep these notes separate from your homework to avoid confusion when you read them over later.

2. Don't hesitate to ask questions in class. It is not a sign of weakness, but of strength. There are always other students with the same question who are too shy to ask.

3. *Read your text carefully.* Many students read only enough to get by, usually only the examples. Reading the complete section will help you to be successful with the homework problems. Most exercises are keyed to specific examples or objectives that will explain the procedures for working them.

4. Before you start on your homework assignment, rework the problems the instructor worked in class. This will reinforce what you have learned. Many students say, "I understand it perfectly when you do it, but I get stuck when I try to work the problem myself."

5. Do your homework assignment only *after* reading the text and reviewing your notes from class. Check your work with the answers in the back of the book. If you get a problem wrong and are unable to see why, mark that problem and ask your instructor about it. Then practice working additional problems of the same type to reinforce what you have learned.

6. Work as neatly as you can. Write your symbols clearly, and make sure the problems are clearly separated from each other. Working neatly will help you to think clearly and also make it easier to review the homework before a test.

7. After you have completed a homework assignment, look over the text again. Try to decide what the main ideas are in the lesson. Often they are clearly highlighted or boxed in the text.

8. Use the chapter test at the end of each chapter as a practice test. Work through the problems under test conditions, without referring to the text or the answers until you are finished. You may want to time yourself to see how long it takes you. When you have finished, check your answers against those in the back of the book and study those problems that you missed. Answers are referenced to the appropriate sections of the text.

9. Keep any quizzes and tests that are returned to you and use them when you study for future tests and the final exam. These quizzes and tests indicate what your instructor considers most important. Be sure to correct any problems on these tests that you missed, so you will have the corrected work to study.

10. Don't worry if you do not understand a new topic right away. As you read more about it and work through the problems, you will gain understanding. Each time you look back at a topic you will understand it a little better. No one understands each topic completely right from the start.

SOLUTIONS TO ODD-NUMBERED EXERCISES

Chapter R

ALGEBRA REFERENCE

R.1 Polynomials

1. $(2x^2 - 6x + 11) + (-3x^2 + 7x - 2)$
$= 2x^2 - 6x + 11 - 3x^2 + 7x - 2$
$= (2 - 3)x^2 + (7 - 6)x + (11 - 2)$
$= -x^2 + x + 9$

2. $(-4y^2 - 3y + 8) - (2y^2 - 6y - 2)$
$= (-4y^2 - 3y + 8) + (-2y^2 + 6y + 2)$
$= -4y^2 - 3y + 8 - 2y^2 + 6y + 2$
$= (-4y^2 - 2y^2) + (-3y + 6y) + (8 + 2)$
$= -6y^2 + 3y + 10$

3. $-6(2q^2 + 4q - 3) + 4(-q^2 + 7q - 3)$
$= (-12q^2 - 24q + 18) + (-4q^2 + 28q - 12)$
$= (-12q^2 - 4q^2) + (-24q + 28q) + (18 - 12)$
$= -16q^2 + 4q + 6$

4. $2(3r^2 + 4r + 2) - 3(-r^2 + 4r - 5)$
$= (6r^2 + 8r + 4) + (3r^2 - 12r + 15)$
$= (6r^2 + 3r^2) + (8r - 12r) + (4 + 15)$
$= 9r^2 - 4r + 19$

5. $(0.613x^2 - 4.215x + 0.892) - 0.47(2x^2 - 3x + 5)$
$= 0.613x^2 - 4.215x + 0.892 - 0.94x^2 + 1.41x - 2.35$
$= -0.327x^2 - 2.805x - 1.458$

6. $0.5(5r^2 + 3.2r - 6) - (1.7r^2 - 2r - 1.5)$
$= (2.5r^2 + 1.6r - 3) + (-1.7r^2 + 2r + 1.5)$
$= (2.5r^2 - 1.7r^2) + (1.6r + 2r) + (-3 + 1.5)$
$= 0.8r^2 + 3.6r - 1.5$

7. $-9m(2m^2 + 3m - 1)$
$= -9m(2m^2) - 9m(3m) - 9m(-1)$
$= -18m^3 - 27m^2 + 9m$

8. $(6k - 1)(2k - 3)$
$= (6k)(2k) + (6k)(-3) + (-1)(2k) + (-1)(-3)$
$= 12k^2 - 18k - 2k + 3$
$= 12k^2 - 20k + 3$

9. $(3t - 2y)(3t + 5y)$
$= (3t)(3t) + (3t)(5y) + (-2y)(3t) + (-2y)(5y)$
$= 9t^2 + 15ty - 6ty - 10y^2$
$= 9t^2 + 9ty - 10y^2$

10. $(9k + q)(2k - q)$
$= (9k)(2k) + (9k)(-q) + (q)(2k) + (q)(-q)$
$= 18k^2 - 9kq + 2kq - q^2$
$= 18k^2 - 7kq - q^2$

11. $\left(\frac{2}{5}y + \frac{1}{8}z\right)\left(\frac{3}{5}y + \frac{1}{2}z\right)$
$= \left(\frac{2}{5}y\right)\left(\frac{3}{5}y\right) + \left(\frac{2}{5}y\right)\left(\frac{1}{2}z\right) + \left(\frac{1}{8}z\right)\left(\frac{3}{5}y\right) + \left(\frac{1}{8}z\right)\left(\frac{1}{2}z\right)$
$= \frac{6}{25}y^2 + \frac{1}{5}yz + \frac{3}{40}yz + \frac{1}{16}z^2$
$= \frac{6}{25}y^2 + \left(\frac{8}{40} + \frac{3}{40}\right)yz + \frac{1}{16}z^2$
$= \frac{6}{25}y^2 + \frac{11}{40}yz + \frac{1}{16}z^2$

12. $\left(\frac{3}{4}r - \frac{2}{3}s\right)\left(\frac{5}{4}r + \frac{1}{3}s\right)$
$= \left(\frac{3}{4}r\right)\left(\frac{5}{4}r\right) + \left(\frac{3}{4}r\right)\left(\frac{1}{3}s\right) + \left(-\frac{2}{3}s\right)\left(\frac{5}{4}r\right) + \left(-\frac{2}{3}s\right)\left(\frac{1}{3}s\right)$
$= \frac{15}{16}r^2 + \frac{1}{4}rs - \frac{5}{6}rs - \frac{2}{9}s^2$
$= \frac{15}{16}r^2 - \frac{7}{12}rs - \frac{2}{9}s^2$

13. $(2 - 3x)(2 + 3x)$
$= (2)(2) + (2)(3x) + (-3x)(2) + (-3x)(3x)$
$= 4 + 6x - 6x - 9x^2$
$= 4 - 9x^2$

14. $(6m + 5)(6m - 5)$
$= (6m)(6m) + (6m)(-5) + (5)(6m) + (5)(-5)$
$= 36m^2 - 30m + 30m - 25$
$= 36m^2 - 25$

15. $(3p-1)(9p^2+3p+1)$
$= (3p-1)(9p^2)+(3p-1)(3p)$
$\quad + (3p-1)(1)$
$= 3p(9p^2)-1(9p^2)+3p(3p)$
$\quad -1(3p)+3p(1)-1(1)$
$= 27p^3-9p^2+9p^2-3p+3p-1$
$= 27p^3-1$

16. $(3p+2)(5p^2+p-4)$
$= (3p)(5p^2)+(3p)(p)+(3p)(-4)$
$\quad +(2)(5p^2)+(2)(p)+(2)(-4)$
$= 15p^3+3p^2-12p+10p^2+2p-8$
$= 15p^3+13p^2-10p-8$

17. $(2m+1)(4m^2-2m+1)$
$= 2m(4m^2-2m+1)+1(4m^2-2m+1)$
$= 8m^3-4m^2+2m+4m^2-2m+1$
$= 8m^3+1$

18. $(k+2)(12k^3-3k^2+k+1)$
$= k(12k^3)+k(-3k^2)+k(k)+k(1)$
$\quad +2(12k^3)+2(-3k^2)+2(k)+2(1)$
$= 12k^4-3k^3+k^2+k+24k^3-6k^2$
$\quad +2k+2$
$= 12k^4+21k^3-5k^2+3k+2$

19. $(x+y+z)(3x-2y-z)$
$= x(3x)+x(-2y)+x(-z)+y(3x)+y(-2y)$
$\quad +y(-z)+z(3x)+z(-2y)+z(-z)$
$= 3x^2-2xy-xz+3xy-2y^2-yz+3xz-2yz-z^2$
$= 3x^2+xy+2xz-2y^2-3yz-z^2$

20. $(r+2s-3t)(2r-2s+t)$
$= r(2r)+r(-2s)+r(t)+2s(2r)$
$\quad +2s(-2s)+2s(t)-3t(2r)-3t(-2s)-3t(t)$
$= 2r^2-2rs+rt+4rs+2st-6rt+6st-3t^2$
$\quad +2rt-st+t^2$
$= 2r^2+2rs-5rt-4s^2+8st-3t^2$

21. $(x+1)(x+2)(x+3)$
$= [x(x+2)+1(x+2)](x+3)$
$= \left[x^2+2x+x+2\right](x+3)$
$= \left[x^2+3x+2\right](x+3)$
$= x^2(x+3)+3x(x+3)+2(x+3)$
$= x^3+3x^2+3x^2+9x+2x+6$
$= x^3+6x^2+11x+6$

22. $(x-1)(x+2)(x-3)$
$= [x(x+2)+(-1)(x+2)](x-3)$
$= (x^2+2x-x-2)(x-3)$
$= (x^2+x-2)(x-3)$
$= x^2(x-3)+x(x-3)+(-2)(x-3)$
$= x^3-3x^2+x^2-3x-2x+6$
$= x^3-2x^2-5x+6$

23. $(x+2)^2 = (x+2)(x+2)$
$= x(x+2)+2(x+2)$
$= x^2+2x+2x+4$
$= x^2+4x+4$

24. $(2a-4b)^2 = (2a-4b)(2a-4b)$
$= 2a(2a-4b)-4b(2a-4b)$
$= 4a^2-8ab-8ab+16b^2$
$= 4a^2-16ab+16b^2$

25. $(x-2y)^3$
$= [(x-2y)(x-2y)](x-2y)$
$= (x^2-2xy-2xy+4y^2)(x-2y)$
$= (x^2-4xy+4y^2)(x-2y)$
$= (x^2-4xy+4y^2)x+(x^2-4xy+4y^2)(-2y)$
$= x^3-4x^2y+4xy^2-2x^2y+8xy^2-8y^3$
$= x^3-6x^2y+12xy^2-8y^3$

R.2 Factoring

1. $7a^3+14a^2 = 7a^2\cdot a+7a^2\cdot 2$
$= 7a^2(a+2)$

2. $3y^3+24y^2+9y = 3y\cdot y^2+3y\cdot 8y+3y\cdot 3$
$= 3y(y^2+8y+3)$

3. $13p^4q^2-39p^3q+26p^2q^2$
$= 13p^2q\cdot p^2q-13p^2q\cdot 3p+13p^2q\cdot 2q$
$= 13p^2q(p^2q-3p+2q)$

4. $60m^4-120m^3n+50m^2n^2$
$= 10m^2\cdot 6m^2-10m^2\cdot 12mn$
$\quad +10m^2\cdot 5n^2$
$= 10m^2(6m^2-12mn+5n^2)$

5. $m^2-5m-14 = (m-7)(m+2)$

since $(-7)(2)=-14$ and $-7+2=-5$.

6. $x^2+4x-5 = (x+5)(x-1)$

since $5(-1)=-5$ and $-1+5=4$.

7. $z^2+9z+20 = (z+4)(z+5)$

since $4\cdot 5=20$ and $4+5=9$.

8. $b^2 - 8b + 7 = (b-7)(b-1)$

since $(-7)(-1) = 7$ and $-7 + (-1) = -8$.

9. $a^2 - 6ab + 5b^2 = (a-b)(a-5b)$

since $(-b)(-5b) = 5b^2$ and $-b + (-5b) = -6b$.

10. $s^2 + 2st - 35t^2 = (s-5t)(s+7t)$

since $(-5t)(7t) = -35t^2$ and $7t + (-5t) = 2t$.

11. $y^2 - 4yz - 21z^2 = (y+3z)(y-7z)$

since $(3z)(-7z) = -21z^2$ and $3z + (-7z) = -4z$.

12. $6a^2 - 48a - 120 = 6(a^2 - 8a - 20)$
$= 6(a-10)(a+2)$

13. $3m^3 + 12m^2 + 9m = 3m(m^2 + 4m + 3)$
$= 3m(m+1)(m+3)$

14. $3x^2 + 4x - 7$

The possible factors of $3x^2$ are $3x$ and x and the possible factors of -7 are -7 and 1, or 7 and -1. Try various combinations until one works.

$$3x^2 + 4x - 7 = (3x+7)(x-1)$$

15. $3a^2 + 10a + 7$

The possible factors of $3a^2$ are $3a$ and a and the possible factors of 7 are 7 and 1. Try various combinations until one works.

$$3a^2 + 10a + 7 = (a+1)(3a+7)$$

16. $4a^2 + 10a + 6 = 2(2a^2 + 5a + 3)$
$= 2(2a+3)(a+1)$

17. $15y^2 + y - 2 = (5y+2)(3y-1)$

18. $21m^2 + 13mn + 2n^2$
$= (7m+2n)(3m+n)$

19. $24a^4 + 10a^3b - 4a^2b^2$
$= 2a^2(12a^2 + 5ab - 2b^2)$
$= 2a^2(4a-b)(3a+2b)$

20. $24x^4 + 36x^3y - 60x^2y^2$
$= 12x^2(2x^2 + 3xy - 5y^2)$
$= 12x^2(x-y)(2x+5y)$

21. $x^2 - 64 = x^2 - 8^2$
$= (x+8)(x-8)$

22. $9m^2 - 25 = (3m)^2 - (5)^2$
$= (3m+5)(3m-5)$

23. $10x^2 - 160 = 10(x^2 - 16)$
$= 10(x^2 - 4^2)$
$= 10(x+4)(x-4)$

24. $9x^2 + 64$ is the *sum* of two perfect squares. It cannot be factored. It is prime.

25. $z^2 + 14zy + 49y^2$
$= z^2 + 2 \cdot 7zy + 7^2y^2$
$= (z+7y)^2$

26. $s^2 - 10st + 25t^2$
$= s^2 - 2 \cdot 5st + (5t)^2$
$= (s-5t)^2$

27. $9p^2 - 24p + 16$
$= (3p)^2 - 2 \cdot 3p \cdot 4 + 4^2$
$= (3p-4)^2$

28. $a^3 - 216$
$= a^3 - 6^3$
$= (a-6)[(a)^2 + (a)(6) + (6)^2]$
$= (a-6)(a^2 + 6a + 36)$

29. $27r^3 - 64s^3$
$= (3r)^3 - (4s)^3$
$= (3r-4s)(9r^2 + 12rs + 16s^2)$

30. $3m^3 + 375$
$= 3(m^3 + 125)$
$= 3(m^3 + 5^3)$
$= 3(m+5)(m^2 - 5m + 25)$

31. $x^4 - y^4 = (x^2)^2 - (y^2)^2$
$= (x^2 + y^2)(x^2 - y^2)$
$= (x^2 + y^2)(x+y)(x-y)$

32. $16a^4 - 81b^4$
$= (4a^2)^2 - (9b^2)^2$
$= (4a^2 + 9b^2)(4a^2 - 9b^2)$
$= (4a^2 + 9b^2)[(2a)^2 - (3b)^2]$
$= (4a^2 + 9b^2)(2a+3b)(2a-3b)$

R.3 Rational Expressions

1. $\dfrac{5v^2}{35v} = \dfrac{5 \cdot v \cdot v}{5 \cdot 7 \cdot v} = \dfrac{v}{7}$

2. $\dfrac{25p^3}{10p^2} = \dfrac{5 \cdot 5 \cdot p \cdot p \cdot p}{2 \cdot 5 \cdot p \cdot p} = \dfrac{5p}{2}$

3. $\dfrac{8k+16}{9k+18} = \dfrac{8(k+2)}{9(k+2)} = \dfrac{8}{9}$

4. $\dfrac{2(t-15)}{(t-15)(t+2)} = \dfrac{2}{(t+2)}$

5. $\dfrac{4x^3-8x^2}{4x^2} = \dfrac{4x^2(x-2)}{4x^2}$

$= x-2$

6. $\dfrac{36y^2+72y}{9y} = \dfrac{36y(y+2)}{9y}$

$= \dfrac{9\cdot 4\cdot y(y+2)}{9\cdot y}$

$= 4(y+2)$

7. $\dfrac{m^2-4m+4}{m^2+m-6} = \dfrac{(m-2)(m-2)}{(m-2)(m+3)}$

$= \dfrac{m-2}{m+3}$

8. $\dfrac{r^2-r-6}{r^2+r-12} = \dfrac{(r-3)(r+2)}{(r+4)(r-3)}$

$= \dfrac{r+2}{r+4}$

9. $\dfrac{3x^2+3x-6}{x^2-4} = \dfrac{3(x+2)(x-1)}{(x+2)(x-2)}$

$= \dfrac{3(x-1)}{x-2}$

10. $\dfrac{z^2-5z+6}{z^2-4} = \dfrac{(z-3)(z-2)}{(z+2)(z-2)}$

$= \dfrac{z-3}{z+2}$

11. $\dfrac{m^4-16}{4m^2-16} = \dfrac{(m^2+4)(m+2)(m-2)}{4(m+2)(m-2)}$

$= \dfrac{m^2+4}{4}$

12. $\dfrac{6y^2+11y+4}{3y^2+7y+4} = \dfrac{(3y+4)(2y+1)}{(3y+4)(y+1)}$

$= \dfrac{2y+1}{y+1}$

13. $\dfrac{9k^2}{25}\cdot\dfrac{5}{3k} = \dfrac{3\cdot 3\cdot 5k^2}{5\cdot 5\cdot 3k} = \dfrac{3k^2}{5k} = \dfrac{3k}{5}$

14. $\dfrac{15p^3}{9p^2} \div \dfrac{6p}{10p^2}$

$= \dfrac{15p^3}{9p^2}\cdot\dfrac{10p^2}{6p}$

$= \dfrac{150p^5}{54p^3}$

$= \dfrac{25\cdot 6p^5}{9\cdot 6p^3}$

$= \dfrac{25p^2}{9}$

15. $\dfrac{3a+3b}{4c}\cdot\dfrac{12}{5(a+b)} = \dfrac{3(a+b)}{4c}\cdot\dfrac{3\cdot 4}{5(a+b)}$

$= \dfrac{3\cdot 3}{c\cdot 5}$

$= \dfrac{9}{5c}$

16. $\dfrac{a-3}{16} \div \dfrac{a-3}{32} = \dfrac{a-3}{16}\cdot\dfrac{32}{a-3}$

$= \dfrac{a-3}{16}\cdot\dfrac{16\cdot 2}{a-3}$

$= \dfrac{2}{1} = 2$

17. $\dfrac{2k-16}{6} \div \dfrac{4k-32}{3}$

$= \dfrac{2k-16}{6}\cdot\dfrac{3}{4k-32}$

$= \dfrac{2(k-8)}{6}\cdot\dfrac{3}{4(k-8)}$

$= \dfrac{1}{4}$

18. $\dfrac{9y-18}{6y+12}\cdot\dfrac{3y+6}{15y-30}$

$= \dfrac{9(y-2)}{6(y+2)}\cdot\dfrac{3(y+2)}{15(y-2)}$

$= \dfrac{27}{90} = \dfrac{3\cdot 3}{10\cdot 3} = \dfrac{3}{10}$

19. $\dfrac{4a+12}{2a-10} \div \dfrac{a^2-9}{a^2-a-20}$

$= \dfrac{4(a+3)}{2(a-5)}\cdot\dfrac{(a-5)(a+4)}{(a-3)(a+3)}$

$= \dfrac{2(a+4)}{a-3}$

20. $\dfrac{6r-18}{9r^2+6r-24}\cdot\dfrac{12r-16}{4r-12}$

$$=\frac{6(r-3)}{3(3r^2+2r-8)}\cdot\frac{4(3r-4)}{4(r-3)}$$
$$=\frac{6(r-3)}{3(3r-4)(r+2)}\cdot\frac{4(3r-4)}{4(r-3)}$$
$$=\frac{6}{3(r+2)}$$
$$=\frac{2}{r+2}$$

21. $\dfrac{k^2+4k-12}{k^2+10k+24}\cdot\dfrac{k^2+k-12}{k^2-9}$

$$=\frac{(k+6)(k-2)}{(k+6)(k+4)}\cdot\frac{(k+4)(k-3)}{(k+3)(k-3)}$$
$$=\frac{k-2}{k+3}$$

22. $\dfrac{m^2+3m+2}{m^2+5m+4}\div\dfrac{m^2+5m+6}{m^2+10m+24}$

$$=\frac{m^2+3m+2}{m^2+5m+4}\cdot\frac{m^2+10m+24}{m^2+5m+6}$$
$$=\frac{(m+1)(m+2)}{(m+4)(m+1)}\cdot\frac{(m+6)(m+4)}{(m+3)(m+2)}$$
$$=\frac{m+6}{m+3}$$

23. $\dfrac{2m^2-5m-12}{m^2-10m+24}\div\dfrac{4m^2-9}{m^2-9m+18}$

$$=\frac{2m^2-5m-12}{m^2-10m+24}\cdot\frac{m^2-9m+18}{4m^2-9}$$
$$=\frac{(2m+3)(m-4)(m-6)(m-3)}{(m-6)(m-4)(2m-3)(2m+3)}$$
$$=\frac{m-3}{2m-3}$$

24. $\dfrac{4n^2+4n-3}{6n^2-n-15}\cdot\dfrac{8n^2+32n+30}{4n^2+16n+15}$

$$=\frac{(2n+3)(2n-1)}{(2n+3)(3n-5)}\cdot\frac{2(2n+3)(2n+5)}{(2n+3)(2n+5)}$$
$$=\frac{2(2n-1)}{3n-5}$$

25. $\dfrac{a+1}{2}-\dfrac{a-1}{2}$

$$=\frac{(a+1)-(a-1)}{2}$$
$$=\frac{a+1-a+1}{2}$$
$$=\frac{2}{2}=1$$

26. $\dfrac{3}{p}+\dfrac{1}{2}$

Multiply the first term by $\frac{2}{2}$ and the second by $\frac{p}{p}$.

$$\frac{2\cdot 3}{2\cdot p}+\frac{p\cdot 1}{p\cdot 2}=\frac{6}{2p}+\frac{p}{2p}$$
$$=\frac{6+p}{2p}$$

27. $\dfrac{6}{5y}-\dfrac{3}{2}=\dfrac{6\cdot 2}{5y\cdot 2}-\dfrac{3\cdot 5y}{2\cdot 5y}$

$$=\frac{12-15y}{10y}$$

28. $\dfrac{1}{6m}+\dfrac{2}{5m}+\dfrac{4}{m}$

$$=\frac{5\cdot 1}{5\cdot 6m}+\frac{6\cdot 2}{6\cdot 5m}+\frac{30\cdot 4}{30\cdot m}$$
$$=\frac{5}{30m}+\frac{12}{30m}+\frac{120}{30m}$$
$$=\frac{5+12+120}{30m}$$
$$=\frac{137}{30m}$$

29. $\dfrac{1}{m-1}+\dfrac{2}{m}$

$$=\frac{m}{m}\left(\frac{1}{m-1}\right)+\frac{m-1}{m-1}\left(\frac{2}{m}\right)$$
$$=\frac{m+2m-2}{m(m-1)}$$
$$=\frac{3m-2}{m(m-1)}$$

30. $\dfrac{5}{2r+3}-\dfrac{2}{r}$

$$=\frac{5r}{r(2r+3)}-\frac{2(2r+3)}{r(2r+3)}$$
$$=\frac{5r-2(2r+3)}{r(2r+3)}$$
$$=\frac{5r-4r-6}{r(2r+3)}$$
$$=\frac{r-6}{r(2r+3)}$$

31. $\dfrac{8}{3(a-1)}+\dfrac{2}{a-1}$

$$=\frac{8}{3(a-1)}+\frac{3}{3}\left(\frac{2}{a-1}\right)$$
$$=\frac{8+6}{3(a-1)}$$
$$=\frac{14}{3(a-1)}$$

32. $\frac{2}{5(k-2)}+\frac{3}{4(k-2)}=\frac{4\cdot 2}{4\cdot 5(k-2)}+\frac{5\cdot 3}{5\cdot 4(k-2)}$

$=\frac{8}{20(k-2)}+\frac{15}{20(k-2)}$

$=\frac{8+15}{20(k-2)}$

$=\frac{23}{20(k-2)}$

33. $\frac{4}{x^2+4x+3}+\frac{3}{x^2-x-2}$

$=\frac{4}{(x+3)(x+1)}+\frac{3}{(x-2)(x+1)}$

$=\frac{4(x-2)}{(x-2)(x+3)(x+1)}$

$+\frac{3(x+3)}{(x-2)(x+3)(x+1)}$

$=\frac{4(x-2)+3(x+3)}{(x-2)(x+3)(x+1)}$

$=\frac{4x-8+3x+9}{(x-2)(x+3)(x+1)}$

$=\frac{7x+1}{(x-2)(x+3)(x+1)}$

34. $\frac{y}{y^2+2y-3}-\frac{1}{y^2+4y+3}$

$=\frac{y}{(y+3)(y-1)}-\frac{1}{(y+3)(y+1)}$

$=\frac{y(y+1)}{(y+3)(y+1)(y-1)}$

$-\frac{1(y-1)}{(y+3)(y+1)(y-1)}$

$=\frac{y(y+1)-(y-1)}{(y+3)(y+1)(y-1)}$

$=\frac{y^2+y-y+1}{(y+3)(y+1)(y-1)}$

$=\frac{y^2+1}{(y+3)(y+1)(y-1)}$

35. $\frac{3k}{2k^2+3k-2}-\frac{2k}{2k^2-7k+3}$

$=\frac{3k}{(2k-1)(k+2)}-\frac{2k}{(2k-1)(k-3)}$

$=\left(\frac{k-3}{k-3}\right)\frac{3k}{(2k-1)(k+2)}$

$-\left(\frac{k+2}{k+2}\right)\frac{2k}{(2k-1)(k-3)}$

$=\frac{(3k^2-9k)-(2k^2+4k)}{(2k-1)(k+2)(k-3)}$

$=\frac{k^2-13k}{(2k-1)(k+2)(k-3)}$

$=\frac{k(k-13)}{(2k-1)(k+2)(k-3)}$

36. $\frac{4m}{3m^2+7m-6}-\frac{m}{3m^2-14m+8}$

$=\frac{4m}{(3m-2)(m+3)}-\frac{m}{(3m-2)(m-4)}$

$=\frac{4m(m-4)}{(3m-2)(m+3)(m-4)}$

$-\frac{m(m+3)}{(3m-2)(m-4)(m+3)}$

$=\frac{4m(m-4)-m(m+3)}{(3m-2)(m-4)(m+3)}$

$=\frac{4m^2-16m-m^2-3m}{(3m-2)(m+3)(m-4)}$

$=\frac{3m^2-19m}{(3m-2)(m+3)(m-4)}$

$=\frac{m(3m-19)}{(3m-2)(m+3)(m-4)}$

37. $\frac{2}{a+2}+\frac{1}{a}+\frac{a-1}{a^2+2a}$

$=\frac{2}{a+2}+\frac{1}{a}+\frac{a-1}{a(a+2)}$

$=\left(\frac{a}{a}\right)\frac{2}{a+2}+\left(\frac{a+2}{a+2}\right)\frac{1}{a}+\frac{a-1}{a(a+2)}$

$=\frac{2a+a+2+a-1}{a(a+2)}$

$=\frac{4a+1}{a(a+2)}$

38. $\frac{5x+2}{x^2-1}+\frac{3}{x^2+x}-\frac{1}{x^2-x}$

$=\frac{5x+2}{(x+1)(x-1)}+\frac{3}{x(x+1)}-\frac{1}{x(x-1)}$

$=\left(\frac{x}{x}\right)\left(\frac{5x+2}{(x+1)(x-1)}\right)+\left(\frac{x-1}{x-1}\right)\left(\frac{3}{x(x+1)}\right)$

$-\left(\frac{x+1}{x+1}\right)\left(\frac{1}{x(x-1)}\right)$

$=\frac{x(5x+2)+(x-1)(3)-(x+1)(1)}{x(x+1)(x-1)}$

$=\frac{5x^2+2x+3x-3-x-1}{x(x+1)(x-1)}$

$=\frac{5x^2+4x-4}{x(x+1)(x-1)}$

R.4 Equations

1.
$$\begin{aligned} 0.2m - 0.5 &= 0.1m + 0.7 \\ 10(0.2m - 0.5) &= 10(0.1m + 0.7) \\ 2m - 5 &= m + 7 \\ m - 5 &= 7 \\ m &= 12 \end{aligned}$$

The solution is 12.

2. $\frac{2}{3}k - k + \frac{3}{8} = \frac{1}{2}$

Multiply both sides of the equation by 24.

$$\begin{aligned} 24\left(\frac{2}{3}k\right) - 24(k) + 24\left(\frac{3}{8}\right) &= 24\left(\frac{1}{2}\right) \\ 16k - 24k + 9 &= 12 \\ -8k + 9 &= 12 \\ -8k &= 3 \\ k &= -\frac{3}{8} \end{aligned}$$

The solution is $-\frac{3}{8}$.

3.
$$\begin{aligned} 2x + 8 &= x - 4 \\ x + 8 &= -4 \\ x &= -12 \end{aligned}$$

The solution is -12.

4.
$$\begin{aligned} 5x + 2 &= 8 - 3x \\ 8x + 2 &= 8 \\ 8x &= 6 \\ x &= \frac{3}{4} \end{aligned}$$

The solution is $\frac{3}{4}$.

5.
$$\begin{aligned} 3r + 2 - 5(r + 1) &= 6r + 4 \\ 3r + 2 - 5r - 5 &= 6r + 4 \\ -3 - 2r &= 6r + 4 \\ -3 &= 8r + 4 \\ -7 &= 8r \\ -\frac{7}{8} &= r \end{aligned}$$

The solution is $-\frac{7}{8}$.

6.
$$\begin{aligned} 5(a + 3) + 4a - 5 &= -(2a - 4) \\ 5a + 15 + 4a - 5 &= -2a + 4 \\ 9a + 10 &= -2a + 4 \\ 11a + 10 &= 4 \\ 11a &= -6 \\ a &= -\frac{6}{11} \end{aligned}$$

The solution is $-\frac{6}{11}$.

7.
$$\begin{aligned} 2[3m - 2(3 - m) - 4] &= 6m - 4 \\ 2[3m - 6 + 2m - 4] &= 6m - 4 \\ 2[5m - 10] &= 6m - 4 \\ 10m - 20 &= 6m - 4 \\ 4m - 20 &= -4 \\ 4m &= 16 \\ m &= 4 \end{aligned}$$

The solution is 4.

8.
$$\begin{aligned} 4[2p - (3 - p) + 5] &= -7p - 2 \\ 4[2p - 3 + p + 5] &= -7p - 2 \\ 4[3p + 2] &= -7p - 2 \\ 12p + 8 &= -7p - 2 \\ 19p + 8 &= -2 \\ 19p &= -10 \\ p &= -\frac{10}{19} \end{aligned}$$

The solution is $-\frac{10}{19}$.

9.
$$\begin{aligned} x^2 + 5x + 6 &= 0 \\ (x + 3)(x + 2) &= 0 \\ x + 3 = 0 \quad &\text{or} \quad x + 2 = 0 \\ x = -3 \quad &\text{or} \quad x = -2 \end{aligned}$$

The solutions are -3 and -2.

10.
$$\begin{aligned} x^2 &= 3 + 2x \\ x^2 - 2x - 3 &= 0 \\ (x - 3)(x + 1) &= 0 \\ x - 3 = 0 \quad &\text{or} \quad x + 1 = 0 \\ x = 3 \quad &\text{or} \quad x = -1 \end{aligned}$$

The solutions are 3 and -1.

11.
$$\begin{aligned} m^2 &= 14m - 49 \\ m^2 - 14m + 49 &= 0 \\ (m)^2 - 2(7m) + (7)^2 &= 0 \\ (m - 7)^2 &= 0 \\ m - 7 &= 0 \\ m &= 7 \end{aligned}$$

The solution is 7.

12.
$$\begin{aligned} 2k^2 - k &= 10 \\ 2k^2 - k - 10 &= 0 \\ (2k - 5)(k + 2) &= 0 \\ 2k - 5 = 0 \quad &\text{or} \quad k + 2 = 0 \\ k = \frac{5}{2} \quad &\text{or} \quad k = -2 \end{aligned}$$

The solutions are $\frac{5}{2}$ and -2.

13.
$$12x^2 - 5x = 2$$
$$12x^2 - 5x - 2 = 0$$
$$(4x+1)(3x-2) = 0$$
$$4x+1=0 \quad \text{or} \quad 3x-2=0$$
$$4x=-1 \quad \text{or} \quad 3x=2$$
$$x=-\frac{1}{4} \quad \text{or} \quad x=\frac{2}{3}$$

The solutions are $-\frac{1}{4}$ and $\frac{2}{3}$.

14.
$$m(m-7) = -10$$
$$m^2 - 7m + 10 = 0$$
$$(m-5)(m-2) = 0$$
$$m-5=0 \quad \text{or} \quad m-2=0$$
$$m=5 \quad \text{or} \quad m=2$$

The solutions are 5 and 2.

15. $4x^2 - 36 = 0$

Divide both sides of the equation by 4.

$$x^2 - 9 = 0$$
$$(x+3)(x-3) = 0$$
$$x+3=0 \quad \text{or} \quad x-3=0$$
$$x=-3 \quad \text{or} \quad x=3$$

The solutions are -3 and 3.

16.
$$z(2z+7) = 4$$
$$2z^2 + 7z - 4 = 0$$
$$(2z-1)(z+4) = 0$$
$$2z-1=0 \quad \text{or} \quad z+4=0$$
$$z=\frac{1}{2} \quad \text{or} \quad z=-4$$

The solutions are $\frac{1}{2}$ and -4.

17.
$$12y^2 - 48y = 0$$
$$12y(y) - 12y(4) = 0$$
$$12y(y-4) = 0$$
$$12y=0 \quad \text{or} \quad y-4=0$$
$$y=0 \quad \text{or} \quad y=4$$

The solutions are 0 and 4.

18. $3x^2 - 5x + 1 = 0$

Use the quadratic formula.

$$x = \frac{-(-5) \pm \sqrt{(-5)^2 - 4(3)(1)}}{2(3)}$$
$$= \frac{5 \pm \sqrt{25-12}}{6}$$
$$x = \frac{5+\sqrt{13}}{6} \quad \text{or} \quad x = \frac{5-\sqrt{13}}{6}$$
$$\approx 1.4343 \qquad \approx 0.2324$$

The solutions are $\frac{5+\sqrt{13}}{6} \approx 1.4343$ and $\frac{5-\sqrt{13}}{6} \approx 0.2324$.

19.
$$2m^2 - 4m = 3$$
$$2m^2 - 4m - 3 = 0$$
$$m = \frac{-(-4) \pm \sqrt{(-4)^2 - 4(2)(-3)}}{2(2)}$$
$$= \frac{4 \pm \sqrt{40}}{4} = \frac{4 \pm \sqrt{4 \cdot 10}}{4}$$
$$= \frac{4 \pm \sqrt{4}\sqrt{10}}{4}$$
$$= \frac{4 \pm 2\sqrt{10}}{4} = \frac{2 \pm \sqrt{10}}{2}$$

The solutions are $\frac{2+\sqrt{10}}{2} \approx 2.5811$ and $\frac{2-\sqrt{10}}{2} \approx -0.5811$.

20. $p^2 + p - 1 = 0$

$$p = \frac{-1 \pm \sqrt{1^2 - 4(1)(-1)}}{2(1)}$$
$$= \frac{-1 \pm \sqrt{5}}{2}$$

The solutions are $\frac{-1+\sqrt{5}}{2} \approx 0.6180$ and $\frac{-1-\sqrt{5}}{2} \approx -1.6180$.

21.
$$k^2 - 10k = -20$$
$$k^2 - 10k + 20 = 0$$
$$k = \frac{-(-10) \pm \sqrt{(-10)^2 - 4(1)(20)}}{2(1)}$$
$$k = \frac{10 \pm \sqrt{100-80}}{2}$$
$$k = \frac{10 \pm \sqrt{20}}{2}$$
$$k = \frac{10 \pm \sqrt{4}\sqrt{5}}{2}$$
$$k = \frac{10 \pm 2\sqrt{5}}{2}$$
$$k = \frac{2(5 \pm \sqrt{5})}{2}$$
$$k = 5 \pm \sqrt{5}$$

The solutions are $5 + \sqrt{5} \approx 7.2361$ and $5 - \sqrt{5} \approx 2.7639$.

22. $5x^2 - 8x + 2 = 0$

$$x = \frac{-(-8) \pm \sqrt{(-8)^2 - 4(5)(2)}}{2(5)}$$
$$= \frac{8 \pm \sqrt{24}}{10} = \frac{8 \pm \sqrt{4 \cdot 6}}{10}$$
$$= \frac{8 \pm \sqrt{4}\sqrt{6}}{10} = \frac{8 \pm 2\sqrt{6}}{10}$$
$$= \frac{4 \pm \sqrt{6}}{5}$$

The solutions are $\frac{4+\sqrt{6}}{5} \approx 1.2899$ and $\frac{4-\sqrt{6}}{5} \approx 0.3101$.

23. $2r^2 - 7r + 5 = 0$

$$(2r - 5)(r - 1) = 0$$
$$2r - 5 = 0 \quad \text{or} \quad r - 1 = 0$$
$$2r = 5$$
$$r = \frac{5}{2} \quad \text{or} \quad r = 1$$

The solutions are $\frac{5}{2}$ and 1.

24. $2x^2 - 7x + 30 = 0$

$$x = \frac{-(-7) \pm \sqrt{(-7)^2 - 4(2)(30)}}{2(2)}$$
$$x = \frac{7 \pm \sqrt{49 - 240}}{4}$$
$$x = \frac{7 \pm \sqrt{-191}}{4}$$

Since there is a negative number under the radical sign, $\sqrt{-191}$ is not a real number. Thus, there are no real number solutions.

25. $3k^2 + k = 6$

$$3k^2 + k - 6 = 0$$
$$k = \frac{-1 \pm \sqrt{1 - 4(3)(-6)}}{2(3)}$$
$$= \frac{-1 \pm \sqrt{73}}{6}$$

The solutions are $\frac{-1+\sqrt{73}}{6} \approx 1.2573$ and $\frac{-1-\sqrt{73}}{6} \approx -1.5907$.

26. $5m^2 + 5m = 0$

$$5m(m + 1) = 0$$
$$5m = 0 \quad \text{or} \quad m + 1 = 0$$
$$m = 0 \quad \text{or} \quad m = -1$$

The solutions are 0 and -1.

27. $\dfrac{3x - 2}{7} = \dfrac{x + 2}{5}$

$$35\left(\frac{3x - 2}{7}\right) = 35\left(\frac{x + 2}{5}\right)$$
$$5(3x - 2) = 7(x + 2)$$
$$15x - 10 = 7x + 14$$
$$8x = 24$$
$$x = 3$$

The solution is $x = 3$.

28. $\dfrac{x}{3} - 7 = 6 - \dfrac{3x}{4}$

Multiply both sides by 12, the least common denominator of 3 and 4.

$$12\left(\frac{x}{3} - 7\right) = 12\left(6 - \frac{3x}{4}\right)$$
$$12\left(\frac{x}{3}\right) - (12)(7) = (12)(6) - (12)\left(\frac{3x}{4}\right)$$
$$4x - 84 = 72 - 9x$$
$$13x - 84 = 72$$
$$13x = 156$$
$$x = 12$$

The solution is 12.

29. $\dfrac{4}{x - 3} - \dfrac{8}{2x + 5} + \dfrac{3}{x - 3} = 0$

$$\frac{4}{x - 3} + \frac{3}{x - 3} - \frac{8}{2x + 5} = 0$$
$$\frac{7}{x - 3} - \frac{8}{2x + 5} = 0$$

Multiply both sides by $(x - 3)(2x + 5)$. Note that $x \neq 3$ and $x \neq -\frac{5}{2}$.

$$(x-3)(2x+5)\left(\frac{7}{x - 3} - \frac{8}{2x + 5}\right) = (x-3)(2x+5)(0)$$
$$7(2x + 5) - 8(x - 3) = 0$$
$$14x + 35 - 8x + 24 = 0$$
$$6x + 59 = 0$$
$$6x = -59$$
$$x = -\frac{59}{6}$$

Note: It is especially important to check solutions of equations that involve rational expressions. Here, a check shows that $-\frac{59}{6}$ is a solution.

30. $\frac{5}{p-2} - \frac{7}{p+2} = \frac{12}{p^2-4}$

$$\frac{5}{p-2} - \frac{7}{p+2} = \frac{12}{(p-2)(p+2)}$$

Multiply both sides by $(p-2)(p+2)$. Note that $p \neq 2$ and $p \neq -2$.

$$(p-2)(p+2)\left(\frac{5}{p-2} - \frac{7}{p+2}\right) = (p-2)(p+2)\left(\frac{12}{(p-2)(p+2)}\right)$$

$$(p-2)(p+2)\left(\frac{5}{p-2}\right) -$$

$$(p-2)(p+2)\left(\frac{7}{p+2}\right) = (p-2)(p+2)\left(\frac{12}{(p-2)(p+2)}\right)$$

$$(p+2)(5) - (p-2)(7) = 12$$

$$5p + 10 - 7p + 14 = 12$$

$$-2p + 24 = 12$$

$$-2p = -12$$

$$p = 6$$

The solution is 6.

31. $\frac{2m}{m-2} - \frac{6}{m} = \frac{12}{m^2-2m}$

$$\frac{2m}{m-2} - \frac{6}{m} = \frac{12}{m(m-2)}$$

Multiply both sides by $m(m-2)$. Note that $m \neq 0$ and $m \neq 2$.

$$m(m-2)\left(\frac{2m}{m-2} - \frac{6}{m}\right) = m(m-2)\left(\frac{12}{m(m-2)}\right)$$

$$m(2m) - 6(m-2) = 12$$

$$2m^2 - 6m + 12 = 12$$

$$2m^2 - 6m = 0$$

$$2m(m-3) = 0$$

$$2m = 0 \quad \text{or} \quad m - 3 = 0$$

$$m = 0 \quad \text{or} \quad m = 3$$

Since $m \neq 0$, 0 is not a solution. The solution is 3.

32. $\frac{2y}{y-1} = \frac{5}{y} + \frac{10-8y}{y^2-y}$

$$\frac{2y}{y-1} = \frac{5}{y} + \frac{10-8y}{y(y-1)}$$

Multiply both sides by $y(y-1)$. Note that $y \neq 0$ and $y \neq 1$.

$$y(y-1)\left(\frac{2y}{y-1}\right) = y(y-1)\left[\frac{5}{y} + \frac{10-8y}{y(y-1)}\right]$$

$$y(y-1)\left(\frac{2y}{y-1}\right) = y(y-1)\left(\frac{5}{y}\right) + y(y-1)\left[\frac{10-8y}{y(y-1)}\right]$$

$$y(2y) = (y-1)(5) + (10-8y)$$

$$2y^2 = 5y - 5 + 10 - 8y$$

$$2y^2 = 5 - 3y$$

$$2y^2 + 3y - 5 = 0$$

$$(2y+5)(y-1) = 0$$

$$2y + 5 = 0 \quad \text{or} \quad y - 1 = 0$$

$$y = -\frac{5}{2} \quad \text{or} \quad y = 1$$

Since $y \neq 1$, 1 is not a solution. The solution is $-\frac{5}{2}$.

33. $\frac{1}{x-2} - \frac{3x}{x-1} = \frac{2x+1}{x^2-3x+2}$

$$\frac{1}{x-2} - \frac{3x}{x-1} = \frac{2x+1}{(x-2)(x-1)}$$

Multiply both sides by $(x-2)(x-1)$. Note that $x \neq 2$ and $x \neq 1$.

$$(x-2)(x-1)\left(\frac{1}{x-2} - \frac{3x}{x-1}\right) = (x-2)(x-1) \cdot \left[\frac{2x+1}{(x-2)(x-1)}\right]$$

$$(x-2)(x-1)\left(\frac{1}{x-2}\right) - (x-2)(x-1)\cdot\left(\frac{3x}{x-1}\right) = \frac{(x-2)(x-1)(2x+1)}{(x-2)(x-1)}$$

$$(x-1) - (x-2)(3x) = 2x + 1$$

$$x - 1 - 3x^2 + 6x = 2x + 1$$

$$-3x^2 + 7x - 1 = 2x + 1$$

$$-3x^2 + 5x - 2 = 0$$

$$3x^2 - 5x + 2 = 0$$

$$(3x-2)(x-1) = 0$$

$$3x - 2 = 0 \quad \text{or} \quad x - 1 = 0$$

$$x = \frac{2}{3} \quad \text{or} \quad x = 1$$

1 is not a solution since $x \neq 1$. The solution is $\frac{2}{3}$.

34. $$\frac{5}{a}+\frac{-7}{a+1}=\frac{a^2-2a+4}{a^2+a}$$

$$a(a+1)\left(\frac{5}{a}+\frac{-7}{a+1}\right)=a(a+1)\left(\frac{a^2-2a+4}{a^2+a}\right)$$

Note that $a \neq 0$ and $a \neq -1$.

$$\begin{aligned}5(a+1)+(-7)(a)&=a^2-2a+4\\ 5a+5-7a&=a^2-2a+4\\ 5-2a&=a^2-2a+4\\ 5&=a^2+4\\ 0&=a^2-1\\ 0&=(a+1)(a-1)\end{aligned}$$

$$a+1=0 \quad\text{or}\quad a-1=0$$
$$a=-1 \quad\text{or}\quad a=1$$

Since -1 would make two denominators zero, 1 is the only solution.

35. $$\frac{5}{b+5}-\frac{4}{b^2+2b}=\frac{6}{b^2+7b+10}$$

$$\frac{5}{b+5}-\frac{4}{b(b+2)}=\frac{6}{(b+5)(b+2)}$$

Multiply both sides by $b(b+5)(b+2)$.
Note that $b \neq 0$, $b \neq -5$, and $b \neq -2$.

$$b(b+5)(b+2)\left(\frac{5}{b+5}-\frac{4}{b(b+2)}\right)$$
$$=b(b+5)(b+2)\left(\frac{6}{(b+5)(b+2)}\right)$$

$$\begin{aligned}5b(b+2)-4(b+5)&=6b\\ 5b^2+10b-4b-20&=6b\\ 5b^2-20&=0\\ b^2-4&=0\\ (b+2)(b-2)&=0\end{aligned}$$

$$b+2=0 \quad\text{or}\quad b-2=0$$
$$b=-2 \quad\text{or}\quad b=2$$

Since $b \neq -2, -2$ is not a solution. The solution is 2.

36. $$\frac{2}{x^2-2x-3}+\frac{5}{x^2-x-6}=\frac{1}{x^2+3x+2}$$

$$\frac{2}{(x-3)(x+1)}+\frac{5}{(x-3)(x+2)}=\frac{1}{(x+2)(x+1)}$$

Multiply both sides by $(x-3)(x+1)(x+2)$.
Note that $x \neq 3$, $x \neq -1$, and $x \neq -2$.

$$(x-3)(x+1)(x+2)\left(\frac{2}{(x-3)(x+1)}\right)$$
$$+(x-3)(x+1)(x+2)\left(\frac{5}{(x-3)(x+2)}\right)$$
$$=(x-3)(x+1)(x+2)\left(\frac{1}{(x+2)(x+1)}\right)$$

$$\begin{aligned}2(x+2)+5(x+1)&=x-3\\ 2x+4+5x+5&=x-3\\ 7x+9&=x-3\\ 6x+9&=-3\\ 6x&=-12\\ x&=-2\end{aligned}$$

However, $x \neq -2$. Therefore there is no solution.

37. $$\frac{4}{2x^2+3x-9}+\frac{2}{2x^2-x-3}=\frac{3}{x^2+4x+3}$$

$$\frac{4}{(2x-3)(x+3)}+\frac{2}{(2x-3)(x+1)}=\frac{3}{(x+3)(x+1)}$$

Multiply both sides by $(2x-3)(x+3)(x+1)$.
Note that $x \neq \frac{3}{2}$, $x \neq -3$, and $x \neq -1$.

$$(2x-3)(x+3)(x+1)$$
$$\cdot\left(\frac{4}{(2x-3)(x+3)}+\frac{2}{(2x-3)(x+1)}\right)$$
$$=(2x-3)(x+3)(x+1)\left(\frac{3}{(x+3)(x+1)}\right)$$

$$\begin{aligned}4(x+1)+2(x+3)&=3(2x-3)\\ 4x+4+2x+6&=6x-9\\ 6x+10&=6x-9\\ 10&=-9\end{aligned}$$

This is a false statement. Therefore, there is no solution.

R.5 Inequalities

1. $x < 4$

Because the inequality symbol means "less than," the endpoint at 4 is not included. This inequality is written in interval notation as $(-\infty, 4)$. To graph this interval on a number line, place an open circle at 4 and draw a heavy arrow pointing to the left.

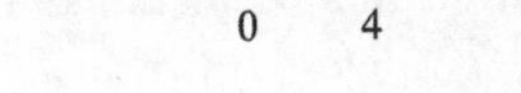

2. $x \geq -3$

Because the inequality sign means "greater than or equal to," the endpoint at -3 is included. This inequality is written in interval notation as $[-3, \infty)$. To graph this interval on a number line, place a closed circle at -3 and draw a heavy arrow pointing to the right.

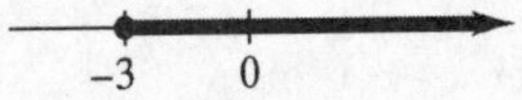

3. $1 \le x < 2$

The endpoint at 1 is included, but the endpoint at 2 is not. This inequality is written in interval notation as $[1, 2)$. To graph this interval, place a closed circle at 1 and an open circle at 2; then draw a heavy line segment between them.

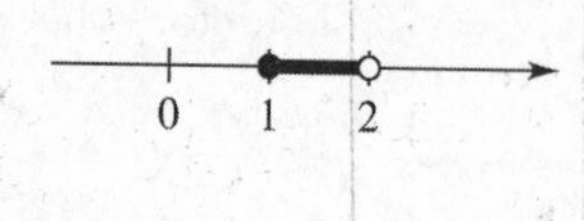

4. $-2 \le x \le 3$

The endpoints at -2 and 3 are both included. This inequality is written in interval notation as $[-2, 3]$. To graph this interval, place an open circle at -2 and another at 3 and draw a heavy line segment between them.

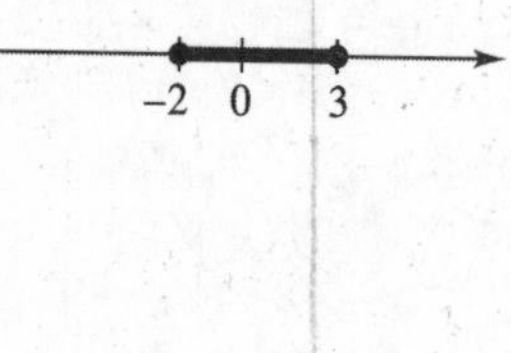

5. $-9 > x$

This inequality may be rewritten as $x < -9$, and is written in interval notation as $(-\infty, -9)$. Note that the endpoint at -9 is not included. To graph this interval, place an open circle at -9 and draw a heavy arrow pointing to the left.

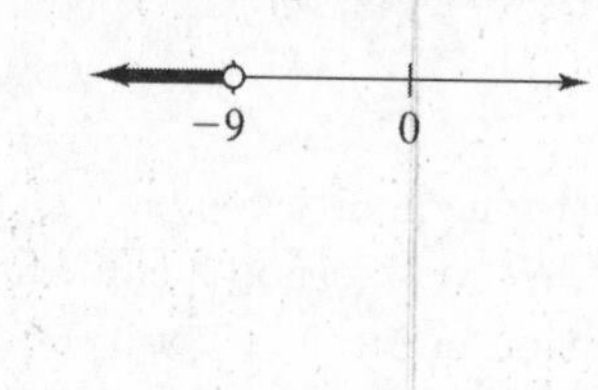

6. $6 \le x$

This inequality may be written as $x \ge 6$, and is written in interval notation as $[6, \infty)$. Note that the endpoint at 6 is included. To graph this interval, place a closed circle at 6 and draw a heavy arrow pointing to the right.

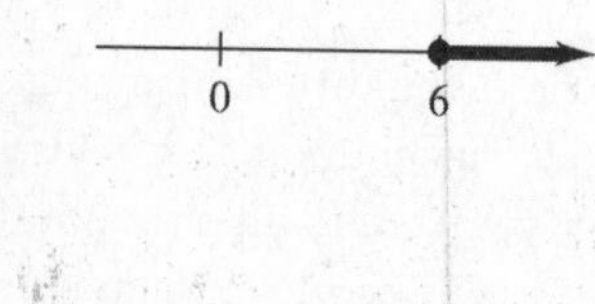

7. $[-7, -3]$

This represents all the numbers between -7 and -3, including both endpoints. This interval can be written as the inequality $-7 \le x \le -3$.

8. $[4, 10)$

This represents all the numbers between 4 and 10, including 4 but not including 10. This interval can be written as the inequality $4 \le x < 10$.

9. $(-\infty, -1]$

This represents all the numbers to the left of -1 on the number line and includes the endpoint. This interval can be written as the inequality $x \le -1$.

10. $(3, \infty)$

This represents all the numbers to the right of 3, and does not include the endpoint. This interval can be written as the inequality $x > 3$.

11. Notice that the endpoint -2 is included, but 6 is not. The interval shows in the graph can be written as the inequality $-2 \le x < 6$.

12. Notice that neither endpoint is included. The interval shown in the graph can be written as $0 < x < 8$.

13. Notice that both endpoints are included. The interval shown in the graph can be written as $x \le -4$ or $x \ge 4$.

14. Notice that the endpoint 0 is not included, but 3 is included. The interval shown in the graph can be written as $x < 0$ or $x \ge 3$.

15.
$$6p + 7 \le 19$$
$$6p \le 12$$
$$\left(\frac{1}{6}\right)(6p) \le \left(\frac{1}{6}\right)(12)$$
$$p \le 2$$

The solution in interval notation is $(-\infty, 2]$.

0 2

16.
$$6k - 4 < 3k - 1$$
$$6k < 3k + 3$$
$$3k < 3$$
$$k < 1$$

The solution in interval notation is $(-\infty, 1)$.

0 1

17. $m-(3m-2)+6<7m-19$

$$m-3m+2+6<7m-19$$
$$-2m+8<7m-19$$
$$-9m+8<-19$$
$$-9m<-27$$
$$-\frac{1}{9}(-9m)>-\frac{1}{9}(-27)$$
$$m>3$$

The solution is $(3,\ \infty)$.

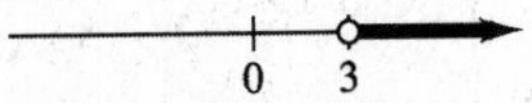

18. $-2(3y-8)\geq 5(4y-2)$

$$-6y+16\geq 20y-10$$
$$-6y+16+(-16)\geq 20y-10+(-16)$$
$$-6y\geq 20y-26$$
$$-6y+(-20y)\geq 20y+(-20y)-26$$
$$-26y\geq -26$$
$$-\frac{1}{26}(-26)y\leq -\frac{1}{26}(-26)$$
$$y\leq 1$$

The solution is $(-\infty,1]$.

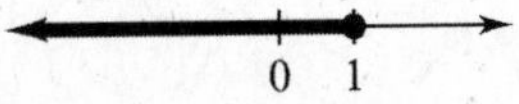

19. $3p-1<6p+2(p-1)$

$$3p-1<6p+2p-2$$
$$3p-1<8p-2$$
$$-5p-1<-2$$
$$-5p<-1$$
$$-\frac{1}{5}(-5p)>-\frac{1}{5}(-1)$$
$$p>\frac{1}{5}$$

The solution is $\left(\frac{1}{5},\ \infty\right)$.

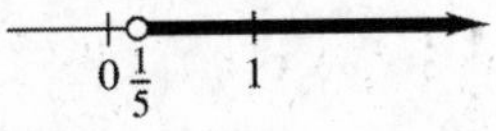

20. $x+5(x+1)>4(2-x)+x$

$$x+5x+5>8-4x+x$$
$$6x+5>8-3x$$
$$6x>3-3x$$
$$9x>3$$
$$x>\frac{1}{3}$$

The solution is $\left(\frac{1}{3},\infty\right)$.

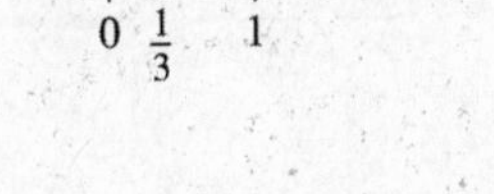

21. $-11<y-7<-1$

$$-11+7<y-7+7<-1+7$$
$$-4<y<6$$

The solution is $(-4,6)$.

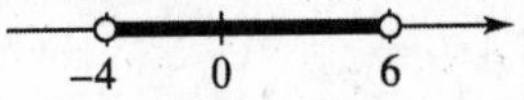

22. $8\leq 3r+1\leq 13$

$$8+(-1)\leq 3r+1+(-1)\leq 13+(-1)$$
$$7\leq 3r\leq 12$$
$$\frac{1}{3}(7)\leq\frac{1}{3}(3r)\leq\frac{1}{3}(12)$$
$$\frac{7}{3}\leq r\leq 4$$

The solution is $\left[\frac{7}{3},4\right]$.

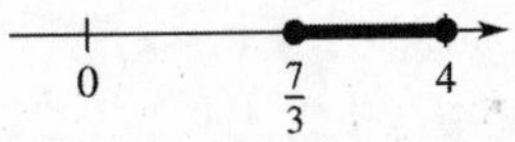

23. $-2<\frac{1-3k}{4}\leq 4$

$$4(-2)<4\left(\frac{1-3k}{4}\right)\leq 4(4)$$
$$-8<1-3k\leq 16$$
$$-9<-3k\leq 15$$
$$-\frac{1}{3}(-9)>-\frac{1}{3}(-3k)\geq-\frac{1}{3}(15)$$

Rewrite the inequalities in the proper order.

$$-5\leq k<3$$

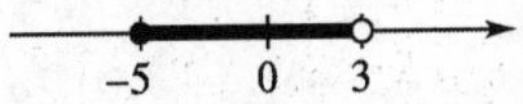

24. $-1\leq\frac{5y+2}{3}\leq 4$

$$3(-1)\leq 3\left(\frac{5y+2}{3}\right)\leq 3(4)$$
$$-3\leq 5y+2\leq 12$$
$$-5\leq 5y\leq 10$$
$$-1\leq y\leq 2$$

The solution is $[-1,\ 2]$.

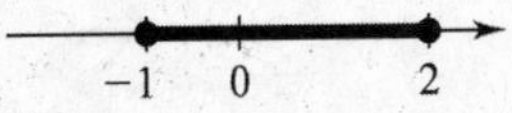

25. $\frac{3}{5}(2p+3) \geq \frac{1}{10}(5p+1)$

$$10\left(\frac{3}{5}\right)(2p+3) \geq 10\left(\frac{1}{10}\right)(5p+1)$$

$$6(2p+3) \geq 5p+1$$
$$12p+18 \geq 5p+1$$
$$7p \geq -17$$
$$p \geq -\frac{17}{7}$$

The solution is $[-\frac{17}{7}, \infty)$.

$-\frac{17}{7}$ 0

26. $\frac{8}{3}(z-4) \leq \frac{2}{9}(3z+2)$

$$(9)\frac{8}{3}(z-4) \leq (9)\frac{2}{9}(3z+2)$$

$$24(z-4) \leq 2(3z+2)$$
$$24z-96 \leq 6z+4$$
$$24z \leq 6z+100$$
$$18z \leq 100$$
$$z \leq \frac{100}{18}$$
$$z \leq \frac{50}{9}$$

The solution is $(-\infty, \frac{50}{9}]$.

0 $\frac{50}{9}$

27. $(m-3)(m+5) < 0$

Solve $(m-3)(m+5) = 0$.

$$(m-3)(m+5) = 0$$

$$m = 3 \quad \text{or} \quad m = -5$$

Intervals: $(-\infty, -5)$, $(-5, 3)$, $(3, \infty)$

For $(-\infty, -5)$, choose -6 to test for m.

$$(-6-3)(-6+5) = -9(-1) = 9 \not< 0$$

For $(-5, 3)$, choose 0.

$$(0-3)(0+5) = -3(5) = -15 < 0$$

For $(3, \infty)$, choose 4.

$$(4-3)(4+5) = 1(9) = 9 \not< 0$$

The solution is $(-5, 3)$.

-5 0 3

28. $(t+6)(t-1) \geq 0$

Solve $(t+6)(t-1) = 0$.

$$(t+6)(t-1) = 0$$
$$t = -6 \quad \text{or} \quad t = 1$$

Intervals: $(-\infty, -6)$, $(-6, 1)$, $(1, \infty)$

For $(-\infty, -6)$, choose -7 to test for t.

$$(-7+6)(-7-1) = (-1)(-8) = 8 \geq 0$$

For $(-6, 1)$, choose 0.

$$(0+6)(0-1) = (6)(-1) = -6 \not\geq 0$$

For $(1, \infty)$, choose 2.

$$(2+6)(2-1) = (8)(1) = 8 \geq 0$$

Because the symbol $\geq$ is used, the endpoints -6 and 1 are included in the solution, $(-\infty, -6] \cup [1, \infty)$.

−6 0 1

29. $y^2 - 3y + 2 < 0$

$(y-2)(y-1) < 0$

Solve $(y-2)(y-1) = 0$.

$$y = 2 \quad \text{or} \quad y = 1$$

Intervals: $(-\infty, 1)$, $(1, 2)$, $(2, \infty)$

For $(-\infty, 1)$, choose $y = 0$.

$$0^2 - 3(0) + 2 = 2 \not< 0$$

For $(1, 2)$, choose $y = \frac{3}{2}$.

$$\left(\frac{3}{2}\right)^2 - 3\left(\frac{3}{2}\right) + 2 = \frac{9}{4} - \frac{9}{2} + 2$$
$$= \frac{9-18+8}{4}$$
$$= -\frac{1}{4} < 0$$

For $(2, \infty)$, choose 3.

$$3^2 - 3(3) + 2 = 2 \not< 0$$

The solution is $(1, 2)$.

0 1 2

30. $2k^2 + 7k - 4 > 0$

Solve $2k^2 + 7k - 4 = 0$.

$$2k^2 + 7k - 4 = 0$$
$$(2k - 1)(k + 4) = 0$$
$$k = \frac{1}{2} \quad \text{or} \quad k = -4$$

Intervals: $(-\infty, -4), \left(-4, \frac{1}{2}\right), \left(\frac{1}{2}, \infty\right)$

For $(-\infty, -4)$, choose -5.

$$2(-5)^2 + 7(-5) - 4 = 11 > 0$$

For $\left(-4, \frac{1}{2}\right)$, choose 0.

$$2(0)^2 + 7(0) - 4 = -4 \not> 0$$

For $\left(\frac{1}{2}, \infty\right)$, choose 1.

$$2(1)^2 + 7(1) - 4 = 5 > 0$$

The solution is $(-\infty, -4) \cup \left(\frac{1}{2}, \infty\right)$.

31. $x^2 - 16 > 0$

Solve $x^2 - 16 = 0$.

$$x^2 - 16 = 0$$
$$(x + 4)(x - 4) = 0$$
$$x = -4 \quad \text{or} \quad x = 4$$

Intervals: $(-\infty, -4)$, $(-4, 4)$, $(4, \infty)$

For $(-\infty, -4)$, choose -5.

$$(-5)^2 - 16 = 9 > 0$$

For $(-4, 4)$, choose 0.

$$0^2 - 16 = -16 \not> 0$$

For $(4, \infty)$, choose 5.

$$5^2 - 16 = 9 > 0$$

The solution is $(-\infty, -4) \cup (4, \infty)$.

32. $2k^2 - 7k - 15 \leq 0$

Solve $2k^2 - 7k - 15 = 0$.

$$2k^2 - 7k - 15 = 0$$
$$(2k + 3)(k - 5) = 0$$
$$k = -\frac{3}{2} \quad \text{or} \quad k = 5$$

Intervals: $\left(-\infty, -\frac{3}{2}\right), \left(-\frac{3}{2}, 5\right), (5, \infty)$

For $\left(-\infty, -\frac{3}{2}\right)$, choose -2.

$$2(-2)^2 - 7(-2) - 15 = 7 \not\leq 0$$

For $\left(-\frac{3}{2}, 5\right)$, choose 0.

$$2(0)^2 - 7(0) - 15 = -15 \leq 0$$

For $(5, \infty)$, choose 6.

$$2(6)^2 - 7(6) - 15 \not\leq 0$$

The solution is $\left[-\frac{3}{2}, 5\right]$.

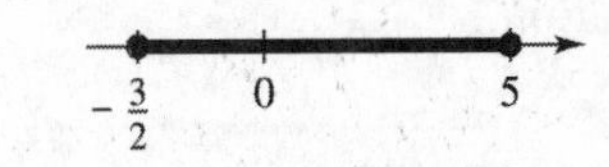

33. $x^2 - 4x \geq 5$

Solve $x^2 - 4x = 5$.

$$x^2 - 4x = 5$$
$$x^2 - 4x - 5 = 0$$
$$(x + 1)(x - 5) = 0$$
$$x + 1 = 0 \quad \text{or} \quad x - 5 = 0$$
$$x = -1 \quad \text{or} \quad x = 5$$

Intervals: $(-\infty, -1), (-1, 5), (5, \infty)$

For $(-\infty, -1)$, choose -2.

$$(-2)^2 - 4(-2) = 12 \geq 5$$

For $(-1, 5)$, choose 0.

$$0^2 - 4(0) = 0 \not\geq 5$$

For $(5, \infty)$, choose 6.

$$(6)^2 - 4(6) = 12 \geq 5$$

The solution is $(-\infty, -1] \cup [5, \infty)$.

34. $10r^2 + r \le 2$

Solve $10r^2 + r = 2$.

$$\begin{aligned} 10r^2 + r &= 2 \\ 10r^2 + r - 2 &= 0 \\ (5r - 2)(2r + 1) &= 0 \end{aligned}$$

$$r = \frac{2}{5} \quad \text{or} \quad r = -\frac{1}{2}$$

Intervals: $\left(-\infty, -\frac{1}{2}\right), \left(-\frac{1}{2}, \frac{2}{5}\right), \left(\frac{2}{5}, \infty\right)$

For $\left(-\infty, -\frac{1}{2}\right)$, choose -1.

$$10(-1)^2 + (-1) = 9 \not\le 2$$

For $\left(-\frac{1}{2}, \frac{2}{5}\right)$, choose 0.

$$10(0)^2 + 0 = 0 \le 2$$

For $\left(\frac{2}{5}, \infty\right)$, choose 1.

$$10(1)^2 + 1 = 11 \not\le 2$$

The solution is $\left[-\frac{1}{2}, \frac{2}{5}\right]$.

$-\frac{1}{2}$ 0 $\frac{2}{5}$

35. $3x^2 + 2x > 1$

Solve $3x^2 + 2x = 1$.

$$\begin{aligned} 3x^2 + 2x &= 1 \\ 3x^2 + 2x - 1 &= 0 \\ (3x - 1)(x + 1) &= 0 \end{aligned}$$

$$x = \frac{1}{3} \quad \text{or} \quad x = -1$$

Intervals: $(-\infty, -1)$, $\left(-1, \frac{1}{3}\right)$, $\left(\frac{1}{3}, \infty\right)$

For $(-\infty, -1)$, choose -2.

$$3(-2)^2 + 2(-2) = 8 > 1$$

For $\left(-1, \frac{1}{3}\right)$, choose 0.

$$3(0)^2 + 2(0) = 0 \not> 1$$

For $\left(\frac{1}{3}, \infty\right)$, choose 1.

$$3(1)^2 + 2(1) = 5 > 1$$

The solution is $(-\infty, -1) \cup \left(\frac{1}{3}, \infty\right)$

-1 0 $\frac{1}{3}$

36. $3a^2 + a > 10$

Solve $3a^2 + a = 10$.

$$\begin{aligned} 3a^2 + a &= 10 \\ 3a^2 + a - 10 &= 0 \\ (3a - 5)(a + 2) &= 0 \end{aligned}$$

$$a = \frac{5}{3} \quad \text{or} \quad a = -2$$

Intervals: $(-\infty, -2), \left(-2, \frac{5}{3}\right), \left(\frac{5}{3}, \infty\right)$

For $(-\infty, -2)$, choose -3.

$$3(-3)^2 + (-3) = 24 > 10$$

For $\left(-2, \frac{5}{3}\right)$, choose 0.

$$3(0)^2 + 0 = 0 \not> 10$$

For $\left(\frac{5}{3}, \infty\right)$, choose 2.

$$3(2)^2 + 2 = 14 > 10$$

The solution is $(-\infty, -2) \cup \left(\frac{5}{3}, \infty\right)$.

-2 0 $\frac{5}{3}$

37. $9 - x^2 \le 0$

Solve $9 - x^2 = 0$.

$$\begin{aligned} 9 - x^2 &= 0 \\ (3 + x)(3 - x) &= 0 \\ x = -3 \quad &\text{or} \quad x = 3 \end{aligned}$$

Intervals: $(-\infty, -3)$, $(-3, 3)$, $(3, \infty)$

For $(-\infty, -3)$, choose -4.

$$9 - (-4)^2 = -7 \le 0$$

For $(-3, 3)$, choose 0.

$$9 - (0)^2 = 9 \not\le 0$$

For $(3, \infty)$, choose 4.

$$9 - (4)^2 = -7 \le 0$$

The solution is $(-\infty, -3] \cup [3, \infty)$.

-3 0 3

38. $p^2 - 16p > 0$

Solve $p^2 - 16p = 0$.

$$p^2 - 16p = 0$$
$$p(p - 16) = 0$$
$$p = 0 \quad \text{or} \quad p = 16$$

Intervals: $(-\infty, 0)$, $(0, 16)$, $(16, \infty)$

For $(-\infty, 0)$, choose -1.

$$(-1)^2 - 16(-1) = 17 > 0$$

For $(0, 16)$, choose 1.

$$(1)^2 - 16(1) = -15 \not> 0$$

For $(16, \infty)$, choose 17.

$$(17)^2 - 16(17) = 17 > 0$$

The solution is $(-\infty, 0) \cup (16, \infty)$.

0 16

39. $\dfrac{m-3}{m+5} \le 0$

Solve $\dfrac{m-3}{m+5} = 0$.

$$(m+5)\frac{m-3}{m+5} = (m+5)(0)$$
$$m - 3 = 0$$
$$m = 3$$

Set the denominator equal to 0 and solve.

$$m + 5 = 0$$
$$m = -5$$

Intervals: $(-\infty, -5)$, $(-5, 3)$, $(3, \infty)$

For $(-\infty, -5)$, choose -6.

$$\frac{-6-3}{-6+5} = 9 \not\le 0$$

For $(-5, 3)$, choose 0.

$$\frac{0-3}{0+5} = -\frac{3}{5} \le 0$$

For $(3, \infty)$, choose 4.

$$\frac{4-3}{4+5} = \frac{1}{9} \not\le 0$$

Although the $\le$ symbol is used, including -5 in the solution would cause the denominator to be zero.
The solution is $(-5, 3]$.

40. $\dfrac{r+1}{r-1} > 0$

Solve the equation $\dfrac{r+1}{r-1} = 0$.

$$\frac{r+1}{r-1} = 0$$
$$(r-1)\frac{r+1}{r-1} = (r-1)(0)$$
$$r + 1 = 0$$
$$r = -1$$

Find the value for which the denominator equals zero.

$$r - 1 = 0$$
$$r = 1$$

Intervals: $(-\infty, -1)$, $(-1, 1)$, $(1, \infty)$

For $(-\infty, -1)$, choose -2.

$$\frac{-2+1}{-2-1} = \frac{-1}{-3} = \frac{1}{3} > 0$$

For $(-1, 1)$, choose 0.

$$\frac{0+1}{0-1} = \frac{1}{-1} = -1 \not> 0$$

For $(1, \infty)$, choose 2.

$$\frac{2+1}{2-1} = \frac{3}{1} = 3 > 0$$

The solution is $(-\infty, -1) \cup (1, \infty)$.

41. $\dfrac{k-1}{k+2} > 1$

Solve $\dfrac{k-1}{k+2} = 1$.

$$k - 1 = k + 2$$
$$-1 \ne 2$$

The equation has no solution.
Solve $k + 2 = 0$.

$$k = -2$$

Intervals: $(-\infty, -2)$, $(-2, \infty)$

For $(-\infty, -2)$, choose -3.

$$\frac{-3-1}{-3+2} = 4 > 1$$

For $(-2, \infty)$, choose 0.

$$\frac{0-1}{0+2} = -\frac{1}{2} \not> 1$$

The solution is $(-\infty, -2)$.

42. $\dfrac{a-5}{a+2} < -1$

Solve the equation $\dfrac{a-5}{a+2} = -1$.

$$\frac{a-5}{a+2} = -1$$
$$a-5 = -1(a+2)$$
$$a-5 = -a-2$$
$$2a = 3$$
$$a = \frac{3}{2}$$

Set the denominator equal to zero and solve for a.

$$a+2 = 0$$
$$a = -2$$

Intervals: $(-\infty, -2), \left(-2, \frac{3}{2}\right), \left(\frac{3}{2}, \infty\right)$

For $(-\infty, -2)$, choose -3.

$$\frac{-3-5}{-3+2} = \frac{-8}{-1} = 8 \nless -1$$

For $\left(-2, \frac{3}{2}\right)$, choose 0.

$$\frac{0-5}{0+2} = \frac{-5}{2} = -\frac{5}{2} < -1$$

For $\left(\frac{3}{2}, \infty\right)$, choose 2.

$$\frac{2-5}{2+2} = \frac{-3}{4} = -\frac{3}{4} \nless -1$$

The solution is $\left(-2, \frac{3}{2}\right)$.

43. $\dfrac{2y+3}{y-5} \le 1$

Solve $\dfrac{2y+3}{y-5} = 1$.

$$2y+3 = y-5$$
$$y = -8$$

Solve $y-5=0$.

$$y = 5$$

Intervals: $(-\infty, -8)$, $(-8, 5)$, $(5, \infty)$

For $(-\infty, -8)$, choose $y = -10$.

$$\frac{2(-10)+3}{-10-5} = \frac{17}{15} \nleq 1$$

For $(-8, 5)$, choose $y = 0$.

$$\frac{2(0)+3}{0-5} = -\frac{3}{5} \le 1$$

For $(5, \infty)$, choose $y = 6$.

$$\frac{2(6)+3}{6-5} = \frac{15}{1} \nleq 1$$

The solution is $[-8, 5)$.

44. $\dfrac{a+2}{3+2a} \le 5$

For the equation $\dfrac{a+2}{3+2a} = 5$.

$$\frac{a+2}{3+2a} = 5$$
$$a+2 = 5(3+2a)$$
$$a+2 = 15+10a$$
$$-9a = 13$$
$$a = -\frac{13}{9}$$

Set the denominator equal to zero and solve for a.

$$3+2a = 0$$
$$2a = -3$$
$$a = -\frac{3}{2}$$

Intervals: $\left(-\infty, -\frac{3}{2}\right), \left(-\frac{3}{2}, -\frac{13}{9}\right), \left(-\frac{13}{9}, \infty\right)$

For $\left(-\infty, -\frac{3}{2}\right)$, choose -2.

$$\frac{-2+2}{3+2(-2)} = \frac{0}{-1} = 0 \le 5$$

For $\left(-\frac{3}{2}, -\frac{13}{9}\right)$, choose -1.46.

$$\frac{-1.46+2}{3+2(-1.46)} = \frac{0.54}{0.08} = 6.75 \nleq 5$$

For $\left(-\frac{13}{9}, \infty\right)$, choose 0.

$$\frac{0+2}{3+2(0)} = \frac{2}{3} \le 5$$

The value $-\frac{3}{2}$ cannot be included in the solution since it would make the denominator zero. The solution is $\left(-\infty, -\frac{3}{2}\right) \cup \left[-\frac{13}{9}, \infty\right)$.

45. $\frac{2k}{k-3} \le \frac{4}{k-3}$

Solve $\frac{2k}{k-3} = \frac{4}{k-3}$.

$$\frac{2k}{k-3} = \frac{4}{k-3}$$
$$\frac{2k}{k-3} - \frac{4}{k-3} = 0$$
$$\frac{2k-4}{k-3} = 0$$
$$2k - 4 = 0$$
$$k = 2$$

Set the denominator equal to 0 and solve for k.

$$k - 3 = 0$$
$$k = 3$$

Intervals: $(-\infty, 2)$, $(2, 3)$, $(3, \infty)$

For $(-\infty, 2)$, choose 0.

$$\frac{2(0)}{0-3} = 0 \text{ and } \frac{4}{0-3} = -\frac{4}{3}, \text{ so}$$
$$\frac{2(0)}{0-3} \not\le \frac{4}{0-3}$$

For $(2, 3)$, choose $\frac{5}{2}$.

$$\frac{2(\frac{5}{2})}{\frac{5}{2}-3} = \frac{5}{-\frac{1}{2}} = -10 \text{ and } \frac{4}{\frac{5}{2}-3} = \frac{4}{-\frac{1}{2}} = -8, \text{ so}$$
$$\frac{2(\frac{5}{2})}{\frac{5}{2}-3} \le \frac{4}{\frac{5}{2}-3}$$

For $(3, \infty)$, choose 4.

$$\frac{2(4)}{4-3} = 8 \text{ and } \frac{4}{4-3} = 4, \text{ so}$$
$$\frac{2(4)}{4-3} \not\le \frac{4}{4-3}$$

The solution is $[2, 3)$.

46. $\frac{5}{p+1} > \frac{12}{p+1}$

Solve the equation $\frac{5}{p+1} = \frac{12}{p+1}$.

$$\frac{5}{p+1} = \frac{12}{p+1}$$
$$5 = 12$$

The equation has no solution.
Set the denominator equal to zero and solve for p.

$$p + 1 = 0$$
$$p = -1$$

Intervals: $(-\infty, -1)$, $(-1, \infty)$

For $(-\infty, -1)$, choose -2.

$$\frac{5}{-2+1} = -5 \text{ and } \frac{12}{-2+1} = -12, \text{ so}$$
$$\frac{5}{-2+1} > \frac{12}{-2+1}.$$

For $(-1, \infty)$, choose 0.

$$\frac{5}{0+1} = 5 \text{ and } \frac{12}{0+1} = 12, \text{ so}$$
$$\frac{5}{0+1} \not> \frac{12}{0+1}.$$

The solution is $(-\infty, -1)$.

47. $\frac{2x}{x^2 - x - 6} \ge 0$

Solve $\frac{2x}{x^2 - x - 6} = 0$.

$$\frac{2x}{x^2 - x - 6} = 0$$
$$2x = 0$$
$$x = 0$$

Set the denominator equal to 0 and solve for x.

$$x^2 - x - 6 = 0$$
$$(x + 2)(x - 3) = 0$$

$$x + 2 = 0 \quad \text{or} \quad x - 3 = 0$$
$$x = -2 \quad \text{or} \quad x = 3$$

Intervals: $(-\infty, -2)$, $(-2, 0)$, $(0, 3)$, $(3, \infty)$

For $(-\infty, -2)$, choose -3.

$$\frac{2(-3)}{(-3)^2 - (-3) - 6} = -1 \not\geq 0$$

For $(-2, 0)$, choose -1.

$$\frac{2(-1)}{(-1)^2 - (-1) - 6} = \frac{1}{2} \geq 0$$

For $(0, 3)$, choose 2.

$$\frac{2(2)}{2^2 - 2 - 6} = -1 \not\geq 0$$

For $(3, \infty)$, choose 4.

$$\frac{2(4)}{4^2 - 4 - 6} = \frac{4}{3} \geq 0$$

The solution is $(-2, 0] \cup (3, \infty)$.

48. $\frac{8}{p^2 + 2p} > 1$

Solve the equation $\frac{8}{p^2 + 2p} = 1$.

$$\begin{aligned} \frac{8}{p^2 + 2p} &= 1 \\ 8 &= p^2 + 2p \\ 0 &= p^2 + 2p - 8 \\ 0 &= (p + 4)(p - 2) \end{aligned}$$

$$p + 4 = 0 \quad \text{or} \quad p - 2 = 0$$
$$p = -4 \quad \text{or} \quad p = 2$$

Set the denominator equal to zero and solve for p.

$$p^2 + 2p = 0$$
$$p(p + 2) = 0$$
$$p = 0 \quad \text{or} \quad p + 2 = 0$$
$$p = -2$$

Intervals: $(-\infty, -4)$, $(-4, -2)$, $(-2, 0)$, $(0, 2)$, $(2, \infty)$

For $(-\infty, -4)$, choose -5.

$$\frac{8}{(-5)^2 + 2(-5)} = \frac{8}{15} \not> 1$$

For $(-4, -2)$, choose -3.

$$\frac{8}{(-3)^2 + 2(-3)} = \frac{8}{9 - 6} = \frac{8}{3} > 1$$

For $(-2, 0)$, choose -1.

$$\frac{8}{(-1)^2 + 2(-1)} = \frac{8}{-1} = -8 \not> 1$$

For $(0, 2)$, choose 1.

$$\frac{8}{(1)^2 + 2(1)} = \frac{8}{3} > 1$$

For $(2, \infty)$, choose 3.

$$\frac{8}{(3)^2 + (2)(3)} = \frac{8}{15} \not> 1$$

The solution is $(-4, -2) \cup (0, 2)$.

49. $\frac{z^2 + z}{z^2 - 1} \geq 3$

Solve

$$\begin{aligned} \frac{z^2 + z}{z^2 - 1} &= 3. \\ z^2 + z &= 3z^2 - 3 \\ -2z^2 + z + 3 &= 0 \\ -1(2z^2 - z - 3) &= 0 \\ -1(z + 1)(2z - 3) &= 0 \end{aligned}$$

$$z = -1 \quad \text{or} \quad z = \frac{3}{2}$$

Set $z^2 - 1 = 0$.

$$z^2 = 1$$
$$z = -1 \quad \text{or} \quad z = 1$$

Intervals: $(-\infty, -1)$, $(-1, 1)$, $\left(1, \frac{3}{2}\right)$, $\left(\frac{3}{2}, \infty\right)$

For $(-\infty, -1)$, choose $x = -2$.

$$\frac{(-2)^2 + 3}{(-2)^2 - 1} = \frac{7}{3} \not\geq 3$$

For $(-1, 1)$, choose $x = 0$.

$$\frac{0^2 + 3}{0^2 - 1} = -3 \not\geq 3$$

For $\left(1, \frac{3}{2}\right)$, choose $x = \frac{3}{2}$.

$$\frac{\left(\frac{3}{2}\right)^2 + 3}{\left(\frac{3}{2}\right)^2 - 1} = \frac{21}{5} \geq 3$$

For $\left(\frac{3}{2}, \infty\right)$, choose $x = 2$.

$$\frac{2^2 + 3}{2^2 - 1} = \frac{7}{3} \not\geq 3$$

The solution is $\left(1, \frac{3}{2}\right]$.

50. $\dfrac{a^2+2a}{a^2-4} \le 2$

Solve the equation $\dfrac{a^2+2a}{a^2-4} = 2$.

$$\begin{aligned} \frac{a^2+2a}{a^2-4} &= 2 \\ a^2+2a &= 2(a^2-4) \\ a^2+2a &= 2a^2-8 \\ 0 &= a^2-2a-8 \\ 0 &= (a-4)(a+2) \\ a-4=0 \quad &\text{or} \quad a+2=0 \\ a=4 \quad &\text{or} \quad a=-2 \end{aligned}$$

But -2 is not a possible solution.
Set the denominator equal to zero and solve for a.

$$\begin{aligned} a^2-4 &= 0 \\ (a+2)(a-2) &= 0 \\ a+2=0 \quad &\text{or} \quad a-2=0 \\ a=-2 \quad &\text{or} \quad a=2 \end{aligned}$$

Intervals: $(-\infty,-2)$, $(-2,2)$, $(2,4)$, $(4,\infty)$

For $(-\infty,-2)$, choose -3.

$$\frac{(-3)^2+2(-3)}{(-3)^2-4} = \frac{9-6}{9-4} = \frac{3}{5} \le 2$$

For $(-2,2)$, choose 0.

$$\frac{(0)^2+2(0)}{0-4} = \frac{0}{-4} = 0 \le 2$$

For $(2,4)$, choose 3.

$$\frac{(3)^2+2(3)}{(3)^2-4} = \frac{9+6}{9-5} = \frac{15}{4} \not\le 2$$

For $(4,\infty)$, choose 5.

$$\frac{(5)^2+2(5)}{(5)^2-4} = \frac{25+10}{25-4} = \frac{35}{21} \le 2$$

The value 4 will satisfy the original inequality, but the values -2 and 2 will not since they make the denominator zero. The solution is $(-\infty,-2)\cup(-2,2)\cup[4,\infty)$.

R.6 Exponents

1. $8^{-2} = \dfrac{1}{8^2} = \dfrac{1}{64}$

2. $3^{-4} = \dfrac{1}{3^4} = \dfrac{1}{81}$

3. $5^0 = 1$, by definition.

4. $\left(-\dfrac{3}{4}\right)^0 = 1$, by definition.

5. $-(-3)^{-2} = -\dfrac{1}{(-3)^2} = -\dfrac{1}{9}$

6. $-(-3^{-2}) = -\left(-\dfrac{1}{3^2}\right) = -\left(-\dfrac{1}{9}\right) = \dfrac{1}{9}$

7. $\left(\dfrac{1}{6}\right)^{-2} = \dfrac{1}{\left(\frac{1}{6}\right)^2} = \dfrac{1}{\frac{1}{36}} = 36$

8. $\left(\dfrac{4}{3}\right)^{-3} = \dfrac{1}{\left(\frac{4}{3}\right)^3} = \dfrac{1}{\frac{64}{27}} = \dfrac{27}{64}$

9. $\dfrac{4^{-2}}{4} = 4^{-2-1} = 4^{-3} = \dfrac{1}{4^3} = \dfrac{1}{64}$

10. $\dfrac{8^9 \cdot 8^{-7}}{8^{-3}} = 8^{9+(-7)-(-3)} = 8^{9-7+3} = 8^5$

11. $$\begin{aligned} \frac{10^8 \cdot 10^{-10}}{10^4 \cdot 10^2} &= \frac{10^{8+(-10)}}{10^{4+2}} = \frac{10^{-2}}{10^6} \\ &= 10^{-2-6} = 10^{-8} \\ &= \frac{1}{10^8} \end{aligned}$$

12. $$\begin{aligned} \left(\frac{7^{-12} \cdot 7^3}{7^{-8}}\right)^{-1} &= (7^{-12+3-(-8)})^{-1} \\ &= (7^{-12+3+8})^{-1} = (7^{-1})^{-1} \\ &= 7^{(-1)(-1)} = 7^1 = 7 \end{aligned}$$

13. $\dfrac{x^4 \cdot x^3}{x^5} = \dfrac{x^{4+3}}{x^5} = \dfrac{x^7}{x^5} = x^{7-5} = x^2$

14. $\dfrac{y^{10} \cdot y^{-4}}{y^6} = y^{10-4-6} = y^0 = 1$

15. $$\begin{aligned} \frac{(4k^{-1})^2}{2k^{-5}} &= \frac{4^2k^{-2}}{2k^{-5}} = \frac{16k^{-2-(-5)}}{2} \\ &= 8k^{-2+5} = 8k^3 \\ &= 2^3k^3 \end{aligned}$$

16. $$\frac{(3z^2)^{-1}}{z^5} = \frac{3^{-1}(z^2)^{-1}}{z^5} = \frac{3^{-1}z^{2(-1)}}{z^5}$$
$$= \frac{3^{-1}z^{-2}}{z^5} = 3^{-1}z^{-2-5}$$
$$= 3^{-1}z^{-7} = \frac{1}{3} \cdot \frac{1}{z^7} = \frac{1}{3z^7}$$

17. $$\frac{3^{-1} \cdot x \cdot y^2}{x^{-4} \cdot y^5} = 3^{-1} \cdot x^{1-(-4)} \cdot y^{2-5}$$
$$= 3^{-1} \cdot x^{1+4} \cdot y^{-3}$$
$$= \frac{1}{3} \cdot x^5 \cdot \frac{1}{y^3}$$
$$= \frac{x^5}{3y^3}$$

18. $$\frac{5^{-2}m^2y^{-2}}{5^2m^{-1}y^{-2}} = \frac{5^{-2}}{5^2} \cdot \frac{m^2}{m^{-1}} \cdot \frac{y^{-2}}{y^{-2}}$$
$$= 5^{-2-2}m^{2-(-1)}y^{-2-(-2)}$$
$$= 5^{-2-2}m^{2+1}y^{-2+2}$$
$$= 5^{-4}m^3y^0 = \frac{1}{5^4} \cdot m^3 \cdot 1$$
$$= \frac{m^3}{5^4}$$

19. $$\left(\frac{a^{-1}}{b^2}\right)^{-3} = \frac{(a^{-1})^{-3}}{(b^2)^{-3}} = \frac{a^{(-1)(-3)}}{b^{2(-3)}}$$
$$= \frac{a^3}{b^{-6}} = a^3b^6$$

20. $$\left(\frac{c^3}{7d^{-1/2}}\right)^{-2} = \frac{(c^3)^{-2}}{7^{-2}(d^{-1/2})^{-2}}$$
$$= \frac{c^{(3)(-2)}}{7^{-2}d^{(-1/2)(-2)}} = \frac{c^{-6}}{7^{-2}d^1}$$
$$= \frac{7^2}{c^6d} = \frac{49}{c^6d}$$

21. $$\left(\frac{x^6y^{-3}}{x^{-2}y^5}\right)^{1/2} = (x^{6-(-2)}y^{-3-5})^{1/2}$$
$$= (x^8y^{-8})^{1/2}$$
$$= (x^8)^{1/2}(y^{-8})^{1/2}$$
$$= x^4y^{-4}$$
$$= \frac{x^4}{y^4}$$

22. $$\left(\frac{a^{-7}b^{-1}}{b^{-4}a^2}\right)^{1/3} = \left(a^{-7-2}b^{-1-(-4)}\right)^{1/3}$$
$$= \left(a^{-9}b^3\right)^{1/3}$$
$$= \left(a^{-9}\right)^{1/3}\left(b^3\right)^{1/3}$$
$$= a^{-3}b^1$$
$$= \frac{b}{a^3}$$

23. $$a^{-1} + b^{-1} = \frac{1}{a} + \frac{1}{b}$$
$$= \left(\frac{b}{b}\right)\left(\frac{1}{a}\right) + \left(\frac{a}{a}\right)\left(\frac{1}{b}\right)$$
$$= \frac{b}{ab} + \frac{a}{ab}$$
$$= \frac{b+a}{ab}$$
$$= \frac{a+b}{ab}$$

24. $$b^{-2} - a = \frac{1}{b^2} - a$$
$$= \frac{1}{b^2} - a\left(\frac{b^2}{b^2}\right)$$
$$= \frac{1}{b^2} - \frac{ab^2}{b^2}$$
$$= \frac{1-ab^2}{b^2}$$

25. $$\frac{2n^{-1} - 2m^{-1}}{m+n^2} = \frac{\frac{2}{n} - \frac{2}{m}}{m+n^2}$$
$$= \frac{\frac{2}{n} \cdot \frac{m}{m} - \frac{2}{m} \cdot \frac{n}{n}}{(m+n^2)}$$
$$= \frac{2m-2n}{mn(m+n^2)} \quad \text{or} \quad \frac{2(m-n)}{mn(m+n^2)}$$

26. $$\left(\frac{m}{3}\right)^{-1} + \left(\frac{n}{2}\right)^{-2} = \left(\frac{3}{m}\right)^1 + \left(\frac{2}{n}\right)^2$$
$$= \frac{3}{m} + \frac{4}{n^2}$$
$$= \left(\frac{3}{m}\right)\left(\frac{n^2}{n^2}\right) + \left(\frac{4}{n^2}\right)\left(\frac{m}{m}\right)$$
$$= \frac{3n^2}{mn^2} + \frac{4m}{mn^2}$$
$$= \frac{3n^2+4m}{mn^2}$$

27. $$(x^{-1} - y^{-1})^{-1} = \frac{1}{\frac{1}{x} - \frac{1}{y}}$$
$$= \frac{1}{\frac{1}{x} \cdot \frac{y}{y} - \frac{1}{y} \cdot \frac{x}{x}}$$
$$= \frac{1}{\frac{y}{xy} - \frac{x}{xy}}$$
$$= \frac{1}{\frac{y-x}{xy}}$$
$$= \frac{xy}{y-x}$$

28. $(x \cdot y^{-1} - y^{-2})^{-2} = \left(\frac{x}{y} - \frac{1}{y^2}\right)^{-2}$

$$= \left[\left(\frac{x}{y}\right)\left(\frac{y}{y}\right) - \frac{1}{y^2}\right]^{-2}$$

$$= \left(\frac{xy}{y^2} - \frac{1}{y^2}\right)^{-2}$$

$$= \left(\frac{xy - 1}{y^2}\right)^{-2}$$

$$= \left(\frac{y^2}{xy - 1}\right)^{2}$$

$$= \frac{(y^2)^2}{(xy-1)^2}$$

$$= \frac{y^4}{(xy-1)^2}$$

29. $121^{1/2} = (11^2)^{1/2} = 11^{2(1/2)} = 11^1 = 11$

30. $27^{1/3} = \sqrt[3]{27} = 3$

31. $32^{2/5} = (32^{1/5})^2 = 2^2 = 4$

32. $-125^{2/3} = -(125^{1/3})^2 = -5^2 = -25$

33. $\left(\frac{36}{144}\right)^{1/2} = \frac{36^{1/2}}{144^{1/2}} = \frac{6}{12} = \frac{1}{2}$

This can also be solved by reducing the fraction first.

$$\left(\frac{36}{144}\right)^{1/2} = \left(\frac{1}{4}\right)^{1/2} = \frac{1^{1/2}}{4^{1/2}} = \frac{1}{2}$$

34. $\left(\frac{64}{27}\right)^{1/3} = \frac{64^{1/3}}{27^{1/3}} = \frac{4}{3}$

35. $8^{-4/3} = (8^{1/3})^{-4} = 2^{-4} = \frac{1}{2^4} = \frac{1}{16}$

36. $625^{-1/4} = \frac{1}{625^{1/4}} = \frac{1}{5}$

37. $\left(\frac{27}{64}\right)^{-1/3} = \frac{27^{-1/3}}{64^{-1/3}} = \frac{64^{1/3}}{27^{1/3}} = \frac{4}{3}$

38. $\left(\frac{121}{100}\right)^{-3/2} = \frac{1}{\left(\frac{121}{100}\right)^{3/2}} = \frac{1}{\left[\left(\frac{121}{100}\right)^{1/2}\right]^3}$

$$= \frac{1}{\left(\frac{11}{10}\right)^3} = \frac{1}{\frac{1331}{1000}} = \frac{1000}{1331}$$

39. $3^{2/3} \cdot 3^{4/3} = 3^{(2/3)+(4/3)} = 3^{6/3} = 3^2 = 9$

40. $27^{2/3} \cdot 27^{-1/3} = 27^{(2/3)+(-1/3)}$

$$= 27^{2/3-1/3}$$

$$= 27^{1/3}$$

41. $\frac{4^{9/4} \cdot 4^{-7/4}}{4^{-10/4}} = 4^{9/4-7/4-(-10/4)}$

$$= 4^{12/4} = 4^3 = 64$$

42. $\frac{3^{-5/2} \cdot 3^{3/2}}{3^{7/2} \cdot 3^{-9/2}}$

$$= 3^{(-5/2)+(3/2)-(7/2)-(-9/2)}$$

$$= 3^{-5/2+3/2-7/2+9/2}$$

$$= 3^0 = 1$$

43. $\frac{7^{-1/3} \cdot 7r^{-3}}{7^{2/3} \cdot (r^{-2})^2}$

$$= \frac{7^{-1/3+1}r^{-3}}{7^{2/3} \cdot r^{-4}}$$

$$= 7^{-1/3+3/3-2/3}r^{-3-(-4)}$$

$$= 7^0 r^{-3+4} = 1 \cdot r^1 = r$$

44. $\frac{12^{3/4} \cdot 12^{5/4} \cdot y^{-2}}{12^{-1} \cdot (y^{-3})^{-2}}$

$$= \frac{12^{3/4+5/4} \cdot y^{-2}}{12^{-1} \cdot y^{(-3)(-2)}} = \frac{12^{8/4} \cdot y^{-2}}{12^{-1} \cdot y^6}$$

$$= \frac{12^2 \cdot y^{-2}}{12^{-1}y^6}$$

$$= 12^{2-(-1)} \cdot y^{-2-6} = 12^3 y^{-8}$$

$$= \frac{12^3}{y^8}$$

45. $\frac{3k^2 \cdot (4k^{-3})^{-1}}{4^{1/2} \cdot k^{7/2}}$

$$= \frac{3k^2 \cdot 4^{-1}k^3}{2 \cdot k^{7/2}}$$

$$= 3 \cdot 2^{-1} \cdot 4^{-1} k^{2+3-(7/2)}$$

$$= \frac{3}{8} \cdot k^{3/2}$$

$$= \frac{3k^{3/2}}{8}$$

46. $\frac{8p^{-3}(4p^2)^{-2}}{p^{-5}} = \frac{8p^{-3}\cdot 4^{-2}p^{(2)(-2)}}{p^{-5}}$

$= \frac{8p^{-3}4^{-2}p^{-4}}{p^{-5}}$

$= 8\cdot 4^{-2}p^{(-3)+(-4)-(-5)}$

$= 8\cdot 4^{-2}p^{-3-4+5}$

$= 8\cdot 4^{-2}p^{-2}$

$= 8\cdot \frac{1}{4^2}\cdot\frac{1}{p^2}$

$= 8\cdot \frac{1}{16}\cdot\frac{1}{p^2}$

$= \frac{8}{16p^2} = \frac{1}{2p^2}$

47. $\frac{a^{4/3}}{a^{2/3}}\cdot\frac{b^{1/2}}{b^{-3/2}} = a^{4/3-2/3}b^{1/2-(-3/2)}$

$= a^{2/3}b^2$

48. $\frac{x^{3/2}\cdot y^{4/5}\cdot z^{-3/4}}{x^{5/3}\cdot y^{-6/5}\cdot z^{1/2}}$

$= x^{3/2-(5/3)}\cdot y^{4/5-(-6/5)}\cdot z^{-3/4-(1/2)}$

$= x^{-1/6}\cdot y^2\cdot z^{-5/4}$

$= \frac{y^2}{x^{1/6}z^{5/4}}$

49. $\frac{k^{-3/5}\cdot h^{-1/3}\cdot t^{2/5}}{k^{-1/5}\cdot h^{-2/3}\cdot t^{1/5}}$

$= k^{-3/5-(-1/5)}h^{-1/3-(-2/3)}t^{2/5-1/5}$

$= k^{-3/5+1/5}h^{-1/3+2/3}t^{2/5-1/5}$

$= k^{-2/5}h^{1/3}t^{1/5}$

$= \frac{h^{1/3}t^{1/5}}{k^{2/5}}$

50. $\frac{m^{7/3}\cdot n^{-2/5}\cdot p^{3/8}}{m^{-2/3}\cdot n^{3/5}\cdot p^{-5/8}}$

$= m^{7/3-(-2/3)}n^{-2/5-(3/5)}p^{3/8-(-5/8)}$

$= m^{7/3+2/3}n^{-2/5-3/5}p^{3/8+5/8}$

$= m^{9/3}n^{-5/5}p^{8/8}$

$= m^3n^{-1}p^1$

$= \frac{m^3p}{n}$

51. $3x^3(x^2+3x)^2 - 15x(x^2+3x)^2$

$= 3x\cdot x^2(x^2+3x)^2 - 3x\cdot 5(x^2+3x)^2$

$= 3x(x^2+3x)^2(x^2-5)$

52. $6x(x^3+7)^2 - 6x^2(3x^2+5)(x^3+7)$

$= 6x(x^3+7)(x^3+7) - 6x(x)(3x^2+5)(x^3+7)$

$= 6x(x^3+7)[(x^3+7) - x(3x^2+5)]$

$= 6x(x^3+7)(x^3+7-3x^3-5x)$

$= 6x(x^3+7)(-2x^3-5x+7)$

53. $10x^3(x^2-1)^{-1/2} - 5x(x^2-1)^{1/2}$

$= 5x\cdot 2x^2(x^2-1)^{-1/2} - 5x(x^2-1)^{-1/2}(x^2-1)^1$

$= 5x(x^2-1)^{-1/2}[2x^2-(x^2-1)]$

$= 5x(x^2-1)^{-1/2}(x^2+1)$

54. $9(6x+2)^{1/2} + 3(9x-1)(6x+2)^{-1/2}$

$= 3\cdot 3(6x+2)^{-1/2}(6x+2)^1$

$+ 3(9x-1)(6x+2)^{-1/2}$

$= 3(6x+2)^{-1/2}[3(6x+2)+(9x-1)]$

$= 3(6x+2)^{-1/2}(18x+6+9x-1)$

$= 3(6x+2)^{-1/2}(27x+5)$

55. $x(2x+5)^2(x^2-4)^{-1/2} + 2(x^2-4)^{1/2}(2x+5)$

$= (2x+5)^2(x^2-4)^{-1/2}(x)$

$+ (x^2-4)^1(x^2-4)^{-1/2}(2)(2x+5)$

$= (2x+5)(x^2-4)^{-1/2}$

$\cdot [(2x+5)(x)+(x^2-4)(2)]$

$= (2x+5)(x^2-4)^{-1/2}$

$\cdot (2x^2+5x+2x^2-8)$

$= (2x+5)(x^2-4)^{-1/2}(4x^2+5x-8)$

56. $(4x^2+1)^2(2x-1)^{-1/2} + 16x(4x^2+1)(2x-1)^{1/2}$

$= (4x^2+1)(4x^2+1)(2x-1)^{-1/2}$

$+ 16x(4x^2+1)(2x-1)^{-1/2}(2x-1)$

$= (4x^2+1)(2x-1)^{-1/2}$

$\cdot [(4x^2+1)+16x(2x-1)]$

$= (4x^2+1)(2x-1)^{-1/2}(4x^2+1+32x^2-16x)$

$= (4x^2+1)(2x-1)^{-1/2}(36x^2-16x+1)$

R.7 Radicals

1. $\sqrt[3]{125} = 5$ because $5^3 = 125$.

2. $\sqrt[4]{1296} = \sqrt[4]{6^4} = 6$

3. $\sqrt[5]{-3125} = -5$ because $(-5)^5 = -3125$.

4. $\sqrt{50} = \sqrt{25 \cdot 2} = \sqrt{25}\sqrt{2} = 5\sqrt{2}$

5. $\sqrt{2000} = \sqrt{4 \cdot 100 \cdot 5}$
$= 2 \cdot 10\sqrt{5}$
$= 20\sqrt{5}$

6. $\sqrt{32y^5} = \sqrt{(16y^4)(2y)}$
$= \sqrt{16y^4}\sqrt{2y}$
$= 4y^2\sqrt{2y}$

7. $\sqrt{27} \cdot \sqrt{3} = \sqrt{27 \cdot 3} = \sqrt{81} = 9$

8. $\sqrt{2} \cdot \sqrt{32} = \sqrt{2 \cdot 32} = \sqrt{64} = 8$

9. $7\sqrt{2} - 8\sqrt{18} + 4\sqrt{72}$
$= 7\sqrt{2} - 8\sqrt{9 \cdot 2} + 4\sqrt{36 \cdot 2}$
$= 7\sqrt{2} - 8(3)\sqrt{2} + 4(6)\sqrt{2}$
$= 7\sqrt{2} - 24\sqrt{2} + 24\sqrt{2}$
$= 7\sqrt{2}$

10. $4\sqrt{3} - 5\sqrt{12} + 3\sqrt{75}$
$= 4\sqrt{3} - 5(\sqrt{4}\sqrt{3}) + 3(\sqrt{25}\sqrt{3})$
$= 4\sqrt{3} - 5(2\sqrt{3}) + 3(5\sqrt{3})$
$= 4\sqrt{3} - 10\sqrt{3} + 15\sqrt{3}$
$= (4 - 10 + 15)\sqrt{3} = 9\sqrt{3}$

11. $4\sqrt{7} - \sqrt{28} + \sqrt{343}$
$= 4\sqrt{7} - \sqrt{4}\sqrt{7} + \sqrt{49}\sqrt{7}$
$= 4\sqrt{7} - 2\sqrt{7} + 7\sqrt{7}$
$= (4 - 2 + 7)\sqrt{7}$
$= 9\sqrt{7}$

12. $3\sqrt{28} - 4\sqrt{63} + \sqrt{112}$
$= 3(\sqrt{4}\sqrt{7}) - 4(\sqrt{9}\sqrt{7}) + (\sqrt{16}\sqrt{7})$
$= 3(2\sqrt{7}) - 4(3\sqrt{7}) + (4\sqrt{7})$
$= 6\sqrt{7} - 12\sqrt{7} + 4\sqrt{7}$
$= (6 - 12 + 4)\sqrt{7}$
$= -2\sqrt{7}$

13. $\sqrt[3]{2} - \sqrt[3]{16} + 2\sqrt[3]{54}$
$= \sqrt[3]{2} - (\sqrt[3]{8 \cdot 2}) + 2(\sqrt[3]{27 \cdot 2})$
$= \sqrt[3]{2} - \sqrt[3]{8}\sqrt[3]{2} + 2(\sqrt[3]{27}\sqrt[3]{2})$
$= \sqrt[3]{2} - 2\sqrt[3]{2} + 2(3\sqrt[3]{2})$
$= \sqrt[3]{2} - 2\sqrt[3]{2} + 6\sqrt[3]{2}$
$= 5\sqrt[3]{2}$

14. $2\sqrt[3]{5} - 4\sqrt[3]{40} + 3\sqrt[3]{135}$
$= 2\sqrt[3]{5} - 4\sqrt[3]{8 \cdot 5} + 3\sqrt[3]{27 \cdot 5}$
$= 2\sqrt[3]{5} - 4(2)\sqrt[3]{5} + 3(3)\sqrt[3]{5}$
$= 2\sqrt[3]{5} - 8\sqrt[3]{5} + 9\sqrt[3]{5}$
$= 3\sqrt[3]{5}$

15. $\sqrt{2x^3y^2z^4} = \sqrt{x^2y^2z^4 \cdot 2x}$
$= xyz^2\sqrt{2x}$

16. $\sqrt{160r^7s^9t^{12}}$
$= \sqrt{(16 \cdot 10)(r^6 \cdot r)(s^8 \cdot s)(t^{12})}$
$= \sqrt{(16r^6s^8t^{12})(10rs)}$
$= \sqrt{16r^6s^8t^{12}}\sqrt{10rs}$
$= 4r^3s^4t^6\sqrt{10rs}$

17. $\sqrt[3]{128x^3y^8z^9} = \sqrt[3]{64x^3y^6z^9 \cdot 2y^2}$
$= \sqrt[3]{64x^3y^6z^9}\sqrt[3]{2y^2}$
$= 4xy^2z^3\sqrt[3]{2y^2}$

18. $\sqrt[4]{x^8y^7z^{11}} = \sqrt[4]{(x^8)(y^4 \cdot y^3)(z^8z^3)}$
$= \sqrt[4]{(x^8y^4z^8)(y^3z^3)}$
$= \sqrt[4]{x^8y^4z^8}\sqrt[4]{y^3z^3}$
$= x^2yz^2\sqrt[4]{y^3z^3}$

19. $\sqrt{a^3b^5} - 2\sqrt{a^7b^3} + \sqrt{a^3b^9}$
$= \sqrt{a^2b^4ab} - 2\sqrt{a^6b^2ab} + \sqrt{a^2b^8ab}$
$= ab^2\sqrt{ab} - 2a^3b\sqrt{ab} + ab^4\sqrt{ab}$
$= (ab^2 - 2a^3b + ab^4)\sqrt{ab}$
$= ab\sqrt{ab}(b - 2a^2 + b^3)$

20. $\sqrt{p^7q^3} - \sqrt{p^5q^9} + \sqrt{p^9q}$
$= \sqrt{(p^6p)(q^2q)} - \sqrt{(p^4p)(q^8q)}$
$+ \sqrt{(p^8p)q}$
$= \sqrt{(p^6q^2)(pq)} - \sqrt{(p^4q^8)(pq)}$
$+ \sqrt{(p^8)(pq)}$
$= \sqrt{p^6q^2}\sqrt{pq} - \sqrt{p^4q^8}\sqrt{pq} + \sqrt{p^8}\sqrt{pq}$
$= p^3q\sqrt{pq} - p^2q^4\sqrt{pq} + p^4\sqrt{pq}$
$= p^2pq\sqrt{pq} - p^2q^4\sqrt{pq} + p^2p^2\sqrt{pq}$
$= p^2\sqrt{pq}(pq - q^4 + p^2)$

21. $\sqrt{a} \cdot \sqrt[3]{a} = a^{1/2} \cdot a^{1/3} = a^{1/2+(1/3)} = a^{5/6} = \sqrt[6]{a^5}$

22. $\sqrt{b^3} \cdot \sqrt[4]{b^3} = b^{3/2} \cdot b^{3/4}$
$= b^{3/2+(3/4)} = b^{9/4}$
$= \sqrt[4]{b^9} = \sqrt[4]{b^8 \cdot b}$
$= \sqrt[4]{b^8}\sqrt[4]{b} = b^2\sqrt[4]{b}$

23. $\dfrac{5}{\sqrt{7}} = \dfrac{5}{\sqrt{7}} \cdot \dfrac{\sqrt{7}}{\sqrt{7}} = \dfrac{5\sqrt{7}}{7}$

24. $\dfrac{5}{\sqrt{10}} = \dfrac{5}{\sqrt{10}} \cdot \dfrac{\sqrt{10}}{\sqrt{10}} = \dfrac{5\sqrt{10}}{\sqrt{100}} = \dfrac{5\sqrt{10}}{10} = \dfrac{\sqrt{10}}{2}$

25. $\dfrac{-3}{\sqrt{12}} = \dfrac{-3}{\sqrt{4 \cdot 3}}$

$= \dfrac{-3}{2\sqrt{3}} \cdot \dfrac{\sqrt{3}}{\sqrt{3}}$

$= \dfrac{-3\sqrt{3}}{6}$

$= -\dfrac{\sqrt{3}}{2}$

26. $\dfrac{4}{\sqrt{8}} = \dfrac{4}{\sqrt{8}} \cdot \dfrac{\sqrt{2}}{\sqrt{2}} = \dfrac{4\sqrt{2}}{\sqrt{16}} = \dfrac{4\sqrt{2}}{4} = \sqrt{2}$

27. $\dfrac{3}{1-\sqrt{2}} = \dfrac{3}{1-\sqrt{2}} \cdot \dfrac{1+\sqrt{2}}{1+\sqrt{2}}$

$= \dfrac{3(1+\sqrt{2})}{1-2}$

$= \dfrac{-3(1+\sqrt{2})}{4}$

28. $\dfrac{5}{2-\sqrt{6}} = \dfrac{5}{2-\sqrt{6}} \cdot \dfrac{2+\sqrt{6}}{2+\sqrt{6}}$

$= \dfrac{5(2+\sqrt{6})}{4+2\sqrt{6}-2\sqrt{6}-\sqrt{36}}$

$= \dfrac{5(2+\sqrt{6})}{4-\sqrt{36}}$

$= \dfrac{5(2+\sqrt{6})}{4-6}$

$= \dfrac{5(2+\sqrt{6})}{-2}$

$= -\dfrac{5(2+\sqrt{6})}{2}$

29. $\dfrac{6}{2+\sqrt{2}} = \dfrac{6}{2+\sqrt{2}} \cdot \dfrac{2-\sqrt{2}}{2-\sqrt{2}}$

$= \dfrac{6(2-\sqrt{2})}{4-2\sqrt{2}+2\sqrt{2}-\sqrt{4}}$

$= \dfrac{6(2-\sqrt{2})}{4-2}$

$= \dfrac{6(2-\sqrt{2})}{2}$

$= 3(2-\sqrt{2})$

30. $\dfrac{\sqrt{5}}{\sqrt{5}+\sqrt{2}} = \dfrac{\sqrt{5}}{\sqrt{5}+\sqrt{2}} \cdot \dfrac{\sqrt{5}-\sqrt{2}}{\sqrt{5}-\sqrt{2}}$

$= \dfrac{\sqrt{5}(\sqrt{5}-\sqrt{2})}{\sqrt{25}-\sqrt{10}+\sqrt{10}-\sqrt{4}}$

$= \dfrac{5-\sqrt{10}}{5-2}$

$= \dfrac{5-\sqrt{10}}{3}$

31. $\dfrac{1}{\sqrt{r}-\sqrt{3}} = \dfrac{1}{\sqrt{r}-\sqrt{3}} \cdot \dfrac{\sqrt{r}+\sqrt{3}}{\sqrt{r}+\sqrt{3}}$

$= \dfrac{\sqrt{r}+\sqrt{3}}{r-3}$

32. $\dfrac{5}{\sqrt{m}-\sqrt{5}} = \dfrac{5}{\sqrt{m}-\sqrt{5}} \cdot \dfrac{\sqrt{m}+\sqrt{5}}{\sqrt{m}+\sqrt{5}}$

$= \dfrac{5(\sqrt{m}+\sqrt{5})}{\sqrt{m^2}+\sqrt{5m}-\sqrt{5m}-\sqrt{25}}$

$= \dfrac{5(\sqrt{m}+\sqrt{5})}{\sqrt{m^2}-\sqrt{25}} = \dfrac{5(\sqrt{m}+\sqrt{5})}{m-5}$

33. $\dfrac{y-5}{\sqrt{y}-\sqrt{5}} = \dfrac{y-5}{\sqrt{y}-\sqrt{5}} \cdot \dfrac{\sqrt{y}+\sqrt{5}}{\sqrt{y}+\sqrt{5}}$

$= \dfrac{(y-5)(\sqrt{y}+\sqrt{5})}{y-5}$

$= \sqrt{y}+\sqrt{5}$

34. $\dfrac{\sqrt{z}-1}{\sqrt{z}-\sqrt{5}} = \dfrac{\sqrt{z}-1}{\sqrt{z}-\sqrt{5}} \cdot \dfrac{\sqrt{z}+\sqrt{5}}{\sqrt{z}+\sqrt{5}}$

$= \dfrac{\sqrt{z^2}+\sqrt{5z}-\sqrt{z}-\sqrt{5}}{\sqrt{z^2}+\sqrt{5z}-\sqrt{5z}-\sqrt{25}}$

$= \dfrac{z+\sqrt{5z}-\sqrt{z}-\sqrt{5}}{z-5}$

35. $\dfrac{\sqrt{x}+\sqrt{x+1}}{\sqrt{x}-\sqrt{x+1}} = \dfrac{\sqrt{x}+\sqrt{x+1}}{\sqrt{x}-\sqrt{x+1}} \cdot \dfrac{\sqrt{x}+\sqrt{x+1}}{\sqrt{x}+\sqrt{x+1}}$

$= \dfrac{x+2\sqrt{x(x+1)}+(x+1)}{x-(x+1)}$

$= \dfrac{2x+2\sqrt{x(x+1)}+1}{-1}$

$= -2x-2\sqrt{x(x+1)}-1$

36. $\dfrac{\sqrt{p}+\sqrt{p^2-1}}{\sqrt{p}-\sqrt{p^2-1}}$

$= \dfrac{\sqrt{p}+\sqrt{p^2-1}}{\sqrt{p}-\sqrt{p^2-1}} \cdot \dfrac{\sqrt{p}+\sqrt{p^2-1}}{\sqrt{p}+\sqrt{p^2-1}}$

$= \dfrac{(\sqrt{p})^2+2\sqrt{p}\sqrt{p^2-1}+(\sqrt{p^2-1}))^2}{\sqrt{p^2}+\sqrt{p}\sqrt{p^2-1}-\sqrt{p}\sqrt{p^2-1}-(\sqrt{p^2-1})^2}$

$= \dfrac{p+2\sqrt{p}\sqrt{p^2-1}+(p^2-1)}{p-(p^2-1)}$

$= \dfrac{p^2+p+2\sqrt{p(p^2-1)}-1}{-p^2+p+1}$

37. $$\frac{1+\sqrt{2}}{2} = \frac{(1+\sqrt{2})(1-\sqrt{2})}{2(1-\sqrt{2})} = \frac{1-2}{2(1-\sqrt{2})} = -\frac{1}{2(1-\sqrt{2})}$$

38. $$\frac{3-\sqrt{3}}{6} = \frac{3-\sqrt{3}}{6}\cdot\frac{3+\sqrt{3}}{3+\sqrt{3}} = \frac{9+3\sqrt{3}-3\sqrt{3}-\sqrt{9}}{6(3+\sqrt{3})} = \frac{9-3}{6(3+\sqrt{3})} = \frac{6}{6(3+\sqrt{3})} = \frac{1}{3+\sqrt{3}}$$

39. $$\frac{\sqrt{x}+\sqrt{x+1}}{\sqrt{x}-\sqrt{x+1}} = \frac{\sqrt{x}+\sqrt{x+1}}{\sqrt{x}-\sqrt{x+1}}\cdot\frac{\sqrt{x}-\sqrt{x+1}}{\sqrt{x}-\sqrt{x+1}} = \frac{x-(x+1)}{x-2\sqrt{x}\cdot\sqrt{x+1}+(x+1)} = \frac{-1}{2x-2\sqrt{x(x+1)}+1}$$

40. $$\frac{\sqrt{p}-\sqrt{p-2}}{\sqrt{p}} = \frac{\sqrt{p}-\sqrt{p-2}}{\sqrt{p}}\cdot\frac{\sqrt{p}+\sqrt{p-2}}{\sqrt{p}+\sqrt{p-2}} = \frac{\sqrt{p^2}+\sqrt{p}\sqrt{p-2}-\sqrt{p}\sqrt{p-2}-\sqrt{(p-2)^2}}{\sqrt{p^2}+\sqrt{p}\sqrt{p-2}} = \frac{p-(p-2)}{p+\sqrt{p(p-2)}} = \frac{2}{p+\sqrt{p(p-2)}}$$

41. $$\sqrt{16-8x+x^2} = \sqrt{(4-x)^2} = |4-x|$$

Since $\sqrt{\ }$ denotes the nonnegative root, we must have $4-x\geq 0$.

42. $\sqrt{9y^2+30y+25} = \sqrt{(3y+5)^2} = |3y+5|$

Since $\sqrt{\ }$ denotes the nonnegative root, we must have $3y+5\geq 0$.

43. $\sqrt{4-25z^2} = \sqrt{(2+5z)(2-5z)}$

This factorization does not produce a perfect square, so the expression $\sqrt{4-25z^2}$ cannot be simplified.

44. $\sqrt{9k^2+h^2}$

The expression $9k^2+h^2$ is the sum of two squares and cannot be factored. Therefore, $\sqrt{9k^2+h^2}$ cannot be simplified.

Chapter 1

LINEAR FUNCTIONS

1.1 Slopes and Equations of Lines

1. Find the slope of the line through $(4, 5)$ and $(-1, 2)$.

$$\begin{aligned} m &= \frac{5-2}{4-(-1)} \\ &= \frac{3}{5} \end{aligned}$$

3. Find the slope of the line through $(8, 4)$ and $(8, -7)$.

$$\begin{aligned} m &= \frac{4-(-7)}{8-8} \\ &= \frac{11}{0} \end{aligned}$$

The slope is undefined; the line is vertical.

5. $y = x$

Using the slope-intercept form, $y = mx + b$, we see that the slope is 1.

7. $5x - 9y = 11$

Rewrite the equation in slope-intercept form.

$$\begin{aligned} 9y &= 5x - 11 \\ y &= \frac{5}{9}x - \frac{11}{9} \end{aligned}$$

The slope is $\frac{5}{9}$.

9. $x = 5$

This is a vertical line. The slope is undefined.

11. $y = 8$

This is a horizontal line, which has a slope of 0.

13. Find the slope of a line parallel to $6x - 3y = 12$.

Rewrite the equation in slope-intercept form.

$$\begin{aligned} -3y &= -6x + 12 \\ y &= 2x - 4 \end{aligned}$$

The slope is 2, so a parallel line will also have slope 2.

15. The line goes through $(1, 3)$, with slope $m = -2$. Use point-slope form.

$$\begin{aligned} y - 3 &= -2(x - 1) \\ y &= -2x + 2 + 3 \\ y &= -2x + 5 \end{aligned}$$

17. The line goes through $(-5, -7)$ with slope $m = 0$. Use point-slope form.

$$\begin{aligned} y - (-7) &= 0[x - (-5)] \\ y + 7 &= 0 \\ y &= -7 \end{aligned}$$

19. The line goes through $(4, 2)$ and $(1, 3)$. Find the slope, then use point-slope form with either of the two given points.

$$\begin{aligned} m &= \frac{3-2}{1-4} \\ &= -\frac{1}{3} \end{aligned}$$

$$\begin{aligned} y - 3 &= -\frac{1}{3}(x - 1) \\ y &= -\frac{1}{3}x + \frac{1}{3} + 3 \\ y &= -\frac{1}{3}x + \frac{10}{3} \end{aligned}$$

21. The line goes through $\left(\frac{2}{3}, \frac{1}{2}\right)$ and $\left(\frac{1}{4}, -2\right)$.

$$m = \frac{-2 - \frac{1}{2}}{\frac{1}{4} - \frac{2}{3}} = \frac{-\frac{4}{2} - \frac{1}{2}}{\frac{3}{12} - \frac{8}{12}}$$

$$m = \frac{-\frac{5}{2}}{-\frac{5}{12}} = \frac{60}{10} = 6$$

$$\begin{aligned} y - (-2) &= 6\left(x - \frac{1}{4}\right) \\ y + 2 &= 6x - \frac{3}{2} \\ y &= 6x - \frac{3}{2} - 2 \\ y &= 6x - \frac{3}{2} - \frac{4}{2} \\ y &= 6x - \frac{7}{2} \end{aligned}$$

23. The line goes through $(-8, 4)$ and $(-8, 6)$.

$$m = \frac{4-6}{-8-(-8)} = \frac{-2}{0};$$

which is undefined.
This is a vertical line; the value of x is always -8.
The equation of this line is $x = -8$.

25. The line has x-intercept -6 and y-intercept -3. Two points on the line are $(-6, 0)$ and $(0, -3)$. Find the slope; then use slope-intercept form.

$$m = \frac{-3-0}{0-(-6)} = \frac{-3}{6} = -\frac{1}{2}$$
$$b = -3$$
$$y = -\frac{1}{2}x - 3$$

27. The vertical line through $(-6, 5)$ goes through the point $(-6, 0)$, so the equation is $x = -6$.

29. Write an equation of the line through $(-4, 6)$, parallel to $3x + 2y = 13$.
Rewrite the equation of the given line in slope-intercept form.

$$3x + 2y = 13$$
$$2y = -3x + 13$$
$$y = -\frac{3}{2}x + \frac{13}{2}$$

The slope is $-\frac{3}{2}$.
Use $m = -\frac{3}{2}$ and the point $(-4, 6)$ in the point-slope form.

$$y - 6 = -\frac{3}{2}[x - (-4)]$$
$$y = -\frac{3}{2}(x + 4) + 6$$
$$y = -\frac{3}{2}x - 6 + 6$$
$$y = -\frac{3}{2}x$$

31. Write an equation of the line through $(3, -4)$, perpendicular to $x + y = 4$.
Rewrite the equation of the given line as

$$y = -x + 4.$$

The slope of this line is -1. To find the slope of a perpendicular line, solve

$$-1m = -1.$$
$$m = 1$$

Use $m = 1$ and $(3, -4)$ in the point-slope form.

$$y - (-4) = 1(x - 3)$$
$$y = x - 3 - 4$$
$$y = x - 7$$

33. Write an equation of the line with y-intercept 4, perpendicular to $x + 5y = 7$.
Find the slope of the given line.

$$x + 5y = 7$$
$$5y = -x + 7$$
$$y = -\frac{1}{5}x + \frac{7}{5}$$

The slope is $-\frac{1}{5}$, so the slope of the perpendicular line will be 5. If the y-intercept is 4, then using the slope-intercept form we have

$$y = mx + b$$
$$y = 5x + 4.$$

35. Do the points $(4, 3), (2, 0)$, and $(-18, -12)$ lie on the same line?
Find the slope between $(4, 3)$ and $(2, 0)$.

$$m = \frac{0-3}{2-4} = \frac{-3}{-2} = \frac{3}{2}$$

Find the slope between $(4, 3)$ and $(-18, -12)$.

$$m = \frac{-12-3}{-18-4} = \frac{-15}{-22} = \frac{15}{22}$$

Since these slopes are not the same, the points do not lie on the same line.

37. A parallelogram has 4 sides, with opposite sides parallel. The slope of the line through $(1, 3)$ and $(2, 1)$ is

$$m = \frac{3-1}{1-2} = \frac{2}{-1} = -2.$$

The slope of the line through $(-\frac{5}{2}, 2)$ and $(-\frac{7}{2}, 4)$ is

$$m = \frac{2-4}{-\frac{5}{2} - (-\frac{7}{2})} = \frac{-2}{1} = -2.$$

Since these slopes are equal, these two sides are parallel.
The slope of the line through $(-\frac{7}{2}, 4)$ and $(1, 3)$ is

$$m = \frac{4-3}{-\frac{7}{2} - 1} = \frac{1}{-\frac{9}{2}} = -\frac{2}{9}.$$

Slope of the line through $(-\frac{5}{2}, 2)$ and $(2, 1)$ is

$$m = \frac{2-1}{-\frac{5}{2} - 2} = \frac{1}{-\frac{9}{2}} = -\frac{2}{9}.$$

Since these slopes are equal, these two sides are parallel.
Since both pairs of opposite sides are parallel, the quadrilateral is a parallelogram.

39. The line goes through $(0, 2)$ and $(-2, 0)$

$$m = \frac{2-0}{0-(-2)} = \frac{2}{2} = 1$$

The correct choice is (a).

41. The line appears to go through $(0, 0)$ and $(-1, 4)$.

$$m = \frac{4-0}{-1-0} = \frac{4}{-1} = -4$$

43. **(a)** See the figure in the textbook.
Segment MN is drawn perpendicular to segment PQ. Recall that MQ is the length of segment MQ.

$$m_1 = \frac{\Delta y}{\Delta x} = \frac{MQ}{PQ}$$

From the diagram, we know that $PQ = 1$. Thus, $m_1 = \frac{MQ}{1}$, so MQ has length m_1.

(b) $m_2 = \frac{\Delta y}{\Delta x} = \frac{-QN}{PQ} = \frac{-QN}{1}$

$QN = -m_2$

(c) Triangles MPQ, PNQ, and MNP are right triangles by construction. In triangles MPQ and MNP,

$$\text{angle } M = \text{angle } M,$$

and in the right triangles PNQ and MNP,

$$\text{angle } N = \text{angle } N.$$

Since all right angles are equal, and since triangles with two equal angles are similar, triangle MPQ is similar to triangle MNP and triangle PNQ is similar to triangle MNP.

Therefore, triangles MPQ and PNQ are similar to each other.

(d) Since corresponding sides in similar triangles are proportional,

$$MQ = k \cdot PQ \quad \text{and} \quad PQ = k \cdot QN.$$

$$\frac{MQ}{PQ} = \frac{k \cdot PQ}{k \cdot QN}$$

$$\frac{MQ}{PQ} = \frac{PQ}{QN}$$

From the diagram, we know that $PQ = 1$.

$$MQ = \frac{1}{QN}$$

From (a) and (b), $m_1 = MQ$ and $-m_2 = QN$.

Substituting, we get

$$m_1 = \frac{1}{-m_2}.$$

Multiplying both sides by m_2, we have

$$m_1 m_2 = -1.$$

45. $y = 4x + 5$

Three ordered pairs that satisfy this equation are $(-2, -3)$, $(-1, 1)$, and $(0, 5)$. Plot these points and draw a line through them.

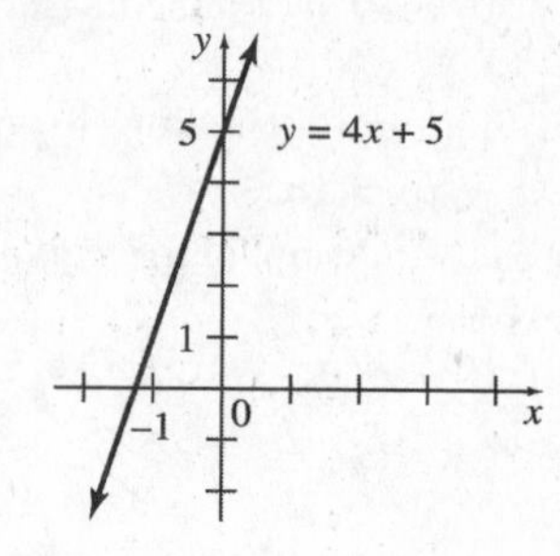

47. $y = -6x + 12$

Three ordered pairs that satisfy this equation are $(0, 12)$, $(1, 6)$, and $(2, 0)$. Plot these points and draw a line through them.

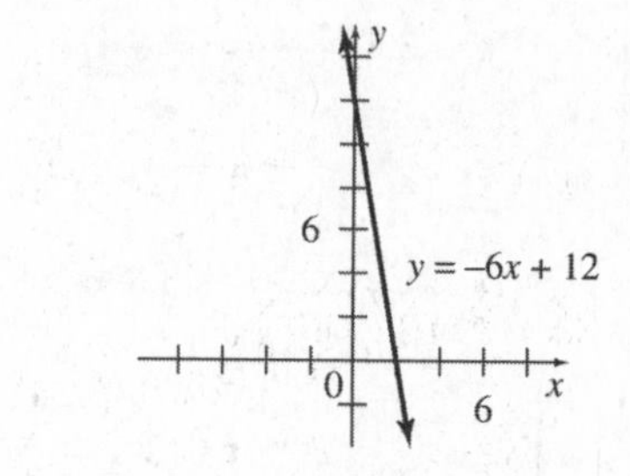

49. $3x - y = -9$

Find the intercepts.
If $y = 0$, then

$$3x - 0 = -9$$
$$3x = -9$$
$$x = -3$$

If $x = 0$, then

$$3(0) - y = -9$$
$$-y = -9$$
$$y = 9$$

so the y-intercept is 9.

Plot the ordered pairs $(-3, 0)$ and $(0, 9)$ and draw a line through these points. (A third point may be used as a check.)

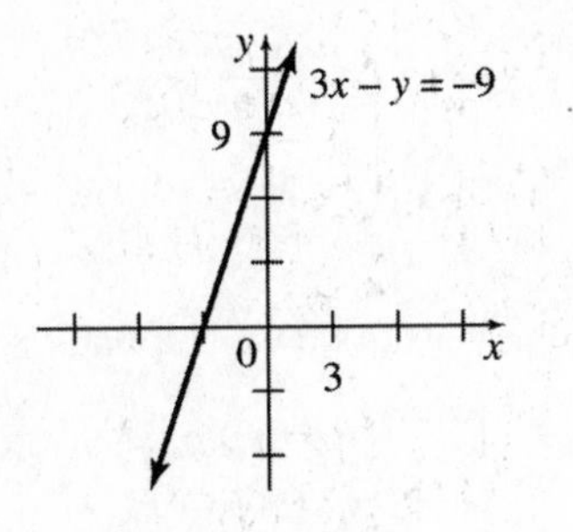

51. $5y + 6x = 11$

Find the intercepts.
If $y = 0$, then

$$\begin{aligned} 5(0) + 6x &= 11 \\ 6x &= 11 \\ x &= \frac{11}{6} \end{aligned}$$

so the x-intercept is $\frac{11}{6}$.
If $x = 0$, then

$$\begin{aligned} 5y + 6(0) &= 11 \\ 5y &= 11 \\ y &= \frac{11}{5} \end{aligned}$$

so the y-intercept is $\frac{11}{5}$.

Plot the ordered pairs $\left(\frac{11}{6}, 0\right)$ and $\left(0, \frac{11}{5}\right)$ and draw a line through these points. (A third point may be used as a check.)

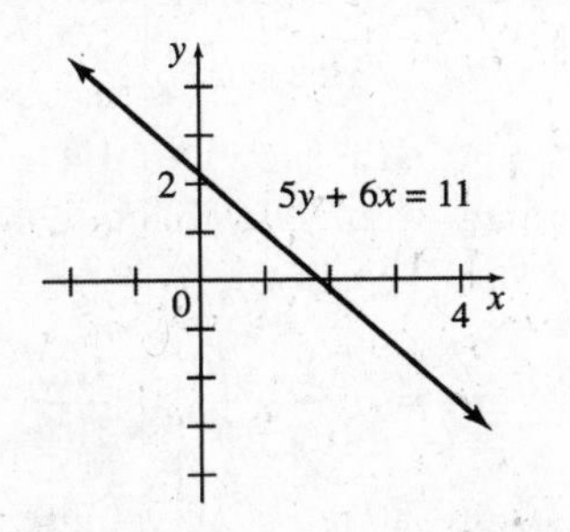

53. $x = 4$

For any value of y, the x-value is 4. Because all ordered pairs that satisfy this equation have the same first number, this equation does not represent a function. The graph is the vertical line with x-intercept 4.

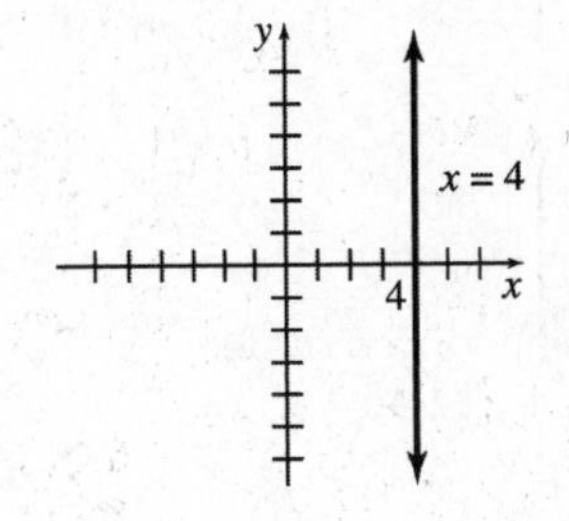

55. $y + 8 = 0$

This equation may be rewritten as $y = -8$, or, equivalently, $y = 0x + -8$. The y-value is -8 for any value of x. The graph is the horizontal line with y-intercept -8.

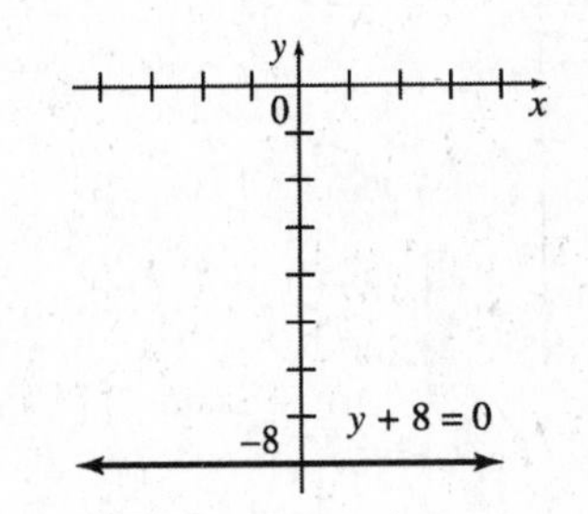

57. $y = -5x$

Three ordered pairs that satisfy this equation are $(0, 0)$, $(-1, 5)$, and $(1, -5)$. Use these points to draw the graph.

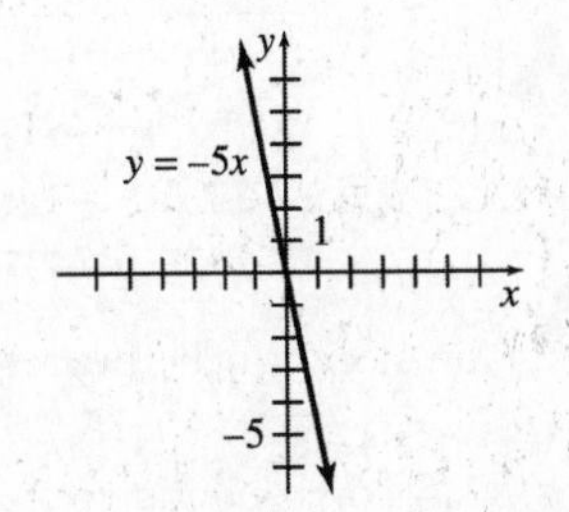

59. $3x - 5y = 0$

If $y = 0$, then $x = 0$, so the x-intercept is 0. If $x = 0$, then $y = 0$, so the y-intercept is 0. Both intercepts give the same ordered pair $(0, 0)$.
To get a second point, choose some other value of x (or y). For example, if $x = 5$, then

$$\begin{aligned} 3x - 5y &= 0 \\ 3(5) - 5y &= 0 \\ 15 - 5y &= 0 \\ -5y &= -15 \\ y &= 3 \end{aligned}$$

giving the ordered pair $(5, 3)$. Graph the line through $(0, 0)$ and $(5, 3)$.

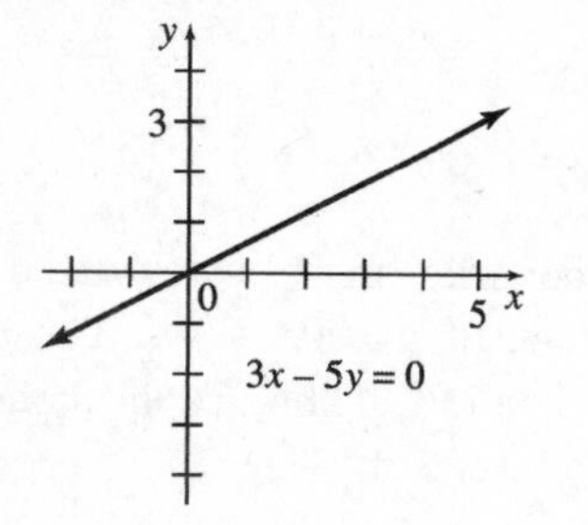

61. **(a)**

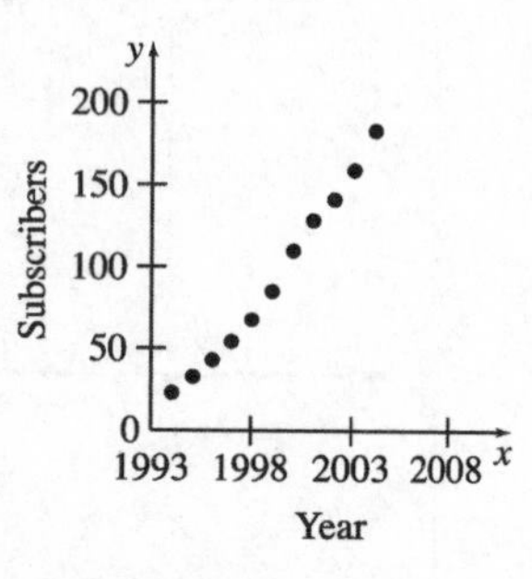

The number of subscribers is increasing and the data appear to be nearly linear.

(b) Find the slope using $(3, 44.04)$ and $(11, 182.14)$.

(c) $$m = \frac{182.14 - 44.04}{11 - 3} = \frac{138.1}{8} = 17.2625$$

$$\begin{aligned} y - 182.14 &= 17.2625(x - 11) \\ y - 182.14 &= 17.2625x - 189.89 \\ y &= 17.2625x - 7.75 \end{aligned}$$

Rounding the slope to the nearest hundredth, the equation is $y = 17.26x - 7.75$.
The year 2005 corresponds to $x = 2005 - 1993 = 12$.

$$\begin{aligned} y &= 17.26(12) - 7.75 \\ y &= 199.37 \end{aligned}$$

The approximation using the equation is less than the actual number of subscribers.

63. **(a)** The line goes through $(0, 100)$ and $(24, 201.6)$.

$$\begin{aligned} m &= \frac{201.6 - 100}{24 - 0} \approx 4.23 \\ b &= 100 \\ y &= 4.23x + 100 \end{aligned}$$

(b) The year 2000 corresponds to $x = 18$.

$$\begin{aligned} y &= 4.23(18) + 100 \\ y &= 176.14 \end{aligned}$$

The estimate is more than, but close to, the actual CPI.

(c) It is increasing at a rate of 4.23 per year.

65. **(a)** Let $x =$ age.

$$\begin{aligned} u &= 0.85(220 - x) = 187 - 0.85x \\ l &= 0.7(220 - x) = 154 - 0.7x \end{aligned}$$

(b) $$\begin{aligned} u &= 187 - 0.85(20) = 170 \\ l &= 154 - 0.7(20) = 140 \end{aligned}$$

The target heart rate zone is 140 to 170 beats per minute.

(c) $$\begin{aligned} u &= 187 - 0.85(40) = 153 \\ l &= 154 - 0.7(40) = 126 \end{aligned}$$

The target heart rate zone is 126 to 153 beats per minute.

(d) $$\begin{aligned} 154 - 0.7x &= 187 - 0.85(x + 36) \\ 154 - 0.7x &= 187 - 0.85x - 30.6 \\ 154 - 0.7x &= 156.4 - 0.85x \\ 0.15x &= 2.4 \\ x &= 16 \end{aligned}$$

The younger woman is 16; the older woman is $16 + 36 = 52$. $l = 0.7(220 - 16) \approx 143$ beats per minute.

67. Let $x = 0$ correspond to 1900. Then the "life expectancy from birth" line contains the points $(0, 46)$ and $(104, 77.8)$.

$$m = \frac{77.8 - 46}{104 - 0} = \frac{31.3}{102} = 0.306$$

Since $(0, 46)$ is one of the points, the line is given by the equation

$$y = 0.306x + 46.$$

The "life expectancy from age 65" line contains the points $(0, 76)$ and $(104, 83.7)$.

$$m = \frac{83.7 - 76}{104 - 0} = \frac{7.7}{104} \approx 0.074$$

Since $(0, 76)$ is one of the points, the line is given by the equation

$$y = 0.074x + 76.$$

Set the two equations equal to determine where the lines intersect. At this point, life expectancy should increase no further.

$$\begin{aligned} 0.306x + 46 &= 0.074x + 76 \\ 0.232x &= 30 \\ x &\approx 129 \end{aligned}$$

Determine the y-value when $x = 129$. Use the first equation.

$$\begin{aligned} y &= 0.306(129) + 46 \\ &= 39.474 + 46 \\ &= 85.474 \end{aligned}$$

Thus, the maximum life expectancy for humans is about 86 years.

69. (a) $$m = \frac{27.4 - 22.8}{45 - 5} = \frac{4.6}{40} = 0.115$$

$$\begin{aligned} y - 22.8 &= 0.115(x - 5) \\ y - 22.8 &= 0.115x - 0.575 \\ y &= 0.115x + 22.2 \end{aligned}$$

(b) $$m = \frac{25.8 - 20.6}{45 - 5} = \frac{5.2}{40} = 0.13$$

$$\begin{aligned} y - 20.6 &= 0.13(x - 5) \\ y - 20.6 &= 0.13x - 0.65 \\ y &= 0.13x + 19.95 \end{aligned}$$

(c) Since $0.13 > 0.115$, women have the faster increase.

(d) Let $y = 30$ and use the equation from part (a) to solve for x.

$$\begin{aligned} 30 &= 0.115x + 22.2 \\ 7.8 &= 0.115x \\ 68 &\approx x \end{aligned}$$

68 years after 1960, or in the year 2028, men's median age at first marriage will reach 30.

(e) Let $x = 68$ and use the equation from part (b) to find y.

$$\begin{aligned} y &= 0.13(68) + 19.95 \\ y &= 8.84 + 19.95 \\ y &= 28.79 \end{aligned}$$

The median age for women at first marriage will be about 28.8 years.

71. (a) The line goes through $(0, 1.59)$ and $(24, 5.08)$.

$$m = \frac{5.08 - 1.59}{24 - 0} = \frac{3.49}{24} \approx 0.145$$

$$\begin{aligned} b &= 1.59 \\ y &= 0.145x + 1.59 \end{aligned}$$

(b) The year 2014 corresponds to $x = 30$.

$$\begin{aligned} y &= 0.145(30) + 1.59 \\ y &= 5.94 \end{aligned}$$

In 2010, the number of cohabitating adults will be about 5.94 million.

73. (a) Plot the points (15, 1600), (200, 15,000), (290, 24,000), and (520, 40,000).

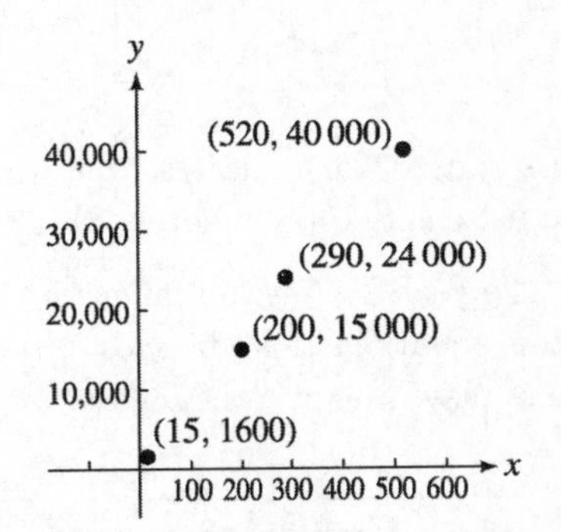

The points lie approximately on a line, so there appears to be a linear relationship between distance and time.

(b) The graph of any equation of the form $y = mx$ goes through the origin, so the line goes through (520, 40,000) and $(0, 0)$.

$$m = \frac{40{,}000 - 0}{520 - 0} \approx 76.9$$

$$\begin{aligned} b &= 0 \\ y &= 76.9x + 0 \\ y &= 76.9x \end{aligned}$$

(c) Let $y = 60{,}000$; solve for x.

$$60{,}000 = 76.9x$$
$$780.23 \approx x$$

Hydra is about 780 megaparsecs from earth.

(d) $A = \dfrac{9.5 \times 10^{11}}{m}, m = 76.9$

$$A = \frac{9.5 \times 10^{11}}{76.9}$$
$$= 12.4 \text{ billion years}$$

75. (a)

Tuition and fees (y): 16,000–22,000; Year (x): 2000, 2002, 2004, 2006

Yes, the data appear to lie roughly along a straight line.

(b) $m = \dfrac{21{,}235 - 16{,}072}{5 - 0} = \dfrac{5163}{5} = 1032.6$

$$b = 16{,}072$$
$$y = 1032.6x + 16{,}072$$

The slope 1032.6 indicates that tuition and fees have increased approximately \$1033 per year.

(c) The year 2025 is too far in the future to rely on this equation to predict costs; too many other factors may influence these costs by then.

1.2 Linear Functions and Applications

1. $f(2) = 7 - 5(2) = 7 - 10 = -3$

3. $f(-3) = 7 - 5(-3) = 7 + 15 = 22$

5. $g(1.5) = 2(1.5) - 3 = 3 - 3 = 0$

7. $g\left(-\frac{1}{2}\right) = 2\left(-\frac{1}{2}\right) - 3 = -1 - 3 = -4$

9. $f(t) = 7 - 5(t) = 7 - 5t$

11. This statement is true.
When we solve $y = f(x) = 0$, we are finding the value of x when $y = 0$, which is the x-intercept.
When we evaluate $f(0)$, we are finding the value of y when $x = 0$, which is the y-intercept.

13. This statement is true.
Only a vertical line has an undefined slope, but a vertical line is not the graph of a function. Therefore, the slope of a linear function cannot be defined.

15. The fixed cost is constant for a particular product and does not change as more items are made. The marginal cost is the rate of change of cost at a specific level of production and is equal to the slope of the cost function at that specific value; it approximates the cost of producing one additional item.

19. \$10 is the fixed cost and \$2.25 is the cost per hour.

Let $x =$ number of hours;
$R(x) =$ cost of renting a snowboard for x hours.

Thus,

$$R(x) = \text{fixed cost} + (\text{cost per hour}) \cdot (\text{number of hours})$$
$$R(x) = 10 + (2.25)(x)$$
$$= 2.25x + 10$$

21. 50¢ is the fixed cost and 35¢ is the cost per half-hour.

Let $x =$ the number of half-hours;
$C(x) =$ the cost of parking a car for x half-hours.

Thus,

$$C(x) = 50 + 35x$$
$$= 35x + 50.$$

23. Fixed cost, \$100; 50 items cost \$1600 to produce.

Let $C(x) =$ cost of producing x items.
$C(x) = mx + b$, where b is the fixed cost.

$$C(x) = mx + 100$$

Now,

$C(x) = 1600$ when $x = 50$, so

$$1600 = m(50) + 100$$
$$1500 = 50m$$
$$30 = m.$$

Thus, $C(x) = 30x + 100$.

25. Marginal cost: \$75; 50 items cost \$4300.

$$C(x) = 75x + b$$

Now, $C(x) = 4300$ when $x = 50$.

$$4300 = 75(50) + b$$
$$4300 = 3750 + b$$
$$550 = b$$

Thus, $C(x) = 75x + 550$.

27. $D(q) = 16 - 1.25q$

(a) $D(0) = 16 - 1.25(0) = 16 - 0 = 16$

When 0 watches are demanded, the price is \$16.

(b) $D(4) = 16 - 1.25(4) = 16 - 5 = 11$

When 400 watches are demanded, the price is \$11.

(c) $D(8) = 16 - 1.25(8) = 16 - 10 = 6$

When 800 watches are demanded, the price is \$6.

(d) Let $D(q) = 8$. Find q.

$$8 = 16 - 1.25q$$
$$\frac{5}{4}q = 8$$
$$q = 6.4$$

When the price is \$8, 640 watches are demanded.

(e) Let $D(q) = 10$. Find q.

$$10 = 16 - 1.25q$$
$$\frac{5}{4}q = 6$$
$$q = 4.8$$

When the price is \$10, 480 watches are demanded.

(f) Let $D(q) = 12$. Find q.

$$12 = 16 - 1.25q$$
$$\frac{5}{4}q = 4$$
$$q = 3.2$$

When the price is \$12, 320 watches are demanded.

(g)

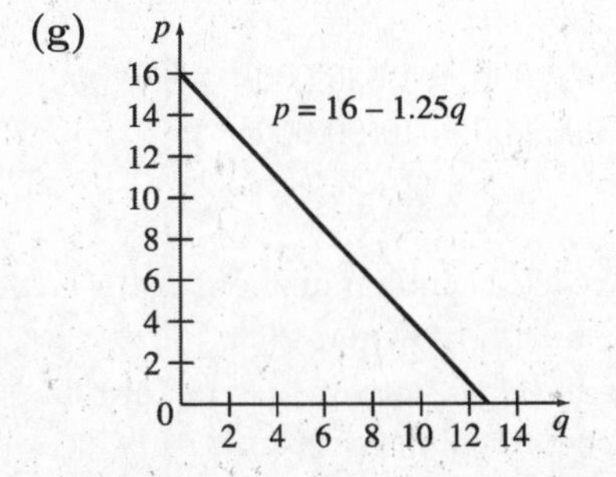

(h) $S(q) = 0.75q$

Let $S(q) = 0$. Find q.

$$0 = 0.75q$$
$$0 = q$$

When the price is \$0, 0 watches are supplied.

(i) Let $S(q) = 10$. Find q.

$$10 = 0.75q$$
$$\frac{40}{3} = q$$
$$q = 13.\overline{3}$$

When the price is \$10, about 1333 watches are supplied.

(j) Let $S(q) = 20$. Find q.

$$20 = 0.75q$$
$$\frac{80}{3} = q$$
$$q = 26.\overline{6}$$

When the price is \$20, about 2667 watches are demanded.

(k)

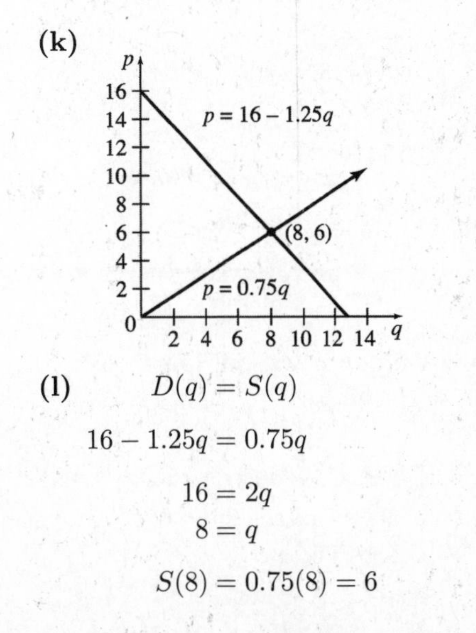

(l)

$$D(q) = S(q)$$
$$16 - 1.25q = 0.75q$$
$$16 = 2q$$
$$8 = q$$
$$S(8) = 0.75(8) = 6$$

The equilibrium quantity is 800 watches, and the equilibrium price is \$6.

29. $p = S(q) = \frac{2}{5}q$; $p = D(q) = 100 - \frac{2}{5}q$

(a)

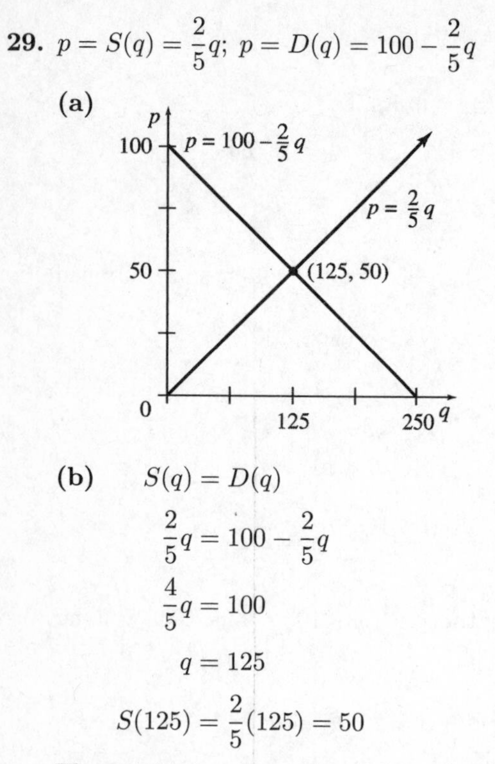

(b)
$$S(q) = D(q)$$
$$\frac{2}{5}q = 100 - \frac{2}{5}q$$
$$\frac{4}{5}q = 100$$
$$q = 125$$
$$S(125) = \frac{2}{5}(125) = 50$$

The equilibrium quantity is 125, the equilibrium price is \$50

31. **(a)** $C(x) = mx + b$; $m = 3.50$; $C(60) = 300$

$$C(x) = 3.50x + b$$

Find b.

$$300 = 3.50(60) + b$$
$$300 = 210 + b$$
$$90 = b$$
$$C(x) = 3.50x + 90$$

(b) $R(x) = 9x$
$C(x) = R(x)$

$$3.50x + 90 = 9x$$
$$90 = 5.5x$$
$$16.36 = x$$

Joanne must produce and sell 17 shirts.

(c) $P(x) = R(x) - C(x)$; $P(x) = 500$

$$500 = 9x - (3.50x + 90)$$
$$500 = 5.5x - 90$$
$$590 = 5.5x$$
$$107.27 = x$$

To make a profit of \$500, Joanne must produce and sell 108 shirts.

33. **(a)** Using the points $(100, 11.02)$ and $(400, 40.12)$,

$$m = \frac{40.12 - 11.02}{400 - 100} = \frac{29.1}{300} = 0.097.$$
$$y - 11.02 = 0.097(x - 100)$$
$$y - 11.02 = 0.097x - 9.7$$
$$y = 0.097x + 1.32$$
$$C(x) = 0.097x + 1.32$$

(b) The fixed cost is given by the constant in $C(x)$. It is \$1.32.

(c) $C(1000) = 0.097(1000) + 1.32 = 97 + 1.32$
$= 98.32$

The total cost of producing 1000 cups is \$98.32.

(d) $C(1001) = 0.097(1001) + 1.32 = 97.097 + 1.32$
$= 98.417$

The total cost of producing 1001 cups is \$98.417.

(e) Marginal cost $= 98.417 - 98.32$
$= \$0.097$ or $9.7¢$

(f) The marginal cost for *any* cup is the slope, \$0.097 or 9.7¢. This means the cost of producing one additional cup of coffee would be 9.7¢.

35. **(a)** $(100{,}000)(50) = 5{,}000{,}000$

Sales in 1996 would be 100,000 + 5,000,000 = 5,100,000.

(b) The ordered pairs are (1, 100,000) and (6, 5,100,000).

$$m = \frac{5{,}100{,}000 - 100{,}000}{6 - 1} = \frac{5{,}000{,}000}{5} = 1{,}000{,}000$$

$$y - 100{,}000 = 1{,}000{,}000(x - 1)$$
$$y - 100{,}000 = 1{,}000{,}000x - 1{,}000{,}000$$
$$y = 1{,}000{,}000x - 900{,}000$$
$$S(x) = 1{,}000{,}000x - 900{,}000$$

(c) Let $S(x) = 1{,}000{,}000{,}000$. Find x.

$$1{,}000{,}000{,}000 = 1{,}000{,}000x - 900{,}000$$
$$1{,}000{,}900{,}000 = 1{,}000{,}000x$$
$$x = 1000.9$$

Sales would reach \$1 billion in about 1991 + 1000.9 = 2991.9, or during the year 2991.
Sales would have to grow much faster than linearly to reach \$1 billion by 2003.

(d) Use ordered pairs (13, 356,000,000) and (14, 479,000,000).

$$m = \frac{479{,}000{,}000 - 356{,}000{,}000}{14 - 13} = 123{,}000{,}000$$

$$\begin{aligned} S(x) - 356{,}000{,}000 &= 123{,}000{,}000(x - 13) \\ S(x) - 356{,}000{,}000 &= 123{,}000{,}000x - 1{,}599{,}000000 \\ S(x) &= 123{,}000{,}000x - 1{,}243{,}000{,}000 \end{aligned}$$

(e) The year 2005 corresponds to $x = 2005 - 1990 = 15$.

$$\begin{aligned} S(15) &= 123{,}000{,}000(15) - 1{,}243{,}000{,}000 \\ S(15) &= 602{,}000{,}000 \end{aligned}$$

The estimated sales are \$602,000,000, which is less than the actual sales.

(f) Let $S(x) = 1{,}000{,}000{,}000$. Find x.

$$\begin{aligned} 1,000,000,000 &= 123{,}000{,}000x - 1{,}243{,}000{,}000 \\ 2{,}243{,}000{,}000 &= 123{,}000{,}000x \\ x &\approx 18.2 \end{aligned}$$

Sales would reach \$1 billion in about 1990 + 18.2 = 2008.2, or during the year 2009.

37. $C(x) = 12x + 39$; $R(x) = 25x$

(a)
$$\begin{aligned} C(x) &= R(x) \\ 12x + 39 &= 25x \\ 39 &= 13x \\ 3 &= x \end{aligned}$$

The break-even quantity is 3 units.

(b)
$$\begin{aligned} P(x) &= R(x) - C(x) \\ P(x) &= 25x - (12x + 39) \\ P(x) &= 13x - 39 \\ P(250) &= 13(250) - 39 \\ &= 3250 - 39 \\ &= 3211 \end{aligned}$$

The profit from 250 units is \$3211.

(c) $P(x) = \$130$; find x.

$$\begin{aligned} 130 &= 13x - 39 \\ 169 &= 13x \\ 13 &= x \end{aligned}$$

For a profit of \$130, 13 units must be produced.

39. $C(x) = 105x + 6000$
$R(x) = 250x$

Set $C(x) = R(x)$ to find the break-even quantity.

$$\begin{aligned} 105x + 6000 &= 250x \\ 6000 &= 145x \\ 41.38 &\approx x \end{aligned}$$

The break-even quantity is about 41 units, so you should decide to produce.

$$\begin{aligned} P(x) &= R(x) - C(x) \\ &= 250x - (105x + 6000) \\ &= 145x - 6000 \end{aligned}$$

The profit function is $P(x) = 145x - 6000$.

41. $C(x) = 1000x + 5000$
$R(x) = 900x$

$$\begin{aligned} 900x &= 1000x + 5000 \\ -5000 &= 100x \\ -50 &= x \end{aligned}$$

It is impossible to make a profit when the break-even quantity is negative. Cost will always be greater than revenue.

$$\begin{aligned} P(x) &= R(x) - C(x) \\ &= 900x - (1000x + 5000) \\ &= -100x - 5000 \end{aligned}$$

The profit function is $P(x) = -100x - 5000$ (always a loss).

43. Use the formula derived in Example 7 in this section of the textbook.

$$F = \frac{9}{5}C + 32$$
$$C = \frac{5}{9}(F - 32)$$

(a) $C = 37$; find F.

$$F = \frac{9}{5}(37) + 32$$
$$F = \frac{333}{5} + 32$$
$$F = 98.6$$

The Fahrenheit equivalent of 37°C is 98.6°F.

(b) $C = 36.5$; find F.

$$F = \frac{9}{5}(36.5) + 32$$
$$F = 65.7 + 32$$
$$F = 97.7$$

$C = 37.5$; find F.

$$F = \frac{9}{5}(37.5) + 32$$
$$= 67.5 + 32 = 99.5$$

The range is between 97.7°F and 99.5°F.

1.3 The Least Squares Line

3. (a)

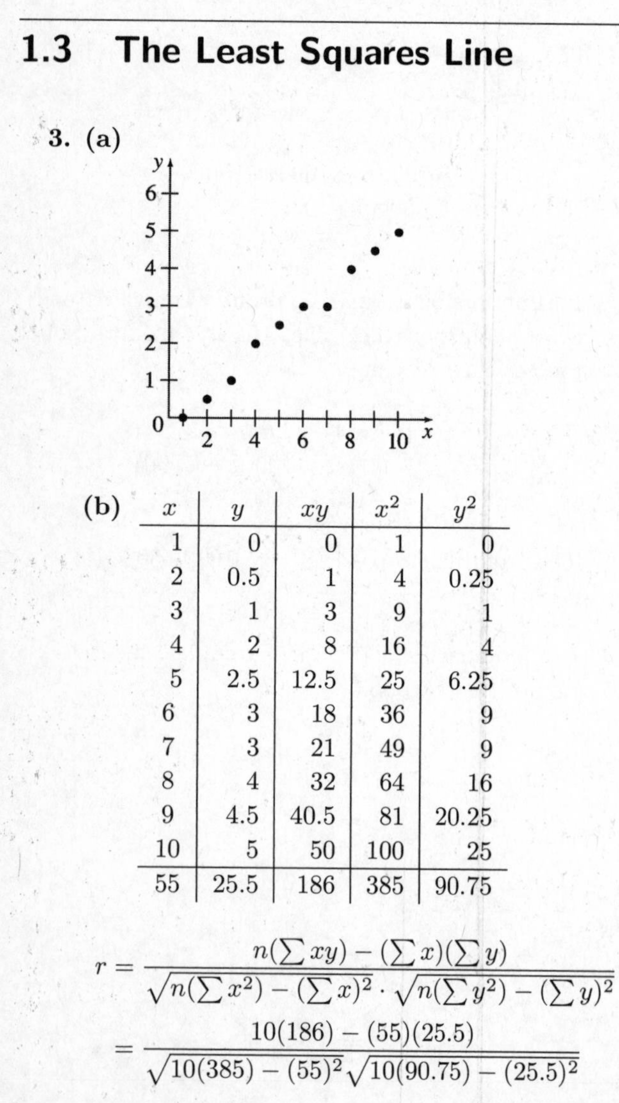

(b)

x	y	xy	x^2	y^2
1	0	0	1	0
2	0.5	1	4	0.25
3	1	3	9	1
4	2	8	16	4
5	2.5	12.5	25	6.25
6	3	18	36	9
7	3	21	49	9
8	4	32	64	16
9	4.5	40.5	81	20.25
10	5	50	100	25
55	25.5	186	385	90.75

$$r = \frac{n(\sum xy) - (\sum x)(\sum y)}{\sqrt{n(\sum x^2) - (\sum x)^2} \cdot \sqrt{n(\sum y^2) - (\sum y)^2}}$$
$$= \frac{10(186) - (55)(25.5)}{\sqrt{10(385) - (55)^2}\sqrt{10(90.75) - (25.5)^2}}$$
$$\approx 0.993$$

(c) The least squares line is of the form $Y = mx + b$. First solve for m.

$$m = \frac{n(\sum xy) - (\sum x)(\sum y)}{n(\sum x^2) - (\sum x)^2}$$
$$= \frac{10(186) - (55)(25.5)}{10(385) - (55)^2}$$
$$= 0.5545454545 \approx 0.55$$

Now find b.

$$b = \frac{\sum y - m(\sum x)}{n}$$
$$= \frac{25.5 - 0.5545454545(55)}{10}$$
$$= -0.5$$

Thus, $Y = 0.55x - 0.5$.

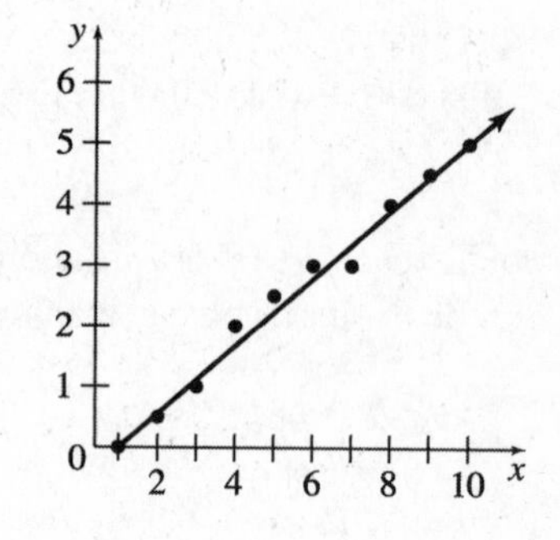

(d) Let $x = 11$. Find Y.

$$Y = 0.55(11) - 0.5 = 5.55$$

5.

$$nb + (\sum x)m = \sum y$$
$$(\sum x)b + (\sum x^2)m = \sum xy$$
$$nb + (\sum x)m = \sum y$$
$$nb = (\sum y) - (\sum x)m$$
$$b = \frac{\sum y - m(\sum x)}{n}$$
$$(\sum x)\left(\frac{\sum y - m(\sum x)}{n}\right) + (\sum x^2)m = \sum xy$$
$$(\sum x)[(\sum y) - m(\sum x)] + nm(\sum x^2) = n(\sum xy)$$
$$(\sum x)(\sum y) - m(\sum x)^2 + nm(\sum x^2) = n(\sum xy)$$
$$nm(\sum x^2) - m(\sum x)^2 = n(\sum xy) - (\sum x)(\sum y)$$
$$m\left[n(\sum x^2) - (\sum x)^2\right] = n(\sum xy) - (\sum x)(\sum y)$$
$$m = \frac{n(\sum xy) - (\sum x)(\sum y)}{n(\sum x^2) - (\sum x)^2}$$

7. (a) $m = \dfrac{n(\sum xy) - (\sum x)(\sum y)}{n(\sum x^2) - (\sum y)}$

$$m = \frac{10(8501.39) - (995)(85.65)}{10(99{,}085) - 995^2}$$

$$m = -0.2519393939 \approx -0.2519$$

$$b = \frac{\sum y - m(\sum x)}{n}$$

$$b = \frac{85.65 - (-0.2519393939)(995)}{10} \approx 33.6330$$

$$Y = -0.2519x + 33.6330$$

(b) The year 2010 corresponds to $x = 110$.

$$Y = -0.2519(110) + 33.6330 \approx 5.924 \text{ (in thousands)}$$

If the trend continues, there will be about 5924 banks in 2010.

(c) $r = \dfrac{10(8501.39) - (995)(85.65)}{\sqrt{10(99{,}085) - 995^2} \cdot \sqrt{10(739.08) - 85.65^2}} \approx -0.977$

This means that the least squares line fits the data points very well. The negative sign indicates that the number of banks is decreasing as the years increase.

9.

x	y	xy	x^2	y^2
97	6247	605,959	9409	39,025,009
98	6618	648,565	9604	43,797,924
99	7031	696,069	9801	49,434,961
100	7842	784,200	10,000	61,496,964
101	8234	831,634	10,201	67,798,756
102	8940	911,880	10,404	79,923,600
103	9205	948,115	10,609	84,732,025
104	9312	968,448	10,816	86,713,344
804	63,429	6,394,869	80,844	512,922,583

(a)

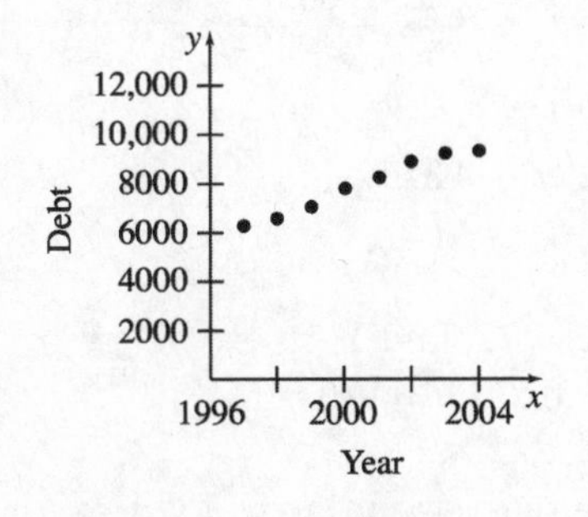

Yes, the pattern is linear.

(b) $m = \dfrac{n(\sum xy) - (\sum x)(\sum y)}{n(\sum x^2) - (\sum x)^2}$

$m = \dfrac{8(6{,}394{,}869) - (804)(63{,}429)}{8(80{,}844) - 804^2}$

$m = 482.25$

$b = \dfrac{\sum y - m(\sum x)}{n}$

$b = \dfrac{63{,}429 - 482.25(804)}{8} = -40{,}537.5$

$Y = 482.25x - 40{,}537.5$

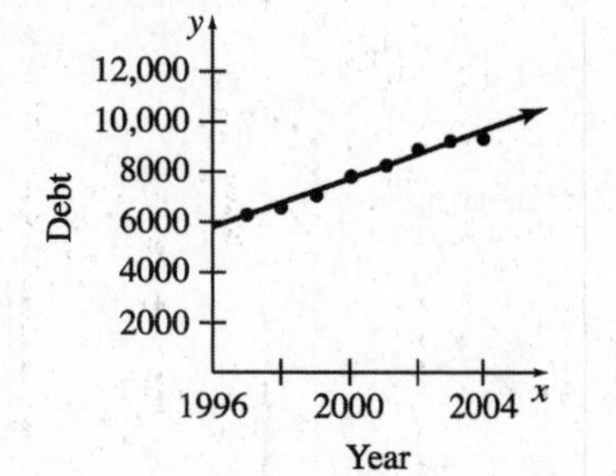

The least squares line seems to be a good fit.

(c) $r = \dfrac{8(6{,}394{,}869) - (804)(63{,}429)}{\sqrt{8(80{,}844) - 804^2} \cdot \sqrt{8(512{,}922{,}583) - 63{,}429^2}} \approx 0.987$

This confirms the least squares line is a good fit.

(d) Let $Y = 12{,}000$ and solve for x.

$$12{,}000 = 482.25x - 40{,}537.5$$
$$52{,}537.5 = 482.25x$$
$$x \approx 109$$

If the trend continues, credit card debt will reach \$12,000 in 1900 + 109, or the year 2009.

11. (a)

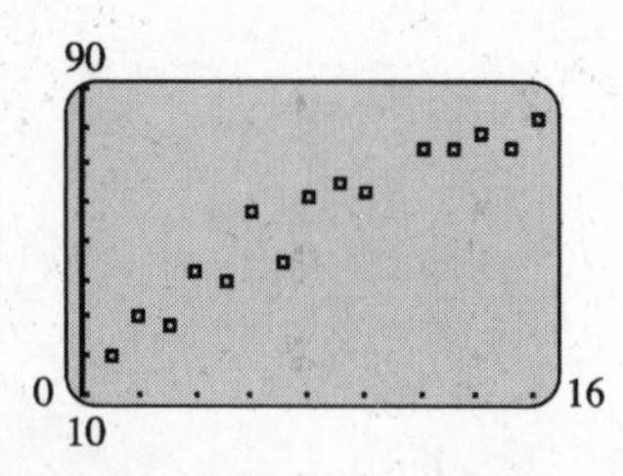

Yes, the points lie in a linear pattern.

(b) Using a calculator's STAT feature, the correlation coefficient is found to be $r \approx 0.959$. This indicates that the percentage of successful hunts does trend to increase with the size of the hunting party.

(c) $Y = 3.98x + 22.7$

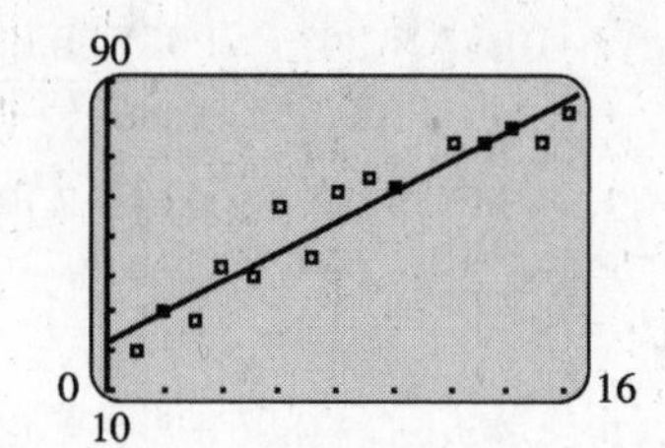

13. (a)

x	y	xy	x^2	y^2
88.6	20.0	1772	7849.96	400.0
71.6	16.0	1145.6	5126.56	256.0
93.3	19.8	1847.34	8704.89	392.04
84.3	18.4	1551.12	7106.49	338.56
80.6	17.1	1378.26	6496.36	292.41
75.2	15.5	1165.6	5655.04	240.25
69.7	14.7	1024.59	4858.09	216.09
82.0	17.1	1402.2	6724	292.41
69.4	15.4	1068.76	4816.36	237.16
83.3	16.2	1349.46	6938.89	262.44
79.6	15.0	1194	6336.16	225
82.6	17.2	1420.72	6822.76	295.84
80.6	16.0	1289.6	6496.36	256.0
83.5	17.0	1419.5	6972.25	289.0
76.3	14.4	1098.72	5821.69	207.36
1200.6	249.8	20,127.47	96,725.86	4200.56

$$\begin{aligned} m &= \frac{n(\sum xy) - (\sum x)(\sum y)}{n(\sum x^2) - (\sum x)^2} \\ &= \frac{15(20{,}127.47) - (1200.6)(249.8)}{15(96{,}725.86) - 1200.6^2} \\ &= 0.211925009 \approx 0.212 \\ b &= \frac{\sum y - m(\sum x)}{n} \\ &= \frac{249.8 - 0.212(1200.6)}{15} \\ &\approx -0.315 \\ Y &= 0.212x - 0.315 \end{aligned}$$

(b) Let $x = 73$; find Y.

$$Y = 0.212(73) - 0.315$$
$$\approx 15.2$$

If the temperature were 73°F, you would expect to hear 15.2 chirps per second.

(c) Let $Y = 18$; find x.

$$18 = 0.212x - 0.315$$
$$18.315 = 0.212x$$
$$86.4 \approx x$$

When the crickets are chirping 18 times per second, the temperature is 86.4°F.

(d)

$$r = \frac{15(20{,}127) - (1200.6)(249.8)}{\sqrt{15(96{,}725.86) - (1200.6)^2} \cdot \sqrt{15(4200.56) - (249.8)^2}}$$
$$= 0.835$$

15. (a)

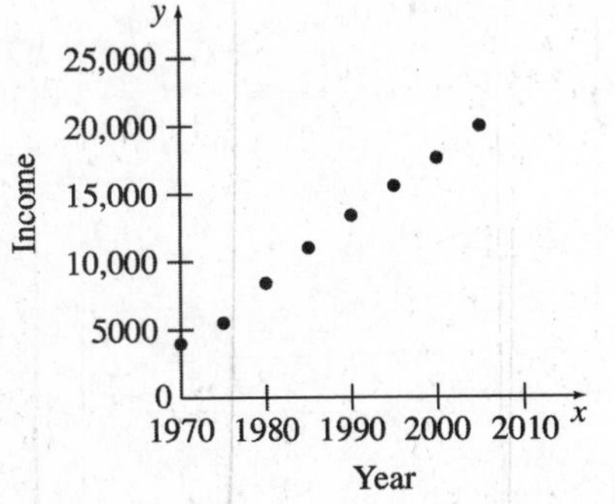

Yes, the data appear to lie along a straight line.

$$r = \frac{8(2159.635) - (140)(95.364)}{\sqrt{8(3500) - 140^2} \cdot \sqrt{8(1366.748) - 95.364^2}}$$
$$\approx 0.999$$

Yes, there is a strong positive linear correlation between the income and the year.

(c) $m = \dfrac{n(\sum xy) - (\sum x)(\sum y)}{n(\sum x^2) - (\sum x)^2}$

$$m = \frac{8(2159.635) - (140)(95.364)}{8(3500) - 140^2}$$

$$m = 0.4673952381 \approx 0.467$$

$$b = \frac{\sum y - m(\sum x)}{n}$$

$$b = \frac{95.364 - 0.4673952381(140)}{8}$$
$$\approx 3.74$$
$$Y = 0.467x + 3.74$$

(d) The year 2020 corresponds to $x = 50$.

$$Y = 0.467(50) + 3.74 = 27.09$$

The predicted poverty level in the year 2020 is $27,090.

17. (a)

x	y	xy	x^2	y^2
150	5000	750,000	22,500	25,000,000
175	5500	962,500	30,625	30,250,000
215	6000	1,290,000	46,225	36,000,000
250	6500	1,625,000	62,500	42,250,000
280	7000	1,960,000	78,400	49,000,000
310	7500	2,325,000	96,100	56,250,000
350	8000	2,800,000	122,500	64,000,000
370	8500	3,145,000	136,900	72,250,000
420	9000	3,780,000	176,400	81,000,000
450	9500	4,275,000	202,500	90,250,000
2970	72,500	22,912,500	974,650	546,250,000

$$m = \frac{n(\sum xy) - (\sum x)(\sum y)}{n(\sum x^2) - (\sum x)^2}$$

$$m = \frac{10(22{,}912{,}500) - (2970)(72{,}500)}{10(974{,}650) - 2970^2}$$

$$m = 14.90924806 \approx 14.9$$

$$b = \frac{\sum y - m(\sum x)}{n}$$

$$b = \frac{72{,}500 - 14.9(2970)}{10}$$
$$\approx 2820$$
$$Y = 14.9x + 2820$$

(b) Let $x = 150$; find Y.

$$Y = 14.9(150) + 2820$$
$Y \approx 5060$, compared to actual 5000

Let $x = 280$; find Y.

$$Y = 14.9(280) + 2820$$
≈ 6990, compared to actual 7000

Let $x = 420$; find Y.

$$Y = 14.9(420) + 2820$$
≈ 9080, compared to actual 9000

(c) Let $x = 230$; find Y.

$$Y = 14.9(230) + 2820$$
$$\approx 6250$$

Adam would need to buy a 6500 BTU air conditioner.

19. (a) Use a calculator's statistical features to obtain the least squares line.

$$y = -0.1358x + 113.94$$

(b) $y = -0.3913x + 148.98$

(c) Set the two equations equal and solve for x.

$$\begin{aligned} -0.1358x + 113.94 &= -0.3913x + 148.98 \\ 0.2555x &= 35.04 \\ x &\approx 137 \end{aligned}$$

The women's record will catch up with the men's record in 1900 + 137, or in the year 2037.

(d) $r_{men} \approx -0.9823$
$r_{women} \approx -0.9487$

Both sets of data points closely fit a line with negative slope.

(e)

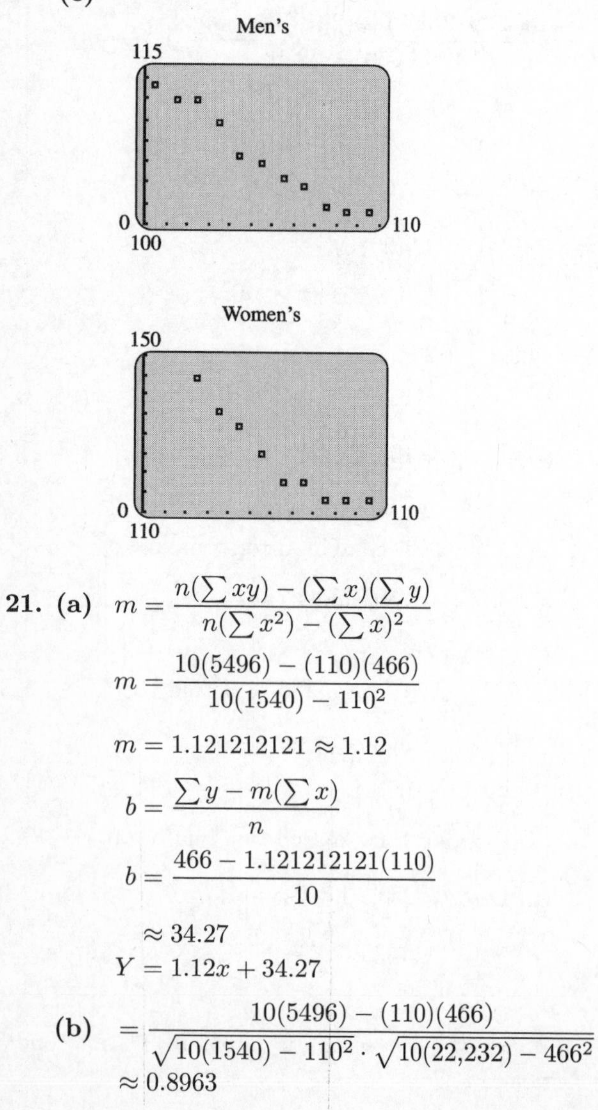

21. (a) $m = \dfrac{n(\sum xy) - (\sum x)(\sum y)}{n(\sum x^2) - (\sum x)^2}$

$$m = \frac{10(5496) - (110)(466)}{10(1540) - 110^2}$$

$$m = 1.121212121 \approx 1.12$$

$$b = \frac{\sum y - m(\sum x)}{n}$$

$$b = \frac{466 - 1.121212121(110)}{10}$$

$$\approx 34.27$$

$$Y = 1.12x + 34.27$$

(b) $= \dfrac{10(5496) - (110)(466)}{\sqrt{10(1540) - 110^2} \cdot \sqrt{10(22{,}232) - 466^2}}$
≈ 0.8963

Yes, the value indicates a good fit of the least squares line to the data.

(c) The year 2005 corresponds to $x = 25$.

$$Y = 1.12(25) + 34.27 = 62.27 \approx 62$$

The predicted length of a game in 2005 is 2 hours + 62 minutes, or 3:02.

Chapter 1 Review Exercises

3. Through $(-3, 7)$ and $(2, 12)$

$$m = \frac{12 - 7}{2 - (-3)} = \frac{5}{5} = 1$$

5. Through the origin and $(11, -2)$

$$m = \frac{-2 - 0}{11 - 0} = -\frac{2}{11}$$

7. $$\begin{aligned} 4x + 3y &= 6 \\ 3y &= -4x + 6 \\ y &= -\frac{4}{3}x + 2 \end{aligned}$$

Therefore, the slope is $m = -\frac{4}{3}$.

9. $$\begin{aligned} y + 4 &= 9 \\ y &= 5 \\ y &= 0x + 5 \\ m &= 0 \end{aligned}$$

11. $$\begin{aligned} y &= 5x + 4 \\ m &= 5 \end{aligned}$$

13. Through $(5, -1)$; slope $\frac{2}{3}$

Use point-slope form.

$$\begin{aligned} y - (-1) &= \frac{2}{3}(x - 5) \\ y + 1 &= \frac{2}{3}(x - 5) \\ 3(y + 1) &= 2(x - 5) \\ 3y + 3 &= 2x - 10 \\ 3y &= 2x - 13 \\ y &= \frac{2}{3}x - \frac{13}{3} \end{aligned}$$

15. Through $(-6, 3)$ and $(2, -5)$

$$m = \frac{-5-3}{2-(-6)} = \frac{-8}{8} = -1$$

Use point-slope form.

$$y - 3 = -1[x - (-6)]$$
$$y - 3 = -x - 6$$
$$y = -x - 3$$

17. Through $(-1, 4)$; undefined slope

Undefined slope means the line is vertical. The equation of the vertical line through $(-1, 4)$ is $x = -1$.

19. Through $(3, -4)$ parallel to $4x - 2y = 9$
Solve $4x - 2y = 9$ for y.

$$-2y = -4x + 9$$
$$y = 2x - \frac{9}{2}$$
$$m = 2$$

The desired line has the same slope. Use the point-slope form.

$$y - (-4) = 2(x - 3)$$
$$y + 4 = 2x - 6$$
$$y = 2x - 10$$

21. Through $(2, -10)$, perpendicular to a line with undefined slope
A line with undefined slope is a vertical line. A line perpendicular to a vertical line is a horizontal line with equation of the form $y = k$. The desired line passed through $(2, -10)$, so $k = -10$. Thus, an equation of the desired line is $y = -10$.

23. Through $(-3, 5)$, perpendicular to $y = -2$
The given line, $y = -2$, is a horizontal line. A line perpendicular to a horizontal line is a vertical line with equation of the form $x = h$.
The desired line passes through $(-3, 5)$, so $h = -3$. Thus, an equation of the desired line is $x = -3$.

25. $y = 6 - 2x$

Find the intercepts.
Let $x = 0$.

$$y = 6 - 2(0) = 6$$

The y-intercept is 6.
Let $y = 0$.

$$0 = 6 - 2x$$
$$2x = 6$$
$$x = 3$$

The x-intercept is 3.
Draw the line through $(0, 6)$ and $(3, 0)$.

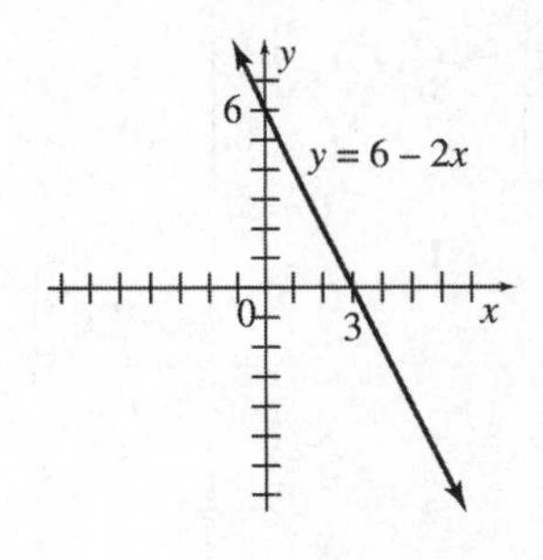

27. $4x + 6y = 12$

Find the intercepts.
When $x = 0$, $y = 2$, so the y-intercept is 2.
When $y = 0$, $x = 3$, so the x-intercept is 3.
Draw the line through $(0, 2)$ and $(3, 0)$.

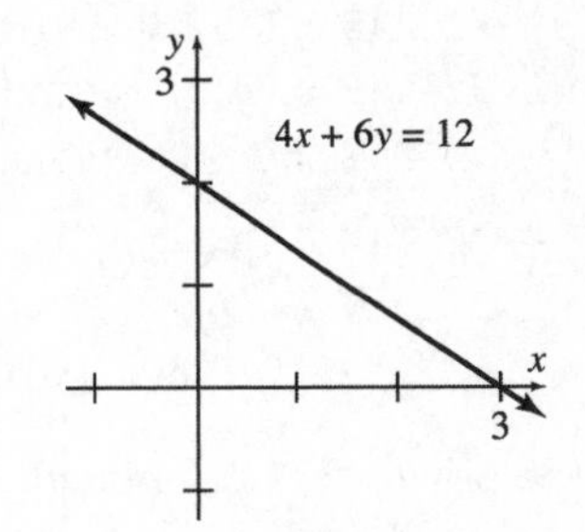

29. $y = 1$

This is the horizontal line passing through $(0, 1)$.

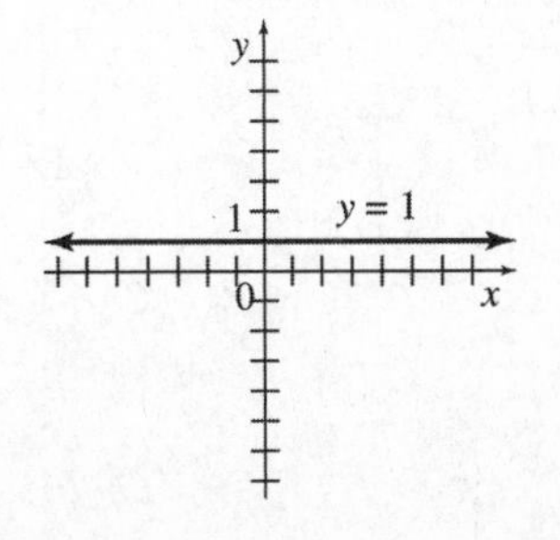

31. $x + 3y = 0$

When $x = 0,\ y = 0.$
When $x = 3,\ y = -1.$
Draw the line through $(0, 0)$ and $(3, -1)$.

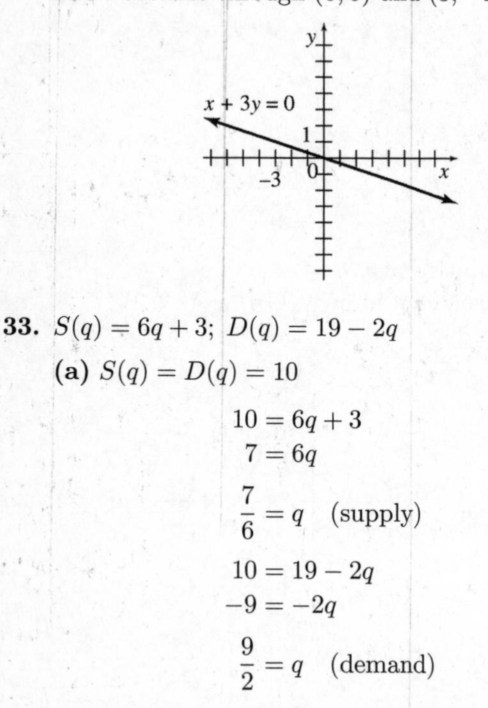

33. $S(q) = 6q + 3;\ D(q) = 19 - 2q$

(a) $S(q) = D(q) = 10$

$$\begin{aligned} 10 &= 6q + 3 \\ 7 &= 6q \\ \frac{7}{6} &= q \quad \text{(supply)} \end{aligned}$$

$$\begin{aligned} 10 &= 19 - 2q \\ -9 &= -2q \\ \frac{9}{2} &= q \quad \text{(demand)} \end{aligned}$$

When the price is \$10 per pound, the supply is $\frac{7}{6}$ pounds per day, and the demand is $\frac{9}{2}$ pounds per day.

(b) $S(q) = D(q) = 15$

$$\begin{aligned} 15 &= 6q + 3 \\ 12 &= 6q \\ 2 &= q \quad \text{(supply)} \end{aligned}$$

$$\begin{aligned} 15 &= 19 - 2q \\ -4 &= -2q \\ 2 &= q \quad \text{(demand)} \end{aligned}$$

When the price is \$15 per pound, the supply is 2 pounds per day, and the demand is 2 pounds per day.

(c) $S(q) = D(q) = \$18$

$$\begin{aligned} 18 &= 6q + 3 \\ 15 &= 6q \\ \frac{5}{2} &= q \quad \text{(supply)} \end{aligned}$$

$$\begin{aligned} 18 &= 19 - 2q \\ -1 &= -2q \\ \frac{1}{2} &= q \quad \text{(demand)} \end{aligned}$$

When the price is $\frac{5}{2}$ pounds per day, the demand is $\frac{1}{2}$ pound per day.

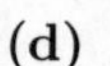

(d)

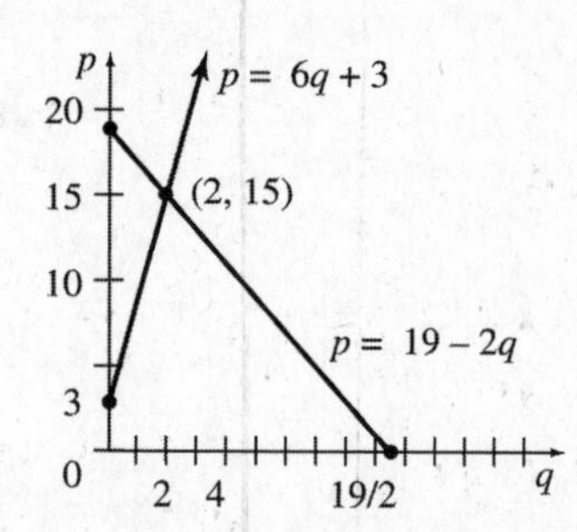

(e) The graph shows that the lines representing the supply and demand functions intersect at the point $(2, 15)$. The y-coordinate of this point gives the equilibrium price. Thus, the equilibrium price is \$15.

(f) The x-coordinate of the intersection point gives the equilibrium quantity. Thus, the equilibrium quantity is 2, representing 2 pounds of crabmeat per day.

35. Using the points $(50, 47.50)$ and $(80, 32.50)$,

$$m = \frac{47.50 - 32.50}{50 - 80} = \frac{15}{-30} = \frac{-1}{2} = -0.5.$$

$$\begin{aligned} p - 47.50 &= -0.5(q - 50) \\ p - 47.50 &= -0.5q + 25 \\ p &= -0.5q + 72.50 \\ D(q) &= -0.5q + 72.50 \end{aligned}$$

37. Eight units cost \$300; fixed cost is \$60.
The fixed cost is the cost if zero units are made.
$(8, 300)$ and $(0, 60)$ are points on the line.

$$m = \frac{60 - 300}{0 - 8} = 30$$

Use slope-intercept form.

$$\begin{aligned} y &= 30x + 60 \\ C(x) &= 30x + 60 \end{aligned}$$

39. Twelve units cost \$445; 50 units cost \$1585. Points on the line are $(12, 445)$ and $(50, 1585)$.

$$m = \frac{1585 - 445}{50 - 12} = 30$$

Use point-slope form.

$$\begin{aligned} y - 445 &= 30(x - 12) \\ y - 445 &= 30x - 360 \\ y &= 30x + 85 \\ C(x) &= 30x + 85 \end{aligned}$$

41. $C(x) = 200x + 1000$
$R(x) = 400x$

(a) $C(x) = R(x)$

$$\begin{aligned} 200x + 1000 &= 400x \\ 1000 &= 200x \\ 5 &= x \end{aligned}$$

The break-even quantity is 5 cartons.

(b) $R(5) = 400(5) = 2000$

The revenue from 5 cartons of CD's is $2000.

43. Let y represent imports from China in billions of dollars. Using the points $(1, 102)$ and $(5, 243)$,

$$m = \frac{243 - 102}{5 - 1} = \frac{141}{4} = 35.25$$

$$\begin{aligned} y - 102 &= 35.25(x - 1) \\ y - 102 &= 35.25x - 35.25 \\ y &= 35.25x + 66.75. \end{aligned}$$

45. Using the points (97, 44,883) and (105, 46,326),

$$m = \frac{46{,}326 - 44{,}883}{105 - 97} = \frac{1443}{8} \approx 180.4$$

$$\begin{aligned} I - 44{,}883 &= 180.4(x - 97) \\ I - 44{,}883 &= 180.4x - 17{,}498.8 \\ I(x) &= 180.4x + 27{,}384.2. \end{aligned}$$

Rounded to the nearest dollar,

$$I(x) = 180.4x + 27{,}384.$$

47. (a)

x	y
1960	43
2840	74
2060	54
3630	79
2420	63
3160	74
3220	78
2550	70
3140	80
3790	77

Using a graphing calculator, $r = 0.881$. Yes, the data seem to fit a straight line.

(b)

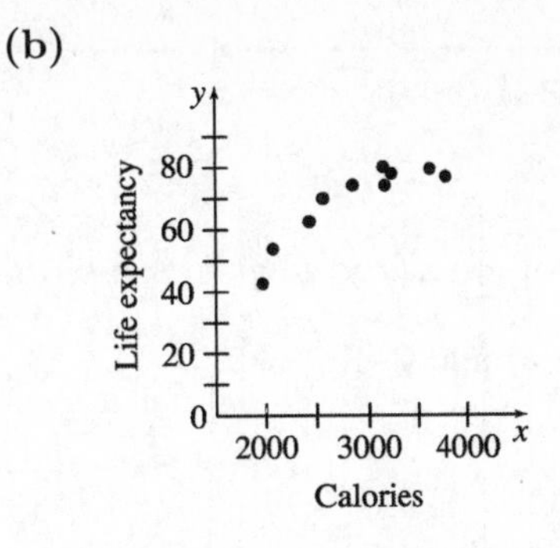

The data somewhat fit a straight line, but there is also a nonlinear trend.

(c) Using a graphing calculator,
$Y = 0.0173x + 19.3$.

(d) Let $x = 3400$. Find Y.

$$Y = 0.0173(3400) + 19.3 \approx 78.1$$

The predicted life expectancy in the United Kingdom, with a daily calorie supply of 3400, is about 78.1 years. This agrees with the actual value of 78 years.

49. Using the points $(74, 142.3)$ and $(104, 118.4)$,

$$m = \frac{118.4 - 142.3}{104 - 74} = \frac{-23.9}{30} = -0.797$$

$$\begin{aligned} y - 142.3 &= -0.797(x - 74) \\ y - 142.3 &= -0.797x + 59 \\ y &= -0.797x + 201.3. \end{aligned}$$

51. (a) Using a graphing calculator, $r = 0.749$.

The data seem to fit a line but the fit is not very good.

(b)

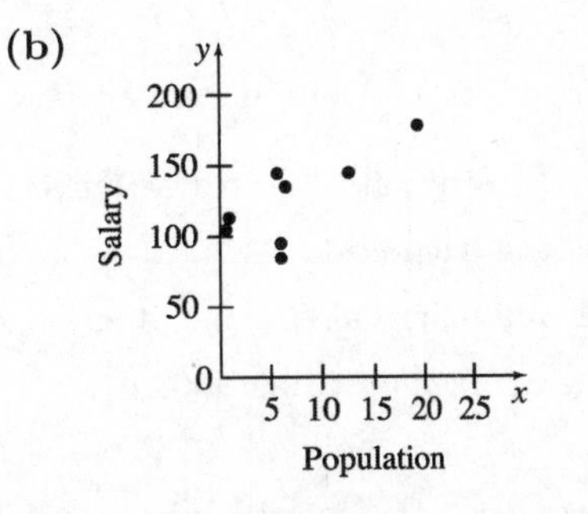

(c) Using a graphing calculator,

$$Y = 3.81x + 98.24$$

(d) The slope is 3.81 thousand (or 3810). On average, the governor's salary increases $3810 for each additional million in population.

Chapter 1 Test

[1.1]

Find the slope of each line that has a slope.

1. Through $(2, -5)$ and $(-1, 7)$

2. Through $(9, 5)$ and $(9, 2)$

3. $3x - 7y = 9$

4. Perpendicular to $2x + 5y = 7$

Find an equation in the form $ax + by = c$ for each line.

5. Through $(-1, 6)$ and $(5, -3)$

6. x–intercept 6, y–intercept -5

7. Through $(0, 3)$, parallel to $2x - 4y = 1$

8. Through $(1, -4)$, perpendicular to $3x + y = 1$

9. Through $(2, 5)$, perpendicular to $x = 5$

Graph each of the following.

10. $3x + 5y = 15$

11. $2x + y = 0$

12. $x - 3 = 0$

[1.2]

13. Let the supply and demand functions for a certain product be given by the following equations.

Supply: $p = 0.20q - 5$ Demand: $p = 100 - 0.15q$,

where p represents the price (in dollars) at a supply or demand, respectively, of q units.

(a) Graph these equations on the same axes.
(b) Find the equilibrium price.
(c) Find the equilibrium quantity.

14. For a given product, eight units cost \$450, while forty units cost \$770.

(a) Find the appropriate linear cost function.
(b) What is the fixed cost?
(c) What is the marginal cost per item?
(d) Find the average cost function.

15. For a given product, the variable cost is \$100, while 150 items cost \$16,000 to produce.

(a) Find the appropriate linear cost function.
(b) What is the fixed cost?
(c) What is the marginal cost per item?
(d) Find the average cost function.

16. Producing x hundred units of widgets costs $C(x) = 4x + 16$; revenue is $12x$, where revenue and $C(x)$ are in thousands of dollars.

(a) What is the break-even quantity?

(b) What is the profit from 300 units?

(c) How many units will produce a profit of $40,000?

[1.3]

17. An electronics firm was planning to expand its product line and wanted to get an idea of the salary picture for technicians it would hire in this field. The following data was collected.

$$n = 12 \qquad \sum x^2 = 3162$$
$$\sum x = 176 \qquad \sum y^2 = 10{,}870$$
$$\sum y = 356$$
$$\sum xy = 5629$$

(a) Find an equation for the least squares line.

(b) Find the coefficient of correlation.

18. An economist was interested in the production costs for companies supplying chemicals for use in fertilizers. The data below represents the relationship between the number of tons produced during a given year (x) and the production cost per ton (y) for seven companies.

Number of Tons (in thousands)	Cost per Ton (in dollars)
3.0	40
4.0	50
2.4	50
5.0	35
2.6	55
4.0	35
5.5	30

(a) Find the equation for the least squares line.

(b) Find the coefficient of correlation.

Chapter 1 Test Answers

1. -4

2. Undefined

3. $\frac{3}{7}$

4. $\frac{5}{2}$

5. $3x + 2y = 9$

6. $5x - 6y = 30$

7. $x - 2y = -6$

8. $x - 3y = 13$

9. $y = 5$

10.

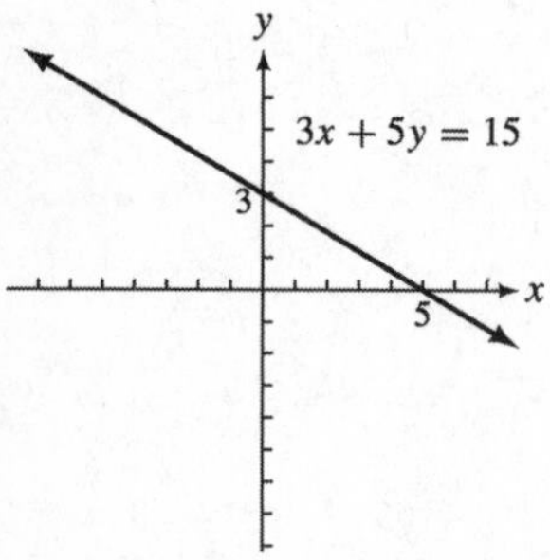

11.

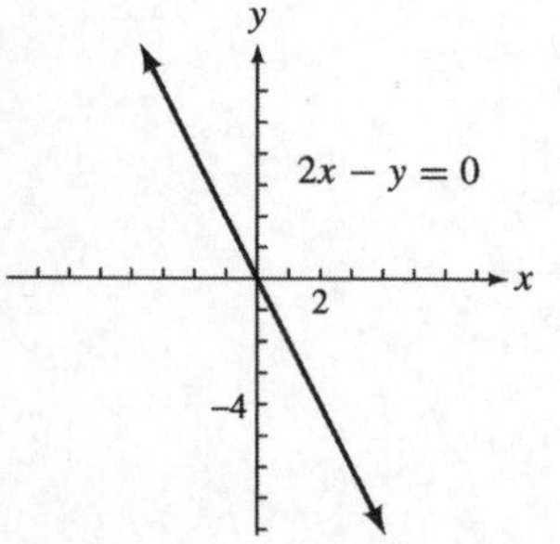

12.

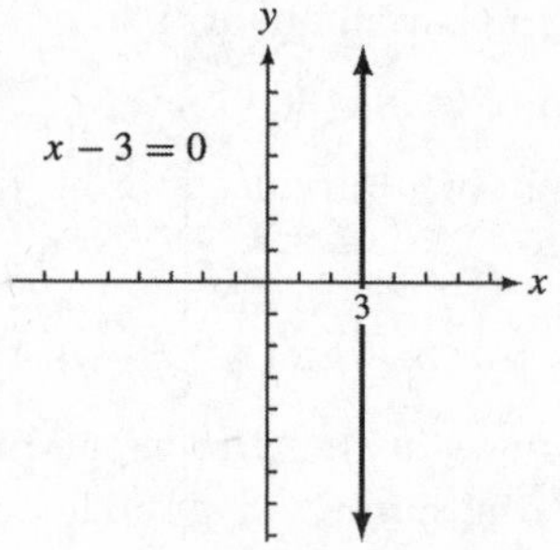

13. (a)

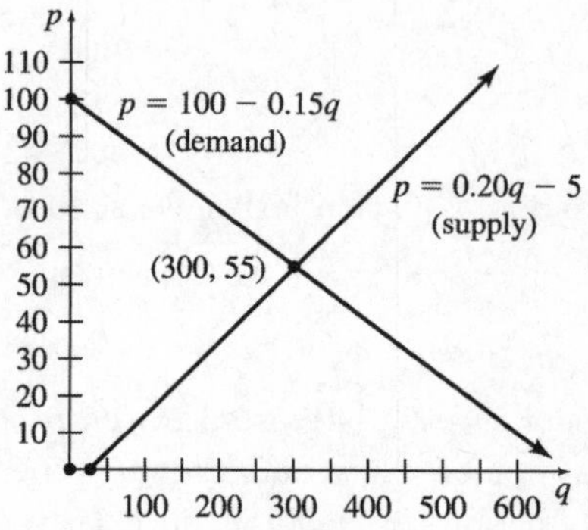

(b) \$55 **(c)** 300 units

14. (a) $y = 10x + 370$ **(b)** \$370
(c) \$10 **(d)** $\overline{C}(x) = 10 + \frac{370}{x}$

15. (a) $y = 100x + 1000$ **(b)** \$1000
(c) \$100 **(d)** $\overline{C}(x) = 100 + \frac{1000}{x}$

16. (a) 200 units **(b)** \$8000 **(c)** 700 units

17. (a) $Y = 0.70x + 19.37$ **(b)** $r = 0.96$

18. (a) $Y = -6.37x + 66.24$ **(b)** $r = -0.79$

Chapter 2

NONLINEAR FUNCTIONS

2.1 Properties of Functions

1. The x-value of 82 corresponds to two y-values, 93 and 14. In a function, each value of x must correspond to exactly one value of y.
The rule is not a function.

3. Each x-value corresponds to exactly one y-value.
The rule is a function.

5. $y = x^3 + 2$

Each x-value corresponds to exactly one y-value.
The rule is a function.

7. $x = |y|$

Each value of x (except 0) corresponds to two y-values.
The rule is not a function.

9. $y = 2x + 3$

x	-2	-1	0	1	2	3
y	-1	1	3	5	7	9

Pairs: $(-2, -1)$, $(-1, 1)$, $(0, 3)$,
$(1, 5)$, $(2, 7)$, $(3, 9)$

Range: $\{-1,\ 1,\ 3,\ 5,\ 7,\ 9\}$

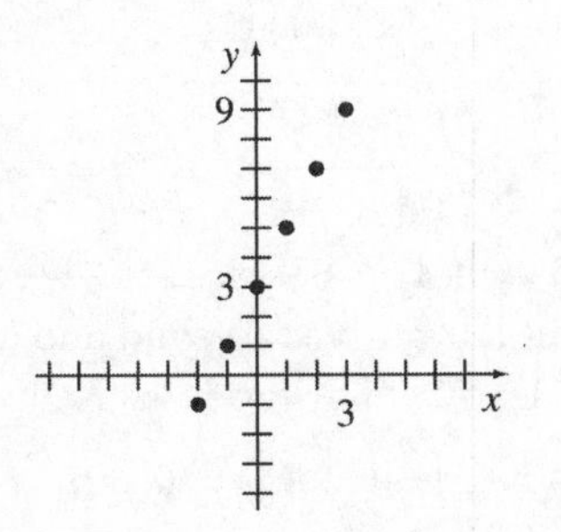

11. $2y - x = 5$

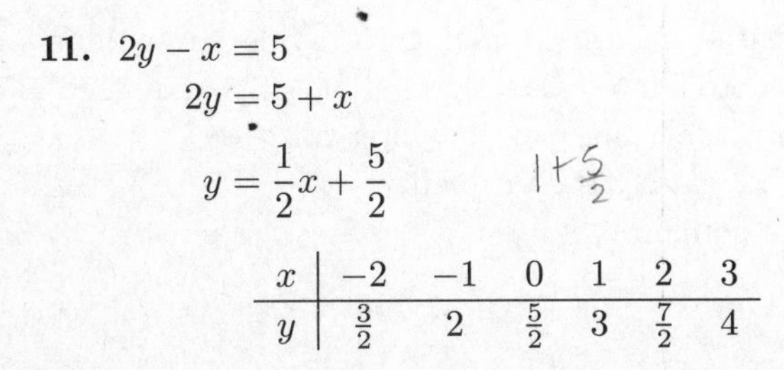

$$2y = 5 + x$$

$$y = \frac{1}{2}x + \frac{5}{2}$$

x	-2	-1	0	1	2	3
y	$\frac{3}{2}$	2	$\frac{5}{2}$	3	$\frac{7}{2}$	4

Pairs:$(-2, \frac{3}{2})$, $(-1, 2)$, $(0, \frac{5}{2})$, $(1, 3)$, $(2, \frac{7}{2})$, $(3, 4)$

Range: $\{\frac{3}{2},\ 2,\ \frac{5}{2}\ 3,\ \frac{7}{2},\ 4\}$

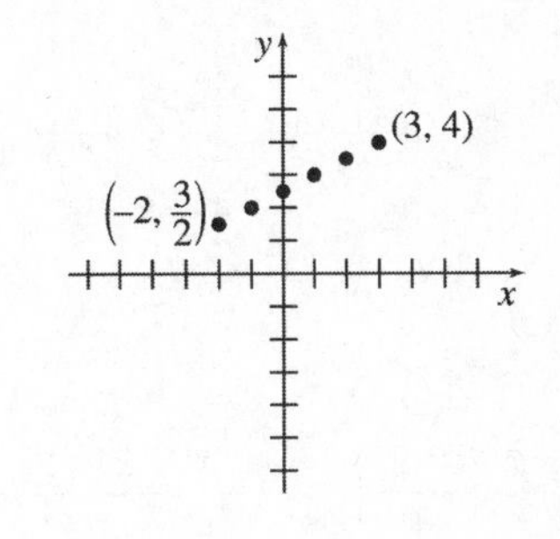

13. $y = x(x + 2)$

x	-2	-1	0	1	2	3
y	0	-1	0	3	8	15

Pairs: $(-2, 0)$, $(-1, -1)$, $(0, 0)$,
$(1, 3)$, $(2, 8)$, $(3, 15)$

Range: $\{-1,\ 0,\ 3,\ 8,\ 15\}$

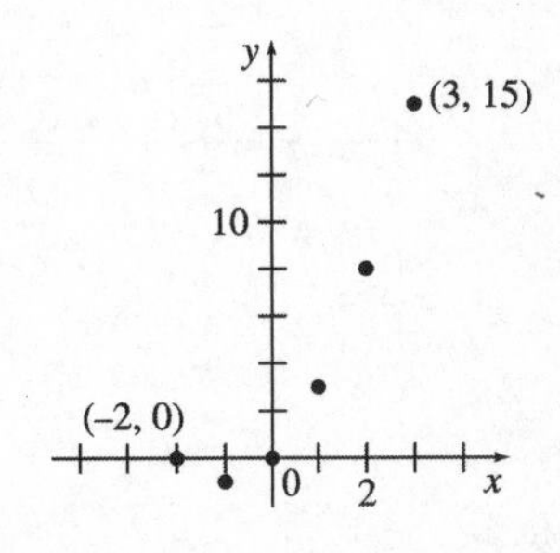

15. $y = x^2$

x	-2	-1	0	1	2	3
y	4	1	0	1	4	9

Pairs: $(-2,4)$, $(-1,1)$, $(0,0)$, $(1,1)$, $(2,4)$, $(3,9)$

Range: $\{0,\ 1,\ 4,\ 9\}$

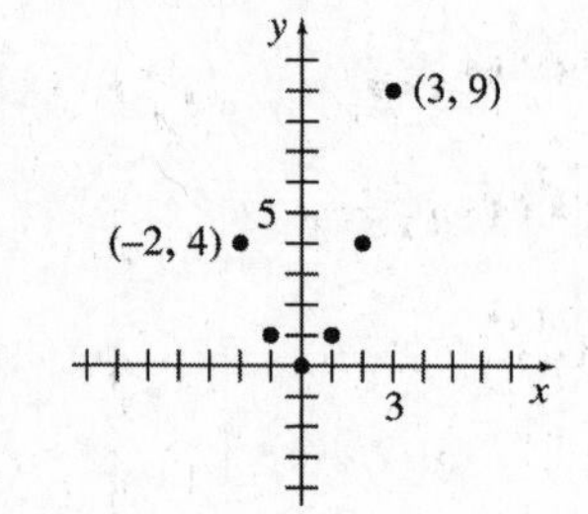

17. $y = \dfrac{1}{x+3}$

x	-2	-1	0	1	2	3
y	1	$\frac{1}{2}$	$\frac{1}{3}$	$\frac{1}{4}$	$\frac{1}{5}$	$\frac{1}{6}$

Pairs: $(-2,1), (-1,\frac{1}{2}), (0,\frac{1}{3}), (1,\frac{1}{4}), (2,\frac{1}{5}), (3,\frac{1}{6})$

Range: $\{1,\ \frac{1}{2},\ \frac{1}{3},\ \frac{1}{4},\ \frac{1}{5},\ \frac{1}{6}\}$

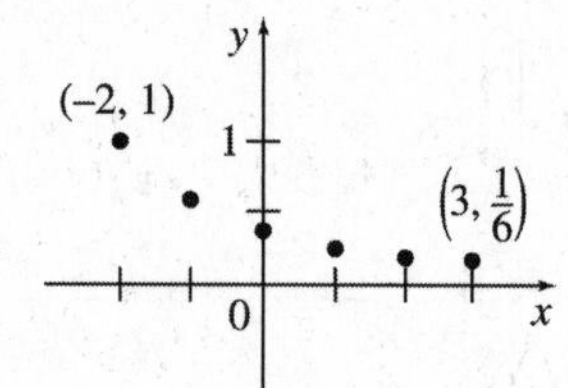

19. $y = \dfrac{2x-2}{x+4}$

x	-2	-1	0	1	2	3
y	-3	$-\frac{4}{3}$	$-\frac{1}{2}$	0	$\frac{1}{3}$	$\frac{4}{7}$

Pairs:$(-2,-3)$, $(-1,-\frac{4}{3})$, $(0,-\frac{1}{2}), (1,0)$, $(2,\frac{1}{3})$, $(3,\frac{4}{7})$

Range: $\{-3,\ -\frac{4}{3},\ -\frac{1}{2},\ 0,\ \frac{1}{3},\ \frac{4}{7}\}$

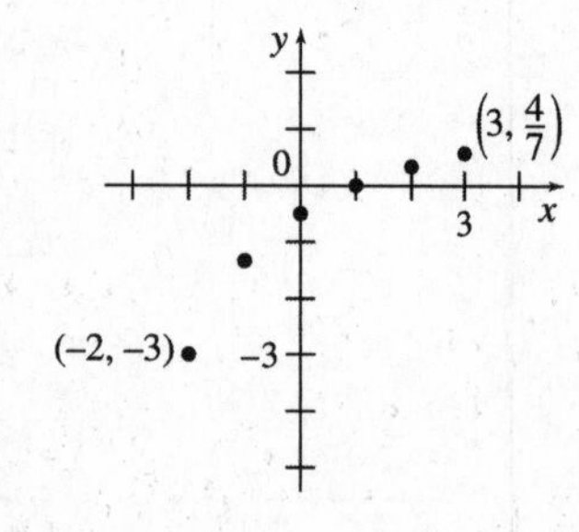

21. $f(x) = 2x$

x can take on any value, so the domain is the set of real numbers, $(-\infty, \infty)$.

23. $f(x) = x^4$

x can take on any value, so the domain is the set of real numbers, $(-\infty, \infty)$.

25. $f(x) = \sqrt{4-x^2}$

For $f(x)$ to be a real number, $4 - x^2 \geq 0$.
Solve $4 - x^2 = 0$.

$$(2-x)(2+x) = 0$$
$$x = 2 \quad \text{or} \quad x = -2$$

The numbers form the intervals $(-\infty,-2)$, $(-2,2)$, and $(2,\infty)$.
Only values in the interval $(-2,2)$ satisfy the inequality. The domain is $[-2,2]$.

27. $f(x) = (x-3)^{1/2} = \sqrt{x-3}$

For $f(x)$ to be a real number,

$$x - 3 \geq 0$$
$$x \geq 3.$$

The domain is $[3,\infty)$.

29. $f(x) = \dfrac{2}{1-x^2} = \dfrac{2}{(1-x)(1+x)}$

Since division by zero is not defined, $(1-x)\cdot(1+x) \neq 0$.
When $(1-x)(1+x) = 0$,

$$1 - x = 0 \quad \text{or} \quad 1 + x = 0$$
$$x = 1 \quad \text{or} \quad x = -1.$$

Thus, x can be any real number except ± 1.
The domain is

$$(-\infty,-1) \cup (-1,1) \cup (1,\infty).$$

31. $f(x) = -\sqrt{\dfrac{2}{x^2-16}} = -\sqrt{\dfrac{2}{(x-4)(x+4)}}.$

$(x-4)\cdot(x+4) > 0$, since $(x-4)\cdot(x+4) < 0$ would produce a negative radicand and $(x-4)\cdot(x+4) = 0$ would lead to division by zero.

Solve $(x-4)\cdot(x+4) = 0$.

$$x - 4 = 0 \quad \text{or} \quad x + 4 = 0$$
$$x = 4 \quad \text{or} \quad x = -4$$

Use the values -4 and 4 to divide the number line into 3 intervals, $(-\infty,-4)$, $(-4,4)$ and $(4,\infty)$. Only the values in the intervals $(-\infty,-4)$ and $(4,\infty)$ satisfy the inequality.
The domain is

$$(-\infty,-4) \cup (4,\infty).$$

33. $f(x) = \sqrt{x^2 - 4x - 5} = \sqrt{(x-5)(x+1)}$

See the method used in Exercise 25.

$$(x-5)(x+1) \geq 0$$

when $x \geq 5$ and when $x \leq -1$.
The domain is $(-\infty, -1] \cup [5, \infty)$.

35. $f(x) = \dfrac{1}{\sqrt{3x^2 + 2x - 1}} = \dfrac{1}{\sqrt{(3x-1)(x+1)}}$

$(3-2)(x+1) > 0$, since the radicand cannot be negative and the denominator of the function cannot be zero.
Solve $(3-1)(x+1) = 0$.

$$3 - 1 = 0 \quad \text{or} \quad x + 1 = 0$$
$$x = \tfrac{1}{3} \quad \text{or} \quad x = -1$$

Use the values -1 and $\frac{1}{3}$ to divide the number line into 3 intervals, $(-\infty, -1)$, $(-1, 4)$ and $(\frac{1}{3}, \infty)$.
Only the values in the intervals $(-\infty, -\frac{1}{3})$ and $(\frac{1}{3}, \infty)$ satisfy the inequality.
The domain is $(-\infty, -1) \cup (\frac{1}{3}, \infty)$.

37. By reading the graph, the domain is all numbers greater than or equal to -5 and less than 4. The range is all numbers greater than or equal to -2 and less than or equal to 6.
Domain: $[-5, 4)$; range: $[-2, 6]$

39. By reading the graph, x can take on any value, but y is less than or equal to 12.
Domain: $(-\infty, \infty)$; range: $(-\infty, 12]$

41. $f(x) = 3x^2 - 4x + 1$

(a) $f(4) = 3(4)^2 - 4(4) + 1$
$= 48 - 16 + 1$
$= 33$

(b) $f\left(-\dfrac{1}{2}\right) = 3\left(-\dfrac{1}{2}\right)^2 - 4\left(-\dfrac{1}{2}\right) + 1$
$= \dfrac{3}{4} + 2 + 1$
$= \dfrac{15}{4}$

(c) $f(a) = 3(a)^2 - 4(a) + 1$
$= 3a^2 - 4a + 1$

(d) $f\left(\dfrac{2}{m}\right) = 3\left(\dfrac{2}{m}\right)^2 - 4\left(\dfrac{2}{m}\right) + 1$
$= \dfrac{12}{m^2} - \dfrac{8}{m} + 1$
or $\dfrac{12 - 8m + m^2}{m^2}$

(e) $f(x) = 1$
$3x^2 - 4x + 1 = 1$
$3x^2 - 4x = 0$
$x(3x - 4) = 0$
$x = 0 \quad \text{or} \quad x = \dfrac{4}{3}$

43. $f(x) = \dfrac{2x + 1}{x - 2}$

(a) $f(4) = \dfrac{2(4) + 1}{4 - 2} = \dfrac{9}{2}$

(b) $f\left(-\dfrac{1}{2}\right) = \dfrac{2(-\frac{1}{2}) + 1}{-\frac{1}{2} - 2}$
$= \dfrac{-1 + 1}{\frac{5}{2}}$
$= \dfrac{0}{\frac{5}{2}} = 0$

(c) $f(a) = \dfrac{2(a) + 1}{(a) - 2} = \dfrac{2a + 1}{a - 2}$

(d) $f\left(\dfrac{2}{m}\right) = \dfrac{2\left(\frac{2}{m}\right) + 1}{\frac{2}{m} - 2}$
$= \dfrac{\frac{4}{m} + \frac{m}{m}}{\frac{2}{m} - \frac{2m}{m}}$
$= \dfrac{\frac{4+m}{m}}{\frac{2-2m}{m}}$
$= \dfrac{4 + m}{m} \cdot \dfrac{m}{2 - 2m}$
$= \dfrac{4 + m}{2 - 2m}$

(e) $f(x) = 1$
$\dfrac{2x + 1}{x - 2} = 1$
$2x + 1 = x - 2$
$x = -3$

45. The domain is all real numbers between the endpoints of the curve, or $[-2, 4]$.
The range is all real numbers between the minimum and maximum values of the function or $[0, 4]$.

(a) $f(-2) = 0$

(b) $f(0) = 4$

(c) $f\left(\dfrac{1}{2}\right) = 3$

(d) From the graph, $f(x) = 1$ when $x = -1.5$, 1.5, or 2.5.

47. The domain is all real numbers between the endpoints of the curve, or $[-2, 4]$.
The range is all real numbers between the minimum and maximum values of the function or $[-3, 2]$.

(a) $f(-2) = -3$

(b) $f(0) = -2$

(c) $f\left(\frac{1}{2}\right) = -1$

(d) From the graph, $f(x) = 1$ when $x = 2.5$.

49. $f(x) = 6x^2 - 2$

$$\begin{aligned} f(t+1) &= 6(t+1)^2 - 2 \\ &= 6(t^2 + 2t + 1) - 2 \\ &= 6t^2 + 12t + 6 - 2 \\ &= 6t^2 + 12t + 4 \end{aligned}$$

51. $g(r+h)$
$= (r+h)^2 - 2(r+h) + 5$
$= r^2 + 2hr + h^2 - 2r - 2h + 5$

53. $g\left(\frac{3}{q}\right) = \left(\frac{3}{q}\right)^2 - 2\left(\frac{3}{q}\right) + 5$
$= \frac{9}{q^2} - \frac{6}{q} + 5$
or $\frac{9 - 6q + 5q^2}{q^2}$

55. A vertical line drawn anywhere through the graph will intersect the graph in only one place. The graph represents a function.

57. A vertical line drawn through the graph may intersect the graph in two places. The graph does not represent a function.

59. A vertical line drawn anywhere through the graph will intersect the graph in only one place. The graph represents a function.

61. $f(x) = 2x + 1$

(a) $f(x+h) = 2(x+h) + 1$
$= 2x + 2h + 1$

(b) $f(x+h) - f(x)$
$= \frac{2x + 2h + 1}{2x + 1}$
$= 2x + 2h + 1 - 2x - 1$
$= 2h$

(c) $\frac{f(x+h) - f(x)}{h}$
$= \frac{\frac{2x+2h+1}{2x+1}}{h}$
$= \frac{2x + 2h + 1 - 2x - 1}{h}$
$= \frac{2h}{h}$
$= 2$

63. $f(x) = 2x^2 - 4x - 5$

(a) $f(x+h)$
$= 2(x+h)^2 - 4(x+h) - 5$
$= 2(x^2 + 2hx + h^2) - 4x - 4h - 5$
$= 2x^2 + 4hx + 2h^2 - 4x - 4h - 5$

(b) $f(x+h) - f(x)$
$= 2x^2 + 4hx + 2h^2 - 4x - 4h - 5$
$- (2x^2 - 4x - 5)$
$= 2x^2 + 4hx + 2h^2 - 4x - 4h - 5$
$- 2x^2 + 4x + 5$
$= 4hx + 2h^2 - 4h$

(c) $\frac{f(x+h) - f(x)}{h}$
$= \frac{4hx + 2h^2 - 4h}{h}$
$= \frac{h(4x + 2h - 4)}{h}$
$= 4x + 2h - 4$

65. $f(x) = \frac{1}{x}$

(a) $f(x+h) = \frac{1}{x+h}$

(b) $f(x+h) - f(x)$
$= \frac{1}{x+h} - \frac{1}{x}$
$= \left(\frac{x}{x}\right)\frac{1}{x+h} - \frac{1}{x}\left(\frac{x+h}{x+h}\right)$
$= \frac{x - (x+h)}{x(x+h)}$
$= \frac{-h}{x(x+h)}$

(c) $\frac{f(x+h)-f(x)}{h}$

$$= \frac{\frac{1}{x+h}-\frac{1}{x}}{h}$$

$$= \frac{\frac{1}{x+h}\left(\frac{x}{x}\right)-\frac{1}{x}\left(\frac{x+h}{x+h}\right)}{h}$$

$$= \frac{\frac{x-(x+h)}{(x+h)x}}{h}$$

$$= \frac{1}{h}\left[\frac{x-x-h}{(x+h)x}\right]$$

$$= \frac{1}{h}\left[\frac{-h}{(x+h)x}\right]$$

$$= \frac{-1}{x(x+h)}$$

67.
$$\begin{aligned} f(x) &= 3x \\ f(-x) &= 3(-x) \\ &= -(3x) \\ &= -f(x) \end{aligned}$$

The function is odd.

69.
$$\begin{aligned} f(x) &= 2x^2 \\ f(-x) &= 2(-x)^2 \\ &= 2x^2 \\ &= f(x) \end{aligned}$$

The function is even.

71.
$$\begin{aligned} f(x) &= \frac{1}{x^2+4} \\ f(-x) &= \frac{1}{(-x)^2+4} \\ &= \frac{1}{x^2+4} \\ &= f(x) \end{aligned}$$

The function is even.

73.
$$\begin{aligned} f(x) &= \frac{x}{x^2-9} \\ f(-x) &= \frac{-x}{(-x)^2-9} \\ &= -\frac{x}{x^2-9} \\ &= -f(x) \end{aligned}$$

The function is odd.

75. **(a)** The independent variable is the years.

(b) The dependent variable is the number of Internet users.

(c) $f(2003) = 719$ million users.

(d) The domain is $1995 \le x \le 2006$.
The range is $16{,}000{,}000 \le y \le 1{,}043{,}000{,}000$.

77. If x is a whole number of days, the cost of renting a car is given by

$$C(x) = 54x + 44.$$

For x in whole days plus a fraction of a day, substitute the next whole number for x in $54x + 44$ because a fraction of a day is charged as a whole day.

(a) $C\left(\frac{3}{4}\right) = C(1) = 54(1) + 44 = \$98$

(b) $C\left(\frac{9}{10}\right) = C(1) = \$98$

(c) $C(1) = \$98$

(d) $C\left(1\frac{5}{8}\right) = C(2) = 54(2) + 44 = \$152$

(e) $C(2.4) = C(3) = 54(3) + 44 = \$206$

(f)

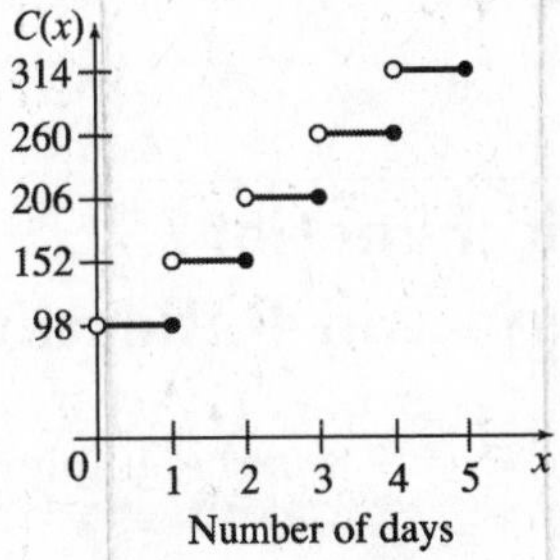

(g) Yes, C is a function.

(h) No, C is not a linear function.

79. **(a)** (i) $y = f(5) = 19.7(5)^{0.753} \approx 66$ kcal/day

(ii) $y = f(25) = 19.7(25)^{0.753} \approx 222$ kcal/day

(b) Since 1 pound equals 0.454 kg, then $x = g(z) = 0.454z$ is the number of kilograms equal in weight to z pounds.

81. **(a)** In the graph, the curves representing wood and coal intersect approximately at the point $(1880, 50)$. So, in 1880 use of wood and coal were both about 50% of the global energy consumption.

(b) In the graph, the curves representing oil and coal intersect approximately at the point $(1965, 35)$. So, in 1965 use of oil and coal were both about 35% of the global energy consumption.

83. **(a)** Let $w =$ the width of the field;
$l =$ the length.

The perimeter of the field is 6000 ft, so

$$\begin{aligned} 2l + 2w &= 6000 \\ l + w &= 3000 \\ l &= 3000 - w. \end{aligned}$$

Thus, the area of the field is given by

$$\begin{aligned} A &= lw \\ A &= (3000 - w)w. \end{aligned}$$

(b) Since $l = 3000 - w$ and w cannot be negative, $0 \le w \le 3000$.
The domain of A is $0 \le w \le 3000$.

(c)

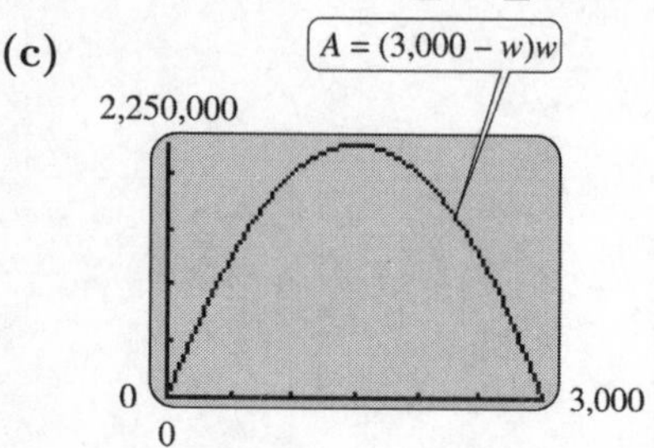

2.2 Quadratic Functions; Translation and Reflection

3. The graph of $y = x^2 - 3$ is the graph of $y = x^2$ translated 3 units downward.
This is graph d.

5. The graph of $y = (x - 3)^2 + 2$ is the graph of $y = x^2$ translated 3 units to the right and 2 units upward.
This is graph a.

7. The graph of $y = -(3 - x)^2 + 2$ is the same as the graph of $y = -(x - 3)^2 + 2$. This is the graph of $y = x^2$ reflected in the x-axis, translated 3 units to the right, and translated 2 units upward.
This is graph c.

9. $y = x^2 + 5x + 6$
$y = (x + 3)(x + 2)$
Set $y = 0$ to find the x-intercepts.

$$\begin{aligned} 0 &= (x + 3)(x + 2) \\ x &= -3, \ x = -2 \end{aligned}$$

The x-intercepts are -3 and -2.
Set $x = 0$ to find the y-intercept.

$$\begin{aligned} y &= 0^2 + 5(0) + 6 \\ y &= 6 \end{aligned}$$

The y-intercept is 6.
The x-coordinate of the vertex is

$$x = \frac{-b}{2a} = \frac{-5}{2} = -\frac{5}{2}.$$

Substitute to find the y-coordinate.

$$y = (-\frac{5}{2})^2 + 5(-\frac{5}{2}) + 6 = \frac{25}{4} - \frac{25}{2} + 6 = -\frac{1}{4}$$

The vertex is $(-\frac{5}{2}, -\frac{1}{4})$.
The axis is $x = -\frac{5}{2}$, the vertical line through the vertex.

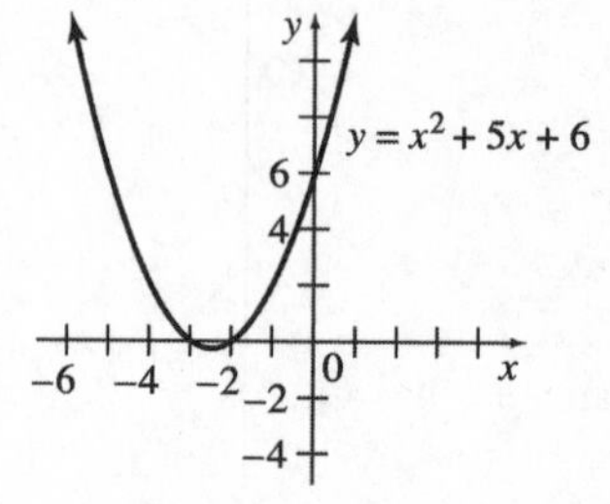

11. $y = -2x^2 - 12x - 16$
$= -2(x^2 + 6x + 8)$
$= -2(x + 4)(x + 2)$

Let $y = 0$.

$$\begin{aligned} 0 &= -2(x + 4)(x + 2) \\ x &= -4, \ x = -2 \end{aligned}$$

-4 and -2 are the x-intercepts.
Let $x = 0$.

$$y = -2(0)^2 + 12(0) - 16$$

-16 is the y-intercept.

Vertex: $x = \dfrac{-b}{2a} = \dfrac{12}{-4} = -3$

$$\begin{aligned} y &= -2(-3)^2 - 12(-3) - 16 \\ &= -18 + 36 - 16 = 2 \end{aligned}$$

The vertex is $(-3, 2)$.

The axis is $x = -3$, the vertical line through the vertex.

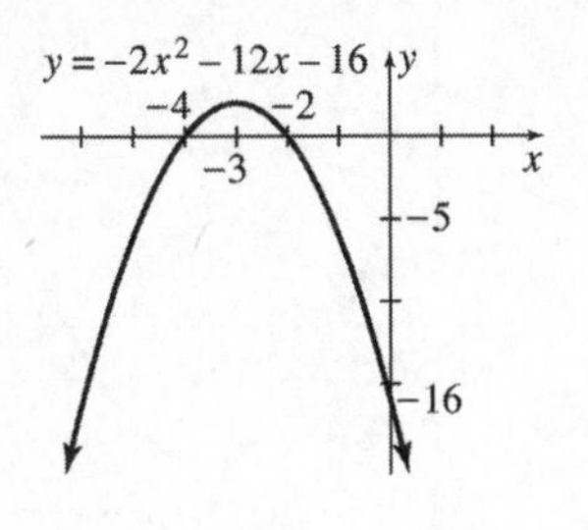

13. $y = 2x^2 + 8x - 8$

Let $y = 0$.

$$2x^2 + 8x - 8 = 0$$
$$x^2 + 4x - 4 = 0$$

$$x = \frac{-4 \pm \sqrt{4^2 - 4(1)(-4)}}{2(1)}$$
$$= \frac{-4 \pm \sqrt{32}}{2} = \frac{-4 \pm 4\sqrt{2}}{2}$$
$$= -2 \pm 2\sqrt{2}$$

The x-intercepts are $-2 \pm 2\sqrt{2} \approx 0.83$ or -4.83.
Let $x = 0$.

$$y = 2(0)^2 + 8(0) - 8 = -8$$

The y-intercept is -8.

The x-coordinate of the vertex is

$$x = \frac{-b}{2a} = -\frac{8}{4} = -2.$$

If $x = -2$,

$$y = 2(-2)^2 + 8(-2) - 8 = 8 - 16 - 8 = -16.$$

The vertex is $(-2, -16)$.
The axis is $x = -2$.

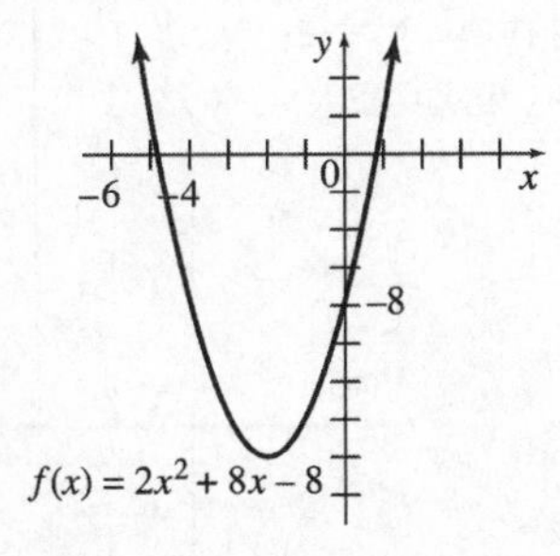

15. $f(x) = 2x^2 - 4x + 5$

Let $f(x) = 0$.

$$0 = 2x^2 - 4x + 5$$

$$x = \frac{-(-4) \pm \sqrt{(-4)^2 - 4(2)(5)}}{2(2)}$$
$$= \frac{4 \pm \sqrt{16 - 40}}{4}$$
$$= \frac{4 \pm \sqrt{-24}}{4}$$

Since the radicand is negative, there are no x-intercepts.

Let $x = 0$.

$$y = 2(0)^2 - 4(0) + 5$$
$$y = 5$$

5 is the y-intercept.

Vertex: $x = \dfrac{-b}{2a} = \dfrac{-(-4)}{2(2)} = \dfrac{4}{4} = 1$

$$y = 2(1)^2 - 4(1) + 5 = 2 - 4 + 5 = 3$$

The vertex is $(1, 3)$.
The axis is $x = 1$.

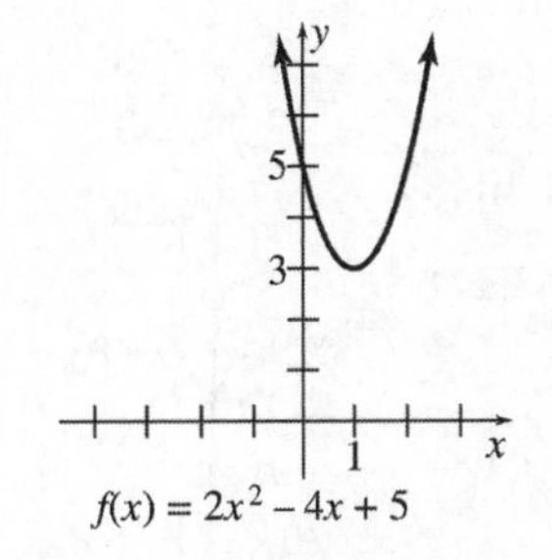

17. $f(x) = -2x^2 + 16x - 21$

Let $f(x) = 0$
Use the quadratic formula.

$$x = \frac{-16 \pm \sqrt{16^2 - 4(-2)(-21)}}{2(-2)}$$
$$= \frac{-16 \pm \sqrt{88}}{-4}$$
$$= \frac{-16 \pm 2\sqrt{22}}{-4}$$
$$= 4 \pm \frac{\sqrt{22}}{2}$$

The x-intercepts are $4 + \frac{\sqrt{22}}{2} \approx 6.35$ and

$4 - \frac{\sqrt{22}}{2} \approx 1.65$.

Let $x = 0$.

$$y = -2(0)^2 + 16(0) - 21$$

$$y = -21$$

-21 is the y-intercept.

Vertex: $x = \frac{-b}{2a} = \frac{-16}{2(-2)} = \frac{-16}{-4} = 4$

$$y = -2(4)^2 + 16(4) - 21$$

$$= -32 + 64 - 21 = 11$$

The vertex is $(4, 11)$.
The axis is $x = 4$.

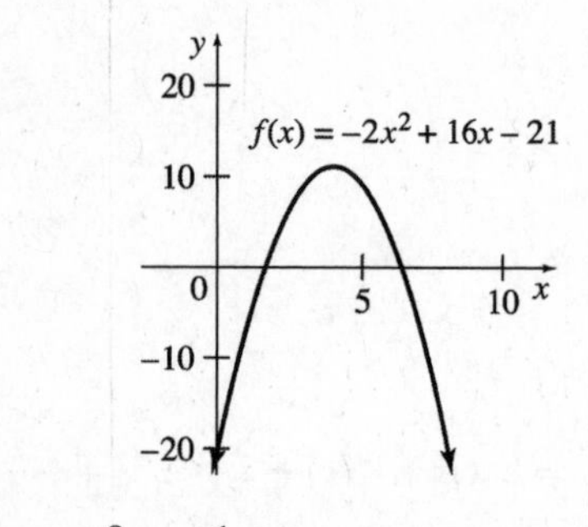

19. $y = \frac{1}{3}x^2 - \frac{8}{3}x + \frac{1}{3}$
Let $y = 0$.

$$0 = \frac{1}{3}x^2 - \frac{8}{3}x + \frac{1}{3}$$

Multiply by 3.

$0 = x^2 - 8x + 1$

$$x = \frac{-(-8) \pm \sqrt{(-8)^2 - 4(1)(1)}}{2(1)}$$

$$= \frac{8 \pm \sqrt{64 - 4}}{2} = \frac{8 \pm \sqrt{60}}{2}$$

$$= \frac{8 \pm 2\sqrt{15}}{2} = 4 \pm \sqrt{15}$$

The x-intercepts are $4+\sqrt{15} \approx 7.87$ and $4-\sqrt{15} \approx 0.13$.
Let $x = 0$.

$$y = \frac{1}{3}(0)^2 - \frac{8}{3}(0) + \frac{1}{3}$$

$\frac{1}{3}$ is the y-intercept.

Vertex: $x = \frac{-b}{2a} = \frac{-\left(-\frac{8}{3}\right)}{2\left(\frac{1}{3}\right)} = \frac{\frac{8}{3}}{\frac{2}{3}} = 4$

$$y = \frac{1}{3}(4)^2 - \frac{8}{3}(4) + \frac{1}{3}$$

$$= \frac{16}{3} - \frac{32}{3} + \frac{1}{3} = -\frac{15}{3} = -5$$

The vertex is $(4, -5)$.
The axis is $x = 4$.

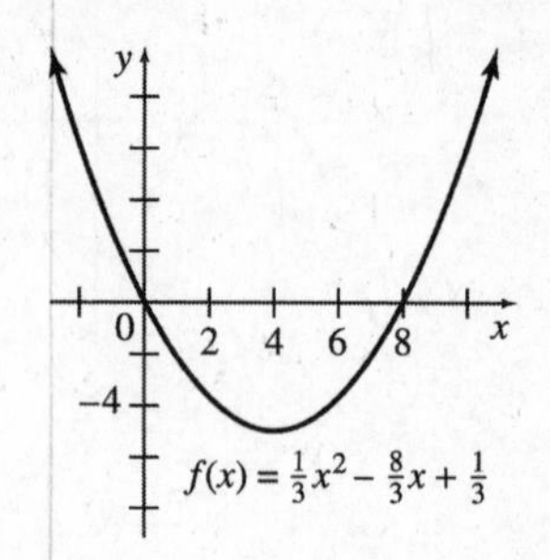

21. The graph of $y = \sqrt{x+2} - 4$ is the graph of $y = \sqrt{x}$ translated 2 units to the left and 4 units downward.
This is graph d.

23. The graph of $y = \sqrt{-x+2} - 4$ is the graph of $y = \sqrt{-(x-2)} - 4$, which is the graph of $y = \sqrt{x}$ reflected in the y-axis, translated 2 units to the right, and translated 4 units downward.
This is graph c.

25. The graph of $y = -\sqrt{x+2} - 4$ is the graph of $y = \sqrt{x}$ reflected in the x-axis, translated 2 units to the left, and translated 4 units downward.
This is graph e.

27. The graph of $y = -f(x)$ is the graph of $y = f(x)$ reflected in the x-axis.

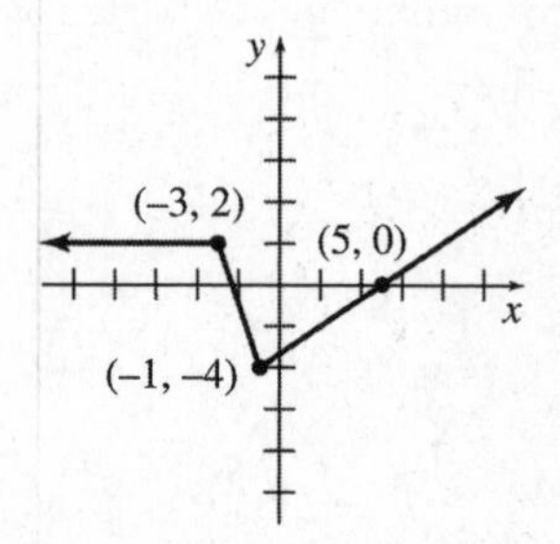

29. The graph of $y = f(-x)$ is the graph of $y = f(x)$ reflected in the y-axis.

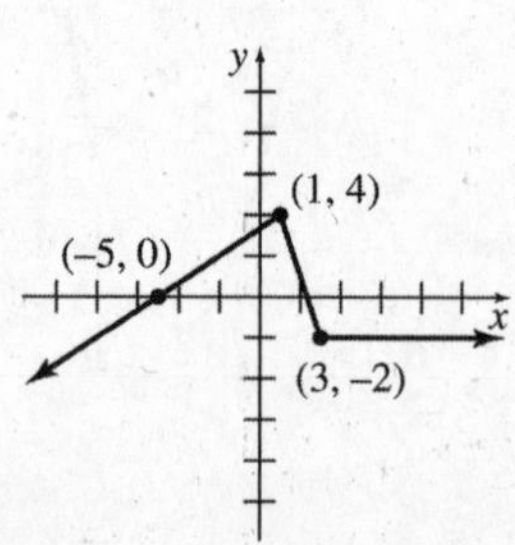

31. $f(x) = \sqrt{x-2} + 2$

Translate the graph of $f(x) = \sqrt{x}$ 2 units right and 2 units up.

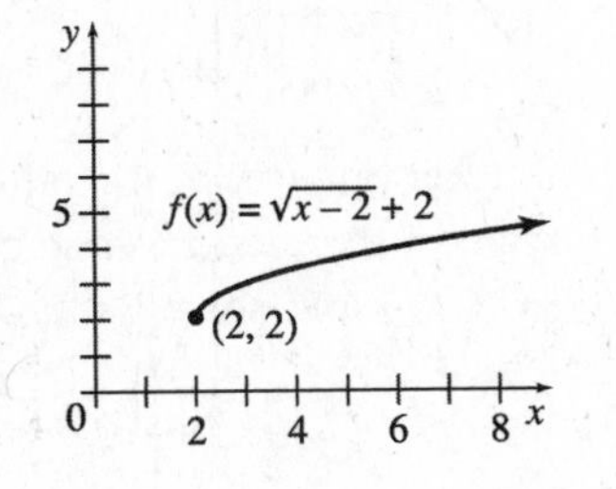

33. $f(x) = -\sqrt{2-x} - 2$
$= -\sqrt{-(x-2)} - 2$

Reflect the graph of $f(x)$ vertically and horizontally.
Translate the graph 2 units right and 2 units down.

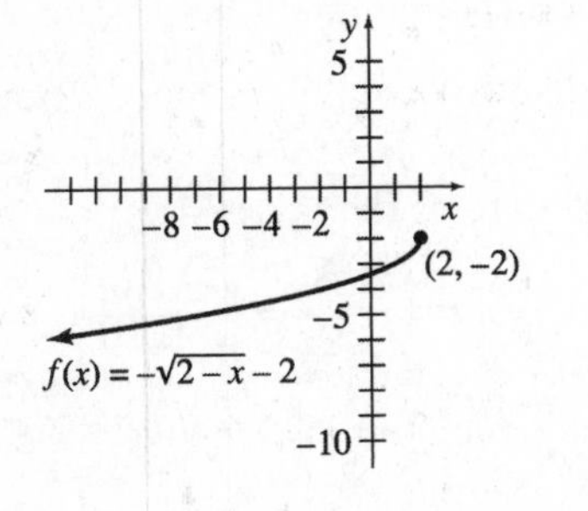

35. If $0 < a < 1$, the graph of $f(ax)$ will be flatter and wider than the graph of $f(x)$.
Multiplying x by a fraction makes the y-values less than the original y-values.

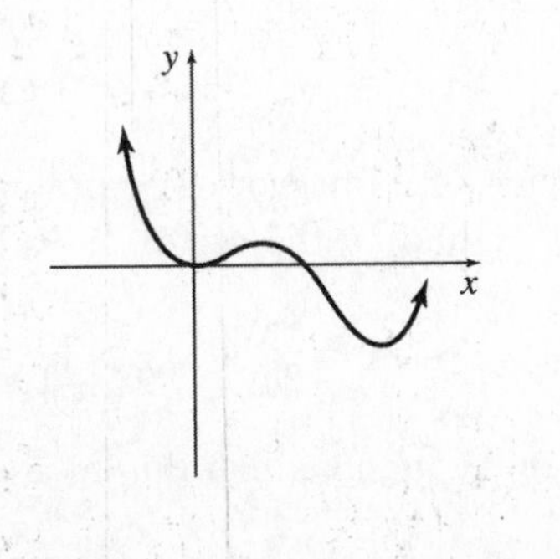

37. If $-1 < a < 0$, the graph of $f(ax)$ will be reflected horizontally, since a is negative. It will be flatter because multiplying x by a fraction decreases the corresponding y-values.

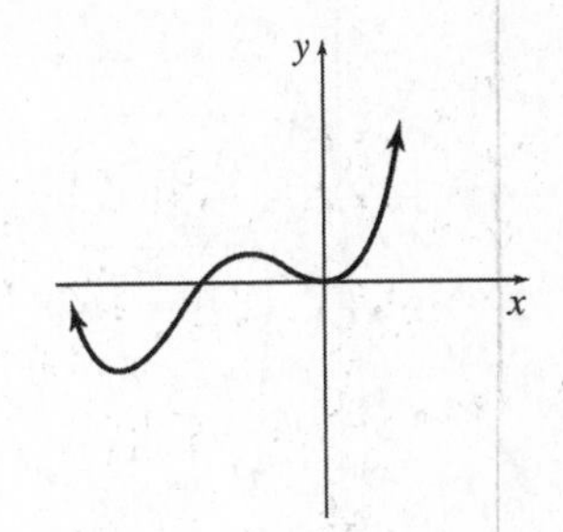

39. If $0 < a < 1$, the graph of $af(x)$ will be flatter and wider than the graph of $f(x)$. Each y-value is only a fraction of the height of the original y-values.

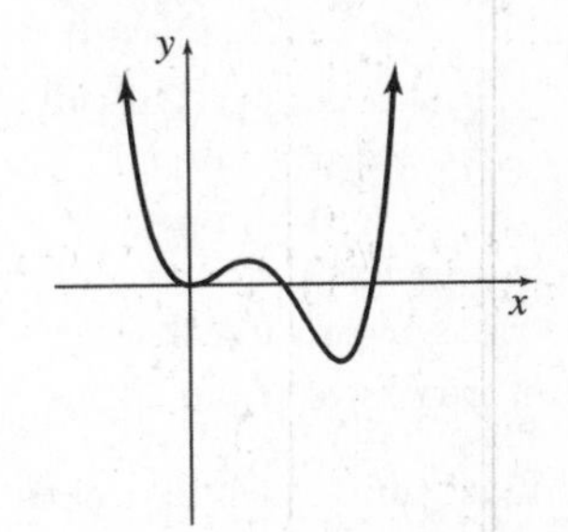

41. If $-1 < a < 0$, the graph will be reflected vertically, since a will be negative. Also, because a is a fraction, the graph will be flatter because each y-value will only be a fraction of its original height.

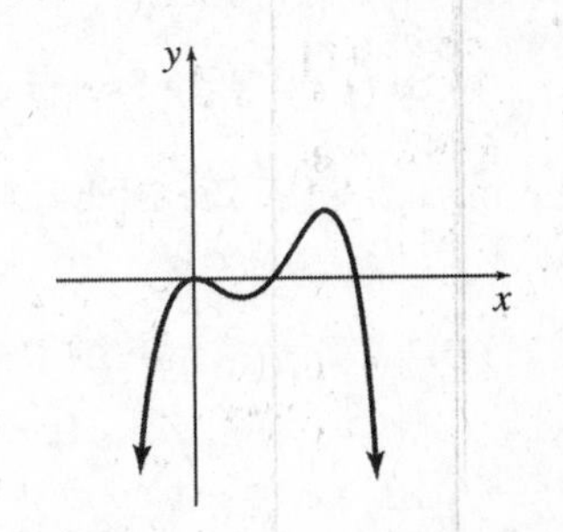

43. **(a)** Since the graph of $y = f(x)$ is reflected vertically to obtain the graph of $y = -f(x)$, the x-intercept is unchanged. The x-intercept of the graph of $y = f(x)$ is r.

(b) Since the graph of $y = f(x)$ is reflected horizontally to obtain the graph of $y = f(-x)$, the x-intercept of the graph of $y = f(-x)$ is $-r$.

(c) Since the graph of $y = f(x)$ is reflected both horizontally and vertically to obtain the graph of $y = -f(-x)$, the x-intercept of the graph of $y = -f(-x)$ is $-r$.

45. (a)

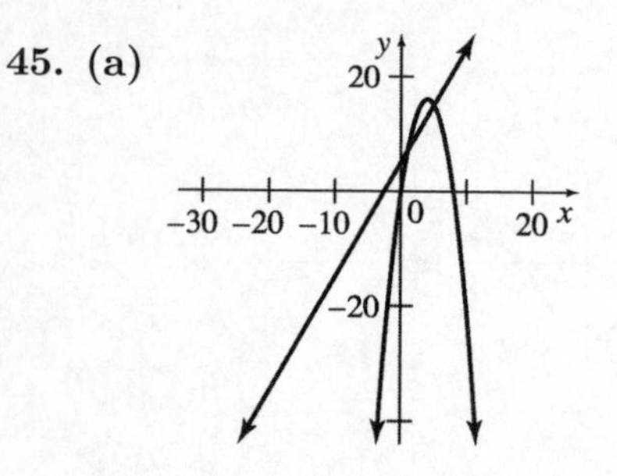

(b) Break-even quantities are values of x = number of widgets for which revenue and cost are equal. Set $R(x) = C(x)$ and solve for x.

$$\begin{aligned} -x^2 + 8x &= 2x + 5 \\ x^2 - 6x + 5 &= 0 \\ (x-5)(x-1) &= 0 \\ x - 5 = 0 \quad \text{or} \quad x - 1 &= 0 \\ x = 5 \quad \text{or} \quad x &= 1 \end{aligned}$$

So, the break-even quantities are 1 and 5. The minimum break-even quantity is $x = 1$.

(c) The maximum revenue occurs at the vertex of R. Since $R(x) = -x^2 + 8x$, then the x-coordinate of the vertex is

$$x = -\frac{b}{2a} = -\frac{8}{2(-1)} = 4.$$

So, the maximum revenue is

$$R(4) = -4^2 + 8(4) = 16.$$

(d) The maximum profit is the maximum difference $R(x) - C(x)$. Since

$$\begin{aligned} P(x) &= R(x) - C(x) \\ &= -x^2 + 8x - (2x + 5) \\ &= -x^2 + 6x - 5 \end{aligned}$$

is a quadratic function, we can find the maximum profit by finding the vertex of P. This occurs at

$$x = -\frac{b}{2a} = \frac{-6}{2(-1)} = 3.$$

Therefore, the maximum profit is

$$P(3) = -(3)^2 + 6(3) - 5 = 4.$$

47. (a)

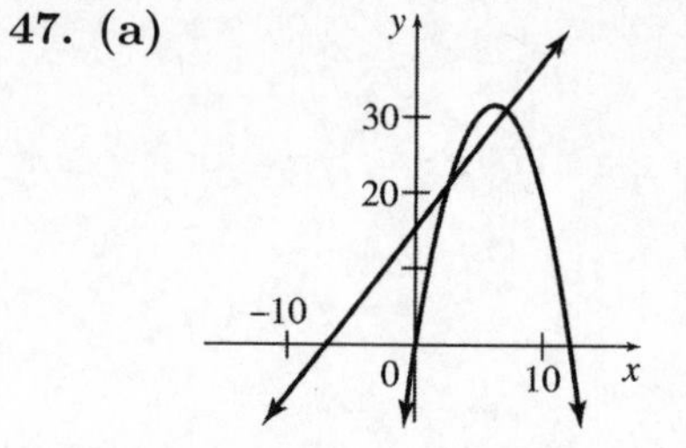

(b) Break-even quantities are values of x = number of widgets for which revenue equals cost. Set $R(x) = C(x)$ and solve for x.

$$\begin{aligned} -\frac{4}{5}x^2 + 10x &= 2x + 15 \\ \frac{4}{5}x^2 - 8x + 15 &= 0 \\ 4x^2 - 40x + 75 &= 0 \\ 4x^2 - 10x - 30x + 75 &= 0 \\ 2x(2x-5) - 15(2x-5) &= 0 \\ (2x-5)(2x-15) &= 0 \\ 2x - 5 = 0 \quad \text{or} \quad 2x - 15 &= 0 \\ x = 2.5 \quad \text{or} \quad x &= 7.5 \end{aligned}$$

So, the break-even quantities are 2.5 and 7.5 with $x = 2.5$ the minimum break-even quantity.

(c) The maximum revenue occurs at the vertex of R. Since $R(x) = -\frac{4}{5}x^2 + 10x$, then the x-coordinate of the vertex is

$$x = -\frac{b}{2a} = -\frac{10}{2\left(\frac{-4}{5}\right)} = 6.25.$$

So, the maximum revenue is

$$R(6.25) = 31.25.$$

(d) The maximum profit is the maximum difference $R(x) - C(x)$. Since

$$\begin{aligned} P(x) &= R(x) - C(x) \\ &= -\frac{4}{5}x^2 + 10x - (2x + 15) \\ &= -\frac{4}{5}x^2 + 8x - 15 \end{aligned}$$

is a quadratic function, we can find the maximum profit by finding the vertex of P. This occurs at

$$x = -\frac{b}{2a} = -\frac{8}{2\left(\frac{-4}{5}\right)} = 5.$$

Therefore, the maximum profit is

$$P(5) = -\frac{4}{5}5^2 + 8(5) - 15 = 5.$$

49. $R(x) = 8000 + 70x - x^2$
$= -x^2 + 70x + 8000$

The maximum revenue occurs at the vertex.

$$x = \frac{-b}{2a} = \frac{-70}{2(-1)} = 35$$

$$\begin{aligned} y &= 8000 + 70(35) - (35)^2 \\ &= 8000 + 2450 - 1225 \\ &= 9225 \end{aligned}$$

The vertex is $(35, 9225)$.
The maximum revenue of \$9225 is realized when 35 seats are left unsold.

51. $p = 500 - x$

(a) The revenue is

$$\begin{aligned} R(x) &= px \\ &= (500 - x)(x) \\ &= 500x - x^2. \end{aligned}$$

(b)

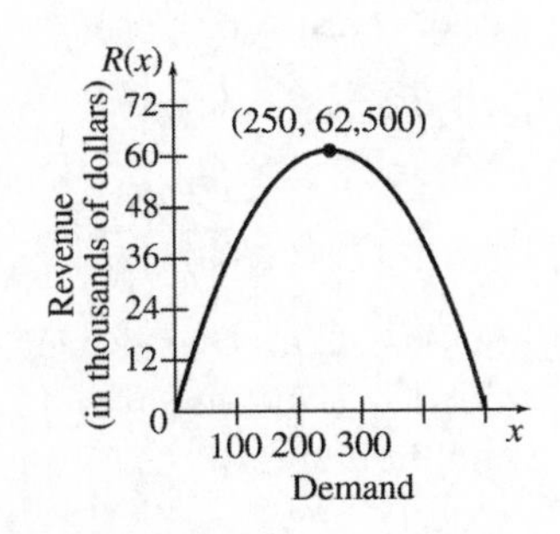

(c) From the graph, the vertex is halfway between $x = 0$ and $x = 500$, so $x = 250$ units corresponds to maximum revenue. Then the price is

$$\begin{aligned} p &= 500 - x \\ &= 500 - 250 = \$250. \end{aligned}$$

Note that price, p, cannot be read directly from the graph of

$$R(x) = 500x - x^2.$$

(d) $R(x) = 500x - x^2$
$= -x^2 + 500x$

Find the vertex.

$$x = \frac{-b}{2a} = \frac{-500}{2(-1)} = 250$$

$$\begin{aligned} y &= -(250)^2 + 500(250) \\ &= 62,500 \end{aligned}$$

The vertex is (250, 62,500).
The maximum revenue is \$62,500.

53. Let $x =$ the number of \$25 increases.

(a) Rent per apartment: $800 + 25x$

(b) Number of apartments rented: $80 - x$

(c) Revenue:

$$\begin{aligned} R(x) &= (\text{number of apartments rented}) \\ &\quad \times (\text{rent per apartment}) \\ &= (80 - x)(800 + 25x) \\ &= -25x^2 + 1200x + 64,000 \end{aligned}$$

(d) Find the vertex:

$$x = \frac{-b}{2a} = \frac{-1200}{2(-25)} = 24$$

$$\begin{aligned} y &= -25(24)^2 + 1200(24) + 64,000 \\ &= 78,400 \end{aligned}$$

The vertex is (24, 78,400). The maximum revenue occurs when $x = 24$.

(e) The maximum revenue is the y-coordinate of the vertex, or \$78,400.

55. $S(x) = 1 - 0.058x - 0.076x^2$

(a)

$$\begin{aligned} 0.50 &= 1 - 0.058x - 0.076x^2 \\ 0.076x^2 + 0.058x - 0.50 &= 0 \\ 76x^2 + 58x - 500 &= 0 \\ 38x^2 + 29x - 250 &= 0 \end{aligned}$$

$$\begin{aligned} x &= \frac{-29 \pm \sqrt{(29)^2 - 4(38)(-250)}}{2(38)} \\ &= \frac{-29 \pm \sqrt{38,841}}{76} \end{aligned}$$

$\frac{-29-\sqrt{38,841}}{76} \approx -2.97$ and $\frac{-29+\sqrt{38,841}}{76} \approx 2.21$

We ignore the negative value.
The value $x = 2.2$ represents 2.2 decades or 22 years, and 22 years after 65 is 87.
The median length of life is 87 years.

(b) If nobody lives, $S(x) = 0$.

$$\begin{aligned} 1 - 0.058x - 0.076x^2 &= 0 \\ 76x^2 + 58x - 1000 &= 0 \\ 38x^2 + 29x - 500 &= 0 \end{aligned}$$

$$\begin{aligned} x &= \frac{-29 \pm \sqrt{(29)^2 - 4(38)(-500)}}{2(38)} \\ &= \frac{-29 \pm \sqrt{76,841}}{76} \end{aligned}$$

$\frac{-29-\sqrt{76,841}}{76} \approx -4.03$ and $\frac{-29+\sqrt{76,841}}{76} \approx 3.27$

We ignore the negative value.

The value $x = 3.3$ represents 3.3 decades or 33 years, and 33 years after 65 is 98.
Virtually nobody lives beyond 98 years.

57. **(a)** The vertex of the quadratic function $y = 0.057x - 0.001x^2$ is at

$$x = -\frac{b}{2a} = -\frac{0.057}{2(-0.001)} = 28.5.$$

Since the coefficient of the leading term, -0.001, is negative, then the graph of the function opens downward, so a maximum is reached at 28.5 weeks of gestation.

(b) The maximum splenic artery resistance reached at the vertex is

$$\begin{aligned} y &= 0.057(28.5) - 0.001(28.5)^2 \\ &\approx 0.81. \end{aligned}$$

(c) The splenic artery resistance equals 0, when $y = 0$.

$$\begin{aligned} 0.057x - 0.001x^2 &= 0 && \text{Substitute in the expression in } x \text{ for } y. \\ x(0.057 - 0.001x) &= 0 && \text{Factor.} \\ x = 0 \text{ or } 0.057 - 0.001x &= 0 && \text{Set each factor equal to 0.} \\ x &= \frac{0.057}{0.001} = 57 \end{aligned}$$

So, the splenic artery resistance equals 0 at 0 weeks or 57 weeks of gestation.
No, this is not reasonable because at $x = 0$ or 57 weeks, the fetus does not exist.

59. **(a)**

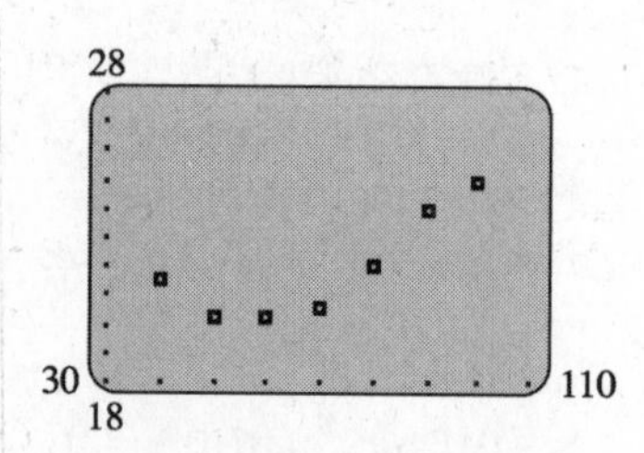

(c) $y = 0.002726x^2 - 0.3113x + 29.33$

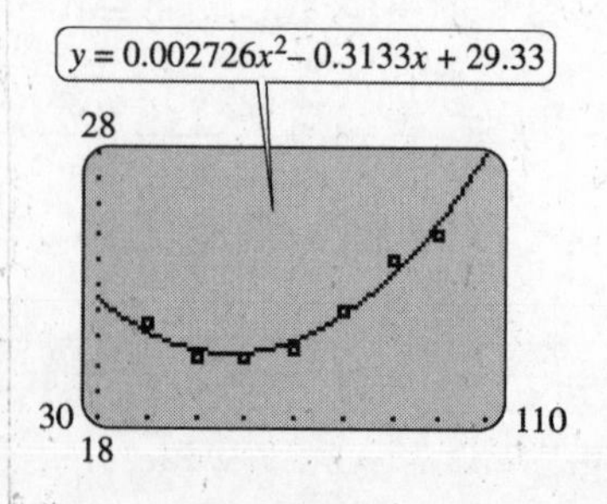

(d) Given that $(h, k) = (60, 20.3)$, the equation has the form

$$y = a(x - 60)^2 + 20.3.$$

Since $(100, 25.1)$ is also on the curve.

$$\begin{aligned} 25.1 &= a(100 - 60)^2 + 20.3 \\ 4.8 &= 1600a \\ a &= 0.003 \end{aligned}$$

(e)

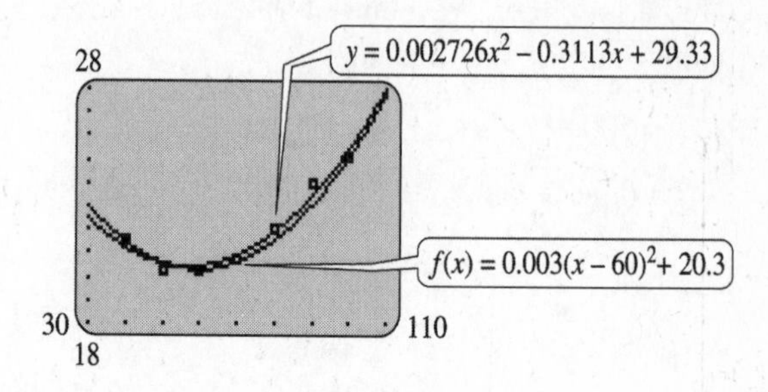

The two graphs are very close.

61. $f(x) = 60.0 - 2.28x + 0.0232x^2$

$$\frac{-b}{2a} = -\frac{-2.28}{2(0.0232)} \approx 49.1$$

The minimum value occurs when $x \approx 49.1$. The age at which the accident rate is a minimum is 49 years. The minimum rate is

$$\begin{aligned} f(49.1) &= 60.0 - 2.28(49.1) + 0.0232(49.1)^2 \\ &= 60.0 - 111.948 + 55.930792 \\ &\approx 3.98. \end{aligned}$$

63. $y = 0.056057x^2 + 1.06657x$

(a) If $x = 25$ mph,

$$\begin{aligned} y &= 0.056057(25)^2 + 1.06657(25) \\ y &\approx 61.7. \end{aligned}$$

At 25 mph, the stopping distance is approximately 61.7 ft.

(b)

$$\begin{aligned} 0.056057x^2 + 1.06657x &= 150 \\ 0.056057x^2 + 1.06657x - 150 &= 0 \end{aligned}$$

$$x = \frac{-1.06657 \pm \sqrt{(1.06657)^2 - 4(0.056057)(-150)}}{2(0.056057)}$$

$x \approx 43.08$ or $x \approx -62.11$

We ignore the negative value.
To stop within 150 ft, the fastest speed you can drive is 43 mph.

65. Let $x =$ the length of the lot and
$y =$ the width of the lot.

The perimeter is given by

$$\begin{aligned} P &= 2x + 2y \\ 380 &= 2x + 2y \\ 190 &= x + y \\ 190 - x &= y. \end{aligned}$$

Area $= xy$ (quantity to be maximized)

$$\begin{aligned} A &= x(190 - x) \\ &= 190x - x^2 \\ &= -x^2 + 190x \end{aligned}$$

Find the vertex: $\frac{-b}{2a} = \frac{-190}{-2} = 95$

$$\begin{aligned} y &= -(95)^2 + 190(95) \\ &= 9025 \end{aligned}$$

This is a parabola with vertex $(95, 9025)$ that opens downward. The maximum area is the value of A at the vertex, or 9025 sq ft.

67. Sketch the culvert on the xy-axes as a parabola that opens upward with vertex at $(0, 0)$.

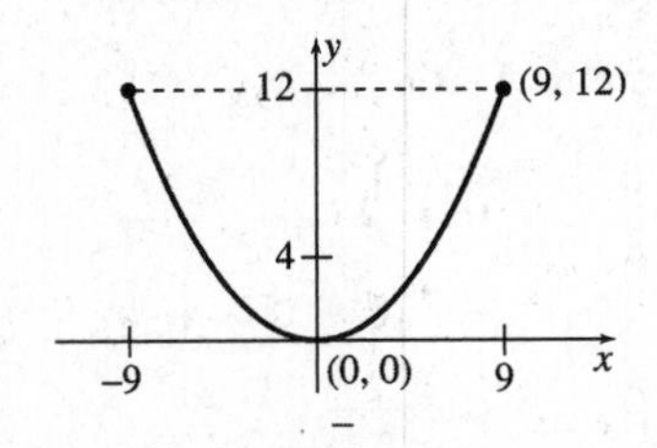

The equation is of the form $y = ax^2$. Since the culvert is 18 ft wide at 12 ft from its vertex, the points $(9, 12)$ and $(-9, 12)$ are on the parabola. Use $(9, 12)$ as one point on the parabola.

$$\begin{aligned} 12 &= a(9)^2 \\ 12 &= 81a \\ \frac{12}{81} &= a \\ \frac{4}{27} &= a \end{aligned}$$

Thus,

$$y = \frac{4}{27}x^2.$$

To find the width 8 feet from the top, find the points with

$$y\text{-value } = 12 - 8 = 4.$$

Thus,

$$\begin{aligned} 4 &= \frac{4}{27}x^2 \\ 108 &= 4x^2 \\ 27 &= x^2 \\ x^2 &= 27 \\ x &= \pm\sqrt{27} \\ x &= \pm 3\sqrt{3}. \end{aligned}$$

The width of the culvert is $3\sqrt{3} + |-3\sqrt{3}| = 6\sqrt{3}$ ft ≈ 10.39 ft.

2.3 Polynomial and Rational Functions

3. The graph of $f(x) = (x - 2)^3 + 3$ is the graph of $y = x^3$ translated 2 units to the right and 3 units upward.

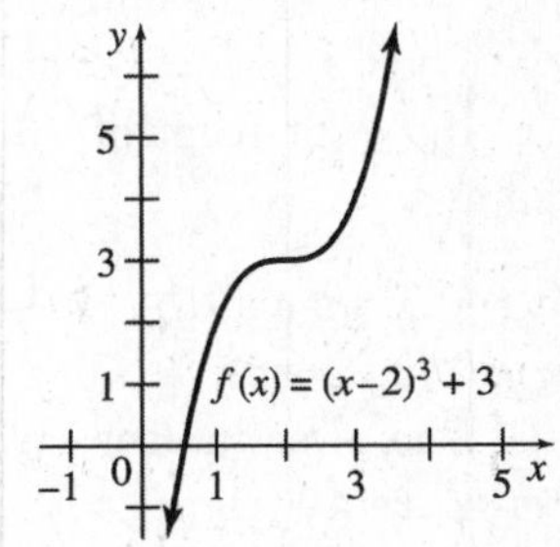

5. The graph of $f(x) = -(x + 3)^4 + 1$ is the graph of $y = x^4$ reflected horizontally, translated 3 units to the left, and translated 1 unit upward.

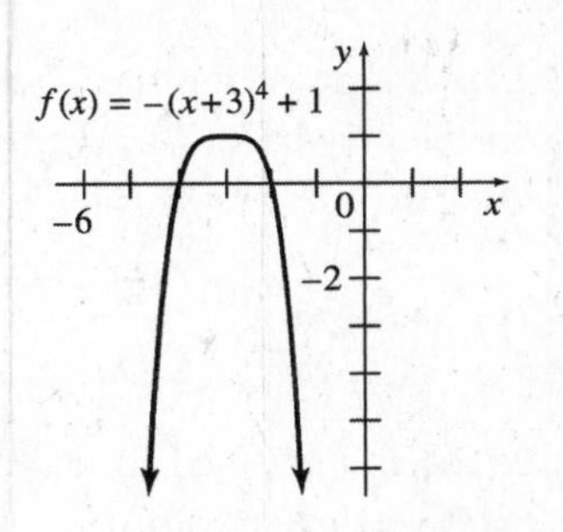

7. The graph of $y = x^3 - 7x - 9$ has the right end up, the left end down, at most two turning points, and a y-intercept of -9.
This is graph d.

9. The graph of $y = -x^3 - 4x^2 + x + 6$ has the right end down, the left end up, at most two turning points, and a y-intercept of 6.
This is graph e.

11. The graph of $y = x^4 - 5x^2 + 7$ has both ends up, at most three turning points, and a y-intercept of 7.
This is graph i.

13. The graph of $y = -x^4 + 2x^3 + 10x + 15$ has both ends down, at most three turning points, and a y-intercept of 15.
This is graph g.

15. The graph of $y = -x^5 + 4x^4 + x^3 - 16x^2 + 12x + 5$ has the right end down, the left end up, at most four turning points, and a y-intercept of 5.
This is graph a.

17. The graph of $y = \frac{2x^2+3}{x^2+1}$ has no vertical asymptote, the line with equation $y = 2$ as a horizontal asymptote, and a y-intercept of 3.
This is graph d.

19. The graph $y = \frac{-2x^2-3}{x^2+1}$ has no vertical asymptote, the line with equation $y = -2$ as a horizontal asymptote, and a y-intercept of -3.
This is graph e.

21. The right end is up and the left end is up. There are three turning points.
The degree is an even integer equal to 4 or more.
The x^n term has a $+$ sign.

23. The right end is up and the left end is down. There are four turning points. The degree is an odd integer equal to 5 or more. The x^n term has a $+$ sign.

25. The right end is down and the left end is up. There are six turning points. The degree is an odd integer equal to 7 or more. The x^n term has a $-$ sign.

27. $y = \dfrac{-4}{x+2}$

The function is undefined for $x = -2$, so the line $x = -2$ is a vertical asymptote.

x	-102	-12	-7	-5	-3	-1	8	98
$x+2$	-100	-10	-5	-3	-1	1	10	100
y	0.04	0.4	0.8	1.3	4	-4	-0.4	-0.04

The graph approaches $y = 0$, so the line $y = 0$ (the x-axis) is a horizontal asymptote.
Asymptotes: $y = 0$, $x = -2$

x-intercept:
none, because the x-axis is an asymptote

y-intercept:

-2, the value when $x = 0$

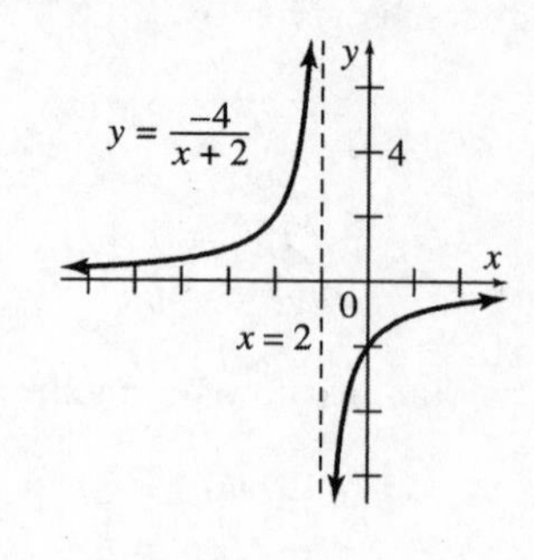

29. $y = \dfrac{2}{3+2x}$

$3 + 2x = 0$ when $2x = -3$ or $x = -\frac{3}{2}$, so the line $x = -\frac{3}{2}$ is a vertical asymptote.

x	-51.5	-6.5	-2	-1	3.5	48.5
$3+2x$	-100	-10	-1	1	10	100
y	-0.02	-0.2	-2	2	0.2	0.02

The graph approaches $y = 0$, so the line $y = 0$ (the x-axis) is a horizontal asymptote.
Asymptote: $y = 0$, $x = -\frac{3}{2}$

x-intercept:
none, since the x-axis is an asymptote

y-intercept:

$\frac{2}{3}$, the value when $x = 0$

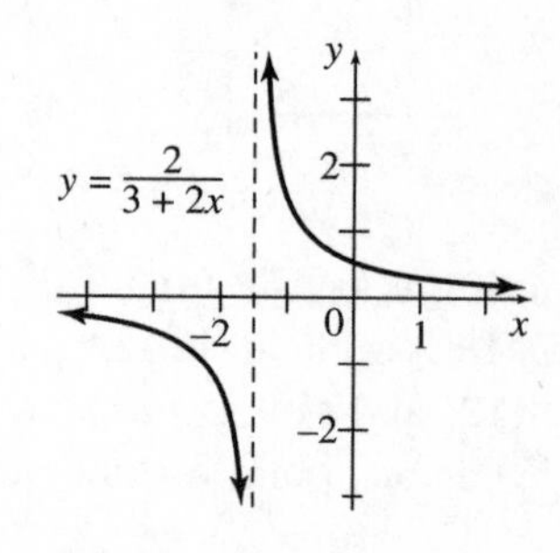

31. $y = \dfrac{2x}{x-3}$

$x - 3 = 0$ when $x = 3$, so the line $x = 3$ is a vertical asymptote.

x	-97	-7	-1	1	2	2.5
$2x$	-194	-14	-2	2	4	5
$x-3$	-100	-10	-4	-2	-1	-0.5
y	1.94	1.4	0.5	-1	-4	-10

x	3.5	4	5	7	11	103
$2x$	7	8	10	14	22	206
$x-3$	0.5	1	2	4	8	100
y	14	8	5	3.5	2.75	2.06

As x gets larger,

$$\frac{2x}{x-3} \approx \frac{2x}{x} = 2.$$

Thus, $y = 2$ is a horizontal asymptote.
Asymptotes: $y = 2$, $x = 3$
x-intercept:
0, the value when $y = 0$

y-intercept:
0, the value when $x = 0$

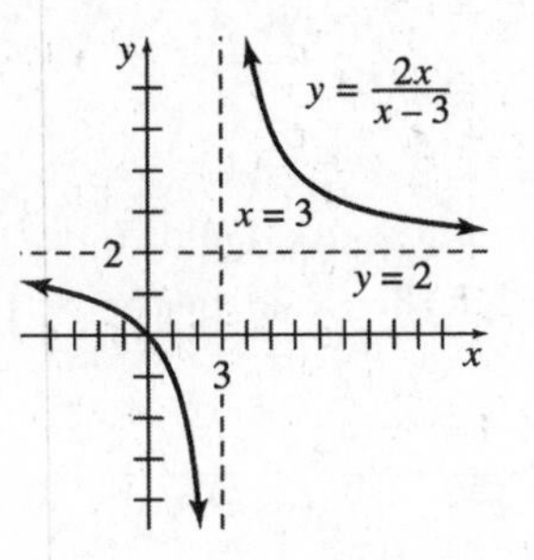

33. $y = \dfrac{x+1}{x-4}$

$x - 4 = 0$ when $x = 4$, so $x = 4$ is a vertical asymptote.

x	−96	−6	−1	0	3
$x+1$	−95	−5	0	1	4
$x-4$	−100	−10	−5	−4	−1
y	0.95	0.5	0	−0.25	−4

x	3.5	4.5	5	14	104
$x+1$	4.5	5.5	6	15	105
$x-4$	−0.5	0.5	1	10	100
y	−9	11	6	1.5	1.05

As x gets larger,

$$\frac{x+1}{x-4} \approx \frac{x}{x} = 1.$$

Thus, $y = 1$ is a horizontal asymptote.
Asymptotes: $y = 1$, $x = 4$

x-intercept:
−1, the value when $y = 0$

y-intercept:
$-\frac{1}{4}$, the value when $x = 0$

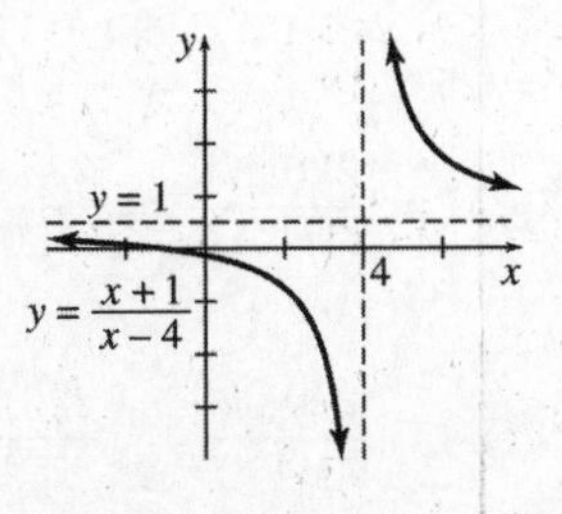

35. $y = \dfrac{3-2x}{4x+20}$

$4x + 20 = 0$ when $4x = -20$ or $x = -5$, so the line $x = -5$ is a vertical asymptote.

x	−8	−7	−6	−4	−3	−2
$3-2x$	−26	−23	−20	−14	−11	−8
$4x+20$	−12	−8	−4	4	8	12
y	2.17	2.88	5	−3.5	−1.38	−0.67

As x gets larger,

$$\frac{3-2x}{4x+20} \approx \frac{-2x}{4x} = -\frac{1}{2}.$$

Thus, the line $y = -\frac{1}{2}$ is a horizontal asymptote.

Asymptotes: $x = -5$, $y = -\frac{1}{2}$

x-intercept:

$\frac{3}{2}$, the value when $y = 0$

y-intercept:

$\frac{3}{20}$, the value when $x = 0$

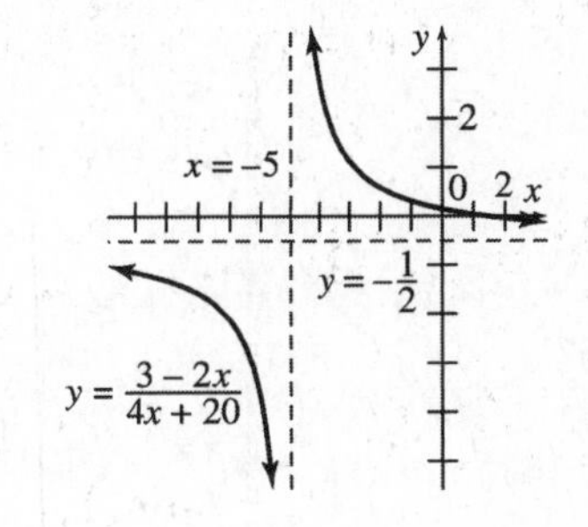

37. $y = \dfrac{-x-4}{3x+6}$

$3x + 6 = 0$ when $3x = -6$ or $x = -2$, so the line $x = -2$ is a vertical asymptote.

x	−5	−4	−3	−1	0	1
$-x-4$	1	0	−1	−3	−4	−5
$3x+6$	−9	−6	−3	3	6	9
y	−0.11	0	0.33	−1	−0.67	−0.56

As x gets larger,

$$\frac{-x-4}{3x+6} \approx \frac{-x}{3x} = -\frac{1}{3}.$$

The line $y = -\frac{1}{3}$ is a horizontal asymptote.
Asymptotes: $y = -\frac{1}{3}$, $x = -2$

x-intercept:
−4, the value when $y = 0$

y-intercept:

$-\frac{2}{3}$, the value when $x = 0$

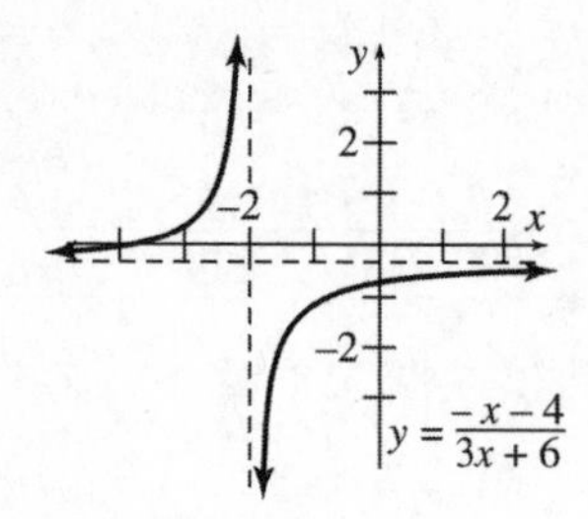

39. $y = \dfrac{x^2 + 7x + 12}{x + 4}$
$= \dfrac{(x + 3)(x + 4)}{x + 4}$
$= x + 3,\ x \neq -4$

There are no asymptotes, but there is a hole at $x = -4$.
x-intercept: -3, the value when $y = 0$.
y-intercept: 3, the value when $x = 0$.

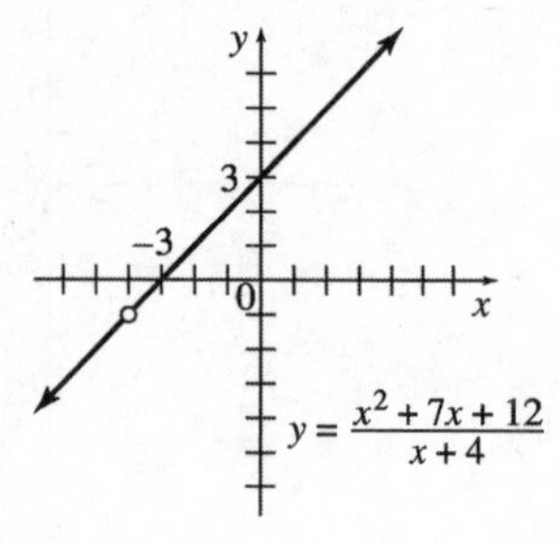

41. For a vertical asymptote at $x = 1$, put $x - 1$ in the denominator. For a horizontal asymptote at $y = 2$, the degree of the numerator must equal the degree of the denominator and the quotient of their leading terms must equal 2. So, $2x$ in the numerator would cause y to approach 2 as x gets larger.
So, one possible answer is $y = \dfrac{2x}{x - 1}$.

43. $f(x) = (x - 1)(x - 2)(x + 3)$,
$g(x) = x^3 + 2x^2 - x - 2$,
$h(x) = 3x^3 + 6x^2 - 3x - 6$

(a) $f(1) = (0)(-1)(4) = 0$

(b) $f(x)$ is zero when $x = 2$ and when $x = -3$.

(c)
$$\begin{aligned} g(-1) &= (-1)^3 + 2(-1)^2 - (-1) - 2 \\ &= -1 + 2 + 1 - 2 = 0 \\ g(1) &= (1)^3 + 2(1)^2 - (1) - 2 \\ &= 1 + 2 - 1 - 2 \\ &= 0 \\ g(-2) &= (-2)^3 + 2(-2)^2 - (-2) - 2 \\ &= -8 + 8 + 2 - 2 \\ &= 0 \end{aligned}$$

(d)
$$\begin{aligned} g(x) &= [x - (-1)](x - 1)[x - (-2)] \\ g(x) &= (x + 1)(x - 1)(x + 2) \end{aligned}$$

(e)
$$\begin{aligned} h(x) &= 3g(x) \\ &= 3(x + 1)(x - 1)(x + 2) \end{aligned}$$

(f) If f is a polynomial and $f(a) = 0$ for some number a, then one factor of the polynomial is $x - a$.

45. $f(x) = \dfrac{1}{x^5 - 2x^3 - 3x^2 + 6}$

(a) Two vertical asymptotes appear, one at $x = -1.4$ and one at $x = 1.4$.

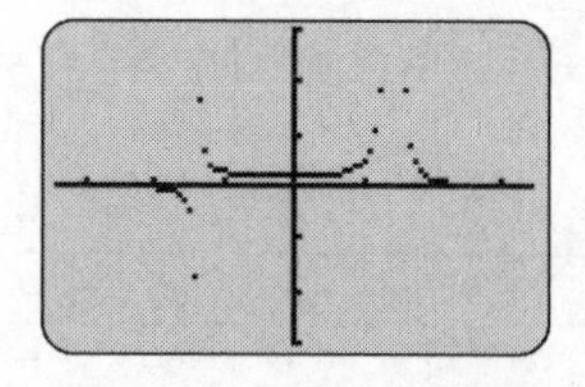

(b) Three vertical asymptotes appear, one at $x = -1.414$, one at $x = 1.414$, and one at $x = 1.442$.

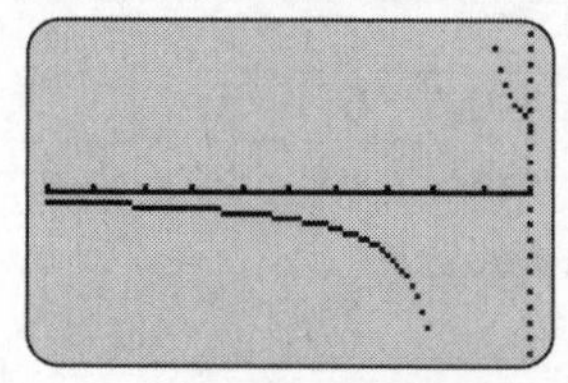

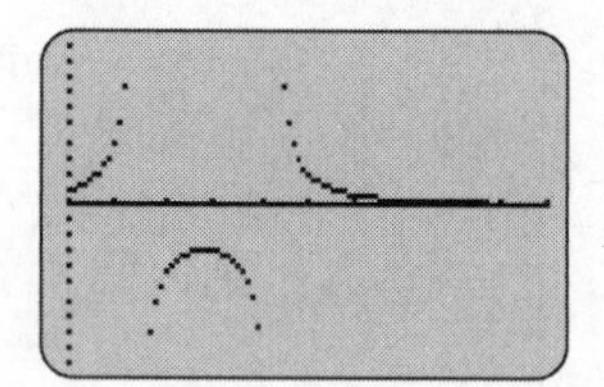

47. $\overline{C}(x) = \dfrac{220{,}000}{x + 475}$

(a) If $x = 25$,

$$\overline{C}(25) = \frac{220{,}000}{25 + 475} = \frac{220{,}000}{500} = 440.$$

If $x = 50$,

$$\overline{C}(50) = \frac{220{,}000}{50 + 475} = \frac{220{,}000}{525} \approx 419.$$

If $x = 100$,

$$\overline{C}(100) = \frac{220{,}000}{100 + 475} = \frac{220{,}000}{575} \approx 383.$$

If $x = 200$,

$$\overline{C}(200) = \frac{220{,}000}{200 + 475} = \frac{220{,}000}{675} \approx 326.$$

If $x = 300$,

$$\overline{C}(300) = \frac{220{,}000}{300 + 475} = \frac{220{,}000}{775} \approx 284.$$

If $x = 400$,

$$\overline{C}(400) = \frac{220{,}000}{400 + 475} = \frac{220{,}000}{875} \approx 251.$$

(b) A vertical asymptote occurs when the denominator is 0.

$$x + 475 = 0$$
$$x = -475$$

A horizontal asymptote occurs when $\overline{C}(x)$ approaches a value as x gets larger. In this case, $\overline{C}(x)$ approaches 0.
The asymptotes are $x = -475$ and $y = 0$.

(c) x-intercepts:

$$0 = \frac{220{,}000}{x + 475};\text{ no such } x\text{, so no } x\text{-intercepts}$$

y-intercepts:

$$\overline{C}(0) = \frac{220{,}000}{0 + 475} \approx 463.2$$

(d) Use the following ordered pairs:
$(25, 440)$, $(50, 419)$, $(100, 383)$,
$(200, 326)$, $(300, 284)$, $(400, 251)$.

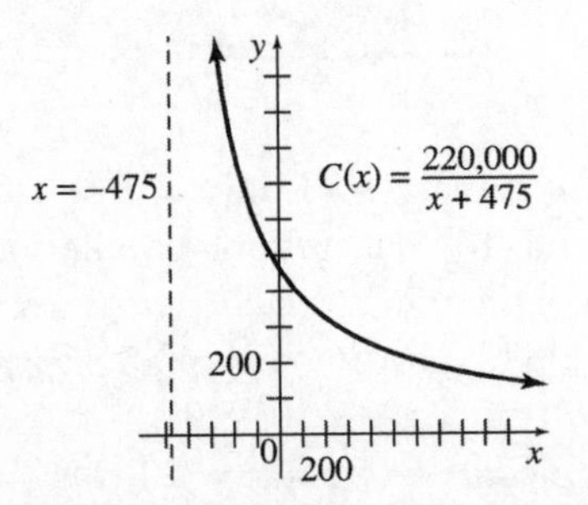

49. Quadratic functions with roots at $x = 0$ and $x = 100$ are of the form $f(x) = ax(100 - x)$.
$f_1(x)$ has a maximum of 100, which occurs at the vertex. The x-coordinate of the vertex lies between the two roots.
The vertex is $(50, 100)$.

$$100 = a(50)(100 - 50)$$
$$100 = a(50)(50)$$
$$\frac{100}{2500} = a$$
$$\frac{1}{25} = a$$

$$f_1(x) = \frac{1}{25}x(100 - x) \text{ or } \frac{x(100 - x)}{25}$$

$f_2(x)$ has a maximum of 250, occurring at $(50, 250)$.

$$250 = a(50)(100 - 50)$$
$$250 = a(50)(50)$$
$$\frac{250}{2500} = a$$
$$\frac{1}{10} = a$$

$$f_2(x) = \frac{1}{10}x(100 - x) \text{ or } \frac{x(100 - x)}{10}$$

$$f_1(x) \cdot f_2(x) = \left[\frac{x(100 - x)}{25}\right] \cdot \left[\frac{x(100 - x)}{10}\right]$$
$$= \frac{x^2(100 - x)^2}{250}$$

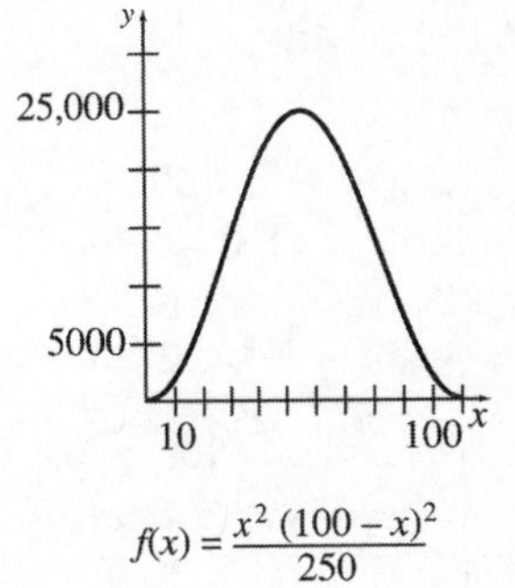

51. $y = \dfrac{6.7x}{100 - x}$,

Let x = percent of pollutant;
y = cost in thousands.

(a) $x = 50$

$$y = \frac{6.7(50)}{100 - 50} = 6.7$$

The cost is \$6700.

$x = 70$

$$y = \frac{6.7(70)}{100 - 70} \approx 15.6$$

The cost is \$15,600.

$x = 80$

$$y = \frac{6.7(80)}{100 - 80} = 26.8$$

The cost is \$26,800.

$x = 90$

$$y = \frac{6.7(90)}{100 - 90} = 60.3$$

The cost is \$60,300.

$x = 95$

$$y = \frac{6.7(95)}{100 - 95}$$

The cost is \$127,300.

$x = 98$

$$y = \frac{6.7(98)}{100 - 98} = 328.3$$

The cost is \$328,300.

$x = 99$

$$y = \frac{6.7(99)}{100 - 99} = 663.3$$

The cost is \$663,300.

(b) No, because $x = 100$ makes the denominator zero, so $x = 100$ is a vertical asymptote.

(c)

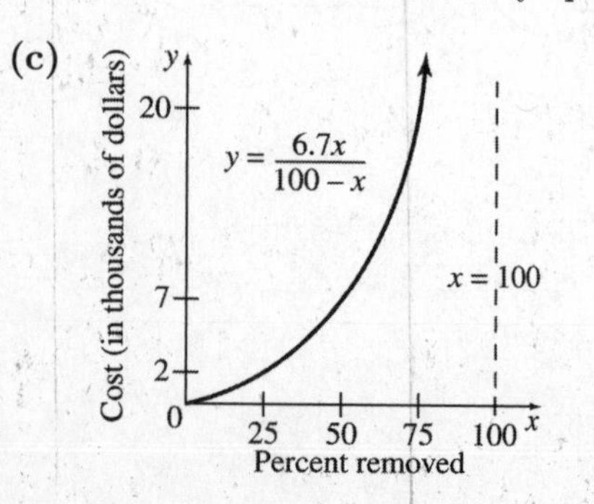

53. (a) $a = \frac{k}{d}$

$k = ad$

d	a	$k = ad$
36.000	9.37	337.32
36.125	9.34	337.4075
36.250	9.31	337.4875
36.375	9.27	337.19625
36.500	9.24	337.26
36.625	9.21	337.31625
36.750	9.18	337.365
36.875	9.15	337.40625
37.000	9.12	337.44

We find the average of the nine values of k by adding them and dividing by 9. This gives 337.35, or, rounding to the nearest integer, $k = 337$. Therefore,

$$a = \frac{337}{d}.$$

(b) When $d = 40.50$,

$$a = \frac{337}{40.50} \approx 8.32.$$

The strength for 40.50 diopter lenses is 8.32 mm of arc.

55. $D(x) = -0.125x^5 + 3.125x^4 + 4000$

(a)

x	0	5	10	15
$D(x)$	4000	5563	22,750	67,281

x	20	25
$D(x)$	104,000	4000

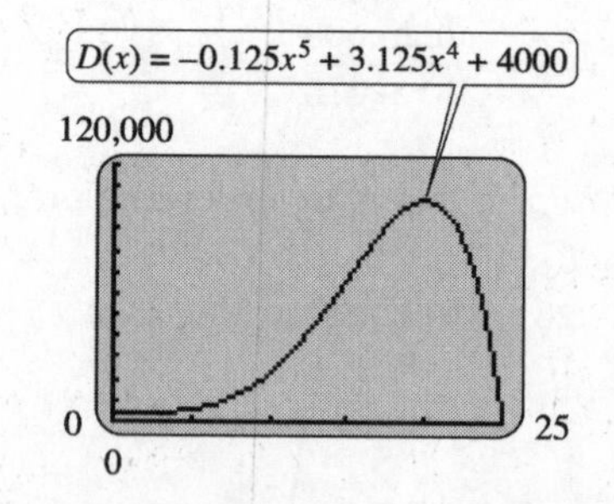

(b) $D(x)$ increases from $x = 0$ to $x = 20$. This corresponds to an increasing population from 1905 to 1925. $D(x)$ does not change much from $x = 0$ to $x = 5$. This corresponds to a relatively stable population from 1905 to 1910.

$D(x)$ decreases from $x = 20$ to $x = 25$. This corresponds to a decreasing population from 1925 to 1930.

57. $f(t) = -0.0014t^3 + 0.092t^2 - 0.67t + 11.89$

(a)

x	9	18	27	36	45
$A(x)$	12.29	21.47	33.31	41.68	40.47

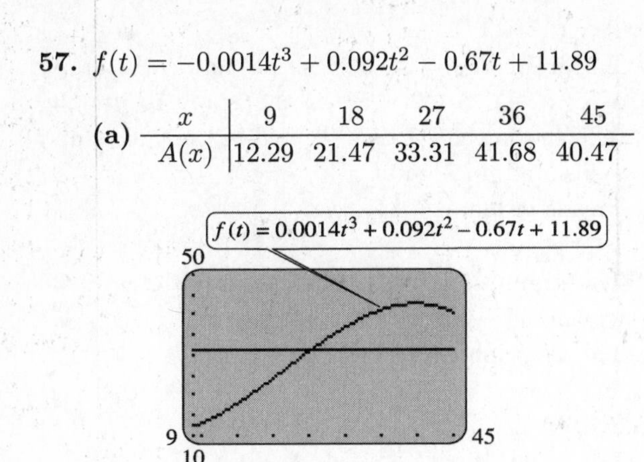

(b) The graph of $f(t)$ intersects the graph of $g(t) = 31$ at about $t = 25$, which corresponds to the year 1985.

59. (a) A reasonable domain for the function is $[0, \infty)$. Populations are not measured using negative numbers, and they may get extremely large.

(b) $f(x) = \dfrac{Kx}{A+x}$

When $K = 5$ and $A = 2$,

$$f(x) = \frac{5x}{2+x}.$$

The graph has a horizontal asymptote at $y = 5$ since

$$\frac{5x}{2+x} \approx \frac{5x}{x} = 5$$

as x gets larger.

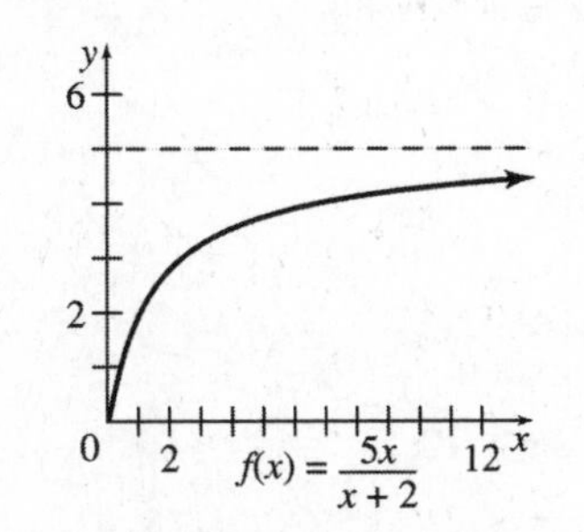

(c) $f(x) = \dfrac{Kx}{A+x}$

As x gets larger,

$$\frac{Kx}{A+x} \approx \frac{Kx}{x} = K.$$

Thus, $y = K$ will always be a horizontal asymptote for this function.

(d) K represents the maximum growth rate. The function approaches this value asymptotically, showing that although the growth rate can get very close to K, it can never reach the maximum, K.

(e) $f(x) = \dfrac{Kx}{A+x}$

Let $A = x$, the quantity of food present.

$$f(x) = \frac{Kx}{A+x} = \frac{Kx}{2x} = \frac{K}{2}$$

K is the maximum growth rate, so $\frac{K}{2}$ is half the maximum. Thus, A represents the quantity of food for which the growth rate is half of its maximum.

61. (a)

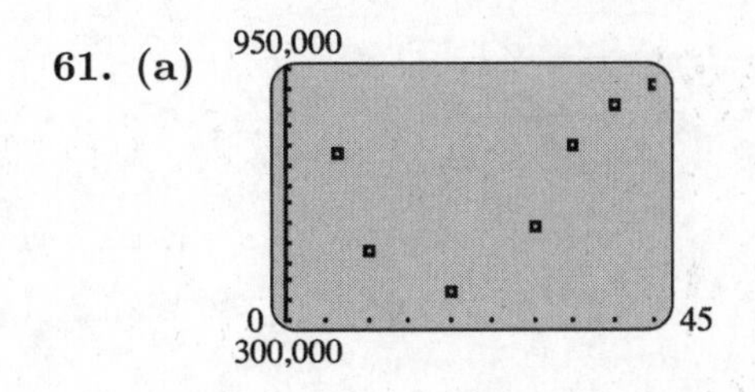

(b) $y = 890.37x^2 - 36{,}370x + 830{,}144$

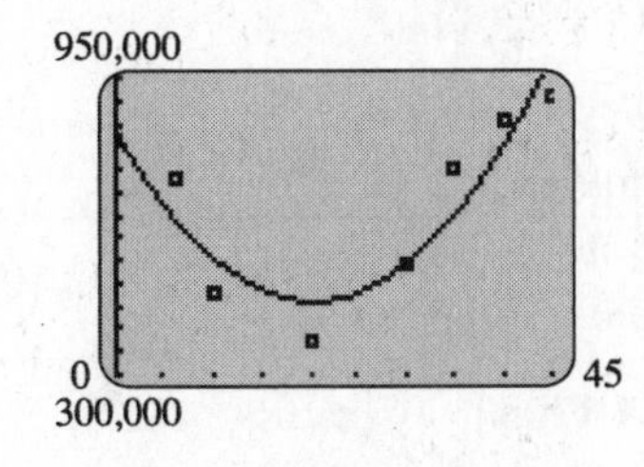

(c) $y = -52.954x^3 + 5017.88x^2 - 127{,}714x + 1{,}322{,}606$

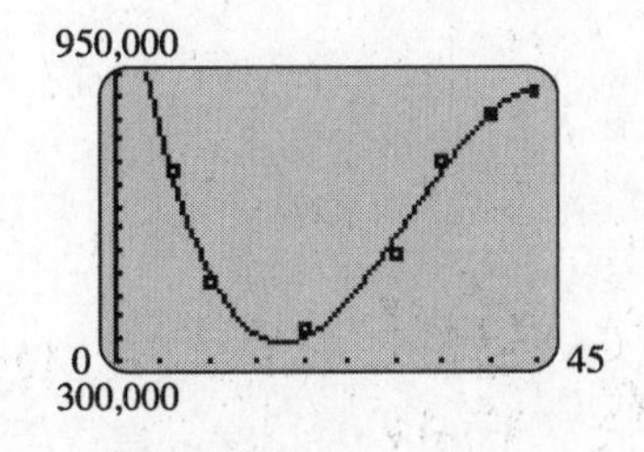

63. (a)

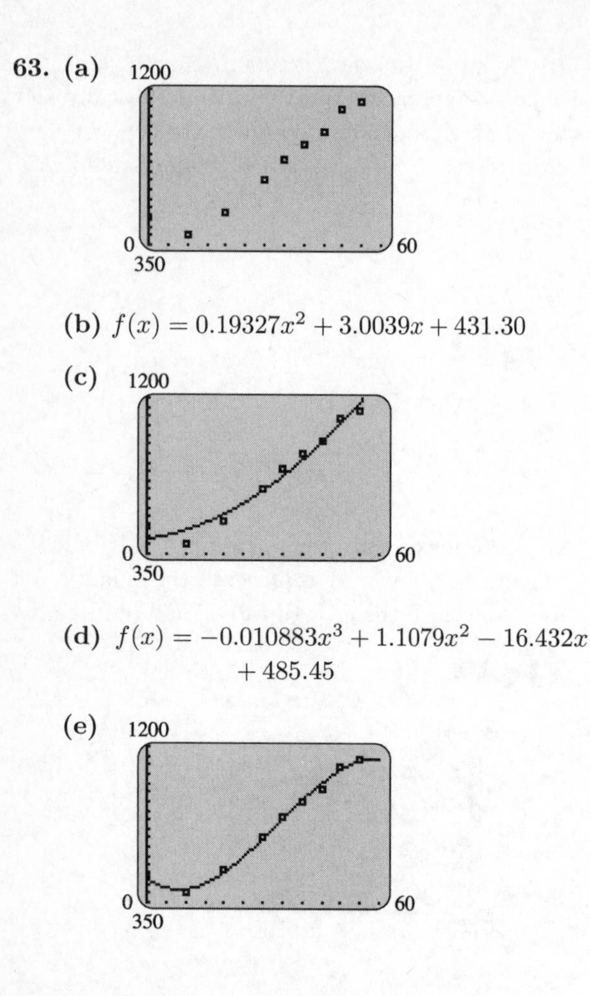

(b) $f(x) = 0.19327x^2 + 3.0039x + 431.30$

(c)

(d) $f(x) = -0.010883x^3 + 1.1079x^2 - 16.432x + 485.45$

(e)

2.4 Exponential Functions

1.

number of folds	1 2 3 4 5 ... 10 ... 50
layers of paper	2 4 8 16 32 ... 1024 ... 2^{50}

$2^{50} = 1.125899907 \times 10^{15}$

3. The graph of $y = 3^x$ is the graph of an exponential function $y = a^x$ with $a > 1$.
This is graph e.

5. The graph of $y = \left(\frac{1}{3}\right)^{1-x}$ is the graph of $y = (3^{-1})^{1-x}$ or $y = 3^{x-1}$. This is the graph of $y = 3^x$ translated 1 unit to the right.
This is graph c.

7. The graph of $y = 3(3)^x$ is the same as the graph of $y = 3^{x+1}$. This is the graph of $y = 3^x$ translated 1 unit to the left.
This is graph f.

9. The graph of $y = 2 - 3^{-x}$ is the same as the graph of $y = -3^{-x} + 2$. This is the graph of $y = 3^x$ reflected in the x-axis, reflected in the y-axis, and translated up 2 units.
This is graph a.

11. The graph of $y = 3^{x-1}$ is the graph of $y = 3^x$ translated 1 unit to the right.
This is graph c.

13.
$$\begin{aligned} 2^x &= 32 \\ 2^x &= 2^5 \\ x &= 5 \end{aligned}$$

15.
$$\begin{aligned} 3^x &= \frac{1}{81} \\ 3^x &= \frac{1}{3^4} \\ 3^x &= 3^{-4} \\ x &= -4 \end{aligned}$$

17.
$$\begin{aligned} 4^x &= 8^{x+1} \\ (2^2)^x &= (2^3)^{x+1} \\ 2^{2x} &= 2^{3x+3} \\ 2x &= 3x + 3 \\ -x &= 3 \\ x &= -3 \end{aligned}$$

19.
$$\begin{aligned} 16^{x+3} &= 64^{2x-5} \\ (2^4)^{x+3} &= (2^6)^{2x-5} \\ 2^{4x+12} &= 2^{12x-30} \\ 4x + 12 &= 12x - 30 \\ 42 &= 8x \\ \frac{21}{4} &= x \end{aligned}$$

21.
$$\begin{aligned} e^{-x} &= (e^4)^{x+3} \\ e^{-x} &= e^{4x+12} \\ -x &= 4x + 12 \\ -5x &= 12 \\ x &= -\frac{12}{5} \end{aligned}$$

23.
$$\begin{aligned} 5^{-|x|} &= \frac{1}{25} \\ 5^{-|x|} &= 5^{-2} \\ -|x| &= -2 \\ |x| &= 2 \\ x = 2 \quad &\text{or} \quad x = -2 \end{aligned}$$

25.
$$\begin{aligned} 5^{x^2+x} &= 1 \\ 5^{x^2+x} &= 5^0 \\ x^2+x &= 0 \\ x(x+1) &= 0 \\ x = 0 \quad \text{or} \quad x+1 &= 0 \\ x = 0 \quad \text{or} \quad x &= -1 \end{aligned}$$

27.
$$\begin{aligned} 27^x &= 9^{x^2+x} \\ (3^3)^x &= (3^2)^{x^2+x} \\ 3^{3x} &= 3^{2x^2+2x} \\ 3x &= 2x^2+2x \\ 0 &= 2x^2-x \\ 0 &= x(2x-1) \\ x = 0 \quad \text{or} \quad 2x-1 &= 0 \\ x = 0 \quad \text{or} \quad x &= \frac{1}{2} \end{aligned}$$

33. $A = P\left(1+\frac{r}{m}\right)^{tm}$, $P = 10{,}000$, $r = 0.04$, $t = 5$

(a) annually, $m = 1$

$$\begin{aligned} A &= 10{,}000\left(1+\frac{0.04}{1}\right)^{5(1)} \\ &= 10{,}000(1.04)^5 \\ &= \$12{,}166.53 \end{aligned}$$

$$\begin{aligned} \text{Interest} &= \$12{,}166.53 - \$10{,}000 \\ &= \$2166.53 \end{aligned}$$

(b) semiannually, $m = 2$

$$\begin{aligned} A &= 10{,}000\left(1+\frac{0.04}{2}\right)^{5(2)} \\ &= 10{,}000(1.02)^{10} \\ &= \$12{,}189.94 \end{aligned}$$

$$\begin{aligned} \text{Interest} &= \$12{,}189.94 - \$10{,}000 \\ &= \$2189.94 \end{aligned}$$

(c) quarterly, $m = 4$

$$\begin{aligned} A &= 10{,}000\left(1+\frac{0.04}{4}\right)^{5(4)} \\ &= 10{,}000(1.01)^{20} \\ &= \$12{,}201.90 \end{aligned}$$

$$\begin{aligned} \text{Interest} &= \$12{,}201.90 - \$10{,}000 \\ &= \$2201.90 \end{aligned}$$

(d) monthly, $m = 12$

$$\begin{aligned} A &= 10{,}000\left(1+\frac{0.04}{12}\right)^{5(12)} \\ &= 10{,}000(1.00\overline{3})^{60} \\ &= \$12{,}209.97 \end{aligned}$$

$$\begin{aligned} \text{Interest} &= \$12{,}209.97 - \$10{,}000 \\ &= \$2209.97 \end{aligned}$$

35. For 6% compounded annually for 2 years,

$$\begin{aligned} A &= 18{,}000(1+0.06)^2 \\ &= 18{,}000(1.06)^2 \\ &= 20{,}224.80 \end{aligned}$$

For 5.9% compounded monthly for 2 years,

$$\begin{aligned} A &= 18{,}000\left(1+\frac{0.059}{12}\right)^{12(2)} \\ &= 18{,}000\left(\frac{12.059}{12}\right)^{24} \\ &= 20{,}248.54 \end{aligned}$$

The 5.9% investment is better. The additional interest is

$$\$20{,}248.54 - \$20{,}224.80 = \$23.74.$$

37. $A = Pe^{rt}$

(a) $r = 3\%$

$$A = 10e^{0.03(3)} = \$10.94$$

(b) $r = 4\%$

$$A = 10e^{0.04(3)} = \$11.27$$

(c) $r = 5\%$

$$A = 10e^{0.05(3)} = \$11.62$$

39.
$$\begin{aligned} 1200 &= 500\left(1+\frac{r}{4}\right)^{(14)(4)} \\ \frac{1200}{500} &= \left(1+\frac{r}{4}\right)^{56} \\ 2.4 &= \left(1+\frac{r}{4}\right)^{56} \\ 1+\frac{r}{4} &= (2.4)^{1/56} \\ 4+r &= 4(2.4)^{1/56} \\ r &= 4(2.4)^{1/56} - 4 \\ r &\approx 0.0630 \end{aligned}$$

The required interest rate is 6.30%.

41. $y = (0.92)^t$

(a)

t	y
0	$(0.92)^0 = 1$
1	$(0.92)^1 = 0.92$
2	$(0.92)^2 \approx 0.85$
3	$(0.92)^3 \approx 0.78$
4	$(0.92)^4 \approx 0.72$
5	$(0.92)^5 \approx 0.66$
6	$(0.92)^6 \approx 0.61$
7	$(0.92)^7 \approx 0.56$
8	$(0.92)^8 \approx 0.51$
9	$(0.92)^9 \approx 0.47$
10	$(0.92)^{10} \approx 0.43$

(b)

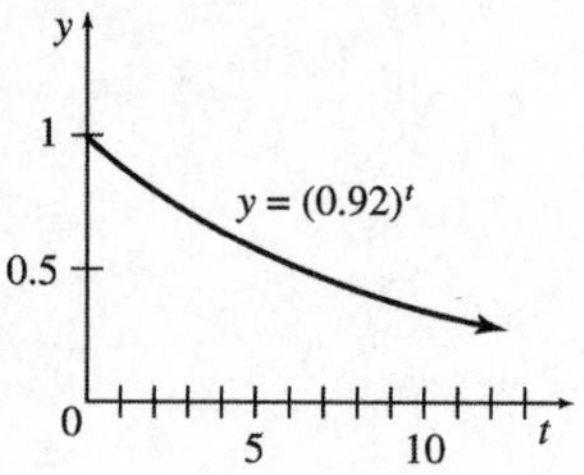

(c) Let $x =$ the cost of the house in 10 years.

$$\text{Then, } 0.43x = 165{,}000$$
$$x \approx 383{,}721.$$

In 10 years, the house will cost about \$384,000.

(d) Let $x =$ the cost of the book in 8 years.

$$\text{Then, } 0.51x = 50$$
$$x \approx 98$$

In 8 years, the textbook will cost about \$98.

43. $A = P\left(1+\frac{r}{m}\right)^{tm}$

$$A = 1000\left(1+\frac{j}{2}\right)^{5(2)} = 1000\left(1+\frac{j}{2}\right)^{10}$$

This represents the amount in Bank X on January 1, 1985.

$$A = P\left(1+\frac{r}{m}\right)^{tm}$$
$$= \left[1000\left(1+\frac{j}{2}\right)^{10}\right]\left(1+\frac{k}{4}\right)^{3(4)}$$
$$= 1000\left(1+\frac{j}{2}\right)^{10}\left(1+\frac{k}{4}\right)^{12}$$

This represents the amount in Bank Y on January 1, 1988, \$1990.76.

$$A = P\left(1+\frac{r}{m}\right)^{tm} = 1000\left(1+\frac{k}{4}\right)^{8\cdot 4}$$
$$= 1000\left(1+\frac{k}{4}\right)^{32}$$

This represents the amount he could have had from January 1, 1980, to January 1, 1988, at a rate of k per annum compounded quarterly, \$2203.76.

So,

$$1000\left(1+\frac{j}{2}\right)^{10}\left(1+\frac{k}{4}\right)^{12} = 1990.76$$

$$\text{and} \qquad 1000\left(1+\frac{k}{4}\right)^{32} = 2203.76.$$
$$\left(1+\frac{k}{4}\right)^{32} = 2.20376$$
$$1+\frac{k}{4} = (2.20376)^{1/32}$$
$$1+\frac{k}{4} = 1.025$$
$$\frac{k}{4} = 0.025$$
$$k = 0.1 \quad \text{or} \quad 10\%$$

Substituting, we have

$$1000\left(1+\frac{j}{2}\right)^{10}\left(1+\frac{1.0}{4}\right)^{12} = 1990.76$$
$$1000\left(1+\frac{j}{2}\right)^{10}(1.025)^{12} = 1990.76$$
$$\left(1+\frac{j}{2}\right)^{10} = 1.480$$
$$1+\frac{j}{2} = (1.480)^{1/10}$$
$$1+\frac{j}{2} = 1.04$$
$$\frac{j}{2} = 0.04$$
$$j = 0.08 \quad \text{or} \quad 8\%.$$

The ratio $\frac{k}{j} = \frac{0.1}{0.08} = 1.25$, is choice (a).

45. $f(x) = 500 \cdot 2^{3x}$

(a) After 1 hour:

$$f(1) = 500 \cdot 2^{3(1)} = 500 \cdot 8 = 4000 \text{ bacteria}$$

(b) initially:

$$f(0) = 500 \cdot 2^{3(0)} = 500 \cdot 1 = 500 \text{ bacteria}$$

(c) The bacteria double every $3x = 1$ hour, or every $\frac{1}{3}$ hour.

(d) When does $f(x) = 32{,}000$?

$$\begin{aligned} 32{,}000 &= 500 \cdot 2^{3x} \\ 64 &= 2^{3x} \\ 2^6 &= 2^{3x} \\ 6 &= 3x \\ x &= 2 \end{aligned}$$

The number of bacteria will increase to 32,000 in 2 hours.

47. (a)

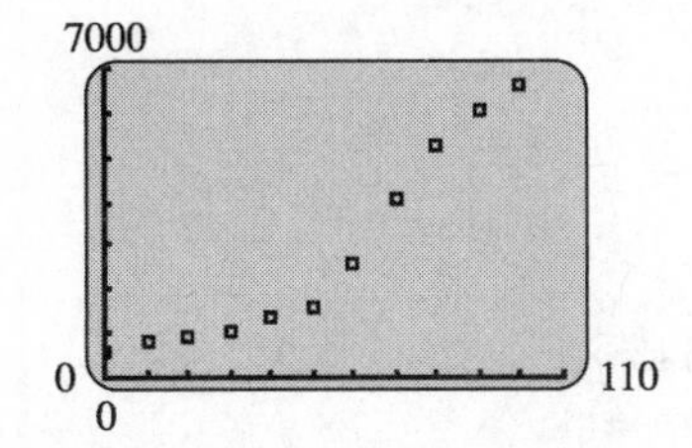

The emissions appear to grow exponentially.

(b) $f(x) = f_0 a^x$

$f_0 = 534$

Use the point $(100, 6672)$ to find a.

$$\begin{aligned} 6672 &= 534a^{100} \\ a^{100} &= \frac{6672}{534} \\ a &= \sqrt[100]{\frac{6672}{534}} \\ &\approx 1.026 \\ f(x) &= 534(1.026)^x \end{aligned}$$

(c) $1.026 - 1 = 0.026 = 2.6\%$

(d) Double the 2000 value is $2(6672) = 13{,}344$.

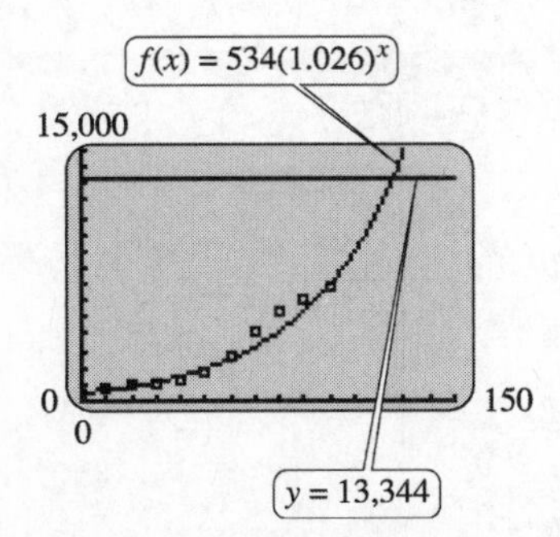

The doubling point is reached when $x \approx 125.4$. The first year in which emissions equal or exceed that threshold is 2026.

49. (a) When $x = 0$, $P = 1013$.
When $x = 10{,}000$, $P = 265$.
First we fit $P = ae^{kx}$.

$$\begin{aligned} 1013 &= ae^0 \\ a &= 1013 \\ P &= 1013e^{kx} \\ 265 &= 1013e^{k(10{,}000)} \\ \frac{265}{1013} &= e^{10{,}000k} \\ 10{,}000k &= \ln\left(\frac{265}{1013}\right) \\ k &= \frac{\ln\left(\frac{265}{1013}\right)}{10{,}000} \approx -1.34 \times 10^{-4} \end{aligned}$$

Therefore $P = 1013e^{(-1.34 \times 10^{-4})x}$.

Next we fit $P = mx + b$.
We use the points $(0, 1013)$ and $(10{,}000, 265)$.

$$\begin{aligned} m &= \frac{265 - 1013}{10{,}000 - 0} = -0.0748 \\ b &= 1013 \end{aligned}$$

Therefore $P = -0.0748x + 1013$.

Finally, we fit $P = \frac{1}{ax+b}$.

$$\begin{aligned} 1013 &= \frac{1}{a(0) + b} \\ b &= \frac{1}{1013} \approx 9.87 \times 10^{-4} \\ P &= \frac{1}{ax + \frac{1}{1013}} \\ 265 &= \frac{1}{10{,}000a + \frac{1}{1013}} \\ \frac{1}{265} &= 10{,}000a + \frac{1}{1013} \\ 10{,}000a &= \frac{1}{265} - \frac{1}{1013} \\ a &= \frac{\frac{1}{265} - \frac{1}{1013}}{10{,}000} \approx 2.79 \times 10^{-7} \end{aligned}$$

Therefore,

$$P = \frac{1}{(2.79 \times 10^{-7})x + (9.87 \times 10^{-4})}.$$

(b)

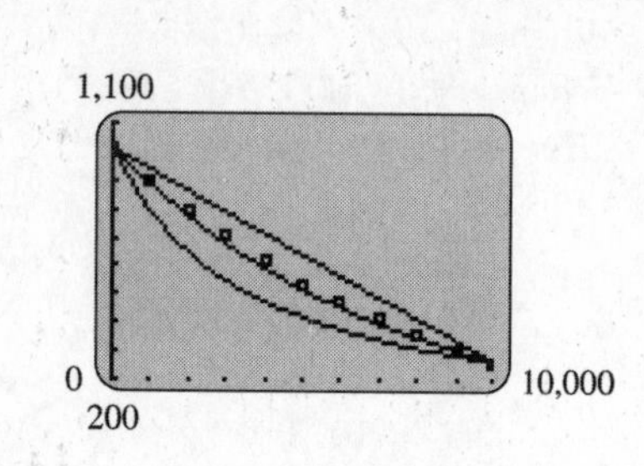

$P = 1013e^{(-1.34\times10^{-4})x}$ is the best fit.

(c) $P(1500) = 1013e^{-1.34\times10^{-4}(1500)} \approx 829$
$P(11,000) = 1013e^{-1.34\times10^{-4}(11,000)} \approx 232$

We predict that the pressure at 1500 meters will be 829 millibars, and at 11,000 meters will be 232 millibars.

(d) Using exponential regression, we obtain $P = 1038(0.99998661)^x$ which differs slightly from the function found in part (b) which can be rewritten as

$$P = 1013(0.99998660)^x.$$

2.5 Logarithmic Functions

1. $5^3 = 125$

Since $a^y = x$ means $y = \log_a x$, the equation in logarithmic form is

$$\log_5 125 = 3.$$

3. $3^4 = 81$

The equation in logarithmic form is

$$\log_3 81 = 4.$$

5. $3^{-2} = \frac{1}{9}$

The equation in logarithmic form is

$$\log_3 \frac{1}{9} = -2.$$

7. $\log_2 32 = 5$

Since $y = \log_a x$ means $a^y = x$, the equation in exponential form is

$$2^5 = 32.$$

9. $\ln \frac{1}{e} = -1$

The equation in exponential form is

$$e^{-1} = \frac{1}{e}.$$

11.
$$\log 100,000 = 5$$
$$\log_{10} 100,000 = 5$$
$$10^5 = 100,000$$

When no base is written, $\log_{10}$ is understood.

13. Let $\log_8 64 = x$.

Then, $8^x = 64$
$8^x = 8^2$
$x = 2.$

Thus, $\log_8 64 = 2.$

15. $\log_4 64 = x$
$4^x = 64$
$4^x = 4^3$
$x = 3$

17. $\log_2 \frac{1}{16} = x$
$2^x = \frac{1}{16}$
$2^x = 2^{-4}$
$x = -4$

19. $\log_2 \sqrt[3]{\frac{1}{4}} = x$
$2^x = \left(\frac{1}{4}\right)^{1/3}$
$2^x = \left(\frac{1}{2^2}\right)^{1/3}$
$2^x = 2^{-2/3}$
$x = -\frac{2}{3}$

21. $\ln e = x$

Recall that ln means $\log_e$.

$$e^x = e$$
$$x = 1$$

23. $\ln e^{5/3} = x$
$e^x = e^{5/3}$
$x = \frac{5}{3}$

25. The logarithm to the base 3 of 4 is written $\log_3 4$. The subscript denotes the base.

27. $\log_5 (3k) = \log_5 3 + \log_5 k$

29. $\log_3 \dfrac{3p}{5k}$

$$= \log_3 3p - \log_3 5k$$
$$= (\log_3 3 + \log_3 p) - (\log_3 5 + \log_3 k)$$
$$= 1 + \log_3 p - \log_3 5 - \log_3 k$$

31. $\ln \dfrac{3\sqrt{5}}{\sqrt[3]{6}}$

$$= \ln 3\sqrt{5} - \ln \sqrt[3]{6}$$
$$= \ln 3 \cdot 5^{1/2} - \ln 6^{1/3}$$
$$= \ln 3 + \ln 5^{1/2} - \ln 6^{1/3}$$
$$= \ln 3 + \frac{1}{2}\ln 5 - \frac{1}{3}\ln 6$$

33. $\log_b 32 = \log_b 2^5$

$$= 5 \log_b 2$$
$$= 5a$$

35. $\log_b 72b = \log_b 72 + \log_b b$

$$= \log_b 72 + 1$$
$$= \log_b 2^3 \cdot 3^3 + 1$$
$$= \log_b 2^3 + \log_b 3^2 + 1$$
$$= 3 \log_b 2 + 2 \log_b 3 + 1$$
$$= 3a + 2c + 1$$

37. $\log_5 30 = \dfrac{\ln 30}{\ln 5}$

$$\approx \frac{3.4012}{1.6094}$$
$$\approx 2.113$$

39. $\log_{1.2} 0.95 = \dfrac{\ln 0.95}{\ln 1.2}$

$$\approx -0.281$$

41. $\log_x 36 = -2$

$$x^{-2} = 36$$
$$(x^{-2})^{-1/2} = 36^{-1/2}$$
$$x = \frac{1}{6}$$

43. $\log_8 16 = z$

$$8^z = 16$$
$$(2^3)^z = 2^4$$
$$2^{3z} = 2^4$$
$$3z = 4$$
$$z = \frac{4}{3}$$

45. $\log_r 5 = \dfrac{1}{2}$

$$r^{1/2} = 5$$
$$(r^{1/2})^2 = 5^2$$
$$r = 25$$

47. $\log_5 (9x - 4) = 1$

$$5^1 = 9x - 4$$
$$9 = 9x$$
$$1 = x$$

49. $\log_9 m - \log_9 (m - 4) = -2$

$$\log_9 \frac{m}{m-4} = -2$$
$$9^{-2} = \frac{m}{m-4}$$
$$\frac{1}{81} = \frac{m}{m-4}$$
$$m - 4 = 81m$$
$$-4 = 80m$$
$$-0.05 = m$$

This value is not possible since $\log_9 (-0.05)$ does not exist.
Thus, there is no solution to the original equation.

51. $\log_3 (x - 2) + \log_3 (x + 6) = 2$

$$\log_3 [(x-2)(x+6)] = 2$$
$$(x-2)(x+6) = 3^2$$
$$x^2 + 4x - 12 = 9$$
$$x^2 + 4x - 21 = 0$$
$$(x+7)(x-3) = 0$$
$$x = -7 \quad \text{or} \quad x = 3$$

$x = -7$ does not check in the original equation. The only solution is 3.

53. $\log_2(x^2 - 1) - \log_2(x + 1) = 2$

$$\log_2 \frac{x^2-1}{x+1} = 2$$
$$2^2 = \frac{x^2-1}{x+1}$$
$$4 = \frac{(x-1)(x+1)}{x+1}$$
$$4 = x - 1$$
$$x = 5$$

55. $5^x = 12$

$$x \log 5 = \log 12$$
$$x = \frac{\log 12}{\log 5}$$
$$\approx 1.544$$

57.
$$\begin{aligned} e^{2y} &= 15 \\ \ln e^{2y} &= \ln 15 \\ 2y \ln e &= \ln 15 \\ 2y(1) &= \ln 15 \\ y &= \frac{\ln 15}{2} \\ &\approx 1.354 \end{aligned}$$

59.
$$\begin{aligned} 10e^{3z-7} &= 100 \\ \ln 10e^{3z-7} &= \ln 100 \\ \ln 10 + \ln e^{3z-7} &= \ln 100 \\ \ln 10 + (3z-7) \ln e &= \ln 100 \\ 3z - 7 &= \ln 100 - \ln 10 \\ 3z &= \ln 100 - \ln 10 + 7 \\ z &= \frac{\ln 100 - \ln 10 + 7}{3} \\ &\approx 3.101 \end{aligned}$$

61.
$$\begin{aligned} 1.5(1.05)^x &= 2(1.01)^x \\ \ln[1.5(1.05)^x] &= \ln[2(1.01)^x] \\ \ln 1.5 + x \ln 1.05 &= \ln 2 + x \ln 1.01 \\ x(\ln 1.05 - \ln 1.01) &= \ln 2 - \ln 1.5 \\ x &= \frac{\ln 2 - \ln 1.5}{\ln 1.05 - \ln 1.01} \\ &\approx 7.407 \end{aligned}$$

63. $f(x) = \ln (x^2 - 9)$

Since the domain of $f(x) = \ln x$ is $(0, \infty)$, the domain of $f(x) = \ln (x^2 - 9)$ is the set of all real numbers x for which

$$x^2 - 9 > 0.$$

To solve this quadratic inequality, first solve the corresponding quadratic equation.

$$\begin{aligned} x^2 - 9 &= 0 \\ (x+3)(x-3) &= 0 \\ x + 3 = 0 \quad &\text{or} \quad x - 3 = 0 \\ x = -3 \quad &\text{or} \quad x = 3 \end{aligned}$$

These two solutions determine three intervals on the number line: $(-\infty, -3), (-3, 3)$, and $(3, \infty)$.

If $x = -4$, $(-4+2)(-4-2) > 0$.
If $x = 0$, $(0+2)(0-2) \not> 0$.
If $x = 4$, $(4+2)(4-2) > 0$.

The domain is $x < -3$ or $x > 3$, which is written in interval notation as $(-\infty, -3) \cup (3, \infty)$.

65. Let $m = \log_a \frac{x}{y}$, $n = \log_a x$, and $p = \log_a y$.
Then $a^m = \frac{x}{y}$, $a^n = x$, and $a^p = y$.
Substituting gives

$$a^m = \frac{x}{y} = \frac{a^n}{a^p} = a^{n-p}.$$

So $m = n - p$.
Therefore,

$$\log_a \frac{x}{y} = \log_a x - \log_a y.$$

67. From Example 7, the doubling time t in years when $m = 1$ is given by

$$t = \frac{\ln 2}{\ln (1+r)}.$$

(a) Let $r = 0.03$.

$$\begin{aligned} t &= \frac{\ln 2}{\ln 1.03} \\ &= 23.4 \text{ years} \end{aligned}$$

(b) Let $r = 0.06$.

$$\begin{aligned} t &= \frac{\ln 2}{\ln 1.06} \\ &= 11.9 \text{ years} \end{aligned}$$

(c) Let $r = 0.08$.

$$\begin{aligned} t &= \frac{\ln 2}{\ln 1.08} \\ &= 9.0 \text{ years} \end{aligned}$$

(d) Since $0.001 \le 0.03 \le 0.05$, for $r = 0.03$, we use the rule of 70.

$$\frac{70}{100r} = \frac{70}{100(0.03)} = 23.3 \text{ years}$$

Since $0.05 \le 0.06 \le 0.12$, for $r = 0.06$, we use the rule of 72.

$$\frac{72}{100r} = \frac{72}{100(0.06)} = 12 \text{ years}$$

For $r = 0.08$, we use the rule of 72.

$$\frac{72}{100(0.08)} = 9 \text{ years}$$

69. $A = Pe^{rt}$

$$\begin{aligned} 1200 &= 500e^{r\cdot 14} \\ 2.4 &= e^{14r} \\ \ln(2.4) &= \ln e^{14r} \\ \ln(2.4) &= 14r \\ \frac{\ln(2.4)}{14} &= r \\ 0.0625 &\approx r \end{aligned}$$

The interest rate should be 6.25%.

71. After x years at Humongous Enterprises, your salary would be 45,000 $(1 + 0.04)^x$ or 45,000 $(1.04)^x$. After x years at Crabapple Inc., your salary would be 30,000 $(1 + 0.06)^x$ or 30,000 $(1.06)^x$. First we find when the salaries would be equal.

$$\begin{aligned} 45{,}000(1.04)^x &= 30{,}000(1.06)^x \\ \frac{(1.04)^x}{(1.06)^x} &= \frac{30{,}000}{45{,}000} \\ \left(\frac{1.04}{1.06}\right)^x &= \frac{2}{3} \\ \log\left(\frac{1.04}{1.06}\right)^x &= \log\left(\frac{2}{3}\right) \\ x\log\left(\frac{1.04}{1.06}\right) &= \log\left(\frac{2}{3}\right) \end{aligned}$$

$$x = \frac{\log\left(\frac{2}{3}\right)}{\log\left(\frac{1.04}{1.06}\right)}$$

$x \approx 21.29$

$2009 + 21.29 = 2030.29$

Therefore, on July 1, 2031, the job at Crabapple, Inc., will pay more.

73. (a) The total number of individuals in the community is 50 + 50, or 100.

Let $P_1 = \frac{50}{100} = 0.5,\ P_2 = 0.5.$

$$\begin{aligned} H &= -1[P_1 \ln P_1 + P_2 \ln P_2] \\ &= -1[0.5 \ln 0.5 + 0.5 \ln 0.5] \\ &\approx 0.693 \end{aligned}$$

(b) For 2 species, the maximum diversity is $\ln 2$.

(c) Yes, $\ln 2 \approx 0.693$.

75. (a) 3 species, $\frac{1}{3}$ each:

$$P_1 = P_2 = P_3 = \frac{1}{3}$$

$$\begin{aligned} H &= -(P_1 \ln P_1 + P_2 \ln P_2 + P_3 \ln P_3) \\ &= -3\left(\frac{1}{3}\ln\frac{1}{3}\right) \\ &= -\ln\frac{1}{3} \\ &\approx 1.099 \end{aligned}$$

(b) 4 species, $\frac{1}{4}$ each:

$$P_1 = P_2 = P_3 = P_4 = \frac{1}{4}$$

$$\begin{aligned} H &= (P_1 \ln P_1 + P_2 \ln P_2 + P_3 \ln P_3 + P_4 \ln P_4) \\ &= -4\left(\frac{1}{4}\ln\frac{1}{4}\right) \\ &= -\ln\frac{1}{4} \\ &\approx 1.386 \end{aligned}$$

(c) Notice that

$$-\ln\frac{1}{3} = \ln(3^{-1})^{-1} = \ln 3 \approx 1.099$$

and

$$-\ln\frac{1}{4} = \ln(4^{-1})^{-1} = \ln 4 \approx 1.386$$

by Property c of logarithms, so the populations are at a maximum index of diversity.

77. $C(t) = C_0e^{-kt}$

When $t = 0$, $C(t) = 2$, and when $t = 3$, $C(t) = 1$.

$$\begin{aligned} 2 &= C_0e^{-k(0)} \\ C_0 &= 2 \\ 1 &= 2e^{-3k} \\ \frac{1}{2} &= e^{-3k} \\ -3k &= \ln\frac{1}{2} = \ln 2^{-1} = -\ln 2 \\ k &= \frac{\ln 2}{3} \\ T &= \frac{1}{k}\ln\frac{C_2}{C_1} \\ T &= \frac{1}{\frac{\ln 2}{3}}\ln\frac{5\,C_1}{C_1} \\ T &= \frac{3\ln 5}{\ln 2} \\ T &\approx 7.0 \end{aligned}$$

The drug should be given about every 7 hours.

79. (a) $h(t) = 37.79(1.021)^t$

Double the 2005 population is $2(42.69) = 85.38$ million

$$85.38 = 37.79(1.021)^t$$
$$\frac{85.38}{37.79} = (1.021)^t$$
$$\log_{1.021}\left(\frac{85.38}{37.79}\right) = t$$
$$t = \frac{\ln\left(\frac{85.38}{37.79}\right)}{\ln 1.021}$$
$$\approx 39.22$$

The Hispanic population is estimated to double their 2005 population in 2039.

(b) $h(t) = 11.14(1.023)^t$

Double the 2005 population is $2(12.69) = 25.38$ million

$$25.38 = 11.14(1.023)^t$$
$$\frac{25.38}{11.14} = (1.023)^t$$
$$\log_{1.023}\left(\frac{25.38}{11.14}\right) = t$$
$$t = \frac{\ln\left(\frac{25.38}{11.14}\right)}{\ln 1.023}$$
$$\approx 36.21$$

The Asian population is estimated to double their 2005 population in 2036.

81. $$C = B\log_2\left(\frac{s}{n} + 1\right)$$
$$\frac{C}{B} = \log_2\left(\frac{s}{n} + 1\right)$$
$$2^{C/B} = \frac{s}{n} + 1$$
$$\frac{s}{n} = 2^{C/B} - 1$$

83. Let I_1 be the intensity of the sound whose decibel rating is 85.

(a) $$10\log\frac{I_1}{I_0} = 85$$
$$\log\frac{I_1}{I_0} = 8.5$$
$$\log I_1 - \log I_0 = 8.5$$
$$\log I_1 = 8.5 + \log I_0$$

Let I_2 be the intensity of the sound whose decimal rating is 75.

$$10\log\frac{I_2}{I_0} = 75$$
$$\log\frac{I_2}{I_0} = 7.5$$
$$\log I_2 - \log I_0 = 7.5$$
$$\log I_0 = \log I_2 - 7.5$$

Substitute for I_0 in the equation for $\log I_1$.

$$\log I_1 = 8.5 + \log I_0$$
$$= 8.5 + \log I_2 - 7.5$$
$$= 1 + \log I_2$$
$$\log I_1 - \log I_2 = 1$$
$$\log\frac{I_1}{I_2} = 1$$

Then $\frac{I_1}{I_2} = 10$, so $I_2 = \frac{1}{10}I_1$. This means the intensity of the sound that had a rating of 75 decibels is $\frac{1}{10}$ as intense as the sound that had a rating of 85 decibels.

85. $\text{pH} = -\log\,[\text{H}^+]$

(a) For pure water:

$$7 = -\log\,[\text{H}^+]$$
$$-7 = \log\,[\text{H}^+]$$
$$10^{-7} = [\text{H}^+]$$

For acid rain:

$$4 = -\log\,[\text{H}^+]$$
$$-4 = \log\,[\text{H}^+]$$
$$10^{-4} = [\text{H}^+]$$
$$\frac{10^{-4}}{10^{-7}} = 10^3 = 1000$$

The acid rain has a hydrogen ion concentration 1000 times greater than pure water.

(b) For laundry solution:

$$11 = -\log\,[\text{H}^+]$$
$$10^{-11} = [\text{H}^+]$$

For black coffee:

$$5 = -\log\,[\text{H}^+]$$
$$10^{-5} = [\text{H}^+]$$
$$\frac{10^{-5}}{10^{-11}} = 10^6 = 1{,}000{,}000$$

The coffee has a hydrogen ion concentration 1,000,000 times greater than the laundry mixture.

2.6 Applications: Growth and Decay; Mathematics of Finance

5. Assume that $y = y_0e^{kt}$ represents the amount remaining of a radioactive substance decaying with a half-life of T. Since $y = y_0$ is the amount of the substance at time $t = 0$, then $y = \frac{y_0}{2}$ is the amount at time $t = T$. Therefore, $\frac{y_0}{2} = y_0e^{kT}$, and solving for k yields

$$\begin{aligned} \frac{1}{2} &= e^{kT} \\ \ln\left(\frac{1}{2}\right) &= kT \\ k &= \frac{\ln\left(\frac{1}{2}\right)}{T} \\ &= \frac{\ln(2^{-1})}{T} \\ &= -\frac{\ln 2}{T}. \end{aligned}$$

7. $r = 4\%$ compounded quarterly, $m = 4$

$$\begin{aligned} r_E &= \left(1+\frac{r}{m}\right)^m - 1 \\ &= \left(1+\frac{0.04}{4}\right)^4 - 1 \\ &\approx 0.0406 \\ &\approx 4.06\% \end{aligned}$$

9. $r = 8\%$ compounded continuously

$$\begin{aligned} r_E &= e^r - 1 \\ &= e^{0.08} - 1 \\ &= 0.0833 \\ &= 8.33\% \end{aligned}$$

11. $A = \$10,000$, $r = 6\%$, $m = 4$, $t = 8$

$$\begin{aligned} P &= A\left(1+\frac{r}{m}\right)^{-tm} \\ &= 10,000\left(1+\frac{0.06}{4}\right)^{-8(4)} \\ &\approx \$6209.93 \end{aligned}$$

13. $A = \$7300$, $r = 5\%$ compounded continuously, $t = 3$

$$\begin{aligned} A &= Pe^{rt} \\ P &= \frac{A}{e^{rt}} \\ &= \frac{7300}{e^{0.5(3)}} \\ &\approx \$6283.17 \end{aligned}$$

15. $r = 9\%$ compounded semiannually

$$\begin{aligned} r_E &= \left(1+\frac{0.09}{2}\right)^2 - 1 \\ &\approx 0.0920 \\ &\approx 9.20\% \end{aligned}$$

17. $r = 6\%$ compounded monthly

$$\begin{aligned} r_E &= \left(1+\frac{0.06}{12}\right)^{12} - 1 \\ &\approx 0.0617 \\ &\approx 6.17\% \end{aligned}$$

19. (a) $A = \$307,000$, $t = 3$, $r = 6\%$, $m = 2$

$$\begin{aligned} A &= P\left(1+\frac{r}{m}\right)^{mt} \\ 307,000 &= P\left(1+\frac{0.06}{2}\right)^{3(2)} \\ 307,000 &= P(1.03)^6 \\ \frac{307,000}{(1.03)^6} &= P \\ \$257,107.67 &= P \end{aligned}$$

(b) $\begin{aligned} \text{Interest} &= 307,000 - 257,107.67 \\ &= \$49,892.33 \end{aligned}$

(c) $\begin{aligned} P &= \$200,000 \\ A &= 200,000(1.03)^6 \\ &= 238,810.46 \end{aligned}$

The additional amount needed is

$$\begin{aligned} &307,000 - 238,810.46 \\ &= \$68,189.54. \end{aligned}$$

21. $P = \$60{,}000$

(a) $r = 8\%$ compounded quarterly:

$$\begin{aligned} A &= P\left(1+\frac{r}{m}\right)^{tm} \\ &= 60,000\left(1+\frac{0.08}{4}\right)^{5(4)} \\ &\approx \$89{,}156.84 \end{aligned}$$

$r = 7.75\%$ compounded continuously

$$\begin{aligned} A &= Pe^{rt} \\ &= 60,000e^{0.0775(5)} \\ &\approx \$88,397.58 \end{aligned}$$

Linda will earn more money at 8% compounded quarterly.

(b) She will earn \$759.26 more.

(c) $r = 8\%,\ m = 4:$

$$\begin{aligned} r_E &= \left(1+\frac{r}{m}\right)^m - 1 \\ &= \left(1+\frac{0.08}{4}\right)^4 - 1 \\ &\approx 0.0824 \\ &= 8.24\% \end{aligned}$$

$r = 7.75\%$ compounded continuously:

$$\begin{aligned} r_E &= e^r - 1 \\ &= e^{0.0775} - 1 \\ &\approx 0.0806 \\ &= 8.06\% \end{aligned}$$

(d) $A = \$80{,}000$

$$\begin{aligned} A &= Pe^{rt} \\ 80{,}000 &= 60{,}000e^{0.0775t} \\ \frac{4}{3} &= e^{0.0775t} \\ \ln\frac{4}{3} &= \ln e^{0.0775t} \\ \ln 4 - \ln 3 &= 0.0775t \\ \frac{\ln 4 - \ln 3}{0.0775} &= t \\ 3.71 &= t \end{aligned}$$

\$60,000 will grow to \$80,000 in about 3.71 years.

(e) $60{,}000\left(1+\frac{0.08}{4}\right)^{4x} \geq 80{,}000$

$$\begin{aligned} (1.02)^{4x} &\geq \frac{80{,}000}{60{,}000} \\ (1.02)^{4x} &\geq \frac{4}{3} \\ \log\,(1.02)^{4x} &\geq \log\left(\frac{4}{3}\right) \\ 4x\,\log\,(1.02) &\geq \log\left(\frac{4}{3}\right) \\ x &\geq \frac{\log\,\left(\frac{4}{3}\right)}{4\,\log\,(1.02)} \approx 3.63 \end{aligned}$$

It will take about 3.63 years.

23. The figure is not correct.

$(1 + 0.09)(1 + 0.08)(1 + 0.07) = 1.2596$

This is a 25.96% increase.

25. $S(x) = 1000 - 800e^{-x}$

(a)
$$\begin{aligned} S(0) &= 1000 - 800e^0 \\ &= 1000 - 800 \\ &= 200 \end{aligned}$$

(b)
$$\begin{aligned} S(x) &= 500 \\ 500 &= 1000 - 800e^{-x} \\ -500 &= -800e^{-x} \\ \frac{5}{8} &= e^{-x} \\ \ln\frac{5}{8} &= \ln e^{-x} \\ -\ln\frac{5}{8} &= x \\ 0.47 &\approx x \end{aligned}$$

Sales reach 500 in about $\frac{1}{2}$ year.

(c) Since $800e^{-x}$ will never actually be zero, $S(x) = 1000 - 800e^{-x}$ will never be 1000.

(d) Graphing the function $y = S(x)$ on a graphing calculator will show that there is a horizontal asymptote at $y = 1000$. This indicates that the limit on sales is 1000 units.

27. (a) $P = P_0 e^{kt}$

When $t = 1650$, $P = 470$.
When $t = 2005$, $P = 6451$.

$$470 = P_0 e^{1650k}$$
$$6451 = P_0 e^{2005k}$$
$$\frac{6451}{470} = \frac{P_0 e^{2005k}}{P_0 e^{1650k}}$$
$$\frac{6451}{470} = e^{355k}$$
$$355k = \ln\left(\frac{6451}{470}\right)$$
$$k = \frac{\ln\left(\frac{6451}{470}\right)}{355}$$
$$k \approx 0.007378$$

Substitute this value into $470 = P_0 e^{1650k}$ to find P_0.

$$470 = P_0 e^{1650(0.007378)}$$
$$P_0 = \frac{470}{e^{1650(0.007378)}}$$
$$P_0 \approx 0.002427$$

Therefore, $P(t) = 0.002427e^{0.007378t}$.

(b) $P(1) = 0.002427e^{0.007378} \approx 0.002445$ million or 2445

The exponential equation gives a world population of only 2445 in the year 1.

(c) No, the answer in part (b) is too small. Exponential growth does not accurately describe population growth for the world over a long period of time.

29. From 1960 to 2005 is an interval of $t = 45$ years.

$$P = P_0 e^{rt}$$
$$1500 = 59e^{45r}$$
$$\frac{1500}{59} = e^{45r}$$
$$45r = \ln\frac{1500}{59}$$
$$r = \frac{1}{45}\ln\frac{1500}{59}$$
$$\approx 0.0719$$

This is an annual increase of 7.19%.

31. $y = y_0 e^{kt}$

(a) $y = 20{,}000,\ y_0 = 50{,}000,\ t = 9$

$$20{,}000 = 50{,}000e^{9k}$$
$$0.4 = e^{9k}$$
$$\ln 0.4 = 9k$$
$$-0.102 = k$$

The equation is

$$y = 50{,}000e^{-0.102t}.$$

(b) $\frac{1}{2}(50{,}000) = 25{,}000$

$$25{,}000 = 50{,}000e^{-0.102t}$$
$$0.5 = e^{-0.102t}$$
$$\ln 0.5 = -0.102t$$
$$6.8 = t$$

Half the bacteria remain after about 6.8 hours.

33. Use $y = y_0 e^{-kt}$.

When $t = 5$, $y = 0.37y_0$.

$$0.37y_0 = y_0 e^{-5k}$$
$$0.37 = e^{-5k}$$
$$-5k = \ln(0.37)$$
$$k = \frac{\ln(0.37)}{-5}$$
$$k \approx 0.1989$$

35.

$$A(t) = A_0 e^{kt}$$
$$0.60\ A_0 = A_0 e^{(-\ln 2/5600)t}$$
$$0.60 = e^{(-\ln 2/5600)t}$$
$$\ln 0.60 = -\frac{\ln 2}{5600}t$$
$$\frac{5600(\ln 0.60)}{-\ln 2} = t$$
$$4127 \approx t$$

The sample was about 4100 years old.

37.

$$\frac{1}{2}A_0 = A_0 e^{-0.00043t}$$
$$\frac{1}{2} = e^{-0.00043t}$$
$$\ln\frac{1}{2} = -0.00043t$$
$$\ln 1 - \ln 2 = -0.00043t$$
$$\frac{0 - \ln 2}{-0.00043} = t$$
$$1612 \approx t$$

The half-life of radium 226 is about 1600 years.

39. (a) $A(t) = A_0 \left(\frac{1}{2}\right)^{t/1620}$

$$A(100) = 4.0 \left(\frac{1}{2}\right)^{100/1620}$$
$$A(100) \approx 3.8$$

After 100 years, about 3.8 grams will remain.

(b)
$$0.1 = 4.0 \left(\frac{1}{2}\right)^{t/1620}$$
$$\frac{0.1}{4} = \left(\frac{1}{2}\right)^{t/1620}$$
$$\ln 0.025 = \frac{t}{1620} \ln \frac{1}{2}$$
$$t = \frac{1620 \ln 0.025}{\ln \left(\frac{1}{2}\right)}$$
$$t \approx 8600$$

The half-life is about 8600 years.

41. (a) $y = y_0 e^{kt}$

When $t = 0$, $y = 25.0$, so $y_0 = 25.0$.
When $t = 50$, $y = 19.5$.

$$19.5 = 25.0 e^{50k}$$
$$\frac{19.5}{25.0} = e^{50k}$$
$$50k = \ln \left(\frac{19.5}{25.0}\right)$$
$$k = \frac{\ln \left(\frac{19.5}{25.0}\right)}{50}$$
$$k \approx -0.00497$$
$$y = 25.0 e^{-0.00497t}$$

(b)
$$\frac{1}{2} y_0 = y_0 e^{-0.00497t}$$
$$\frac{1}{2} = e^{-0.00497t}$$
$$-0.00497t = \ln \left(\frac{1}{2}\right)$$
$$t = \frac{\ln \left(\frac{1}{2}\right)}{-0.00497}$$
$$t \approx 139$$

The half-life is about 139 days.

43. $A(t) = A_0 \left(\frac{1}{2}\right)^{t/5600}$

$$A(43{,}000) = A_0 \left(\frac{1}{2}\right)^{43{,}000/5600}$$
$$\approx 0.005 A_0$$

About 0.5% of the original carbon 14 was present.

45. (a) Let $t =$ the number of degrees Celsius.

$y = y_0 \cdot e^{kt}$
$y_0 = 10$ when $t = 0°$.
To find k, let $y = 11$ when $t = 10°$.

$$11 = 10 e^{10k}$$
$$e^{10k} = \frac{11}{10}$$
$$10k = \ln 1.1$$
$$k = \frac{\ln 1.1}{10}$$
$$\approx 0.0095$$

The equation is

$$y = 10 e^{0.0095t}.$$

(b) Let $y = 15$; solve for t.

$$15 = 10 e^{0.0095t}$$
$$\ln 1.5 = 0.0095t$$
$$t = \frac{\ln 1.5}{0.0095}$$
$$\approx 42.7$$

15 grams will dissolve at 42.7°C.

47. $f(t) = T_0 + Ce^{-kt}$

$$25 = 20 + 100 e^{-0.1t}$$
$$5 = 100 e^{-0.1t}$$
$$e^{-0.1t} = 0.05$$
$$-0.1t = \ln 0.05$$
$$t = \frac{\ln 0.05}{-0.1}$$
$$\approx 30$$

It will take about 30 min.

Chapter 2 Review Exercises

5. $y = (2x-1)(x+1)$
$= 2x^2 + x - 1$

x	-3	-2	-1	0	1	2	3
y	14	5	0	-1	2	9	20

Pairs: $(-3,14), (-2,5), (-1,0), (0,-1), (1,2), (2,9), (3, 20)$
Range: $\{-1, 0, 2, 5, 9, 14, 20\}$

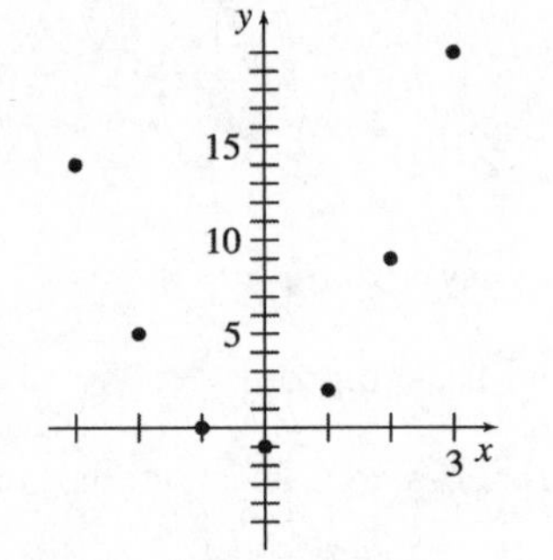

7. $f(x) = 5x^2 - 3$ and $g(x) = -x^2 + 4x + 1$

(a) $f(-2) = 5(-2)^2 - 3 = 17$

(b) $g(3) = -(3)^2 + 4(3) + 1 = 4$

(c) $f(-k) = 5(-k)^2 - 3 = 5k^2 - 3$

(d) $g(3m) = -(3m)^2 + 4(3m) + 1$
$= -9m^2 + 12m + 1$

(e) $f(x+h) = 5(x+h)^2 - 3$
$= 5(x^2 + 2xh + h^2) - 3$
$= 5x^2 + 10xh + 5h^2 - 3$

(f) $g(x+h) = -(x+h)^2 + 4(x+h) + 1$
$= -(x^2 + 2xh + h^2) + 4x + 4h + 1$
$= -x^2 - 2xh - h^2 + 4x + 4h + 1$

(g) $\dfrac{f(x+h) - f(x)}{h}$

$$= \frac{5(x+h)^2 - 3 - (5x^2 - 3)}{h}$$
$$= \frac{5(x^2 + 2hx + h^2) - 3 - 5x^2 + 3}{h}$$
$$= \frac{5x^2 + 10hx + 5h^2 - 5x^2}{h}$$
$$= \frac{10hx + 5h^2}{h}$$
$$= 10x + 5h$$

(h) $\dfrac{g(x+h) - g(x)}{h}$

$$= \frac{-(x+h)^2 + 4(x+h) + 1 - (-x^2 + 4x + 1)}{h}$$
$$= \frac{-(x^2 + 2xh + h^2) + 4x + 4h + 1 + x^2 - 4x - 1}{h}$$
$$= \frac{-x^2 - 2xh - h^2 + 4h + x^2}{h}$$
$$= \frac{-2xh - h^2 + 4h}{h}$$
$$= -2x - h + 4$$

9. $y = \ln(x+7)$

$$x + 7 > 0$$
$$x > -7$$

Domain: $(-7, \infty)$.

11. $y = \dfrac{3x-4}{x}$

$x \neq 0$

Domain: $(-\infty, 0) \cup (0, \infty)$

13. $y = 2x^2 + 3x - 1$

The graph is a parabola.
Let $y = 0$.

$$0 = 2x^2 + 3x - 1$$
$$x = \frac{-3 \pm \sqrt{3^2 - 4(2)(-1)}}{2(2)}$$
$$= \frac{-3 \pm \sqrt{9+8}}{4}$$
$$= \frac{-3 \pm \sqrt{17}}{4}$$

The x-intercepts are $\frac{-3+\sqrt{17}}{4} \approx 0.28$ and $\frac{-3-\sqrt{17}}{4} \approx -1.48$.

Let $x = 0$.

$$y = 2(0)^2 + 3(0) - 1$$

-1 is the y-intercept.

Vertex: $x = \dfrac{-b}{2a} = \dfrac{-3}{2(2)} = -\dfrac{3}{4}$

$$y = 2\left(-\frac{3}{4}\right)^2 + 3\left(-\frac{3}{4}\right) - 1$$
$$= \frac{9}{8} - \frac{9}{4} - 1$$
$$= -\frac{17}{8}$$

The vertex is $\left(-\frac{3}{4}, -\frac{17}{8}\right)$.

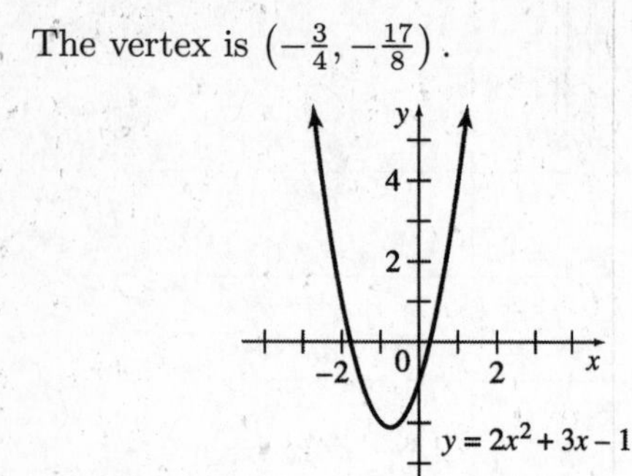

15. $y = -x^2 + 4x + 2$

Let $y = 0$.

$$\begin{aligned} 0 &= -x^2 + 4x + 2 \\ x &= \frac{-4 \pm \sqrt{4^2 - 4(-1)(2)}}{2(-1)} \\ &= \frac{-4 \pm \sqrt{24}}{-2} \\ &= 2 \pm \sqrt{6} \end{aligned}$$

The x-intercepts are $2 + \sqrt{6} \approx 4.45$ and $2 - \sqrt{6} \approx -0.45$.
Let $x = 0$.

$$y = -0^2 + 4(0) + 2$$

2 is the y-intercept.

Vertex: $x = \dfrac{-b}{2a} = \dfrac{-4}{2(-1)} = \dfrac{-4}{-2} = 2$

$$y = -2^2 + 4(2) + 2 = 6$$

The vertex is $(2, 6)$.

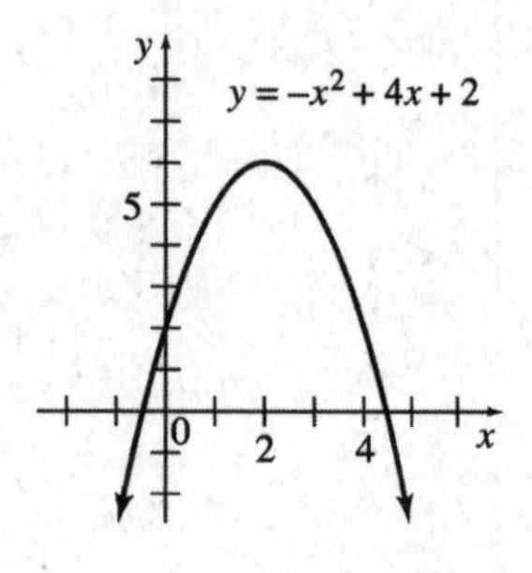

17. $f(x) = x^3 - 3$

Translate the graph of $f(x) = x^3$ 3 units down.

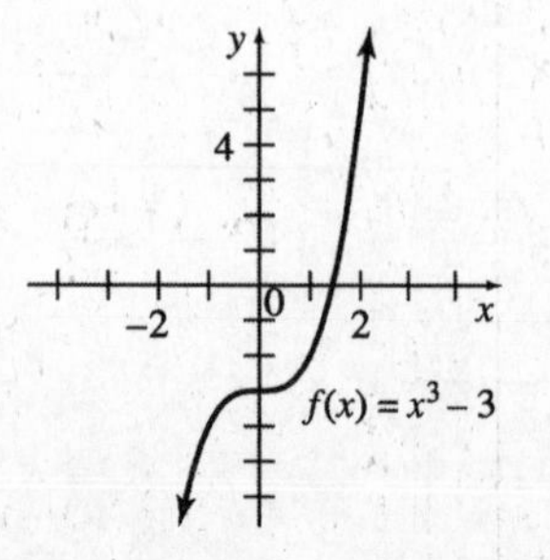

19. $y = -(x - 1)^4 + 4$

Translate the graph of $y = x^4$ 1 unit to the right and reflect vertically. Translate 4 units upward.

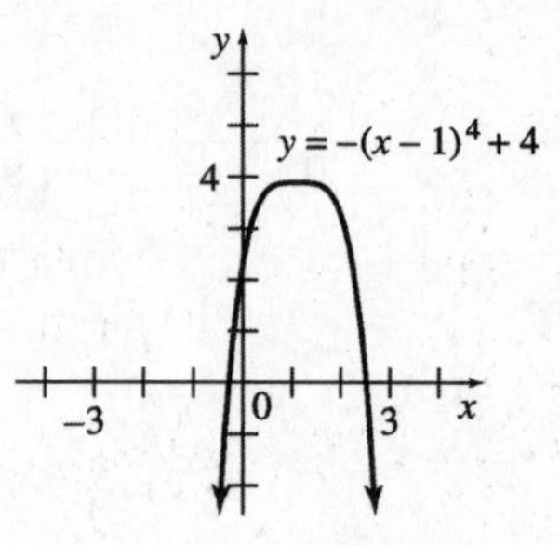

21. $f(x) = \dfrac{8}{x}$

Vertical asymptote: $x = 0$

Horizontal asymptote:

$\frac{8}{x}$ approaches zero as x gets larger.

$y = 0$ is an asymptote.

x	−4	−3	−2	−1	1	2	3	4
y	−2	−2.7	−4	−8	8	4	2.7	2

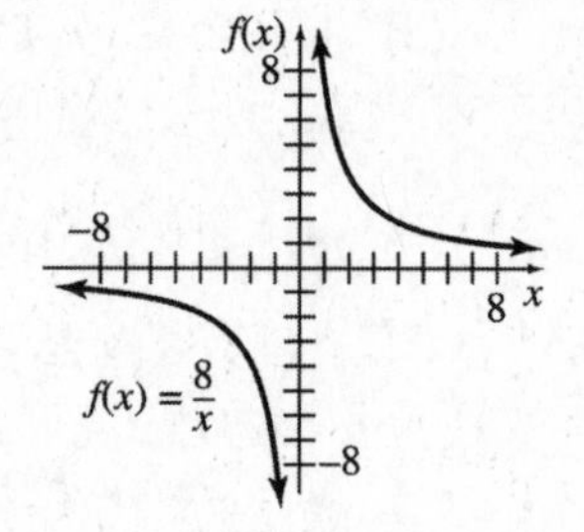

23. $f(x) = \dfrac{4x - 2}{3x + 1}$

Vertical asymptote:

$$\begin{aligned} 3x + 1 &= 0 \\ x &= -\frac{1}{3} \end{aligned}$$

Horizontal asymptote:

As x gets larger,

$$\frac{4x - 2}{3x - 1} \approx \frac{4x}{3x} = \frac{4}{3}.$$

$y = \frac{4}{3}$ is an asymptote.

x	-3	-2	-1	0	1	2	3
y	1.75	2	3	-2	0.5	0.86	1

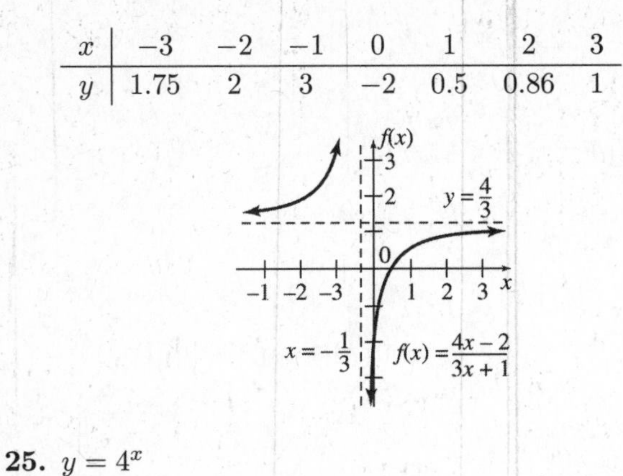

25. $y = 4^x$

x	-2	-1	0	1	2
y	$\frac{1}{16}$	$\frac{1}{4}$	1	4	16

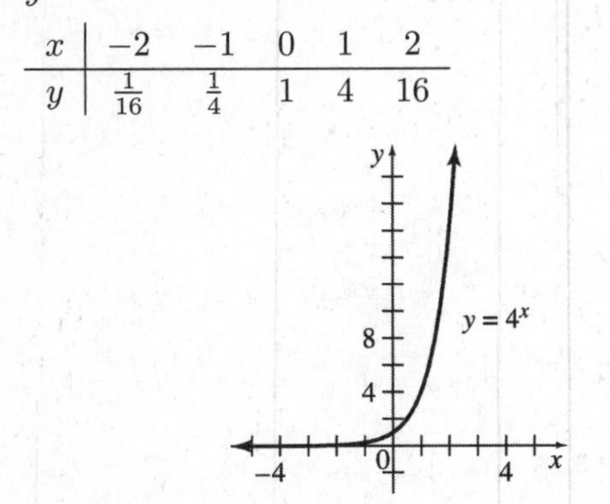

27. $y = \left(\frac{1}{5}\right)^{2x-3}$

x	0	1	2
y	125	5	$\frac{1}{5}$

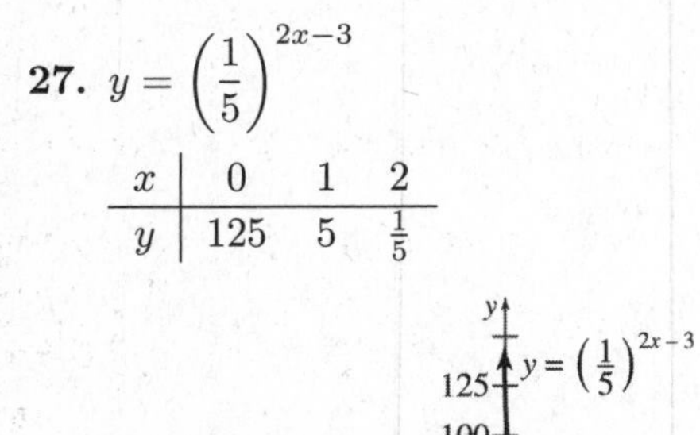

29. $y = \log_2 (x-1)$

$2^y = x - 1$

$x = 1 + 2^y$

x	2	3	5	9
y	0	1	2	3

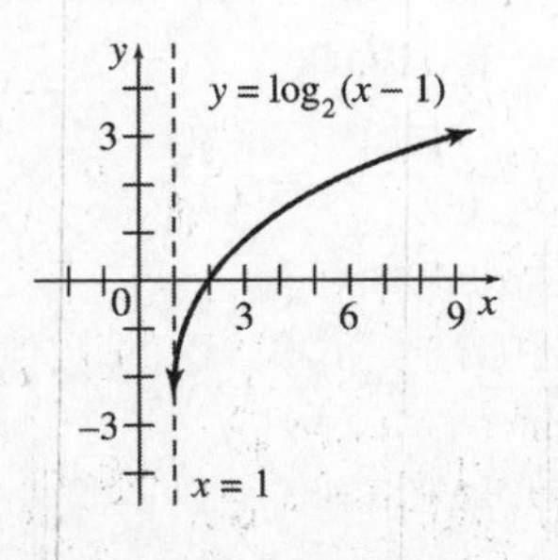

31.

$$\begin{aligned} y &= -\ln (x+3) \\ -y &= \ln (x+3) \\ e^{-y} &= x+3 \\ e^{-y} - 3 &= x \end{aligned}$$

x	-2.63	-2	-0.28	4.39
y	1	0	-1	-2

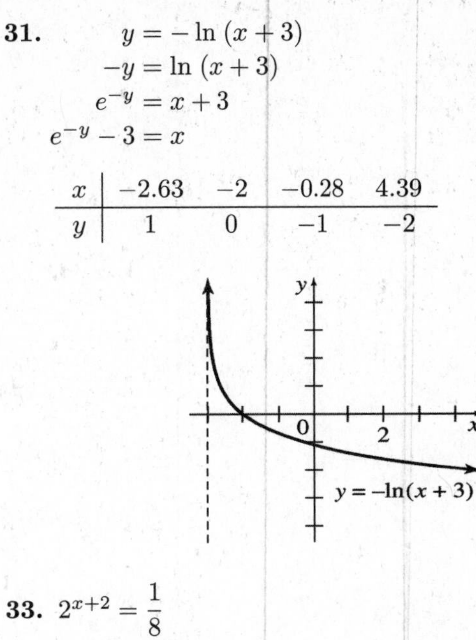

33.

$$\begin{aligned} 2^{x+2} &= \frac{1}{8} \\ 2^{x+2} &= \frac{1}{2^3} \\ 2^{x+2} &= 2^{-3} \\ x+2 &= -3 \\ x &= -5 \end{aligned}$$

35.

$$\begin{aligned} 9^{2y+3} &= 27^y \\ (3^2)^{2y+3} &= (3^3)^y \\ 3^{4y+6} &= 3^{3y} \\ 4y+6 &= 3y \\ y &= -6 \end{aligned}$$

37. $3^5 = 243$

The equation in logarithmic form is

$$\log_3 243 = 5.$$

39. $e^{0.8} = 2.22554$

The equation in logarithmic form is

$$\ln 2.22554 = 0.8.$$

41. $\log_2 32 = 5$

The equation in exponential form is

$$2^5 = 32.$$

43. $\ln 82.9 = 4.41763$

The equation in exponential form is

$$e^{4.41763} = 82.9.$$

45. $\log_3 81 = x$

$$3^x = 81$$
$$3^x = 3^4$$
$$x = 4$$

47. $\log_4 8 = x$

$$4^x = 8$$
$$(2^2)^x = 2^3$$
$$2x = 3$$
$$x = \frac{3}{2}$$

49. $\log_5 3k + \log_5 7k^3$

$$= \log_5 3k(7k^3)$$
$$= \log_5 (21k^4)$$

51. $4 \log_3 y - 2 \log_3 x$

$$= \log_3 y^4 - \log_3 x^2$$
$$= \log_3 \left(\frac{y^4}{x^2}\right)$$

53.

$$6^p = 17$$
$$\ln 6^p = \ln 17$$
$$p \ln 6 = \ln 17$$
$$p = \frac{\ln 17}{\ln 6}$$
$$\approx 1.581$$

55.

$$2^{1-m} = 7$$
$$\ln 2^{1-m} = \ln 7$$
$$(1-m) \ln 2 = \ln 7$$
$$1 - m = \frac{\ln 7}{\ln 2}$$
$$-m = \frac{\ln 7}{\ln 2} - 1$$
$$m = 1 - \frac{\ln 7}{\ln 2}$$
$$\approx -1.807$$

57.

$$e^{-5-2x} = 5$$
$$\ln e^{-5-2x} = \ln 5$$
$$(-5 - 2x) \ln e = \ln 5$$
$$(-5 - 2x) \cdot 1 = \ln 5$$
$$-2x = \ln 5 + 5$$
$$x = \frac{\ln 5 + 5}{-2}$$
$$\approx -3.305$$

59.

$$\left(1 + \frac{m}{3}\right)^5 = 15$$
$$\left[\left(1 + \frac{m}{3}\right)^5\right]^{1/5} = 15^{1/5}$$
$$1 + \frac{m}{3} = 15^{1/5}$$
$$\frac{m}{3} = 15^{1/5} - 1$$
$$m = 3(15^{1/5} - 1)$$
$$\approx 2.156$$

61. $\log_k 64 = 6$

$$k^6 = 64$$
$$k^6 = 2^6$$
$$k = 2$$

63. $\log(4p + 1) + \log p = \log 3$

$$\log[p(4p + 1)] = \log 3$$
$$\log(4p^2 + p) = \log 3$$
$$4p^2 + p = 3$$
$$4p^2 + p - 3 = 0$$
$$(4p - 3)(p + 1) = 0$$
$$4p - 3 = 0 \quad \text{or} \quad p + 1 = 0$$
$$p = \frac{3}{4} \qquad\qquad p = -1$$

p cannot be negative, so $p = \frac{3}{4}$.

65. $f(x) = a^x; a > 0, a \neq 1$

(a) The domain is $(-\infty, \infty)$.

(b) The range is $(0, \infty)$.

(c) The y-intercept is 1.

(d) The graph has no discontinuities.

(e) The x-axis, $y = 0$, is a horizontal asymptote.

(f) The function is increasing if $a > 1$.

(g) The function is decreasing if $0 < a < 1$.

69. $y = \dfrac{7x}{100 - x}$

(a) $y = \dfrac{7(80)}{100 - 80} = \dfrac{560}{20} = 28$

The cost is \$28,000.

(b) $y = \dfrac{7(50)}{100 - 50} = \dfrac{350}{50} = 7$

The cost is \$7000.

(c) $\dfrac{7(90)}{100 - 90} = \dfrac{630}{10} = 63$

The cost is \$63,000.

(d) Plot the points $(80, 28)$, $(50, 7)$, and $(90, 63)$.

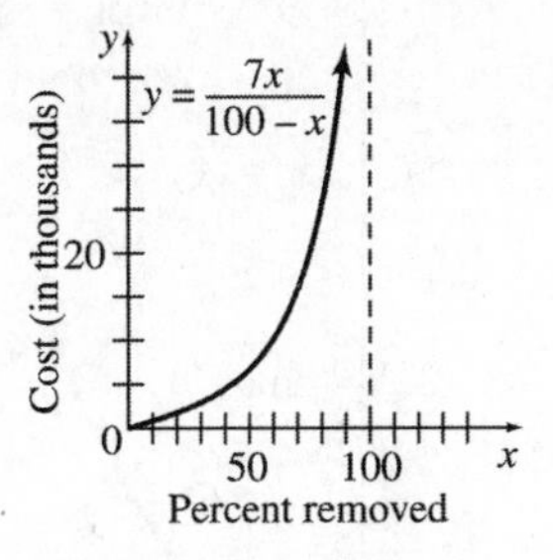

(e) No, because all of the pollutant would be removed when $x = 100$, at which point the denominator of the function would be zero.

71. $P = \$2781.36,\ r = 4.8\%,\ t = 6,\ m = 4$

$$A = P\left(1 + \frac{r}{m}\right)^{tm}$$

$$\begin{aligned} A &= 2781.36\left(1 + \frac{0.048}{4}\right)^{(6)(4)} \\ &= 2781.36(1.012)^{24} \\ &= \$3703.31 \end{aligned}$$

$$\begin{aligned} \text{Interest} &= \$3703.31 - \$2781.36 \\ &= \$921.95 \end{aligned}$$

73. \$2100 deposited at 4% compounded quarterly.

$$A = P\left(1 + \frac{r}{m}\right)^{tm}$$

To double:

$$\begin{aligned} 2(2100) &= 2100\left(1 + \frac{0.04}{4}\right)^{t \cdot 4} \\ 2 &= 1.01^{4t} \\ \ln 2 &= 4t \ln 1.01 \\ t &= \frac{\ln 2}{4 \ln 1.01} \\ &\approx 17.4 \end{aligned}$$

Because interest is compounded quarterly, round the result up to the nearest quarter, which is 17.5 years or 70 quarters.

To triple:

$$\begin{aligned} 3(2100) &= 2100\left(1 + \frac{0.04}{4}\right)^{t \cdot 4} \\ 3 &= 1.01^{4t} \\ \ln 3 &= 4t \ln 1.01 \\ t &= \frac{\ln 3}{4 \ln 1.01} \\ &\approx 27.6 \end{aligned}$$

Because interest is compounded quarterly, round the result up to the nearest quarter, which is 27.75 years or 111 quarters.

75. $P = \$12{,}104,\ r = 6.2\%,\ t = 4$

$$\begin{aligned} A &= Pe^{rt} \\ A &= 12{,}104e^{0.062(4)} \\ &= 12{,}104e^{0.248} \\ &= \$15{,}510.79 \end{aligned}$$

77. $P = \$12{,}000,\ r = 0.05,\ t = 8$

$$\begin{aligned} A &= 12{,}000e^{0.05(8)} \\ &= 12{,}000e^{0.40} \\ &= \$17{,}901.90 \end{aligned}$$

79. $r = 6\%,\ m = 12$

$$\begin{aligned} r_E &= \left(1 + \frac{r}{m}\right)^m - 1 \\ &= \left(1 + \frac{0.06}{12}\right)^{12} - 1 \\ &= 0.0617 = 6.17\% \end{aligned}$$

81. $A = \$2000,\ r = 6\%,\ t = 5,\ m = 1$

$$\begin{aligned} P &= A\left(1 + \frac{r}{m}\right)^{-tm} \\ &= 2000\left(1 + \frac{0.06}{1}\right)^{-5(1)} \\ &= 2000(1.06)^{-5} \\ &= \$1494.52 \end{aligned}$$

83. $r = 7\%,\ t = 8,\ m = 2,\ P = 10{,}000$

$$\begin{aligned} A &= P\left(1 + \frac{r}{m}\right)^{tm} \\ &= 10{,}000\left(1 + \frac{0.07}{2}\right)^{8(2)} \\ &= 10{,}000(1.035)^{16} \\ &= \$17{,}339.86 \end{aligned}$$

85. $P = \$6000,\ A = \$8000,\ t = 3$

$$\begin{aligned} A &= Pe^{rt} \\ 8000 &= 6000e^{3r} \\ \frac{4}{3} &= e^{3r} \\ \ln 4 - \ln 3 &= 3r \\ r &= \frac{\ln 4 - \ln 3}{3} \\ r &\approx 0.0959 \text{ or about } 9.59\% \end{aligned}$$

87. **(a)** $n = 1000 - (p - 50)(10),\ p \geq 50$
$= 1000 - 10p + 500$
$= 1500 - 10p$

(b) $R = pn$
$R = p(1500 - 10p)$

(c) $p \geq 50$

Since n cannot be negative,

$$1500 - 10p \geq 0$$
$$-10p \geq -1500$$
$$p \leq 150.$$

Therefore, $50 \leq p \leq 150$.

(d) Since $n = 1500 - 10p$,

$$10p = 1500 - n$$
$$p = 150 - \frac{n}{10}.$$
$$R = pn$$
$$R = \left(150 - \frac{n}{10}\right)n$$

(e) Since she can sell at most 1000 tickets, $0 \leq n \leq 1000$.

(f) $R = -10p^2 + 1500p$

$$\frac{-b}{2a} = \frac{-1500}{2(-10)} = 75$$

The price producing maximum revenue is \$75.

(g) $R = -\frac{1}{10}n^2 + 150n$

$$\frac{-b}{2a} = \frac{-150}{2\left(-\frac{1}{10}\right)} = 750$$

The number of tickets producing maximum revenue is 750.

(h) $R(p) = -10p^2 + 1500p$
$R(75) = -10(75)^2 + 1500(75)$
$= -56{,}250 + 112{,}500$
$= 56{,}250$

The maximum revenue is \$56,250.

(i)

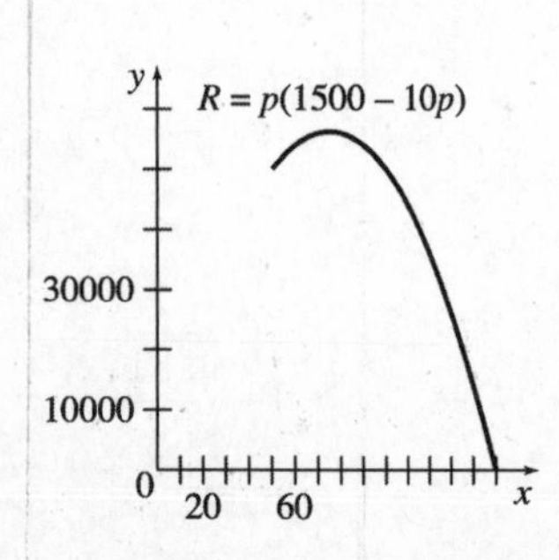

(j) The revenue starts at \$50,000 when the price is \$50, rises to a maximum of \$56,250 when the price is \$75, and falls to 0 when the price is \$150.

89. $C(x) = x^2 + 4x + 7$

(a)

y
30
5
C(x) = x² + 4x + 7
3
5
x
Production cost (in hundreds of dollars)
Hundreds of nails

(b) $C(x+1) - C(x)$
$= (x+1)^2 + 4(x+1) + 7$
$\quad - (x^2 + 4x + 7)$
$= x^2 + 2x + 1 + 4x + 4 + 7$
$\quad - x^2 - 4x - 7$
$= 2x + 5$

(c) $A(x) = \frac{C(x)}{x} = \frac{x^2 + 4x + 7}{x}$
$= x + 4 + \frac{7}{x}$

(d) $A(x+1) - A(x)$
$= (x+1) + 4 + \frac{7}{x+1}$
$\quad - \left(x + 4 + \frac{7}{x}\right)$
$= x + 1 + 4 + \frac{7}{x+1} - x - 4 - \frac{7}{x}$
$= 1 + \frac{7}{x+1} - \frac{7}{x}$
$= 1 + \frac{7x - 7(x+1)}{x(x+1)}$
$= 1 + \frac{7x - 7x - 7}{x(x+1)}$
$= 1 - \frac{7}{x(x+1)}$

91. $F(x) = -\frac{2}{3}x^2 + \frac{14}{3}x + 96$

The maximum fever occurs at the vertex of the parabola.

$$x = \frac{-b}{2a} = \frac{-\frac{14}{3}}{-\frac{4}{3}} = \frac{7}{2}$$

$$\begin{aligned} y &= -\frac{2}{3}\left(\frac{7}{2}\right)^2 + \frac{14}{3}\left(\frac{7}{2}\right) + 96 \\ &= -\frac{2}{3}\left(\frac{49}{4}\right) + \frac{49}{3} + 96 \\ &= -\frac{49}{6} + \frac{49}{3} + 96 \\ &= -\frac{49}{6} + \frac{98}{6} + \frac{576}{6} = \frac{625}{6} \approx 104.2 \end{aligned}$$

The maximum fever occurs on the third day. It is about 104.2°F.

93. **(a)**

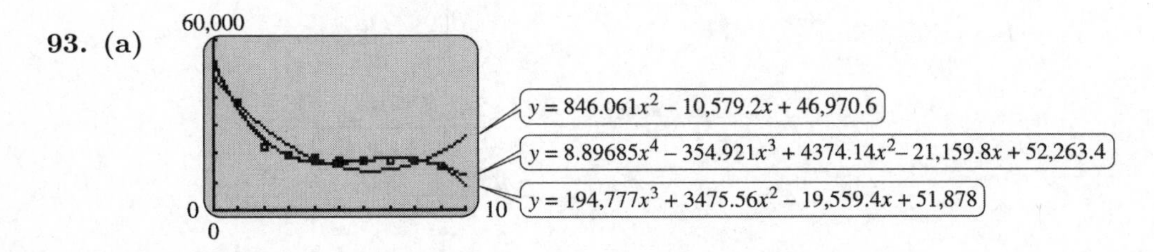

(b) Quadratic:

$y = 846.061x^2 - 10{,}579.2x + 46{,}970.6$

Cubic:

$y = -194.777x^3 + 3475.56x^2 - 19{,}558.4x + 51{,}879$

Quartic:

$y = 8.89685x^4 - 354.921x^3 + 4374.14x^2 - 21{,}159.8x + 52{,}263.4$

(c)

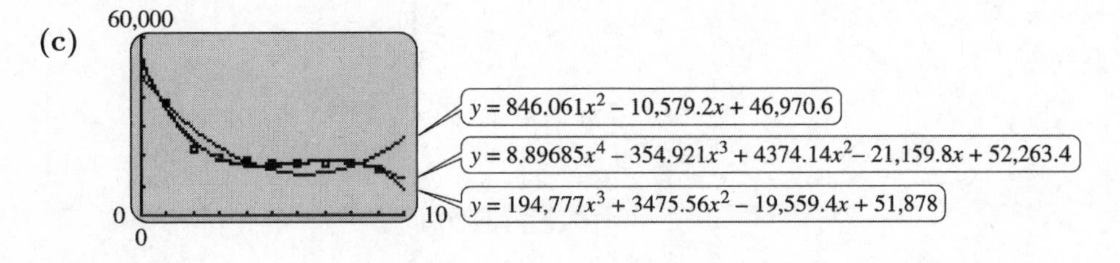

(d) Use the functions found in part (b), with $x = 11$.

Quadratic: 32,973
Cubic: -1969
Quartic: 6635

95.

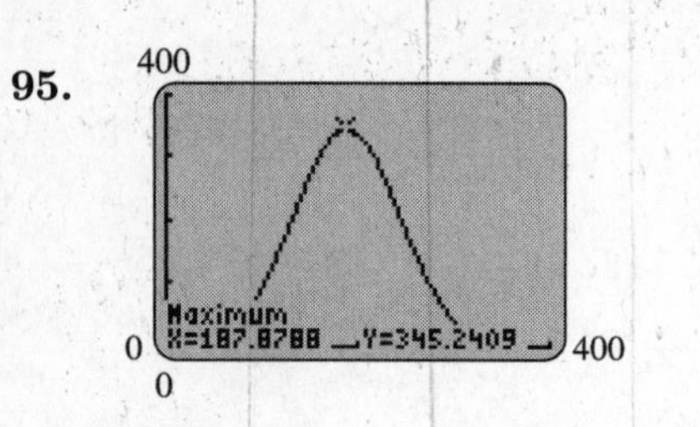

This function has a maximum value at $x \approx 187.9$. At $x \approx 187.9, y \approx 345$. The largest girth for which this formula gives a reasonable answer is 187.9 cm. The predicted mass of a polar bear with this girth is 345 kg.

97. $p(t) = \dfrac{1.79 \cdot 10^{11}}{(2026.87 - t)^{0.99}}$

(a) $p(2005) \approx 8.441$ billion

This is about 1.964 billion more than the estimate of 6.477 billion.

(b) $p(2020) \approx 26.56$ billion
$p(2025) \approx 96.32$ billion

99. Graph

$$y = c(t) = e^{-t} - e^{-2t}$$

on a graphing calculator and locate the maximum point. A calculator shows that the x-coordinate of the maximum point is about 0.69, and the y-coordinate is exactly 0.25. Thus, the maximum concentration of 0.25 occurs at about 0.69 minutes.

101. $y = y_0 e^{-kt}$

(a)
$$\begin{aligned} 100{,}000 &= 128{,}000e^{-k(5)} \\ 128{,}000 &= 100{,}000e^{5k} \\ \frac{128}{100} &= e^{5k} \\ \ln\left(\frac{128}{100}\right) &= 5k \\ 0.05 &\approx k \end{aligned}$$

$y = 100{,}000e^{-0.05t}$

(b)
$$\begin{aligned} 70{,}000 &= 100{,}000e^{-0.05t} \\ \frac{7}{10} &= e^{-0.05t} \\ \ln\frac{7}{10} &= -0.05t \\ 7.1 &\approx t \end{aligned}$$

It will take about 7.1 years.

103. (a) Since the speed in one direction is $v + w$ and in the other direction is $v - w$, the time in one direction is $\frac{d}{v+w}$ and in the other direction is $\frac{d}{v-w}$. So the total time is $\frac{d}{v+w} + \frac{d}{v-w}$.

(b) The average speed is the total distance divided by the total time. So

$$v_{aver} = \frac{2d}{\frac{d}{v+w} + \frac{d}{v-w}}.$$

(c)
$$\begin{aligned} &\frac{2d}{\frac{d}{v+w} + \frac{d}{v-w}} \\ &= \frac{2d}{\frac{d}{v+w} + \frac{d}{v-w}} \cdot \frac{(v+w)(v-w)}{(v+w)(v-w)} \\ &= \frac{2d(v^2 - w^2)}{d(v-w) + d(v+w)} \\ &= \frac{2d(v^2 - w^2)}{dv - dw + dv + dw} \\ &= \frac{2d(v^2 - w^2)}{2dv} \\ &= \frac{v^2 - w^2}{v} = v - \frac{w^2}{v} \end{aligned}$$

(d) $v_{aver} = v - \dfrac{w^2}{v}$

v_{aver} will be greatest when $w = 0$.

105. (a)
$$\begin{aligned} P &= kD^1 \\ 164.8 &= k(30.1) \\ k &= \frac{164.8}{30.1} \approx 5.48 \end{aligned}$$

For $n = 1,\ P = 5.48D$.

$$\begin{aligned} P &= kD^{1.5} \\ 164.8 &= k(30.1)^{1.5} \\ k &= \frac{164.8}{(30.1)^{1.5}} \approx 1.00 \end{aligned}$$

For $n = 1.5,\ P = 1.00D^{1.5}$.

$$\begin{aligned} P &= kD^2 \\ 164.8 &= k(30.1)^2 \\ k &= \frac{164.8}{(30.1)^2} \approx 0.182 \end{aligned}$$

For $n = 2,\ P = 0.182D^2$.

(b)

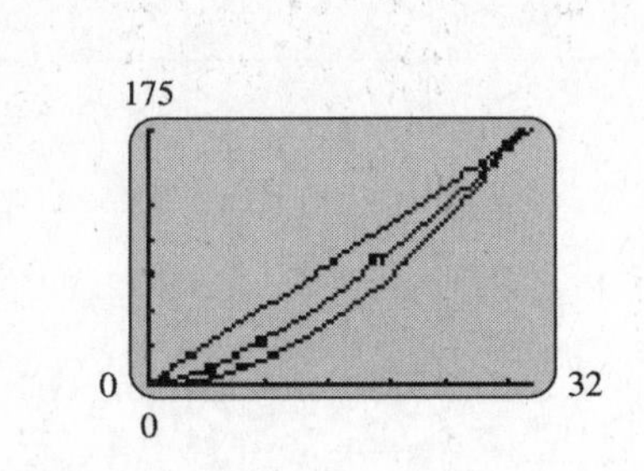

$P = 1.00D^{1.5}$ appears to be the best fit.

(c) $P = 1.00(39.5)^{1.5} \approx 248.3$ years

(d) We obtain

$$P = 1.00D^{1.5}.$$

This is the same as the function found in part (b).

Chapter 2 Test

[2.1]

1. List the ordered pairs obtained from each equation, given $\{-3, -2, -1, 0, 1, 2, 3\}$ as the domain. Give the range.

(a) $2x + 3y = 6$ (b) $y = \dfrac{1}{x^2 - 2}$

2. Let $f(x) = 3x - 4$ and $g(x) = -x^2 + 5x$. Find each of the following.

(a) $f(-3)$ (b) $g(-2)$ (c) $f(2m)$ (d) $g(k-1)$ (e) $f(x+h)$

[2.2]

3. Graph the parabola

$$y = 2x^2 - 4x - 2$$

and give its vertex, axis, x–intercept, and y–intercept.

4. The manufacturer of a certain product has determined that his profit in dollars for making x units of this product is given by the equation

$$p = -2x^2 + 120x + 3000.$$

(a) Find the number of units that will maximize the profit.

(b) What is the maximum profit for making this product?

Use translations and reflections to graph the following functions.

5. $y = 4 - x^2$

6. $f(x) = \dfrac{1}{2}x^3 + 1$

[2.3]

Graph each function.

7. $f(x) = x^3 - 2x^2 - x + 2$

8. $f(x) = \dfrac{3}{2x - 1}$

9. $f(x) = \dfrac{x - 1}{3x + 6}$

10. Find the horizontal and vertical asymptotes and $x-$ and y–intercepts, if any, for the following function.

$$f(x) = \frac{3x - 5}{5x + 10}$$

11. Suppose a cost-benefit model is given by the equation

$$y = \frac{20x}{110 - x},$$

where y is the cost in thousands of dollars of removing x percent of a certain pollutant. Find the cost in thousands of dollars of removing each of the following percents of pollution.

(a) 70% **(b)** 90% **(c)** 100%

[2.4]

12. Solve the equation $8^{2y-1} = 4^{y+1}$.

13. Graph the function $f(x) = 4^{x-1}$.

14. $315 is deposited in an account paying 6% compounded quarterly for 3 years. Find the following.

(a) The amount in the account after 3 years.

(b) The amount of interest earned by this deposit.

[2.5]

15. Evaluate each logarithm without using a calculator.

(a) $\log_8 16$ **(b)** $\ln e^{-3/4}$

16. Use properties of logarithms to simplify the following.

(a) $\log_2 3k + \log_2 4k^2$ **(b)** $4 \log_3 r - 3 \log_3 m$

17. Solve each equation. Round to the nearest thousandth if necessary.

(a) $2^{x+1} = 10$ **(b)** $\log_2 (x + 3) + \log_2 (x - 3) = 4$

[2.6]

18. Find the interest rate needed for $5000 to grow to $10,000 in 8 years with continuous compounding.

19. How long will it take for $1 to triple at an average rate of 7% compounded continuously?

20. Suppose sales of a certain item are given by $\mathrm{S}(x) = 2500 - 1500e^{-2x}$, where x represents the number of years that the item has been on the market and $S(x)$ represents sales in thousands. Find the limit on sales.

21. Find the effective rate for the account described in Problem 14.

22. The population of Smalltown has grown exponentially from 14,000 in 1994 to 16,500 in 1997. At this rate, in what year will the population reach 17,400?

23. Potassium 42 decays exponentially. A sample which contained 1000 grams 5 hours ago has decreased to 758 grams at present.

(a) Write an exponential equation to express the amount, y, present after t hours.

(b) What is the half-life of potassium 42?

24. Find the present value of \$15,000 at 6% compounded quarterly for 4 years.

25. Mr. Jones needs \$20,000 for a down payment on a house in 5 years. How much must he deposit now at 5.8% compounded quarterly in order to have \$20,000 in 5 years?

Chapter 2 Test Answers

1. **(a)** $(-3,4)$, $\left(-2,\frac{10}{3}\right)$, $\left(-1,\frac{8}{3}\right)$, $(0,2)$, $\left(1,\frac{4}{3}\right)$, $\left(2,\frac{2}{3}\right)$, $(3,0)$; range: $\left\{0,\frac{2}{3},\frac{4}{3},2,\frac{8}{3},\frac{10}{3},4\right\}$

(b) $\left(-3,\frac{1}{7}\right)$, $\left(-2,\frac{1}{2}\right)$, $(-1,-1)$, $\left(0,-\frac{1}{2}\right)$, $(1,-1)$, $\left(2,\frac{1}{2}\right)$, $\left(3,\frac{1}{7}\right)$; range: $\left\{-1,-\frac{1}{2},\frac{1}{7},\frac{1}{2}\right\}$

2. **(a)** -13 **(b)** -14 **(c)** $6m-4$
(d) $-k^2+7k-6$ **(e)** $3x+3h-4$

3. Vertex: $(1,-4)$; axis: $x=1$;
x–intercepts: $1-\sqrt{2}\approx -0.414$,
$1+\sqrt{2}\approx 2.414$; y–intercept: -2

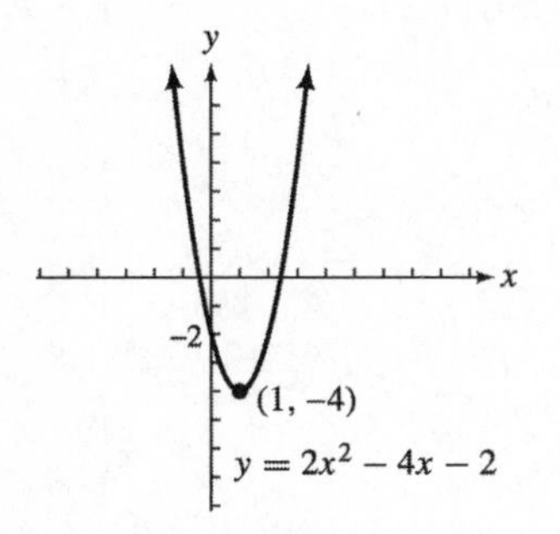

4. **(a)** 30 **(b)** \$4800

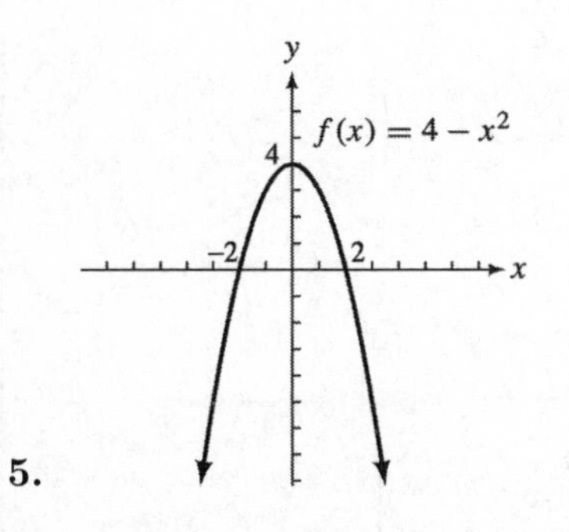

5.

6.

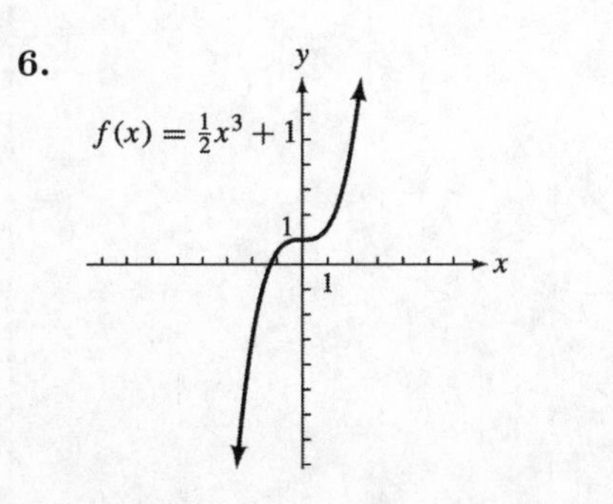

7.

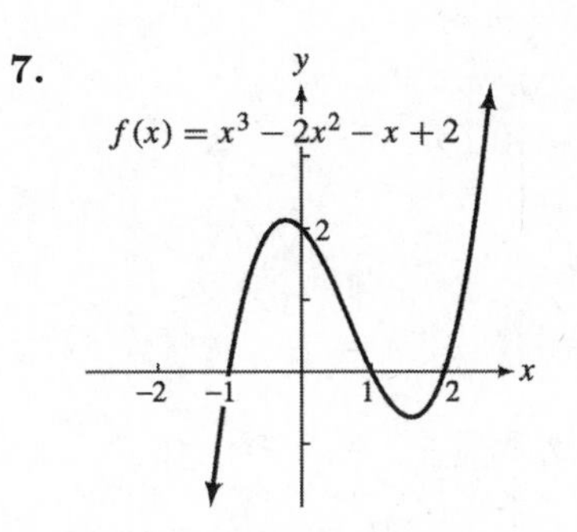

8.

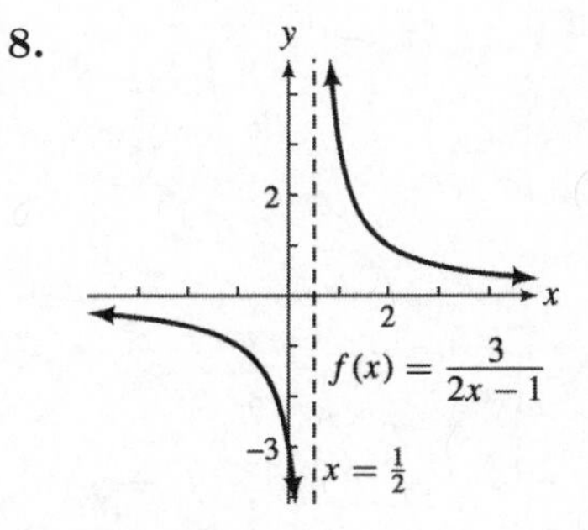

9.

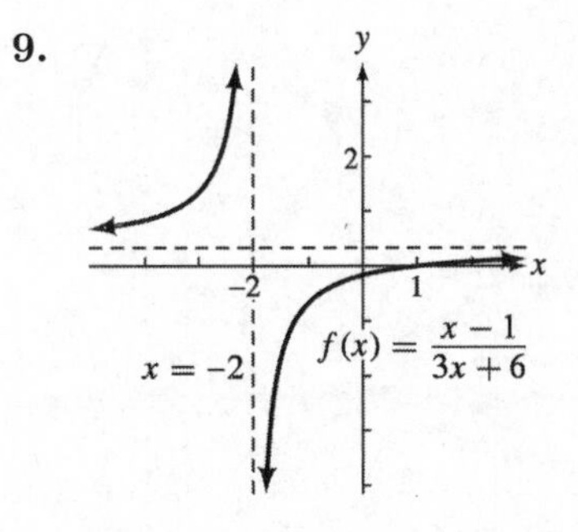

10. Horizontal: $y=\frac{3}{5}$; vertical: $x=-2$;

x–intercept: $\frac{5}{3}$; y–intercept: $-\frac{1}{2}$

11. **(a)** \$35,000 **(b)** \$90,000 **(c)** \$200,000

12. $\frac{5}{4}$

13.

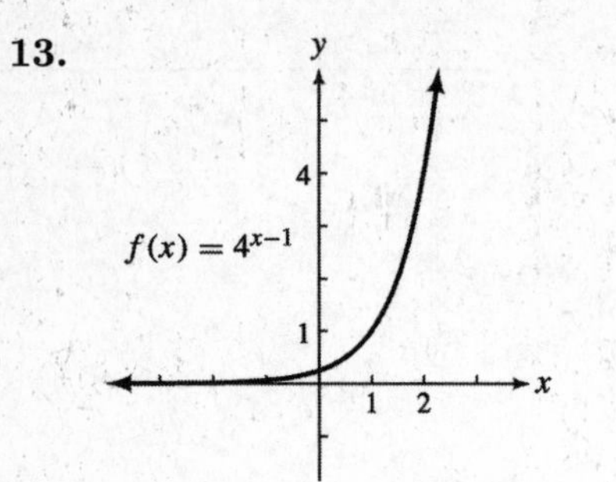

14. **(a)** \$376.62 **(b)** \$61.62

15. **(a)** $\frac{4}{3}$ **(b)** $-\frac{3}{4}$

16. **(a)** $\log_2 12k^3$ **(b)** $\log_3 \frac{r^4}{m^3}$

17. **(a)** 2.322 **(b)** 5

18. 8.66%

19. About 15.7 years

20. 2,500,000

21. 6.14%

22. 1998

23. **(a)** $y = 1000e^{-0.0554t}$ **(b)** About 12.5 hours

24. \$11,820.47

25. \$14,996.47

Chapter 3

THE DERIVATIVE

3.1 Limits

1. Since $\lim_{x\to 2^-} f(x)$ does not equal $\lim_{x\to 2^+} f(x)$, $\lim_{x\to 2} f(x)$ does not exist. The answer is c.

3. Since $\lim_{x\to 4^-} f(x) = \lim_{x\to 4^+} f(x) = 6$,

$\lim_{x\to 4} f(x) = 6$. The answer is b.

5. (a) By reading the graph, as x gets closer to 3 from the left or right, $f(x)$ gets closer to 3.

$$\lim_{x\to 3} f(x) = 3$$

(b) By reading the graph, as x gets closer to 0 from the left or right, $f(x)$ gets closer to 1.

$$\lim_{x\to 0} f(x) = 1$$

7. (a) By reading the graph, as x gets closer to 0 from the left or right, $f(x)$ gets closer to 0.

$$\lim_{x\to 0} f(x) = 0$$

(b) By reading the graph, as x gets closer to 2 from the left, $f(x)$ gets closer to -2, but as x gets closer to 2 from the right, $f(x)$ gets closer to 1.

$\lim_{x\to 2} f(x)$ does not exist.

9. (a) (i) By reading the graph, as x gets closer to -2 from the left, $f(x)$ gets closer to -1.

$$\lim_{x\to -2^-} f(x) = -1$$

(ii) By reading the graph, as x gets closer to -2 from the right, $f(x)$ gets closer to $-\frac{1}{2}$.

$$\lim_{x\to -2^+} f(x) = -\frac{1}{2}$$

(iii) Since $\lim_{x\to -2^-} f(x) = -1$ and $\lim_{x\to -2^+} f(x) = -\frac{1}{2}$, $\lim_{x\to -2} f(x)$ does not exist.

(iv) $f(-2)$ does not exist since there is no point on the graph with an x-coordinate of -2.

(b) (i) By reading the graph, as x gets closer to -1 from the left, $f(x)$ gets closer to $-\frac{1}{2}$.

$$\lim_{x\to -1^-} f(x) = -\frac{1}{2}$$

(ii) By reading the graph, as x gets closer to -1 from the right, $f(x)$ gets closer to $-\frac{1}{2}$.

$$\lim_{x\to -1^+} f(x) = -\frac{1}{2}$$

(iii) Since $\lim_{x\to -1^-} f(x) = -\frac{1}{2}$ and

$\lim_{x\to -1^+} f(x) = -\frac{1}{2}$, $\lim_{x\to -1} f(x) = -\frac{1}{2}$.

(iv) $f(-1) = -\frac{1}{2}$ since $\left(-1, -\frac{1}{2}\right)$ is a point of the graph.

11. By reading the graph, as x moves further to the right, $f(x)$ gets closer to 3. Therefore, $\lim_{x\to\infty} f(x) = 3$.

13. $\lim_{x\to 2} F(x)$ in Exercise 6 exists because $\lim_{x\to 2^-} F(x) = 4$ and $\lim_{x\to 2^+} F(x) = 4$.
$\lim_{x\to -2} f(x)$ in Exercise 9 does not exist since $\lim_{x\to -2^-} f(x) = -1$, but $\lim_{x\to -2^+} f(x) = -\frac{1}{2}$.

15. From the table, as x approaches 1 from the left or the right, $f(x)$ approaches 4.

$$\lim_{x\to 1} f(x) = 4$$

17. $k(x) = \dfrac{x^3 - 2x - 4}{x - 2}$; find $\lim_{x\to 2} k(x)$.

x	1.9	1.99	1.999
$k(x)$	9.41	9.9401	9.9941

x	2.001	2.01	2.1
$k(x)$	10.006	10.0601	10.61

As x approaches 2 from the left or the right, $k(x)$ approaches 10.

$$\lim_{x\to 2} k(x) = 10$$

19. $h(x) = \frac{\sqrt{x}-2}{x-1}$; find $\lim_{x\to 1} h(x)$.

x	0.9	0.99	0.999
$h(x)$	10.51317	100.50126	1000.50013

x	1.001	1.01	1.1
$h(x)$	−999.50012	−99.50124	−9.51191

$\lim_{x\to 1^-} = -\infty$

$\lim_{x\to 1^+} = -\infty$

Thus, $\lim_{x\to 1} h(x)$ does not exist.

21. $\lim_{x\to 4} [f(x) - g(x)] = \lim_{x\to 4} f(x) - \lim_{x\to 4} g(x) = 9 - 27 = -18$

23. $\lim_{x\to 4} \frac{f(x)}{g(x)} = \frac{\lim_{x\to 4} f(x)}{\lim_{x\to 4} g(x)} = \frac{9}{27} = \frac{1}{3}$

25.
$$\begin{aligned}\lim_{x\to 4} \sqrt{f(x)} &= \lim_{x\to 4} [f(x)]^{1/2}\\ &= [\lim_{x\to 4} f(x)]^{1/2}\\ &= 9^{1/2} = 3\end{aligned}$$

27.
$$\begin{aligned}\lim_{x\to 4} 2^{f(x)} &= 2^{\lim_{x\to 4} f(x)}\\ &= 2^9\\ &= 512\end{aligned}$$

29.
$$\begin{aligned}&\lim_{x\to 4} \frac{f(x)+g(x)}{2g(x)}\\ &= \frac{\lim_{x\to 4} [f(x)+g(x)]}{\lim_{x\to 4} 2g(x)}\\ &= \frac{\lim_{x\to 4} f(x) + \lim_{x\to 4} g(x)}{2\lim_{x\to 4} g(x)}\\ &= \frac{9+27}{2(27)} = \frac{36}{54} = \frac{2}{3}\end{aligned}$$

31.
$$\begin{aligned}\lim_{x\to 3} \frac{x^2-9}{x-3} &= \lim_{x\to 3} \frac{(x-3)(x+3)}{x-3}\\ &= \lim_{x\to 3} (x+3)\\ &= \lim_{x\to 3} x + \lim_{x\to 3} 3\\ &= 3+3\\ &= 6\end{aligned}$$

33.
$$\begin{aligned}\lim_{x\to 1} \frac{5x^2-7x+2}{x^2-1} &= \lim_{x\to 1} \frac{(5x-2)(x-1)}{(x+1)(x-1)}\\ &= \lim_{x\to 1} \frac{5x-2}{x+1}\\ &= \frac{5-2}{2}\\ &= \frac{3}{2}\end{aligned}$$

35.
$$\begin{aligned}\lim_{x\to -2} \frac{x^2-x-6}{x+2} &= \lim_{x\to -2} \frac{(x-3)(x+2)}{x+2}\\ &= \lim_{x\to -2} (x-3)\\ &= \lim_{x\to -2} x + \lim_{x\to -2} (-3)\\ &= -2-3\\ &= -5\end{aligned}$$

37.
$$\begin{aligned}&\lim_{x\to 0} \frac{\frac{1}{x+3} - \frac{1}{3}}{x}\\ &= \lim_{x\to 0} \left(\frac{1}{x+3} - \frac{1}{3}\right)\left(\frac{1}{x}\right)\\ &= \lim_{x\to 0} \left[\frac{3}{3(x+3)} - \frac{x+3}{3(x+3)}\right]\left(\frac{1}{x}\right)\\ &= \lim_{x\to 0} \frac{3-x-3}{3(x+3)(x)}\\ &= \lim_{x\to 0} \frac{-x}{3(x+3)x}\\ &= \lim_{x\to 0} \frac{-1}{3(x+3)}\\ &= \frac{-1}{3(0+3)}\\ &= -\frac{1}{9}\end{aligned}$$

39.
$$\begin{aligned}&\lim_{x\to 25} \frac{\sqrt{x}-5}{x-25}\\ &= \lim_{x\to 25} \frac{\sqrt{x}-5}{x-25} \cdot \frac{\sqrt{x}+5}{\sqrt{x}+5}\\ &= \lim_{x\to 25} \frac{x-25}{(x-25)(\sqrt{x}+5)}\\ &= \lim_{x\to 25} \frac{1}{\sqrt{x}+5}\\ &= \frac{1}{\sqrt{25}+5}\\ &= \frac{1}{10}\end{aligned}$$

41. $\lim_{h\to 0} \frac{(x+h)^2 - x^2}{h}$

$= \lim_{h\to 0} \frac{x^2 + 2hx + h^2 - x^2}{h}$

$= \lim_{h\to 0} \frac{2hx + h^2}{h}$

$= \lim_{h\to 0} \frac{h(2x+h)}{h}$

$= \lim_{h\to 0} (2x+h)$

$= 2x + 0 = 2x$

43. $\lim_{x\to\infty} \frac{3x}{7x-1} = \lim_{x\to\infty} \frac{\frac{3x}{x}}{\frac{7x}{x} - \frac{1}{x}}$

$= \lim_{x\to\infty} \frac{3}{7 - \frac{1}{x}}$

$= \frac{3}{7-0} = \frac{3}{7}$

45. $\lim_{x\to-\infty} \frac{-x^2 + 2x}{2x^2 - 2x + 1}$

$= \lim_{x\to-\infty} \frac{\frac{3x^2}{x^2} + \frac{2x}{x^2}}{\frac{2x^2}{x^2} - \frac{2x}{x^2} + \frac{1}{x^2}}$

$= \lim_{x\to-\infty} \frac{3 + \frac{2}{x}}{2 - \frac{2}{x} + \frac{1}{x^2}}$

$= \frac{3-0}{2+0+0} = \frac{3}{2}$

47. $\lim_{x\to\infty} \frac{3x^3 + 2x - 1}{2x^4 - 3x^3 - 2}$

$= \lim_{x\to\infty} \frac{\frac{3x^3}{x^4} + \frac{2x}{x^4} - \frac{1}{x^4}}{\frac{2x^4}{x^4} - \frac{3x^3}{x^4} - \frac{2}{x^4}}$

$= \lim_{x\to\infty} \frac{\frac{3}{x} + \frac{2}{x^3} - \frac{1}{x^4}}{2 - \frac{3}{x} - \frac{2}{x^4}}$

$= \frac{0+0-0}{2-0-0} = 0$

49. $\lim_{x\to\infty} \frac{2x^3 - x - 3}{6x^2 - x - 1}$

$= \lim_{x\to\infty} \frac{\frac{2x^3}{x^3} - \frac{x}{x^3} - \frac{3}{x^3}}{\frac{6x^2}{x^3} - \frac{x}{x^3} - \frac{1}{x^3}}$

$= \lim_{x\to\infty} \frac{2 - \frac{1}{x^2} - \frac{3}{x^3}}{\frac{6}{x} - \frac{1}{x^2} - \frac{1}{x^3}}$

$= \frac{2-0-0}{0-0-0} = \frac{2}{0} = \infty$

Therefore $\lim_{x\to\infty} \frac{2x^3 - x - 3}{6x^2 - x - 1}$ does not exist.

51. $\lim_{x\to\infty} \frac{2x^2 - 7x^4}{9x^2 + 5x - 6} = \lim_{x\to\infty} \frac{\frac{2x^2}{x^2} - \frac{7x^4}{x^2}}{\frac{9x^2}{x^2} + \frac{5x}{x^2} - \frac{6}{x^2}}$

$= \lim_{x\to\infty} \frac{2 - 7x^2}{9 + \frac{5}{x} - \frac{6}{x^2}}$

The denominator approaches 9, while the numerator becomes a negative number that is larger and larger in magnitude, so

$$\lim_{x\to\infty} \frac{2x^2 - 7x^4}{9x^2 + 5x - 6} = -\infty.$$

53. Find $\lim_{x\to 3} f(x)$, where $f(x) = \frac{x^2-9}{x-3}$.

x	2.9	2.99	2.999	3.001	3.01	3.1
$f(x)$	5.9	5.99	5.999	6.001	6.01	6.1

$\lim_{x\to 3} f(x) = \lim_{x\to 3} \frac{x^2 - 9}{x - 3} = 6.$

55. Find $\lim_{x\to 1} f(x)$, where $f(x) = \frac{5x^2-7x+2}{x^2-1}$.

x	0.9	0.99	0.999	1.001	1.01	1.1
$f(x)$	1.316	1.482	1.498	1.502	1.517	1.667

$\lim_{x\to 1} f(x) = \lim_{x\to 1} \frac{5x^2 - 7x + 2}{x^2 - 1} = 1.5 = \frac{3}{2}.$

57. (a) $\lim_{x\to-2} \frac{3x}{(x+2)^3}$ does not exist since $\lim_{x\to-2^+} \frac{3x}{(x+2)^3} = -\infty$ and $\lim_{x\to-2^-} \frac{3x}{(x+2)^3} = \infty$.

(b) Since $(x+2)^3 = 0$ when $x = -2$, $x = -2$ is the vertical asymptote of the graph of $F(x)$.

(c) The two answers are related. Since $x = -2$ is a vertical asymptote, we know that $\lim_{x\to-2} F(x)$ does not exist.

61. (a) $\lim_{x\to-\infty} e^x = 0$ since, as the graph goes further to the left, e^x gets closer to 0.

(b) The graph of e^x has a horizontal asymptote at $y = 0$ since $\lim_{x\to-\infty} e^x = 0$.

63. (a) $\lim_{x\to 0^+} \ln x = -\infty$ since, as the graph gets closer to $x = 0$, the value of $\ln x$ get smaller.

(b) The graph of $y = \ln x$ has a vertical asymptote at $x = 0$ since $\lim_{x\to 0^+} \ln x = -\infty$.

67. $\lim_{x\to 1} \dfrac{x^4+4x^3-9x^2+7x-3}{x-1}$

(a)

x	1.01	1.001	1.0001	0.99	0.999	0.9999
$f(x)$	5.0908	5.009	5.0009	4.9108	4.991	4.9991

As $x\to 1^-$ and as $x\to 1^+$, we see that $f(x)\to 5$.

(b) Graph

$$y=\frac{x^4+4x^3-9x^2+7x-3}{x-1}$$

on a graphing calculator. One suitable choice for the viewing window is $[-6,6]$ by $[-10,40]$ with Xscl $=1$, Yscl $=10$.

Because $x-1=0$ when $x=1$, we know that the function is undefined at this x-value. The graph does not show an asymptote at $x=1$. This indicates that the rational expression that defines this function is not written in lowest terms, and that the graph should have an open circle to show a "hole" in the graph at $x=1$. The graphing calculator doesn't show the hole, but if we try to find the value of the function at $x=1$, we see that it is undefined. (Using the TABLE feature on a TI-83, we see that for $x=1$, the y-value is listed as "ERROR.")
By viewing the function near $x=1$ and using the ZOOM feature, we see that as x gets close to 1 from the left or the right, y gets close to 5, suggesting that

$$\lim_{x\to 1}\frac{x^4+4x^3-9x^2+7x-3}{x-1}=5.$$

69. $\lim_{x\to -1}\dfrac{x^{1/3}+1}{x+1}$

(a)

x	-1.01	-1.001	-1.0001
$f(x)$	0.33223	0.33322	0.33332

x	-0.99	-0.999	-0.9999
$f(x)$	0.33445	0.33344	0.33334

We see that as $x\to -1^-$ and as $x\to -1^+$,

$f(x)\to 0.3333$ or $\frac{1}{3}$.

(b) Graph

$$y=\frac{x^{1/3}+1}{x+1}.$$

One suitable choice for the viewing window is $[-5,5]$ by $[-2,2]$.

Because $x+1=0$ when $x=-1$, we know that the function is undefined at this x-value. The graph does not show an asymptote at $x=-1$. This indicates that the rational expression that defined this function is not written lowest terms, and that the graph should have an open circle to show a "hole" in the graph at $x=-1$. The graphing calculator doesn't show the hole, but if we try to find the value of the function at $x=-1$, we see that it is undefined. (Using the TABLE feature on a TI-83, we see that for $x=-1$, the y-value is listed as "ERROR.")
By viewing the function near $x=-1$ and using the ZOOM feature, we see that as x gets close to -1 from the left or right, y gets close to 0.3333, suggesting that

$$\lim_{x\to -1}\frac{x^{1/3}+1}{x+1}=0.3333 \text{ or } \frac{1}{3}.$$

71. $\lim_{x\to\infty}\dfrac{\sqrt{9x^2+5}}{2x}$

Graph the functions on a graphing calculator. A good choice for the viewing window is $[-10,10]$ by $[-5,5]$.

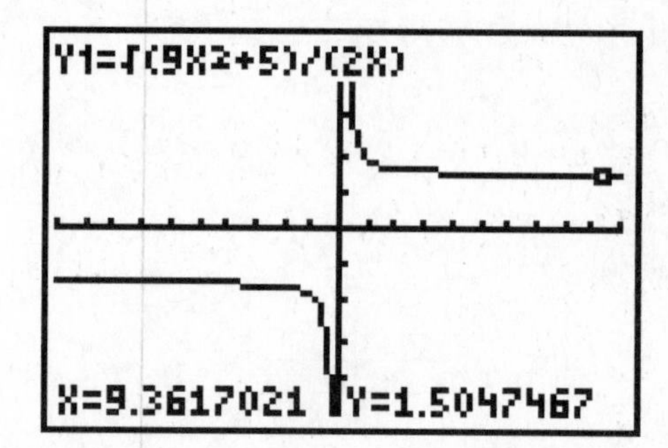

(a) The graph appears to have horizontal asymptotes at $y=\pm 1.5$. We see that as $x\to\infty$, $y\to 1.5$, so we determine that

$$\lim_{x\to\infty}\frac{\sqrt{9x^2+5}}{2x}=1.5.$$

(b) As $x\to\infty$,

$$\sqrt{9x^2+5}\to\sqrt{9x^2}=3\,|x|,$$

and

$$\frac{\sqrt{9x^2+5}}{2x}\to\frac{3\,|x|}{2x}.$$

Since $x>0$, $|x|=x$, so

$$\frac{3\,|x|}{2x}=\frac{3x}{2x}=\frac{3}{2}.$$

Thus,

$$\lim_{x\to\infty} \frac{\sqrt{9x^2+5}}{2x} = \frac{3}{2} \text{ or } 1.5.$$

73. $\lim_{x\to-\infty} \frac{\sqrt{36x^2+2x+7}}{3x}$

Graph this function on a graphing calculator. A good choice for the viewing window is $[-10, 10]$ by $[-5, 5]$.

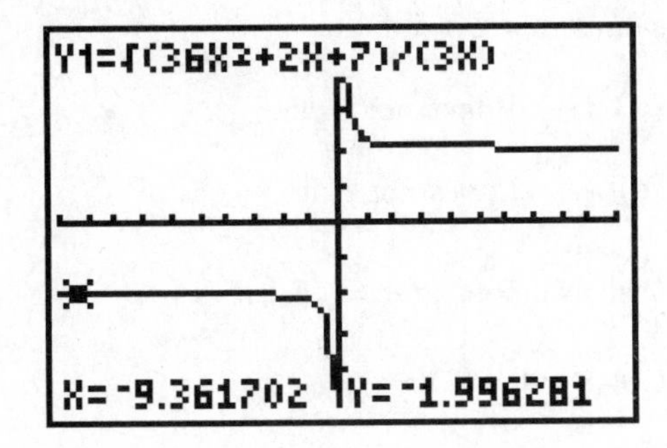

(a) The graph appears to have horizontal asymptotes at $y = \pm 2$. We see that as $x \to -\infty$, $y \to -2$, so we determine that

$$\lim_{x\to-\infty} \frac{\sqrt{36x^2+2x+7}}{3x} = -2.$$

(b) As $x \to -\infty$,

$$\sqrt{36x^2+2x+7} \to \sqrt{36x^2} = 6\,|x|$$

and

$$\frac{\sqrt{36x^2+2x+7}}{3x} \to \frac{6\,|x|}{3x}.$$

Since $x < 0$, $|x| = -x$, so

$$\frac{6\,|x|}{3x} = \frac{6(-x)}{3x} = -2.$$

Thus,

$$\lim_{x\to-\infty} \frac{\sqrt{36x^2+2x+7}}{3x} = -2.$$

75. $\lim_{x\to\infty} \frac{\left(1+5x^{1/3}+2x^{5/3}\right)^3}{x^5}$

Graph this function on a graphing calculator. A good choice for the viewing window is $[-20, 20]$ by $[0, 20]$ with Xscl = 5, Yscl = 5.

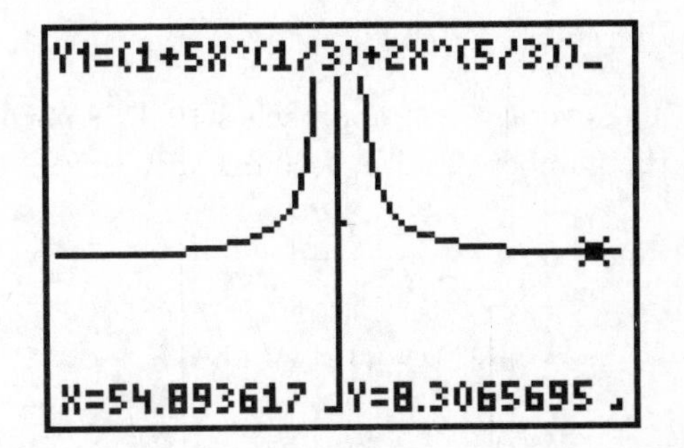

(a) The graph appears to have a horizontal asymptote at $y = 8$. We see that as $x \to \infty$, $y \to 8$, so we determine that

$$\lim_{x\to\infty} \frac{\left(1+5x^{1/3}+2x^{5/3}\right)^3}{x^5} = 8.$$

(b) As $x \to \infty$, the highest power term dominates in the numerator, so

$$\begin{aligned}(1+5x^{1/3}+2x^{5/3})^3 \to (2x^{5/3})^3 &= 2^3x^5 \\ &= 8x^5.\end{aligned}$$

and

$$\frac{\left(1+5x^{1/3}+2x^{5/3}\right)^3}{x^5} \to \frac{8x^5}{x^5} = 8.$$

Thus,

$$\lim_{x\to\infty} \frac{\left(1+5x^{1/3}+2x^{5/3}\right)^3}{x^5} = 8.$$

79. **(a)** $\lim_{x\to 98} T(x) = 7.25$ cents because $T(x)$ is constant at 7.25 cents as x approaches 98 from the left or the right.

(b) $\lim_{x\to 02^-} T(x) = 7$ cents because as x approaches 02 from the left, $T(x)$ is constant at 7 cents.

(c) $\lim_{x\to 02^+} T(x) = 7.25$ cents because as x approaches 02 from the right, $T(x)$ is constant at 7.25 cents.

(d) $\lim_{x\to 02} T(x)$ does not exist because

$$\lim_{x\to 02^-} T(x) \neq \lim_{x\to 02^+} T(x).$$

(e) $T(02) = 7.25$ cents since $(02, 7.25)$ is a point on the graph.

81. $\overline{C}(x) = \frac{C(x)}{x}$
$= \frac{0.0738x + 111.83}{x}$
$= 0.0738 + \frac{111.83}{x}$

$\lim_{x\to\infty} C(x) = 0.0738$

The average cost approaches $0.0738 per mile as the number of miles becomes very large.

83. $\lim_{n\to\infty}\left[R\left[\frac{1-(1+i)^{-n}}{i}\right]\right]$
$= \frac{R}{i}\lim_{n\to\infty}[1-(1+i)^{-n}]$
$= \frac{R}{i}\left[\lim_{n\to\infty} 1 - \lim_{n\to\infty}(1+i)^{-n}\right]$
$= \frac{R}{i}[1-0] = \frac{R}{i}$

85. **(a)** $N(65) = 71.8e^{-8.96e^{(-0.0685(65))}}$
≈ 64.68

To the nearest whole number, this species of alligator has approximately 65 teeth after 65 days of incubation by this formula.

(b) Since $\lim_{t\to\infty}(-8.96e^{-0.0685t}) = -8.96\cdot 0 = 0$, it follows that

$$\lim_{t\to\infty} 71.8e^{-8.96e^{(-0.0685t)}} = 71.8e^0 = 71.8\cdot 1 = 71.8$$

So, to the nearest whole number, $\lim_{t\to\infty} N(t) \approx 72$. Therefore, by this model a newborn alligator of this species will have about 72 teeth.

87. $A(h) = \frac{0.17h}{h^2+2}$

$$\lim_{x\to\infty} A(h) = \lim_{x\to\infty}\frac{0.17h}{h^2+2} = \lim_{x\to\infty}\frac{\frac{0.17h}{h^2}}{\frac{h^2}{h^2}+\frac{2}{h^2}} = \lim_{x\to\infty}\frac{\frac{0.17}{h}}{1+\frac{2}{h^2}} = \frac{0}{1+0} = 0$$

This means that the concentration of the drug in the bloodstream approaches 0 as the number of hours after injection increases.

3.2 Continuity

1. Discontinuous at $x = -1$

(a) $\lim_{x\to-1^-} f(x) = \frac{1}{2}$

(b) $\lim_{x\to-1^+} f(x) = \frac{1}{2}$

(c) $\lim_{x\to-1} f(x) = \frac{1}{2}$ (since (a) and (b) have the same answers)

(d) $f(-1)$ does not exist.

(e) $f(-1)$ does not exist.

3. Discontinuous at $x = 1$

(a) $\lim_{x\to1^-} f(x) = -2$

(b) $\lim_{x\to1^+} f(x) = -2$

(c) $\lim_{x\to1} f(x) = -2$ (since (a) and (b) have the same answers)

(d) $f(1) = 2$

(e) $\lim_{x\to1} f(x) \neq f(1)$

5. Discontinuous at $x = -5$ and $x = 0$

(a) $\lim_{x\to-5^-} f(x) = \infty$ (limit does not exist)
$= \lim_{x\to0^-} f(x) = 0$

(b) $\lim_{x\to-5^+} f(x) = -\infty$ (limit does not exist)
$= \lim_{x\to0^+} f(x) = 0$

(c) $\lim_{x\to-5} f(x)$ does not exist, since the answers to (a) and (b) are different.
$\lim_{x\to0} f(x) = 0$, since the answers to (a) and (b) are the same.

(d) $f(-5)$ does not exist. $f(0)$ does not exist.

(e) $f(-5)$ does not exist and $\lim_{x\to-5} f(x)$ does not exist. $f(0)$ does not exist.

7. $f(x) = \dfrac{5+x}{x(x-2)}$

$f(x)$ is discontinuous at $x = 0$ and $x = 2$ since the denominator equals 0 at these two values.
$\lim_{x\to 0} f(x)$ does not exist since $\lim_{x\to 0^-} f(x) = \infty$ and $\lim_{x\to 0+} f(x) = -\infty$.
$\lim_{x\to 2} f(x)$ does not exist since $\lim_{x\to 2^-} f(x) = -\infty$ and $\lim_{x\to 2+} f(x) = \infty$.

9. $f(x) = \dfrac{x^2-4}{x-2}$

$f(x)$ is discontinuous at $x = 2$ since the denominator equals zero at that value.

Since

$$\frac{x^2-4}{x-2} = \frac{(x+2)(x-2)}{x-2} = x+2,$$

$\lim_{x\to 2} f(x) = 2 + 2 = 4.$

11. $p(x) = x^2 - 4x + 11$

Since $p(x)$ is a polynomial function, it is continuous everywhere and thus discontinuous nowhere.

13. $p(x) = \dfrac{|x+2|}{x+2}$

$p(x)$ is discontinuous at $x = -2$ since the denominator is undefined at that value.
Since $\lim_{x\to -2^-} p(x) = -1$ and $\lim_{x\to -2+} p(x) = 1$, $\lim_{x\to -2} p(x)$ does not exist.

15. $k(x) = e^{\sqrt{x-1}}$

The function is undefined for $x < 1$, so the function is discontinuous for $a < 1$. The limit as x approaches any $a < 1$ does not exist because the function is undefined for $x < 1$.

17. As x approaches 0 from the left or the right, $\left|\frac{x}{x-1}\right|$ approaches 0 and $r(x) = \ln\left|\frac{x}{x-1}\right|$ goes to $-\infty$. So $\lim_{x\to 0} r(x)$ does not exist. As x approaches 1 from the left or the right, $\left|\frac{x}{x-1}\right|$ goes to ∞ and so does $r(x) = \ln\left|\frac{x}{x-1}\right|$. So $\lim_{x\to 1} r(x)$ does not exist.

19. $f(x) = \begin{cases} 1 & \text{if } x < 2 \\ x+3 & \text{if } 2 \le x \le 4 \\ 7 & \text{if } x > 4 \end{cases}$

(a)

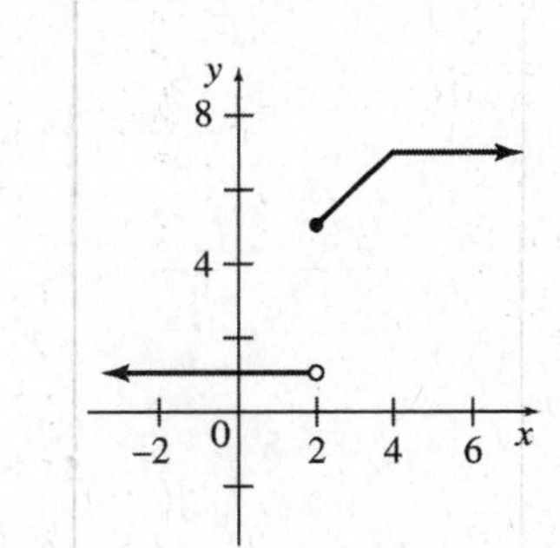

(b) $f(x)$ is discontinuous at $x = 2$.

(c) $\lim_{x\to 2^-} f(x) = 1$ $\lim_{x\to 2+} f(x) = 5$

21. $g(x) = \begin{cases} 11 & \text{if } x < -1 \\ x^2+2 & \text{if } -1 \le x \le 3 \\ 11 & \text{if } x > 3 \end{cases}$

(a)

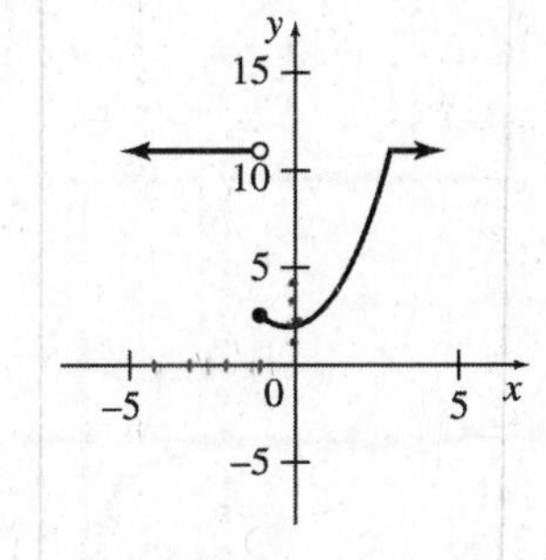

(b) $g(x)$ is discontinuous at $x = -1$.

(c) $\lim_{x\to -1^-} g(x) = 11$
$\lim_{x\to -1^+} g(x) = (-1)^2 + 2 = 3$

23. $h(x) = \begin{cases} 4x+4 & \text{if } x \le 0 \\ x^2 - 4x + 4 & \text{if } x > 0 \end{cases}$

(a)

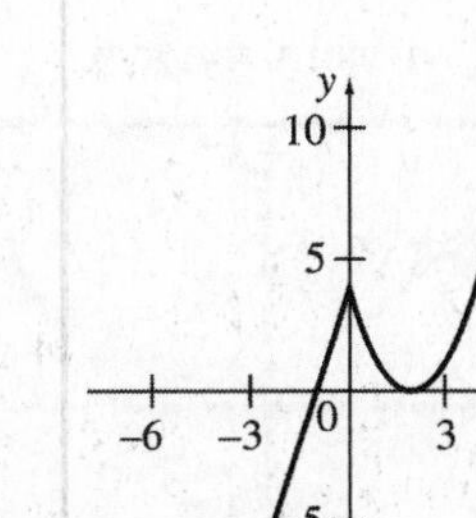

(b) There are no points of discontinuity.

25. Find k so that $kx^2 = x + k$ for $x = 2$.

$$\begin{aligned} k(2)^2 &= 2 + k \\ 4k &= 2 + k \\ 3k &= 2 \\ k &= \frac{2}{3} \end{aligned}$$

27. $\dfrac{2x^2 - x - 15}{x - 3} = \dfrac{(2x+5)(x-3)}{x-3} = 2x + 5$

Find k so that $2x + 5 = kx - 1$ for $x = 3$.

$$\begin{aligned} 2(3) + 5 &= k(3) - 1 \\ 6 + 5 &= 3k - 1 \\ 11 &= 3k - 1 \\ 12 &= 3k \\ 4 &= k \end{aligned}$$

31. $f(x) = \dfrac{x^2 + x + 2}{x^3 - 0.9x^2 + 4.14x - 5.4} = \dfrac{P(x)}{Q(x)}$

(a) Graph

$$Y_1 = \frac{P(x)}{Q(x)} = \frac{x^2 + x + 2}{x^3 - 0.9x^2 + 4.14x - 5.4}$$

on a graphing calculator. A good choice for the viewing window is $[-3, 3]$ by $[-10, 10]$.

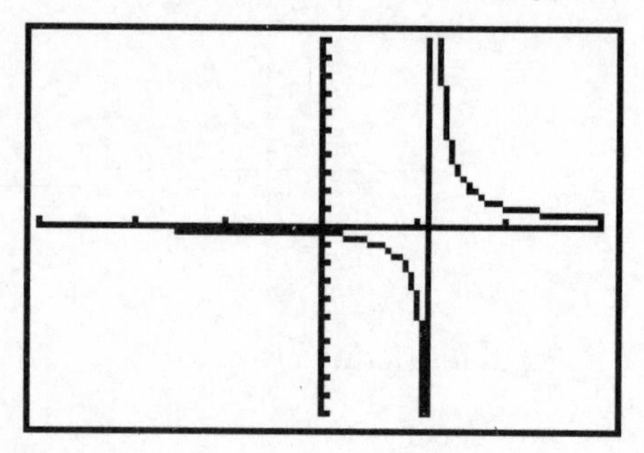

The graph has a vertical asymptote at $x = 1.2$, which indicates that f is discontinuous at $x = 1.2$.

(b) Graph

$$Y_2 = Q(x) = x^3 - 0.9x^2 + 4.14x - 5.4$$

using the same viewing window.

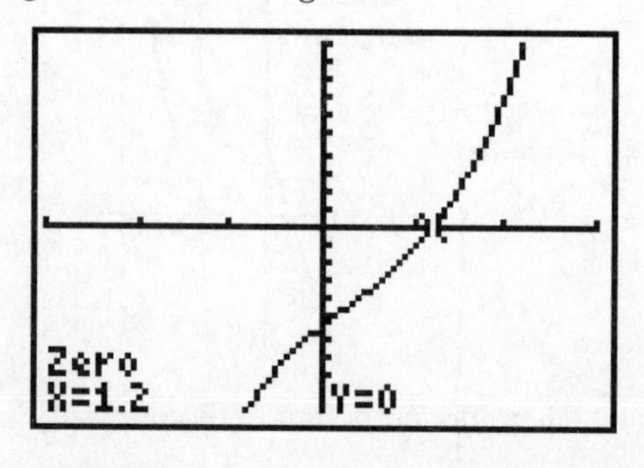

We see that this graph has one x-intercept, 1.2. This indicates that 1.2 is the only real solution of the equation $Q(x) = 0$.

This result verifies our answer from part (a) because a rational function of the form

$$f(x) = \frac{P(x)}{Q(x)}$$

will be discontinuous wherever $Q(x) = 0$.

33.
$$\begin{aligned} g(x) &= \frac{x+4}{x^2 + 2x - 8} \\ &= \frac{x+4}{(x-2)(x+4)} \\ &= \frac{1}{x-2}, \ x \neq -4 \end{aligned}$$

If $g(x)$ is defined so that $g(-4) = \dfrac{1}{-4-2} = -\dfrac{1}{6}$, then the function becomes continuous at -4. It cannot be made continuous at 2. The correct answer is (a).

35. In dollars,

$C(x) = 4x$ if $0 < x \leq 150$
$C(x) = 3x$ if $150 < x \leq 400$
$C(x) = 2.5x$ if $400 < x$.

(a) $C(130) = 4(130) = \$520$

(b) $C(150) = 4(150) = \$600$

(c) $C(210) = 3(210) = \$630$

(d) $C(400) = 3(400) = \$1200$

(e) $C(500) = 2.5(500) = \$1250$

(f) C is discontinuous at $x = 150$ and $x = 400$ because those represent points of price change.

37. In dollars,

$C(t) = 36t$ if $0 < t \leq 5$
$C(t) = 36(5) = 180$ if $t = 6$ or $t = 7$
$C(t) = 180 + 36(t - 7)$ if $7 < t \leq 12$.

The average cost per day is

$$A(t) = \frac{C(t)}{t}.$$

(a) $A(4) = \dfrac{36(4)}{4} = \$36$

(b) $A(5) = \dfrac{36(5)}{5} = \$36$

(c) $A(6) = \frac{180}{6} = \$30$

(d) $A(7) = \frac{180}{7} \approx \$25.71$

(e) $A(8) = \frac{180 + 36(8-7)}{8}$

$= \frac{216}{8} = \$27$

(f) $\lim_{t \to 5^-} A(t) = 36$ because as t approaches 5 from the left, $A(t)$ approaches 36 (think of the graph for $t = 1, 2, ..., 5$).

(g) $\lim_{t \to 5^+} A(t) = 30$ because as t approaches 5 from the right, $A(t)$ approaches 30. 1, 2, 3, 4, 7, 8, 9, 10

(h) A is discontinuous at $t = 11$, and because the average cost will differ for each different rental length.

39. **(a)** Since $t = 0$ weeks the woman weighs 120 lbs. and at $t = 40$ weeks she weighs 147 lbs., graph the line beginning at coordinate $(0, 120)$ and ending at $(40, 147)$, with closed circles at these points. Since immediately after giving birth, she loses 14 lbs. and continues to lose 13 more lbs. over the following 20 weeks, graph the line between the points $(40, 133)$ and $(60, 120)$ with an open circle at $(40, 133)$ and a closed circle at $(60, 120)$.

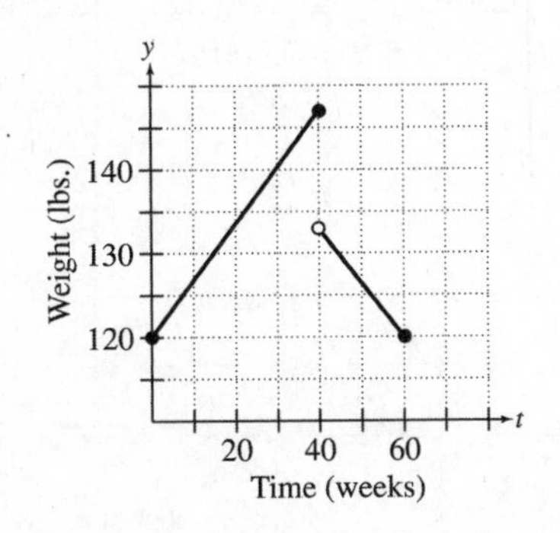

(b) From the graph, we see that

$$\lim_{t \to 40^-} w(t) = 147 \neq 133$$
$$= \lim_{t \to 40^+} w(t),$$

where $w(t)$ is the weight in pounds t weeks after conception. Therefore, w is discontinuous at $t = 40$.

3.3 Rates of Change

1. $y = x^2 + 2x = f(x)$ between $x = 1$ and $x = 3$

Average rate of change

$= \frac{f(3) - f(1)}{3 - 1}$

$= \frac{15 - 3}{2}$

$= 6$

3. $y = -3x^3 + 2x^2 - 4x + 1 = f(x)$ between $x = -2$ and $x = 1$

$$\text{Average rate of change} = \frac{f(1) - f(-2)}{1 - (-2)}$$
$$= \frac{(-4) - (41)}{1 - (-2)}$$
$$= \frac{-45}{3} = -15$$

5. $y = \sqrt{x} = f(x)$ between $x = 1$ and $x = 4$

Average rate of change

$= \frac{f(4) - f(1)}{4 - 1}$

$= \frac{2 - 1}{3}$

$= \frac{1}{3}$

7. $y = e^x = f(x)$ between $x = -2$ and $x = 0$

Average rate of change

$= \frac{f(0) - f(-2)}{0 - (-2)}$

$= \frac{1 - e^{-2}}{2}$

≈ 0.4323

9. $\lim_{h \to 0} \frac{s(6+h) - s(6)}{h}$

$$= \lim_{h \to 0} \frac{(6+h)^2 + 5(6+h) + 2 - [6^2 + 5(6) + 2]}{h}$$

$$= \lim_{h \to 0} \frac{h^2 + 17h + 68 - 68}{h} = \lim_{h \to 0} \frac{h^2 + 17h}{h}$$

$$= \lim_{h \to 0} \frac{h(h+17)}{h} = \lim_{h \to 0} (h + 17) = 17$$

The instantaneous velocity at $t = 6$ is 17.

11. $s(t) = 5t^2 - 2t - 7$

$$\lim_{h\to 0} \frac{s(2+h)-s(2)}{h}$$

$$= \lim_{h\to 0} \frac{[5(2+h)^2 - 2(2+h) - 7] - [5(2)^2 - 2(2) - 7]}{h}$$

$$= \lim_{h\to 0} \frac{[20 + 20h + 5h^2 - 4 - 2h - 7] - [20 - 4 - 7]}{h}$$

$$= \lim_{h\to 0} \frac{9 + 18h + 5h^2 - 9}{h}$$

$$= \lim_{h\to 0} \frac{18h + 5h^2}{h}$$

$$= \lim_{h\to 0} \frac{h(18+5h)}{h}$$

$$= \lim_{h\to 0} (18 + 5h) = 18$$

The instantaneous velocity at $t = 2$ is 18.

13. $s(t) = t^3 + 2t + 9$

$$\lim_{h\to 0} \frac{s(1+h)-s(1)}{h}$$

$$= \lim_{h\to 0} \frac{[(1+h)^3+2(1+h)+9]-[(1)^3+2(1)+9]}{h}$$

$$= \lim_{h\to 0} \frac{[1+3h+3h^2+h^3+2+2h+9]-[1+2+9]}{h}$$

$$= \lim_{h\to 0} \frac{h^3 + 3h^2 + 5h + 12 - 12}{h}$$

$$= \lim_{h\to 0} \frac{h^3 + 3h^2 + 5h}{h} = \lim_{h\to 0} \frac{h(h^2 + 3h + 5)}{h}$$

$$= \lim_{h\to 0} (h^2 + 3h + 5) = 5$$

The instantaneous velocity at $t = 1$ is 5.

15. $f(x) = x^2 + 2x$ at $x = 0$

$$\lim_{h\to 0} \frac{f(0+h)-f(0)}{h}$$

$$= \lim_{h\to 0} \frac{(0+h)^2 + 2(0+h) - [0^2 + 2(0)]}{h}$$

$$= \lim_{h\to 0} \frac{h^2 + 2h}{h} = \lim_{h\to 0} \frac{h(h+2)}{h}$$

$$= \lim_{h\to 0} h + 2 = 2$$

The instantaneous rate of change at $x = 0$ is 2.

17. $g(t) = 1 - t^2$ at $t = -1$

$$\lim_{h\to 0} \frac{g(-1+h)-g(-1)}{h}$$

$$= \lim_{h\to 0} \frac{1-(-1+h)^2 - [1-(-1)^2]}{h}$$

$$= \lim_{h\to 0} \frac{1-(1-2h+h^2) - 1 + 1}{h}$$

$$= \lim_{h\to 0} \frac{2h - h^2}{h} = \lim_{h\to 0} \frac{h(2-h)}{h}$$

$$= \lim_{h\to 0} (2-h) = 2$$

The instantaneous rate of change at $t = -1$ is 2.

19. $f(x) = x^x$ at $x = 2$

h	
0.01	$\frac{f(2+0.01)-f(2)}{0.01} = \frac{2.01^{2.01} - 2^2}{0.01} = 6.84$
0.001	$\frac{f(2+0.001)-f(2)}{0.001} = \frac{2.001^{2.001} - 2^2}{0.001} = 6.779$
0.0001	$\frac{f(2+0.0001)-f(2)}{0.0001} = \frac{2.0001^{2.0001} - 2^2}{0.0001} = 6.773$
0.00001	$\frac{f(2+0.00001)-f(2)}{0.00001} = \frac{2.00001^{2.00001} - 2^2}{0.00001} = 6.7727$
0.000001	$\frac{f(2+0.000001)-f(2)}{0.000001} = \frac{2.000001^{2.000001} - 2^2}{0.000001} = 6.7726$

The instantaneous rate of change at $x = 2$ is 6.7726.

21. $f(x) = x^{\ln x}$ at $x = 2$

h	
0.01	$\frac{f(2+0.01)-f(2)}{0.01} = \frac{2.01^{\ln 2.01} - 2^{\ln 2}}{0.01} = 1.1258$
0.001	$\frac{f(2+0.001)-f(2)}{0.001} = \frac{2.001^{\ln 2.001} - 22^{\ln 2}}{0.001} = 1.1212$
0.0001	$\frac{f(2+0.0001)-f(2)}{0.0001} = \frac{2.0001^{\ln 2.0001} - 2^{\ln 2}}{0.0001} = 1.1207$
0.00001	$\frac{f(2+0.00001)-f(2)}{0.00001} = \frac{2.00001^{\ln 2.00001} - 2^{\ln 2}}{0.00001} = 1.1207$

The instantaneous rate of change at $x = 2$ is 1.1207.

25. Let $B(t)$ = the amount in the Medicare Trust Fund for year t.

(a) $B(1994) = 152$
$B(1998) = 125$

Average change in fund

$$= \frac{125 - 152}{1998 - 1994} = \frac{-27}{4} = -6.75$$

On average, the amount in the fund decreased approximately \$6.75 billion per year from 1994 – 1998.

(b) $B(1998) = 125$
$B(2010) = 300$

Average change in fund

$$= \frac{300 - 125}{2010 - 1998} = \frac{175}{12} \approx 14.58$$

On average, the amount in the fund increases approximately \$14.58 billion per year from 1998 – 2010.

(c) $B(1990) = 125$
$B(1998) = 125$

Average change in fund

$$= \frac{125 - 125}{1998 - 1990} = \frac{0}{8} = 0$$

On average, the amount in the fund changes approximately \$0 per year from 1990 to 1998.

27. (a) From June 2005 to December 2005 is 6 months. During that time, the number of single family housing starts went from 1,724,000 to 1,633,000—a drop of 91,000. The average monthly rate of change is

$$\frac{-91{,}000}{6} = -15{,}167 \text{ starts per month}$$

(b) From December 2005 to June 2006 is 6 months. During that time, the number of single family housing starts went from 1,633,000 to 1,486,000—a drop of 147,000. The average monthly rate of change is

$$\frac{-147{,}000}{6} = -24{,}500 \text{ starts per month}$$

(c) From June 2005 to June 2006 is 12 months. During that time, the number of single family housing starts went from 1,724,000 to 1,486,000—a drop of 238,000. The average monthly rate of change is

$$\frac{-238{,}000}{12} = -19{,}833 \text{ starts per month}$$

(d) $\frac{-15{,}167 + (-24{,}500)}{2} = -19{,}833$ (allowing for rounding)

They are equal. This will not be true for all time periods (only for time periods of equal length).

29. $P(x) = 2x^2 - 5x + 6$

(a) $P(4) = 18$
$P(2) = 4$

Average rate of change of profit

$$= \frac{P(4) - P(2)}{4 - 2}$$
$$= \frac{18 - 4}{2}$$
$$= \frac{14}{2} = 7,$$

which is $700 per item.

(b) $P(3) = 9$
$P(2) = 4$

Average rate of change of profit

$$= \frac{P(3) - P(2)}{3 - 2} = \frac{9 - 4}{1} = 5$$

which is $500 per item.

(c) $$\lim_{h \to 0} \frac{P(2+h) - P(2)}{h}$$
$$= \lim_{h \to 0} \frac{2(2+h)^2 - 5(2+h) + 6 - 4}{h}$$
$$= \lim_{h \to 0} \frac{8 + 8h + 2h^2 - 10 - 5h + 2}{h}$$
$$= \lim_{h \to 0} \frac{2h^2 + 3h}{h}$$
$$= \lim_{h \to 0} \frac{h(2h + 3)}{h}$$
$$= \lim_{h \to 0} (2h + 3) = 3,$$

which is $300 per item.

(d) $$\lim_{h \to 0} \frac{P(4+h) - P(4)}{h}$$
$$= \lim_{h \to 0} \frac{2(4+h)^2 - 5(4+h) + 6 - 18}{h}$$
$$= \lim_{h \to 0} \frac{32 + 16h + 2h^2 - 20 - 5h - 12}{h}$$
$$= \lim_{h \to 0} \frac{2h^2 + 11h}{h}$$
$$= \lim_{h \to 0} \frac{h(2h + 11)}{h}$$
$$= \lim_{h \to 0} 2h + 11 = 11,$$

which is $1100 per item.

31. $N(p) = 80 - 5p^2,\ 1 \le p \le 4$

(a) Average rate of change of demand is

$$\frac{N(3) - N(2)}{3 - 2} = \frac{35 - 60}{1}$$
$$= -25 \text{ boxes per dollar.}$$

(b) Instantaneous rate of change when p is 2 is

$$\lim_{h \to 0} \frac{N(2+h) - N(2)}{h}$$
$$= \lim_{h \to 0} \frac{80 - 5(2+h)^2 - [80 - 5(2)^2]}{h}$$
$$= \lim_{h \to 0} \frac{80 - 20 - 20h - 5h^2 - (80 - 20)}{h}$$
$$= \lim_{h \to 0} \frac{-5h^2 - 20h}{h}$$

$= -20$ boxes per dollar. Around the $2 point, a $1 price increase (say, from $1.50 to $2.50) causes a drop in demand of about 20 boxes.

(c) Instantaneous rate of change when p is 3 is

$$\lim_{h \to 0} \frac{80 - 5(3+h)^2 - [80 - 5(3)^2]}{h}$$
$$= \lim_{h \to 0} \frac{80 - 45 - 30h - 5h^2 - 80 + 45}{h}$$
$$= \lim_{h \to 0} \frac{-30h - 5h^2}{h}$$
$$= -30 \text{ boxes per dollar.}$$

(d) As the price increases, the demand decreases; this is an expected change.

33. Let $P(t)$ = world population estimated in billions for year t.

(a) $P(1990) = 5.3$

If replacement-level fertility is reached in 2010, $P(2050) = 8.6$.

$$\text{Average rate of change} = \frac{P(2050) - P(1990)}{2050 - 1990}$$
$$= \frac{8.6 - 5.3}{60}$$
$$= 0.055$$

On average, the population will increase 55 million per year.

If replacement-level fertility is reached in 2030, $P(2050) = 9.2$.

$$\text{Average rate of change} = \frac{P(2050) - P(1990)}{2050 - 1990} = \frac{9.2 - 5.3}{60} = 0.065$$

On average, the population will increase 65 million per year.
If replacement-level fertility is reached in 2050, $P(2050) = 9.8$.

$$\text{Average rate of change} = \frac{P(2050) - P(1990)}{2050 - 1990} = \frac{9.8 - 5.3}{60} = 0.075$$

On average, the population will increase 75 million per year.
The projection for replacement-level fertility by 2010 predicts the smallest rate of change in world population.

(b) If replacement-level fertility is reached in 2010

$$P(2090) = 9.3$$
$$P(2130) = 9.6$$

$$\text{Average rate of change} = \frac{P(2130) - P(2090)}{2130 - 2090} = \frac{9.6 - 9.3}{40} = 0.0075$$

On average, the population will increase 7.5 million per year.
If replacement-level fertility is reached in 2030,

$$P(2090) = 10.3$$
$$P(2130) = 10.6$$

$$\text{Average rate of change} = \frac{P(2130) - P(2090)}{2130 - 2090} = \frac{10.6 - 10.3}{40} = 0.0075$$

On average, the population will increase 7.5 million per year.
If replacement-level fertility is reached in 2050,

$$P(2090) = 11.35$$
$$P(2130) = 11.75$$

$$\text{Average rate of change} = \frac{P(2130) - P(2090)}{2130 - 2090} = \frac{11.75 - 11.35}{40} = 0.01$$

On average, the population will increase 10 million per year.
From 2090 – 2130 the three projections show almost the same rate of change in world population.

35. $L(t) = -0.01t^2 + 0.788t - 7.048$

(a) $$\frac{L(28) - L(22)}{28 - 22} = \frac{7.176 - 5.448}{6} = 0.288$$

The average rate of growth during weeks 22 through 28 is 0.288 mm per week.

(b)
$$\lim_{h\to 0} \frac{L(t+h) - L(t)}{h}$$
$$= \lim_{h\to 0} \frac{L(22+h) - L(22)}{h}$$
$$= \lim_{h\to 0} \frac{[-0.01(22+h)^2 + 0.788(22+h) - 7.048] - 5.448}{h}$$
$$= \lim_{h\to 0} \frac{-0.01(h^2 + 44h + 484) + 17.336 + 0.788h - 12.496}{h}$$
$$= \lim_{h\to 0} \frac{-0.01h^2 + 0.348h}{h}$$
$$= \lim_{h\to 0} (-0.01h + 0.348)$$
$$= 0.348$$

The instantaneous rate of growth at exactly 22 weeks is 0.348 mm per week.

(c)

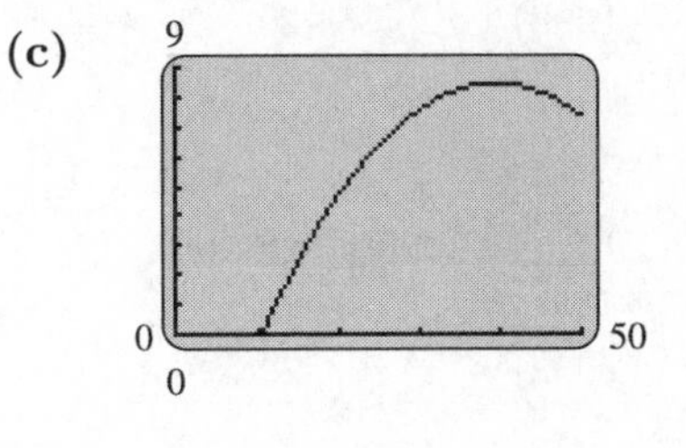

37. **(a)** The average rate of change of $M(t)$ on the interval $[105, 115]$ is

$$\frac{M(115) - M(105)}{115 - 105} = \frac{0.8}{10} = 0.08$$

kilograms per day.

(b) Calculate $\lim_{h \to 0} \frac{M(105 + h) - M(105)}{h}$

$$\begin{aligned} M(105 + h) &= 27.5 + 0.3(105 + h) \\ &\quad - 0.001(105 + h)^2 \\ &= 27.5 + 31.5 + 0.3h \\ &\quad - (11.025 + 0.21h + 0.001h^2) \\ &= 47.975 + 0.09h - 0.001h^2 \\ M(105) &= 47.975 \end{aligned}$$

So, the instantaneous rate of change of $M(t)$ at $t = 105$ is

$$\begin{aligned} &\lim_{h \to 0} \left(\frac{47.975 + 0.09h - 0.001h^2 - 47.975}{h} \right) \\ &= \lim_{h \to 0} \left(\frac{0.09h - 0.001h^2}{h} \right) \\ &= \lim_{h \to 0} (0.09 - 0.001h) \\ &= 0.09 \text{ kilograms per day.} \end{aligned}$$

(c)

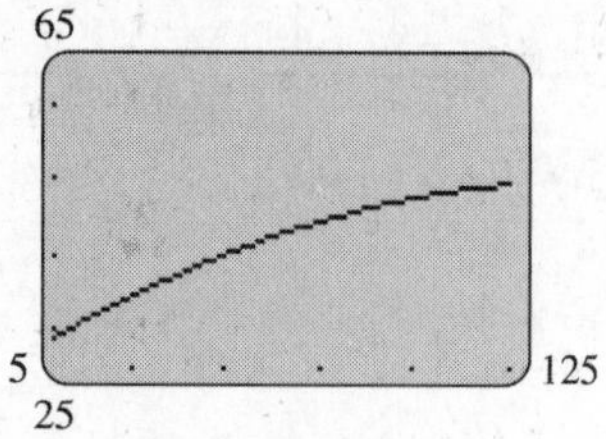

65

5 125

25

39. **(a)** Let $D(t)$ represent percent of kids who have used drugs by grade 8 in year t.

$$\frac{D(2001) - D(1998)}{2001 - 1998} = \frac{26.8 - 29}{3} \approx -0.73$$

The average rate of change from 1998 to 2001 is -0.73 percent per year.

$$\frac{D(2005) - D(2002)}{2005 - 2002} = \frac{21.4 - 24.5}{3} \approx -1.03$$

The average rate of change from 2002 to 2005 is -1.03 percent per year.

(b) Let $D(t)$ represent percent of kids who have used drugs by grade 10 in year t.

$$\frac{D(2001) - D(1998)}{2001 - 1998} = \frac{45.6 - 44.9}{3} \approx 0.23$$

The average rate of change from 1998 to 2001 is 0.23 percent per year.

$$\frac{D(2005) - D(2002)}{2005 - 2002} = \frac{38.2 - 44.6}{3} \approx -2.13$$

The average rate of change from 2002 to 2005 is -2.13 percent per year.

(c) Let $D(t)$ represent percent of kids who have used drugs by grade 12 in year t.

$$\frac{D(2001) - D(1998)}{2001 - 1998} = \frac{53.9 - 54.1}{3} \approx -0.07$$

The average rate of change from 1998 to 2001 is -0.07 percent per year.

$$\frac{D(2005) - D(2002)}{2005 - 2002} = \frac{50.4 - 53}{3} \approx -0.87$$

The average rate of change from 2002 to 2005 is -0.87 percent per year.

41. **(a)** $\frac{s(2) - s(0)}{2 - 0} = \frac{10 - 0}{2} = 5$ ft/sec

(b) $\frac{s(4) - s(2)}{4 - 2} = \frac{14 - 10}{2} = 2$ ft/sec

(c) $\frac{s(6) - s(4)}{6 - 4} = \frac{20 - 14}{2} = 3$ ft/sec

(d) $\frac{s(8) - s(6)}{8 - 6} = \frac{30 - 20}{2} = 5$ ft/sec

(e) (i)

$$\begin{aligned} &\frac{f(x_0 + h) - f(x_0 - h)}{2h} \\ &= \frac{f(4 + 2) - f(4 - 2)}{(2)(2)} \\ &= \frac{f(6) - f(2)}{4} \\ &= \frac{20 - 10}{4} \\ &= \frac{10}{4} = 2.5 \text{ ft/sec} \end{aligned}$$

(ii) $\frac{2 + 3}{2} = 2.5$ ft/sec

(f) (i) $\dfrac{f(x_0+h)-f(x_0-h)}{2h}$

$$= \frac{f(6+2)-f(6-2)}{(2)(2)}$$

$$= \frac{f(8)-f(4)}{4}$$

$$= \frac{30-14}{4}$$

$$= \frac{16}{4} = 4 \text{ ft/sec}$$

(ii) $\dfrac{3+5}{2} = 4 \text{ ft/sec}$

43. $s(t) = t^2 + 5t + 2$

(a) Average velocity $= \dfrac{s(6)-s(4)}{6-4}$

$$= \frac{68-38}{6-4}$$

$$= \frac{30}{2} = 15 \text{ ft/sec}$$

(b) Average velocity $= \dfrac{s(5)-s(4)}{5-4}$

$$= \frac{52-38}{5-4}$$

$$= \frac{14}{1} = 14 ft/sec$$

(c) $\lim\limits_{h\to 0} \dfrac{s(4+h)-s(4)}{h}$

$$= \lim_{h\to 0} \frac{(4+h)^2+5(4+h)+2-38}{h}$$

$$= \lim_{h\to 0} \frac{16+8h+h^2+20+5h+2-38}{h}$$

$$= \lim_{h\to 0} \frac{h^2+13h}{h} = \lim_{h\to 0} \frac{h(h+13)}{h}$$

$$= \lim_{h\to 0} (h+13) = 13 \text{ ft/sec}$$

3.4 Definition of the Derivative

1. **(a)** $f(x) = 5$ is a horizontal line and has slope 0; the derivative is 0.

(b) $f(x) = x$ has slope 1; the derivative is 1.

(c) $f(x) = -x$ has slope of -1; the derivative is -1.

(d) $x = 3$ is vertical and has undefined slope; the derivative does not exist.

(e) $y = mx + b$ has slope m; the derivative is m.

3. $f(x) = \frac{x^2-1}{x+2}$ is not differentiable when $x + 2 = 0$ or $x = -2$ because the function is undefined and a vertical asymptote occurs there.

5. Using the points $(5, 3)$ and $(6, 5)$, we have

$$m = \frac{5-3}{6-5} = \frac{2}{1} = 2.$$

7. Using the points $(-2, 2)$ and $(2, 3)$, we have

$$m = \frac{3-2}{2-(-2)} = \frac{1}{4}.$$

9. Using the points $(-3, -3)$ and $(0, -3)$, we have

$$m = \frac{-3-(-3)}{0-3} = \frac{0}{-3} = 0.$$

11. $f(x) = -4x^2 + 9x + 2$

Step 1 $f(x+h) = -4(x+h)^2 + 9(x+h) + 2$
$= -4(x^2 + 2xh + h^2) + 9x + 9h + 2$
$= -4x^2 - 8xh - 4h^2 + 9x + 9h + 2$

Step 2 $f(x+h) - f(x)$
$= -4x^2 - 8xh - 4h^2 + 9x + 9h + 2$
$\quad - (-4x^2 + 9x + 2)$
$= -8xh - 4h^2 + 9h$
$= h(-8x - 4h + 9)$

Step 3 $\dfrac{f(x+h)-f(x)}{h} = \dfrac{h(-8x-4h+9)}{h}$
$= -8x - 4h + 9$

Step 4 $f'(x) = \lim\limits_{h\to 0} \dfrac{f(x+h)-f(x)}{h}$
$= \lim\limits_{h\to 0} (-8x - 4h + 9)$
$= -8x + 9$

$f'(-2) = -8(-2) + 9 = 25$
$f'(0) = -8(0) + 9 = 9$
$f'(3) = -8(3) + 9 = -15$

13. $f(x) = \dfrac{12}{x}$

$$f(x+h) = \frac{12}{x+h}$$

$$f(x+h) - f(x) = \frac{12}{x+h} - \frac{12}{x}$$

$$= \frac{12x - 12(x+h)}{x(x+h)}$$

$$= \frac{12x - 12x - 12h}{x(x+h)}$$

$$= \frac{-12h}{x(x+h)}$$

$$\frac{f(x+h)-f(x)}{h}=\frac{-12h}{hx(x+h)}$$
$$=\frac{-12}{x(x+h)}$$
$$=\frac{-12}{x^2+xh}$$

$$f'(x)=\lim_{h\to 0}\frac{f(x+h)-f(x)}{h}$$
$$=\lim_{h\to 0}\frac{-12}{x^2+xh}$$
$$=\frac{-12}{x^2}$$

$$f'(-2)=\frac{-12}{(-2)^2}=\frac{-12}{4}=-3$$

$f'(0)=\frac{-12}{0^2}$ which is undefined so $f'(0)$ does not exist.

$$f'(3)=\frac{-12}{3^2}=\frac{-12}{9}=-\frac{4}{3}$$

15. $f(x)=\sqrt{x}$

Steps 1-3 are combined.

$$\frac{f(x+h)-f(x)}{h}$$
$$=\frac{\sqrt{x+h}-\sqrt{x}}{h}$$
$$=\frac{\sqrt{x+h}-\sqrt{x}}{h}\cdot\frac{\sqrt{x+h}+\sqrt{x}}{\sqrt{x+h}+\sqrt{x}}$$
$$=\frac{x+h-x}{h(\sqrt{x+h}+\sqrt{x})}$$
$$=\frac{1}{\sqrt{x+h}+\sqrt{x}}$$

$$f'(x)=\lim_{h\to 0}\frac{f(x+h)-f(x)}{h}$$
$$=\lim_{h\to 0}\frac{1}{\sqrt{x+h}+\sqrt{x}}$$
$$=\frac{1}{2\sqrt{x}}$$

$f'(-2)=\frac{1}{2\sqrt{-2}}$ which is undefined so $f'(-2)$ does not exist.

$f'(0)=\frac{1}{2\sqrt{0}}=\frac{1}{0}$ which is undefined so $f'(0)$ does not exist.

$$f'(3)=\frac{1}{2\sqrt{3}}$$

17. $f(x)=2x^3+5$

Steps 1-3 are combined.

$$\frac{f(x+h)-f(x)}{h}$$
$$=\frac{2(x+h)^3+5-(2x^3+5)}{h}$$
$$=\frac{2(x^3+3x^2h+3xh^2+h^3)+5-2x^3-5}{h}$$
$$=\frac{2x^3+6x^2h+6xh^2+2h^3+5-2x^3-5}{h}$$
$$=\frac{6x^2h+6xh^2+2h^3}{h}$$
$$=\frac{h(6x^2+6xh+2h^2)}{h}$$
$$=6x^2+6xh+2h^2$$

$$f'(x)=\lim_{h\to 0}(6x^2+6xh+2h^2)$$
$$=6x^2$$
$$f'(-2)=6(-2)^2=24$$
$$f'(0)=6(0)^2=0$$
$$f'(3)=6(3)^2=54$$

19. (a) $f(x)=x^2+2x;x=3,x=5$

$$\text{Slope of secant line}=\frac{f(5)-f(3)}{5-3}$$
$$=\frac{(5)^2+2(5)-[(3)^2+2(3)]}{2}$$
$$=\frac{35-15}{2}$$
$$=10$$

Now use $m=10$ and $(3,f(3))=(3,15)$ in the point-slope form.

$$y-15=10(x-3)$$
$$y-15=10x-30$$
$$y=10x-30+15$$
$$y=10x-15$$

(b) $f(x)=x^2+2x;\ x=3$

$$\frac{f(x+h)-f(x)}{h}$$
$$=\frac{[(x+h)^2+2(x+h)]-(x^2+2x)}{h}$$
$$=\frac{(x^2+2hx+h^2+2x+2h)-(x^2+2x)}{h}$$
$$=\frac{2hx+h^2+2h}{h}=2x+h+2$$

$$f'(x)=\lim_{h\to 0}(2x+h+2)=2x+2$$

$f'(3) = 2(3) + 2 = 8$ is the slope of the tangent line at $x = 3$.
Use $m = 8$ and $(3, 15)$ in the point-slope form.

$$y - 15 = 8(x - 3)$$
$$y = 8x - 9$$

21. **(a)** $f(x) = \frac{5}{x}; x = 2, x = 5$

$$\text{Slope of secant line} = \frac{f(5) - f(2)}{5 - 2} = \frac{\frac{5}{5} - \frac{5}{2}}{3} = \frac{1 - \frac{5}{2}}{3} = -\frac{1}{2}$$

Now use $m = -\frac{1}{2}$ and $(5, f(5)) = (5, 1)$ in the point-slope form.

$$y - 1 = -\frac{1}{2}[x - 5]$$
$$y - 1 = -\frac{1}{2}x + \frac{5}{2}$$
$$y = -\frac{1}{2}x + \frac{5}{2} + 1$$
$$y = -\frac{1}{2}x + \frac{7}{2}$$

(b) $f(x) = \dfrac{5}{x};\ x = 2$

$$\frac{f(x+h) - f(x)}{h} = \frac{\frac{5}{x+h} - \frac{5}{x}}{h} = \frac{\frac{5x - 5(x+h)}{(x+h)x}}{h} = \frac{5x - 5x - 5h}{h(x+h)(x)} = \frac{-5h}{h(x+h)x} = \frac{-5}{(x+h)x}$$

$$f'(x) = \lim_{h \to 0} \frac{-5}{(x+h)(x)} = -\frac{5}{x^2}$$

$f'(2) = \frac{-5}{2^2} = -\frac{5}{4}$ is the slope of the tangent line at $x = 2$.

Now use $m = -\frac{5}{4}$ and $\left(2, \frac{5}{2}\right)$ in the point-slope form.

$$y - \frac{5}{2} = -\frac{5}{4}(x - 2)$$
$$y - \frac{5}{2} = -\frac{5}{4}x + \frac{10}{4}$$
$$y = -\frac{5}{4}x + 5$$
$$5x + 4y = 20$$

23. **(a)** $f(x) = 4\sqrt{x};\ x = 9, x = 16$

$$\text{Slope of secant line} = \frac{f(16) - f(9)}{16 - 9} = \frac{4\sqrt{16} - 4\sqrt{9}}{7} = \frac{16 - 12}{7} = \frac{4}{7}$$

Now use $m = \frac{4}{7}$ and $(9, f(9)) = (9, 12)$ in the point-slope form.

$$y - 12 = \frac{4}{7}(x - 9)$$
$$y - 12 = \frac{4}{7}x - \frac{36}{7}$$
$$y = \frac{4}{7}x - \frac{36}{7} + 12$$
$$y = \frac{4}{7}x + \frac{48}{7}$$

(b) $f(x) = 4\sqrt{x};\ x = 9$

$$\frac{f(x+h) - f(x)}{h} = \frac{4\sqrt{x+h} - 4\sqrt{x}}{h} \cdot \frac{4\sqrt{x+h} + 4\sqrt{x}}{4\sqrt{x+h} + 4\sqrt{x}} = \frac{16(x+h) - 16x}{h(4\sqrt{x+h} + 4\sqrt{x})}$$

$$f'(x) = \lim_{h \to 0} \frac{16(x+h) - 16x}{h(4\sqrt{x+h} + 4\sqrt{x})} = \lim_{h \to 0} \frac{16h}{h(4\sqrt{x+h} + 4\sqrt{x})} = \lim_{h \to 0} \frac{4}{(\sqrt{x+h} + \sqrt{x})} = \frac{4}{2\sqrt{x}} = \frac{2}{\sqrt{x}}$$

$f'(9) = \frac{2}{\sqrt{9}} = \frac{2}{3}$ is the slope of the tangent line at $x = 9$.

Use $m = \frac{2}{3}$ and $(9, 12)$ in the point-slope form.

$$y - 12 = \frac{2}{3}(x - 9)$$
$$y = \frac{2}{3}x + 6$$
$$3y = 2x + 18$$

25. $f(x) = -4x^2 + 11x$

$$\frac{f(x+h) - f(x)}{h}$$
$$= \frac{-4(x+h)^2 + 11(x+h) - (-4x^2 + 11x)}{h}$$
$$= \frac{-8xh - 4h^2 + 11h}{h}$$

$$f'(x) = \lim_{h \to 0} (-8x - 4h + 11)$$
$$= -8x + 11$$
$$f'(2) = -8(2) + 11 = -5$$
$$f'(16) = -8(16) + 11 = -117$$
$$f'(-3) = -8(-3) + 11 = 35$$

27. $f(x) = e^x$

$$\frac{f(x+h) - f(x)}{h} = \frac{e^{x+h} - e^x}{h}$$
$$f'(x) = \lim_{h \to 0} \frac{e^{x+h} - e^x}{h}$$

$f'(2) \approx 7.3891$; $f'(16) \approx 8,886,111$; $f'(-3) \approx 0.0498$

29. $f(x) = -\frac{2}{x}$

$$\frac{f(x+h) - f(x)}{h} = \frac{\frac{-2}{x+h} - \left(\frac{-2}{x}\right)}{h}$$
$$= \frac{\frac{-2x + 2(x+h)}{(x+h)x}}{h}$$
$$= \frac{2h}{h(x+h)x}$$
$$= \frac{2}{(x+h)x}$$

$$f'(x) = \lim_{h \to 0} \frac{2}{(x+h)x}$$
$$= \frac{2}{x^2}$$
$$f'(2) = \frac{2}{2^2} = \frac{1}{2}$$

$$f'(16) = \frac{2}{16^2}$$
$$= \frac{2}{256}$$
$$\frac{1}{128}.$$
$$f'(-3) = \frac{2}{(-3)^2}$$
$$= \frac{2}{9}$$

31. $f(x) = \sqrt{x}$

$$\frac{f(x+h) - f(x)}{h}$$
$$= \frac{\sqrt{x+h} - \sqrt{x}}{h} \cdot \frac{\sqrt{x+h} + \sqrt{x}}{\sqrt{x+h} + \sqrt{x}}$$
$$= \frac{(x+h) - x}{h(\sqrt{x+h} + \sqrt{x})}$$
$$= \frac{h}{h(\sqrt{x+h} + \sqrt{x})}$$
$$= \frac{1}{\sqrt{x+h} + \sqrt{x}}$$

$$f'(x) = \lim_{h \to 0} \frac{1}{\sqrt{x+h} + \sqrt{x}} = \frac{1}{2\sqrt{x}}$$
$$f'(2) = \frac{1}{2\sqrt{2}}$$
$$f'(16) = \frac{1}{2\sqrt{16}} = \frac{1}{8}$$

$f'(-3) = \frac{1}{2\sqrt{-3}}$ is not a real number, so

$f'(-3)$ does not exist.

33. At $x = 0$, the graph of $f(x)$ has a sharp point. Therefore, there is no derivative for $x = 0$.

35. For $x = -3$ and $x = 0$, the tangent to the graph of $f(x)$ is vertical. For $x = -1$, there is a gap in the graph $f(x)$. For $x = 2$, the function $f(x)$ does not exist. For $x = 3$ and $x = 5$, the graph $f(x)$ has sharp points. Therefore, no derivative exists for $x = -3$, $x = -1, x = 0, x = 2, x = 3$, and $x = 5$.

37. **(a)** The rate of change of $f(x)$ is positive when $f(x)$ is increasing, that is, on $(a, 0)$ and (b, c).

(b) The rate of change of $f(x)$ is negative when $f(x)$ is decreasing, that is, on $(0, b)$.

(c) The rate of change is zero when the tangent to the graph is horizontal, that is, at $x = 0$ and $x = b$.

39. The zeros of graph (b) correspond to the turning points of graph (a), the points where the derivative is zero. Graph (a) gives the distance, while graph (b) gives the velocity.

41. $f(x) = x^x, a = 3$

(a)

h	
0.01	$\frac{f(3+0.01)-f(3)}{0.01} = \frac{3.01^{3.01}-3^3}{0.01} = 57.3072$
0.001	$\frac{f(3+0.001)-f(3)}{0.001} = \frac{3.001^{3.001}-3^3}{0.001} = 56.7265$
0.00001	$\frac{f(3+.00001)-f(3)}{0.00001} = \frac{3.00001^{3.00001}-3^3}{0.00001} = 56.6632$
0.000001	$\frac{f(3+0.000001)-f(3)}{0.000001} = \frac{3.000001^{3.000001}-3^3}{0.000001} = 56.6626$
0.0000001	$\frac{f(3+0.0000001)-f(3)}{0.0000001} = \frac{3.0000001^{3.0000001}-3^3}{0.0000001} = 56.6625$

It appears that $f'(3) = 56.6625$.

(b) Graph the function on a graphing calculator and move the cursor to an x-value near $x = 3$. A good choice for the initial viewing window is $[0, 4]$ by $[0, 60]$ with Xscl = 1, Yscl = 10.

Now zoom in on the function several times. Each time you zoom in, the graph will look less like a curve and more like a straight line. Use the TRACE feature to select two points on the graph, and record their coordinates. Use these two points to compute the slope. The result will be close to the most accurate value found in part (a), which is 56.6625.

Note: In this exercise, the method used in part (a) gives more accurate results than the method used in part (b).

43. $f(x) = x^{1/x},\ a = 3$

(a)

h	
0.01	$\frac{f(3+0.01)-f(3)}{0.01} = \frac{3.01^{1/3.01}-3^{1/3}}{0.01} = -0.0160$
0.001	$\frac{f(3+0.001)-f(3)}{0.001} = \frac{3.001^{1/3.001}-3^{1/3}}{0.001} = -0.0158$
0.0001	$\frac{f(3+0.0001)-f(3)}{0.0001} = \frac{3.0001^{1/3.0001}-3^{1/3}}{0.0001} = -0.0158$

It appears that $f'(3) = -0.0158$.

(b) Graph the function on a graphing calculator and move the cursor to an x-value near $x = 3$. A good choice for the initial viewing window is $[0, 5]$ by $[0, 3]$.

Follow the procedure outlined in the solution for Exercise 39, part (b). Note that near $x = 3$, the graph is very close to a horizontal line, so we expect that it slope will be close to 0. The final result will be close to the value found in part (a) of this exercise, which is -0.0158.

47. $D(p) = -2p^2 - 4p + 300$

D is demand; p is price.

(a) Given that $D'(p) = -4p-4$, the rate of change of demand with respect to price is $-4p - 4$, the derivative of the function $D(p)$.

(b) $D'(10) = -4(10) - 4$
$= -44$

The demand is decreasing at the rate of about 44 items for each increase in price of $1.

49. $R(x) = 20x - \frac{x^2}{500}$

(a) $R'(x) = 20 - \frac{1}{250}x$

At $y = 1000$,

$$R'(1000) = 20 - \frac{1}{250}(1000)$$
$$= \$16 \text{ per table.}$$

(b) The marginal revenue for the 1001st table is approximately $R'(1000)$. From (a), this is about \$16.

(c) The actual revenue is

$$R(1001) - R(1000) = 20(1001) - \frac{1001^2}{500} - \left[20(1000) - \frac{1000^2}{500}\right]$$
$$= 18{,}015.998 - 18{,}000$$
$$= \$15.998 \text{ or } \$16.$$

(d) The marginal revenue gives a good approximation of the actual revenue from the sale of the 1001st table.

51. (a) $f(x) = -0.0142x^4 + 0.6698x^3 - 6.113x^2 + 84.05x + 203.9$

$f(10) = 961$
$f(20) = 2526$
$f(30) = 3806$

(b) $Y_1 = -0.0142x^4 + 0.6698x^3 - 6.113x^2 + 84.05x + 203.9$

$$nDeriv(Y_1, x, 10) \approx 106$$
$$nDeriv(Y_1, x, 20) \approx 189$$
$$nDeriv(Y_1, x, 30) \approx -8$$
$$nDeriv(Y_1, x, 35) \approx -318$$

53. The derivative at $(2, 4000)$ can be approximated by the slope of the line through $(0, 2000)$ and $(2, 4000)$. The derivative is approximately

$$\frac{4000 - 2000}{2 - 0} = \frac{2000}{2} = 1000.$$

Thus the shellfish population is increasing at a rate of 1000 shellfish per unit time.
The derivative at about $(10, 10{,}300)$ can be approximated by the slope of the line through $(10, 10{,}300)$ and $(13, 12{,}000)$. The derivative is approximately

$$\frac{12{,}000 - 10{,}300}{13 - 10} = \frac{1700}{3} \approx 570.$$

The shellfish population is increasing at a rate of about 583 shellfish per unit time. The derivative at about $(13, 11{,}250)$ can be approximated by the slope of the line through $(13, 11{,}250)$ and $(16, 12{,}000)$. The derivative is approximately

$$\frac{12{,}000 - 11{,}250}{16 - 13} = \frac{750}{3} = 250.$$

The shellfish population is increasing at a rate of 250 shellfish per unit time.

55. (a) Set $M(v) = 150$ and solve for v.

$$0.0312443v^2 - 101.39v + 82{,}264 = 150$$
$$0.0312443v^2 - 101.39v + 82{,}114 = 0$$

Solve using the quadratic formula.
Let D equal the discriminant.

$$D = b^2 - 4ac$$
$$= (-101.39)^2 - 4(0.0312443)(82{,}114)$$
$$\approx 17.55$$

$$v = \frac{101.39 \pm \sqrt{D}}{2(0.0312443)}$$

$v \approx 1690$ meter per second or
$v \approx 1560$ meters per second.
Since the functions is defined only for $v \geq 1620$, the only solution is 1690 meters per second.

(b) Calculate $\lim_{h \to 0} \frac{M(1700 + h) - M(1700)}{h}$

$$M(1700 + h)$$
$$= 0.0312443(1700 + h)^2 - 101.39(1700 + h) + 82{,}264$$
$$= 90{,}296.027 + 106.23062h + 0.0312443h^2 - 172{,}363 - 101.39h + 82{,}264$$
$$= 0.01312443h^2 + 4.84062h + 197.027$$

$M(1700) = 197.027$, so the derivative of $M(v)$ at $v = 1700$ is

$$\lim_{h \to 0} \left(\frac{0.0312443h^2 + 4.84062h + 197.027 - 197.027}{h}\right)$$
$$= \lim_{h \to 0} \left(\frac{0.0312443h^2 + 4.84062h}{h}\right)$$
$$= \lim_{h \to 0} (0.0312443h + 4.84062)$$
$$= 4.84062$$
$$\approx 4.84 \text{ days per meter per second}$$

The increase in velocity for this cheese from 1700 m/s to 1701 m/s indicates that the approximate age of the cheese has increased by 4.84 days.

57. (a) The derivative does not exist at the two "corners" or "sharp points." The x-values of these points are 0.75 and 3.

(b) To find $T'(0.5)$, calculate the slope of the line segment with positive slope, since this portion of the graph includes the value $x = 0.5$ (marked as $\frac{1}{2}$ hr on the graph.) To find this slope, use the points $(0, 100)$ (the starting point) and $(0.75, 875)$ (the beginning point of the cleaning cycle).

$$\begin{aligned} m = T'(0.5) &= \frac{875 - 100}{0.75 - 0} \\ &= \frac{775}{0.75} \\ &\approx 1033 \end{aligned}$$

The oven temperature is increasing at 1033° per hour.

(c) To find $T'(2)$, find the slope of the line segment containing the value $x = 2$. This segment is horizontal, so

$$m = T'(2) = 0.$$

The oven temperature is not changing.

(d) To find $T'(3.5)$, calculate the slope of the line segment with negative slope, since this portion of the graph includes the value $x = 3.5$. To find this slope, use the points $(3, 875)$ (the end of the cleaning cycle) and $(3.75, 100)$ (the stopping point).

$$\begin{aligned} m = T'(3.5) &= \frac{100 - 875}{3.75 - 3} \\ &= \frac{-775}{0.75} \\ &\approx -1033 \end{aligned}$$

The oven temperature is decreasing at 1033° per hour.

59. The velocities are equal at approximately $t = 0.13$. The acceleration for the hands is approximately 0 mph per sec. The acceleration for the bat is approximately 640 mph per sec.

3.5 Graphical Differentiation

3. Since the x-intercepts of the graph of f' occur whenever the graph of f has a horizontal tangent line, Y_1 is the derivative of Y_2. Notice that Y_1 has 2 x-intercepts; each occurs at an x-value where the tangent line to Y_2 is horizontal.
Note also that Y_1 is positive whenever Y_2 is increasing, and that Y_1 is negative whenever Y_2 is decreasing.

5. Since the x-intercepts of the graph of f' occur whenever the graph of f has a horizontal tangent line, Y_2 is the derivative of Y_1. Notice that Y_2 has 1 x-intercept which occurs at the x-value where the tangent line to Y_1 is horizontal. Also notice that the range on which Y_1 is increasing, Y_2 is positive and the range on which it is decreasing, Y_2 is negative.

7. To graph f', observe the intervals where the slopes of tangent lines are positive and where they are negative to determine where the derivative is positive and where it is negative. Also, whenever f has a horizontal tangent, f' will be 0, so the graph of f' will have an x-intercept. The x-values of the three turning point on the graph of f become the three x-intercepts of the graph of f.

Estimate the magnitude of the slope at several points by drawing tangents to the graph of f.

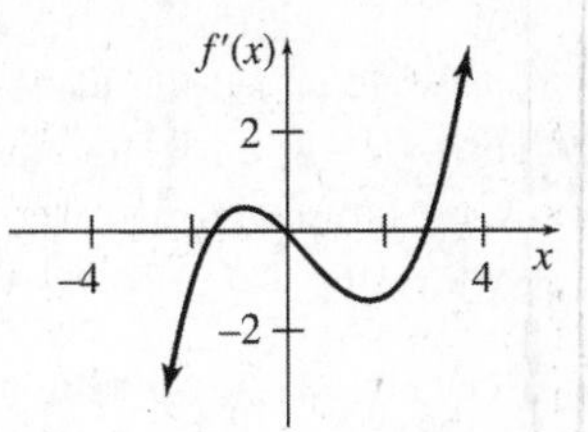

9. On the interval $(-\infty, -2)$, the graph of f is a horizontal line, so its slope is 0. Thus, on this interval, the graph of f' is $y = 0$ on $(-\infty, -2)$. On the interval $(-2, 0)$, the graph of f is a straight line, so its slope is constant. To find this slope, use the points $(-2, 2)$ and $(0, 0)$.

$$m = \frac{2 - 0}{-2 - 0} = \frac{2}{-2} = -1$$

On the interval $(0, 1)$, the slope is also constant. To find this slope, use the points $(0, 0)$ and $(1, 1)$.

$$m = \frac{1 - 0}{1 - 0} = 1$$

On the interval $(1, \infty)$, the graph is again a horizontal line, so $m = 0$. The graph of f' will be made up of portions of the y-axis and the lines $y = -1$ and $y = 1$.

Because the graph of f has "sharp points" or "corners" at $x = -2, x = 0$, and $x = 1$, we know that $f'(-2), f'(0)$, and $f'(1)$ do not exist. We show this on the graph of f' by using open circles at the endpoints of the portions of the graph.

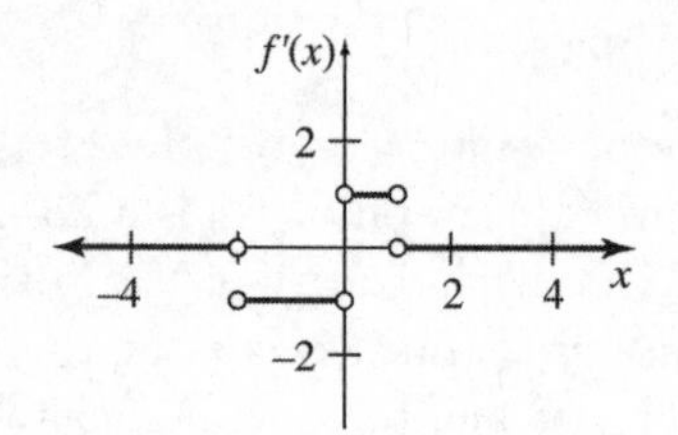

11. On the interval $(-\infty, -2)$, the graph of f is a straight line, so its slope is constant. To find this slope, use the points $(-4, 2)$ and $(-2, 0)$.

$$m = \frac{0-2}{-2-(-4)} = \frac{-2}{2} = -1$$

On the interval $(2, \infty)$, the slope of f is also constant. To find this slope, use the points $(2, 0)$ and $(3, 2)$.

$$m = \frac{2-0}{3-2} = \frac{2}{1} = 2$$

Thus, we have $f'(x) = -1$ on $(-\infty, -2)$ and $f'(x) = 2$ on $(2, \infty)$.

Because f is discontinuous at $x = -2$ and $x = 2$, we know that $f'(-2)$ and $f'(2)$ do not exist, which we indicate with open circles at $(-2, -1)$ and $(2, 2)$ on the graph of f'.

On the interval $(-2, 2)$, all tangent lines have positive slopes, so the graph of f' will be above the y-axis. Notice that the slope of f (and thus the y-value of f') decreases on $(-2, 0)$ and increases on $(0, 2)$, with a minimum value on this interval of 1 at $x = 0$.

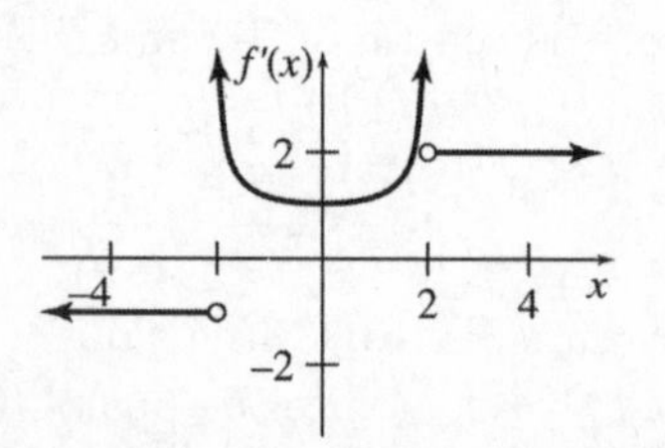

13. We observe that the slopes of tangent lines are positive on the interval $(-\infty, 0)$ and negative on the interval $(0, \infty)$, so the value of f' will be positive on $(-\infty, 0)$ and negative on $(0, \infty)$. Since f is undefined at $x = 0$, $f'(0)$ does not exist.

Notice that the graph of f becomes very flat when $|x| \to \infty$. The *value* of f approaches 0 and also the *slope* approaches 0. Thus, $y = 0$ (the x-axis) is a horizontal asymptote for both the graph of f and the graph of f'.

As $x \to 0^-$ and $x \to 0^+$, the graph of f gets very steep, so $|f'(x)| \to \infty$. Thus, $x = 0$ (the y-axis) is a vertical asymptote for both the graph of f and the graph of f'.

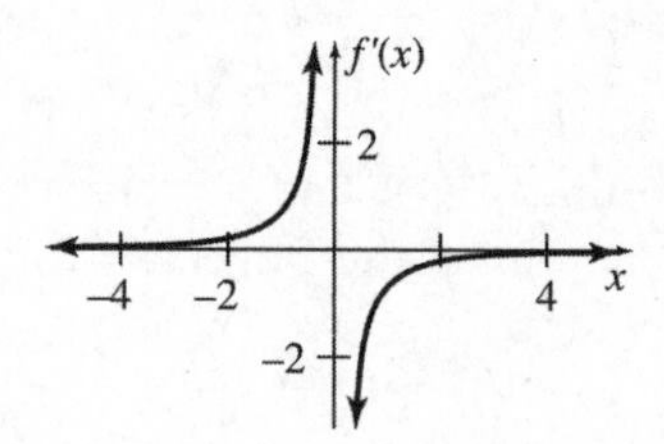

15. The slope of $f(x)$ is undefined at $x = -2, -1, 0, 1$, and 2, and the graph approaches vertical (unbounded slope) as x approaches those values. Accordingly, the graph of $f'(x)$ has vertical asymptotes at $x = -2, -1, 0, 1$, and 2. $f(x)$ has turning points (zero slope) at $x = -1.5, -0.5, 0.5$, and 1.5, so the graph of $f'(x)$ crosses the x-axis at those values. Elsewhere, the graph of $f'(x)$ is negative where $f(x)$ is decreasing and positive where $f(x)$ is increasing.

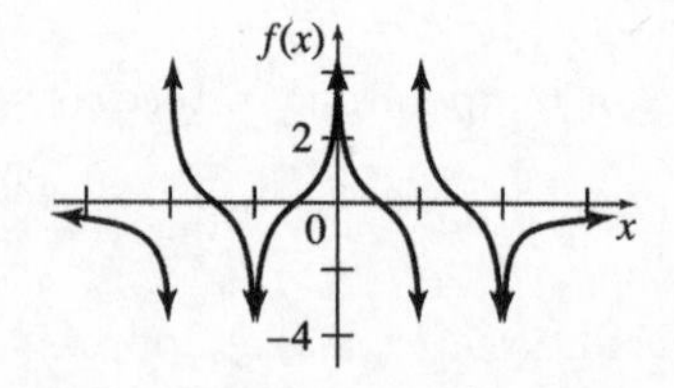

17. The graph rises steadily, with varying degrees of steepness. The graph is steepest around 1976 and nearly flat around 1950 and 1980. Accordingly, the rate of change is always positive, with a maximum value around 1976 and values near zero around 1950 and 1980.

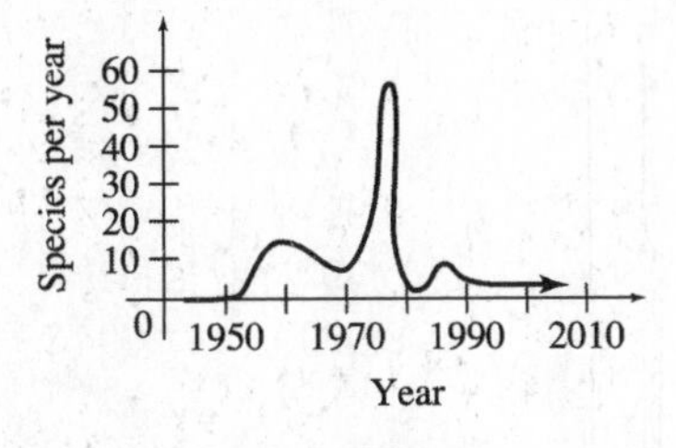

19. The growth rate of the function $y = f(x)$ is given by the derivative of this function $y' = f(x)$. We use the graph of f to sketch the graph of f'. First, notice as x increase, y increases throughout the domain of f, but at a slower and slower rate. The slope of f is positive but always decreasing, and approaches 0 as t gets large. Thus, y' will always be positive and decreasing. It will approach but never reach 0.

To plot point on the graph of f', we need to estimate the slope of f at several points. From the graph of f, we obtain the values given in the following table.

t	y'
2	1000
10	700
13	250

Use these points to sketch the graph.

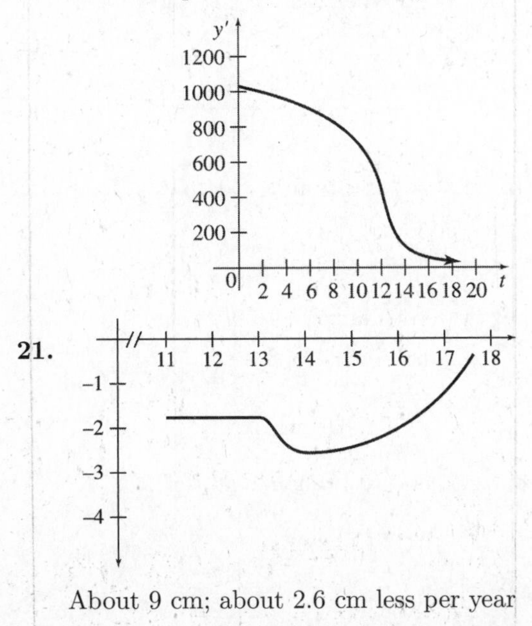

About 9 cm; about 2.6 cm less per year

Chapter 3 Review Exercises

5. (a) $\lim_{x \to -3^-} = 4$

(b) $\lim_{x \to -3^+} = 4$

(c) $\lim_{x \to -3} = 4$ (since parts (a) and (b) have the same answer)

(d) $f(-3) = 4$, since $(-3, 4)$ is a point of the graph.

7. (a) $\lim_{x \to 4^-} f(x) = \infty$

(b) $\lim_{x \to 4^+} f(x) = -\infty$

(c) $\lim_{x \to 4} f(x)$ does not exist since the answers to parts (a) and (b) are different.

(d) $f(4)$ does not exist since the graph has no point with an x-value of 4.

9. $\lim_{x \to -\infty} g(x) = \infty$ since the y-value gets very large as the x-value gets very small.

11.
$$\lim_{x \to 6} \frac{2x+7}{x+3} = \frac{2(6)+7}{6+3} = \frac{19}{9}$$

13.
$$\lim_{x \to 4} \frac{x^2-16}{x-4} = \lim_{x \to 4} \frac{(x-4)(x+4)}{x-4} = \lim_{x \to 4} (x+4) = 4+4 = 8$$

15.
$$\lim_{x \to -4} \frac{2x^2+3x-20}{x+4} = \lim_{x \to -4} \frac{(2x-5)(x+4)}{x+4} = \lim_{x \to -4} (2x-5) = 2(-4)-5 = -13$$

17.
$$\lim_{x \to 9} \frac{\sqrt{x}-3}{x-9} = \lim_{x \to 9} \frac{\sqrt{x}-3}{x-9} \cdot \frac{\sqrt{x}+3}{\sqrt{x}+3} = \lim_{x \to 9} \frac{x-9}{(x-9)(\sqrt{x}+3)} = \lim_{x \to 9} \frac{1}{\sqrt{x}+3} = \frac{1}{\sqrt{9}+3} = \frac{1}{6}$$

19. $\lim\limits_{x\to\infty} \dfrac{2x^2+5}{5x^2-1} = \lim\limits_{x\to\infty} \dfrac{\frac{2x^2}{x^2}+\frac{5}{x^2}}{\frac{5x^2}{x^2}-\frac{1}{x^2}}$

$= \lim\limits_{x\to\infty} \dfrac{2+\frac{5}{x^2}}{5-\frac{1}{x^2}}$

$= \dfrac{2+0}{5-0}$

$= \dfrac{2}{5}$

21. $\lim\limits_{x\to-\infty} \left(\dfrac{3}{8}+\dfrac{3}{x}-\dfrac{6}{x^2}\right)$

$= \lim\limits_{x\to-\infty} \dfrac{3}{8} + \lim\limits_{x\to-\infty} \dfrac{3}{x} - \lim\limits_{x\to-\infty} \dfrac{6}{x^2}$

$= \dfrac{3}{8}+0-0$

$= \dfrac{3}{8}$

23. As shown on the graph, $f(x)$ is discontinuous at x_2 and x_4.

25. $f(x)$ is discontinuous at $x=0$ and $x=-\frac{1}{3}$ since that is where the denominator of $f(x)$ equals 0. $f(0)$ and $f\left(-\frac{1}{3}\right)$ do not exist.
$\lim\limits_{x\to 0} f(x)$ does not exist since $\lim\limits_{x\to 0^+} f(x) = -\infty$, but $\lim\limits_{x\to 0^-} f(x) = \infty$. $\lim\limits_{x\to -\frac{1}{3}} f(x)$ does not exist since $\lim\limits_{x\to -\frac{1}{3}^-} = -\infty$, but $\lim\limits_{x\to -\frac{1}{3}^+} f(x) = \infty$.

27. $f(x)$ is discontinuous at $x=-5$ since that is where the denominator of $f(x)$ equals 0.
$f(-5)$ does not exist.
$\lim\limits_{x\to -5} f(x)$ does not exist since $\lim\limits_{x\to -5^-} f(x) = \infty$, but $\lim\limits_{x\to -5^+} f(x) = -\infty$.

29. $f(x) = x^2+3x-4$ is continuous everywhere since f is a polynomial function.

31. **(a)**

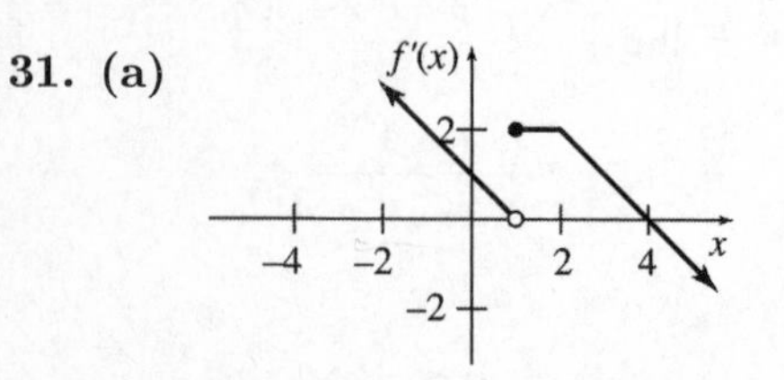

(b) The graph is discontinuous at $x=1$.

(c) $\lim\limits_{x\to 1^-} f(x) = 0;\ \lim\limits_{x\to 1^+} f(x) = 2$

33. $f(x) = \dfrac{x^4+2x^3+2x^2-10x+5}{x^2-1}$

(a) Find the values of $f(x)$ when x is close to 1.

x	y
1.1	2.6005
1.01	2.06
1.001	2.006
1.0001	2.0006
0.99	1.94
0.999	1.994
0.9999	1.9994

It appears that $\lim\limits_{x\to 1} f(x) = 2$.

(b) Graph

$$y = \frac{x^4+2x^3+2x^2-10x+5}{x^2-1}$$

on a graphing calculator. One suitable choice for the viewing window is $[-2, 6]$ by $[-10, 10]$.

Because $x^2-1=0$ when $x=-1$ or $x=1$, this function is discontinuous at these two x-values. The graph shows a vertical asymptote at $x=-1$ but not at $x=1$. The graph should have an open circle to show a "hole" in the graph at $x=1$. The graphing calculator doesn't show the hole, but trying to find the value of the function of $x=1$ will show that this value is undefined.

By viewing the function near $x=1$ and using the ZOOM feature, we see that as x gets close to 1 from the left or the right, y gets close to 2, suggesting that

$$\lim_{x\to 1} \frac{x^4+2x^3+2x^2-10x+5}{x^2-1} = 2.$$

35. $y = 6x^3+2 = f(x)$; from $x=1$ to $x=4$

$f(4) = 6(4)^3+2 = 386$
$f(1) = 6(1)^3+2 = 8$

Average rate of change:

$$= \frac{386-8}{4-1} = \frac{378}{3} = 126$$

$y' = 18x$

Instantaneous rate of change at $x=1$:

$$f'(1) = 18(1) = 18$$

37. $y = \dfrac{-6}{3x-5} = f(x)$; from $x = 4$ to $x = 9$

$$f(9) = \frac{-6}{3(9)-5} = \frac{-6}{22} = -\frac{3}{11}$$

$$f(4) = \frac{-6}{3(4)-5} = -\frac{6}{7}$$

Average rate of change:

$$= \frac{\frac{-3}{11} - \left(-\frac{6}{7}\right)}{9-4}$$

$$= \frac{\frac{-21+66}{77}}{5}$$

$$= \frac{45}{5(77)} = \frac{9}{77}$$

$$y' = \frac{(3x-5)(0) - (-6)(3)}{(3x-5)^2}$$

$$= \frac{18}{(3x-5)^2}$$

Instantaneous rate of change at $x = 4$:

$$f'(4) = \frac{18}{(3 \cdot 4 - 5)^2} = \frac{18}{7^2} = \frac{18}{49}$$

39. **(a)** $f(x) = 3x^2 - 5x + 7$; $x = 2, x = 4$

Slope of secant line

$$= \frac{f(4) - f(2)}{4-2}$$

$$= \frac{[3(4)^2 - 5(4) + 7] - [3(2)^2 - 5(2) + 7]}{2}$$

$$= \frac{35 - 9}{2}$$

$$= 13$$

Now use $m = 13$ and $(2, f(2)) = (2, 9)$ in the point-slope form.

$$y - 9 = 13(x-2)$$
$$y - 9 = 13x - 26$$
$$y = 13x - 26 + 9$$
$$y = 13x - 17$$

(b) $f(x) = 3x^2 - 5x + 7$; $x = 2$

$$\frac{f(x+h) - f(x)}{h}$$

$$= \frac{[3(x+h)^2 - 5(x+h) + 7] - [3x^2 - 5x + 7]}{h}$$

$$= \frac{3x^2 + 6xh + 3h^2 - 5x - 5h + 7 - 3x^2 + 5x - 7}{h}$$

$$= \frac{6xh + 3h^2 - 5h}{h}$$

$$= 6x + 3h - 5$$

$$f'(x) = \lim_{h \to 0} 6x + 3h - 5$$
$$= 6x - 5$$
$$f'(2) = 6(2) - 5$$
$$= 7$$

Now use $m = 7$ and $(2, f(2)) = (2, 9)$ in the point-slope form.

$$y - 9 = 7(x-2)$$
$$y - 9 = 7x - 14$$
$$y = 7x - 14 + 9$$
$$y = 7x - 5$$

41. **(a)** $f(x) = \dfrac{12}{x-1}$; $x = 3, x = 7$

$$\text{Slope of secant line} = \frac{f(7) - f(3)}{7-3}$$

$$= \frac{\frac{12}{7-1} - \frac{12}{3-1}}{4}$$

$$= \frac{2-6}{4}$$

$$= -1$$

Now use $m = -1$ and $(3, f(x)) = (3, 6)$ in the point-slope form.

$$y - 6 = -1(x-3)$$
$$y - 6 = -x + 3$$
$$y = -x + 3 + 6$$
$$y = -x + 9$$

(b) $f(x) = \dfrac{12}{x-1}$; $x = 3$

$$\frac{f(x+h) - f(x)}{h} = \frac{\frac{12}{x+h-1} - \frac{12}{x-1}}{h}$$

$$= \frac{12(x-1) - 12(x+h-1)}{h(x-1)(x+h-1)}$$

$$= \frac{-12h}{h(x-1)(x+h-1)}$$

$$= -\frac{12}{(x-1)(x+h-1)}$$

$$f'(x) = \lim_{h \to 0} -\frac{12}{(x-1)(x+h-1)} = -\frac{12}{(x-1)^2}$$
$$f'(3) = -\frac{12}{(3-1)^2} = -3$$

Now use $m = -3$ and $(3, f(x)) = (3, 6)$ in the point-slope form.

$$\begin{aligned} y - 6 &= -3(x-3) \\ y - 6 &= -3x + 9 \\ y &= -3x + 9 + 6 \\ y &= -3x + 15 \end{aligned}$$

43. $y = 4x^2 + 3x - 2 = f(x)$

$$y' = \lim_{h \to 0} \frac{f(x+h) - f(x)}{h} = \lim_{h \to 0} \frac{[4(x+h)^2 + 3(x+h) - 2] - [4x^2 + 3x - 2]}{h}$$
$$= \lim_{h \to 0} \frac{4(x^2 + 2xh + h^2) + 3x + 3h - 2 - 4x^2 - 3x + 2}{h} = \lim_{h \to 0} \frac{4x^2 + 8xh + 4h^2 + 3x + 3h - 2 - 4x^2 - 3x + 2}{h}$$
$$= \lim_{h \to 0} \frac{8xh + 4h^2 + 3h}{h} = \lim_{h \to 0} \frac{h(8x + 4h + 3)}{h} = \lim_{h \to 0} (8x + 4h + 3) = 8x + 3$$

45. $f(x) = (\ln x)^x, x_0 = 3$

(a)

h	
0.01	$\frac{f(3+0.01) - f(3)}{0.01} = \frac{(\ln 3.01)^{3.01} - (\ln 3)3}{0.01} = 1.3385$
0.001	$\frac{f(3+0.001) - f(3)}{0.001} = \frac{(\ln 3.001)^{3.001} - (\ln 3)^3}{0.001} = 1.3323$
0.0001	$\frac{f(3+0.0001) - f(3)}{0.0001} = \frac{(\ln 3.0001)^{3.0001} - (\ln 3)^3}{0.0001} = 1.3317$
0.00001	$\frac{f(3+0.00001) - f(3)}{0.00001} = \frac{(\ln 3.00001)^{3.00001} - (\ln 3)^3}{0.00001} = 1.3317$

(b) Using a graphing calculator will confirm this result.

47. On the interval $(-\infty, 0)$, the graph of f is a straight line, so its slope is constant. To find this slope, use the points $(-2, 2)$ and $(0, 0)$.

$$m = \frac{0-2}{0-(-2)} = \frac{-2}{2} = -1$$

Thus, the value of f' will be -1 on this interval. The graph of f has a sharp point at 0, so $f'(0)$ does not exist. To show this, we use an open circle on the graph of f' at $(0, -1)$.
We also observe that the slope of f is positive but decreasing from $x = 0$ to about $x = 1$, and then negative from there on. As $x \to \infty$, $f(x) \to 0$ and also $f'(x) = 0$.

Use this information to complete the graph of f'.

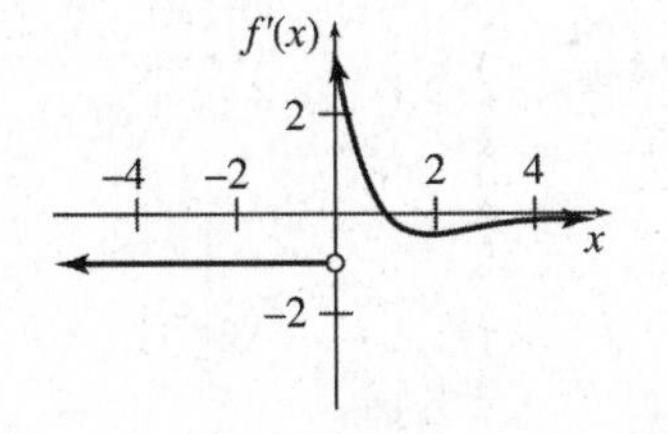

49. $$\lim_{x\to\infty} \frac{cf(x) - dg(x)}{f(x) - g(x)} = \frac{\lim_{x\to\infty} [cf(x) - dg(x)]}{\lim_{x\to\infty} [f(x) - g(x)]}$$
$$= \frac{\lim_{x\to\infty} [cf(x)] - \lim_{x\to\infty} [dg(x)]}{\lim_{x\to\infty} [f(x)] - \lim_{x\to\infty} [g(x)}$$
$$= \frac{c \lim_{x\to\infty} [f(x)] - d \lim_{x\to\infty} [g(x)]}{\lim_{x\to\infty} [f(x)] - \lim_{x\to\infty} [g(x)]}$$
$$= \frac{c \cdot c - d \cdot d}{c - d}$$
$$= \frac{(c+d)(c-d)}{c-d}$$
$$= c + d$$

The answer is (e).

51. $R(x) = 5000 + 16x - 3x^2$

(a) $R'(x) = 16 - 6x$

(b) Since x is in hundreds of dollars, \$1000 corresponds to $x = 10$.

$$R'(10) = 16 - 6(10)$$
$$= 16 - 60 = -44$$

An increase of \$100 spent on advertising when advertising expenditures are \$1000 will result in the revenue decreasing by \$44.

53. $P(x) = 15x + 25x^2$

(a) $P(6) = 15(6) + 25(6)^2$
$= 90 + 900 = 990$
$P(7) = 15(7) + 25(7)^2$
$= 105 + 1225 = 1330$

Average rate of change:

$$= \frac{P(7) - P(6)}{7 - 6} = \frac{1330 - 990}{1}$$
$= 340$ cents or \$3.40

(b) $P(6) = 990$
$P(6.5) = 15(6.5) + 25(6.5)^2$
$= 97.5 + 1056.25$
$= 1153.75$

Average rate of change:

$$= \frac{P(6.5) - P(6)}{6.5 - 6}$$
$$= \frac{1153.75 - 990}{0.5}$$
$= 327.5$ cents or \$3.28

(c) $P(6) = 990$
$P(6.1) = 15(6.1) + 25(6.1)^2$
$= 91.5 + 930.25$
$= 1021.75$

Average rate of change:

$$= \frac{P(6.1) - P(6)}{6.1 - 6}$$
$$= \frac{1021.75 - 990}{0.1}$$
$= 317.5$ cents or \$3.18

(d) $P'(x) = 15 + 50x$
$P'(6) = 15 + 50(6)$
$= 15 + 300$
$= 315$ cents or \$3.15

(e) $P'(20) = 15 + 50(20)$
$= 1015$ cents or \$10.15

(f) $P'(30) = 15 + 50(30)$
$= 1515$ cents or \$15.15

(g) The domain of x is $[0, \infty)$ since pounds cannot be measured with negative numbers.

(h) Since $P'(x) = 15 + 50x$ gives the marginal profit, and $x \geq 0$, $P'(x)$ can never be negative.

(i) $\overline{P}(x) = \frac{P(x)}{x}$

$= \frac{15x + 25x^2}{x}$

$= 15 + 25x$

(j) $\overline{P}'(x) = 25$

(k) The marginal average profit cannot change since $\overline{P}'(x)$ is constant. The profit per pound never changes, no matter now many pounds are sold.

55. (a) $\lim\limits_{x \to 29{,}300^-} T(x) = (29{,}300)(0.15)$

$= \$4395$

(b) $\lim\limits_{x \to 29{,}300^+} T(x) = 4350 + (0.27)(29{,}300 - 29{,}300)$

$= \$4350$

(c) $\lim\limits_{x \to 29{,}300} T(x)$ does not exist since parts (a) and (b) have different answers.

(d)

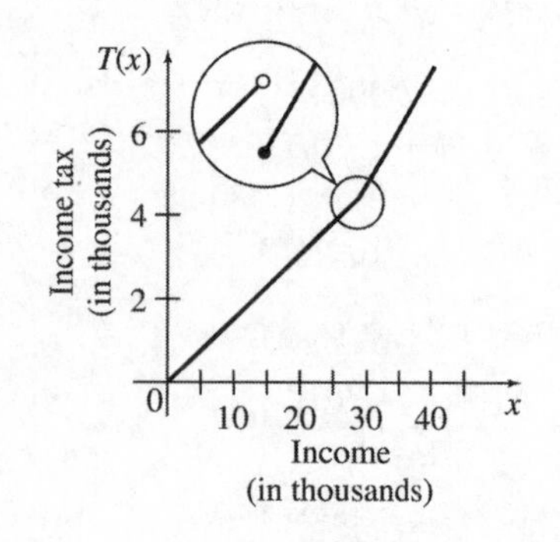

(e) The graph is discontinuous at $x = 29{,}300$.

(f) For $0 \le x \le 29{,}300$,

$$A(x) = \frac{T(x)}{x} = \frac{0.15x}{x} = 0.15.$$

For $x > 29{,}300$,

$$A(x) = \frac{T(x)}{x}$$

$$= \frac{4350 + (0.27)(x - 29{,}300)}{x}$$

$$= \frac{0.27x - 3561}{x}$$

$$= 0.27 - \frac{3561}{x}.$$

(g) $\lim\limits_{x \to 29{,}300^-} A(x) = 0.15$

(h) $\lim\limits_{x \to 29{,}300^+} A(x) = 0.27 - \frac{3561}{29{,}300} = 0.14846$

(i) $\lim\limits_{x \to 29{,}300} A(x)$ does not exist since parts (g) and (h) have different answers.

(j) $\lim\limits_{x \to \infty} A(x) = 0.27 - 0 = 0.27$

(k)

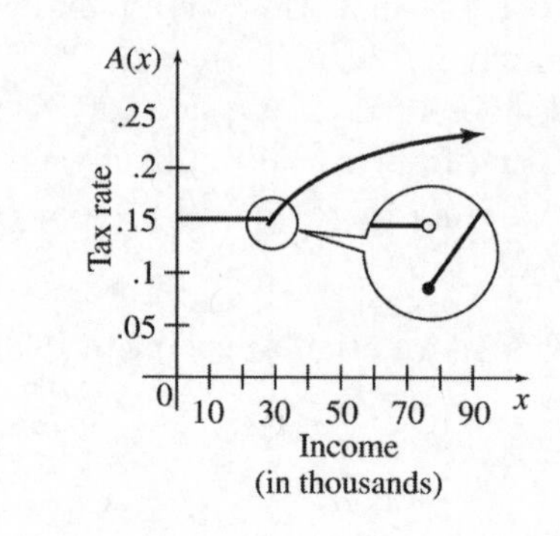

57. $V(t) = -t^2 + 6t - 4$

(a)

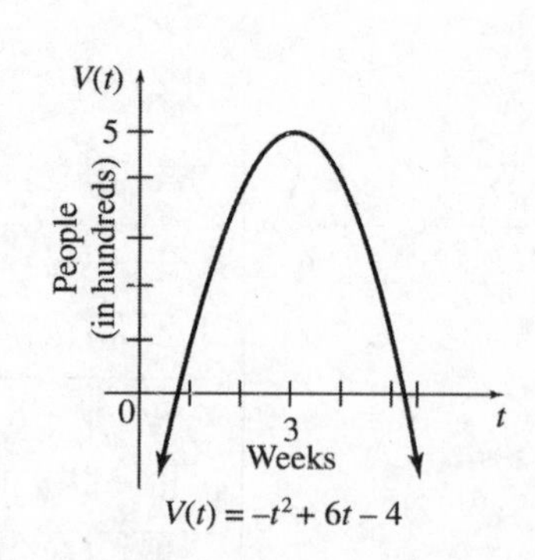

(b) The x-intercepts of the parabola are 0.8 and 5.2, so a reasonable domain would be $[0.8, 5.2]$, which represents the time period from 0.8 to 5.2 weeks.

(c) The number of cases reaches a maximum at the vertex;

$$x = \frac{-b}{2a} = \frac{-6}{-2} = 3$$

$$V(3) = -3^2 + 6(3) - 4 = 5$$

The vertex of the parabola is $(3, 5)$. This represents a maximum at 3 weeks of 500 cases.

(d) The rate of change function is

$$V'(t) = -2t + 6.$$

(e) The rate of change in the number of cases at the maximum is

$$V'(3) = -2(3) + 6 = 0.$$

(f) The sign of the rate of change up to the maximum is + because the function is increasing. The sign of the rate of change after the maximum is − because the function is decreasing.

59. **(a)** The curve slants downward up to about age 4, where it turns and begins to rise. There is a slight decline in steepness between ages 10.5 and 18. Correspondingly, the graph of the rate of change lies below the horizontal axis to the left of 4 years and above the horizontal axis to the right of that point. The graph of the rate of change is declining between ages 10.5 and 18.

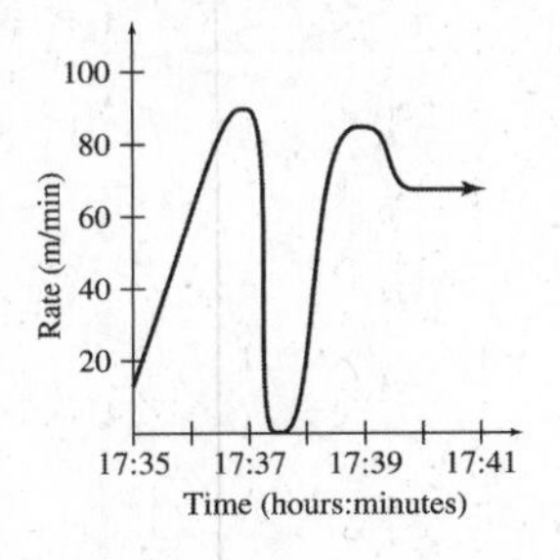

(b) The curve slants downward up to about age 5.25, where it turns and begins to rise. There is a slight decline in steepness between ages 11 and 20. Correspondingly, the graph of the rate of change lies below the horizontal axis to the left of 5.25 years and above the horizontal axis to the right of that point. The graph of the rate of change is declining between ages 11 and 20.

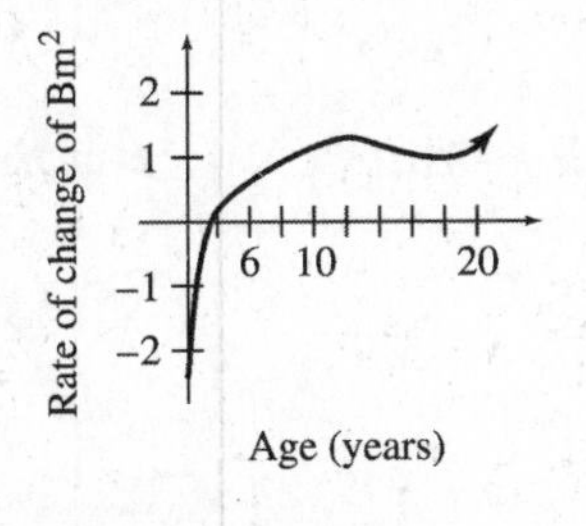

61. **(a)** The slope of the tangent line at $x = 100$ is 1. This means that the ball is rising 1 ft for each foot it travels horizontally.

(b) The slope of the tangent line at $x = 200$ is -2.7. This means that the ball is dropping 2.7 ft for each foot it travels horizontally.

Chapter 3 Test

[3.1]

Decide whether each limit exists. If a limit exists, find its value.

1. $\lim_{x \to 1} f(x)$

2. $\lim_{x \to -1} f(x)$

3. $\lim_{x \to 2} f(x)$

4. $\lim_{x \to -2} f(x)$

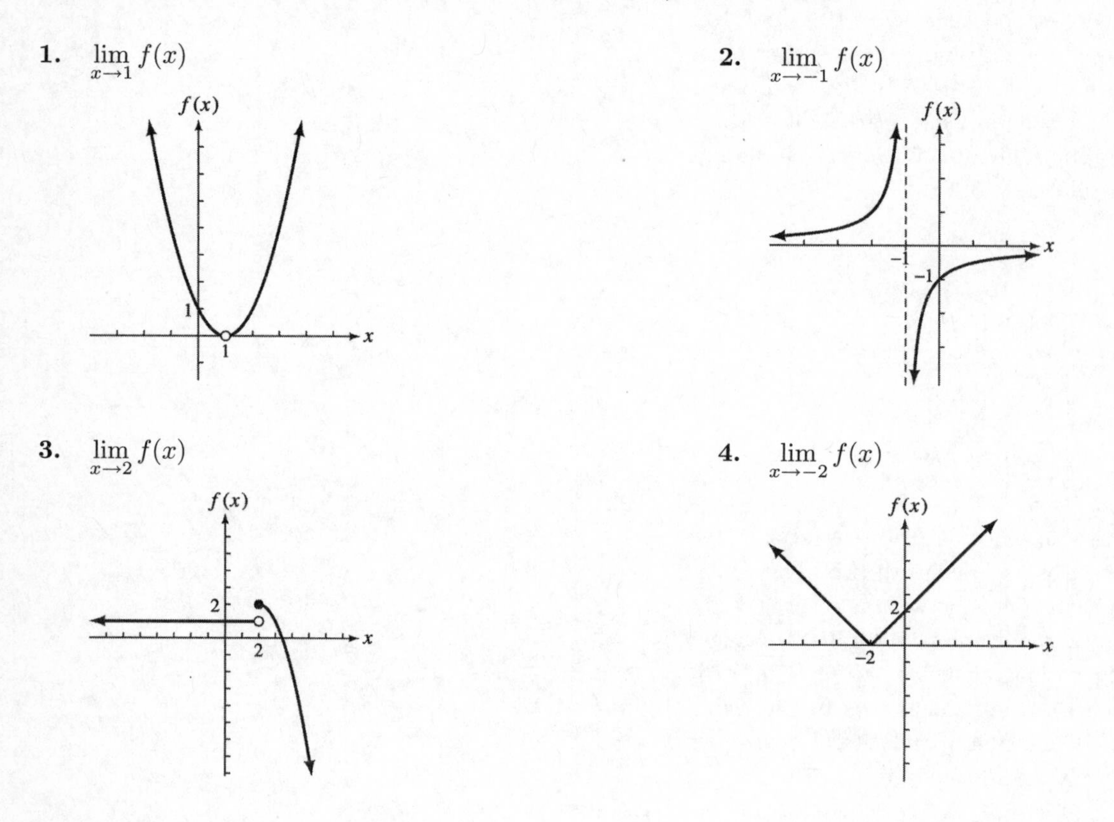

Use the properties of limits to help decide whether the following limits exist. If a limit exists, find its value.

5. $\lim_{x \to 2} \left(\frac{1}{x} + 1\right)(3x - 2)$

6. $\lim_{x \to 0} \frac{3x + 5}{4x}$

7. $\lim_{x \to 2} \frac{x^2 - 5x + 6}{x - 2}$

8. $\lim_{x \to 1} \frac{x - 1}{\sqrt{x} - 1}$

[3.2]

Find all points $x = a$ where the function is discontinuous. For each point of discontinuity, give $\lim_{x \to a} f(x)$ if it exists.

9. $f(x) = \frac{x^2 - 16}{x + 4}$

10. $g(x) = \frac{3 + x}{(2x + 3)(x - 5)}$

11. $h(x) = \frac{3x^4 + 2x^2 - 7}{4}$

12. Is $f(x) = \frac{x-2}{x(3-x)(x+4)}$ continuous at the given values of x?

(a) $x = 2$ **(b)** $x = 0$ **(c)** $x = 5$ **(d)** $x = 3$

13. Use the graph to answer the following questions.

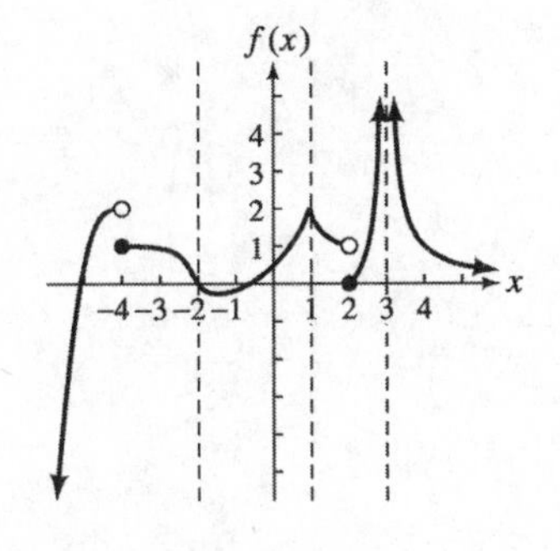

(a) On which of the following intervals is the graph continuous?

$$(-5,-2),\ (-3,2),\ (1,4)$$

(b) Where is the function discontinuous?

(c) Find $\lim_{x\to-4^-} f(x)$.

(d) Find $\lim_{x\to-4^+} f(x)$.

14. Consider the following function.

$$f(x) = \begin{cases} -3 & \text{if } x < -1 \\ 2x-1 & \text{if } -1 \le x \le 2 \\ 6 & \text{if } x > 2 \end{cases}$$

(a) Graph this function.

(b) Find any points of discontinuity.

(c) Find the limit from the left and from the right at any point(s) of discontinuity.

[3.3]

15. Find the average rate of change of $f(x) = x^3 - 5x$ between $x = 1$ and $x = 5$.

16. Use the graph to find the average rate of change of f on the given intervals.

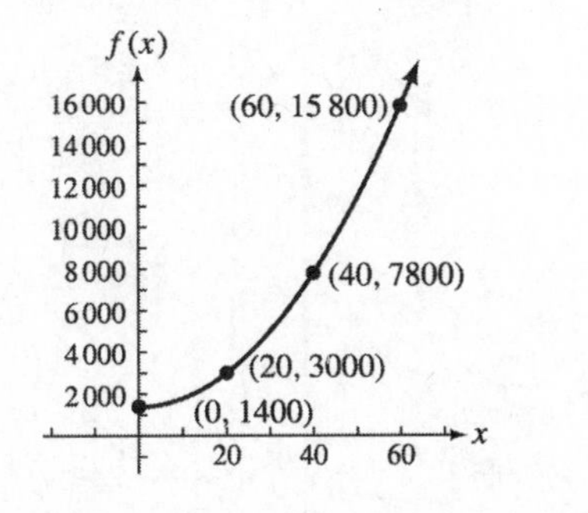

(a) From $x = 0$ to $x = 20$

(b) From $x = 20$ to $x = 60$

(c) From $x = 0$ to $x = 40$

(d) From $x = 0$ to $x = 60$

Find the instantaneous rate of change for each function at the given value.

17. $f(x) = 3x^2 - 7$ at $x = 1$

18. $g(t) = 4 - 2t^2$ at $t = -3$

19. Suppose the total profit in thousands of dollars from selling x units is given by

$$P(x) = 2x^2 - 6x + 9.$$

(a) Find the average rate of change of profit as x increases from 2 to 4.

(b) Find the marginal profit when 10 units are sold.

(c) Find the average rate of change of profit when sales are increased from 100 to 200 units.

[3.4]

Use the definition of the derivative to find the derivative of each function.

20. $y = x^3 - 5x^2$

21. $y = \dfrac{3}{x}$

Find $f'(x)$ and use it to find $f'(3)$, $f'(0)$, $f'(-1)$.

22. $f(x) = \dfrac{-3}{x}$

23. $f(x) = 3x^2 + 4x$

24. $f(x) = -\dfrac{1}{2}\sqrt{x}$

[3.5]

Sketch the graph of the derivative for each function shown.

25.

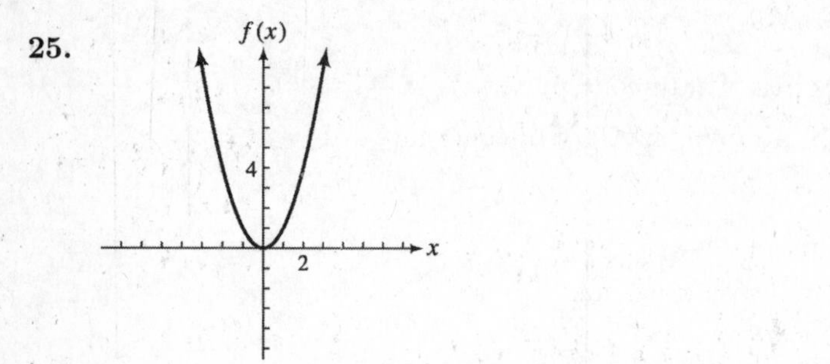

26.

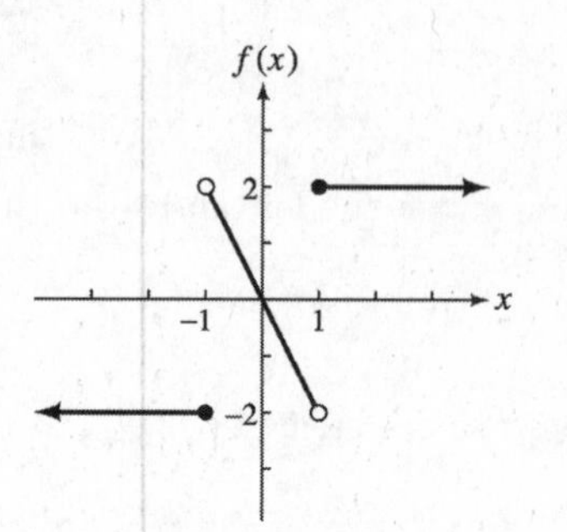

Chapter 3 Test Answers

1. 0

2. Does not exist

3. Does not exist

4. 0

5. 6

6. Does not exist

7. -1

8. 2

9. $a = -4$, $\lim_{x \to -4} f(x) = -8$

10. $a = -\frac{3}{2}$, limit does not exist; $a = 5$, limit does not exist.

11. Discontinuous nowhere

12. **(a)** Yes **(b)** No **(c)** Yes **(d)** No

13. **(a)** $(-3, 2)$ **(b)** At $x = -4$, $x = 2$, and $x = 3$ **(c)** 2 **(d)** 1

14. **(a)**

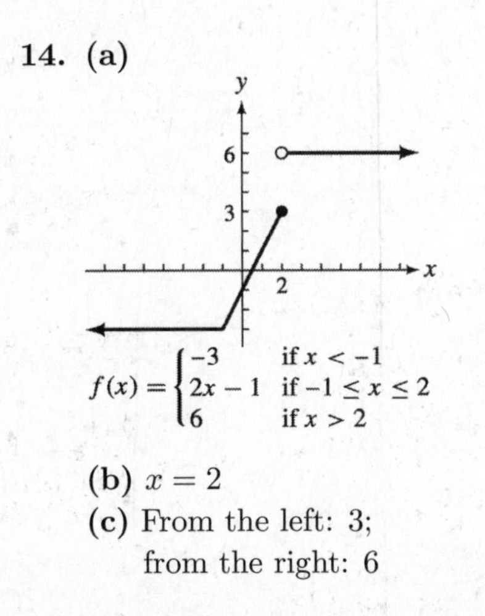

(b) $x = 2$
(c) From the left: 3; from the right: 6

15. 26

16. **(a)** 80 **(b)** 320 **(c)** 160 **(d)** 240

17. 6

18. 12

19. **(a)** \$6000 per unit **(b)** \$34,000 per unit **(c)** \$594,000 per unit

20. $y' = 3x^2 - 10x$

21. $y' = -\frac{3}{x^2}$

22. $f'(x) = \frac{3}{x^2}$, $\frac{1}{3}$, undefined, 3

23. $f'(x) = 6x + 4$, 22, 4, -2

24. $f'(x) = \frac{-1}{4\sqrt{x}}$, $\frac{-1}{4\sqrt{3}}$ or $\frac{-\sqrt{3}}{12}$, undefined, undefined

25.

26.

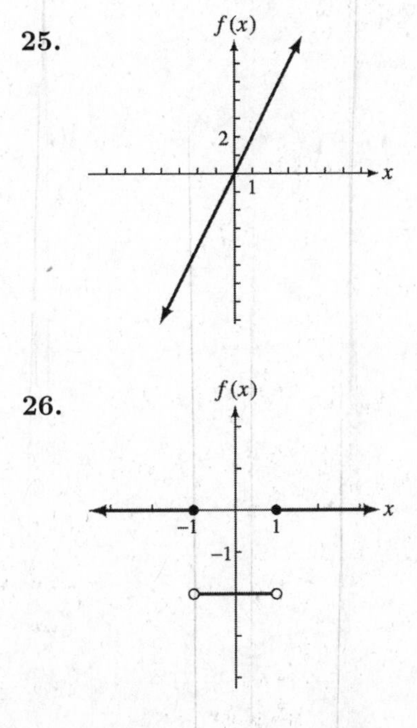

Chapter 4

CALCULATING THE DERIVATIVE

4.1 Techniques for Finding Derivatives

1. $y = 12x^3 - 8x^2 + 7x + 5$

$$\frac{dy}{dx} = 12(3x^{3-1}) - 8(2x^{2-1}) + 7x^{1-1} + 0$$
$$= 36x^2 - 16x + 7$$

3. $y = 3x^4 - 6x^3 + \frac{x^2}{8} + 5$

$$\frac{dy}{dx} = 3(4x^{4-1}) - 6(3x^{3-1}) + \frac{1}{8}(2x^{2-1}) + 0$$
$$= 12x^3 - 18x^2 + \frac{1}{4}x$$

5. $y = 6x^{3.5} - 10x^{0.5}$

$$\frac{dy}{dx} = 6(3.5x^{3.5-1}) - 10(0.5x^{0.5-1})$$
$$= 21x^{2.5} - 5x^{-0.5} \text{ or } 21x^{2.5} - \frac{5}{x^{0.5}}$$

7. $y = 8\sqrt{x} + 6x^{3/4}$

$$= 8x^{1/2} + 6x^{3/4}$$
$$\frac{dy}{dx} = 8\left(\frac{1}{2}x^{1/2-1}\right) + 6\left(\frac{3}{4}x^{3/4-1}\right)$$
$$= 4x^{-1/2} + \frac{9}{2}x^{-1/4}$$
$$\text{or } \frac{4}{x^{1/2}} + \frac{9}{2x^{1/4}}$$

9. $g(x) = 6x^{-5} - x^{-1}$

$$g'(x) = 6(-5)x^{-5-1} - (-1)x^{-1-1}$$
$$= -30x^{-6} + x^{-2}$$
$$\text{or } \frac{-30}{x^6} + \frac{1}{x^2}$$

11. $y = 5x^{-5} - 6x^{-2} + 13x^{-1}$

$$\frac{dy}{dx} = 5(-5x^{-5-1}) - 6(-2x^{-2-1}) + 13(-1x^{-1-1})$$
$$= -25x^{-6} + 12x^{-3} - 13x^{-2}$$
$$\text{or } \frac{-25}{x^6} + \frac{12}{x^3} - \frac{13}{x^2}$$

13. $f(t) = \frac{14}{t} + \frac{12}{t^4} + \sqrt{2}$

$$= 14t^{-1} + 12t^{-4} + \sqrt{2}$$
$$f'(t) = 14(-1t^{-1-1}) + 12(-4t^{-4-1}) + 0$$
$$= -14t^{-2} - 48t^{-5} \text{ or } \frac{-14}{t^2} - \frac{48}{t^5}$$

15. $y = \frac{3}{x^6} + \frac{1}{x^5} - \frac{7}{x^2}$

$$= 3x^{-6} + x^{-5} - 7x^{-2}$$
$$\frac{dy}{dx} = 3(-6x^{-7}) + (-5x^{-6}) - 7(-2x^{-3})$$
$$= -18x^{-7} - 5x^{-6} + 14x^{-3}$$
$$\text{or } \frac{-18}{x^7} - \frac{5}{x^6} + \frac{14}{x^3}$$

17. $h(x) = x^{-1/2} - 14x^{-3/2}$

$$h'(x) = -\frac{1}{2}x^{-3/2} - 14\left(-\frac{3}{2}x^{-5/2}\right)$$
$$= \frac{-x^{-3/2}}{2} + 21x^{-5/2}$$
$$\text{or } \frac{-1}{2x^{3/2}} + \frac{21}{x^{5/2}}$$

19. $y = \frac{-2}{\sqrt[3]{x}}$

$$= \frac{-2}{x^{1/3}} = -2x^{-1/3}$$
$$\frac{dy}{dx} = -2\left(-\frac{1}{3}x^{-4/3}\right)$$
$$= \frac{2x^{-4/3}}{3} \text{ or } \frac{2}{3x^{4/3}}$$

21. $g(x) = \frac{x^3 - 4x}{\sqrt{x}}$

$$= \frac{x^3 - 4x}{x^{1/2}}$$
$$= x^{5/2} - 4x^{1/2}$$
$$g'(x) = \frac{5}{2}x^{5/2-1} - 4\left(\frac{1}{2}x^{1/2-1}\right)$$
$$= \frac{5}{2}x^{3/2} - 2x^{-1/2}$$
$$\text{or } \frac{5}{2}x^{3/2} - \frac{2}{\sqrt{x}}$$

23. $h(x) = (x^2 - 1)^3$
$= x^6 - 3x^4 + 3x^2 - 1$
$h'(x) = 6x^{6-1} - 3(4x^{4-1}) + 3(2x^{2-1}) - 0$
$= 6x^5 - 12x^3 + 6x$

27. $D_x\left[9x^{-1/2} + \dfrac{2}{x^{3/2}}\right]$

$= D_x[9x^{-1/2} + 2x^{-3/2}]$

$= 9\left(-\dfrac{1}{2}x^{-3/2}\right) + 2\left(-\dfrac{3}{2}x^{-5/2}\right)$

$= -\dfrac{9}{2}x^{-3/2} - 3x^{-5/2}$

or $\dfrac{-9}{2x^{3/2}} - \dfrac{3}{x^{5/2}}$

29. $f(x) = \dfrac{x^4}{6} - 3x$
$= \dfrac{1}{6}x^4 - 3x$
$f'(x) = \dfrac{1}{6}(4x^3) - 3$
$= \dfrac{2}{3}x^3 - 3$
$f'(-2) = \dfrac{2}{3}(-2)^3 - 3$
$= -\dfrac{16}{3} - 3$
$= -\dfrac{25}{3}$

31. $y = x^4 - 5x^3 + 2;\ x = 2$
$y' = 4x^3 - 15x^2$
$y'(2) = 4(2)^3 - 15(2)^2$
$= -28$

The slope of tangent line at $x = 2$ is -28.
Use $m = -28$ and $(x_1, y_1) = (2, -22)$ to obtain the equation.

$y - (-22) = -28(x - 2)$
$y = -28x + 34$

33. $y = -2x^{1/2} + x^{3/2}$
$y' = -2\left(\dfrac{1}{2}x^{-1/2}\right) + \dfrac{3}{2}x^{1/2}$
$= -x^{-1/2} + \dfrac{3}{2}x^{1/2}$
$= -\dfrac{1}{x^{1/2}} + \dfrac{3x^{1/2}}{2}$
$y'(9) = -\dfrac{1}{(9)^{1/2}} + \dfrac{3(9)^{1/2}}{2}$
$= -\dfrac{1}{3} + \dfrac{9}{2}$
$= \dfrac{25}{6}$

The slope of the tangent line at $x = 9$ is $\frac{25}{6}$.

35. $f(x) = 9x^2 - 8x + 4$
$f'(x) = 18x - 8$

Let $f'(x) = 0$ to find the point where the slope of the tangent line is zero.

$18x - 8 = 0$
$18x = 8$
$x = \dfrac{8}{18} = \dfrac{4}{9}$

Find the y-coordinate.

$f(x) = 9x^2 - 8x + 4$
$f\left(\dfrac{4}{9}\right) = 9\left(\dfrac{4}{9}\right)^2 - 8\left(\dfrac{4}{9}\right) + 4$
$= 9\left(\dfrac{16}{81}\right) - \dfrac{32}{9} + 4$
$= \dfrac{16}{9} - \dfrac{32}{9} + \dfrac{36}{9} = \dfrac{20}{9}$

The slope of the tangent line is zero at one point, $\left(\frac{4}{9}, \frac{20}{9}\right)$.

37. $f(x) = 2x^3 + 9x^2 - 60x + 4$
$f'(x) = 6x^2 + 18x - 60$

If the tangent line is horizontal, then its slope is zero and $f'(x) = 0$.

$6x^2 + 18x - 60 = 0$
$6(x^2 + 3x - 10) = 0$
$6(x + 5)(x - 2) = 0$
$x = -5$ or $x = 2$

Thus, the tangent line is horizontal at $x = -5$ and $x = 2$.

39. $f(x) = x^3 - 4x^2 - 7x + 8$
$f'(x) = 3x^2 - 8x - 7$

If the tangent line is horizontal, then its slope is zero and $f'(x) = 0$.

$$\begin{aligned} 3x^2 - 8x - 7 &= 0 \\ x &= \frac{8 \pm \sqrt{64 + 84}}{6} \\ x &= \frac{8 \pm \sqrt{148}}{6} \\ x &= \frac{8 \pm 2\sqrt{37}}{6} \\ x &= \frac{2(4 \pm \sqrt{37})}{6} \\ x &= \frac{4 \pm \sqrt{37}}{3} \end{aligned}$$

Thus, the tangent line is horizontal at $x = \frac{4 \pm \sqrt{37}}{3}$.

41. $f(x) = 6x^2 + 4x - 9$
$f'(x) = 12x + 4$

If the slope of the tangent line is -2, $f'(x) = -2$.

$$\begin{aligned} 12x + 4 &= -2 \\ 12x &= -6 \\ x &= -\frac{1}{2} \\ f\left(-\frac{1}{2}\right) &= -\frac{19}{2} \end{aligned}$$

The slope of the tangent line is -2 at $\left(-\frac{1}{2}, -\frac{19}{2}\right)$.

43. $f(x) = x^3 + 6x^2 + 21x + 2$
$f'(x) = 3x^2 + 12x + 21$

If the slope of the tangent line is 9, $f'(x) = 9$.

$$\begin{aligned} 3x^2 + 12x + 21 &= 9 \\ 3x^2 + 12x + 12 &= 0 \\ 3(x^2 + 4x + 4) &= 0 \\ 3(x + 2)^2 &= 0 \\ x &= -2 \\ f(-2) &= -24 \end{aligned}$$

The slope of the tangent line is 9 at $(-2, -24)$.

45.
$$\begin{aligned} f(x) &= \frac{1}{2}g(x) + \frac{1}{4}h(x) \\ f'(x) &= \frac{1}{2}g'(x) + \frac{1}{4}h'(x) \\ f'(2) &= \frac{1}{2}g'(2) + \frac{1}{4}h'(2) \\ &= \frac{1}{2}(7) + \frac{1}{4}(14) = 7 \end{aligned}$$

49. $\frac{f(x)}{k} = \frac{1}{k} \cdot f(x)$

Use the rule for the derivative of a constant times a function.

$$\begin{aligned} \frac{d}{dx}\left[\frac{f(x)}{k}\right] &= \frac{d}{dx}\left[\frac{1}{k} \cdot f(x)\right] \\ &= \frac{1}{k}f'(x) \\ &= \frac{f'(x)}{k} \end{aligned}$$

51. The demand is given by $q = 5000 - 100p$.

Solve for p.

$$\begin{aligned} p &= \frac{5000 - q}{100} \\ R(q) &= q\left(\frac{5000 - q}{100}\right) \\ &= \frac{5000q - q^2}{100} \\ R'(q) &= \frac{5000 - 2q}{100} \end{aligned}$$

(a) $R'(1000) = \frac{5000 - 2(1000)}{100}$
$= 30$

(b) $R'(2500) = \frac{5000 - 2(2500)}{100}$
$= 0$

(c) $R'(3000) = \frac{5000 - 2(3000)}{100}$
$= -10$

53.
$$\begin{aligned} S(t) &= 100 - 100t^{-1} \\ S'(t) &= -100(-1t^{-2}) \\ &= 100t^{-2} \\ &= \frac{100}{t^2} \end{aligned}$$

(a) $S'(1) = \frac{100}{(1)^2} = \frac{100}{1} = 100$

(b) $S'(10) = \frac{100}{(10)^2} = \frac{100}{100} = 1$

55. Profit = Revenue − Cost

$$P(q) = qp(q) - C(q)$$

$$P(q) = q\left(\frac{1000}{q^2} + 1000\right) - (0.2q^2 + 6q + 50)$$

$$= \frac{1000}{q} + 1000q - 0.2q^2 - 6q - 50$$

$$= 1000q^{-1} + 994q - 0.2x^2 - 50$$

$$P'(q) = -1000q^{-2} + 994 - 0.4q$$

$$= 994 - 0.4q - \frac{1000}{q^2}$$

$$P'(10) = 994 - 0.4(10) - \frac{1000}{(10)^2}$$

$$= 994 - 4 - 10$$

$$= 980$$

The marginal profit is \$980.

57. (a) 1982 when $t = 50$:

$$C(50) = 0.00875(50)^2 - 0.108(50) + 1.42$$
$$= 17.895 \approx 17.9 \text{ cents}$$

2002 when $t = 70$:

$$C(70) = 0.00875(70)^2 - 0.108(70) + 1.42$$
$$= 36.735 \approx 36.7 \text{ cents}$$

(b) $C'(t) = 0.00875(2t) - 0.108(1)$
$= 0.0175t - 0.108$

1982 when $t = 50$:

$$C'(50) = 0.0175(50) - 0.108$$
$$= 0.767 \text{ cents/year}$$

2002 when $t = 70$:

$$C'(70) = 0.0175(70) - 0.108$$
$$\approx 1.12 \text{ cents/year}$$

(c) Using a graphing calculator, a cubic function that models the data is

$$C(t) = (-1.790 \times 10^{-4})t^3 + 0.02947t^2$$
$$- 0.7105t + 3.291.$$

Using the values from the calculator, the rate of change in 1982 is $C'(50) \approx 0.894$ cents/year. Using the values from the calculator, the rate of change in 2002 is $C'(70) \approx 0.784$ cents/year.

59. $N(t) = 0.00437t^{3.2}$
$N'(t) = 0.013984t^{2.2}$

(a) $N'(5) \approx 0.4824$

(b) $N'(10) \approx 2.216$

61. $V(t) = -2159 + 1313t - 60.82t^2$

(a) $V(3) = -2159 + 1313(3) - 60.82(3)^2$
$= 1232.62 \text{ cm}^3$

(b) $V'(t) = 1313 - 121.64t$
$V'(3) = 1313 - 121.64(3)$
$= 948.08 \text{ cm}^3/\text{yr}$

63. $v = 2.69l^{1.86}$

$$\frac{dv}{dl} = (1.86)2.69l^{1.86-1} \approx 5.00l^{0.86}$$

65. $t = 0.0588s^{1.125}$

(a) When $s = 1609, t \approx 238.1$ seconds, or 3 minutes, 58.1 seconds.

(b) $\frac{dt}{ds} = 0.0588(1.125s^{1.125-1})$
$= 0.06615s^{0.125}$

When $s = 100$, $\frac{dt}{ds} \approx 0.118$ sec/m. At 100 meters, the fastest possible time increases by 0.118 seconds for each additional meter.

(c) Yes, they have been surpassed. In 2000, the world record in the mile stood at 3:43.13. (Ref: www.runnersworld.com)

67. $\text{BMI} = \frac{703w}{h^2}$

(a) $6'2'' = 74$ in.

$$\text{BMI} = \frac{703(220)}{74^2} \approx 28$$

(b) $\text{BMI} = \frac{703w}{74^2} = 24.9$ implies

$$w = \frac{24.9(74)^2}{703} \approx 194.$$

A 220-lb person needs to lose 26 pounds to get down to 194 lbs.

(c) If $f(h) = \frac{703(125)}{h^2} = 87{,}875h^{-2}$, then

$$f'(h) = 87{,}875(-2h^{-2-1})$$
$$= -175{,}750h^{-3} = -\frac{175{,}750}{h^3}$$

(d) $f'(65) = -\frac{175{,}750}{65^3} \approx -0.64$

For a 125-lb female with a height of 65 in. (5′5″), the BMI decreases by 0.64 for each additional inch of height.

(e) Sample Chart

ht/wt	140	160	180	200
60	27	31	35	39
65	23	27	30	33
70	20	23	26	29
75	17	20	22	25

69. $s(t) = 18t^2 - 13t + 8$

(a) $$v(t) = s'(t) = 18(2t) - 13 + 0 \\ = 36t - 13$$

(b) $$v(0) = 36(0) - 13 = -13 \\ v(5) = 36(5) - 13 = 167 \\ v(10) = 36(10) - 13 = 347$$

71. $s(t) = -3t^3 + 4t^2 - 10t + 5$

(a) $$v(t) = s'(t) = -3(3t^2) + 4(2t) - 10 + 0 \\ = -9t^2 + 8t - 10$$

(b) $$v(0) = -9(0)^2 + 8(0) - 10 = -10 \\ v(5) = -9(5)^2 + 8(5) - 10 \\ = -225 + 40 - 10 = -195 \\ v(10) = -9(10)^2 + 8(10) - 10 \\ = -900 + 80 - 10 = -830$$

73. $s(t) = -16t^2 + 64t$

(a) $$v(t) = s'(t) = -16(2t) + 64 \\ = -32t + 64$$

$$v(2) = -32(2) + 64 = -64 + 64 = 0$$

$$v(3) = -32(3) + 64 = -96 + 64 = -32$$

The ball's velocity is 0 ft/sec after 2 seconds and -32 ft/sec after 3 seconds.

(b) As the ball travels upward, its speed decreases because of the force of gravity until, at maximum height, its speed is 0 ft/sec.
In part (a), we found that $v(2) = 0$.

It takes 2 seconds for the ball to reach its maximum height.

(c) $$s(2) = -16(2)^2 + 64(2) \\ = -16(4) + 128 \\ = -64 + 128 \\ = 64$$

It will go 64 ft high.

75. $$y_1 = 4.13x + 14.63 \\ y_2 = -0.033x^2 + 4.647x + 13.347$$

(a) When $x = 5, y_1 \approx 35$ and $y_2 \approx 36$.

(b) $$\frac{dy_1}{dx} = 4.13$$

$$\frac{dy_2}{dx} = 0.033(2x) + 4.647 \\ = -0.066x + 4.647$$

When $x = 5$, $\frac{dy_1}{dx} = 4.13$ and $\frac{dy_2}{dx} \approx 4.32$. These values are fairly close and represent the rate of change of four years for a dog for one year of a human, for a dog that is actually 5 years old.

(c) With the first three points eliminated, the dog age increases in 2-year steps and the human age increases in 8-year steps, for a slope of 4. The equation has the form $y = 4x + b$. A value of 16 for b makes the numbers come out right. $y = 4x + b$. For a dog of age $x = 5$ years or more, the equivalent human age is given by $y = 4x + 16$.

4.2 Derivatives of Products and Quotients

1. $$y = (3x^2 + 2)(2x - 1)$$

$$\frac{dy}{dx} = (3x^2 + 2)(2) + (2x - 1)(6x) \\ = 6x^2 + 4 + 12x^2 - 6x \\ = 18x^2 - 6x + 4$$

3. $$y = (2x - 5)^2 \\ = (2x - 5)(2x - 5)$$

$$\frac{dy}{dx} = (2x - 5)(2) + (2x - 5)(2) \\ = 4x - 10 + 4x - 10 \\ = 8x - 20$$

5. $$k(t) = (t^2 - 1)^2 = (t^2 - 1)(t^2 - 1) \\ k'(t) = (t^2 - 1)(2t) + (t^2 - 1)(2t) \\ = 2t^3 - 2t + 2t^3 - 2t \\ = 4t^3 - 4t$$

7. $$\begin{aligned} y &= (x+1)(\sqrt{x}+2) \\ &= (x+1)(x^{1/2}+2) \\ \frac{dy}{dx} &= (x+1)\left(\frac{1}{2}x^{-1/2}\right) + (x^{1/2}+2)(1) \\ &= \frac{1}{2}x^{1/2} + \frac{1}{2}x^{-1/2} + x^{1/2} + 2 \\ &= \frac{3}{2}x^{1/2} + \frac{1}{2}x^{-1/2} + 2 \\ \text{or } & \frac{3x^{1/2}}{2} + \frac{1}{2x^{1/2}} + 2 \end{aligned}$$

9. $$\begin{aligned} p(y) &= (y^{-1}+y^{-2})(2y^{-3}-5y^{-4}) \\ p'(y) &= (y^{-1}+y^{-2})(-6y^{-4}+20y^{-5}) \\ &\quad + (-y^{2}-2y^{-3})(2y^{-3}-5y^{-4}) \\ &= -6y^{-5}+20y^{-6}-6y^{-6}+20y^{-7} \\ &\quad -2y^{-5}+5y^{-6}-4y^{-6}+10y^{-7} \\ &= -8y^{-5}+15y^{-6}+30y^{-7} \end{aligned}$$

11. $$\begin{aligned} f(x) &= \frac{6x+1}{3x+10} \\ f'(x) &= \frac{(3x+10)(6)-(6x+1)(3)}{(3x+10)^2} \\ &= \frac{18x+60-18x-3}{(3x+10)^2} \\ &= \frac{57}{(3x+10)^2} \end{aligned}$$

13. $$\begin{aligned} y &= \frac{5-3t}{4+t} \\ \frac{dy}{dx} &= \frac{(4+t)(-3)-(5-3t)(1)}{(4+t)^2} \\ &= \frac{-12-3t-5+3t}{(4+t)^2} \\ &= \frac{-17}{(4+t)^2} \end{aligned}$$

15. $$\begin{aligned} y &= \frac{x^2+x}{x-1} \\ \frac{dy}{dx} &= \frac{(x-1)(2x+1)-(x^2+x)(1)}{(x-1)^2} \\ &= \frac{2x^2+x-2x-1-x^2-x}{(x-1)^2} \\ &= \frac{x^2-2x-1}{(x-1)^2} \end{aligned}$$

17. $$\begin{aligned} f(t) &= \frac{4t^2+11}{t^2+3} \\ f'(t) &= \frac{(t^2+3)(8t)-(4t^2+11)(2t)}{(t^2+3)^2} \\ &= \frac{8t^3+24t-8t^3-22t}{(t^2+3)^2} \\ &= \frac{2t}{(t^2+3)^2} \end{aligned}$$

19. $$\begin{aligned} g(x) &= \frac{x^2-4x+2}{x^2+3} \\ g'(x) &= \frac{(x^2+3)(2x-4)-(x^2-4x+2)(2x)}{(x^2+3)^2} \\ &= \frac{2x^3-4x^2+6x-12-2x^3+8x^2-4x}{(x^2+3)^2} \\ &= \frac{4x^2+2x-12}{(x^2+3)^2} \end{aligned}$$

21. $$\begin{aligned} p(t) &= \frac{\sqrt{t}}{t-1} \\ &= \frac{t^{1/2}}{t-1} \\ p'(t) &= \frac{(t-1)\left(\frac{1}{2}t^{-1/2}\right)-t^{1/2}(1)}{(t-1)^2} \\ &= \frac{\frac{1}{2}t^{1/2}-\frac{1}{2}t^{-1/2}-t^{1/2}}{(t-1)^2} \\ &= \frac{-\frac{1}{2}t^{1/2}-\frac{1}{2}t^{-1/2}}{(t-1)^2} \\ &= \frac{-\frac{\sqrt{t}}{2}-\frac{1}{2\sqrt{t}}}{(t-1)^2} \quad \text{or} \quad \frac{-t-1}{2\sqrt{t}(t-1)^2} \end{aligned}$$

23. $$\begin{aligned} y &= \frac{5x+6}{\sqrt{x}} = \frac{5x+6}{x^{1/2}} \\ \frac{dy}{dx} &= \frac{(x^{1/2})(5)-(5x+6)\left(\frac{1}{2}x^{-1/2}\right)}{(x^{1/2})^2} \\ &= \frac{5x^{1/2}-\frac{5}{2}x^{1/2}-3x^{-1/2}}{x} \\ &= \frac{\frac{5}{2}x^{1/2}-3x^{-1/2}}{x} \\ &= \frac{\frac{5\sqrt{x}}{2}-\frac{3}{\sqrt{x}}}{x} \quad \text{or} \quad \frac{5x-6}{2x\sqrt{x}} \end{aligned}$$

25. $g(y) = \dfrac{y^{1.4}+1}{y^{2.5}+2}$

$$g'(y) = \frac{(y^{2.5}+2)(1.4y^{0.4}) - (y^{1.4}+1)(2.5y^{1.5})}{(y^{2.5}+2)^2} = \frac{1.4y^{2.9}+2.8y^{0.4}-2.5y^{2.9}-2.5y^{1.5}}{(y^{2.5}+2)^2}$$

$$= \frac{-1.1y^{2.9}-2.5y^{1.5}+2.8y^{0.4}}{(y^{2.5}+2)^2}$$

27. $g(x) = \dfrac{(2x^2+3)(5x+2)}{6x-7}$

$$g'(x) = \frac{(6x-7)[(2x^2+3)(5)+(4x)(5x+2)] - (2x^2+3)(5x+2)(6)}{(6x-7)^2}$$

$$= \frac{(6x-7)(30x^2+8x+15)-(2x^2+3)(30x+12)}{(6x-7)^2}$$

$$= \frac{180x^3+48x^2+90x-210x^2-56x-105-60x^3-24x^2-90x-36}{(6x-7)^2}$$

$$= \frac{120x^3-186x^2-56x-141}{(6x-7)^2}$$

29. $h(x) = \dfrac{f(x)}{g(x)}$

$$h'(x) = \frac{g(x)f'(x)-f(x)g'(x)}{[g(x)]^2}$$

$$h'(3) = \frac{g(3)f'(3)-f(3)g'(3)}{[g(3)]^2} = \frac{4(8)-9(5)}{4^2} = -\frac{13}{16}$$

31. In the first step, the denominator, $(x^3)^2 = x^6$, was omitted. The correct work follows.

$$D_x\left(\frac{x^2-4}{x^3}\right) = \frac{x^3(2x)-(x^2-4)(3x^2)}{(x^3)^2} = \frac{2x^4-3x^4+12x^2}{x^6} = \frac{-x^4+12x^2}{x^6} = \frac{x^2(-x^2+12)}{x^2(x^4)} = \frac{-x^2+12}{x^4}$$

33. **(a)** $f(x) = \dfrac{3x^3+6}{x^{2/3}}$

$$f'(x) = \frac{(x^{2/3})(9x^2)-(3x^3+6)(\frac{2}{3}x^{-1/3})}{(x^{2/3})^2} = \frac{9x^{8/3}-2x^{8/3}-4x^{-1/3}}{x^{4/3}} = \frac{7x^{8/3}-\frac{4}{x^{1/3}}}{x^{4/3}} = \frac{7x^3-4}{x^{5/3}}$$

(b) $f(x) = 3x^{7/3}+6x^{-2/3}$

$$f'(x) = 3\left(\frac{7}{3}x^{4/3}\right)+6\left(-\frac{2}{3}x^{-5/3}\right) = 7x^{4/3}-4x^{-5/3}$$

(c) The derivatives are equivalent.

35. $f(x) = \dfrac{u(x)}{v(x)}$

$$f'(x) = \lim_{h\to 0}\frac{f(x+h)-f(x)}{h} = \lim_{h\to 0}\frac{\frac{u(x+h)}{v(x+h)}-\frac{u(x)}{v(x)}}{h} = \lim_{h\to 0}\frac{u(x+h)v(x)-u(x)v(x+h)}{hv(x+h)v(x)}$$

$$= \lim_{h\to 0}\frac{u(x+h)v(x)-u(x)v(x)+u(x)v(x)-u(x)v(x+h)}{hv(x+h)v(x)}$$

$$= \lim_{h\to 0}\frac{v(x)[u(x+h)-u(x)]-u(x)[v(x+h)-v(x)]}{hv(x+h)v(x)}$$

$$= \lim_{h\to 0}\frac{v(x)\frac{u(x+h)-u(x)}{h}-u(x)\frac{v(x+h)-v(x)}{h}}{v(x+h)v(x)} = \frac{v(x)\cdot u'(x)-u(x)v'(x)}{[v(x)]^2}$$

37. Graph the numerical derivative of $f(x) = (x^2 - 2)(x^2 - \sqrt{2})$ for x ranging from -2 to 2. The derivative crosses the x-axis at 0 and at approximately -1.307 and 1.307.

39. $C(x) = \dfrac{3x+2}{x+4}$

$\overline{C}(x) = \dfrac{C(x)}{x} = \dfrac{3x+2}{x^2+4x}$

(a) $\overline{C}(10) = \dfrac{3(10)+2}{10^2+4(10)} = \dfrac{32}{140} \approx 0.2286$ hundreds of dollars or \$22.86 per unit

(b) $\overline{C}(20) = \dfrac{3(20)+2}{(20)^2+4(20)} = \dfrac{62}{480} \approx 0.1292$ hundreds of dollars or \$12.92 per unit

(c) $\overline{C}(x) = \dfrac{3x+2}{x^2+4x}$ per unit

(d) $\overline{C}'(x) = \dfrac{(x^2+4x)(3)-(3x+2)(2x+4)}{(x^2+4x)^2} = \dfrac{3x^2+12x-6x^2-12x-4x-8}{(x^2+4x)^2} = \dfrac{-3x^2-4x-8}{(x^2+4x)^2}$

41. $M(d) = \dfrac{100d^2}{3d^2+10}$

(a) $M'(d) = \dfrac{(3d^2+10)(200d)-(100d^2)(6d)}{(3d^2+10)^2} = \dfrac{600d^3+2000d-600d^3}{(3d^2+10)^2} = \dfrac{2000d}{(3d^2+10)^2}$

(b) $M'(2) = \dfrac{2000(2)}{[3(2)^2+10]^2} = \dfrac{4000}{484} \approx 8.3$

This means the new employee can assemble about 8.3 additional bicycles per day after 2 days of training.

$$M'(5) = \frac{2000(5)}{[3(5)^2+10]^2} = \frac{10{,}000}{7225} \approx 1.4$$

This means the new employee can assemble about 1.4 additional bicycles per day after 5 days of training.

43. $\overline{C}(x) = \dfrac{C(x)}{x}$

Let $u(x) = C(x)$, with $u'(x) = C'(x)$.
Let $v(x) = x$ with $v'(x) = 1$. Then, by the quotient rule,

$$\overline{C}(x) = \frac{v(x)\cdot u'(x) - u(x)\cdot v'(x)}{[v(x)]^2} = \frac{x\cdot C'(x) - C(x)\cdot 1}{x^2} = \frac{xC'(x)-C(x)}{x^2}$$

45. Let $C(t)$ be the cost as a function of time and $q(t)$ be the quantity as a function of time. Then $\overline{C}(t) = \frac{C(t)}{q(t)}$ is the revenue as a function of time. Let $t = t_1$ represent last month.

$$\overline{C}'(t) = \frac{q(t)C'(t) - C(t)q'(t)}{[g(t)]^2}$$

$$\overline{C}'(t_1) = \frac{q(t_1)C'(t_1) - C(t_1)q'(t_1)}{[g(t_1)]^2} = \frac{(12{,}500)(1200) - (27{,}000)(350)}{(12{,}500)^2} = 0.03552$$

The average cost is increasing at a rate of \$0.03552 per gallon per month.

47. $f(x) = \dfrac{Kx}{A+x}$

(a) $f'(x) = \dfrac{(A+x)K - Kx(1)}{(A+x)^2}$

$f'(x) = \dfrac{AK}{(A+x)^2}$

(b) $f'(A) = \dfrac{AK}{(A+A)^2} = \dfrac{AK}{4A^2} = \dfrac{K}{4A}$

49. $R(w) = \dfrac{30(w-4)}{w-1.5}$

(a) $R(5) = \dfrac{30(5-4)}{5-1.5} \approx 8.57$ min

(b) $R(7) = \dfrac{30(7-4)}{7-1.5} \approx 16.36$ min

(c)
$$\begin{aligned} R'(w) &= \frac{(w-1.5)(30) - 30(w-4)(1)}{(w-1.5)^2} \\ &= \frac{30w - 45 - 30w + 120}{(w-1.5)^2} \\ &= \frac{75}{(w-1.5)^2} \end{aligned}$$

$R'(5) = \dfrac{75}{(5-1.5)^2} \approx 6.12 \dfrac{\text{min}^2}{\text{kcal}}$

$R'(7) = \dfrac{75}{(7-1.5)^2} \approx 2.48 \dfrac{\text{min}^2}{\text{kcal}}$

51. $f(t) = \dfrac{90t}{99t - 90}$

$f'(t) = \dfrac{(99t-90)(90) - (90t)(99)}{(99t-90)^2} = \dfrac{-8100}{(99t-90)^2}$

(a) $f'(1) = \dfrac{-8100}{(99-90)^2} = \dfrac{-8100}{9^2} = \dfrac{-8100}{81} = -100$

(b)
$$\begin{aligned} f'(10) &= \frac{-8100}{[99(10) - 90]^2} \\ &= \frac{-8100}{(900)^2} \\ &= \frac{-8100}{810,000} \\ &= -\frac{1}{100} \text{ or } -0.01 \end{aligned}$$

4.3 The Chain Rule

In Exercises 1-5, $f(x) = 5x^2 - 2x$ and $g(x) = 8x + 3$.

1.
$$\begin{aligned} g(2) &= 8(2) + 3 = 19 \\ f[g(2)] &= f[19] \\ &= 5(19)^2 - 2(19) \\ &= 1805 - 38 = 1767 \end{aligned}$$

3.
$$\begin{aligned} f(2) &= 5(2)^2 - 2(2) \\ &= 20 - 4 = 16 \\ g[f(2)] &= g[16] \\ &= 8(16) + 3 \\ &= 128 + 3 = 131 \end{aligned}$$

5.
$$\begin{aligned} g(k) &= 8k + 3 \\ f[g(k)] &= f[8k+3] \\ &= 5(8k+3)^2 - 2(8k+3) \\ &= 5(64k^2 + 48k + 9) - 16k - 6 \\ &= 320k^2 + 224k + 39 \end{aligned}$$

7. $f(x) = \dfrac{x}{8} + 7$; $g(x) = 6x - 1$

$$\begin{aligned} f[g(x)] &= \frac{6x-1}{8} + 7 \\ &= \frac{6x-1}{8} + \frac{56}{8} \\ &= \frac{6x+55}{8} \\ g[f(x)] &= 6\left[\frac{x}{8} + 7\right] - 1 \\ &= \frac{6x}{8} + 42 - 1 \\ &= \frac{3x}{4} + 41 \\ &= \frac{3x}{4} + \frac{164}{4} \\ &= \frac{3x + 164}{4} \end{aligned}$$

9. $f(x) = \dfrac{1}{x}$; $g(x) = x^2$

$$\begin{aligned} f[g(x)] &= \frac{1}{x^2} \\ g[f(x)] &= \left(\frac{1}{x}\right)^2 \\ &= \frac{1}{x^2} \end{aligned}$$

11. $f(x) = \sqrt{x+2}$; $g(x) = 8x^2 - 6$

$$\begin{aligned} f[g(x)] &= \sqrt{(8x^2 - 6) + 2} \\ &= \sqrt{8x^2 - 4} \\ g[f(x)] &= 8(\sqrt{x+2})^2 - 6 \\ &= 8x + 16 - 6 \\ &= 8x + 10 \end{aligned}$$

13. $f(x) = \sqrt{x+1};\ g(x) = \dfrac{-1}{x}$

$$f[g(x)] = \sqrt{\frac{-1}{x} + 1} = \sqrt{\frac{x-1}{x}}$$

$$g[f(x)] = \frac{-1}{\sqrt{x+1}}$$

17. $y = (5 - x^2)^{3/5}$

If $f(x) = x^{3/5}$ and $g(x) = 5 - x^2$, then

$$y = f[g(x)] = (5 - x^2)^{3/5}.$$

19. $y = -\sqrt{13 + 7x}$

If $f(x) = -\sqrt{x}$ and
$g(x) = 13 + 7x$,

then $y = f[g(x)] = -\sqrt{13 + 7x}$.

21. $y = (x^2 + 5x)^{1/3} - 2(x^2 + 5x)^{2/3} + 7$

If $f(x) = x^{1/3} - 2x^{2/3} + 7$ and
$g(x) = x^2 + 5x$,

then

$$y = f[g(x)] = (x^2 + 5x)^{1/3} - 2(x^2 + 5x)^{2/3} + 7.$$

23. $y = (8x^4 - 5x^2 + 1)^4$

Let $f(x) = x^4$ and $g(x) = 8x^4 - 5x^2 + 1$. Then

$$(8x^4 - 5x^2 + 1)^4 = f[g(x)].$$

Use the alternate form of the chain rule.

$$\frac{dy}{dx} = f'[g(x)] \cdot g'(x)$$

$$f'(x) = 4x^3$$

$$f'[g(x)] = 4[g(x)]^3 = 4(8x^4 - 5x^2 + 1)^3$$

$$g'(x) = 32x^3 - 10x$$

$$\frac{dy}{dx} = 4(8x^4 - 5x^2 + 1)^3(32x^3 - 10x)$$

25. $f(x) = -2(12x^2 + 5)^{-6}$

Use the generalized power rule with
$u = 12x^2 + 5, n = -6$, and $u' = 24x$.

$$f'(x) = -2[-6(12x^2 + 5)^{-6-1} \cdot 24x] = -2[-144x(12x^2 + 5)^{-7}] = 288x(12x^2 + 5)^{-7}$$

27. $s(t) = 45(3t^3 - 8)^{3/2}$

Use the generalized power rule with
$u = 3t^3 - 8$, $n = \frac{3}{2}$, and $u' = 9t^2$.

$$s'(t) = 45\left[\frac{3}{2}(3t^3 - 8)^{1/2} \cdot 9t^2\right] = 45\left[\frac{27}{2}t^2(3t^3 - 8)^{1/2}\right] = \frac{1215}{2}t^2(3t^3 - 8)^{1/2}$$

29. $g(t) = -3\sqrt{7t^3 - 1} = -3(7t^3 - 1)^{1/2}$

Use generalized power rule with
$u = 7t^3 - 1$, $n = \frac{1}{2}$, and $u' = 21t^2$.

$$g'(t) = -3\left[\frac{1}{2}(7t^3 - 1)^{-1/2} \cdot 21t^2\right] = -3\left[\frac{21}{2}t^2(7t^3 - 1)^{-1/2}\right] = \frac{-63}{2}t^2 \cdot \frac{1}{(7t^3 - 1)^{1/2}} = \frac{-63t^2}{2\sqrt{7t^3 - 1}}$$

31. $m(t) = -6t(5t^4 - 1)^4$

Use the product rule and the power rule.

$$m'(t) = -6t[4(5t^4 - 1)^3 \cdot 20t^3] + (5t^4 - 1)^4(-6) = -480t^4(5t^4 - 1)^3 - 6(5t^4 - 1)^4 = -6(5t^4 - 1)^3[80t^4 + (5t^4 - 1)] = -6(5t^4 - 1)^3(85t^4 - 1)$$

33. $y = (3x^4 + 1)^4(x^3 + 4)$

Use the product rule and the power rule.

$$\frac{dy}{dx} = (3x^4 + 1)^4(3x^2) + (x^3 + 4)[4(3x^4 + 1)^3 \cdot 12x^3]$$
$$= 3x^2(3x^4 + 1)^4 + 48x^3(x^3 + 4)(3x^4 + 1)^3$$
$$= 3x^2(3x^4 + 1)^3[3x^4 + 1 + 16x(x^3 + 4)]$$
$$= 3x^2(3x^4 + 1)^3(3x^4 + 1 + 16x^4 + 64x)$$
$$= 3x^2(3x^4 + 1)^3(19x^4 + 64x + 1)$$

35. $q(y) = 4y^2(y^2 + 1)^{5/4}$

Use the product rule and the power rule.

$$q'(y) = 4y^2 \cdot \frac{5}{4}(y^2 + 1)^{1/4}(2y) + 8y(y^2 + 1)^{5/4}$$
$$= 10y^3(y^2 + 1)^{1/4} + 8y(y^2 + 1)^{5/4}$$
$$= 2y(y^2 + 1)^{1/4}[5y^2 + 4(y^2 + 1)^{4/4}]$$
$$= 2y(y^2 + 1)^{1/4}(9y^2 + 4)$$

37. $y = \dfrac{-5}{(2x^3+1)^2} = -5(2x^3+1)^{-2}$

$$\frac{dy}{dx} = -5[-2(2x^3+1)^{-3} \cdot 6x^2]$$
$$= -5[-12x^2(2x^3+1)^{-3}]$$
$$= 60x^2(2x^3+1)^{-3}$$
$$= \frac{60x^2}{(2x^3+1)^3}$$

39. $r(t) = \dfrac{(5t-6)^4}{3t^2+4}$

$r'(t)$
$$= \frac{(3t^2+4)[4(5t-6)^3 \cdot 5] - (5t-6)^4(6t)}{(3t^2+4)^2}$$
$$= \frac{20(3t^2+4)(5t-6)^3 - 6t(5t-6)^4}{(3t^2+4)^2}$$
$$= \frac{2(5t-6)^3[10(3t^2+4) - 3t(5t-6)]}{(3t^2+4)^2}$$
$$= \frac{2(5t-6)^3(30t^2+40-15t^2+18t)}{(3t^2+4)^2}$$
$$= \frac{2(5t-6)^3(15t^2+18t+40)}{(3t^2+4)^2}$$

41. $y = \dfrac{3x^2-x}{(2x-1)^5}$

$$\frac{dy}{dx} = \frac{(2x-1)^5(6x-1) - (3x^2-x)[5(2x-1)^4 \cdot 2]}{[(2x-1)^5]^2}$$
$$= \frac{(2x-1)^5(6x-1) - 10(3x^2-x)(2x-1)^4}{(2x-1)^{10}}$$
$$= \frac{(2x-1)^4[(2x-1)(6x-1) - 10(3x^2-x)]}{(2x-1)^{10}}$$
$$= \frac{12x^2-2x-6x+1-30x^2+10x}{(2x-1)^6}$$
$$= \frac{-18x^2+2x+1}{(2x-1)^6}$$

43. (a) $D_x(f[g(x)])$ at $x = 1$

$$= f'[g(1)] \cdot g'(1)$$
$$= f'(2) \cdot \left(\frac{2}{7}\right)$$
$$= -7\left(\frac{2}{7}\right)$$
$$= -2$$

(b) $D_x(f[g(x)])$ at $x = 2$

$$= f'[g(2)] \cdot g'(2)$$
$$= f'(3) \cdot \left(\frac{3}{7}\right)$$
$$= -8\left(\frac{3}{7}\right)$$
$$= -\frac{24}{7}$$

45. $f(x) = \sqrt{x^2+16}; x = 3$
$f(x) = (x^2+16)^{1/2}$

$$f'(x) = \frac{1}{2}(x^2+16)^{-1/2}(2x)$$
$$f'(x) = \frac{x}{\sqrt{x^2+16}}$$
$$f'(3) = \frac{3}{\sqrt{3^2+16}} = \frac{3}{5}$$
$$f(3) = \sqrt{3^2+16} = 5$$

We use $m = \frac{3}{5}$ and the point $P(3,5)$ in the point-slope form.

$$y - 5 = \frac{3}{5}(x-3)$$
$$y - 5 = \frac{3}{5}x - \frac{9}{5}$$
$$y = \frac{3}{5}x + \frac{16}{5}$$

47. $f(x) = x(x^2-4x+5)^4; x = 2$

$$f'(x) = x \cdot 4(x^2-4x+5)^3 \cdot (2x-4) + 1 \cdot (x^2-4x+5)^4$$
$$= (x^2-4x+5)^3 \cdot [4x(2x-4) + (x^2-4x+5)]$$
$$= (x^2-4x+5)^3(9x^2-20x+5)$$
$$f'(2) = (1)^3(1) = 1$$
$$f(2) = 2(1)^4 = 2$$

We use $m = 1$ and the point $P(2,2)$.

$$y - 2 = 1(x-2)$$
$$y - 2 = x - 2$$
$$y = x$$

49. $f(x) = \sqrt{x^3 - 6x^2 + 9x + 1}$

$$f(x) = (x^3 - 6x^2 + 9x + 1)^{1/2}$$

$$f'(x) = \frac{1}{2}(x^3 - 6x^2 + 9x + 1)^{-1/2} \cdot (3x^2 - 12x + 9)$$

$$f'(x) = \frac{3(x^2 - 4x + 3)}{2\sqrt{x^3 - 6x^2 + 9x + 1}}$$

If the tangent line is horizontal, its slope is zero and $f'(x) = 0$.

$$\frac{3(x^2 - 4x + 3)}{2\sqrt{x^3 - 6x^2 + 9x + 1}} = 0$$

$$3(x^2 - 4x + 3) = 0$$
$$3(x - 1)(x - 3) = 0$$
$$x = 1 \quad \text{or} \quad x = 3$$

The tangent line is horizontal at $x = 1$ and $x = 3$.

53. $D(p) = \frac{-p^2}{100} + 500$; $p(c) = 2c - 10$

The demand in terms of the cost is

$$D(c) = D[p(c)]$$
$$= \frac{-(2c - 10)^2}{100} + 500$$
$$= \frac{-4(c - 5)^2}{100} + 500$$
$$= \frac{-c^2 + 10c - 25}{25} + 500$$
$$= \frac{-c^2 + 10c - 25 + 12{,}500}{25}$$
$$= \frac{-c^2 + 10c + 12{,}475}{25}.$$

55. $A = 1500\left(1 + \frac{r}{36{,}500}\right)^{1825}$

$\frac{dA}{dr}$ is the rate of change of A with respect to r.

$$\frac{dA}{dr} = 1500(1825)\left(1 + \frac{r}{36{,}500}\right)^{1824}\left(\frac{1}{36{,}500}\right)$$
$$= 75\left(1 + \frac{r}{36{,}500}\right)^{1824}$$

(a) For $r = 6\%$,

$$\frac{dA}{dr} = 75\left(1 + \frac{6}{36{,}500}\right)^{1824} = \$101.22.$$

(b) For $r = 8\%$,

$$\frac{dA}{dr} = 75\left(1 + \frac{8}{36{,}500}\right)^{1824} = \$111.86.$$

(c) For $r = 9\%$,

$$\frac{dA}{dr} = 75\left(1 + \frac{9}{36{,}500}\right)^{1824} = \$117.59.$$

57. $V = \frac{60{,}000}{1 + 0.3t + 0.1t^2}$

The rate of change of the value is

$$V'(t) = \frac{(1 + 0.3t + 0.1t^2)(0) - 60{,}000(0.3 + 0.2t)}{(1 + 0.3t + 0.1t^2)^2}$$
$$= \frac{-60{,}000(0.3 + 0.2t)}{(1 + 0.3t + 0.1t^2)^2}.$$

(a) 2 years after purchase, the rate of change in the value is

$$V'(2) = \frac{-60{,}000[0.3 + 0.2(2)]}{[1 + 0.3(2) + 0.1(2)^2]^2}$$
$$= \frac{-60{,}000(0.3 + 0.4)}{(1 + 0.6 + 0.4)^2}$$
$$= \frac{-42{,}000}{4}$$
$$= -\$10{,}500.$$

(b) 4 years after purchase, the rate of change in the value is

$$V'(4) = \frac{-60{,}000[0.3 + 0.2(4)]}{[1 + 0.3(4) + 0.1(4)^2]^2}$$
$$= \frac{-66{,}000}{14.44}$$
$$= -\$4570.64.$$

59. $P(x) = 2x^2 + 1$; $x = f(a) = 3a + 2$

$$P[f(a)] = 2(3a + 2)^2 + 1$$
$$= 2(9a^2 + 12a + 4) + 1$$
$$= 18a^2 + 24a + 9$$

61. **(a)** $r(t) = 2t$; $A(r) = \pi r^2$

$$A[r(t)] = \pi(2t)^2$$
$$= 4\pi t^2$$

$A = 4\pi t^2$ gives the area of the pollution in terms of the time since the pollutants were first emitted.

(b) $D_t A[r(t)] = 8\pi t$

$$D_t A[r(4)] = 8\pi(4) = 32\pi$$

At 12 P.M., the area of pollution is changing at the rate of 32π mi^2/hr.

63. $C(t) = \frac{1}{2}(2t+1)^{-1/2}$

$$C'(t) = \frac{1}{2}\left(-\frac{1}{2}\right)(2t+1)^{-3/2}(2)$$
$$= -\frac{1}{2}(2t+1)^{-3/2}$$

(a) $C'(0) = -\frac{1}{2}[2(0)+1]^{-3/2}$
$$= -\frac{1}{2}$$
$$= -0.5$$

(b) $C'(4) = -\frac{1}{2}[2(4)+1]^{-3/2}$
$$= -\frac{1}{2}(9)^{-3/2}$$
$$= \frac{-1}{2}\cdot\frac{1}{(\sqrt{9})^3}$$
$$= -\frac{1}{54}$$
$$\approx -0.02$$

(c) $C'(7.5) = -\frac{1}{2}[2(7.5)+1]^{-3/2}$
$$= -\frac{1}{2}(16)^{-3/2}$$
$$= -\frac{1}{2}\left(\frac{1}{(\sqrt{16})^3}\right)$$
$$= -\frac{1}{128}$$
$$\approx -0.008$$

(d) C is always decreasing because

$$C' = -\tfrac{1}{2}(2t+1)^{-3/2}$$

is always negative for $t \geq 0$.
(The amount of calcium in the bloodstream will continue to decrease over time.)

65. $V(r) = \frac{4}{3}\pi r^3, S(r) = 4\pi r^2, r(t) = 6 - \frac{3}{17}t$

(a) $r(t) = 0$ when $6 - \frac{3}{17}t = 0$;

$$t = \frac{17(6)}{3} = 34 \text{ min.}$$

(b) $\frac{dV}{dr} = 4\pi r^2, \frac{dS}{dr} = 8\pi r, \frac{dr}{dt} = -\frac{3}{17}$

$$\frac{dV}{dt} = \frac{dV}{dr}\cdot\frac{dr}{dt} = -\frac{12}{17}\pi r^2$$
$$= -\frac{12}{17}\pi\left(6 - \frac{3}{17}t\right)^2$$
$$\frac{dS}{dt} = \frac{dS}{dr}\cdot\frac{dr}{dt} = -\frac{24}{17}\pi r$$
$$= -\frac{24}{17}\pi\left(6 - \frac{3}{17}t\right)$$

When $t = 17$,

$$\frac{dV}{dt} = -\frac{12}{17}\pi\left[6 - \frac{3}{17}(17)\right]^2$$
$$= -\frac{108}{17}\pi \text{ mm}^3/\text{min}$$
$$\frac{dS}{dt} = -\frac{24}{17}\pi\left[6 - \frac{3}{17}(17)\right]$$
$$= -\frac{72}{17}\pi \text{ mm}^2/\text{min}$$

At $t = 17$ minutes, the volume is decreasing by $\frac{108}{17}\pi$ mm^3 per minute and the surface area is decreasing by $\frac{72}{17}\pi$ mm^2 per minute.

4.4 Derivatives of Exponential Functions

1.
$$y = e^{4x}$$
Let $g(x) = 4x$,
with $g'(x) = 4$.
$$\frac{dy}{dx} = 4e^{4x}$$

3. $y = -8e^{3x}$
$$\frac{dy}{dx} = -8(3e^{3x}) = -24e^{3x}$$

5. $y = -16e^{2x+1}$
$$g(x) = 2x+1$$
$$g'(x) = 2$$
$$\frac{dy}{dx} = -16(2e^{2x+1}) = -32e^{2x+1}$$

7. $y = e^{x^2}$
$$g(x) = x^2$$
$$g'(x) = 2x$$
$$\frac{dy}{dx} = 2xe^{x^2}$$

9. $y = 3e^{2x^2}$

$g(x) = 2x^2$

$g'(x) = 4x$

$$\frac{dy}{dx} = 3(4xe^{2x^2}) = 12xe^{2x^2}$$

11. $y = 4e^{2x^2-4}$

$g(x) = 2x^2 - 4$

$g'(x) = 4x$

$$\frac{dy}{dx} = 4[(4x)e^{2x^2-4}] = 16xe^{2x^2-4}$$

13. $y = xe^x$

Use the product rule.

$$\frac{dy}{dx} = xe^x + e^x \cdot 1 = e^x(x+1)$$

15. $y = (x+3)^2 e^{4x}$

Use the product rule.

$$\begin{aligned}\frac{dy}{dx} &= (x+3)^2(4)e^{4x} + e^{4x} \cdot 2(x+3)\\ &= 4(x+3)^2e^{4x} + 2(x+3)e^{4x}\\ &= 2(x+3)e^{4x}[2(x+3)+1]\\ &= 2(x+3)(2x+7)e^{4x}\end{aligned}$$

17. $y = \dfrac{x^2}{e^x}$

Use the quotient rule.

$$\frac{dy}{dx} = \frac{e^x(2x) - x^2e^x}{(e^x)^2} = \frac{xe^x(2-x)}{e^{2x}} = \frac{x(2-x)}{e^x}$$

19. $y = \dfrac{e^x + e^{-x}}{x}$

$$\frac{dy}{dx} = \frac{x(e^x - e^{-x}) - (e^x + e^{-x})}{x^2}$$

21. $p = \dfrac{10{,}000}{9 + 4e^{-0.2t}}$

$$\frac{dp}{dt} = \frac{(9+4e^{-0.2t}) \cdot 0 - 10{,}000[0 + 4(-0.2)e^{-0.2t}]}{(9+4e^{-0.2t})^2} = \frac{8000e^{-0.2t}}{(9+4e^{-0.2t})^2}$$

23. $f(z) = (2z + e^{-z^2})^2$

$$\begin{aligned}f'(z) &= 2(2z + e^{-z^2})^1(2 - 2ze^{-z^2})\\ &= 4(2z + e^{-z^2})(1 - ze^{-z^2})\end{aligned}$$

25. $y = 4^{-5x+2}$

Let $g(x) = -5x + 2$, with $g'(x) = -5$. Then

$$\frac{dy}{dx} = (\ln 4)(4^{-5x+2}) \cdot (-5) = -5(\ln 4)4^{-5x+2}$$

27. $y = -10^{3x^2-4}$

Let $g(x) = 3x^2 - 4$, with $g'(x) = 6x$.

$$\frac{dy}{dx} = -(\ln 10)10^{3x^2-4} \cdot 6x = -6x(10^{3x^2-4}) \ln 10$$

29. $s = 5 \cdot 2^{\sqrt{t-2}}$

Let $g(t) = t - 2$, with $g'(t) = \frac{1}{2\sqrt{t-2}}$. Then

$$\frac{ds}{dt} = 5(\ln 2)(2^{\sqrt{t-2}}) \cdot \frac{1}{2\sqrt{t-2}} = \frac{(5\ln 2)2^{\sqrt{t-2}}}{2\sqrt{t-2}}$$

31. $y = \dfrac{t^2e^{2t}}{t + e^{3t}}$

Use the quotient rule and product rule.

$$\begin{aligned}\frac{dy}{dt} &= \frac{(t+e^{3t})(2te^{2t} + t^2 \cdot 2e^{2t}) - t^2e^{2t}(1+3e^{3t})}{(t+e^{3t})^2}\\ &= \frac{(t+e^{3t})(2te^{2t} + 2t^2e^{2t}) - t^2e^{2t}(1+3e^{3t})}{(t+e^{3t})^2}\\ &= \frac{(2t^2e^{2t} + 2t^3e^{2t} + 2te^{5t} + 2t^2e^{5t}) - (t^2e^{2t} + 3t^2e^{5t})}{(t+e^{3t})^2}\\ &= \frac{t^2e^{2t} + 2t^3e^{2t} + 2te^{5t} - t^2e^{5t}}{(t+e^{3t})^2}\\ &= \frac{(2t^3+t^2)e^{2t} + (2t-t^2)e^{5t}}{(t+e^{3t})^2}\end{aligned}$$

33. $f(x) = e^{x^2/(x^3+2)}$

Let $g(x) = \dfrac{x^2}{x^3+2}$

$$\begin{aligned}g'(x) &= \frac{(x^3+2)(2x) - x^2(3x^2)}{(x^3+2)^2}\\ &= \frac{2x^4 + 4x - 3x^4}{(x^3+2)^2}\\ &= \frac{4x - x^4}{(x^3+2)^2}\\ &= \frac{x(4-x^3)}{(x^3+2)^2}\end{aligned}$$

$$f'(x) = e^{x^2/(x^3+2)} \cdot \left[\frac{x(4-x^3)}{(x^3+2)^2}\right]$$
$$= \frac{x(4-x^3)e^{x^2/(x^3+2)}}{(x^3+2)^2}$$

35. Graph
$$y = \frac{e^{x+0.0001} - e^x}{0.0001}$$
on a graphing calculator. A good choice for the viewing window is $[-1, 4]$ by $[-1, 16]$ with Xscl = 1, Yscl = 2.

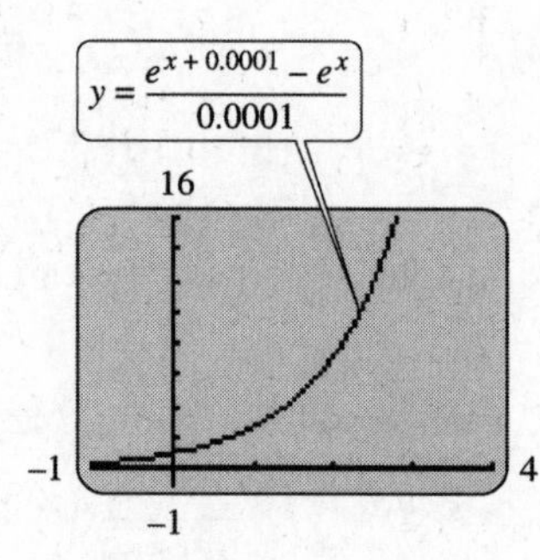

If we graph $y = e^x$ on the same screen, we see that the two graphs coincide. They are close enough to being identical that they are indistinguishable. By the definition of the derivative, if $f(x) = e^x$,
$$f'(x) = \lim_{h\to 0} \frac{f(x+h) - f(x)}{h} = \lim_{h\to 0} \frac{e^{x+h} - e^x}{h},$$
and $h = 0.0001$ is very close to 0.
Comparing the two graphs provides graphical evidence that
$$f'(x) = e^x.$$

37. $S(t) = 100 - 90e^{-0.3t}$
$S'(t) = -90(-0.3)e^{-0.3t}$
$= 27e^{-0.3t}$

(a) $S'(1) = 27e^{-0.3(1)}$
$= 27e^{-0.3}$
≈ 20

(b) $S'(5) = 27e^{-0.3(5)}$
$= 27e^{-1.5}$
≈ 6

(c) As time goes on, the rate of change of sales is decreasing.

(d) $S'(t) = 27e^{-0.3t} \neq 0$, but
$$\lim_{t\to\infty} S'(t) = \lim_{t\to\infty} 27e^{-0.3t} = 0.$$
Although the rate of change of sales never equals zero, it gets closer and closer to zero as t increases.

39. $A(t) = 10t^2\, 2^{-t}$
$A'(t) = 10t^2(\ln 2)2^{-t}(-1) + 20t\, 2^{-t}$
$A'(t) = 10t\, 2^{-t}(-t \ln 2 + 2)$

(a) $A'(2) = 10(2)(2^{-2})(-2 \ln 2 + 2)$
≈ 3.07

(b) $A'(4) = 10(4)(2^{-4})(-4 \ln 2 + 2)$
≈ -1.93

(c) Public awareness increased at first and then decreased.

41. $y = 100e^{-0.03045t}$

(a) For $t = 0$,
$$y = 100e^{-0.03045(0)}$$
$$= 100e^0$$
$$= 100\%.$$

(b) For $t = 2$,
$$y = 100e^{-0.03045(2)}$$
$$= 100e^{-0.0609}$$
$$\approx 94\%.$$

(c) For $t = 4$,
$$y = 100e^{-0.03045(4)}$$
$$\approx 89\%.$$

(d) For $t = 6$,
$$y' = 100e^{-0.03045(6)}$$
$$\approx 83\%.$$

(e) $y' = 100(-0.03045)e^{-0.03045t}$
$= -3.045e^{-0.03045t}$

For $t = 0$,
$$y' = -3.045e^{-0.03045(0)}$$
$$= -3.045.$$

(f) For $t = 2$,
$$y' = -3.045e^{-0.03045(2)}$$
$$\approx -2.865.$$

(g) The percent of these cars on the road is decreasing, but at a slower rate as they age.

43. **(a)** $G_0 = 0.7, m = 10.3$

$$G(t) = \frac{10.3}{1 + \left(\frac{10.3}{0.7} - 1\right) e^{-0.03036(10.3)t}} = \frac{10.3}{1 + 13.71e^{-0.3127t}}$$

(b) $G'(t) = -10.3(1 + 13.71e^{-0.3127t})^{-2} \cdot 13.71e^{-0.3127t}(-0.3127)$

$$= \frac{44.1573051e^{-0.3127t}}{(1 + 13.71e^{-0.3127t})^2}$$

1990 when $t = 5$:

$$G(5) = \frac{10.3}{1 + 13.71e^{-0.3127(5)}} \approx 2.66$$

$$G'(5) = \frac{44.1573051e^{-0.3127(5)}}{[1 + 13.71e^{-0.3127(5)}]^2} \approx 0.617$$

The population in 1990 is 2.66 million and the growth rate is 0.617 million per year.

(c) 1995 when $t = 10$:

$$G(10) = \frac{10.3}{1 + 13.71e^{-0.3127(10)}} \approx 6.43$$

$$G'(10) = \frac{44.1573051e^{-0.3127(10)}}{[1 + 13.71e^{-0.3127(10)}]^2} \approx 0.755$$

The population in 1995 is 6.43 million and the growth rate is 0.755 million per year.

(d) 2000 when $t = 15$:

$$G(15) = \frac{10.3}{1 + 13.71e^{-0.3127(15)}} \approx 9.15$$

$$G'(15) = \frac{44.1573051e^{-0.3127(15)}}{[1 + 13.71e^{-0.3127(15)}]^2} \approx 0.320$$

The population in 2000 is 9.15 million and the growth rate is 0.320 million per year.

(e) The rate of growth over time increases for a while and then gradually decreases to 0.

45. $p(t) = 9.865(1.025)^t$
$p'(t) = 9.865(\ln 1.025)(1.025)^t$

(a) For 1998, $t = 18$.

$$p'(18) = 9.865(\ln 1.025)(1.025)^{18} = 0.380$$

The instantaneous rate of growth is 380,000 people per year.

(b) For 2006, $t = 26$.

$$p'(26) = 9.865(\ln 1.025)(1.025)^{26} = 0.463$$

The instantaneous rate of growth is 463,000 people per year.

47. $G(t) = \dfrac{mG_o}{G_o + (m - G_o)e^{-kmt}}$, where $G_o = 400$; $m = 5200$; and $k = 0.0001$.

(a) $$G(t) = \frac{(5200)(400)}{400 + (5200 - 400)e^{(-0.0001)(5200)t}} = \frac{(400)(5200)}{400 + 4800e^{-0.52t}} = \frac{5200}{1 + 12e^{-0.52t}}$$

(b) $G(t) = 5200(1 + 12e^{-0.52t})^{-1}$
$G'(t) = -5200(1 + 12e^{-0.52t})^{-2}(-6.24e^{-0.52t})$

$$= \frac{32{,}448e^{-0.52t}}{(1 + 12e^{-0.52t})^2}$$

$$G(1) = \frac{5200}{1 + 12e^{-0.52}} \approx 639$$

$$G'(1) = \frac{32{,}448e^{-0.52}}{(1 + 12e^{-0.52})^2} \approx 292$$

(c) $$G(4) = \frac{5200}{1 + 12e^{-2.08}} \approx 2081$$

$$G'(4) = \frac{32{,}448e^{-2.08}}{(1 + 12e^{-2.08})^2} \approx 649$$

(d) $$G(10) = \frac{5200}{1 + 12e^{-5.2}} \approx 4877$$

$$G'(10) = \frac{34{,}448e^{-5.2}}{(1 + 12e^{-5.2})^2} \approx 167$$

(e) It increases for a while and then gradually decreases to 0.

49. $V(t) = 1100[1023e^{-0.02415t} + 1]^{-4}$

(a) $V(240) = 1100[1023e^{-0.02415(240)} + 1]^{-4}$
$\approx 3.857 \text{ cm}^3$

(b) $V = \frac{4}{3}\pi r^3$, so $r(V) = \sqrt[3]{\frac{3V}{4\pi}}$

$$r(3.857) = \sqrt[3]{\frac{3(3.857)}{4\pi}} \approx 0.973 \text{ cm}$$

(c) $V(t) = 1100[1023e^{-0.02415t} + 1]^{-4} = 0.5$

$$[1023e^{-0.02415t} + 1]^{-4} = \frac{1}{2200}$$

$$(1023e^{-0.02415t} + 1)^4 = 2200$$

$$1023e^{-0.02415t} + 1 = 2200^{1/4}$$

$$1023e^{-0.02415t} = 2200^{1/4} - 1$$

$$e^{-0.02415t} = \frac{2200^{1/4} - 1}{1023}$$

$$-0.02415t = \ln\left(\frac{2200^{1/4} - 1}{1023}\right)$$

$$t = \frac{1}{-0.02415} \ln\left(\frac{2200^{1/4} - 1}{1023}\right) \approx 214 \text{ months}$$

The tumor has been growing for almost 18 years.

(d) As t goes to infinity, $e^{-0.02415t}$ goes to zero, and $V(t) = 1100[1023e^{-0.02415t}+1]^{-4}$ goes to 1100 cm^3, which corresponds to a sphere with a radius of $\sqrt[3]{\frac{3(1100)}{4\pi}} \approx 6.4$ cm. It makes sense that a tumor growing in a person's body reaches a maximum volume of this size.

(e) By the chain rule,

$$\frac{dV}{dt} = 1100(-4)[1023e^{-0.02415t} + 1]^{-5} \cdot (1023)(e^{-0.02415t})(-0.02415)$$
$$= 108{,}703.98[1023e^{-0.02415t} + 1]^{-5}e^{-0.02415t}$$

At $t = 240$, $\frac{dV}{dt} \approx 0.282$.

At 240 months old, the tumor is increasing in volume at the instantaneous rate of 0.282 cm^3/month.

51. $URR = 1 - \left\{(0.96)^{0.14t-1} + \frac{8t}{126t+900}[1-(0.96)^{0.14t-1}]\right\}$

(a) When $t = 180$, $URR \approx 0.589$. The patient has not received adequate dialysis.

(b) When $t = 240$, $URR \approx 0.690$. The patient has received adequate dialysis.

(c) $D_t URR$

$$= -\Big\{(\ln 0.96)(0.96)^{0.14t-1}(0.14) + \frac{8t}{126t + 900}(-\ln 0.96)(0.96)^{0.14t-1}(0.14) + \frac{(126t + 900)(8) - 8t(126)}{(126t + 900)^2}[1 - (0.96)^{0.14t-1}]\Big\}$$

When $t = 240$, $D_t URR \approx 0.001$. The URR is increasing instantaneously by 0.001 units per minute when $t = 240$ minutes.
The rate of increase is low, and it will take a significant increase in time on dialysis to increase URR significantly.

53. $M(t) = 3102e^{-e^{-0.022(t-56)}}$

(a) $M(200) = 3102e^{-e^{-0.022(200-56)}} \approx 2974.15$ grams, or about 3 kilograms.

(b) As t gets very large, $-e^{-0.022(t-56)}$ goes to zero, $e^{-e^{-0.022(t-56)}}$ goes to 1, and $M(t)$ approaches 3102 grams or about 3.1 kilograms.

(c) 80% of 3102 is 2481.6.

$$2481.6 = 3102e^{-e^{-0.022(t-56)}}$$

$$-\ln\frac{2481.6}{3102} = e^{-0.022(t-56)}$$

$$\ln\left(\ln\frac{3102}{2481.6}\right) = -0.022(t - 56)$$

$$t = -\frac{1}{0.022}\ln\left(\ln\frac{3102}{2481.6}\right) + 56$$

$$\approx 124 \text{ days}$$

(d) $D_t M(t) = 3102e^{-e^{-0.022(t-56)}} D_t(-e^{-0.022(t-56)})$
$= 3102e^{-e^{-.0022(t-56)}}(-e^{-0.022(t-56)})(-0.02$
$= 68.244e^{-e^{-0.022(t-56)}}e^{-0.022(t-56)}$

When $t = 200$, $D_t M(t) \approx 2.75$ g/day.

(e)

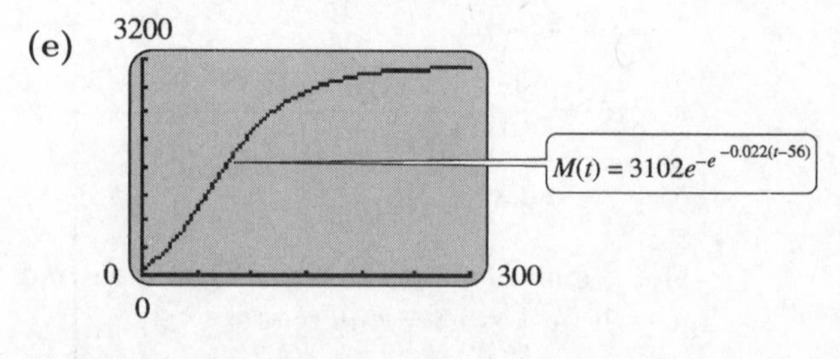

Growth is initially rapid, then tapers off.

(f)

Day	*Weight*	*Rate*
50	991	24.88
100	2122	17.73
150	2734	7.60
200	2974	2.75
250	3059	0.94
300	3088	0.32

55. $W_1(t) = 509.7(1 - 0.941e^{-0.00181t})$
$W_2(t) = 498.4(1 - 0.889e^{-0.00219t})^{1.25}$

(a) Both W_1 and W_2 are strictly increasing functions, so they approach their maximum values as t approaches ∞.

$$\lim_{t\to\infty} W_1(t) = \lim_{t\to\infty} 509.7(1 - 0.941e^{-0.00181t}) = 509.7(1-0) = 509.7$$

$$\lim_{t\to\infty} W_2(t) = \lim_{t\to\infty} 498.4(1 - 0.889e^{-0.00219t})^{1.25} = 498.4(1-0)^{1.25} = 498.4$$

So, the maximum values of W_1 and W_2 are 509.7 kg and 498.4 kg respectively.

(b) $0.9(509.7) = 509.7(1 - 0.941e^{-0.00181t})$

$$0.9 = 1 - 0.941e^{-0.00181t}$$

$$\frac{0.1}{0.941} = e^{-0.00181t}$$

$$1239 \approx t$$

$$0.9(498.4) = 498.4(1 - 0.889e^{-0.00219t})^{1.25}$$

$$0.9 = (1 - 0.889e^{-0.00219t})^{1.25}$$

$$\frac{1-0.9^{0.8}}{0.889} = e^{-0.00219t}$$

$$1095 \approx t$$

Respectively, it will take the average beef cow about 1239 days or 1095 days to reach 90% of its maximum.

(c) $W_1'(t) = (509.7)(-0.941)(-0.00181)e^{-0.00181t}$

$$\approx 0.868126e^{-0.00181t}$$

$$W_1'(750) \approx 0.868126e^{-0.00181(750)} \approx 0.22 \text{ kg/day}$$

$$W_2'(t) = (498.4)(1.25)(1 - 0.889e^{-0.00219t})^{0.25} \cdot (-0.889)(-0.00219)e^{-0.00219t}$$

$$\approx 1.21292e^{-0.00219t}(1 - 0.889e^{-0.00219t})^{0.25}$$

$$W_2'(750) \approx 1.12192e^{-0.00219(750)} \cdot (1 - 0.889e^{-0.00219(750)})^{0.25} \approx 0.22 \text{ kg/day}$$

Both functions yield a rate of change of about 0.22 kg per day.

(d) Looking at the graph, the growth patterns of the two functions are very similar.

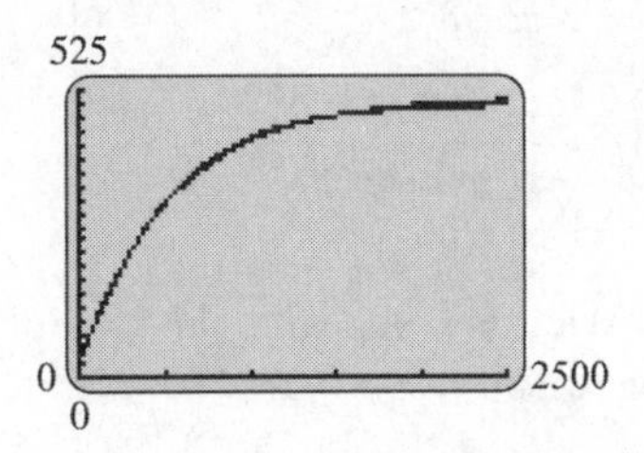

(e) The graphs of the rates of change of the two functions are also very similar.

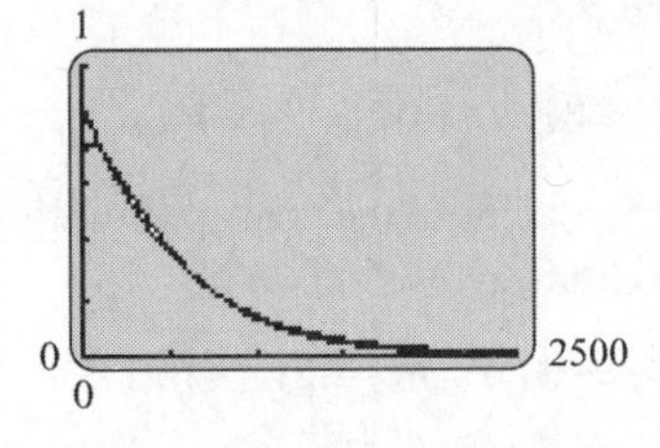

57. (a) $G_0 = 0.00369, m = 1, k = 3.5$

$$G(t) = \frac{1}{1 + \left(\frac{1}{0.00369} - 1\right)e^{-3.5(1)t}} = \frac{1}{1 + 270e^{-3.5t}}$$

(b) $G'(t) = -(1 + 270e^{-3.5t})^{-2} \cdot 270e^{-3.5t}(-3.5)$

$$= \frac{945e^{-3.5t}}{(1 + 270e^{-3.5t})^2}$$

$$G(1) = \frac{1}{1 + 270e^{-3.5(1)}} \approx 0.109$$

$$G'(1) = \frac{945e^{-3.5(1)}}{[1 + 270e^{-3.5(1)}]^2} \approx 0.341$$

The proportion is 0.109 and the rate of growth is 0.341 per century.

(c) $G(2) = \dfrac{1}{1 + 270e^{-3.5(2)}} \approx 0.802$

$$G'(2) = \frac{945e^{-3.5(2)}}{[1 + 270e^{-3.5(2)}]^2} \approx 0.555$$

The proportion is 0.802 and the rate of growth is 0.555 per century.

(d) $$G(3) = \frac{1}{1 + 270e^{-3.5(3)}} \approx 0.993$$

$$G'(3) = \frac{945e^{-3.5(2)}}{[1 + 270e^{-3.5(2)}]^2} \approx 0.0256$$

The proportion is 0.993 and the rate of growth is 0.0256 per century.

(e) The rate of growth increases for a while and then gradually decreases to 0.

59. $P(t) = 37.79(1.012)^t$

(a) $P(10) = 37.79(1.021)^{10} \approx 46.5$

So, the U.S. Latino-American population in 2010 was approximately 46,500,000.

(b) $$P'(t) = 37.79(\ln 1.021)(1.021)^t \approx 0.7854(1.021)^t$$

$$P'(10) = 0.7854(1.021)^{10} \approx 0.967$$

The Latino-American population was increasing at the rate of 0.967 million/year at the end of the year 2010.

61. $Q(t) = CV(1 - e^{-t/RC})$

(a) $$I_c = \frac{dQ}{dt} = CV\left[0 - e^{-t/RC}\left(-\frac{1}{RC}\right)\right] = CV\left(\frac{1}{RC}\right)e^{-t/RC} = \frac{V}{R}e^{-t/RC}$$

(b) When $C = 10^{-5}$ farads, $R = 10^7$ ohms, and $V = 10$ volts, after 200 seconds

$I_c = \frac{10}{10^7}e^{-200/(10^7 \cdot 10^{-5})} \approx 1.35 \times 10^{-7}$ amps

63. $T(h) = 80e^{-0.000065h}$

$$\frac{dT}{dt} = 80e^{-0.000065h}\left(-0.000065\frac{dh}{dt}\right) = -0.0052e^{-0.000065h}\frac{dh}{dt}$$

If $h = 1000$ and $\frac{dh}{dt} = 800$, then

$$\frac{dT}{dt} = -0.0052e^{-0.000065(1000)}(800) \approx -3.90$$

The temperature is decreasing at 3.90 degrees/hr.

4.5 Derivatives of Logarithmic Functions

1. $y = \ln(8x)$

$$\frac{dy}{dx} = \frac{d}{dx}(\ln 8x) = \frac{d}{dx}(\ln 8 + \ln x) = \frac{d}{dx}(\ln 8) + \frac{d}{dx}(\ln x) = 0 + \frac{1}{x} = \frac{1}{x}$$

3. $y = \ln(8 - 3x)$

$g(x) = 8 - 3x$

$g'(x) = -3$

$$\frac{dy}{dx} = \frac{g'(x)}{g(x)} = \frac{-3}{8 - 3x} \text{ or } \frac{3}{3x - 8}$$

5. $y = \ln|4x^2 - 9x|$

$g(x) = 4x^2 - 9x$

$g'(x) = 8x - 9$

$$\frac{dy}{dx} = \frac{g'(x)}{g(x)} = \frac{8x - 9}{4x^2 - 9x}$$

7. $y = \ln\sqrt{x + 5}$

$g(x) = \sqrt{x + 5} = (x + 5)^{1/2}$

$g'(x) = \frac{1}{2}(x + 5)^{-1/2}$

$$\frac{dy}{dx} = \frac{\frac{1}{2}(x + 5)^{-1/2}}{(x + 5)^{1/2}} = \frac{1}{2(x + 5)}$$

9. $y = \ln (x^4 + 5x^2)^{3/2}$

$= \frac{3}{2} \ln (x^4 + 5x^2)$

$\frac{dy}{dx} = \frac{3}{2} D_x [\ln (x^4 + 5x^2)]$

$g(x) = x^4 + 5x^2$

$g'(x) = 4x^3 + 10x$

$\frac{dy}{dx} = \frac{3}{2}\left(\frac{4x^3 + 10x}{x^4 + 5x^2}\right)$

$= \frac{3}{2}\left[\frac{2x(2x^2+5)}{x^2(x^2+5)}\right]$

$= \frac{3(2x^2+5)}{x(x^2+5)}$

11. $y = -5x \ln(3x+2)$

Use the product rule.

$\frac{dy}{dx} = -5x \left[\frac{d}{dx} \ln(3x+2)\right]$

$+ \ln(3x+2)\left[\frac{d}{dx}(-5x)\right]$

$= -5x\left(\frac{3}{3x+2}\right) + [\ln(3x+2)](-5)$

$= -\frac{15x}{3x+2} - 5\ln(3x+2)$

13. $s = t^2 \ln |t|$

$\frac{ds}{dt} = t^2 \cdot \frac{1}{t} + 2t \ln |t|$

$= t + 2t \ln |t|$

$= t(1 + 2 \ln |t|)$

15. $y = \frac{2 \ln (x+3)}{x^2}$

Use the quotient rule.

$\frac{dy}{dx} = \frac{x^2\left(\frac{2}{x+3}\right) - 2 \ln (x+3) \cdot 2x}{(x^2)^2}$

$= \frac{\frac{2x^2}{x+3} - 4x \ln (x+3)}{x^4}$

$= \frac{2x^2 - 4x(x+3) \ln (x+3)}{x^4(x+3)}$

$= \frac{x[2x - 4(x+3) \ln (x+3)]}{x^4(x+3)}$

$= \frac{2x - 4(x+3) \ln (x+3)}{x^3(x+3)}$

17. $y = \frac{\ln x}{4x+7}$

Use the quotient rule.

$\frac{dy}{dx} = \frac{(4x+7)\left(\frac{1}{x}\right) - (\ln x)(4)}{(4x+7)^2}$

$= \frac{\frac{4x+7}{x} - 4 \ln x}{(4x+7)^2}$

$= \frac{4x + 7 - 4x \ln x}{x(4x+7)^2}$

19. $y = \frac{3x^2}{\ln x}$

$\frac{dy}{dx} = \frac{(\ln x)(6x) - 3x^2\left(\frac{1}{x}\right)}{(\ln x)^2}$

$= \frac{6x \ln x - 3x}{(\ln x)^2}$

21. $y = (\ln |x+1|)^4$

$\frac{dy}{dx} = 4(\ln |x+1|^3\left(\frac{1}{x+1}\right)$

$= \frac{4(\ln |x+1|)^3}{x+1}$

23. $y = \ln |\ln x|$

$g(x) = \ln x$

$g'(x) = \frac{1}{x}$

$\frac{dy}{dx} = \frac{g'(x)}{g(x)}$

$= \frac{\frac{1}{x}}{\ln x}$

$= \frac{1}{x \ln x}$

25. $y = e^{x^2} \ln x,\ x > 0$

$\frac{dy}{dx} = e^{x^2}\left(\frac{1}{x}\right) + (\ln x)(2x)e^{x^2}$

$= \frac{e^{x^2}}{x} + 2xe^{x^2} \ln x$

27. $y = \frac{e^x}{\ln x},\ x > 0$

Use the quotient rule.

$\frac{dy}{dx} = \frac{(\ln x)e^x - e^x\left(\frac{1}{x}\right)}{(\ln x)^2} \cdot \frac{x}{x}$

$= \frac{xe^x \ln x - e^x}{x (\ln x)^2}$

29. $g(z) = (e^{2z} + \ln\ z)^3$

$$g'(z) = 3(e^{2z} + \ln\ z)^2\left(e^{2z}\cdot 2 + \frac{1}{z}\right)$$
$$= 3(e^{2z} + \ln\ z)^2\left(\frac{2ze^{2z}+1}{z}\right)$$

31. $y = \log(4x - 3)$

$$g(x) = 4x - 3$$
$$g'(x) = 4$$
$$\frac{dy}{dx} = \frac{1}{\ln 10}\cdot\frac{4}{4x-3}$$
$$= \frac{4}{(\ln 10)(4x-3)}$$

33. $y = \log\ |3x|$

$g(x) = 3x$ and $g'(x) = 3$.

$$\frac{dy}{dx} = \frac{1}{\ln\ 10}\cdot\frac{3}{3x}$$
$$= \frac{1}{x\ \ln\ 10}$$

35. $y = \log_7\ \sqrt{4x-3}$

$$g(x) = \sqrt{4x-3}$$
$$g'(x) = \frac{4}{2\sqrt{4x-3}} = \frac{2}{\sqrt{4x-3}}$$
$$\frac{dy}{dx} = \frac{1}{\ln 7}\cdot\frac{\frac{2}{\sqrt{4x-3}}}{\sqrt{4x-3}}$$
$$= \frac{2}{(\ln 7)(4x-3)}$$

37. $y = \log_2\ (2x^2 - x)^{5/2}$

$g(x) = (2x^2 - x)^{5/2}$ and

$g'(x) = \frac{5}{2}(2x^2-x)^{3/2}\cdot(4x-1)$.

$$\frac{dy}{dx} = \frac{1}{\ln\ 2}\cdot\frac{\frac{5}{2}(2x^2-x)^{3/2}\cdot(4x-1)}{(2x^2-x)^{5/2}}$$
$$= \frac{5(4x-1)}{(2\ln\ 2)(2x^2-x)}$$

39. $z = 10^y\ \log\ y$

$g(y) = 10^y$ and $g'(y) = (\ln 10)10^y$.

$$\frac{dz}{dy} = 10^y\cdot\frac{1}{(\ln\ 10)y} + \log\ y\cdot(\ln 10)10^y$$
$$= \frac{10^y}{(\ln\ 10)y} + (\log\ y)(\ln 10)10^y$$

41. $f(x) = \ln(xe^{\sqrt{x}} + 2)$

$$g(x) = xe^{\sqrt{x}} + 2$$
$$g'(x) = x\left[e^{\sqrt{x}}\left(\frac{1}{2\sqrt{x}}\right)\right] + e^{\sqrt{x}}(1)$$
$$= \frac{e^{\sqrt{x}}\sqrt{x}}{2} + e^{\sqrt{x}}$$
$$= \frac{e^{\sqrt{x}}}{2}(\sqrt{x}+2)$$
$$f'(x) = \frac{g'(x)}{g(x)}$$
$$= \frac{\frac{e^{\sqrt{x}}}{2}(\sqrt{x}+2)}{xe^{\sqrt{x}}+2}$$
$$= \frac{e^{\sqrt{x}}(\sqrt{x}+2)}{2(xe^{\sqrt{x}}+2)}$$

43. $f(t) = \dfrac{2t^{3/2}}{\ln(2t^{3/2}+1)}$

Use the quotient rule.

$$u(t) = 2t^{3/2}, u'(t) = 3t^{1/2}$$
$$v(t) = \ln(2t^{3/2}+1), v'(t) = \frac{3t^{1/2}}{2t^{3/2}+1}$$
$$f'(t) = \frac{\ln(2t^{3/2}+1)(3t^{1/2}) - 2t^{3/2}\left[\frac{3t^{1/2}}{2t^{3/2}+1}\right]}{[\ln(2t^{3/2}+1)]^2}$$
$$= \frac{(3t^{1/2})\ln(2t^{3/2}+1) - \frac{6t^2}{2t^{3/2}+1}}{[\ln(2t^{3/2}+1)]^2}$$
$$= \frac{(6t^2+3t^{1/2})\ln(2t^{3/2}+1) - 6t^2}{(2t^{3/2}+1)[\ln(2t^{3/2}+1)]^2}$$

45. Note that a is a constant.

$$\frac{d}{dx}\ \ln\ |ax| = \frac{d}{dx}(\ln\ |a| + \ln\ |x|)$$
$$= \frac{d}{dx}\ \ln\ |a| + \frac{d}{dx}\ \ln\ x$$
$$= 0 + \frac{d}{dx}\ \ln\ |x|$$
$$= \frac{d}{dx}\ \ln\ |x|$$

Therefore,

$$\frac{d}{dx}\ \ln\ |ax| = \frac{d}{dx}\ \ln\ |x|.$$

47. Graph

$$y = \frac{\ln |x + 0.0001| - \ln |x|}{0.0001}$$

on a graphing calculator. A good choice for the viewing window is $[-3, 3]$.

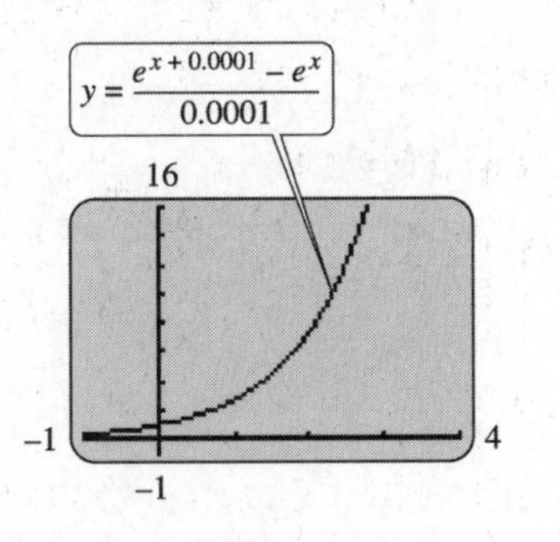

If we graph $y = \frac{1}{x}$ on the same screen, we see that the two graphs coincide.
By the definition of the derivative, if $f(x) = \ln |x|$,

$$f'(x) = \lim_{h\to 0} \frac{f(x+h) - f(x)}{h} = \lim_{h\to 0} \frac{\ln |x+h| - \ln |x|}{h},$$

and $h = 0.0001$ is very close to 0.
Comparing the two graphs provides graphical evidence that

$$f'(x) = \frac{1}{x}.$$

49. Use the derivative of $\ln x$.

$$\frac{d \ln \frac{u(x)}{v(x)}}{dx} = \frac{1}{\frac{u(x)}{v(x)}} \cdot \frac{d\left[\frac{u(x)}{v(x)}\right]}{dx} = \frac{v(x)}{u(x)} \cdot \frac{d\left[\frac{u(x)}{v(x)}\right]}{dx}$$

$$d \ln u(x) = \frac{1}{u(x)} \cdot \frac{d[u(x)]}{dx}$$

$$d \ln v(x) = \frac{1}{v(x)} \cdot \frac{d[v(x)]}{dx}$$

Then, since $\ln \frac{u(x)}{v(x)} = \ln u(x) - \ln v(x)$,

$$\frac{v(x)}{u(x)} \cdot \frac{d\left[\frac{u(x)}{v(x)}\right]}{dx} = \frac{1}{u(x)} \cdot \frac{d[u(x)]}{dx} - \frac{1}{v(x)} \cdot \frac{d[v(x)]}{dx}.$$

Multiply both sides of this equation by $\frac{u(x)}{v(x)}$. Then

$$\frac{d\left[\frac{u(x)}{v(x)}\right]}{dx} = \frac{1}{v(x)} \cdot \frac{d[u(x)]}{dx} - \frac{u(x)}{[v(x)]^2} \cdot \frac{d[v(x)]}{dx}$$

$$= \frac{v(x)}{[v(x)]^2} \cdot \frac{d[u(x)]}{dx} - \frac{u(x)}{[v(x)]^2} \cdot \frac{d[v(x)]}{dx}$$

$$= \frac{v(x) \cdot \frac{d[u(x)]}{dx} - u(x) \cdot \frac{d[v(x)]}{dx}}{[v(x)]^2}$$

This is the quotient rule.

51. The change-of-base theorem for logarithms states $\log_a x = \frac{\ln x}{\ln a}$. Find the derivative of each side.

$$\frac{d \log_a x}{dx} = \frac{\ln a \cdot \frac{d \ln x}{dx} - \ln x \cdot \frac{d \ln a}{dx}}{(\ln a)^2}$$

$$= \frac{\ln a \cdot \frac{1}{x} - \ln x \cdot 0}{(\ln a)^2}$$

$$= \frac{\frac{1}{x}}{\ln a}$$

$$= \frac{\frac{1}{x}}{x \ln a}$$

53.

$$h(x) = x^x$$
$$u(x) = x, u'(x) = 1$$
$$v(x) = x, v'(x) = 1$$

$$h'(x) = x^x \left[\frac{x(1)}{x} + (\ln x)(1)\right]$$
$$= x^x(1 + \ln x)$$

55. $R(x) = 30 \ln (2x + 1)$

$$C(x) = \frac{x}{2}$$

$$P(x) = R(x) - C(x) = 30 \ln (2x + 1) - \frac{x}{2}$$

The profit will be a maximum when the derivative of the profit function is equal to 0.

$$P'(x) = 30\left(\frac{1}{2x+1}\right)(2) - \frac{1}{2} = \frac{60}{2x+1} - \frac{1}{2}$$

Now, $P'(x) = \dfrac{60}{2x+1} - \dfrac{1}{2} = 0$

when

$$\frac{60}{2x+1} = \frac{1}{2}$$
$$120 = 2x + 1$$
$$\frac{119}{2} = x.$$

Thus, a maximum profit occurs when $x = \frac{119}{2}$ or, in a practical sense, when 59 or 60 items are manufactured. (Both 59 and 60 give the same profit.)

57. $C(q) = 100q + 100$

(a) The marginal cost is given by $C'(q)$.

$$C'(q) = 100$$

(b) $P(q) = R(q) - C(q)$

$$= q\left(100 + \frac{50}{\ln q}\right) - (100q + 100)$$

$$= 100q + \frac{50q}{\ln q} - 100q - 100$$

$$= \frac{50q}{\ln q} - 100$$

(c) The profit from one more unit is is $\frac{dP}{dq}$ for $q = 8$.

$$\frac{dP}{dq} = \frac{(\ln q)(50) - 50q\left(\frac{1}{q}\right)}{(\ln q)^2}$$

$$= \frac{50 \ln q - 50}{(\ln q)^2} = \frac{50(\ln q - 10)}{(\ln q)^2}$$

When $q = 8$, the profit from one more unit is

$$\frac{50(\ln 8 - 1)}{(\ln 8)^2} = \$12.48.$$

(d) The manager can use the information from part (c) to decide whether it is profitable to make and sell additional items.

59. $A(w) = 4.688w^{0.8168-0.0154\log_{10} w}$

(a) $A(4000) = 4.688(4000)^{0.8168-0.0154\log_{10} 4000}$
$\approx 2590 \text{ cm}^2$

(b) $\frac{A(w)}{4.688} = w^{0.8166-0.0154\log_{10} w}$

$$\ln A(w) - \ln 4.688 = (\ln w)\left(0.8168 - 0.0154\frac{\ln w}{\ln 10}\right)$$

$$\frac{A'(w)}{A(w)} = \frac{1}{w}\left(0.8168 - 0.0154\frac{\ln w}{\ln 10}\right) + \ln w\left(\frac{-0.0154}{\ln 10}\right)\frac{1}{w}$$

$$= \frac{0.8168}{w} - \frac{0.0308}{\ln 10}\frac{\ln w}{w}$$

$$= \frac{1}{w}\left(0.8168 - \frac{0.0308}{\ln 10}\ln w\right)$$

$$A'(w) = \frac{1}{w}\left(0.8168 - \frac{0.0308}{\ln 10}\ln w\right) \cdot (4.688w^{0.8168-0.0154\log_{10} w})$$

$$A'(4000) \approx 0.4571 \approx 0.46 \text{ g/cm}^2$$

When the infant weighs 4000 g, it is gaining 0.46 square centimeters per gram of weight increase.

(c)

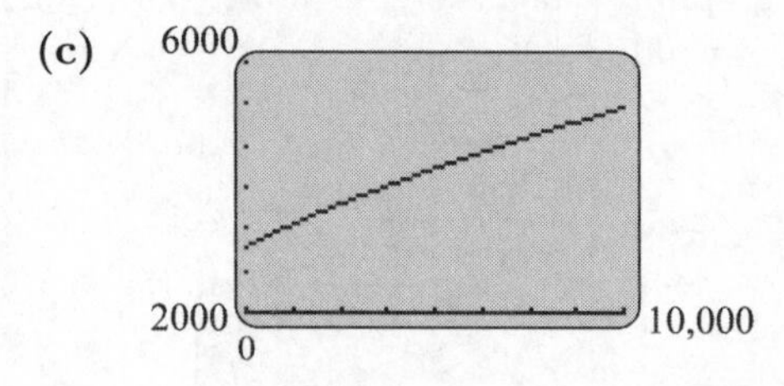

61. $F(x) = 0.774 + 0.727\log(x)$

(a) $F(25{,}000) = 0.774 + 0.727\log(25{,}000)$
$= 3.9713\ldots$
$\approx 4 \text{ kJ/day}$

(b)

$$F'(x) = 0.727\frac{1}{x\ln 10}$$

$$= \frac{0.727}{\ln 10}x^{-1}$$

$$F'(25{,}000) = \frac{0.727}{\ln 10}25{,}000^{-1}$$

$$\approx 0.000012629\ldots$$

$$\approx 1.3 \times 10^{-5}$$

When a fawn is 25 kg in size, the rate of change of the energy expenditure of the fawn is about 1.3×10^{-5} kJ/day per gram.

(c)

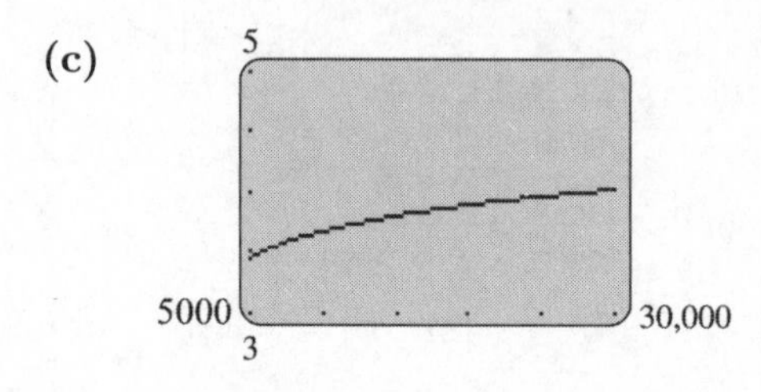

63. $M(t) = (0.1t + 1) \ln \sqrt{t}$

(a) $M(15) = [0.1(15) + 1] \ln \sqrt{15}$
≈ 3.385

When the temperature is 15°C, the number of matings is about 3.

(b) $M(25) = [0.1(25) + 1] \ln \sqrt{25}$
≈ 5.663

When the temperature is 25°C, the number of matings is about 6.

(c) $M(t) = (0.1t + 1) \ln \sqrt{t}$
$= (0.1t + 1) \ln t^{1/2}$

$$M'(t) = (0.1t + 1)\left(\frac{1}{2} \cdot \frac{1}{t}\right) + (\ln t^{1/2})(0.1)$$

$$= 0.1 \ln \sqrt{t} + \frac{1}{2t}(0.1t + 1)$$

$$M'(15) = 0.1 \ln \sqrt{15} + \frac{1}{2 \cdot 15}[(0.1)(15) + 1]$$

$$\approx 0.22$$

When the temperature is 15°C, the rate of change of the number of matings is about 0.22.

65. $M = \frac{2}{3} \log \frac{E}{0.007}$

(a)
$$8.9 = \frac{2}{3} \log \frac{E}{0.007}$$
$$13.35 = \log \frac{E}{0.007}$$
$$10^{13.35} = \frac{E}{0.007}$$
$$E = 0.007(10^{13.35})$$
$$\approx 1.567 \times 10^{11} \text{ kWh}$$

(b) $10{,}000{,}000 \times 247$ kWh/month
$= 2{,}470{,}000{,}000$ kWh/month

$$\frac{1.567 \times 10^{11} \text{ kWh}}{2{,}470{,}000{,}000 \text{ kWh/month}} \approx 63.4 \text{ months}$$

(c) $M = \frac{2}{3} \log E - \frac{2}{3} \log 0.007$

$$\frac{dM}{dE} = \frac{2}{3}\left(\frac{1}{(\ln 10)E}\right)$$
$$= \frac{2}{(3 \ln 10)E}$$

When $E = 70{,}000$,

$$\frac{dM}{dE} = \frac{2}{(3 \ln 10)70{,}000}$$
$$\approx 4.14 \times 10^{-6}$$

(d) $\frac{dM}{dE}$ varies inversely with E, so as E increases, $\frac{dM}{dE}$ decreases and approaches zero.

Chapter 4 Review Exercises

1. $y = 5x^3 - 7x^2 - 9x + \sqrt{5}$

$$\frac{dy}{dx} = 5(3x^2) - 7(2x) - 9 + 0$$
$$= 15x^2 - 14x - 9$$

3. $y = 9x^{8/3}$

$$\frac{dy}{dx} = 9\left(\frac{8}{3}x^{5/3}\right)$$
$$= 24x^{5/3}$$

5. $f(x) = 3x^{-4} + 6\sqrt{x}$
$= 3x^{-4} + 6x^{1/2}$

$$f'(x) = 3(-4x^{-5}) + 6\left(\frac{1}{2}x^{-1/2}\right)$$
$$= -12x^{-5} + 3x^{-1/2} \text{ or } -\frac{12}{x^5} + \frac{3}{x^{1/2}}$$

7. $k(x) = \frac{3x}{4x + 7}$

$$k'(x) = \frac{(4x + 7)(3) - (3x)(4)}{(4x + 7)^2}$$
$$= \frac{12x + 21 - 12x}{(4x + 7)^2}$$
$$= \frac{21}{(4x + 7)^2}$$

9. $y = \frac{x^2 - x + 1}{x - 1}$

$$\frac{dy}{dx} = \frac{(x - 1)(2x - 1) - (x^2 - x + 1)(1)}{(x - 1)^2}$$
$$= \frac{2x^2 - 3x + 1 - x^2 + x - 1}{(x - 1)^2}$$
$$= \frac{x^2 - 2x}{(x - 1)^2}$$

11. $f(x) = (3x^2 - 2)^4$
$f'(x) = 4(3x^2 - 2)^3[3(2x)]$
$= 24x(3x^2 - 2)^3$

13. $y = \sqrt{2t^7 - 5}$
$= (2t^7 - 5)^{1/2}$

$$\frac{dy}{dx} = \frac{1}{2}(2t^7 - 5)^{-1/2}[2(7t^6)]$$
$$= 7t^6(2t^7 - 5)^{-1/2} \text{ or } \frac{7t^6}{(2t^7 - 5)^{1/2}}$$

15. $y = 3x(2x + 1)^3$

$$\frac{dy}{dx} = 3x(3)(2x + 1)^2(2) + (2x + 1)^3(3)$$
$$= (18x)(2x + 1)^2 + 3(2x + 1)^3$$
$$= 3(2x + 1)^2[6x + (2x + 1)]$$
$$= 3(2x + 1)^2(8x + 1)$$

17. $r(t) = \dfrac{5t^2 - 7t}{(3t+1)^3}$

$$r'(t) = \frac{(3t+1)^3(10t-7) - (5t^2-7t)(3)(3t+1)^2(3)}{[(3t+1)^3]^2}$$

$$= \frac{(3t+1)^3(10t-7) - 9(5t^2-7t)(3t+1)^2}{(3t+1)^6}$$

$$= \frac{(3t+1)(10t-7) - 9(5t^2-7t)}{(3t+1)^4}$$

$$= \frac{30t^2 - 11t - 7 - 45t^2 + 63t}{(3t+1)^4}$$

$$= \frac{-15t^2 + 52t - 7}{(3t+1)^4}$$

19. $p(t) = t^2(t^2+1)^{5/2}$

$$p'(t) = t^2 \cdot \frac{5}{2}(t^2+1)^{3/2} \cdot 2t + 2t(t^2+1)^{5/2}$$
$$= 5t^3(t^2+1)^{3/2} + 2t(t^2+1)^{5/2}$$
$$= t(t^2+1)^{3/2}[5t^2 + 2(t^2+1)^1]$$
$$= t(t^2+1)^{3/2}(7t^2+2)$$

21. $y = -6e^{2x}$

$$\frac{dy}{dx} = -6(2e^{2x}) = -12e^{2x}$$

23.
$$y = e^{-2x^3}$$
$$g(x) = -2x^3$$
$$g'(x) = -6x^2$$
$$y' = -6x^2e^{-2x^3}$$

25. $y = 5x \cdot e^{2x}$

Use the product rule.

$$\frac{dy}{dx} = 5x(2e^{2x}) + e^{2x(5)}$$
$$= 10xe^{2x} + 5e^{2x}$$
$$= 5e^{2x}(2x+1)$$

27.
$$y = \ln(2+x^2)$$
$$g(x) = 2 + x^2$$
$$g'(x) = 2x$$
$$\frac{dy}{dx} = \frac{2x}{2+x^2}$$

29. $y = \dfrac{\ln|3x|}{x-3}$

$$\frac{dy}{dx} = \frac{(x-3)\left(\frac{1}{3x}\right)(3) - (\ln|3x|)(1)}{(x-3)^2}$$

$$= \frac{\frac{x-3}{x} - \ln|3x|}{(x-3)^2} \cdot \frac{x}{x}$$

$$= \frac{x - 3 - x\ln|3x|}{x(x-3)^2}$$

31. $y = \dfrac{xe^x}{\ln(x^2-1)}$

$$\frac{dy}{dx} = \frac{\ln(x^2-1)[xe^x + e^x] - xe^x\left(\frac{1}{x^2-1}\right)(2x)}{[\ln(x^2-1)]^2}$$

$$= \frac{e^x(x+1)\ln(x^2-1) - \frac{2x^2e^x}{x^2-1}}{[\ln(x^2-1)]^2} \cdot \frac{x^2-1}{x^2-1}$$

$$= \frac{e^x(x+1)(x^2-1)\ln(x^2-1) - 2x^2e^x}{(x^2-1)[\ln(x^2-1)]^2}$$

33. $s = (t^2 + e^t)^2$

$s' = 2(t^2 + e^t)(2t + e^t)$

35. $y = 3 \cdot 10^{-x^2}$

$$\frac{dy}{dx} = 3 \cdot (\ln 10)10^{-x^2}(-2x)$$
$$= -6x(\ln 10) \cdot 10^{-x^2}$$

37. $g(z) = \log_2(z^3 + z + 1)$

$$g'(z) = \frac{1}{\ln 2} \cdot \frac{3z^2+1}{z^3+z+1}$$
$$= \frac{3z^2+1}{(\ln 2)(z^3+z+1)}$$

39. $f(x) = e^{2x}\ln(xe^x + 1)$

Use the product rule.

$$f'(x) = e^{2x}\left(\frac{e^x + xe^x}{xe^x+1}\right) + [\ln(xe^x+1)](2e^{2x})$$
$$= \frac{(1+x)e^{3x}}{xe^x+1} + 2e^{2x}\ln(xe^x+1)$$

41. (a) $D_x(f[g(x)])$ at $x = 2$

$$= f'[g(2)]g'(2)$$
$$= f'(1)\left(\frac{3}{10}\right)$$
$$= -5\left(\frac{3}{10}\right)$$
$$= -\frac{3}{2}$$

(b) $D_x(f[g(x)])$ at $x = 3$

$$= f'[g(3)]g'(3)$$
$$= f'(2)\left(\frac{4}{11}\right)$$
$$= -6\left(\frac{4}{11}\right)$$
$$= -\frac{24}{11}$$

45. $y = 8 - x^2$; $x = 1$

$$y = 8 - x^2$$
$$\frac{dy}{dx} = -2x$$

slope $= y'(1) = -2(1) = -2$

Use $(1, 7)$ and $m = -2$ in the point-slope form.

$$y - 7 = -2(x - 1)$$
$$y - 7 = -2x + 2$$
$$2x + y = 9$$
$$y = -2x + 9$$

47. $y = \frac{x}{x^2 - 1}$; $x = 2$

$$\frac{dy}{dx} = \frac{(x^2 - 1)\cdot 1 - x(2x)}{(x^2 - 1)^2}$$
$$= \frac{-x^2 - 1}{(x^2 - 1)^2}$$

The value of $\frac{dy}{dx}$ when $x = 2$ is the slope.

$$m = \frac{-(2^2) - 1}{(2^2 - 1)^2} = \frac{-5}{9} = -\frac{5}{9}$$

When $x = 2$,

$$y = \frac{2}{4 - 1} = \frac{2}{3}.$$

Use $m = -\frac{5}{9}$ with $P\left(2, \frac{2}{3}\right)$.

$$y - \frac{2}{3} = -\frac{5}{9}(x - 2)$$
$$y - \frac{6}{9} = -\frac{5}{9}x + \frac{10}{9}$$
$$y = -\frac{5}{9}x + \frac{16}{9}$$

49. $y = -\sqrt{8x + 1}$; $x = 3$

$$y = -(8x + 1)^{1/2}$$
$$\frac{dy}{dx} = -\frac{1}{2}(8x + 1)^{-1/2}(8)$$
$$\frac{dy}{dx} = -\frac{4}{(8x + 1)^{1/2}}$$

The value of $\frac{dy}{dx}$ when $x = 3$ is the slope.

$$m = -\frac{4}{(24 + 1)^{1/2}} = -\frac{4}{5}$$

When $x = 3$,

$$y = -\sqrt{24 + 1} = -5.$$

Use $m = -\frac{4}{5}$ with $P(3, -5)$.

$$y + 5 = -\frac{4}{5}(x - 3)$$
$$y + \frac{25}{5} = -\frac{4}{5}x + \frac{12}{5}$$
$$y = -\frac{4}{5}x - \frac{13}{5}$$

51. $y = xe^x$; $x = 1$

$$\frac{dy}{dx} = xe^x + 1 \cdot e^x$$
$$= e^x(x + 1)$$

The value of $\frac{dy}{dx}$ when $x = 1$ is the slope.

$$m = e^1(1 + 1) = 2e$$

When $x = 1$, $y = 1e^1 = e$. Use $m = 2e$ with $P(1, e)$.

$$y - e = 2e(x - 1)$$
$$y = 2ex - e$$

53. $y = x \ln x$; $x = e$

$$\frac{dy}{dx} = x \cdot \frac{1}{x} + 1 \cdot \ln x$$
$$= 1 + \ln x$$

The value of $\frac{dy}{dx}$ when $x = e$ is the slope

$$m = 1 + \ln e = 1 + 1 = 2.$$

When $x = e$, $y = e \ln e = e \cdot 1 = e$. Use $m = 2$ with $P(e, e)$.

$$y - e = 2(x - e)$$
$$y = 2x - e$$

55. **(a)** Use the chain rule.

Let $g(x) = \ln x$. Then $g'(x) = \frac{1}{x}$.

Let $y = g[f(x)]$. Then $\frac{dy}{dx} = g'[f(x)] \cdot f'(x)$.

so

$$\frac{d \ln f(x)}{dx} = \frac{1}{f(x)} \cdot f'(x) = \frac{f'(x)}{f(x)}.$$

(b) $\hat{f} = \frac{f'(x)}{f(x)}, \hat{g} = f\frac{g'(x)}{g(x)}$

$$\begin{aligned}\hat{fg} &= \frac{(fg)'(x)}{(fg)(x)} \\ &= \frac{f(x) \cdot g'(x) + g(x) \cdot f'(x)}{f(x)g(x)} \\ &= \frac{f(x) \cdot g'(x)}{f(x) \cdot g(x)} + \frac{g(x) \cdot f'(x)}{f(x) \cdot g(x)} \\ &= \frac{g'(x)}{g(x)} + \frac{f'(x)}{f(x)} \\ &= \hat{g} + \hat{f} \text{ or } \hat{f} + \hat{g}\end{aligned}$$

57. $C(x) = \sqrt{x+1}$

$$\begin{aligned}\overline{C}(x) &= \frac{C(x)}{x} = \frac{\sqrt{x+1}}{x} \\ &= \frac{(x+1)^{1/2}}{x} \\ \overline{C}'(x) &= \frac{x\left[\frac{1}{2}(x+1)^{-1/2}\right] - (x+1)^{1/2}(1)}{x^2} \\ &= \frac{\frac{1}{2}x(x+1)^{-1/2} - (x+1)^{1/2}}{x^2} \\ &= \frac{x(x+1)^{-1/2} - 2(x+1)^{1/2}}{2x^2} \\ &= \frac{(x+1)^{-1/2}[x - 2(x+1)]}{2x^2} \\ &= \frac{(x+1)^{-1/2}(-x-2)}{2x^2} \\ &= \frac{-x-2}{2x^2(x+1)^{1/2}}\end{aligned}$$

59. $C(x) = (x^2+3)^3$

$$\begin{aligned}\overline{C}(x) &= \frac{C(x)}{x} = \frac{(x^2+3)^3}{x} \\ \overline{C}'(x) &= \frac{x[3(x^2+3)^2(2x)] - (x^2+3)^3(1)}{x^2} \\ &= \frac{6x^2(x^2+3)^2 - (x^2+3)^3}{x^2} \\ &= \frac{(x^2+3)^2[6x^2 - (x^2+3)]}{x^2} \\ &= \frac{(x^2+3)^2(5x^2-3)}{x^2}\end{aligned}$$

61. $C(x) = 10 - e^{-x}$

$$\begin{aligned}\overline{C}(x) &= \frac{C(x)}{x} \\ \overline{C}(x) &= \frac{10 - e^{-x}}{x} \\ \overline{C}'(x) &= \frac{x(e^{-x}) - (10 - e^{-x}) \cdot 1}{x^2} \\ &= \frac{e^{-x}(x+1) - 10}{x^2}\end{aligned}$$

63. $S(x) = 1000 + 60\sqrt{x} + 12x$

$$\begin{aligned}&= 1000 + 60x^{1/2} + 12x \\ \frac{dS}{dx} &= 60\left(\frac{1}{2}x^{-1/2}\right) + 12 \\ &= 30x^{-1/2} + 12 = \frac{30}{\sqrt{x}} + 12\end{aligned}$$

(a) $\frac{dS}{dx}(9) = \frac{30}{\sqrt{9}} + 12 = \frac{30}{3} + 12 = 22$

Sales will increase by $22 million when $1000 more is spent on research.

(b) $\frac{dS}{dx}(16) = \frac{30}{\sqrt{16}} + 12 = \frac{30}{4} + 12 = 19.5$

Sales will increase by $19.5 million when $1000 more is spent on research.

(c) $\frac{dS}{dx}(25) = \frac{30}{\sqrt{25}} + 12 = \frac{30}{5} + 12 = 18$

Sales will increase by $18 million when $1000 more is spent on research.

(d) As more money is spent on research, the increase in sales is decreasing.

65. $T(x) = \dfrac{1000 + 60x}{4x + 5}$

$$T'(x) = \frac{(4x+5)(60) - (1000+60x)(4)}{(4x+5)^2} = \frac{240x + 300 - 4000 - 240x}{(4x+5)^2} = \frac{-3700}{(4x+5)^2}$$

(a) $T'(9) = \dfrac{-3700}{[4(9)+5]^2} = \dfrac{-3700}{1681} \approx -2.201$

Costs will decrease \$2201 for the next \$100 spent on training.

(b) $T'(19) = \dfrac{-3700}{[4(19)+5]^2} = \dfrac{-3700}{6561} \approx -0.564$

Costs will decrease \$564 for the next \$100 spent on training.

(c) Costs will always decrease because

$$T'(x) = \frac{-3700}{(4x+5)^2}$$

will always be negative.

67. $A(r) = 1000e^{12r/100}$

$A'(r) = 1000e^{12r/100} \cdot \dfrac{12}{100} = 120e^{12r/100}$
$A'(5) = 120e^{0.6} \approx 218.65$

The balance increases by approximately \$218.65 for every 1% increase in the interest rate when the rate is 5%.

69. $f(t) = 1.5207t^4 - 19.166t^3 + 62.91t^2 + 6.0726t + 1026$
$f'(t) = 1.5207(4t^3) - 19.166(3t^2) + 62.91(2t) + 6.0726$
$= 6.0828t^3 - 57.498t^2 + 125.82$

$t = 7$ corresponds to the beginning of 2005.

$f'(7) = 6.0828(7)^3 - 57.498(7)^2 + 125.82(7) + 6.0726$
≈ 156

Rents were increasing at the rate of \$156 per month per year.

71. **(a)** Using the regression feature on a graphing calculator, a cubic function that models the data is

$$y = (2.458 \times 10^{-5})t^3 - (6.767 \times 10^{-4})t^2 - 0.02561t + 2.031.$$

Using the regression feature on a graphing calculator, a quartic function that models the data is

$$y = (-1.314 \times 10^{-6})t^4 + (3.363 \times 10^{-4})t^3 - 0.02565t^2 + 0.7410t - 5.070.$$

(b) Using the cubic function, $\frac{dy}{dx}$ at $x = 95$ is about 0.51 dollar per year. Using the quartic function, $\frac{dy}{dx}$ at $x = 95$ is about 0.47 dollar per year.

73. $G(t) = \dfrac{m\,G_o}{G_o + (m - G_o)e^{-kmt}}$, where $m = 30{,}000$, $G_o = 2000$, and $k = 5 \cdot 10^{-6}$.

(a) $G(t) = \dfrac{(30{,}000)(2000)}{2000 + (30{,}000 - 2000)e^{-5\cdot 10^{-6}(30{,}000)t}} = \dfrac{30{,}000}{1 + 14e^{-0.15t}}$

(b) $G(t) = 30{,}000(1 + 14e^{-0.15t})^{-1}$
$G(6) = 30{,}000(1 + 14e^{-0.90})^{-1} \approx 4483$

$G'(t) = -30{,}000(1 + 14e^{-0.15t})^{-2}(-2.1e^{-0.15t}) = \dfrac{63{,}000e^{-0.15t}}{(1 + 14e^{-0.15t})^2}$

$G'(6) = \dfrac{63{,}000e^{-0.90}}{(1 + 14e^{-0.90})^2} \approx 572$

The population is 4483, and the rate of growth is 572.

75. $M(t) = 3583e^{-e^{-0.020(t-66)}}$

(a) $M(250) = 3583e^{-e^{-0.020(250-66)}}$
≈ 3493.76 grams,
or about 3.5 kilograms

(b) As $t \to \infty$, $-e^{-0.020(t-66)} \to 0$, $e^{-e^{-0.020(t-66)}} \to 1$, and $M(t) \to 3583$ grams or about 3.6 kilograms.

(c) 50% of 3583 is 1791.5.

$$1791.5 = 3583e^{-e^{-0.020(t-66)}}$$
$$\ln\left(\frac{1791.5}{3583}\right) = -e^{-0.020(t-66)}$$
$$\ln\left(\ln\frac{3583}{1791.5}\right) = -0.020(t-66)$$
$$t = -\frac{1}{0.020}\ln\left(\ln\frac{3583}{1791.5}\right) + 66$$
$$\approx 84 \text{ days}$$

(d) $D_tM(t)$
$= 3583e^{-e^{-0.020(t-66)}} D_t(-e^{-0.020(t-66)})$
$= 3583e^{-e^{-.0.020(t-66)}} (-e^{-0.020(t-66)})(-0.020)$
$= 71.66e^{-e^{-0.020(t-66)}} (e^{-0.020(t-66)})$

when $t = 250$, $D_tM(t) \approx 1.76$ g/day.

(e)

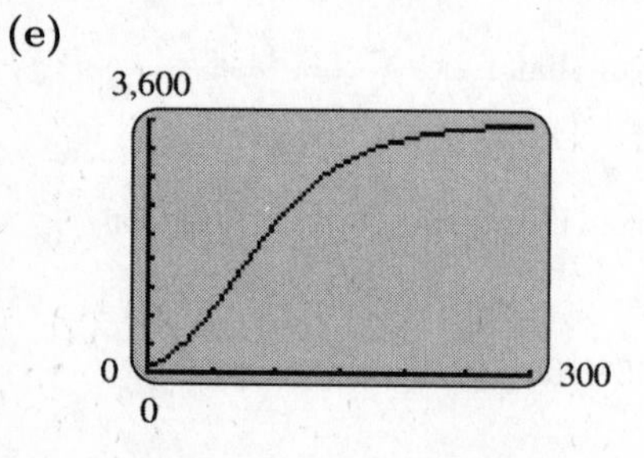

Growth is initially rapid, then tapers off.

(f)

Day	*Weight*	*Rate*
50	904	24.90
100	2159	21.87
150	2974	11.08
200	3346	4.59
250	3494	1.76
300	3550	0.66

77. $f(t) = \dfrac{8}{t+1} + \dfrac{20}{t^2+1}$

(a) The average velocity from $t = 1$ to $t = 3$ is given by

$$\text{average velocity} = \frac{f(3) - f(1)}{3-1}$$
$$= \frac{\left(\frac{8}{4} + \frac{20}{10}\right) - \left(\frac{8}{2} + \frac{20}{2}\right)}{2}$$
$$= \frac{4-14}{2}$$
$$= -5$$

Belmar's average velocity between 1 sec and 3 sec is -5 ft/sec.

(b) $f(t) = 8(t+1)^{-1} + 20(t^2+1)^{-1}$
$f'(t) = -8(t+1)^{-2} \cdot 1 - 20(t^2+1)^{-2} \cdot 2t$
$= -\dfrac{8}{(t+1)^2} - \dfrac{40t}{(t^2+1)^2}$
$f'(3) = -\dfrac{8}{16} - \dfrac{120}{100}$
$= -0.5 - 1.2$
$= -1.7$

Belmar's instantaneous velocity at 3 sec is -1.7 ft/sec.

79. (a) $N(t) = N_0e^{-0.217t}$, where $t = 1$ and $N_0 = 210$

$$N(1) = 210e^{-0.217(1)}$$
$$\approx 169$$

The number of words predicted to be in use in 1950 is 169, and the actual number in use was 167.

(b) $N(2) = 210e^{-0.217(2)}$
≈ 136

In 2050 the will be about 136 words still being used.

(c) $N(t) = 210e^{-0.217t}$
$N'(t) = 210e^{-0.217t} \cdot (-0.217)$
$= -45.57e^{-0.217t}$
$N'(2) = -45.57e^{-0.217(2)}$
≈ -30

In the year 2050 the number of words in use will be decreasing by 30 words per millenium.

Chapter 4 Test

[4.1]

Find the derivative of each function.

1. $y = 2x^4 - 3x^3 + 4x^2 - x + 1$

2. $f(x) = x^{3/4} - x^{-2/3} + x^{-1}$

3. $f(x) = -2x^{-2} - 3\sqrt{x}$

4. Find the slope of the tangent line to $y = -2x + \frac{1}{x} + \sqrt{x}$ at $x = 1$. Find the equation of the tangent line.

5. For the function

$$f(x) = x^3 + \frac{3}{2}x^2 - 60x + 18,$$

find all values of x where the tangent line is horizontal.

6. At the beginning of an experiment, a culture is determined to have 2×10^4 bacteria. Thereafter, the number of bacteria observed at time t (in hours) is given by the equation

$$B(t) = 10^4 \left(2 - 3\sqrt{t} + 2t + t^2\right).$$

How fast is the population growing at the end of 4 hours?

[4.2]

Find the derivative of each function.

7. $f(x) = \left(2x^2 - 5x\right)\left(3x^2 + 1\right)$

8. $y = \dfrac{x^2 + x}{x^3 - 1}$

9. Management has determined that the cost in thousands of dollars for producing x units is given by the equation

$$C(x) = \frac{3x^2}{x^2 + 1} + 200.$$

Find and interpret the marginal cost when $x = 3$.

[4.3]

10. Find $f[g(x)]$ and $g[f(x)]$ for the following pair of functions.

$$f(x) = \frac{3}{x^2};\ g(x) = \frac{1}{x}$$

Find each of the following.

11. $\dfrac{dy}{dx}$ if $y = \dfrac{\sqrt{x - 2}}{x + 1}$

12. $D_x\left[\sqrt{\left(3x^2 - 1\right)^3}\right]$

13. $f'(2)$ if $f(t) = \dfrac{2t - 1}{\sqrt{t + 2}}$

Find the derivative of each function.

14. $y = -2x\sqrt{3x - 1}$

15. $y = \left(5x^3 - 2x\right)^5$

16. Find the equation of the tangent line to the graph of the function $f(x) = \sqrt{x^2 - 9}$ at $x = 5$. (Write the equation in slope-intercept form.)

[4.4]

Find the derivative of each function.

17. $y = 3x^2e^{2x}$

18. $y = \left(e^{2x} - \ln|x|\right)^3$

19. The concentration of a certain drug in the bloodstream at time t in minutes is given by

$$c(t) = e^{-t} - e^{-3t}.$$

Find the rate of change in the concentration at each of the following times. Round to the nearest thousandth.

(a) $t = 0$ **(b)** $t = 1$ **(c)** $t = 2$

[4.5]

Find the derivative of each function.

20. $y = \ln\left(2x^3 + 5\right)^{2/3}$

21. $y = x^2 \ln\left(x^2 + 1\right)$

22. $y = \dfrac{\ln|3x - 1|}{x - 2}.$

Chapter 4 Test Answers

1. $\frac{dy}{dx} = 8x^3 - 9x^2 + 8x - 1$

2. $f'(x) = \frac{3}{4x^{1/4}} + \frac{2}{3x^{5/3}} - \frac{1}{x^2}$

3. $f'(x) = \frac{4}{x^3} - \frac{3}{2\sqrt{x}}$

4. $-\frac{5}{2}$; $5x + 2y = 5$

5. $x = -5$, $x = 4$

6. 9.25×10^4 bacteria per hour

7. $f'(x) = 24x^3 - 45x^2 + 4x - 5$

8. $\frac{dy}{dx} = \frac{-x^4-2x^3-2x-1}{(x^3-1)^2}$

9. $C'(3) = 0.18$; after three units have been produced, the cost to produce one more unit will be approximately 0.18(1000) or \$180.

10. $f[g(x)] = 3x^2$; $g[f(x)] = \frac{x^2}{3}$

11. $\frac{5-x}{2\sqrt{x-2}(x+1)^2}$

12. $9x\sqrt{3x^2-1}$

13. $\frac{13}{16}$

14. $\frac{dy}{dx} = \frac{2-9x}{\sqrt{3x-1}}$

15. $\frac{dy}{dx} = 5\left(5x^3 - 2x\right)^4\left(15x^2 - 2\right)$

16. $y = \frac{5}{4}x - \frac{9}{4}$

17. $6x^2e^{2x} + 6xe^{2x}$

18. $\frac{3\left(e^{2x}-\ln|x|\right)^2\left(2xe^{2x}-1\right)}{x}$

19. **(a)** 2 **(b)** -0.219 **(c)** -0.128

20. $\frac{dy}{dx} = \frac{4x^2}{2x^3+5}$

21. $\frac{dy}{dx} = 2x\ln(x^2+1) + \frac{2x^3}{x^2+1}$

22. $\frac{3x-6-(3x-1)\ln|3x-1|}{(3x-1)(x-2)^2}$

Chapter 5

GRAPHS AND THE DERIVATIVE

5.1 Increasing and Decreasing Functions

1. By reading the graph, f is

(a) increasing on $(1, \infty)$ and
(b) decreasing on $(-\infty, 1)$.

3. By reading the graph, g is

(a) increasing on $(-\infty, -2)$ and
(b) decreasing on $(-2, \infty)$.

5. By reading the graph, h is

(a) increasing on $(-\infty, -4)$ and $(-2, \infty)$ and
(b) decreasing on $(-4, -2)$.

7. By reading the graph, f is

(a) increasing on $(-7, -4)$ and $(-2, \infty)$ and
(b) decreasing on $(-\infty, -7)$ and $(-4, -2)$.

9. (a) Since the graph of the function is positive for $x < -1$ and $x > 3$, the intervals where $f(x)$ is increasing are $(-\infty, -1)$ and $(3, \infty)$.

(b) Since the graph of the function is negative for $-1 < x < 3$, the interval where $f(x)$ is decreasing is $(-1, 3)$.

11. (a) Since the graph of the function is positive for $x < -8$, $-6 < x < -2.5$ and $x > -1.5$, the intervals where $f(x)$ is increasing are $(-\infty, -8), (-6, -2.5)$, and $(-1.5, \infty)$.

(b) Since the graph of the function is negative for $-8 < x < -6$ and $-2.5 < x < -1.5$, the intervals where $f(x)$ is decreasing are $(-8, -6)$ and $(-2.5, -1.5)$.

13. $y = 2.3 + 3.4x - 1.2x^2$

(a) $y' = 3.4 - 2.4x$

y' is zero when

$$3.4 - 2.4x = 0$$
$$x = \frac{3.4}{2.4} = \frac{17}{12}$$

and there are no values of x where y' does not exist, so the only critical number is $x = \frac{17}{12}$.

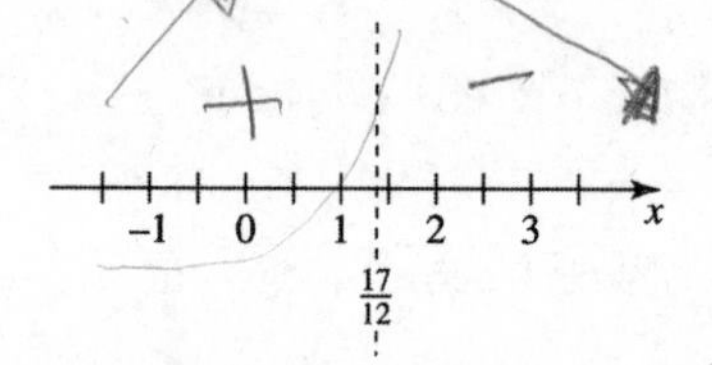

Test a point in each interval.

When $x = 0, y' = 3.4 - 2.4(0) = 3.4 > 0$.
When $x = 2, y' = 3.4 - 2.4(2) = -1.4 < 0$.

(b) The function is increasing on $(-\infty, \frac{17}{12})$.

(c) The function is decreasing on $(\frac{17}{12}, \infty)$.

15. $f(x) = \frac{2}{3}x^3 - x^2 - 24x - 4$

(a)
$$\begin{aligned} f'(x) &= 2x^2 - 2x - 24 \\ &= 2(x^2 - x - 12) \\ &= 2(x + 3)(x - 4) \end{aligned}$$

$f'(x)$ is zero when $x = -3$ or $x = 4$, so the critical numbers are -3 and 4.

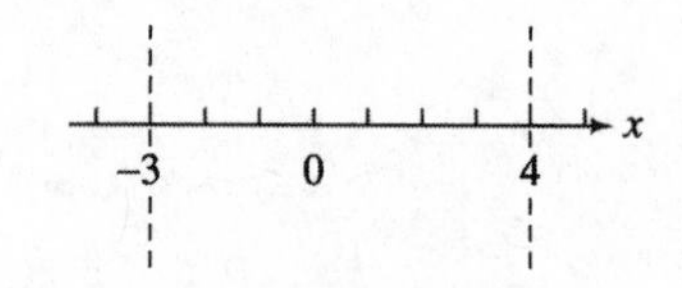

Test a point in each interval.

$$f'(-4) = 16 > 0$$
$$f'(0) = -24 < 0$$
$$f'(5) = 16 > 0$$

(b) f is increasing on $(-\infty, -3)$ and $(4, \infty)$.

(c) f is decreasing on $(-3, 4)$.

17. $f(x) = 4x^3 - 15x^2 - 72x + 5$

(a) $$\begin{aligned} f'(x) &= 12x^2 - 30x - 72 \\ &= 6(2x^2 - 5x - 12) \\ &= 6(2x + 3)(x - 4) \end{aligned}$$

$f'(x)$ is zero when $x = -\frac{3}{2}$ or $x = 4$, so the critical numbers are $-\frac{3}{2}$ and 4.

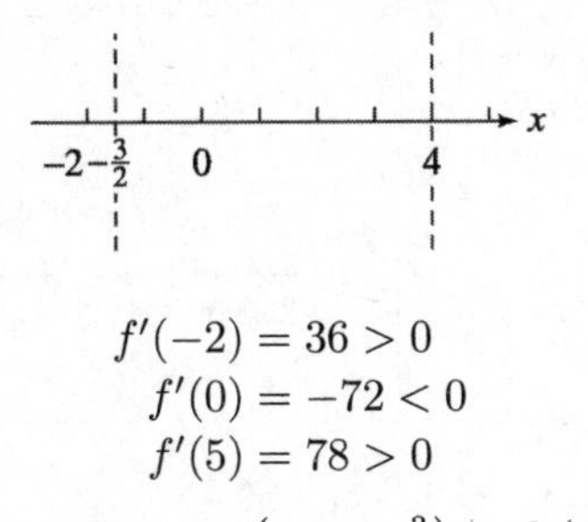

$$\begin{aligned} f'(-2) &= 36 > 0 \\ f'(0) &= -72 < 0 \\ f'(5) &= 78 > 0 \end{aligned}$$

(b) f is increasing on $\left(-\infty, -\frac{3}{2}\right)$ and $(4, \infty)$.

(c) f is decreasing on $\left(-\frac{3}{2}, 4\right)$.

19. $f(x) = x^4 + 4x^3 + 4x^2 + 1$

(a) $$\begin{aligned} f'(x) &= 4x^3 + 12x^2 + 8x \\ &= 4x(x^2 + 3x + 2) \\ &= 4x(x + 2)(x + 1) \end{aligned}$$

$f'(x)$ is zero when $x = 0$, $x = -2$, or $x = -1$, so the critical numbers are 0, -2, and -1.

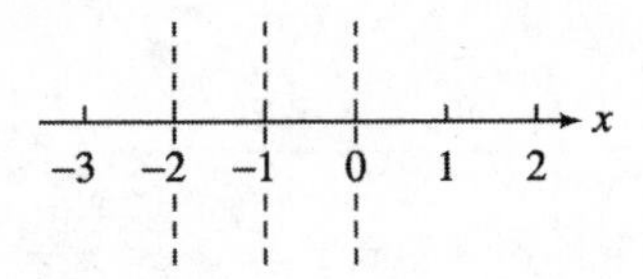

Test a point in each interval.

$$\begin{aligned} f'(-3) &= -12(-1)(-2) = -24 < 0 \\ f'(-1.5) &= -6(.5)(-.5) = 1.5 > 0 \\ f'(-.5) &= -2(1.5)(.5) = -1.5 < 0 \\ f'(1) &= 4(3)(2) = 24 > 0 \end{aligned}$$

(b) f is increasing on $(-2, -1)$ and $(0, \infty)$.

(c) f is decreasing on $(-\infty, -2)$ and $(-1, 0)$.

21. $y = -3x + 6$

(a) $y' = -3 < 0$

There are no critical numbers since y' is never 0 and always exists.

(b) Since y' is always negative, the function is increasing on no interval.

(c) y' is always negative, so the function is decreasing everywhere, or on the interval $(-\infty, \infty)$.

23. $f(x) = \dfrac{x + 2}{x + 1}$

(a) $$\begin{aligned} f'(x) &= \frac{(x + 1)(1) - (x + 2)(1)}{(x + 1)^2} \\ &= \frac{-1}{(x + 1)^2} \end{aligned}$$

The derivative is never 0, but it fails to exist at $x = -1$. Since -1 is not in the domain of f, however, -1 is not a critical number.

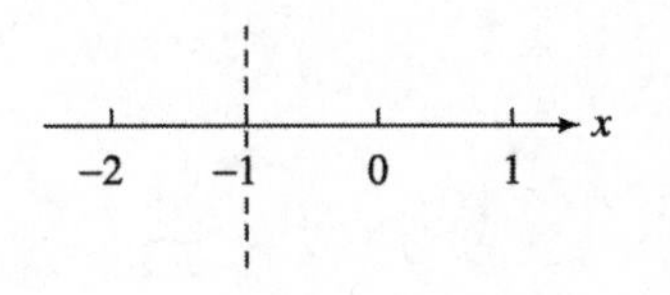

$$\begin{aligned} f'(-2) &= -1 < 0 \\ f'(0) &= -1 < 0 \end{aligned}$$

(b) f is increasing on no interval.

(c) f is decreasing everywhere that it is defined, on $(-\infty, -1)$ and on $(-1, \infty)$.

25. $$\begin{aligned} y &= \sqrt{x^2 + 1} \\ &= (x^2 + 1)^{1/2} \end{aligned}$$

(a) $$\begin{aligned} y' &= \frac{1}{2}(x^2 + 1)^{-1/2}(2x) \\ &= x(x^2 + 1)^{-1/2} \\ &= \frac{x}{\sqrt{x^2 + 1}} \end{aligned}$$

$y' = 0$ when $x = 0$.
Since y does not fail to exist for any x, and since $y' = 0$ when $x = 0$, 0 is the only critical number.

-2 -1 0 1 2 x

$$\begin{aligned} y'(1) &= \frac{1}{\sqrt{2}} > 0 \\ y'(-1) &= \frac{-1}{\sqrt{2}} < 0 \end{aligned}$$

(b) y is increasing on $(0, \infty)$.

(c) y is decreasing on $(-\infty, 0)$.

27. $f(x) = x^{2/3}$

(a) $f'(x) = \frac{2}{3}x^{-1/3} = \frac{2}{3x^{1/3}}$

$f'(x)$ is never zero, but fails to exist when $x = 0$, so 0 is the only critical number.

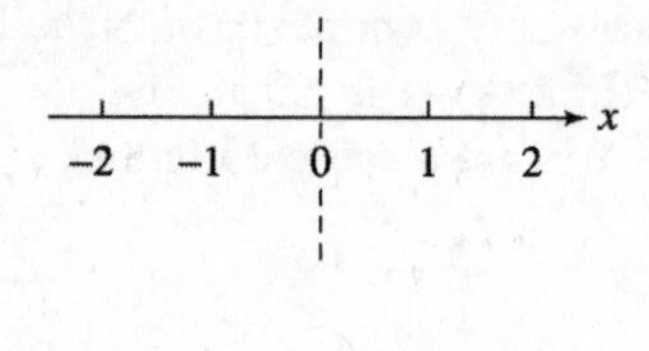

$$f'(-1) = -\frac{2}{3} < 0$$

$$f'(1) = \frac{2}{3} > 0$$

(b) f is increasing on $(0, \infty)$.

(c) f is decreasing on $(-\infty, 0)$.

29. $y = x - 4 \ln (3x - 9)$

(a) $y' = 1 - \frac{12}{3x - 9} = 1 - \frac{4}{x - 3}$

$= \frac{x - 7}{x - 3}$

y' is zero when $x = 7$. The derivative does not exist at $x = 3$, but note that the domain of f is $(3, \infty)$.
Thus, the only critical number is 7.

Choose values in the intervals $(3, 7)$ and $(7, \infty)$.

$$f'(4) = -3 < 0$$

$$f(8) = \frac{1}{5} > 0$$

(b) The function is increasing on $(7, \infty)$.

(c) The function is decreasing on $(3, 7)$.

31. $f(x) = xe^{-3x}$

(a) $f'(x) = e^{-3x} + x(-3e^{-3x})$
$= (1 - 3x)e^{-3x}$
$= \frac{1 - 3x}{e^{3x}}$

$f'(x)$ is zero when $x = \frac{1}{3}$ and there are no values of x where $f'(x)$ does not exist, so the critical number is $\frac{1}{3}$.

Test a point in each interval.

$$f'(0) = \frac{1 - 3(0)}{e^{3(0)}} = 1 > 0$$

$$f'(1) = \frac{1 - 3(1)}{e^{3(1)}} = -\frac{2}{e^3} < 0$$

(b) The function is increasing on $\left(-\infty, \frac{1}{3}\right)$.

(c) The function is decreasing on $\left(\frac{1}{3}, \infty\right)$.

33. $f(x) = x^2 2^{-x}$

(a) $f'(x) = x^2[\ln 2(2^{-x})(-1)] + (2^{-x})2x$
$= 2^{-x}(-x^2 \ln 2 + 2x)$
$= \frac{x(2 - x \ln 2)}{2^x}$

$f'(x)$ is zero when $x = 0$ or $x = \frac{2}{\ln 2}$ and there are no values of x where $f'(x)$ does not exist. The critical numbers are 0 and $\frac{2}{\ln 2}$.

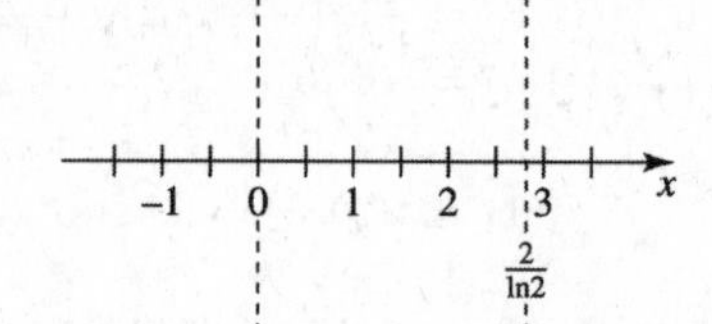

Test a point in each interval.

$$f'(-1) = \frac{(-1)(2 - (-1) \ln 2)}{2^{-1}} = -2(2 + \ln 2) < 0$$

$$f'(1) = \frac{(1)(2 - (1) \ln 2)}{2^1} = \frac{2 - \ln 2}{2} > 0$$

$$f'(3) = \frac{(3)(2 - (3) \ln 2)}{2^3} = \frac{3(2 - 3 \ln 2)}{8} < 0$$

(b) The function is increasing on $\left(0, \frac{2}{\ln 2}\right)$.

(c) The function is decreasing on $(-\infty, 0)$ and $\left(\frac{2}{\ln 2}, \infty\right)$.

35. $y = x^{2/3} - x^{5/3}$

(a) $y' = \frac{2}{3}x^{-1/3} - \frac{5}{3}x^{2/3} = \frac{2-5x}{3x^{1/3}}$

$y' = 0$ when $x = \frac{2}{5}$. The derivative does not exist at $x = 0$. So the critical numbers are 0 and $\frac{2}{5}$.

Test a point in each interval.

$$y'(-1) = \frac{7}{-3} < 0$$

$$y'\left(\frac{1}{5}\right) = \frac{1}{3\left(\frac{1}{5}\right)^{1/3}} = \frac{5^{1/3}}{3} > 0$$

$$y'(1) = \frac{-3}{3} = -1 < 0$$

(b) y is increasing on $(0, \frac{2}{5})$.

(c) y is decreasing on $(-\infty, 0)$ and $(\frac{2}{5}, \infty)$.

39. $f(x) = ax^2 + bx + c,\ a < 0$
$f'(x) = 2ax + b$

Let $f'(x) = 0$ to find the critical number.

$$2ax + b = 0$$
$$2ax = -b$$
$$x = \frac{-b}{2a}$$

Choose a value in the interval $\left(-\infty, -\frac{-b}{2a}\right)$. Since $a < 0$,

$$\frac{-b}{2a} - \frac{-1}{2a} = \frac{-b+1}{2a} < \frac{-b}{2a}.$$

$$f'\left(\frac{-b+1}{2a}\right) = 2a\left(\frac{-b+1}{2a}\right) + b = 1 > 0$$

Choose a value in the interval $\left(\frac{-b}{2a}, \infty\right)$. Since $a < 0$,

$$\frac{-b}{2a} - \frac{-1}{2a} = \frac{-b-1}{2a} < \frac{-b}{2a}.$$

$$f'\left(\frac{-b-1}{2a}\right) = 2a\left(\frac{-b-1}{2a}\right) + b = -1 < 0$$

f' is increasing on $\left(-\infty, \frac{-b}{2a}\right)$ and decreasing on $\left(\frac{-b}{2a}, \infty\right)$.

This tells us that the curve opens downward and $x = \frac{-b}{2a}$ is the x-coordinate of the vertex.

$$f\left(\frac{-b}{2a}\right) = a\left(\frac{-b}{2a}\right)^2 + b\left(\frac{-b}{2a}\right) + c = \frac{ab^2}{4a^2} - \frac{b^2}{2a} + c = \frac{b^2}{4a} - \frac{2b^2}{4a} + \frac{4ac}{4a} = \frac{4ac - b^2}{4a}$$

The vertex is $\left(\frac{-b}{2a}, \frac{4ac-b^2}{4a}\right)$ or $\left(-\frac{b}{2a}, \frac{4ac-b^2}{4a}\right)$.

41. $f(x) = \ln x$

$$f'(x) = \frac{1}{x}$$

$f'(x)$ is undefined at $x = 0$. $f'(x)$ never equals zero. Note that $f(x)$ has a domain of $(0, \infty)$. Pick a value in the interval $(0, \infty)$.

$$f'(2) = \frac{1}{2} > 0$$

$f(x)$ is increasing on $(0, \infty)$.
$f(x)$ is never decreasing.
Since $f(x)$ never equals zero, the tangent line is horizontal nowhere.

43. $f(x) = e^{0.001x} - \ln x$

$$f'(x) = 0.001e^{0.001x} - \frac{1}{x}$$

Note that $f(x)$ is only defined for $x > 0$. Use a graphing calculator to plot $f'(x)$ for $x > 0$.

(a) $f'(x) > 0$ about $(567, \infty)$, so $f(x)$ is increasing about $(567, \infty)$.

(b) $f'(x) < 0$ about $(0, 567)$, so $f(x)$ is decreasing about $(0, 567)$.

45. $H(r) = \frac{300}{1 + 0.03r^2} = 300(1 + 0.03r^2)^{-1}$

$$H'(r) = 300[-1(1 + 0.03r^2)^{-2}(0.06r)] = \frac{-18r}{(1 + 0.03r^2)^2}$$

Since r is a mortgage rate (in percent), it is always positive. Thus, $H'(r)$ is always negative.

(a) H is increasing on nowhere.

(b) H is decreasing on $(0, \infty)$.

47. $C(x) = 0.32x^2 - 0.00004x^3$
$R(x) = 0.848x^2 - 0.0002x^3$
$$\begin{aligned} P(x) &= R(x) - C(x) \\ &= (0.848x^2 - 0.0002x^3) - (0.32x^2 - 0.00004x^3) \\ &= 0.528x^2 - 0.00016x^3 \end{aligned}$$
$P'(x) = 1.056x - 0.00048x^2$

$$\begin{aligned} 1.056x - 0.00048x^2 &= 0 \\ x(1.056 - 0.00048x) &= 0 \\ x = 0 \text{ or } x &= 2200 \end{aligned}$$

Choose $x = 1000$ and $x = 3000$ as test points.

$$\begin{aligned} P'(1000) &= 1.056(1000) - 0.00048(1000)^2 = 576 \\ P'(3000) &= 1.056(3000) - 0.00048(3000)^2 \\ &= -1152 \end{aligned}$$

The function is increasing on $(0, 2200)$.

49. **(a)** These curves are graphs of functions since they all pass the vertical line test.

(b) The graph for particulates increases from April to July; it decreases from July to November; it is constant from January to April and November to December.

(c) All graphs are constant from January to April and November to December. When the temperature is low, as it is during these months, air pollution is greatly reduced.

51. $$\begin{aligned} A(x) &= 0.003631x^3 - 0.03746x^2 \\ &\quad + 0.1012x + 0.009 \\ A'(x) &= 0.010893x^2 - 0.07492x + 0.1012 \end{aligned}$$

Solve for $A'(x) = 0$.

$x \approx 1.85$ or $x \approx 5.03$

Choose $x = 1$ and $x = 4$ as test points.

$$\begin{aligned} A'(1) &= 0.010893(1)^2 - 0.07492(1) \\ &\quad + 0.1012 \\ &= 0.037173 \\ A'(4) &= 0.010893(4)^2 - 0.07492(4) \\ &\quad + 0.1012 \\ &= -0.024192 \end{aligned}$$

(a) The function is increasing on $(0, 1.85)$.

(b) The function is decreasing on $(1.85, 5)$.

53. $$\begin{aligned} K(t) &= \frac{5t}{t^2+1} \\ K'(t) &= \frac{5(t^2+1) - 2t(5t)}{(t^2+1)^2} \\ &= \frac{5t^2 + 5 - 10t^2}{(t^2+1)^2} \\ &= \frac{5 - 5t^2}{(t^2+1)^2} \end{aligned}$$

$K'(t) = 0$ when

$$\begin{aligned} \frac{5-5t^2}{(t^2+1)^2} &= 0 \\ 5 - 5t^2 &= 0 \\ 5t^2 &= 5 \\ t &= \pm 1. \end{aligned}$$

Since t is the time after a drug is administered, the function applies only for $[0, \infty)$, so we discard $t = -1$. Then 1 divides the interval into two intervals.

$$\begin{aligned} K'(0.5) &= 2.4 > 0 \\ K'(2) &= -0.6 < 0 \end{aligned}$$

(a) K is increasing on $(0, 1)$.

(b) K is decreasing on $(1, \infty)$.

55. **(a)** $$\begin{aligned} F(t) &= -10.28 + 175.9te^{-t/1.3} \\ F'(t) &= (175.9)(e^{-t/1.3}) \\ &\quad + (175.9.9t)\left(-\frac{1}{1.3}e^{-t/1.3}\right) \\ &= (175.9)(e^{-t/1.3})\left(1 - \frac{t}{1.3}\right) \\ &\approx 175.9e^{-t/1.3}(1 - 0.769t) \end{aligned}$$

(b) $F'(t)$ is equal to 0 at $t = 1.3$. Therefore, 1.3 is a critical number. Since the domain is $(0, \infty)$, test values in the intervals from $(0, 1.3)$ and $(1.3, \infty)$.

$$F'(1) \approx 18.83 > 0 \text{ and } F'(2) \approx -20.32 < 0$$

$F'(t)$ is increasing on $(0, 1.3)$ and decreasing on $(1.3, \infty)$.

57. $f(x) = \frac{1}{\sqrt{2\pi}} e^{-x^2/2}$

$$f'(x) = \frac{1}{\sqrt{2\pi}} e^{-x^2/2}(-x)$$
$$= \frac{-x}{\sqrt{2\pi}} e^{-x^2/2}$$

$f'(x) = 0$ when $x = 0$.
Choose a value from each of the intervals $(-\infty, 0)$ and $(0, \infty)$.

$$f'(-1) = \frac{1}{\sqrt{2\pi}} e^{-1/2} > 0$$
$$f'(1) = \frac{-1}{\sqrt{2\pi}} e^{-1/2} < 0$$

The function is increasing on $(-\infty, 0)$ and decreasing on $(0, \infty)$.

59. As shown on the graph,

(a) horsepower increases with engine speed on $(1500, 6250)$;

(b) horsepower decreases with engine speed on $(6250, 7200)$;

(c) torque increases with engine speed on $(1500, 2500)$ and $(3500, 4400)$;

(d) torque decreases with engine speed on $(3000, 3500)$ and $(6000, 7200)$.

5.2 Relative Extrema

1. As shown on the graph, the relative minimum of -4 occurs when $x = 1$.

3. As shown on the graph, the relative maximum of 3 occurs when $x = -2$.

5. As shown on the graph, the relative maximum of 3 occurs when $x = -4$ and the relative minimum of 1 occurs when $x = -2$.

7. As shown on the graph, the relative maximum of 3 occurs when $x = -4$; the relative minimum of -2 occurs when $x = -7$ and $x = -2$

9. Since the graph of the function is zero at $x = -1$ and $x = 3$, the critical numbers are -1 and 3.

Since the graph of the function is positive on $(-\infty, -1)$ and negative on $(-1, 3)$, there is a relative maximum at -1. Since the graph of the function is negative on $(-1, 3)$ and positive on $(3, \infty)$, there is a relative minimum at 3.

11. Since the graph of the function is zero at $x = -8, x = -6, x = -2.5$ and $x = -1.5$, the critical numbers are $-8, -6, -2.5$, and -1.5.

Since the graph of the function is positive on $(-\infty, -8)$ and negative on $(-8, -6)$, there is a relative maximum at -8. Since the graph of the function is on $(-8, -6)$ and positive on $(-6, -2.5)$, there is a relative minimum at -6. Since the graph of the function is positive on $(-6, -2.5)$ and negative on $(-2.5, -1.5)$, there is a relative maximum at -2.5. Since the graph of the function is negative on $(-2.5, -1.5)$ and positive on $(-1.5, \infty)$, there is a relative minimum at -1.5.

13. $f(x) = x^2 - 10x + 33$
$f'(x) = 2x - 10$

$f'(x)$ is zero when $x = 5$.

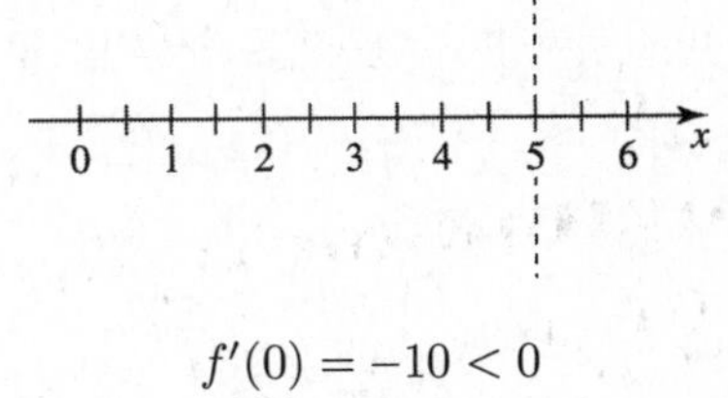

$$f'(0) = -10 < 0$$
$$f'(6) = 2 > 0$$

f is decreasing on $(-\infty, 5)$ and increasing on $(5, \infty)$. Thus, a relative minimum occurs at $x = 5$.

$f(5) = 8$

Relative minimum of 8 at 5

15. $f(x) = x^3 + 6x^2 + 9x - 8$
$f'(x) = 3x^2 + 12x + 9 = 3(x^2 + 4x + 3)$
$= 3(x + 3)(x + 1)$

$f'(x)$ is zero when $x = -1$ or $x = -3$.

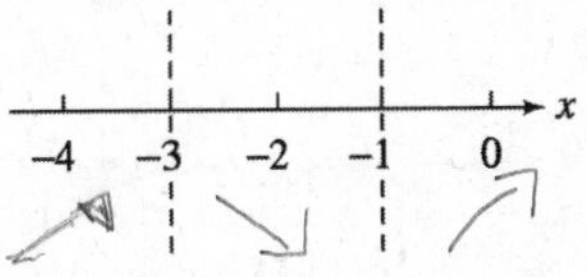

$$f'(-4) = 9 > 0$$
$$f'(-2) = -3 < 0$$
$$f'(0) = 9 > 0$$

Thus, f is increasing on $(-\infty, -3)$, decreasing on $(-3, -1)$, and increasing on $(-1, \infty)$.
f has a relative maximum at -3 and a relative minimum at -1.

$$f(-3) = -8$$
$$f(-1) = -12$$

Relative maximum of -8 at -3; relative minimum of -12 at -1

17. $f(x) = -\frac{4}{3}x^3 - \frac{21}{2}x^2 - 5x + 8$

$$\begin{aligned} f'(x) &= -4x^2 - 21x - 5 \\ &= (-4x - 1)(x + 5) \end{aligned}$$

$f'(x)$ is zero when $x = -5$, or $x = -\frac{1}{4}$.

$$\begin{aligned} f'(-6) &= -23 < 0 \\ f'(-4) &= 15 > 0 \\ f'(0) &= -5 < 0 \end{aligned}$$

f is decreasing on $(-\infty, -5)$, increasing on $\left(-5, -\frac{1}{4}\right)$, and decreasing on $\left(-\frac{1}{4}, \infty\right)$. f has a relative minimum at -5 and a relative maximum at $-\frac{1}{4}$.

$$f(-5) = -\frac{377}{6}$$

$$f\left(-\frac{1}{4}\right) = \frac{827}{96}$$

Relative maximum of $\frac{827}{96}$ at $-\frac{1}{4}$; relative minimum of $-\frac{377}{6}$ at -5

19. $f(x) = x^4 - 18x^2 - 4$

$$\begin{aligned} f'(x) &= 4x^3 - 36x \\ &= 4x(x^2 - 9) \\ &= 4x(x + 3)(x - 3) \end{aligned}$$

$f'(x)$ is zero when $x = 0$ or $x = -3$ or $x = 3$.

$$\begin{aligned} f'(-4) &= 4(-4)^3 - 36(-4) = -112 < 0 \\ f'(-1) &= -4 + 36 = 32 > 0 \\ f'(1) &= 4 - 36 = -32 < 0 \\ f'(4) &= 4(4)^3 - 36(4) = 112 > 0 \end{aligned}$$

f is decreasing on $(-\infty, -3)$ and $(0, 3)$; f is increasing on $(-3, 0)$ and $(3, \infty)$.

$$\begin{aligned} f(-3) &= -85 \\ f(0) &= -4 \\ f(3) &= -85 \end{aligned}$$

Relative maximum of -4 at 0; relative minimum of -85 at 3 and -3

21. $f(x) = 3 - (8 + 3x)^{2/3}$

$$\begin{aligned} f'(x) &= -\frac{2}{3}(8 + 3x)^{-1/3}(3) \\ &= -\frac{2}{(8 + 3x)^{1/3}} \end{aligned}$$

Critical number:

$$\begin{aligned} 8 + 3x &= 0 \\ x &= -\frac{8}{3} \end{aligned}$$

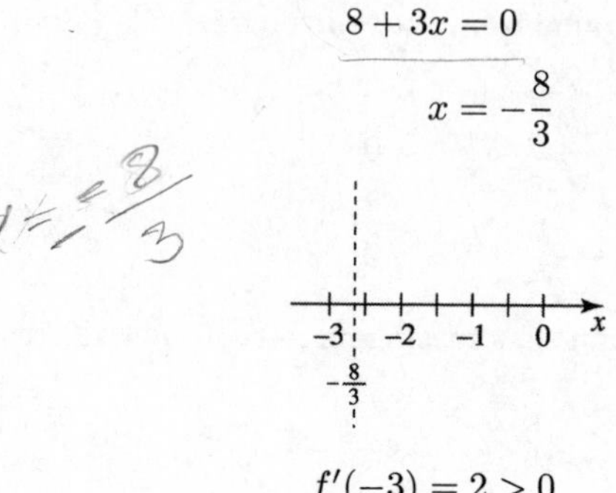

$$\begin{aligned} f'(-3) &= 2 > 0 \\ f'(0) &= -1 < 0 \end{aligned}$$

f is increasing on $\left(-\infty, -\frac{8}{3}\right)$ and decreasing on $\left(-\frac{8}{3}, \infty\right)$.

$$f\left(-\frac{8}{3}\right) = 3$$

Relative maximum of 3 at $-\frac{8}{3}$

23. $f(x) = 2x + 3x^{2/3}$

$$\begin{aligned} f'(x) &= 2 + 2x^{-1/3} \\ &= 2 + \frac{2}{\sqrt[3]{x}} \end{aligned}$$

Find the critical numbers.

$f'(x) = 0$ when

$$\begin{aligned} 2 + \frac{2}{\sqrt[3]{x}} &= 0 \\ \frac{2}{\sqrt[3]{x}} &= -2 \\ \frac{1}{\sqrt[3]{x}} &= -1 \\ x &= (-1)^3 \\ x &= -1. \end{aligned}$$

$f'(x)$ does not exist when

$$\begin{aligned} \sqrt[3]{x} &= 0 \\ x &= 0. \end{aligned}$$

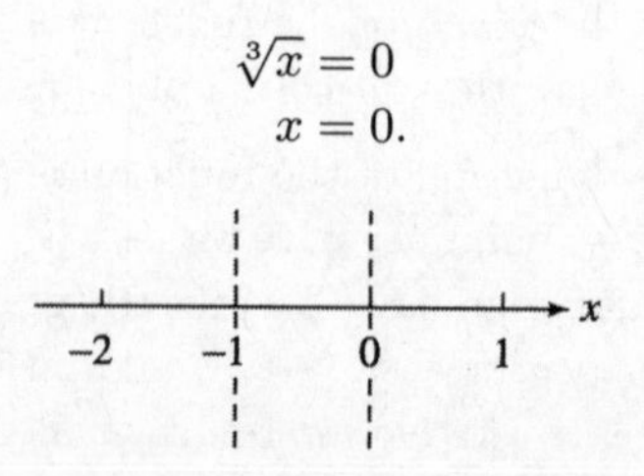

$$f'(-2) = 2 + \frac{2}{\sqrt[3]{-2}} \approx 0.41 > 0$$

$$f'\left(-\frac{1}{2}\right) = 2 + \frac{2}{\sqrt[3]{-\frac{1}{2}}}$$

$$= 2 + \frac{2\sqrt[3]{2}}{-1} \approx -0.52 < 0$$

$$f'(1) = 2 + \frac{2}{\sqrt[3]{1}} = 4 > 0$$

f is increasing on $(-\infty, -1)$ and $(0, \infty)$.
f is decreasing on $(-1, 0)$.

$$f(-1) = 2(-1) + 3(-1)^{2/3} = 1$$
$$f(0) = 0$$

Relative maximum of 1 at -1; relative minimum of 0 at 0

25. $f(x) = x - \frac{1}{x}$

$f'(x) = 1 + \frac{1}{x^2}$ is never zero, but fails to exist at $x = 0$.
Since $f(x)$ also fails to exist at $x = 0$, there are no critical numbers and no relative extrema.

27. $f(x) = \frac{x^2 - 2x + 1}{x - 3}$

$$f'(x) = \frac{(x-3)(2x-2) - (x^2 - 2x + 1)(1)}{(x-3)^2}$$

$$= \frac{x^2 - 6x + 5}{(x-3)^2}$$

Find the critical numbers:

$$x^2 - 6x + 5 = 0$$
$$(x-5)(x-1) = 0$$
$$x = 5 \quad \text{or} \quad x = 1$$

Note that $f(x)$ and $f'(x)$ do not exist at $x = 3$, so the only critical numbers are 1 and 5.

+ − +
0 1 2 3 4 5 6 x

$$f'(0) = \frac{5}{9} > 0$$

$$f'(2) = -3 < 0$$

$$f'(6) = \frac{5}{9} > 0$$

$f(x)$ is increasing on $(-\infty, 1)$ and $(5, \infty)$.
$f(x)$ is decreasing on $(1, 5)$.

$$f(1) = 0$$
$$f(5) = 8$$

Relative maximum of 0 at 1; relative minimum of 8 at 5

29. $f(x) = x^2 e^x - 3$
$f'(x) = x^2 e^x + 2xe^x$
$= xe^x(x + 2)$

$f'(x)$ is zero at $x = 0$ and $x = -2$.

−3 −2 −1 0 1 x

$$f'(-3) = 3e^{-3} = \frac{3}{e^3} > 0$$

$$f'(-1) = -e^{-1} = \frac{-1}{e} < 0$$

$$f'(1) = 3e^1 > 0$$

f is increasing on $(-\infty, -2)$ and $(0, \infty)$.
f is decreasing on $(-2, 0)$.

$$f(0) = 0 \cdot e^0 - 3 = -3$$
$$f(-2) = (-2)^2 e^{-2} - 3$$
$$= \frac{4}{e^2} - 3$$
$$\approx -2.46$$

Relative minimum of -3 at 0; relative maximum of -2.46 at -2

31. $f(x) = 2x + \ln x$

$$f'(x) = 2 + \frac{1}{x} = \frac{2x+1}{x}$$

$f'(x)$ is zero at $x = -\frac{1}{2}$. The domain of $f(x)$ is $(0, \infty)$. Therefore $f'(x)$ is never zero in the domain of $f(x)$.
$f'(1) = 3 > 0$. Since $f(x)$ is always increasing, f has no relative extrema.

33. $f(x) = \frac{2^x}{x}$

$$f'(x) = \frac{(x)\ln 2(2^x) - 2^x(1)}{x^2}$$

$$= \frac{2^x(x \ln 2 - 1)}{x^2}$$

Find the critical numbers:

$$x \ln 2 - 1 = 0 \quad \text{or} \quad x^2 = 0$$

$$x = \frac{1}{\ln 2} \qquad x = 0$$

Since f is not defined for $x = 0$, 0 is not a critical number. $x = \frac{1}{\ln 2} \approx 1.44$ is the only critical number.

$$f'(1) \approx -0.6137 < 0$$
$$f'(2) \approx 0.3863 > 0$$

f is decreasing on $\left(0, \frac{1}{\ln 2}\right)$ and increasing on $\left(\frac{1}{\ln 2}, \infty\right)$.

$$f\left(\frac{1}{\ln 2}\right) = \frac{2^{1/\ln 2}}{\frac{1}{\ln 2}} = e \ln 2$$

Relative minimum of $e \ln 2$ at $\dfrac{1}{\ln 2}$

35.
$$\begin{aligned} y &= -2x^2 + 12x - 5 \\ y' &= -4x + 12 \\ &= -4(x - 3) \end{aligned}$$

The vertex occurs when $y' = 0$ or when

$$\begin{aligned} x - 3 &= 0 \\ x &= 3 \end{aligned}$$

When $x = 3$,

$$y = -2(3)^2 + 12(3) - 5 = 13$$

The vertex is $(3, 13)$.

37. $f(x) = x^5 - x^4 + 4x^3 - 30x^2 + 5x + 6$

$f'(x) = 5x^4 - 4x^3 + 12x^2 - 60x + 5$

Graph f' on a graphing calculator. A suitable choice for the viewing window is $[-4, 4]$ by $[-50, 50]$, Yscl $= 10$.

Use the calculator to estimate the x-intercepts of this graph. These numbers are the solutions of the equation $f'(x) = 0$ and thus the critical numbers for f. Rounded to three decimal places, these x-values are 0.085 and 2.161.

Examine the graph of f' near $x = 0.085$ and $x = 2.161$. Observe that $f'(x) > 0$ to the left of $x = 0.085$ and $f'(x) < 0$ to the right of $x = 0.085$. Also observe that $f'(x) < 0$ to the left of $x = 2.161$ and $f'(x) > 0$ to the right of $x = 2.161$. The first derivative test allows us to conclude that f has a relative maximum at $x = 0.085$ and a relative minimum at $x = 2.161$.

$$f(0.085) \approx 6.211$$
$$f(2.161) \approx -57.607$$

Relative maximum of 6.211 at 0.085; relative minimum of -57.607 at 2.161.

39. $f(x) = 2|x + 1| + 4|x - 5| - 20$

Graph this function in the window $[-10, 10]$ by $[-15, 30]$, Yscl $= 5$.

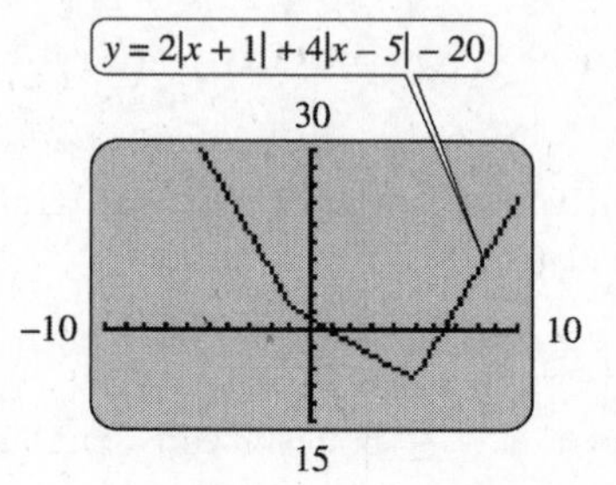

The graph shows that f has no relative maxima, but there is a relative minimum at $x = 5$. (Note that the graph has a sharp point at $(5, -8)$, indicating that $f'(5)$ does not exist.)

41. $C(q) = 80 + 18q$; $p = 70 - 2q$

$$\begin{aligned} P(q) &= R(q) - C(q) = pq - C(q) \\ &= (70 - 2q)q - (80 + 18q) \\ &= -2q^2 + 52q - 80 \end{aligned}$$

(a) Since the graph of P is a parabola that opens downward, we know that its vertex is a maximum point. To find the q-value of this point, we find the critical number.

$P'(q) = -4q + 52$
$P'(q) = 0$ when

$$\begin{aligned} -4q + 52 &= 0 \\ 4q &= 52 \\ q &= 13 \end{aligned}$$

The number of units that produce maximum profit is 13.

(b) If $q = 13$,

$$\begin{aligned} p &= 70 - 2(13) \\ &= 44 \end{aligned}$$

The price that produces maximum profit is \$44.

(c) $P(13) = -2(13)^2 + 52(13) - 80 = 258$

The maximum profit is \$258.

43. $C(q) = 100 + 20qe^{-0.01q}$; $p = 40e^{-0.01q}$

$$\begin{aligned}P(q) &= R(q) - C(q) = pq - C(q)\\ &= (40e^{-0.01q})q - (100 + 20qe^{-0.01q})\\ &= 20qe^{-0.01q} - 100\end{aligned}$$

(a) $$\begin{aligned}P'(q) &= 20e^{-0.01q} + 20qe^{-0.01q}(-0.01)\\ &= (20 - 0.2q)e^{-0.01q}\end{aligned}$$

Solve $P'(q) = 0$.

$$\begin{aligned}(20 - 0.2q)e^{-0.01q} &= 0\\ 20 - 0.2q &= 0\\ q &= 100\end{aligned}$$

Since $e^{-0.01q} > 0$ for all values of q, the sign of $P'(q)$ is the same as the sign of $20 - 0.2q$. For $q < 100, P'(q) > 0$; for $q > 100, P'(q) < 0$. Therefore, the number of units that produces maximum profit is 100.

(b) If $q = 100$,

$$\begin{aligned}p &= 40e^{-0.01(100)}\\ &= 40e^{-1}\\ &\approx 14.72\end{aligned}$$

The price per unit that produces maximum profit is \$14.72.

(c) $$\begin{aligned}P(100) &= 20(100)e^{-0.01(100)} - 100\\ &= 2000e^{-1} - 100\\ &\approx 635.76\end{aligned}$$

The maximum profit is \$635.76.

45. $$\begin{aligned}P(t) &= -0.01432t^3 + 0.3976t^2 - 2.257t + 23.41\\ P'(t) &= -0.04296t^2 + 0.7952t - 2.257\end{aligned}$$

Solve for $P'(t) = 0:$

$t \approx 3.5001$ or $t \approx 15.0101$

Test points $t = 1, t = 4, t = 16$.

$$\begin{aligned}P'(1) &\approx -1.5048\\ P'(4) &\approx 0.2364\\ P'(16) &\approx -0.5316\end{aligned}$$

P is decreasing on $(0, 3.5)$ and $(15, 18)$ and increasing on $(3.5, 15)$.

$$\begin{aligned}P(0) &= 23.41\\ P(3.5) &\approx 19.767\\ P(15) &\approx 30.685\\ P(18) &\approx 28.092\end{aligned}$$

Relative maximum of 23,410 megawatts at midnight; relative minimum of 19,767 megawatts at 3:30 a.m.; relative maximum of 30,685 at 3:00 p.m.; relative minimum of 28,092 at 6:00 p.m.

47. $$\begin{aligned}p &= D(q) = 200e^{-0.1q}\\ R(q) &= pq\\ &= 200qe^{-0.1q}\\ R'(q) &= 200qe^{-0.1q}(-0.1) + 200e^{-0.1q}\\ &= 20e^{-0.1q}(10 - q)\end{aligned}$$

$R'(q) = 0$ when $q = 10$, the only critical number. Use the first derivative test to verify that $q = 10$ gives the maximum revenue.

$$\begin{aligned}R'(9) &= 20e^{-0.9} > 0\\ R'(11) &= -20e^{-1.1} < 0\end{aligned}$$

The maximum revenue results when $q = 10$

$p = D(10) = \frac{200}{e} \approx 73.58$, or when telephones are sold at \$73.58.

49. $$\begin{aligned}C(x) &= 0.002x^3 = 9x + 6912\\ \overline{C}(x) &= \frac{C(x)}{x} = 0.002x^2 + 9 + \frac{6912}{x}\\ \overline{C}'(x) &= 0.004x - \frac{6912}{x^2}\end{aligned}$$

$\overline{C}'(x) = 0$ when

$$\begin{aligned}0.004x - \frac{6912}{x^2} &= 0\\ 0.004x^3 &= 6912\\ x^3 &= 1{,}728{,}000\\ x &= 120\end{aligned}$$

A product level of 120 units will produce the minimum average cost per unit.

51. **(a)** $$\begin{aligned}M(t) &= 6.281t^{0.242}e^{-0.025t}\\ M'(t) &= (1.520002t^{-0.758})(e^{-0.025t})\\ &\quad + (6.281t^{0.242})(-0.025e^{-0.025t})\\ &= e^{-0.025t}(1.520002t^{-0.758} - 0.157025t^{0.242})\end{aligned}$$

$M'(t) = 0$ when

$$\begin{aligned}1.520002t^{-0.758} - 0.157025t^{0.242} &= 0\\ t &= 9.68\end{aligned}$$

Let $t = 9.68$ in $M(t)$.

$$\begin{aligned}M(9.68) &= 6.281(9.68)^{0.242}e^{-0.025(9.68)}\\ &\approx 8.54 \text{ kg}\end{aligned}$$

The maximum daily consumption is 8.54 kg and it occurs at 9.68 weeks.

(b) $M(t) = at^b e^{-ct}$

$$M'(t) = (bat^{b-1})(e^{-ct}) + (at^b)(-ce^{-ct})$$
$$= ae^{-ct}(bt^{b-1} - ct^b)$$

$M'(t) = 0$ when

$$bt^{b-1} - ct^b = 0$$
$$bt^{b-1} = ct^b$$
$$\frac{b}{c} = t$$

Let $t = \frac{b}{c}$ in $M(t)$.

$$M\left(\frac{b}{c}\right) = a\left(\frac{b}{c}\right)^b e^{-c\left(\frac{b}{c}\right)}$$
$$= a\left(\frac{b}{c}\right)^b e^{-b}$$

The maximum daily consumption is $a(b/c)^b e^{-b}$ kg and it occurs at b/c weeks.

53. $F(t) = -10.28 + 175.9te^{-t/1.3}$

$$F'(t) = (175.9t)\left(-\frac{1}{1.3}\right)e^{-t/1.3} + (175.9)(e^{-t/1.3})$$
$$= e^{-t/1.3}\left(-\frac{175.9t}{1.3} + 175.9\right)$$

$F'(t) = 0$ when

$$-\frac{175.9t}{1.3} + 175.9 = 0$$
$$t = 1.3$$

At 1.3 hours, the termal effect of the food is maximized.

55. $R(t) = \frac{20t}{t^2 + 100}$

$$R'(t) = \frac{20(t^2 + 100) - 20t(2t)}{(t^2 + 100)^2} = \frac{2000 - 20t^2}{(t^2 + 100)^2}$$

$R'(t) = 0$ when

$$2000 - 20t^2 = 0$$
$$-20t^2 = -2000$$
$$t^2 = 100$$
$$t = \pm 10.$$

Disregard the negative value.
Use the first derivative test to verify that $t = 10$ gives a maximum rating.

$$R'(9) = 0.0116 > 0$$
$$R'(11) = -0.0086 < 0$$

The film should be 10 minutes long.

5.3 Higher Derivatives, Concavity, and the Second Derivative Test

1. $f(x) = 5x^3 - 7x^2 + 4x + 3$

$$f'(x) = 15x^2 - 14x + 4$$
$$f''(x) = 30x - 14$$
$$f''(0) = 30(0) - 14 = -14$$
$$f''(2) = 30(2) - 14 = 46$$

3. $f(x) = 4x^4 - 3x^3 - 2x^2 + 6$

$$f'(x) = 16x^3 - 9x^2 - 4x$$
$$f''(x) = 48x^2 - 18x - 4$$
$$f''(0) = 48(0)^2 - 18(0) - 4 = -4$$
$$f''(2) = 48(2)^2 - 18(2) - 4 = 152$$

5. $f(x) = 3x^2 - 4x + 8$

$$f'(x) = 6x - 4$$
$$f''(x) = 6$$
$$f''(0) = 6$$
$$f''(2) = 6$$

7. $f(x) = \frac{x^2}{1 + x}$

$$f'(x) = \frac{(1 + x)(2x) - x^2(1)}{(1 + x)^2}$$
$$= \frac{2x + x^2}{(1 + x)^2}$$

$$f''(x) = \frac{(1 + x)^2(2 + 2x) - (2x + x^2)(2)(1 + x)}{(1 + x)^4}$$
$$= \frac{(1 + x)(2 + 2x) - (2x + x^2)(2)}{(1 + x)^3}$$
$$= \frac{2}{(1 + x)^3}$$

$$f''(0) = 2$$
$$f''(2) = \frac{2}{27}$$

9. $f(x) = \sqrt{x^2+4} = (x^2+4)^{1/2}$

$$f'(x) = \frac{1}{2}(x^2+4)^{-1/2} \cdot 2x = \frac{x}{(x^2+4)^{1/2}}$$

$$f''(x) = \frac{(x^2+4)^{1/2}(1) - x\left[\frac{1}{2}(x^2+4)^{-1/2}\right]2x}{x^2+4} = \frac{(x^2+4)^{1/2} - \frac{x^2}{(x^2+4)^{1/2}}}{x^2+4} = \frac{(x^2+4)-x^2}{(x^2+4)^{3/2}} = \frac{4}{(x^2+4)^{3/2}}$$

$$f''(0) = \frac{4}{(0^2+4)^{3/2}} = \frac{4}{4^{3/2}} = \frac{4}{8} = \frac{1}{2}$$

$$f''(2) = \frac{4}{(2^2+4)^{3/2}} = \frac{4}{8^{3/2}} = \frac{4}{16\sqrt{2}} = \frac{1}{4\sqrt{2}}$$

11. $f(x) = 32x^{3/4}$

$f'(x) = 24x^{-1/4}$

$$f''(x) = -6x^{-5/4} = -\frac{6}{x^{5/4}}$$

$f''(0)$ does not exist.

$$f''(2) = -\frac{6}{2^{5/4}} = -\frac{3}{2^{1/4}}$$

13. $f(x) = 5e^{-x^2}$

$f'(x) = 5e^{-x^2}(-2x) = -10xe^{-x^2}$

$f''(x) = -10xe^{-x^2}(-2x) + e^{-x^2}(-10) = 20x^2e^{-x^2} - 10e^{-x^2}$

$f''(0) = 20(0^2)e^{-0^2} - 10e = 0 - 10 = -10$

$f''(2) = 20(2^2)e^{-(2^2)} - 10e^{-(2^2)} = 80e^{-4} - 10e^{-4} = 70e^{-4} \approx 1.282$

15. $f(x) = \dfrac{\ln x}{4x}$

$$f'(x) = \frac{4x\left(\frac{1}{x}\right) - (\ln x)(4)}{(4x)^2} = \frac{4 - 4\ln x}{16x^2} = \frac{1-\ln x}{4x^2}$$

$$f''(x) = \frac{4x^2\left(-\frac{1}{x}\right) - (1-\ln x)8x}{16x^4} = \frac{-4x - 8x + 8x\ln x}{16x^4} = \frac{-12x + 8x\ln x}{16x^4} = \frac{4x(-3+2\ln x)}{16x^4} = \frac{-3+2\ln x}{4x^3}$$

$f''(0)$ does not exist because $\ln 0$ is undefined.

$$f''(2) = \frac{-3+2\ln 2}{4(2)^3} = \frac{-3+2\ln 2}{32} \approx 0.050$$

17. $f(x) = 7x^4 + 6x^3 + 5x^2 + 4x + 3$

$f'(x) = 28x^3 + 18x^2 + 10x + 4$

$f''(x) = 84x^2 + 36x + 10$

$f'''(x) = 168x + 36$

$f^{(4)}(x) = 168$

19. $f(x) = 5x^5 - 3x^4 + 2x^3 + 7x^2 + 4$

$f'(x) = 25x^4 - 12x^3 + 6x^2 + 14x$

$f''(x) = 100x^3 - 36x^2 + 12x + 14$

$f'''(x) = 300x^2 - 72x + 12$

$f^{(4)}(x) = 600x - 72$

21. $f(x) = \dfrac{x-1}{x+2}$

$$f'(x) = \frac{(x+2)-(x-1)}{(x+2)^2} = \frac{3}{(x+2)^2}$$

$$f''(x) = \frac{-3(2)(x+2)}{(x+2)^4} = \frac{-6}{(x+2)^3}$$

$$f'''(x) = \frac{(-6)(-3)(x+2)^2}{(x+2)^6} = 18(x+2)^{-4} \text{ or } \frac{18}{(x+2)^4}$$

$$f^{(4)}(x) = \frac{-18(4)(x+2)^3}{(x+2)^8} = -72(x+2)^{-5} \text{ or } \frac{-72}{(x+2)^5}$$

23. $f(x) = \frac{3x}{x-2}$

$f'(x) = \frac{(x-2)(3) - 3x(1)}{(x-2)^2} = \frac{-6}{(x-2)^2}$

$f''(x) = \frac{-6(-2)(x-2)}{(x-2)^4} = \frac{12}{(x-2)^3}$

$f'''(x) = \frac{-12(3)(x-2)^2}{(x-2)^6} = -36(x-2)^{-4}$

or $\frac{-36}{(x-2)^4}$

$f^{(4)}(x) = \frac{-36(-4)(x-2)^3}{(x-2)^8} = 144(x-2)^{-5}$

or $\frac{144}{(x-2)^5}$

25. $f(x) = \ln x$

(a) $f'(x) = \frac{1}{x} = x^{-1}$

$f''(x) = -x^{-2} = \frac{-1}{x^2}$

$f'''(x) = 2x^{-3} = \frac{2}{x^3}$

$f^{(4)}(x) = -6x^{-4} = \frac{-6}{x^4}$

$f^{(5)}(x) = 24x^{-5} = \frac{24}{x^5}$

(b) $f^{(n)}(x) = \frac{(-1)^{n-1}(n-1)!}{x^n}$

27. Concave upward on $(2, \infty)$
Concave downward on $(-\infty, 2)$
Inflection point at $(2, 3)$

29. Concave upward on $(-\infty, -1)$ and $(8, \infty)$
Concave downward on $(-1, 8)$
Inflection point at $(-1, 7)$ and $(8, 6)$

31. Concave upward on $(2, \infty)$
Concave downward on $(-\infty, 2)$
No points of inflection

33. $f(x) = x^2 + 10x - 9$
$f'(x) = 2x + 10$
$f''(x) = 2 > 0$ for all x.

Always concave upward
No inflection points

35. $f(x) = -2x^3 + 9x^2 + 168x - 3$
$f'(x) = -6x^2 + 18x + 168$
$f''(x) = -12x + 18$
$f''(x) = -12x + 18 > 0$ when
$-6(2x - 3) > 0$
$2x - 3 < 0$
$x < \frac{3}{2}.$

Concave upward on $\left(-\infty, \frac{3}{2}\right)$

$f''(x) = -12x + 18 < 0$ when
$-6(2x - 3) < 0$
$2x - 3 > 0$
$x > \frac{3}{2}.$

Concave downward on $\left(\frac{3}{2}, \infty\right)$

$f''(x) = -12x + 18 = 0$ when
$-6(2x + 3) = 0$
$2x + 3 = 0$
$x = \frac{3}{2}.$

$f\left(\frac{3}{2}\right) = \frac{525}{2}$

Inflection point at $\left(\frac{3}{2}, \frac{525}{2}\right)$

37. $f(x) = \frac{3}{x-5}$

$f'(x) = \frac{-3}{(x-5)^2}$

$f''(x) = \frac{-3(-2)(x-5)}{(x-5)^4} = \frac{6}{(x-5)^3}$

$f''(x) = \frac{6}{(x-5)^3} > 0$ when
$(x-5)^3 > 0$
$x - 5 > 0$
$x > 5.$

Concave upward on $(5, \infty)$

$f''(x) = \frac{6}{(x-5)^3} < 0$ when

$(x-5)^3 < 0$
$x - 5 < 0$
$x < 5.$

Concave downward on $(-\infty, 5)$

$f''(x) \neq 0$ for any value for x; it does not exist when $x = 5$. There is a change of concavity there, but no inflection point since $f(5)$ does not exist.

39. $f(x) = x(x+5)^2$

$f'(x) = x(2)(x+5) + (x+5)^2$
$= (x+5)(2x+x+5)$
$= (x+5)(3x+5)$

$f''(x) = (x+5)(3) + (3x+5)$
$= 3x + 15 + 3x + 5 = 6x + 20$

$f''(x) = 6x + 20 > 0$ when
$2(3x+10) > 0$
$3x > -10$
$x > -\frac{10}{3}.$

Concave upward on $\left(-\frac{10}{3}, \infty\right)$

$f''(x) = 6x + 20 < 0$ when
$2(3x+10) < 0$
$3x < -10$
$x < -\frac{10}{3}.$

Concave downward on $\left(-\infty, -\frac{10}{3}\right)$

$$f\left(-\frac{10}{3}\right) = -\frac{10}{3}\left(-\frac{10}{3}+5\right)^2 = \frac{-10}{3}\left(\frac{-10+15}{3}\right)^2 = -\frac{10}{3}\cdot\frac{25}{9} = -\frac{250}{27}$$

Inflection point at $\left(-\frac{10}{3}, -\frac{250}{27}\right)$

41. $f(x) = 18x - 18e^{-x}$

$f'(x) = 18 - 18e^{-x}(-1) = 18 + 18e^{-x}$

$f''(x) = 18e^{-x}(-1) = -18e^{-x}$

$f''(x) = -18e^{-x} < 0$ for all x

$f(x)$ is never concave upward and always concave downward. There are no points of inflection since $-18e^{-x}$ is never equal to 0.

43. $f(x) = x^{8/3} - 4x^{5/3}$

$$f'(x) = \frac{8}{3}x^{5/3} - \frac{20}{3}x^{2/3}$$

$$f''(x) = \frac{40}{9}x^{2/3} - \frac{40}{9}x^{-1/3} = \frac{40(x-1)}{9x^{1/3}}$$

$f''(x) = 0$ when $x = 1$

$f''(x)$ fails to exist when $x = 0$

Note that both $f(x)$ and $f'(x)$ exist at $x = 0$.

Check the sign of $f''(x)$ in the three intervals determined by $x = 0$ and $x = 1$ using test points.

$$f''(-1) = \frac{40(-2)}{9(-1)} = \frac{80}{9} > 0$$

$$f''\left(\frac{1}{8}\right) = \frac{40\left(-\frac{7}{8}\right)}{9\left(\frac{1}{2}\right)} = -\frac{70}{9} < 0$$

$$f''(8) = \frac{40(7)}{9(2)} = \frac{140}{9} > 0$$

Concave upward on $(-\infty, 0)$ and $(1, \infty)$; concave downward on $(0, 1)$

$f(0) = (0)^{8/3} - 4(0)^{5/3} = 0$
$f(1) = (1)^{8/3} - 4(1)^{5/3} = -3$

Inflection points at $(0, 0)$ and $(1, -3)$

45. $f(x) = \ln(x^2+1)$

$$f'(x) = \frac{2x}{x^2+1}$$

$$f''(x) = \frac{(x^2+1)(2) - (2x)(2x)}{(x^2+1)^2} = \frac{-2x^2+2}{(x^2+1)^2}$$

$$f''(x) = \frac{-2x^2+2}{(x^2+1)^2} > 0 \text{ when}$$

$-2x^2 + 2 > 0$
$-2x^2 > -2$
$x^2 < 1$
$-1 < x < 1$

Concave upward on $(-1, 1)$

$$f''(x) = \frac{-2x^2+2}{(x^2+1)^2} < 0 \text{ when}$$

$-2x^2 + 2 < 0$
$-2x^2 < -2$
$x^2 > 1$
$x > 1$ or $x < -1$

Concave downward on $(-\infty, -1)$ and $(1, \infty)$

$f(1) = \ln[(1)^2 + 1] = \ln 2$
$f(-1) = \ln[(-1)^2 + 1] = \ln 2$

Inflection points at $(-1, \ln 2)$ and $(1, \ln 2)$

47. $f(x) = x^2 \log |x|$

$$f'(x) = 2x \log |x| + x^2 \left(\frac{1}{x \ln 10}\right)$$
$$= 2x \log |x| + \frac{x}{\ln 10}$$
$$f''(x) = 2 \log |x| + 2x \left(\frac{1}{x \ln 10}\right) + \frac{1}{\ln 10}$$
$$= 2 \log |x| + \frac{3}{\ln 10}$$
$$f''(x) = 2 \log |x| + \frac{3}{\ln 10} > 0 \text{ when}$$
$$2 \log |x| + \frac{3}{\ln 10} > 0$$
$$2 \log |x| > -\frac{3}{\ln 10}$$
$$\log |x| > -\frac{3}{2 \ln 10}$$
$$\frac{\ln |x|}{\ln 10} > -\frac{3}{2 \ln 10}$$
$$\ln |x| > -\frac{3}{2}$$
$$|x| > e^{-3/2}$$
$$x > e^{-3/2} \text{ or } x < -e^{-3/2}$$

Concave upward on $(-\infty, -e^{-3/2})$ and $(e^{-3/2}, \infty)$

$$f''(x) = 2 \log |x| + \frac{3}{\ln 10} < 0 \text{ when}$$
$$2 \log |x| + \frac{3}{\ln 10} < 0$$
$$2 \log |x| < -\frac{3}{\ln 10}$$
$$\log |x| < -\frac{3}{2 \ln 10}$$
$$\frac{\ln |x|}{\ln 10} < -\frac{3}{2 \ln 10}$$
$$\ln |x| < -\frac{3}{2}$$
$$|x| < e^{-3/2}$$
$$-e^{-3/2} < x < e^{-3/2}$$

Note that $f(x)$ is not defined at $x = 0$.

Concave downward on $(-e^{-3/2}, 0)$ and $(0, e^{-3/2})$.

$$f(-e^{-3/2}) = (-e^{-3/2})^2 \log |-e^{-3/2}|$$
$$= e^{-3} \log e^{-3/2} = -\frac{3e^{-3}}{2 \ln 10}$$
$$f(e^{-3/2}) = (e^{-3/2})^2 \log |e^{-3/2}|$$
$$= e^{-3} \log e^{-3/2} = -\frac{3e^{-3}}{2 \ln 10}$$

Inflection points at $\left(-e^{-3/2}, -\frac{3e^{-3}}{2 \ln 10}\right)$ and $\left(e^{-3/2}, -\frac{3e^{-3}}{2 \ln 10}\right)$

49. Since the graph of $f'(x)$ is increasing on $(-\infty, 0)$ and $(4, \infty)$, the function is concave upward on $(-\infty, 0)$ and $(4, \infty)$. Since the graph of $f'(x)$ is decreasing on $(0, 4)$, the function is concave downward on $(0, 4)$. The inflection points are at 0 and 4.

51. Since the graph of $f'(x)$ is increasing on $(-7, 3)$ and $(12, \infty)$, the function is concave upward on $(-7, 3)$ and $(12, \infty)$. Since the graph of $f'(x)$ is decreasing on $(-\infty, -7)$ and $(3, 12)$, the function is concave downward on $(-\infty, -7)$ and $(3, 12)$. The inflection points are at $-7, 3$, and 12.

53. Choose $f(x) = x^k$, where $1 < k < 2$.

If $k = \frac{4}{3}$, then

$$f'(x) = \frac{4}{3}x^{1/3} \qquad f''(x) = \frac{4}{9}x^{-2/3} = \frac{4}{9x^{2/3}}$$

Critical number: 0

Since $f'(x)$ is negative when $x < 0$ and positive when $x > 0$, $f(x) = x^{4/3}$ has a relative minimum at $x = 0$.

If $k = \frac{5}{3}$, then

$$f'(x) = \frac{5}{3}x^{2/3} \qquad f''(x) = \frac{10}{9}x^{-1/3} = \frac{10}{9x^{1/3}}$$

$f''(x)$ is never 0, and does not exist when $x = 0$; so, the only candidate for an inflection point is at $x = 0$.

Since $f''(x)$ is negative when $x < 0$ and positive when $x > 0$, $f(x) = x^{5/3}$ has an inflection point at $x = 0$.

55. **(a)** The slope of the tangent line to $f(x) = e^x$ as $x \to -\infty$ is close to 0 since the tangent line is almost horizontal, and a horizontal line has a slope of 0.

(b) The slope of the tangent line to $f(x) = e^x$ as $x \to 0$ is close to 1 since the first derivative represents the slope of the tangent line, $f'(x) = e^x$, and $e^0 = 1$.

57. $f(x) = -x^2 - 10x - 25$

$$f'(x) = -2x - 10$$
$$= -2(x + 5) = 0$$

Critical number: -5

$f''(x) = -2 < 0$ for all x.

The curve is concave downward, which means a relative maximum occurs at $x = -5$.

59. $f(x) = 3x^3 - 3x^2 + 1$
$f'(x) = 9x^2 - 6x$
$= 3x(3x - 2) = 0$

Critical numbers: 0 and $\frac{2}{3}$

$f''(x) = 18x - 6$
$f''(0) = -6 < 0$, which means that a relative maximum occurs at $x = 0$.

$f''\left(\frac{2}{3}\right) = 6 > 0$, which means that a relative minimum occurs at $x = \frac{2}{3}$.

61. $f(x) = (x + 3)^4$
$f'(x) = 4(x + 3)^3 = 0$

Critical number: $x = -3$

$f''(x) = 12(x + 3)^2$
$f''(-3) = 12(-3 + 3)^2 = 0$

The second derivative test fails.
Use the first derivative test.

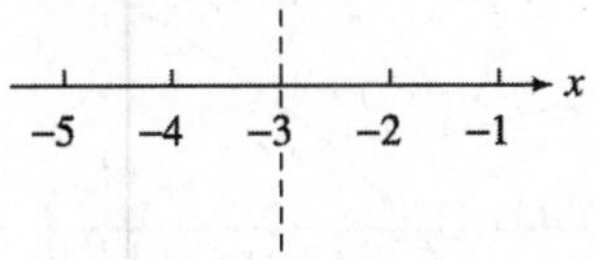

$$f'(-4) = 4(-4 + 3)^2$$
$$= 4(-1)^3 = -4 < 0$$

This indicates that f is decreasing on $(-\infty, -3)$.

$$f'(0) = 4(0 + 3)^3$$
$$= 4(3)^3 = 108 > 0$$

This indicates that f is increasing on $(-3, \infty)$.
A relative minimum occurs at -3.

63. $f(x) = x^{7/3} + x^{4/3}$

$f'(x) = \frac{7}{3}x^{4/3} + \frac{4}{3}x^{1/3}$

$f'(x) = 0$ when

$$\frac{7}{3}x^{4/3} + \frac{4}{3}x^{1/3} = 0$$

$$\frac{x^{1/3}}{3}(7x + 4) = 0$$

$$x = 0 \text{ or } x = -\frac{4}{7}.$$

Critical numbers: $-\frac{4}{7}, 0$

$f''(x) = \frac{28}{9}x^{1/3} + \frac{4}{9}x^{-2/3}$

$f''\left(-\frac{4}{7}\right) = \frac{28}{9}\left(-\frac{4}{7}\right)^{1/3} + \frac{4}{9}\left(-\frac{4}{7}\right)^{-2/3} \approx -1.9363$

Relative maximum occurs at $-\frac{4}{7}$.

$f''(0)$ does not exist, so the second derivative test fails.

Use the first derivative test.

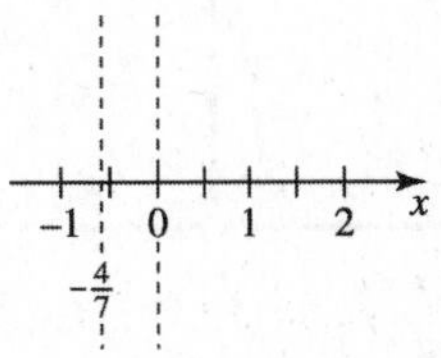

$f'\left(-\frac{1}{2}\right) = \frac{7}{3}\left(-\frac{1}{2}\right)^{4/3} + \frac{4}{3}\left(-\frac{1}{2}\right)^{1/3} \approx -0.1323$

This indicates that f is decreasing on $\left(-\frac{4}{7}, 0\right)$.

$f'(1) = \frac{7}{3}(1)^{4/3} + \frac{4}{3}(1)^{1/3} = \frac{11}{3}$

This indicates that f is increasing on $(0, \infty)$.
Relative minimum occurs at 0.

65. There are many examples. The easiest is $f(x) = \sqrt{x}$. This graph is increasing and concave downward.

$$f'(x) = \tfrac{1}{2}x^{-1/2} = \tfrac{1}{2\sqrt{x}}$$

$f'(0)$ does not exist, while $f'(x) > 0$ for all $x > 0$. (Note that the domain of f is $[0, \infty)$.)
As x increases, the value of $f'(x)$ decreases, but remains positive. It approaches zero, but never becomes zero or negative.

67. $f'(x) = 10x^2(x - 1)(5x - 3)$
$= 10x^2(5x^2 - 8x + 3)$
$= 50x^4 - 80x^3 + 30x^2$
$f''(x) = 200x^3 - 240x^2 + 60x$
$= 20x(10x^2 - 12x + 3)$

Graph f' in the window $[-1, 1.5]$ by $[-2, 2]$, Xscl $= 0.1$.

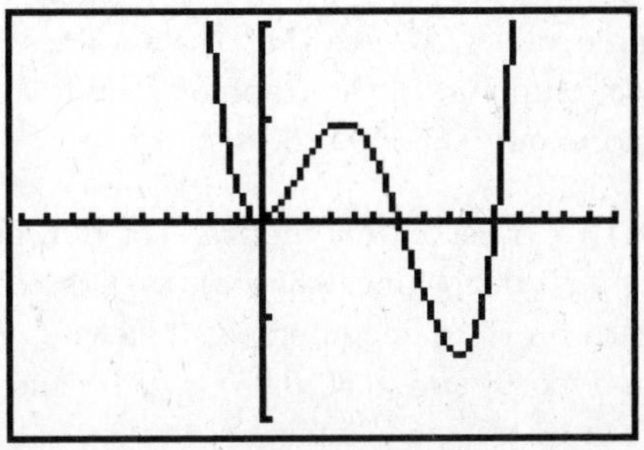

This window does not give a good view of the graph of f'', so we graph f'' in the window $[-1, 1.5]$ by $[-20, 20]$, Xscl $= 0.1$. Yscl $= 5$.

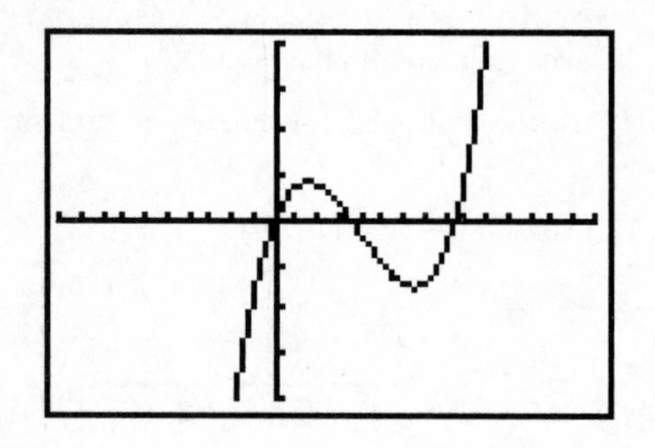

(a) The critical numbers of f are the x-intercepts of the graph of f'. (Note that there are no values where $f'(x)$ does not exist.) From the graph or by examining the factored expression for f', we see that the critical numbers of f are $0, 0.6$, and 1.

By either looking at the graph of f' and applying the first derivative test or by looking at the graph of f'' and applying the second derivative test, we see that f has a relative minimum at 1 and a relative maximum at 0.6.
(At $x = 0$, the second derivative test fails since $f''(0) = 0$, and the first derivative does not change sign, so there is no relative extremum at 0.)

(b) Examine the graph of f' to determine the intervals where the graph lies above and below the x-axis. We see that $f'(x) \geq 0$ on $(-\infty, 0.6)$, $f'(x) < 0$ on $(0.6, 1)$, and $f'(x) > 0$ on $(1, \infty)$. Therefore, f is increasing on $(-\infty, 0.6)$ and $(1, \infty)$ and decreasing on $(0.6, 1)$.

(c) Examine the graph of f''. We see that this graph has three x-intercepts, so there are three values where $f''(x) = 0$. These x-values are 0, about 0.36, and about 0.85. Because the sign of f'' and thus the concavity of f changes at these three values, we see that the x-values of the inflection points of the graph of f are 0, about 0.36, and about 0.85.

(d) We observe from the graph of f'' that $f''(x) > 0$ on $(0, 0.36)$ and $(0.85, \infty)$, so f is concave upward on the same intervals. Likewise, $f''(x) < 0$ on $(-\infty, 0)$ and $(0.36, 0.85)$, so f is concave downward on the same intervals.

69. $f'(x) = x^2 + x \ln x$

$$f''(x) = 2x + (x)\left(\frac{1}{x}\right) + (1)(\ln x)$$
$$= 2x + 1 + \ln x$$

Graph f' and f'' in the window $[0, 1]$ by $[-2, 3]$, Xscl $= 0.1$, Yscl $= 1$.

Graph of f':

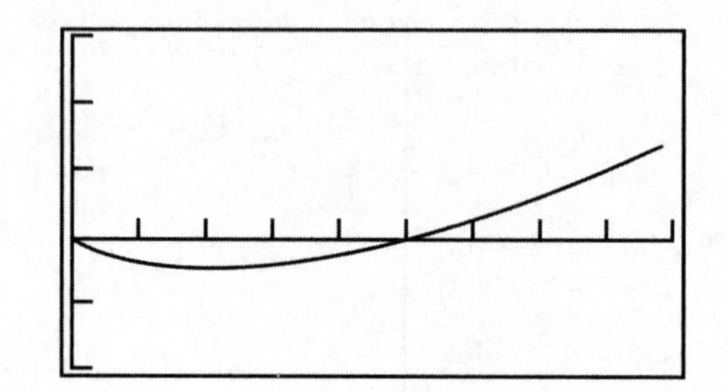

Graph of f'':

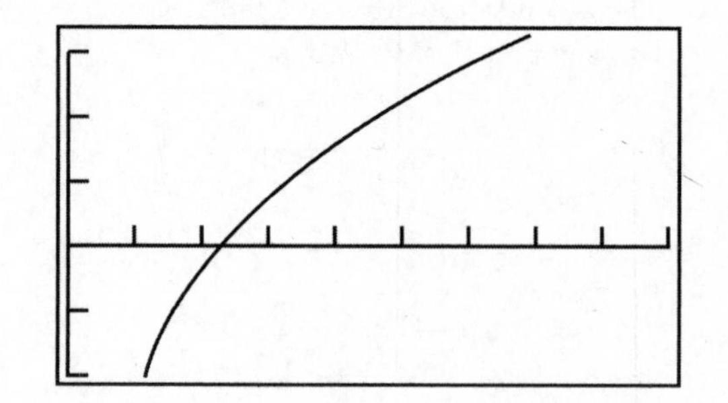

(a) The critical number of f is the x-intercept of the graph of f'. Using the graph, we find a critical number of f is about 0.5671. By looking at the graph of f'' and applying the second derivative test, we see f has a minimum at 0.5671.

(b) Examine the graph of f' to determine the intervals where the graph lies above and below the x-axis. We see that $f'(x) > 0$ on about $(0.5671, \infty)$, indicating that f is increasing on about $(0.5671, \infty)$. We also see that $f'(x) < 0$ on about $(0, 0.5671)$, indicating that f is decreasing on about $(0, 0.5671)$.

(c) Examine the graph of f''. We see that the graph has one x-intercept, so there is one x-value where $f''(x) = 0$. This value is about 0.2315. Because the sign of f'' changes at this value, we see that x-value of the inflection point of the graph of f is about 0.2315.

(d) We observe from the graph f'' that $f'' > 0$ on about $(0.2315, \infty)$, so f is concave upward on about $(0.2315, \infty)$. Likewise, we observe from the graph f'' that $f'' < 0$ on about $(0, 0.2315)$, so f is concave downward on about $(0, 0.2315)$.

71. $f(t) = 1.5207t^4 - 19.166t^3 + 62.91t^2 + 6.0726t + 1026$
$f'(t) = 6.0828t^3 - 57.498t^2 + 125.82t + 6.0726$
$f''(t) = 18.2484t^2 - 114.996t + 125.82$

Solve $f''(x) = 0$ to find the inflection points.

$$18.2484t^2 - 114.996t + 125.82 = 0$$

$$t = \frac{114.996 \pm \sqrt{(-114.996)^2 - 4(18.2484)(125.82)}}{2(18.2484)}$$

$t \approx 1.409$ or $t \approx 4.892$

$$f'(1.409) = 6.0828(1.409)^3 - 57.498(1.409)^2 + 125.82(1.409) + 6.0726 \approx 86.218$$

$$f'(4.892) = 6.0828(4.892)^3 - 57.498(4.892)^2 + 125.82(4.892) + 6.0726 \approx -42.303$$

Rents were increasing most rapidly when $t \approx 1.409$ or about mid 1999.

73. $R(x) = \frac{4}{27}(-x^3 + 66x^2 + 1050x - 400)$

$0 \le x \le 25$

$$R'(x) = \frac{4}{27}(-3x^2 + 132x + 1050)$$

$$R''(x) = \frac{4}{27}(-6x + 132)$$

A point of diminishing returns occurs at a point of inflection, or where $R''(x) = 0$.

$$\frac{4}{27}(-6x + 132) = 0$$
$$-6x + 132 = 0$$
$$6x = 132$$
$$x = 22$$

Test $R''(x)$ to determine whether concavity changes at $x = 22$.

$$R''(20) = \frac{4}{27}(-6 \cdot 20 + 132) = \frac{16}{9} > 0$$

$$R''(24) = \frac{4}{27}(-6 \cdot 24 + 132) = -\frac{16}{9} < 0$$

$R(x)$ is concave upward on $(0, 22)$ and concave downward on $(22, 25)$.

$$R(22) = \frac{4}{27}[-(22)^3 + 66(22)^2 + 1060(22) - 400] \approx 6517.9$$

The point of diminishing returns is $(22, 6517.9)$.

75. $R(x) = -0.6x^3 + 3.7x^2 + 5x$, $0 \le x \le 6$
$R'(x) = -1.8x^2 + 7.4x + 5$
$R''(x) = -3.6x + 7.4$

A point of diminishing returns occurs at a point of inflection or where $R''(x) = 0$.

$$-3.6x + 7.4 = 0$$
$$-3.6x = -7.4$$
$$x = \frac{-7.4}{-3.6} \approx 2.06$$

Test $R''(x)$ to determine whether concavity changes at $x = 2.05$.

$$R''(2) = -3.6(2) + 7.4 = -7.2 + 7.4 = 0.2 > 0$$
$$R''(3) = -3.6(3) + 7.4 = -10.8 + 7.4 = -3.4 < 0$$

$R(x)$ is concave upward on $(0, 2.06)$ and concave downward on $(2.06, 6)$.

$$R(2.06) = -0.6(2.06)^3 + 3.7(2.06)^2 + 5(2.06) \approx 20.8$$

The point of diminishing returns is $(2.06, 20.8)$.

77. Let $D(q)$ represent the demand function. The revenue function, $R(q)$, is $R(q) = qD(q)$. The marginal revenue is given by

$$R'(q) = qD'(q) + D(q)(1) = qD'(q) + D(q).$$

$$R''(q) = qD''(q) + D'(q)(1) + D'(q) = qD''(q) + 2D'(q)$$

gives the rate of decline of marginal revenue.
$D'(q)$ gives the rate of decline of price.
If marginal revenue declines more quickly than price,

$$qD''(q) + 2D'(q) - D'(q) < 0$$
$$\text{or} \quad qD''(q) + D'(q) < 0.$$

79. **(a)** $R(t) = t^2(t-18) + 96t + 1000;\ 0 < t < 8$
$= t^3 - 18t^2 + 96t + 1000$
$R'(t) = 3t^2 - 36t + 96$

Set $R'(t) = 0$.

$$3t^2 - 36t + 96 = 0$$
$$t^2 - 12t + 32 = 0$$
$$(t-8)(t-4) = 0$$
$$t = 8 \quad \text{or} \quad t = 4$$

8 is not in the domain of $R(t)$.
$R''(t) = 6t - 36$
$R''(4) = -12 < 0$ implies that $R(t)$ is maximized at $t = 4$, so the population is maximized at 4 hours.

(b) $R(4) = 16(-14) + 96(4) + 1000$
$= -224 + 384 + 1000$
$= 1160$

The maximum population is 1160 million.

81. $K(x) = \dfrac{3x}{x^2+4}$

(a) $K'(x) = \dfrac{3(x^2+4) - (2x)(3x)}{(x^2+4)^2}$
$= \dfrac{-3x^2+12}{(x^2+4)^2} = 0$

$$-3x^2 + 12 = 0$$
$$x^2 = 4$$
$$x = 2 \quad \text{or} \quad x = -2$$

For this application, the domain of K is $[0, \infty)$, so the only critical number is 2.

$$K''(x) = \frac{(x^2+4)^2(-6x) - (-3x^2+12)(2)(x^2+4)(2x)}{(x^2+4)^4}$$
$$= \frac{-6x(x^2+4) - 4x(-3x^2+12)}{(x^2+4)^3}$$
$$= \frac{6x^3 - 72x}{(x^2+4)^3}$$

$K''(2) = \frac{-96}{512} = -\frac{3}{16} < 0$ implies that $K(x)$ is maximized at $x = 2$.
Thus, the concentration is a maximum after 2 hours.

(b) $K(2) = \dfrac{3(2)}{(2)^2+4} = \dfrac{3}{4}$

The maximum concentration is $\frac{3}{4}\%$.

83. $G(t) = \dfrac{10{,}000}{1+49e^{-0.1t}}$

$$G'(t) = \frac{(1+49e^{-0.1t})(0) - (10{,}000)(-4.9e^{-0.1t})}{(1+49e^{-0.1t})^2}$$
$$= \frac{49{,}000e^{-0.1t}}{(1+49e^{-0.1t})^2}$$

To find $G''(t)$, apply the quotient rule to find the derivative of $G'(t)$.
The numerator of $G''(t)$ will be

$$(1+49e^{-0.1t})^2(-4900e^{-0.1t})$$
$$- (49{,}000e^{-0.1t})(2)(1+49e^{-0.1t})(-4.9e^{-0.1t})$$
$$= (1+49e^{-0.1t})(-4900e^{-0.1t})$$
$$\cdot [(1+49e^{-0.1t}) - 20(4.9e^{-0.1t})]$$
$$= (-4900e^{-0.1t})[1+49e^{-0.1t} - 98e^{-0.1t}]$$
$$= (-4900e^{-0.1t})(1-49e^{-0.1t}).$$

Thus,

$$G''(t) = \frac{(-4900e^{-0.1t})(1-49e^{-0.1t})}{(1+49e^{-0.1t})^4}.$$

$G''(t) = 0$ when $-4900e^{-0.1t} = 0$ or $1 - 49e^{-0.1t} = 0$.
$-4900e^{-0.1t} < 0$, and thus never equals zero.

$$1 - 49e^{-0.1t} = 0$$
$$1 = 49e^{-0.1t}$$
$$\frac{1}{49} = e^{-0.1t}$$
$$\ln\left(\frac{1}{49}\right) = -0.1t$$
$$\ln 1 - \ln 49 = -0.1t$$
$$-\ln 49 = -0.1t$$
$$\ln 49 = 0.1t$$
$$\ln 7^2 = 0.1t$$
$$2 \ln 7 = 0.1t$$
$$20 \ln 7 = t$$
$$38.9182 \approx t$$

The point of inflection is $(38.9182, 5000)$.

85. $L(t) = Be^{-ce^{-kt}}$
$L'(t) = Be^{-ce^{-kt}}(-ce^{-kt})'$
$= Be^{-ce^{-kt}}[-ce^{-kt}(-kt)']$
$= Bcke^{-ce^{-kt}-kt}$
$L''(t) = Bcke^{-ce^{-kt}-kt}(-ce^{-kt} - kt)'$
$= Bcke^{-ce^{-kt}-kt}[-ce^{-kt}(-kt)' - k]$
$= Bcke^{-ce^{-kt}-kt}(cke^{-kt} - k)$
$= Bck^2e^{-ce^{-kt}-kt}(ce^{-kt} - 1)$

$L''(t) = 0$ when $ce^{-kt} - 1 = 0$

$$\begin{aligned} ce^{-kt} - 1 &= 0 \\ \frac{c}{e^{kt}} &= 1 \\ e^{kt} &= c \\ kt &= \ln c \\ t &= \frac{\ln c}{k} \end{aligned}$$

Letting $c = 7.267963$ and $k = 0.670840$

$$t = \frac{\ln 7.267963}{0.670840} \approx 2.96 \text{ years}$$

Verify that there is a point of inflection at $t = \frac{\ln c}{k} \approx 2.96$. For

$$L''(t) = Bck^2 e^{-ce^{-kt}-kt}(ce^{-kt} - 1),$$

we only need to test the factor $ce^{-kt} - 1$ on the intervals determined by $t \approx 2.96$ since the other factors are always positive.
$L''(1)$ has the same sign as

$$7.267963e^{-0.670840(1)} - 1 \approx 2.72 > 0.$$

$L''(3)$ has the same sign as

$$7.267963e^{-0.670840(3)} - 1 \approx -0.029 < 0.$$

Therefore L, is concave up on $\left(0, \frac{\ln c}{k} \approx 2.96\right)$ and concave down on $\left(\frac{\ln c}{k}, \infty\right)$, so there is a point of inflection at $t = \frac{\ln c}{k} \approx 2.96$ years.
This signifies the time when the rate of growth begins to slow down since L changes from concave up to concave down at this inflection point.

87. $v(x) = -35.98 + 12.09x - 0.4450x^2$
$v'(x) = 12.09 - 0.89x$
$v''(x) = -0.89$

Since $-0.89 < 0$, the function is always concave down.

89. Since the rate of violent crimes is decreasing but at a slower rate than in previous years, we know that $f'(t) < 0$ but $f''(t) > 0$. Note that since $f'(t) < 0$, f is decreasing, and since $f''(t) > 0$, the graph of f is concave upward.

91. $s(t) = -16t^2$
$v(t) = s'(t) = -32t$

(a) $v(3) = -32(3) = -96$ ft/sec

(b) $v(5) = -32(5) = -160$ ft/sec

(c) $v(8) = -32(8) = -256$ ft/sec

(d) $a(t) = v'(t) = s''(t)$
$= -32$ ft/sec^2

93. $s(t) = 256t - 16t^2$
$v(t) = s'(t) = 256 - 32t$
$a(t) = v'(t) = s''(t) = -32$

To find when the maximum height occurs, set $s'(t) = 0$.

$$\begin{aligned} 256 - 32t &= 0 \\ t &= 8 \end{aligned}$$

Find the maximum height.

$$\begin{aligned} s(8) &= 256(8) - 16(8^2) \\ &= 1024 \end{aligned}$$

The maximum height of the ball is 1024 ft.
The ball hits the ground when $s = 0$.

$$\begin{aligned} &256t - 16t^2 = 0 \\ &16t(16 - t) = 0 \\ &t = 0 \quad \text{(initial moment)} \\ &t = 16 \ \text{(final moment)} \end{aligned}$$

The ball hits the ground 16 seconds after being thrown.

95. The car was moving most rapidly when $t \approx 6$, because acceleration was positive on $(0, 6)$ and negative after $t = 6$, so velocity was a maximum at $t = 6$.

5.4 Curve Sketching

1. Graph $y = x \ln |x|$ on a graphing calculator. A suitable choice for the viewing window is $[-1, 1]$ by $[-1, 1]$, Xscl = 0.1, Yscl = 0.1.

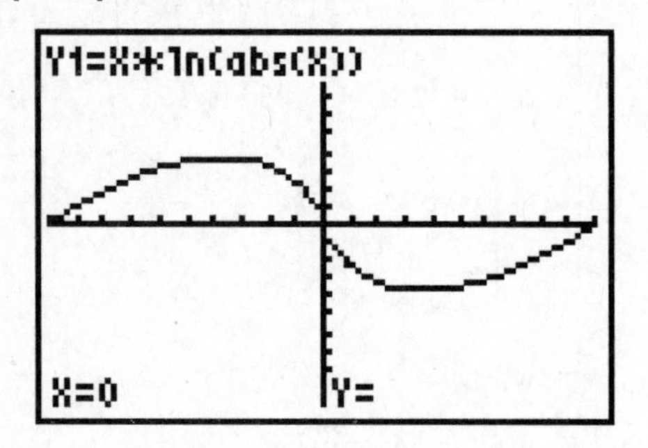

The calculator shows no y-value when $x = 0$ because 0 is not in the domain of this function. However, we see from the graph that

$$\lim_{x \to 0^-} x \ln |x| = 0$$

and

$$\lim_{x \to 0^+} x \ln |x| = 0.$$

Thus,

$$\lim_{x \to 0} x \ln |x| = 0.$$

3. $f(x) = -2x^3 - 9x^2 + 108x - 10$

Domain is $(-\infty, \infty)$.

$$\begin{aligned} f(-x) &= -2(-x)^3 - 9(-x)^2 + 108(-x) - 10 \\ &= 2x^3 - 9x^2 - 108x - 10 \end{aligned}$$

No symmetry

$$\begin{aligned} f'(x) &= -6x^2 - 18x + 108 \\ &= -6(x^2 + 3x - 18) \\ &= -6(x + 6)(x - 3) \end{aligned}$$

$f'(x) = 0$ when $x = -6$ or $x = 3$.
Critical numbers: -6 and 3
Critical points: $(-6, -550)$ and $(3, 179)$

$$\begin{aligned} f''(x) &= -12x - 18 \\ f''(-6) &= 54 > 0 \\ f''(3) &= -54 < 0 \end{aligned}$$

Relative maximum at 3, relative minimum at -6
Increasing on $(-6, 3)$
Decreasing on $(-\infty, -6)$ and $(3, \infty)$

$$\begin{aligned} f''(x) = -12x - 18 &= 0 \\ -6(2x + 3) &= 0 \\ x &= -\frac{3}{2} \end{aligned}$$

Point of inflection at $(-1.5, -185.5)$
Concave upward on $(-\infty, -1.5)$
Concave downward on $(-1.5, \infty)$
y-intercept:
$y = -2(0)^3 - 9(0)^2 + 108(0) - 10 = -10$

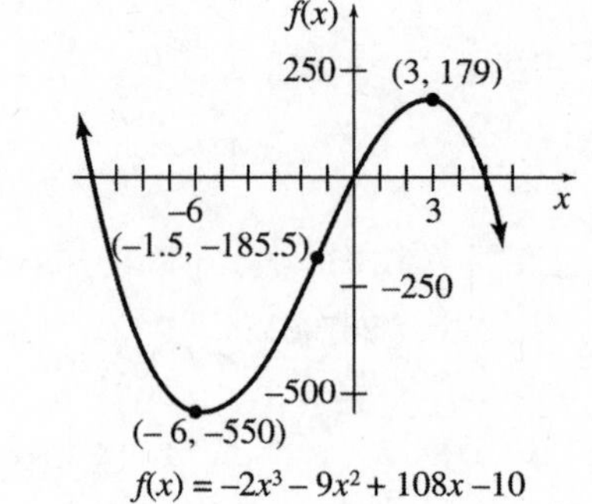

5. $f(x) = -3x^3 + 6x^2 - 4x - 1$

Domain is $(-\infty, \infty)$.

$$\begin{aligned} f(-x) &= -3(-x)^3 + 6(-x)^2 - 4(-x) - 1 \\ &= 3x^3 + 6x^2 + 4x - 1 \end{aligned}$$

No symmetry

$$\begin{aligned} f'(x) &= -9x^2 + 12x - 4 \\ &= -(3x - 2)^2 \end{aligned}$$

$$(3x - 2)^2 = 0$$
$$x = \frac{2}{3}$$

Critical number: $\frac{2}{3}$

$$f\left(\tfrac{2}{3}\right) = -3\left(\tfrac{2}{3}\right)^3 + 6\left(\tfrac{2}{3}\right)^2 - 4\left(\tfrac{2}{3}\right) - 1 = -\tfrac{17}{9}$$

Critical point: $\left(\frac{2}{3}, -\frac{17}{9}\right)$

$$\begin{aligned} f'(0) &= -9(0)^2 + 12(0) - 4 = -4 < 0 \\ f'(1) &= -9(1)^2 + 12(1) - 4 = -1 < 0 \end{aligned}$$

No relative extremum at $\left(\frac{2}{3}, -\frac{17}{9}\right)$

Decreasing on $(-\infty, \infty)$

$$\begin{aligned} f''(x) &= -18x + 12 \\ &= -6(3x - 2) \end{aligned}$$

$$3x - 2 = 0$$
$$x = \frac{2}{3}$$

Point of inflection at $\left(\frac{2}{3}, -\frac{17}{9}\right)$

$$\begin{aligned} f''(0) &= -18(0) + 12 = 12 > 0 \\ f''(1) &= -18(1) + 12 = -6 < 0 \end{aligned}$$

Concave upward on $\left(-\infty, \frac{2}{3}\right)$

Concave downward on $\left(\frac{2}{3}, \infty\right)$

Point of inflection at $\left(\frac{2}{3}, -\frac{17}{9}\right)$

y-intercept: $y = -3(0)^3 + 6(0)^2 - 4(0) - 1 = -1$

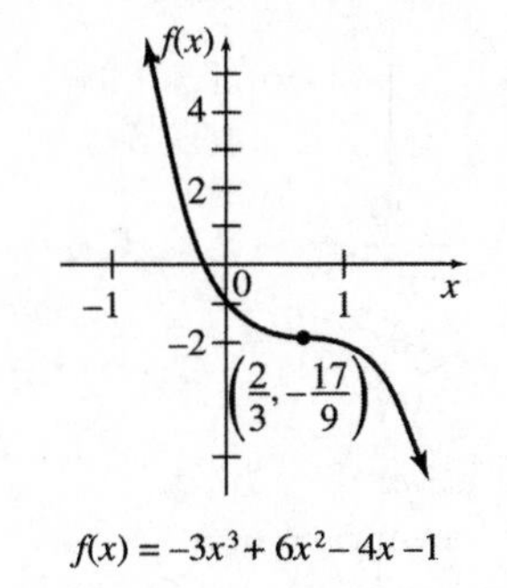

$f(x) = -3x^3 + 6x^2 - 4x - 1$

7. $f(x) = x^4 - 24x^2 + 80$

Domain is $(-\infty, \infty)$.

$$\begin{aligned} f(-x) &= (-x)^4 - 24(-x)^2 + 80 \\ &= x^4 - 24x^2 + 80 = f(x) \end{aligned}$$

The graph is symmetric about the y-axis.

$f'(x) = 4x^3 - 48x$

$$\begin{aligned} 4x^3 - 48x &= 0 \\ 4x(x^2 - 12) &= 0 \\ 4x(x - 2\sqrt{3})(x + 2\sqrt{3}) &= 0 \end{aligned}$$

Critical numbers: $-2\sqrt{3}, 0$, and $2\sqrt{3}$
Critical points: $(-2\sqrt{3}, -64)$, $(0, 80)$, and $(2\sqrt{3}, -64)$

$f''(x) = 12x^2 - 48$

$$\begin{aligned} f''(-2\sqrt{3}) &= 12(-2\sqrt{3})^2 - 48 = 96 > 0 \\ f''(0) &= 12(0)^2 - 48 = -48 < 0 \\ f''(2\sqrt{3}) &= 12(2\sqrt{3})^2 - 48 = 96 > 0 \end{aligned}$$

Relative maximum at 0, relative minima at $-2\sqrt{3}$ and $2\sqrt{3}$
Increasing on $(-2\sqrt{3}, 0)$ and $(2\sqrt{3}, \infty)$
Decreasing on $(-\infty, -2\sqrt{3})$ and $(0, 2\sqrt{3})$

$$\begin{aligned} 12x^2 - 48 &= 0 \\ 12(x^2 - 4) &= 0 \\ x &= \pm 2 \end{aligned}$$

Points of inflection at $(-2, 0)$ and $(2, 0)$
Concave upward on $(-\infty, -2)$ and $(2, \infty)$
Concave downward on $(-2, 2)$

x-intercepts: $0 = x^4 - 24x^2 + 80$

Let $u = x^2$.

$$\begin{aligned} u^2 - 24u + 80 &= 0 \\ (u - 4)(u - 20) &= 0 \end{aligned}$$

$u = 4$ or $u = 20$
$x = \pm 2$ or $x = \pm 2\sqrt{5}$

y-intercept: $y = (0)^4 - 24(0)^2 + 80 = 80$

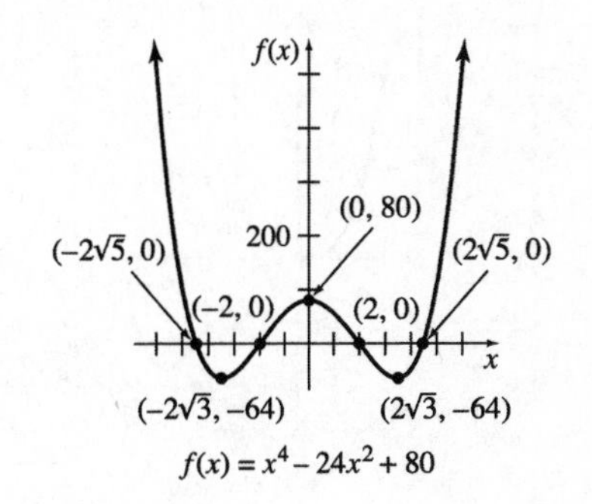

$f(x) = x^4 - 24x^2 + 80$

9. $f(x) = x^4 - 4x^3$

Domain is $(-\infty, \infty)$.

$f(-x) = (-x)^4 - 4(-x)^3 = x^4 + 4x^3 \neq f(x)$ or $-f(x)$

The graph is not symmetric about the y-axis or the origin.

$f'(x) = 4x^3 - 12x^2$

$$\begin{aligned} 4x^3 - 12x^2 &= 0 \\ 4x^2(x - 3) &= 0 \end{aligned}$$

Critical numbers: 0 and 3
Critical points: $(0, 0)$ and $(3, -27)$

$f''(x) = 12x^2 - 24x$

$$\begin{aligned} f''(0) &= 12(0)^2 - 24(0) = 0 \\ f''(3) &= 12(3)^2 - 24(3) = 36 > 0 \end{aligned}$$

Second derivative test fails for 0. Use first derivative test.

$$\begin{aligned} f'(-1) &= 4(-1)^3 - 12(-1)^2 = -16 < 0 \\ f'(1) &= 4(1)^3 - 12(1)^2 = -8 < 0 \end{aligned}$$

Neither a relative minimum nor maximum at 0
Relative minimum at 3
Increasing on $(3, \infty)$
Decreasing on $(-\infty, 3)$

$$\begin{aligned} 12x^2 - 24x &= 0 \\ 12x(x - 2) &= 0 \\ x = 0 &\text{ or } x = 2 \end{aligned}$$

Points of inflection at $(0, 0)$ and $(2, -16)$
Concave upward on $(-\infty, 0)$ and $(2, \infty)$
Concave downward on $(0, 2)$

x-intercepts: $x^4 - 4x^3 = 0$

$$x^3(x-4) = 0$$
$$x = 0 \text{ or } x = 4$$

y-intercept: $y = (0)^4 - 4(0)^3 = 0$

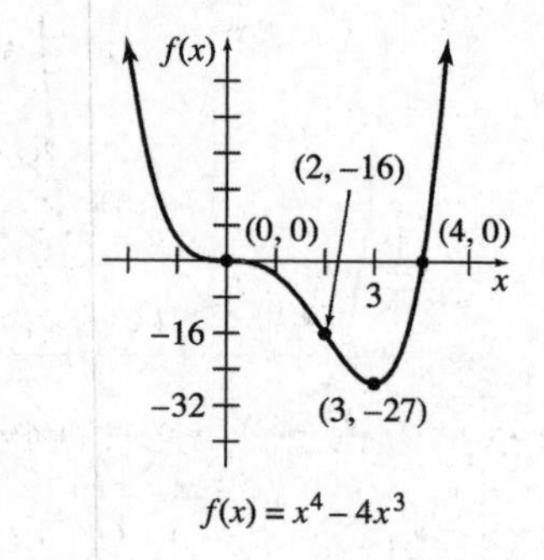

$f(x) = x^4 - 4x^3$

11. $f(x) = 2x + \dfrac{10}{x}$

$$= 2x + 10x^{-1}$$

Since $f(x)$ does not exist when $x = 0$, the domain is $(-\infty, 0) \cup (0, \infty)$.

$$f(-x) = 2(-x) + 10(-x)^{-1}$$
$$= -(2x + 10x^{-1})$$
$$= -f(x)$$

The graph is symmetric about the origin.

$$f'(x) = 2 - 10x^{-2}$$

$$2 - \frac{10}{x^2} = 0$$

$$\frac{2(x^2-5)}{x^2} = 0$$

$$x = \pm\sqrt{5}$$

Critical numbers: $-\sqrt{5}$ and $\sqrt{5}$
Critical points: $(-\sqrt{5}, -4\sqrt{5})$ and $(\sqrt{5}, 4\sqrt{5})$

Test a point in the intervals $(-\infty, -\sqrt{5})$, $(-\sqrt{5}, 0)$, $(0, \sqrt{5})$, and $(\sqrt{5}, \infty)$.

$$f'(-3) = 2 - 10(-3)^{-2} = \frac{8}{9} > 0$$

$$f'(-1) = 2 - 10(-1)^{-2} = -8 < 0$$
$$f'(1) = 2 - 10(1)^{-2} = -8 < 0$$

$$f'(3) = 2 - 10(3)^{-2} = \frac{8}{9} > 0$$

Relative maximum at $-\sqrt{5}$
Relative minimum at $\sqrt{5}$
Increasing on $(-\infty, -\sqrt{5})$ and $(\sqrt{5}, \infty)$
Decreasing on $(-\sqrt{5}, 0)$ and $(0, \sqrt{5})$
(Recall that $f(x)$ does not exist at $x = 0$.)

$$f''(x) = 20x^{-3} = \frac{20}{x^3}$$

$f''(x) = \dfrac{20}{x^3}$ is never equal to zero.

There are no inflection points.
Test a point in the intervals $(-\infty, 0)$ and $(0, \infty)$.

$$f''(-1) = \frac{20}{(-1)^3} = -20 < 0$$

$$f''(1) = \frac{20}{(1)^3} = 20 > 0$$

Concave upward on $(0, \infty)$
Concave downward on $(-\infty, 0)$
$f(x)$ is never zero, so there are no x-intercepts.
$f(x)$ does not exist for $x = 0$, so there is no y-intercept.

Vertical asymptote at $x = 0$
$y = 2x$ is an oblique asymptote.

$$f(-x) = 2(-x) + 10(-x)^{-1}$$
$$= -(2x + 10x^{-1})$$
$$= -f(x)$$

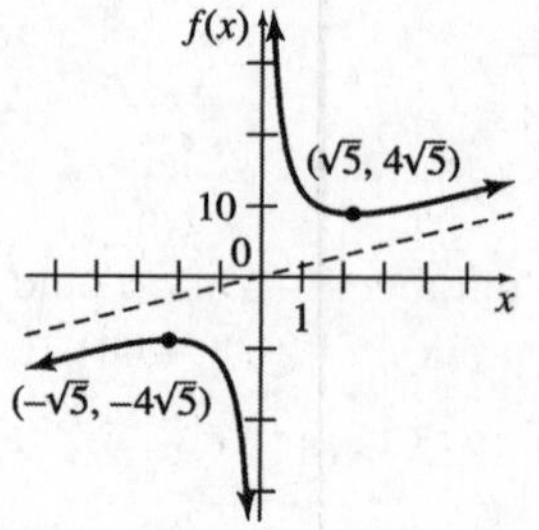

$f(x) = 2x + \frac{10}{x}$

13. $f(x) = \dfrac{-x+4}{x+2}$

Since $f(x)$ does not exist when $x = -2$, the domain is $(-\infty, -2) \cup (-2, \infty)$.

$$f(-x) = \frac{-(-x)+4}{(-x)+2} = \frac{x+4}{-x+2}$$

The graph is not symmetric about the y-axis or the origin.

$$f'(x) = \frac{(x+2)(-1) - (-x+4)(1)}{(x+2)^2}$$
$$= \frac{-6}{(x+2)^2}$$

$f'(x) < 0$ and is never zero. $f'(x)$ fails to exist for $x = -2$.

No critical numbers; no relative extrema

Decreasing on $(-\infty, -2)$ and $(-2, \infty)$

$$f''(x) = \frac{12}{(x+2)^3}$$

$f''(x)$ fails to exist for $x = -2$.
No points of inflection
Test a point in the intervals $(-\infty, -2)$ and $(-2, \infty)$.

$$f''(-3) = -12 < 0$$
$$f''(-1) = 12 > 0$$

Concave upward on $(-2, \infty)$
Concave downward on $(-\infty, -2)$

x-intercept: $\dfrac{-x+4}{x+2} = 0$

$$x = 4$$

y-intercept: $y = \dfrac{-0+4}{0+2} = 2$

Vertical asymptote at $x = -2$
Horizontal asymptote at $y = -1$

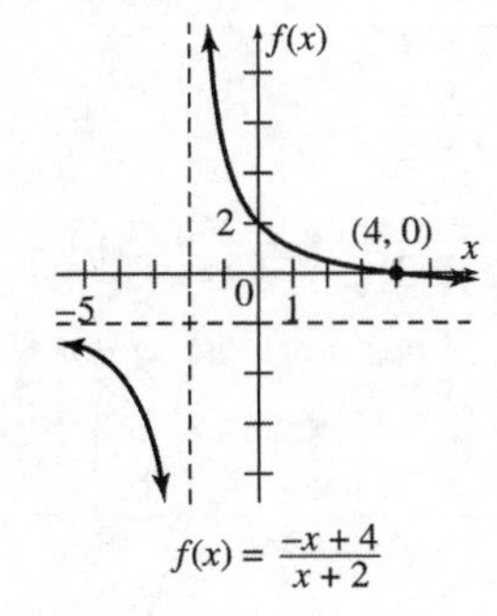

15. $f(x) = \dfrac{1}{x^2+4x+3}$

$$= \frac{1}{(x+3)(x+1)}$$

Since $f(x)$ does not exist when $x = -3$ and $x = -1$, the domain is $(-\infty, -3) \cup (-3, -1) \cup (-1, \infty)$.

$$f(-x) = \frac{1}{(-x)^2 + 4(-x) + 3} = \frac{1}{x^2 - 4x + 3}$$

The graph is not symmetric about the y-axis or the origin.

$$f'(x) = \frac{0 - (2x+4)}{(x^2+4x+3)^2} = \frac{-2(x+2)}{[(x+3)(x+1)]^2}$$

Critical number: -2

Test a point in the intervals $(-\infty, -3), (-3, -2), (-2, -1)$, and $(-1, \infty)$.

$$f'(-4) = \frac{-2(-4+2)}{[(-4+3)(-4+1)]^2} = \frac{4}{9} > 0$$

$$f'\left(-\frac{5}{2}\right) = \frac{-2\left(-\frac{5}{2}+2\right)}{\left[\left(-\frac{5}{2}+3\right)\left(-\frac{5}{2}+1\right)\right]^2} = \frac{16}{9} > 0$$

$$f'\left(-\frac{3}{2}\right) = \frac{-2\left(-\frac{3}{2}+2\right)}{\left[\left(-\frac{3}{2}+3\right)\left(-\frac{3}{2}+1\right)\right]^2} = -\frac{16}{9} < 0$$

$$f'(0) = \frac{-2(0+2)}{[(0+3)(0+1)]^2} = -\frac{4}{9} < 0$$

$$f(-2) = \frac{1}{(-2+3)(-2+1)} = -1$$

Relative maximum at $(-2, -1)$
Increasing on $(-\infty, -3)$ and $(-3, -2)$
Decreasing on $(-2, -1)$ and $(-1, \infty)$

$$f''(x) = \frac{(x^2+4x+3)^2(-2) - (-2x-4)(2)(x^2+4x+3)(2x+4)}{(x^2+4x+3)^4}$$

$$= \frac{-2(x^2+4x+3)[(x^2+4x+3) + (-2x-4)(2x+4)]}{(x^2+4x+3)^4}$$

$$= \frac{-2(x^2+4x+3-4x^2-16x-16)}{(x^2+4x+3)^3}$$

$$= \frac{-2(-3x^2-12x-13)}{(x^2+4x+3)^3}$$

$$= \frac{2(3x^2+12x+13)}{[(x+3)(x+1)]^3}$$

Since $3x^2 + 12x + 13 = 0$ has no real solutions, there are no x-values where $f''(x) = 0$. $f''(x)$ does not exist where $x = -3$ and $x = -1$. Since $f(x)$ does not exist at these x-values, there are no points of inflection.

Test a point in the intervals $(-\infty, -3), (-3, -1)$, and $(-1, \infty)$.

$$f''(-4) = \frac{2[3(-4)^2 + 12(-4) + 13]}{[(-4+3)(-4+1)]^3} = \frac{26}{27} > 0$$

$$f''(-2) = \frac{2[3(-2)^2 + 12(-2) + 13]}{[(-2+3)(-2+1)]^3} = -2 < 0$$

$$f''(0) = \frac{2[3(0)^2 + 12(0) + 13]}{[(0+3)(0+1)]^3} = \frac{26}{27} > 0$$

Concave upward on $(-\infty, -3)$ and $(-1, \infty)$

Concave downward on $(-3, -1)$

$f(x)$ is never zero, so there are no x-intercepts.

y-intercept: $y = \dfrac{1}{(0+3)(0+1)} = \dfrac{1}{3}$

Vertical asymptotes where $f(x)$ is undefined at $x = -3$ and $x = -1$.

Horizontal asymptote at $y = 0$

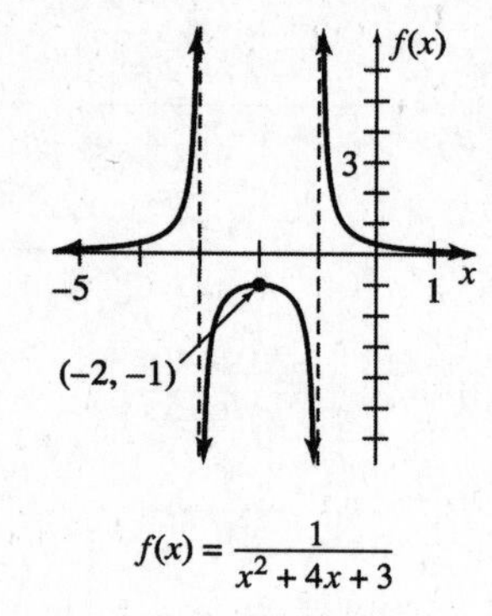

$f(x) = \dfrac{1}{x^2 + 4x + 3}$

17. $f(x) = \dfrac{x}{x^2 + 1}$

Domain is $(-\infty, \infty)$

$$f(-x) = \frac{-x}{(-x)^2 + 1} = -\frac{x}{x^2 + 1} = -f(x)$$

The graph is symmetric about the origin.

$$f'(x) = \frac{(x^2 + 1)(1) - x(2x)}{(x^2 + 1)^2}$$

$$= \frac{1 - x^2}{(x^2 + 1)^2}$$

$$1 - x^2 = 0$$

Critical numbers: 1 and -1

Critical points: $\left(1, \frac{1}{2}\right)$ and $\left(-1, -\frac{1}{2}\right)$

$$f''(x) = \frac{(x^2+1)^2(-2x) - (1-x^2)(2)(x^2+1)(2x)}{(x^2 + 1)^4}$$

$$= \frac{-2x^3 - 2x - 4x + 4x^3}{(x^2 + 1)^3}$$

$$= \frac{2x^3 - 6x}{(x^2 + 1)^3}$$

$$f''(1) = -\frac{1}{2} < 0$$

$$f''(-1) = \frac{1}{2} > 0$$

Relative maximum at 1
Relative minimum at -1
Increasing on $(-1, 1)$

Decreasing on $(-\infty, -1)$ and $(1, \infty)$

$$f''(x) = \frac{2x^3 - 6x}{(x^2 + 1)^3} = 0$$

$$2x^3 - 6x = 0$$

$$2x(x^2 - 3) = 0$$

$$x = 0,\ x = \pm\sqrt{3}$$

Inflection points at $(0, 0)$, $\left(\sqrt{3}, \frac{\sqrt{3}}{4}\right)$ and $\left(-\sqrt{3}, -\frac{\sqrt{3}}{4}\right)$

Concave upward on $(-\sqrt{3}, 0)$ and $(\sqrt{3}, \infty)$
Concave downward on $(-\infty, -\sqrt{3})$ and $(0, \sqrt{3})$

x-intercept: $0 = \dfrac{x}{x^2 + 1}$

$$0 = x$$

y-intercept: $y = \dfrac{0}{0^2 + 1} = 0$

Horizontal asymptote at $y = 0$

$f(x) = \dfrac{x}{x^2 + 1}$

19. $f(x) = \dfrac{1}{x^2 - 9}$

$$= \frac{1}{(x + 3)(x - 3)}$$

Since $f(x)$ does not exist when $x = -3$ and $x = 3$, the domain is $(-\infty, -3) \cup (-3, 3) \cup (3, \infty)$.

$$f(-x) = \frac{1}{(-x)^2 - 9} = \frac{1}{x^2 - 9} = f(x)$$

The graph is symmetric about the y-axis.

$$f'(x) = \frac{-2x}{(x^2 - 9)^2}$$

Critical number: 0

Critical point: $\left(0, -\frac{1}{9}\right)$

Test a point in the intervals $(-\infty, -3)$, $(-3, 0)$, $(0, 3)$, and $(3, \infty)$.

$$f'(-4) = \frac{-2(-4)}{[(-4)^2 - 9]^2} = \frac{8}{49} > 0$$

$$f'(-1) = \frac{-2(-4)}{[(-1)^2 - 9]^2} = \frac{1}{32} > 0$$

$$f'(1) = \frac{-2(1)}{[(1)^2 - 9]^2} = -\frac{1}{32} < 0$$

$$f'(4) = \frac{-2(4)}{[(4)^2 - 9]^2} = -\frac{8}{49} < 0$$

Relative maximum at $\left(0, -\frac{1}{9}\right)$

Increasing on $(-\infty, -3)$ and $(-3, 0)$

Decreasing on $(0, 3)$ and $(3, \infty)$

$$f''(x) = \frac{(x^2-9)^2(-2) - (-2x)(2)(x^2-9)(2x)}{(x^2-9)^4}$$
$$= \frac{-2(x^2-9)[(x^2-9)+(-2x)(2x)]}{(x^2+4)^4}$$
$$= \frac{-2(x^2-9-4x^2)}{(x^2-9)^3}$$
$$= \frac{-2(-3x^2-9)}{(x^2-9)^3}$$
$$= \frac{6(x^2+3)}{[(x+3)(x-3)]^3}$$

Since $x^2 + 3 = 0$ has no solutions, there are no x-values where $f''(x) = 0$. $f''(x)$ does not exist where $x = -3$ and $x = 3$. Since $f(x)$ does not exist at these x-values, there are no points of inflection.

Test a point in the intervals $(-\infty, -3), (-3, 3)$, and $(3, \infty)$.

$$f''(-4) = \frac{6[(-4)^2+3]}{[(-4+3)(-4-3)]^3} = \frac{114}{343} > 0$$
$$f''(0) = \frac{6[(0)^2+3]}{[(0+3)(0-3)]^3} = -\frac{2}{81} < 0$$
$$f''(4) = \frac{6[(4)^2+3]}{[(4+3)(4-3)]^3} = \frac{114}{343} > 0$$

Concave upward on $(-\infty, -3)$ and $(3, \infty)$
Concave downward on $(-3, 3)$

$f(x)$ is never zero, so there are no x-intercepts.

y-intercept: $y = \frac{1}{0^2-9} = -\frac{1}{9}$

Vertical asymptotes where $f(x)$ is undefined at $x = -3$ and $x = 3$.
Horizontal asymptote at $y = 0$

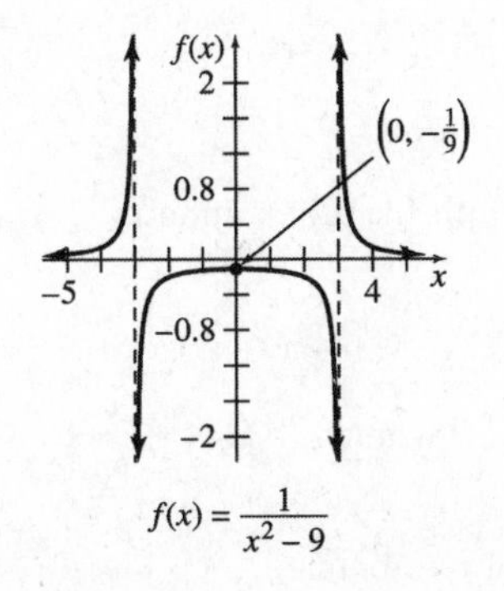

21. $f(x) = x \ln |x|$

The domain of this function is $(-\infty, 0) \cup (0, \infty)$.

$$f(-x) = -x \ln -x\,|-x|$$
$$= x \ln |-x| = -f(x)$$

The graph is symmetric about the origin.

$$f'(x) = x \cdot \frac{1}{x} + \ln |x|$$
$$= 1 + \ln |x|$$

$f'(x) = 0$ when

$$0 = 1 + \ln |x|$$
$$-1 = \ln |x|$$
$$e^{-1} = |x|$$
$$x = \pm\frac{1}{e} \approx \pm 0.37.$$

Critical numbers: $\pm\frac{1}{e} \approx \pm 0.37$.

$$f'(-1) = 1 + \ln |-1| = 1 > 0$$
$$f'(-0.1) = 1 + \ln |-0.1| \approx -1.3 < 0$$
$$f'(0.1) = 1 + \ln |0.1| \approx -1.3 < 0$$
$$f'(1) = 1 + \ln |1| = 1 > 0$$
$$f\left(\frac{1}{e}\right) = \frac{1}{e} \ln \left|\frac{1}{e}\right| = -\frac{1}{e}$$
$$f\left(-\frac{1}{e}\right) = -\frac{1}{e} \ln \left|-\frac{1}{e}\right| = \frac{1}{e}$$

Relative maximum of $(-\frac{1}{e}, \frac{1}{e})$; relative minimum of $(\frac{1}{e}, -\frac{1}{e})$.

Increasing on $\left(-\infty, -\frac{1}{e}\right)$ and $\left(\frac{1}{e}, \infty\right)$ and decreasing on $\left(-\frac{1}{e}, 0\right)$ and $\left(0, \frac{1}{e}\right)$.

$$f''(x) = \frac{1}{x}$$
$$f''(-1) = \frac{1}{-1} = -1 < 0$$
$$f''(1) = \frac{1}{1} = 1 > 0$$

Concave downward on $(-\infty, 0)$;
Concave upward on $(0, \infty)$.
There is no y-intercept.

x-intercept: $0 = x \ln |x|$

$$x = 0 \quad \text{or} \quad \ln |x| = 0$$
$$|x| = e^0 = 1$$
$$x = \pm 1$$

Since 0 is not in the domain, the only x-intercepts are -1 and 1.

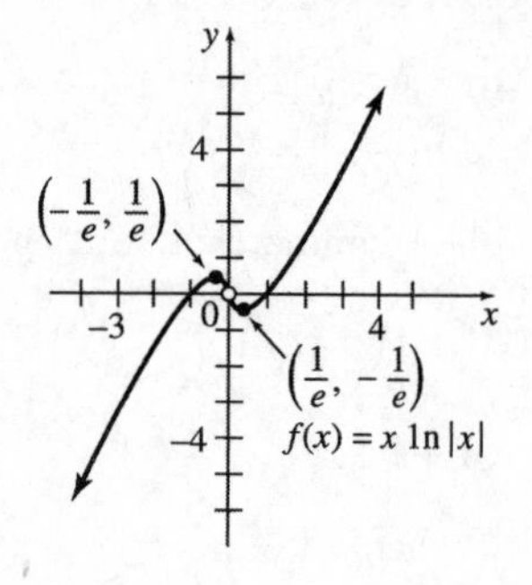

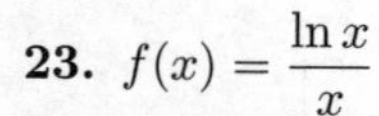

23. $f(x) = \dfrac{\ln x}{x}$

Note that the domain of this function is $(0, \infty)$.

$f(-x) = \dfrac{\ln(-x)}{-x}$ does not exist when $x \geq 0$, no symmetry.

$$f'(x) = \frac{x\left(\frac{1}{x}\right) - \ln x(1)}{x^2} = \frac{1 - \ln x}{x^2}$$

Critical numbers:

$$1 - \ln x = 0$$
$$1 = \ln x$$
$$e^1 = x$$

$$f(e) = \frac{\ln e}{e} = \frac{1}{e}$$

Critical points: $\left(e, \dfrac{1}{e}\right)$

$$f'(1) = \frac{1 - \ln 1}{1^2} = \frac{1}{1} = 1 > 0$$
$$f'(3) = \frac{1 - \ln 3}{3^2} = -0.01 < 0$$

There is a relative maximum at $\left(e, \frac{1}{e}\right)$.
The function is increasing on $(0, e)$ and decreasing on (e, ∞).

$$f''(x) = \frac{x^2\left(-\frac{1}{x}\right) - (1 - \ln x)2x}{x^4} = \frac{-x - 2x(1 - \ln x)}{x^4} = \frac{-x[1 + 2(1 - \ln x)]}{x^4} = \frac{-(1 + 2 - 2\ \ln x)}{x^3} = \frac{-3 + 2\ \ln x}{x^3}$$

$f''(x) = 0$ when $-3 + 2\ \ln x = 0$

$$2\ \ln x = 3$$
$$\ln x = \frac{3}{2} = 1.5$$
$$x = e^{1.5} \approx 4.48.$$

$$f''(1) = \frac{-3 + 2\ln 1}{1^3} = -3 < 0$$
$$f''(5) = \frac{-3 + 2\ln 5}{5^3} \approx 0.0018 > 0$$

Inflection point at $\left(e^{1.5}, \frac{1.5}{e^{1.5}}\right) \approx (4.48, 0.33)$

Concave downward on $(0, e^{1.5})$; concave upward on $(e^{1.5}, \infty)$

$$f(e^{1.5}) = \frac{\ln e^{1.5}}{e^{1.5}} = \frac{1.5}{e^{1.5}} = \frac{3}{2e^{1.5}} \approx 0.33$$

Since $x \neq 0$, there is no y-intercept.

x-intercept: $f(x) = 0$ when $\ln\ x = 0$

$$x = e^0 = 1$$

Vertical asymptote at $x = 0$
Horizontal asymptote at $y = 0$

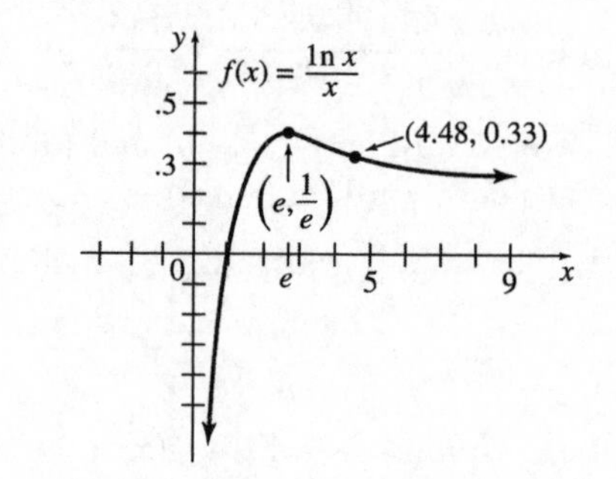

25. $f(x) = xe^{-x}$

Domain is $(-\infty, \infty)$.

$f(-x) = -xe^{x}$

The graph has no symmetry.

$$f'(x) = -xe^{-x} + e^{-x} = e^{-x}(1 - x)$$

$f'(x) = 0$ when $e^{-x}(1 - x) = 0$

$$x = 1$$

Critical numbers: 1

Critical points: $\left(1, \frac{1}{e}\right)$

$$f'(0) = e^{-0}(1-0) = 1 > 0$$

$$f'(2) = e^{-2}(1-2) = \frac{-1}{e^2} < 0$$

Relative maximum at $\left(1, \frac{1}{e}\right)$

Increasing on $(-\infty, 1)$; decreasing on $(1, \infty)$

$$\begin{aligned} f''(x) &= e^{-x}(-1) + (1-x)(-e^{-x}) \\ &= -e^{-x}(1+1-x) \\ &= -e^{-x}(2-x) \end{aligned}$$

$$f'' = 0 \text{ when } -e^{-x}(2-x) = 0$$
$$x = 2.$$

$$f''(0) = -e^{-0}(2-0) = -2 < 0$$

$$f''(3) = -e^{-3}(2-3) = \frac{1}{e^3} > 0$$

Inflection point at $\left(2, \frac{2}{e^2}\right)$

Concave downward on $(-\infty, 2)$, concave upward on $(2, \infty)$

x-intercept: $0 = xe^{-x}$

$x = 0$

y-intercept: $y = 0 \cdot e^{-0} = 0$

Horizontal asymptote at $y = 0$

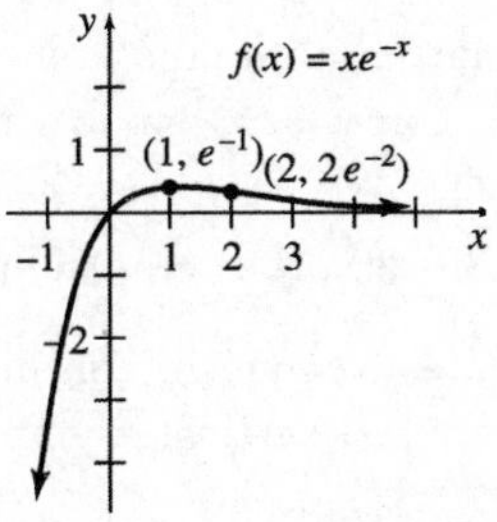

27. $f(x) = (x-1)e^{-x}$

Domain is $(-\infty, \infty)$

$f(-x) = (-x-1)e^{x}$

The graph has no symmetry.

$$\begin{aligned} f'(x) &= -(x-1)e^{-x} + e^{-x}(1) \\ &= e^{-x}[-(x-1)+1] \\ &= e^{-x}(2-x) \end{aligned}$$

$$f'(x) = 0 \text{ when } e^{-x}(2-x) = 0$$
$$x = 2.$$

Critical number: 2

Critical point: $\left(2, \frac{1}{e^2}\right)$

$$\begin{aligned} f''(x) &= -e^{-x} + (2-x)(-e^{-x}) \\ &= -e^{-x}[1 + (2-x)] \\ &= -e^{-x}(3-x) \end{aligned}$$

$$f''(2) = -e^{-2}(3-2) = \frac{-1}{e^2} < 0$$

Relative maximum at $\left(2, \frac{1}{e^2}\right)$

$$f'(0) = e^{-0}(2-0) = 2 > 0$$

$$f'(3) = e^{-3}(2-3) = \frac{-1}{e^3} < 0$$

Increasing on $(-\infty, 2)$; decreasing on $(2, \infty)$.

$$f''(x) = 0 \quad \text{when} \quad -e^{-x}(3-x) = 0$$
$$x = 3.$$

$$f''(0) = -e^{-0}(3-0) = -3 < 0$$

$$f''(4) = -e^{-4}(3-4) = \frac{1}{e^4} > 0$$

Inflection point at $\left(3, \frac{2}{e^3}\right)$

Concave downward on $(-\infty, 3)$; concave upward on $(3, \infty)$

$$f(3) = (3-1)e^{-3} = \frac{2}{e^3}$$

y-intercept: $y = (0-1)e^{-0}$

$= (-1)(1) = -1$

x-intercept: $0 = (x-1)e^{-x}$

$x - 1 = 0$

$x = 1$

Horizontal asymptote at $y = 0$

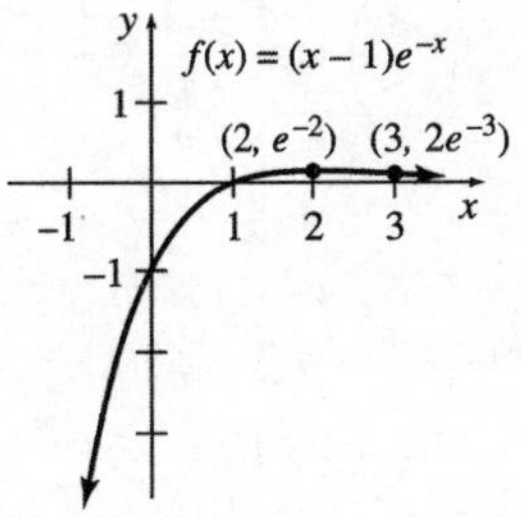

29. $f(x) = x^{2/3} - x^{5/3}$

Domain is $(-\infty, \infty)$.

$f(-x) = x^{2/3} + x^{5/3}$

The graph has no symmetry.

$$\begin{aligned} f'(x) &= \frac{2}{3}x^{-1/3} - \frac{5}{3}x^{2/3} \\ &= \frac{2-5x}{3x^{1/3}} \end{aligned}$$

$$f'(x) = 0 \quad \text{when} \quad 2 - 5x = 0$$

Critical number: $x = \frac{2}{5}$

$$f\left(\frac{2}{5}\right) = \left(\frac{2}{5}\right)^{2/3} - \left(\frac{2}{5}\right)^{5/3}$$

$$= \frac{3 \cdot 2^{2/3}}{5^{5/3}} \approx 0.326$$

Critical point: $(0.4, 0.326)$

$$f''(x) = \frac{3x^{1/3}(-5) - (2-5x)(3)\left(\frac{1}{3}\right)x^{-2/3}}{(3x^{1/3})^2}$$

$$= \frac{-15x^{1/3} - (2-5x)x^{-2/3}}{9x^{2/3}}$$

$$= \frac{-15x - (2-5x)}{9x^{4/3}}$$

$$= \frac{-10x - 2}{9x^{4/3}}$$

$$f''\left(\frac{2}{5}\right) = \frac{-10\left(\frac{2}{5}\right) - 2}{9\left(\frac{2}{5}\right)^{4/3}}$$

$$\approx -2.262 < 0$$

Relative maximum at $\left(\frac{2}{5}, \frac{3 \cdot 2^{2/3}}{5^{5/3}}\right) \approx (0.4, 0.326)$

$f'(x)$ does not exist when $x = 0$

Since $f''(0)$ is undefined, use the first derivative test.

$$f'(-1) = \frac{2-5(-1)}{3(-1)^{1/3}} = \frac{7}{-3} < 0$$

$$f'\left(\frac{1}{8}\right) = \frac{2-5\left(\frac{1}{8}\right)}{3\left(\frac{1}{8}\right)^{1/3}} = \frac{11}{12} > 0$$

$$f'(1) = \frac{2-5}{3 \cdot 1^{1/3}} = -1 < 0$$

Relative minimum at $(0, 0)$

f increases on $\left(0, \frac{2}{5}\right)$.
f decreases on $(-\infty, 0)$ and $\left(\frac{2}{5}, \infty\right)$.

$$f''(x) = 0 \quad \text{when} \quad -10x - 2 = 0$$

$$x = -\frac{1}{5}$$

$$f''(x) \text{ undefined when } 9x^{4/3} = 0$$

$$x = 0$$

$$f''(-1) = \frac{-10(-1) - 2}{9(-1)^{4/3}} = \frac{8}{9} > 0$$

$$f''\left(-\frac{1}{8}\right) = \frac{-10\left(-\frac{1}{8}\right) - 2}{9\left(-\frac{1}{8}\right)^{4/3}} = -\frac{4}{3} < 0$$

$$f''(1) = \frac{-10(1) - 2}{9(1)^{4/3}} = -\frac{4}{3} < 0$$

Concave upward on $\left(-\infty, -\frac{1}{5}\right)$
Concave downward on $\left(-\frac{1}{5}, \infty\right)$
Inflection point at $\left(-\frac{1}{5}, \frac{6}{5^{5/3}}\right) \approx (-0.2, 0.410)$

y-intercept: $y = 0^{2/3} - 0^{5/3} = 0$
x-intercept: $0 = x^{2/3} - x^{5/3}$

$$= x^{2/3}(1 - x)$$

$$x = 0 \text{ or } x = 1$$

31. For Exercises 3, 7, and 9, the relative maxima or minima are outside the vertical window of $-10 \le y \le 10$.

For Exercise 11, the default window shows only a small portion of the graph.

For Exercise 15, the default window does not allow the graph to properly display the vertical asymptotes.

33. For Exercises 17, 19, 23, 25, and 27, the y-coordinate of the relative minimum, relative maximum, or inflection point is so small, it may be hard to distinguish.

For Exercises 35–39 other graphs are possible.

35. **(a)** indicates a smooth, continuous curve except where there is a vertical asymptote.

(b) indicates that the function decreases on both sides of the asymptote, so there are no relative extrema.

(c) gives the horizontal asymptote $y = 2$.

(d) and **(e)** indicate that concavity does not change left of the asymptote, but that the right portion of the graph changes concavity at $x = 2$ and $x = 4$. There are inflection points at 2 and 4.

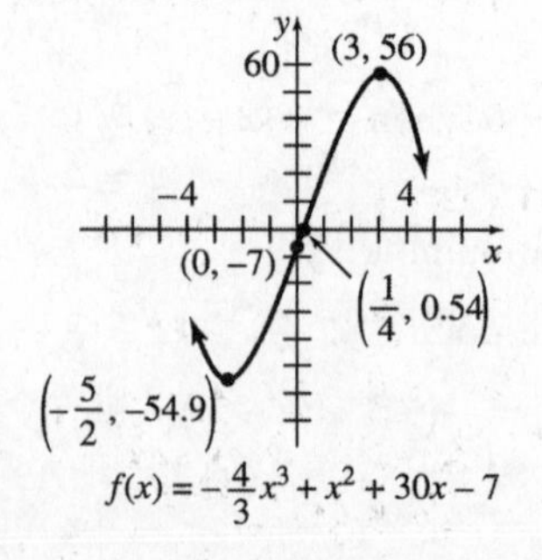

37. **(a)** indicates that there can be no asymptotes, sharp "corners", holes, or jumps. The graph must be one smooth curve.

(b) and **(c)** indicate relative maxima at -3 and 4 and a relative minimum at 1.

(d) and **(e)** are consistent with (g).

(f) indicates turning points at the critical numbers -3 and 4.

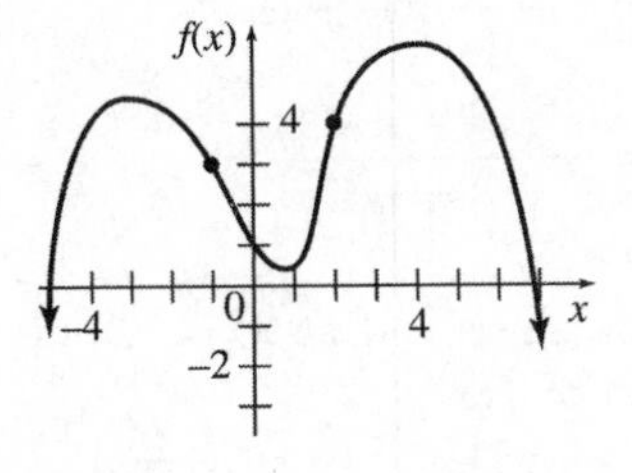

39. **(a)** indicates that the curve may not contain breaks.

(b) indicates that there is a sharp "corner" at 4.

(c) gives a point at $(1, 5)$.

(d) shows critical numbers.

(e) and **(f)** indicate (combined with (c) and (d)) a relative maximum at $(1, 5)$, and (combined with (b)) a relative minimum at 4.

(g) is consistent with (b).

(h) indicates the curve is concave upward on $(2, 3)$.

(i) indicates the curve is concave downward on $(-\infty, 2), (3, 4)$ and $(4, \infty)$.

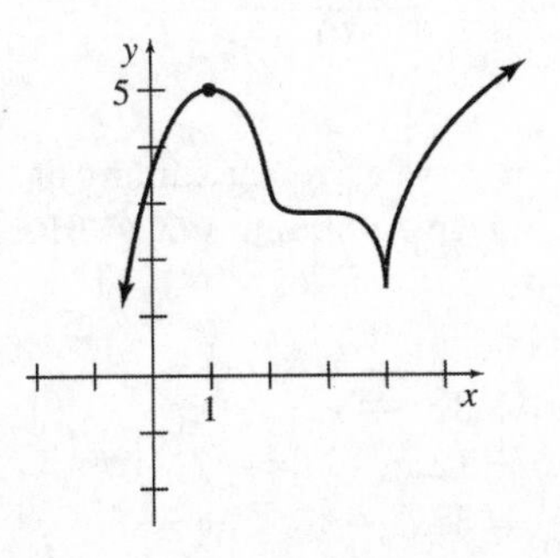

Chapter 5 Review Exercises

5. $f(x) = x^2 + 9x + 8$
$f'(x) = 2x + 9$

$f'(x) = 0$ when $x = -\frac{9}{2}$ and f' exists everywhere.

Critical number: $-\frac{9}{2}$

Test an x-value in the intervals $\left(-\infty, -\frac{9}{2}\right)$ and $\left(-\frac{9}{2}, \infty\right)$.

$f'(-5) = -1 < 0$
$f'(-4) = 1 > 0$

f is increasing on $\left(-\frac{9}{2}, \infty\right)$ and decreasing on $\left(-\infty, -\frac{9}{2}\right)$.

7. $f(x) = -x^3 + 2x^2 + 15x + 16$

$$\begin{aligned} f'(x) &= -3x^2 + 4x + 15 \\ &= -(3x^2 - 4x - 15) \\ &= -(3x + 5)(x - 3) \end{aligned}$$

$f'(x) = 0$ when $x = -\frac{5}{3}$ or $x = 3$ and f' exists everywhere.

Critical numbers: $-\frac{5}{3}$ and 3

Test an x-value in the intervals $\left(-\infty, -\frac{5}{3}\right), \left(-\frac{5}{3}, 3\right)$, and $(3, \infty)$.

$f'(-2) = -5 < 0$
$f'(0) = 15 > 0$
$f'(4) = -17 < 0$

f is increasing on $\left(-\frac{5}{3}, 3\right)$ and decreasing on $\left(-\infty, -\frac{5}{3}\right)$ and $(3, \infty)$.

9. $f(x) = \dfrac{16}{9 - 3x}$

$$f'(x) = \frac{16(-1)(-3)}{(9 - 3x)^2} = \frac{48}{(9 - 3x)^2}$$

$f'(x) > 0$ for all x ($x \neq 3$), and f is not defined for $x = 3$.

f is increasing on $(-\infty, 3)$ and $(3, \infty)$ and never decreasing.

11. $f(x) = \ln|x^2 - 1|$

$$f'(x) = \frac{2x}{x^2 - 1}$$

f is not defined for $x = -1$ and $x = 1$.

$f'(x) = 0$ when $x = 0$.

Test an x-value in the intervals $(-\infty, -1)$, $(-1, 0)$, $(0, 1)$, and $(1, \infty)$.

$$f'(-2) = -\frac{4}{3} < 0$$

$$f'\left(-\frac{1}{2}\right) = \frac{4}{3} > 0$$

$$f'\left(\frac{1}{2}\right) = -\frac{4}{3} < 0$$

$$f'(2) = \frac{4}{3} > 0$$

f is increasing on $(-1, 0)$ and $(1, \infty)$ and decreasing on $(-\infty, -1)$ and $(0, 1)$.

13. $f(x) = -x^2 + 4x - 8$
$f'(x) = -2x + 4 = 0$

Critical number: $x = 2$

$f''(x) = -2 < 0$ for all x, so $f(2)$ is a relative maximum.

$$f(2) = -4$$

Relative maximum of -4 at 2

15. $f(x) = 2x^2 - 8x + 1$
$f'(x) = 4x - 8 = 0$

Critical number: $x = 2$

$f''(x) = 4 > 0$ for all x, so $f(2)$ is a relative minimum.

$$f(2) = -7$$

Relative minimum of -7 at 2

17. $f(x) = 2x^3 + 3x^2 - 36x + 20$

$$\begin{aligned} f'(x) &= 6x^2 + 6x - 36 = 0 \\ 6(x^2 + x - 6) &= 0 \\ (x + 3)(x - 2) &= 0 \end{aligned}$$

Critical numbers: -3 and 2

$f''(x) = 12x + 6$
$f''(-3) = -30 < 0$, so a maximum occurs at $x = -3$.
$f''(2) = 30 > 0$, so a minimum occurs at $x = 2$.
$f(-3) = 101$
$f(2) = -24$

Relative maximum of 101 at -3
Relative minimum of -24 at 2

19. $f(x) = \dfrac{xe^x}{x - 1}$

$$\begin{aligned} f'(x) &= \frac{(x-1)(xe^x + e^x) - xe^x(1)}{(x-1)^2} \\ &= \frac{x^2e^x + xe^x - xe^x - e^x - xe^x}{(x-1)^2} \\ &= \frac{x^2e^x - xe^x - e^x}{(x-1)^2} \\ &= \frac{e^x(x^2 - x - 1)}{(x-1)^2} \end{aligned}$$

$f'(x)$ is undefined at $x = 1$, but 1 is not in the domain of $f(x)$.
$f'(x) = 0$ when $x^2 - x - 1 = 0$

$$\begin{aligned} x &= \frac{1 \pm \sqrt{1 - 4(1)(-1)}}{2} \\ &= \frac{1 \pm \sqrt{5}}{2} \end{aligned}$$

$$\frac{1 + \sqrt{5}}{2} \approx 1.618 \quad \text{or} \quad \frac{1 - \sqrt{5}}{2} = -0.618$$

Critical numbers are -0.618 and 1.618.

$$f'(1.4) = \frac{e^{1.4}(1.4^2 - 1.4 - 1)}{(1.4 - 1)^2} \approx -11.15 < 0$$

$$f'(2) = \frac{e^2(2^2 - 2 - 1)}{(2 - 1)^2} = e^2 \approx 7.39 > 0$$

$$f'(-1) = \frac{e^{-1}[(-1)^2 - (-1) - 1]}{(-1 - 1)^2} \approx 0.09 > 0$$

$$f'(0) = \frac{e^0(0^2 - 0 - 1)}{(0 - 1)^2} = -1 < 0$$

There is a relative maximum at $(-0.618, 0.206)$ and a relative minimum at $(1.618, 13.203)$.

21.

$$\begin{aligned} f(x) &= 3x^4 - 5x^2 - 11x \\ f'(x) &= 12x^3 - 10x - 11 \\ f''(x) &= 36x^2 - 10 \\ f''(1) &= 36(1)^2 - 10 = 26 \\ f''(-3) &= 36(-3)^2 - 10 = 314 \end{aligned}$$

23. $f(x) = \dfrac{4x+2}{3x-6}$

$$f'(x) = \frac{(3x-6)(4)-(4x+2)(3)}{(3x-6)^2}$$
$$= \frac{12x-24-12x-6}{(3x-6)^2}$$
$$= \frac{-30}{(3x-6)^2}$$
$$= -30(3x-6)^{-2}$$

$$f''(x) = -30(-2)(3x-6)^{-3}(3)$$
$$= 180(3x-6)^{-3} \quad \text{or} \quad \frac{180}{(3x-6)^3}$$

$$f''(1) = 180[3(1)-6]^{-3} = -\frac{20}{3}$$

$$f''(-3) = 180[3(-3)-6]^{-3} = -\frac{4}{75}$$

25. $f(t) = \sqrt{t^2+1} = (t^2+1)^{1/2}$

$$f'(t) = \frac{1}{2}(t^2+1)^{-1/2}(2t) = t(t^2+1)^{-1/2}$$
$$f''(t) = (t^2+1)^{-1/2}(1) + t\left[\left(-\frac{1}{2}\right)(t^2+1)^{-3/2}(2t)\right]$$
$$= (t^2+1)^{-1/2} - t^2(t^2+1)^{-3/2}$$
$$= \frac{1}{(t^2+1)^{1/2}} - \frac{t^2}{(t^2+1)^{3/2}} = \frac{t^2+1-t^2}{(t^2+1)^{3/2}}$$
$$= (t^2+1)^{-3/2} \quad \text{or} \quad \frac{1}{(t^2+1)^{3/2}}$$

$$f''(1) = \frac{1}{(1+1)^{3/2}} = \frac{1}{2^{3/2}} \approx 0.354$$

$$f''(-3) = \frac{1}{(9+1)^{3/2}} = \frac{1}{10^{3/2}} \approx 0.032$$

27. $f(x) = -2x^3 - \dfrac{1}{2}x^2 + x - 3$

Domain is $(-\infty, \infty)$
The graph has no symmetry.

$$f'(x) = -6x^2 - x + 1 = 0$$
$$(3x-1)(2x+1) = 0$$

Critical numbers: $\frac{1}{3}$ and $-\frac{1}{2}$

Critical points: $\left(\frac{1}{3}, -2.80\right)$ and $\left(-\frac{1}{2}, -3.375\right)$

$$f''(x) = -12x - 1$$
$$f''\left(\frac{1}{3}\right) = -5 < 0$$
$$f''\left(-\frac{1}{2}\right) = 5 > 0$$

Relative maximum at $\frac{1}{3}$

Relative minimum at $-\frac{1}{2}$

Increasing on $\left(-\frac{1}{2}, \frac{1}{3}\right)$

Decreasing on $\left(-\infty, -\frac{1}{2}\right)$ and $\left(\frac{1}{3}, \infty\right)$

$$f''(x) = -12x - 1 = 0$$
$$x = -\frac{1}{12}$$

Point of inflection at $\left(-\frac{1}{12}, -3.09\right)$

Concave upward on $\left(-\infty, -\frac{1}{12}\right)$

Concave downward on $\left(-\frac{1}{12}, \infty\right)$

y-intercept:

$$y = -2(0)^3 - \frac{1}{2}(0)^2 + (0) - 3 = -3$$

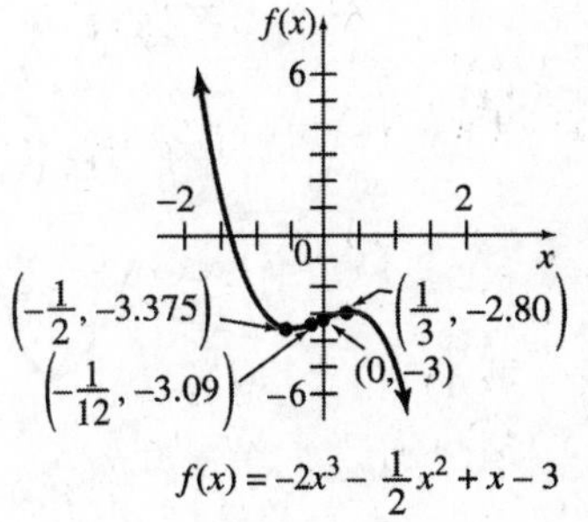

$f(x) = -2x^3 - \frac{1}{2}x^2 + x - 3$

29. $f(x) = x^4 - \dfrac{4}{3}x^3 - 4x^2 + 1$

Domain is $(-\infty, \infty)$
The graph has no symmetry.

$$f'(x) = 4x^3 - 4x^2 - 8x = 0$$
$$4x(x^2 - x - 2) = 0$$
$$4x(x-2)(x+1) = 0$$

Critical numbers: 0, 2, and -1
Critical points: $(0, 1)$, $\left(2, -\frac{29}{3}\right)$ and $\left(-1, -\frac{2}{3}\right)$

$$f''(x) = 12x^2 - 8x - 8$$
$$= 4(3x^2 - 2x - 2)$$
$$f''(-1) = 12 > 0$$
$$f''(0) = -8 < 0$$
$$f''(2) = 24 > 0$$

Relative maximum at 0
Relative minima at -1 and 2
Increasing on $(-1, 0)$ and $(2, \infty)$
Decreasing on $(-\infty, -1)$ and $(0, 2)$

$$f''(x) = 4(3x^2 - 2x - 2) = 0$$

$$x = \frac{2 \pm \sqrt{4 - (-24)}}{6}$$

$$= \frac{1 \pm \sqrt{7}}{3}$$

Points of inflection at $\left(\frac{1+\sqrt{7}}{3}, -5.12\right)$ and $\left(\frac{1-\sqrt{7}}{3}, 0.11\right)$

Concave upward on $\left(-\infty, \frac{1-\sqrt{7}}{3}\right)$ and $\left(\frac{1+\sqrt{7}}{3}, \infty\right)$

Concave downward on $\left(\frac{1-\sqrt{7}}{3}, \frac{1+\sqrt{7}}{3}\right)$

y-intercept:

$$y = (0)^4 - \frac{4}{3}(0)^3 - 4(0)^2 + 1 = 1$$

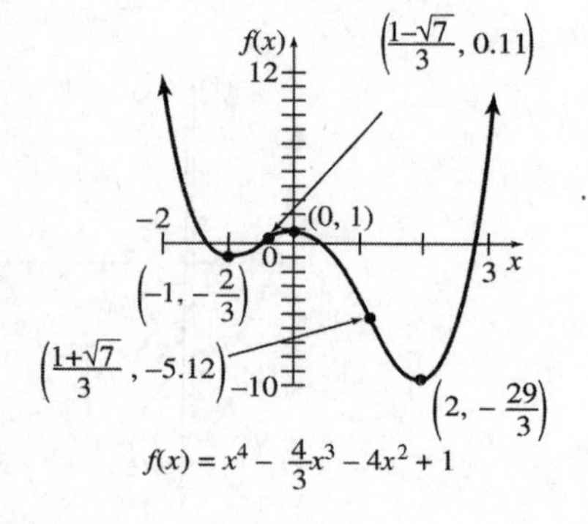

31. $f(x) = \dfrac{x-1}{2x+1}$

Domain is $(-\infty, -\frac{1}{2}) \cup (-\frac{1}{2}, \infty)$
The graph has no symmetry.

$$f'(x) = \frac{(2x+1)(1) - (x-1)(2)}{(2x+1)^2}$$

$$= \frac{3}{(2x+1)^2}$$

f' is never zero.
$f'(-\frac{1}{2})$ does not exist, but $-\frac{1}{2}$ is not a critical number because $-\frac{1}{2}$ is not in the domain of f. Thus, there are no critical numbers, so $f(x)$ has no relative extrema.

Increasing on $(-\infty, \frac{1}{2})$ and $(\frac{1}{2}, \infty)$

$$f''(x) = \frac{-12}{(2x+1)^3}$$

$$f''(0) = -12 < 0$$
$$f''(-1) = 12 > 0$$

No inflection points

Concave upward on $(-\infty, -\frac{1}{2})$

Concave downward on $(-\frac{1}{2}, \infty)$

x-intercept: $\dfrac{x-1}{2x+1} = 0$

$$x = 1$$

y-intercept: $y = \dfrac{0-1}{2(0)+1} = -1$

Vertical asymptote at $x = -\frac{1}{2}$

Horizontal asymptote at $y = \frac{1}{2}$

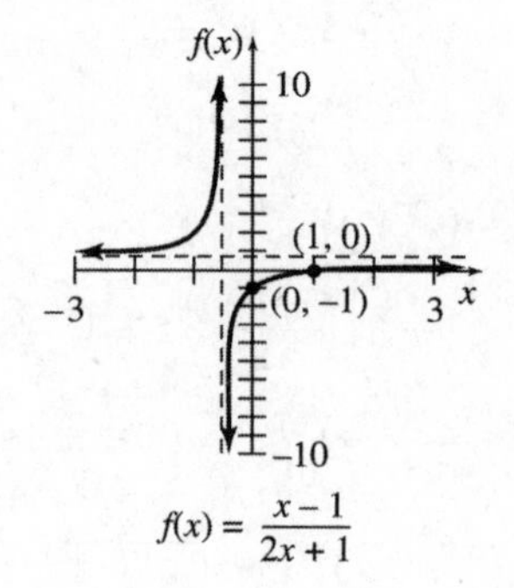

33. $f(x) = -4x^3 - x^2 + 4x + 5$

Domain is $(-\infty, \infty)$
The graph has no symmetry.

$$f'(x) = -12x^2 - 2x + 4$$
$$= -2(6x^2 + x - 2) = 0$$
$$(3x+2)(2x-1) = 0$$

Critical numbers: $-\frac{2}{3}$ and $\frac{1}{2}$

Critical points: $(-\frac{2}{3}, 3.07)$ and $(\frac{1}{2}, 6.25)$

$$f''(x) = -24x - 2$$
$$= -2(12x + 1)$$

$$f''\left(-\frac{2}{3}\right) = 14 > 0$$

$$f''\left(\frac{1}{2}\right) = -14 < 0$$

Relative maximum at $\frac{1}{2}$

Relative minimum at $-\frac{2}{3}$

Increasing on $\left(-\frac{2}{3}, \frac{1}{2}\right)$

Decreasing on $\left(-\infty, -\frac{2}{3}\right)$ and $\left(\frac{1}{2}, \infty\right)$

$$f''(x) = -2(12x + 1) = 0$$
$$x = -\frac{1}{12}$$

Point of inflection at $\left(-\frac{1}{12}, 4.66\right)$

Concave upward on $\left(-\infty, -\frac{1}{12}\right)$

Concave downward on $\left(-\frac{1}{12}, \infty\right)$

y-intercept:

$$y = -4(0)^3 - (0)^2 + 4(0) + 5 = 5$$

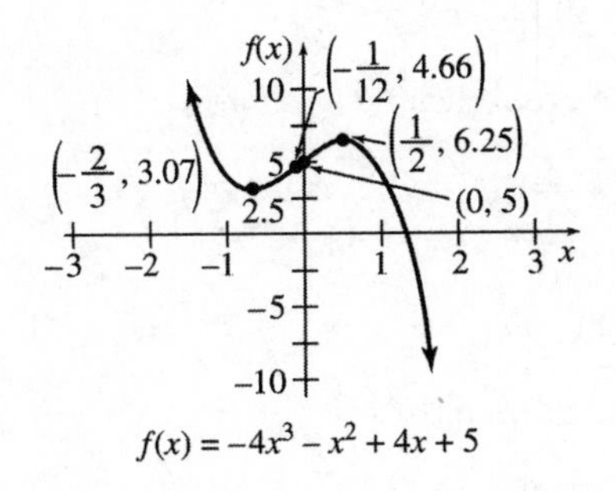

$f(x) = -4x^3 - x^2 + 4x + 5$

35. $f(x) = x^4 + 2x^2$

Domain is $(-\infty, \infty)$

$$f(-x) = (-x)^4 - 2(-x)^2$$
$$= x^4 + 2x^2 = f(x)$$

The graph is symmetric about the y-axis.

$$f'(x) = 4x^3 + 4x$$
$$= 4x(x^2 + 1) = 0$$

Critical number: 0
Critical point: $(0, 0)$

$$f''(x) = 12x^2 + 4 = 4(3x^2 + 1)$$
$$f''(0) = 4 > 0$$

Relative minimum at 0
Increasing on $(0, \infty)$
Decreasing on $(-\infty, 0)$

$$f''(x) = 4(3x^2 + 1) \neq 0 \text{ for any } x$$

No points of inflection

$$f''(-1) = 16 > 0$$
$$f''(1) = 16 > 0$$

Concave upward on $(-\infty, \infty)$
x-intercept: 0; y-intercept: 0

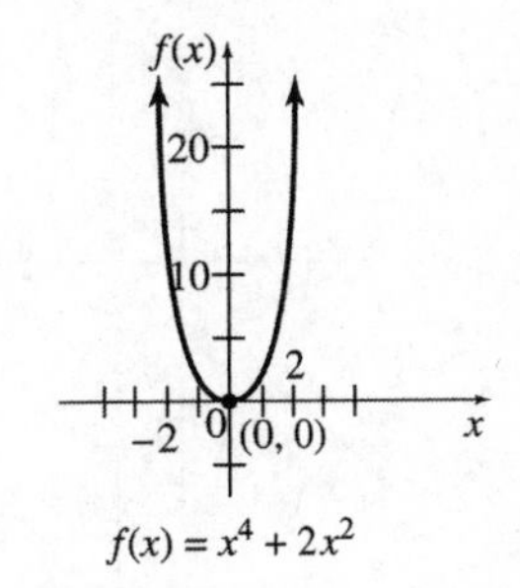

$f(x) = x^4 + 2x^2$

37. $f(x) = \dfrac{x^2 + 4}{x}$

Domain is $(-\infty, 0) \cup (0, \infty)$

$$f(-x) = \frac{(-x)^2 + 4}{-x}$$
$$= \frac{x^2 + 4}{-x} = -f(x)$$

The graph is symmetric about the origin.

$$f'(x) = \frac{x(2x) - (x^2 + 4)}{x^2}$$
$$= \frac{x^2 - 4}{x^2} = 0$$

Critical numbers: -2 and 2
Critical points: $(-2, -4)$ and $(2, 4)$

$$f''(x) = \frac{8}{x^3}$$
$$f''(-2) = -1 < 0$$
$$f''(2) = 1 > 0$$

Relative maximum at -2
Relative minimum at 2
Increasing on $(-\infty, -2)$ and $(2, \infty)$
Decreasing on $(-2, 0)$ and $(0, 2)$

$f''(x) = \frac{8}{x^3} > 0$ for all x.

No inflection points
Concave upward on $(0, \infty)$
Concave downward on $(-\infty, 0)$
No x- or y-intercepts
Vertical asymptote at $x = 0$

Oblique asymptote at $y = x$

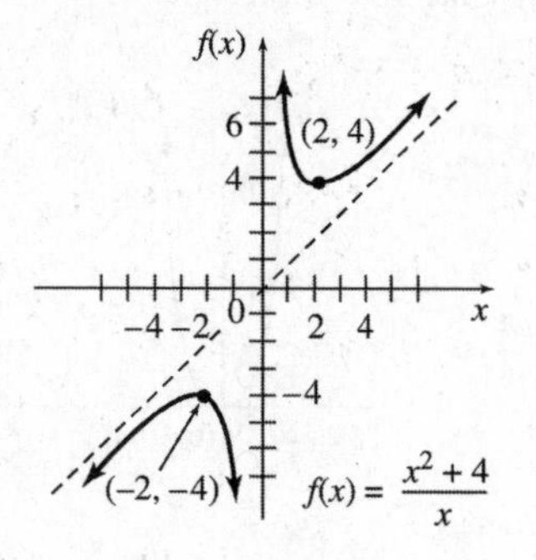

39. $f(x) = \dfrac{2x}{3-x}$

Domain is $(-\infty, 3) \cup (3, \infty)$
The graph has no symmetry.

$$f'(x) = \frac{(3-x)(2) - (2x)(-1)}{(3-x)^2}$$
$$= \frac{6}{(3-x)^2}$$

$f'(x)$ is never zero. $f'(3)$ does not exist, but since 3 is not in the domain of f, it is not a critical number.
No critical numbers, so no relative extrema

$$f'(0) = \frac{2}{3} > 0$$
$$f'(4) = 6 > 0$$

Increasing on $(-\infty, 3)$ and $(3, \infty)$

$$f''(x) = \frac{12}{(3-x)^3}$$

$f''(x)$ is never zero. $f''(3)$ does not exist, but since 3 is not in the domain of f, there is no inflection point at $x = 3$.

$$f''(0) = \frac{12}{27} > 0$$
$$f''(4) = -12 < 0$$

Concave upward on $(-\infty, 3)$
Concave downward on $(3, \infty)$

x-intercept: 0; y-intercept: 0

Vertical asymptote at $x = 3$

Horizontal asymptote at $y = -2$

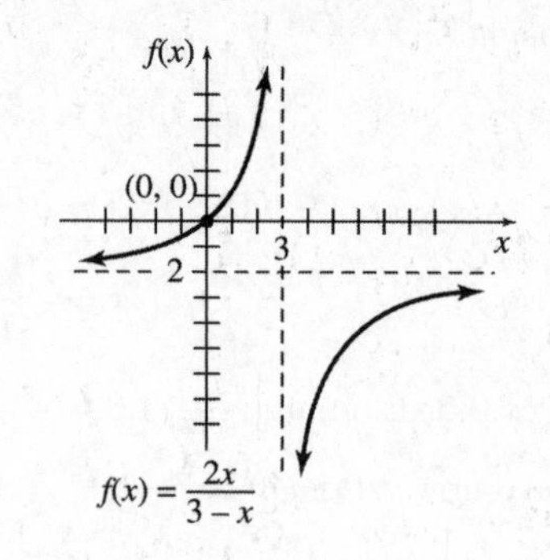

41. $f(x) = xe^{2x}$

Domain is $(-\infty, \infty)$.

$f(-x) = -xe^{-2x}$

The graph has no symmety.

$$f'(x) = (1)(e^{2x}) + (x)(2e^{2x})$$
$$= e^{2x}(2x+1)$$

$f'(x) = 0$ when $x = -\frac{1}{2}$.

Critical number: $-\frac{1}{2}$

Critical point: $\left(-\frac{1}{2}, -\frac{1}{2e}\right)$

$$f'(-1) = e^{2(-1)}[2(-1)+1] = -e^{-2} < 0$$
$$f'(0) = e^{2(0)}[2(0)+1] = 1 > 0$$

No relative maximum

Relative minimum at $\left(-\frac{1}{2}, -\frac{1}{2e}\right)$

Decreasing on $\left(-\infty, -\frac{1}{2}\right)$ and increasing on $\left(-\frac{1}{2}, \infty\right)$

$$f''(x) = 2e^{2x}(2x+1) + e^{2x}(2)$$
$$= 4e^{2x}(x+1)$$

$f''(x) = 0$ when $x = -1$.

$$f''(-2) = 4e^{2(-2)}[(-2)+1] = -4e^{-4} < 0$$
$$f''(0) = 4e^{2(0)}[(0)+1] = 4 > 0$$

Inflection point at $(-1, -e^{-2})$

Concave upward on $(-1, \infty)$
Concave downward on $(-\infty, -1)$

x-intercept: $xe^{2x} = 0$
$x = 0$

y-intercept: $y = (0)e^{2(0)} = 0$

Since $\lim\limits_{x\to-\infty} xe^{2x} = 0$, there is a horizontal asymptote at $y = 0$.

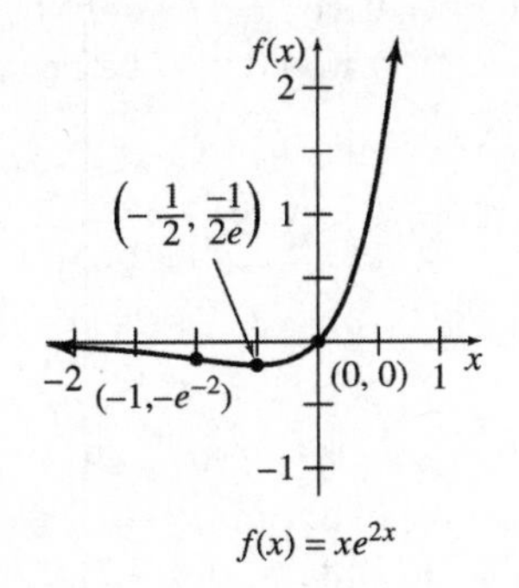

43. $f(x) = \ln(x^2 + 4)$

Domain is $(-\infty, \infty)$.

$f(-x) = \ln[(-x)^2 + 4] = \ln(x^2 + 4) = f(x)$

The graph is symmetric about the y-axis.

$$f'(x) = \frac{2x}{x^2 + 4}$$

$f'(x) = 0$ when $x = 0$.

Critical number: 0

Critical point: $(0, \ln 4)$

$$f'(-1) = \frac{2(-1)}{(-1)^2 + 4} = -\frac{2}{5} < 0$$

$$f'(1) = \frac{2(1)}{(1)^2 + 4} = \frac{2}{5} > 0$$

No relative maximum

Relative minimum at $(0, \ln 4)$

Increasing on $(0, \infty)$

Decreasing on $(-\infty, 0)$

$$f''(x) = \frac{(x^2 + 4)(2) - (2x)(2x)}{(x^2 + 4)^2} = \frac{-2(x^2 - 4)}{(x^2 + 4)^2}$$

$f''(x) = 0$ when

$$x^2 - 4 = 0$$
$$x = \pm 2$$

$$f''(-3) = \frac{-2[(-3)^2 - 4]}{[(-3)^2 + 4]^2} = -\frac{10}{169} < 0$$

$$f''(0) = \frac{-2[(0)^2 - 4]}{[(0)^2 + 4]^2} = \frac{1}{2} > 0$$

$$f''(3) = \frac{-2[(3)^2 - 4]}{[(3)^2 + 4]^2} = -\frac{10}{169} < 0$$

Inflection points at $(-2, \ln 8)$ and $(2, \ln 8)$

Concave upward on $(-2, 2)$

Concave downward on $(-\infty, -2)$ and $(2, \infty)$

Since $f(x)$ never equals zero, there are no x-intercepts.

y-intercept: $y = \ln[(0)^2 + 4] = \ln 4$

No horizontal or vertical asymptotes.

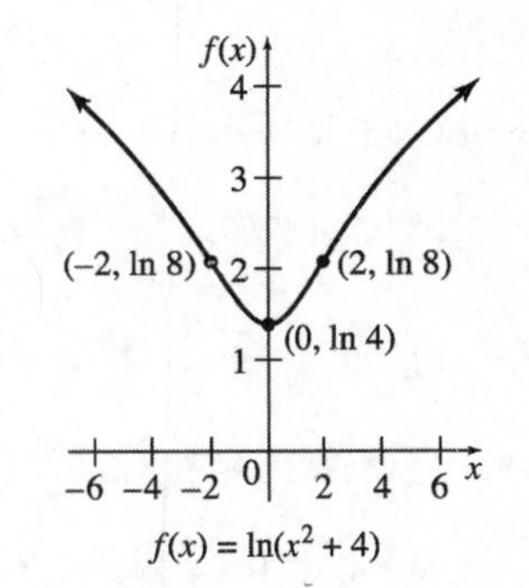

45. $f(x) = 4x^{1/3} + x^{4/3}$

Domain is $(-\infty, \infty)$.

$f(-x) = 4(-x)^{1/3} + (-x)^{4/3} = -4x^{1/3} + x^{4/3}$

The graph is not symmetric about the y-axis or origin.

$$f'(x) = \frac{4}{3}x^{-2/3} + \frac{4}{3}x^{1/3}$$

$f'(x) = 0$ when

$$\frac{4}{3}x^{-2/3} + \frac{4}{3}x^{1/3} = 0$$
$$\frac{4}{3}x^{-2/3}(1 + x) = 0$$
$$x = -1$$

$f'(x)$ is not defined when $x = 0$

Critical numbers: -1 and 0

Critical points: $(-1, -3)$ and $(0, 0)$

$$f'(-8) = \frac{4}{3}(-8)^{-2/3} + \frac{4}{3}(-8)^{1/3} = -\frac{7}{3} < 0$$

$$f'\left(-\frac{1}{8}\right) = \frac{4}{3}\left(-\frac{1}{8}\right)^{-2/3} + \frac{4}{3}\left(-\frac{1}{8}\right)^{1/3} = \frac{14}{3} > 0$$

$$f'(1) = \frac{4}{3}(1)^{-2/3} + \frac{4}{3}(1)^{1/3} = \frac{8}{3} > 0$$

No relative maximum

Relative minimum at $(-1, -3)$

Increasing on $(-1, \infty)$

Decreasing on $(-\infty, -1)$

$$f''(x) = -\frac{8}{9}x^{-5/3} + \frac{4}{9}x^{-2/3}$$

$f''(x) = 0$ when

$$-\frac{8}{9}x^{-5/3} + \frac{4}{9}x^{-2/3} = 0$$
$$\frac{4}{9}x^{-5/3}(-2 + x) = 0$$
$$x = 2$$

$f''(x)$ is not defined when $x = 0$

$$f''(-1) = -\frac{8}{9}(-1)^{-5/3} + \frac{4}{9}(-1)^{-2/3} = \frac{4}{3} > 0$$
$$f''(1) = -\frac{8}{9}(1)^{-5/3} + \frac{4}{9}(1)^{-2/3} = -\frac{4}{9} < 0$$
$$f''(8) = -\frac{8}{9}(8)^{-5/3} + \frac{4}{9}(8)^{-2/3} = \frac{1}{12} > 0$$

Inflection points at $(0, 0)$ and $(2, 6 \cdot 2^{1/3})$

Concave upward on $(-\infty, 0)$ and $(2, \infty)$
Concave downward on $(0, 2)$

x-intercept: $4x^{1/3} + x^{4/3} = 0$
$$x^{1/3}(4 + x) = 0$$
$$x = 0 \text{ or } x = -4$$

y-intercept: $y = 4(0)^{1/3} + (0)^{4/3} = 0$

No horizontal or vertical asymptotes

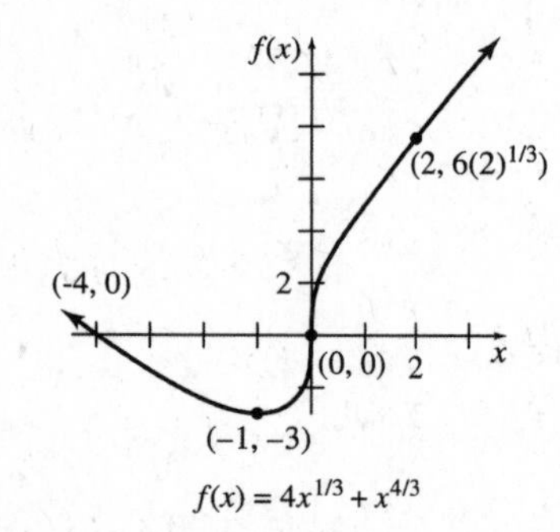

$f(x) = 4x^{1/3} + x^{4/3}$

47.

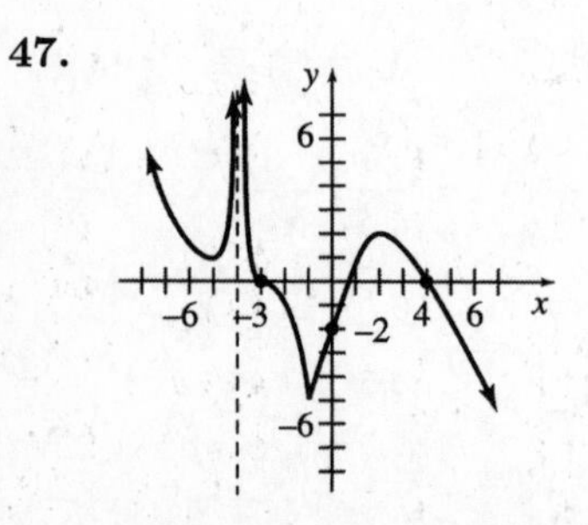

Other graphs are possible.

49. (a)-(b) If the price of the stock is falling faster and faster, $P(t)$ would be decreasing, so $P'(t)$ would be negative. $P(t)$ would be concave downward, so $P''(t)$ would also be negative.

51. (a) Profit = Income − Cost
$$\begin{aligned} P(q) &= qp - C(q) \\ &= q(-q^2 - 3q + 299) - (-10q^2 + 250q) \\ &= -q^3 - 3q^2 + 299q + 10q^2 - 250q \\ &= -q^3 + 7q^2 + 49q \end{aligned}$$

(b)
$$\begin{aligned} P'(q) &= -3q^2 + 14q + 49 \\ &= (-3q - 7)(q - 7) \\ &= -(3q + 7)(q - 7) \end{aligned}$$
$$q = \tfrac{7}{3} \text{ (nonsensical) or } q = 7$$
$$P''(q) = -6q + 14$$
$$P''(7) = -6 \cdot 7 + 14 = -28 < 0$$
(indicates a maximum)

7 brushes would produce the maximum profit.

(c) $p = -7^2 - 3(7) + 299$
$= -49 - 21 + 299 = 229$

\$229 is the price that produces the maximum profit.

(d) $P(7) = -7^3 + 7(7^2) + 49(7) = 343$

The maximum profit is \$343.

(e) $P''(q) = 0$ when $-6q + 14 = 0$
$$q = \frac{7}{3}$$
$$P''(2) = -6(2) + 14 = 2 > 0$$
$$P''(3) = -6 \cdot 3 + 14 = -4 < 0$$

The point of diminishing returns is $q = \frac{7}{3}$ (between 2 and 3 brushes).

53. (a)
$$\begin{aligned} Y(M) &= Y_0M^b \\ Y'(M) &= bY_0M^{b-1} \\ Y''(M) &= b(b-1)Y_0M^{b-2} \end{aligned}$$

When $b > 0$, $Y' > 0$, so metabolic rate and life span are increasing function of mass.
When $b < 0$, $Y' < 0$, so heartbeat is a decreasing function of mass.
When $0 < b < 1$, $b(b-1) < 0$, so metabolic rate and life span have graphs that are concave downward.
When $b < 0, b(b-1) > 0$, so heartbeat has a graph that is concave upward.

(b)
$$\frac{dY}{dM} = bY_0M^{b-1} = \frac{b}{M}Y_0M^b$$
$$= \frac{b}{M}Y$$

55. Let $a = 0.25$.

$$\begin{aligned} f(v) &= v(0.25 - v)(v - 1) \\ f'(v) &= -v^3 + 1.25v^2 - 0.25v \\ &= -3v^2 + 2.5v - 0.25 \end{aligned}$$

By the quadratic formula, $f'(v) = 0$ when $v = 0.12$ and $v \approx 0.72$.
Critical numbers $v \approx 0.12$ and $v \approx 0.72$.

$$\begin{aligned} f'(0.1) &= -0.03 < 0 \\ f'(0.5) &= 0.25 > 0 \\ f'(1) &= -0.75 < 0 \end{aligned}$$

The function is decreasing on $(0, 0.12)$ and $(0.72, 1)$ and increasing on $(0.12, 0.72)$.

$$\begin{aligned} f(0.12) &\approx -0.01 \\ &\approx 0.09 \end{aligned}$$

Relative minimum at $(0.12, -0.01)$, relative maximum at $(0.72, 0.09)$.

$$f''(v) = -6v + 2.5$$

$f''(v) = 0$ when

$$\begin{aligned} -6v + 2.5 &= 0 \\ v &\approx 0.42 \end{aligned}$$

$$\begin{aligned} f''(0.1) &= 1.9 > 0 \\ f''(0.5) &= -0.5 < 0 \end{aligned}$$

Concave upward on $(0, 0.42)$
Concave downward on $(0.42, 1)$

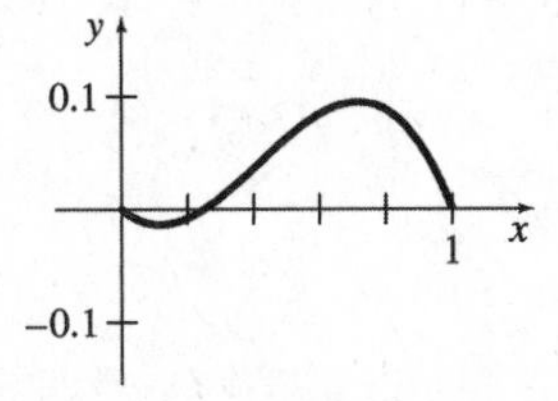

57. (a) Set the two formulas equal to each other.

$$\begin{aligned} 1486S^2 - 4106S + 4514 &= 1486S - 825 \\ 1486S^2 - 5592S + 5339 &= 0 \end{aligned}$$

Take the derivative.

$$2972S - 5592 = 0$$

Solve for S.

$$S \approx 1.88$$

For males with 1.88 square meters of surface area, the red cell volume increases approximately 1486 ml for each additional square meter of surface area.

(b) Set the formulas equal to each other.

$$\begin{aligned} 995e^{0.6085S} &= 1578S \\ 995e^{0.6085S} - 1578S &= 0 \end{aligned}$$

Take the derivative.

$$\begin{aligned} 605.4575e^{0.6085S} - 1578 &= 0 \\ e^{0.6085S} &\approx 2.6063 \\ 0.6085S &\approx \ln 2.6063 \\ S &\approx 1.57 \text{ square meters} \end{aligned}$$

By plugging the exact value of S into the two formulas given for PV, we get about 2593 ml (Hurley) and 2484 for Pearson et al.

(c) For males with 1.57 square meters of surface area, the red cell volume increases approximately 1578 ml for each additional square meter of surface area.

(d) When f and g are closest together, their absolute difference is minimized.

$$\begin{aligned} \frac{d}{dx}|f(x_0) - g(x_0)| &= 0 \\ |f'(x_0) - g'(x_0)| &= 0 \\ f'(x_0) &= g'(x_0) \end{aligned}$$

59. (a) $P(t) = 325 + 7.475(t + 10)e^{-(t+10)/20}$

$$\begin{aligned} P'(t) &= 7.475\left[1 \cdot e^{-(t+10)/20} \right. \\ &\quad \left. + (t+10)e^{-(t+10)/20} \cdot \frac{-1}{20}\right] \\ &= 7.475\left[1 - \frac{1}{20}(t+10)\right]e^{-(t+10)/20} \\ &= 7.475\left(\frac{1}{2} - \frac{t}{20}\right)e^{-(t+10)/20} \end{aligned}$$

$P'(t)$ is zero when

$$\begin{aligned} \frac{1}{2} - \frac{t}{20} &= 0 \\ t &= 10 \end{aligned}$$

$$\begin{aligned} P'(9) &\approx 0.39 \\ P'(11) &\approx -0.36 \end{aligned}$$

$P(t)$ is increasing at 9 and decreasing at 11. So, a relative maximum occurs at $t = 10$. The population is largest in the year 2010.

(b) $P'(t) = 7.475\left(\frac{1}{2} - \frac{t}{20}\right)e^{-(t+10)/20}$

$$\begin{aligned}P''(t) &= 7.475\left[-\frac{1}{20}e^{-(t+10)/20} + \left(\frac{1}{2} - \frac{t}{20}\right)e^{-(t+10)/20} \cdot \frac{-1}{20}\right] \\ &= 7.475\left(-\frac{1}{20} - \frac{1}{40} + \frac{t}{400}\right)e^{-(t+10)/20} \\ &= 7.475\left(\frac{t}{400} - \frac{3}{40}\right)e^{-(t+10)/20}\end{aligned}$$

$P'(t)$ is zero when

$$\begin{aligned}\frac{t}{400} - \frac{3}{40} &= 0 \\ t &= 30 \\ P'(29) &= -0.0027 \\ P'(31) &= 0.0024\end{aligned}$$

P' is decreasing at 29, and increasing at 31. So, a relative minimum occurs at $t = 30$. The population is declining most rapidly in the year 2010.

(c) As time t approaches infinity, the population P approaches

$$\begin{aligned}\lim_{t\to\infty} P(t) &= \lim_{t\to\infty} (325 + 7.475(t+10)e^{-(t+10)/20}) \\ &= 325 + 7.475 \lim_{t\to\infty} \frac{t+10}{e^{(t+10)/20}} \\ &= 325 + 7.475(0) \\ &= 325.\end{aligned}$$

The population is approaching 325 million.

61. (a) $s(t) = 512t - 16t^2$
$v(t) = s'(t) = 512 - 32t$
$a(t) = v'(t) = s''(t) = -32$

(b) The maximum height is attained when $v(t) = 0$.

$$\begin{aligned}512 - 32t &= 0 \\ t &= 16\end{aligned}$$

$$\begin{aligned}v(0) &= 512 > 0 \\ v(20) &= 512 - 640 = -128 < 0\end{aligned}$$

The height reaches a maximum when $t = 16$.

$$s(16) = 512 \cdot 16 - 16(16^2) = 4096$$

The maximum height is 4096 ft.

(c) The projectile hits the ground when $s(t) = 0$.

$$\begin{aligned}512t - 16t^2 &= 0 \\ 16t(32 - t) &= 0 \\ t = 0 &\text{ or } t = 32\end{aligned}$$

$$v(32) = 512 - 32(32) = -512$$

The projectile hits the ground after 32 seconds with a velocity of -512 ft/sec.

Chapter 5 Test

[5.1]

Find the largest open intervals where each of the following functions is (a) increasing or (b) decreasing.

1.

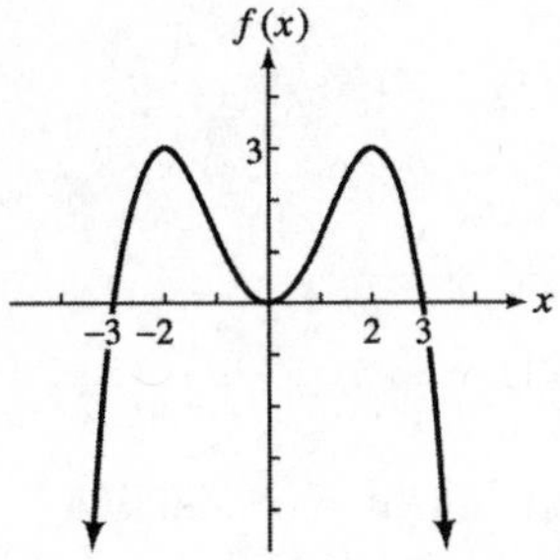

2. $f(x) = 3x^3 - 3x^2 - 3x + 5$

3. $f(x) = \dfrac{x-2}{x+3}$

4. For a certain product, the demand equation is $p = 10 - .05x$. The cost function for this product is $C = 5x - .75$. Over what interval(s) is the profit increasing?

5. What are the critical numbers? Why are they significant?

[5.2]

Find the x-values of all points where the following functions have any relative extrema. Find the value(s) of any relative extrema.

6. $f(x) = 4x^3 - \dfrac{9}{2}x^2 - 3x - 1$

7. $f(x) = 4x^{2/3}$

8. A manufacturer estimates that the cost in dollars per unit for a production run of x thousand units is given by $C = 3x^2 - 60x + 320$. How many thousand units should be produced during each run to minimize the cost per unit, and what is the minimum cost per unit?

9. Use the derivative to find the vertex of the parabola $y = -2x^2 + 6x + 9$.

[5.3]

Find $f'''(x)$, the third derivative of f, and $f^{(4)}(x)$, the fourth derivative of f, for each of the following functions.

10. $f(x) = 2e^{3x}$

11. $f(x) = \dfrac{x}{3x+2}$

12. $f(x) = 4x^6 - 3x^4 + 2x^3 - 7x + 5$

Find any points of inflection for the following functions.

13. $f(x) = 6x^3 - 18x^2 + 12x - 15$

14. $f(x) = (x-1)^3 + 2$

15. $f(x) = \dfrac{2}{1+x^2}$

Find the second derivative of each function; then find $f''(-3)$.

16. $f(x) = -5x^3 + 3x^2 - x + 1$

17. $f(t) = \sqrt{t^2 - 5}$

18. Find the largest open intervals where $f(x) = x^3 - 3x^2 - 9x - 1$ is concave upward or concave downward. Find the location of any points of inflection. Graph the function.

19. The function $s(t) = t^3 - 9t^2 + 15t + 25$ gives the displacement in centimeters at time t (in seconds) of a particle moving along a line. Find the velocity and acceleration functions. Then find the velocity and acceleration at $t = 0$ and $t = 2$.

20. What is the difference between velocity and acceleration?

[5.4]

Find the horizontal asymptotes for the graphs of the following functions.

21. $f(x) = \dfrac{2x}{7x+1}$

22. $f(x) = \dfrac{3x^2 + 2x}{4x^3 - x}$

23. Graph $f(x) = \frac{2}{3}x^3 - \frac{5}{2}x^2 - 3x + 1$. Give critical points, intervals where the function is increasing or decreasing, points of inflection, and intervals where the function is concave upward or concave downward.

24. Graph $f(x) = x + \frac{32}{x^2}$. Give relative extrema, regions where the function is increasing or decreasing, points of inflection, regions where the function is concave upward or downward, intercepts where possible, and asymptotes where applicable.

25. Sketch a graph of a single function that has all the properties listed.

(a) Continuous for all real numbers

(b) Increasing on $(-3, 2)$ and $(4, \infty)$

(c) Decreasing on $(-\infty, -3)$ and $(2, 4)$

(d) Concave upward on $(-\infty, 0)$

(e) Concave downward on $(0, 4)$ and $(4, \infty)$

(f) Differentiable everywhere except $x = 4$

(g) $f'(-3) = f'(2) = 0$

(h) An inflection point at $(0, 0)$

Chapter 5 Test Answers

1. **(a)** $(-\infty, -2)$ and $(0, 2)$ **(b)** $(-2, 0)$ and $(2, \infty)$

2. **(a)** $\left(-\infty, -\frac{1}{3}\right)$ and $(1, \infty)$ **(b)** $\left(-\frac{1}{3}, 1\right)$

3. **(a)** $(-\infty, -3)$ and $(-3, \infty)$ **(b)** Never decreasing

4. $(0, 50)$

5. Critical numbers are x–values for which $f'(x) = 0$ or $f'(x)$ does not exist. They tell where the derivative may change signs, and are used to tell where a function is increasing or decreasing.

6. Relative maximum of $-\frac{19}{32}$ at $-\frac{1}{4}$; relative minimum of $-\frac{9}{2}$ at 1

7. Relative minimum of 0 at 0

8. 10 thousand units; \$20 per unit

9. $\left(\frac{3}{2}, \frac{27}{2}\right)$

10. $f'''(x) = 54e^{3x}$, $f^{(4)}(x) = 162e^{3x}$

11. $f'''(x) = \frac{108}{(3x+2)^4}$, $f^{(4)}(x) = \frac{-1296}{(3x+2)^5}$

12. $f'''(x) = 480x^3 - 72x + 12$, $f^{(4)} = 1440x^2 - 72$

13. $(1, -15)$

14. $(1, 2)$

15. $\left(-\frac{\sqrt{3}}{3}, \frac{3}{2}\right), \left(\frac{\sqrt{3}}{3}, \frac{3}{2}\right)$

16. $f''(x) = -30x + 6$; $f''(-3) = 96$

17. $f''(t) = \frac{-5}{(t^2-5)^{3/2}}$; $f''(-3) = -\frac{5}{8}$

18. Concave upward on $(1, \infty)$; concave downward on $(-\infty, 1)$; point of inflection $(1, -12)$

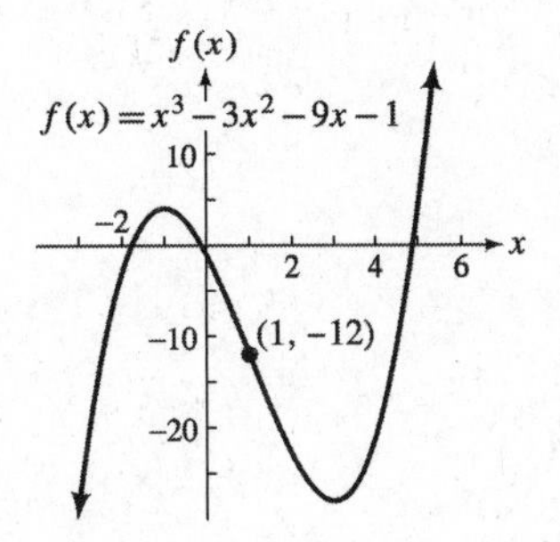

19. $v(t) = 3t^2 - 18t + 15$; $a(t) = 6t - 18$; $v(0) = 15$ cm/sec; $a(0) = -18$ cm/sec^2; $v(2) = -9$ cm/sec; $a(2) = -6$ cm/sec^2

20. Velocity gives the rate of change of position relative to time. Acceleration gives the rate of change of velocity.

21. $y = \frac{2}{7}$

22. $y = 0$

23. Critical points: $\left(-\frac{1}{2}, \frac{43}{24}\right)$ or $(-0.5, 1.79)$ (relative maximum) and $\left(3, -\frac{25}{2}\right)$ or $(3, -12.5)$ (relative minimum); increasing on $\left(-\infty, -\frac{1}{2}\right)$ or $(-\infty, -0.5)$ and $(3, \infty)$; decreasing on $\left(-\frac{1}{2}, 3\right)$ or $(-0.5, 3)$; point of inflection: $\left(\frac{5}{4}, -\frac{257}{48}\right)$ or $(1.25, -5.35)$; concave up on $\left(\frac{5}{4}, \infty\right)$ or $(1.25, \infty)$; concave down on $\left(-\infty, \frac{5}{4}\right)$ or $(-\infty, 1.25)$

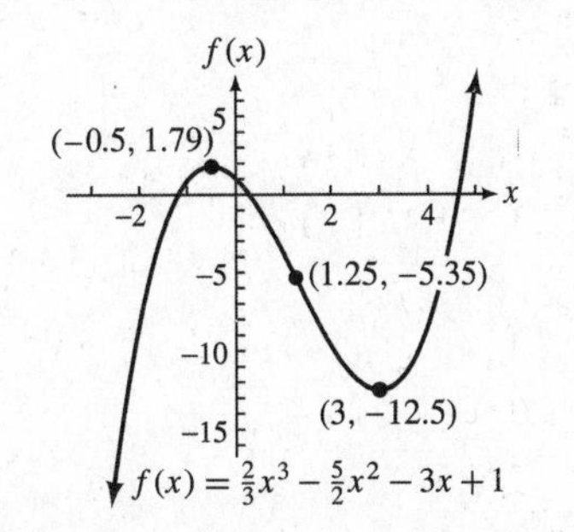

24. Relative extrema: Relative minimum at $(4, 6)$, no relative maxima; increasing on $(-\infty, 0)$ and $(4, \infty)$; decreasing on $(0, 4)$; points of inflection: none; concave upward on $(-\infty, 0)$ and $(0, \infty)$; concave downward nowhere; x–intercept: $-\sqrt[3]{32} \approx -3.17$; y–intercept: none; oblique asymptote: $y = x$

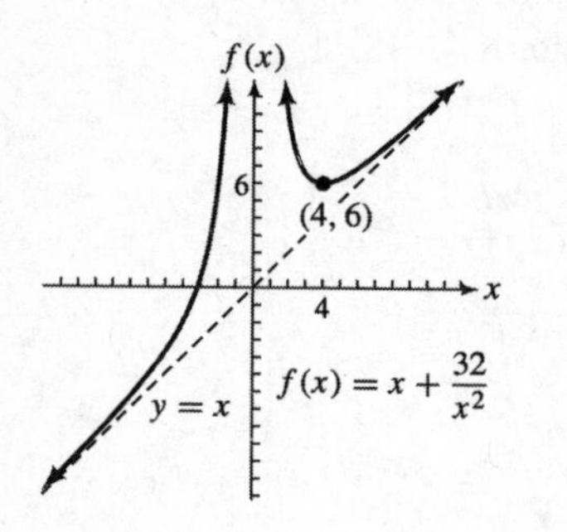

25. One such graph is shown. Other answers are possible.

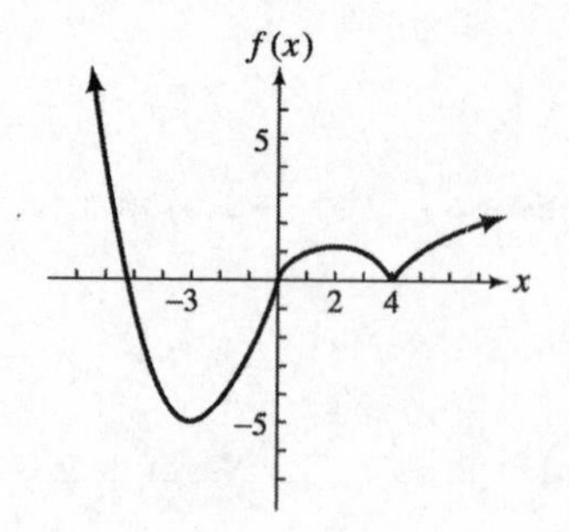

Chapter 6

APPLICATIONS OF THE DERIVATIVE

6.1 Absolute Extrema

1. As shown on the graph, the absolute maximum occurs at x_3; there is no absolute minimum. (There is no functional value that is less than all others.)

3. As shown on the graph, there are no absolute extrema.

5. As shown on the graph, the absolute minimum occurs at x_1; there is no absolute maximum.

7. As shown on the graph, the absolute maximum occurs at x_1; the absolute minimum occurs at x_2.

11. $f(x) = x^3 - 6x^2 + 9x - 8$; $[0, 5]$

Find critical numbers:

$$\begin{aligned} f'(x) &= 3x^2 - 12x + 9 = 0 \\ x^2 - 4x + 3 &= 0 \\ (x-3)(x-1) &= 0 \\ x = 1 \quad &\text{or} \quad x = 3 \end{aligned}$$

x	$f(x)$	
0	-8	Absolute minimum
1	-4	
3	-8	Absolute minimum
5	12	Absolute maximum

13. $f(x) = \frac{1}{3}x^3 + \frac{3}{2}x^2 - 4x + 1$; $[-5, 2]$

Find critical numbers:

$$\begin{aligned} f'(x) &= x^2 + 3x - 4 = 0 \\ (x+4)(x-1) &= 0 \\ x = -4 \quad &\text{or} \quad x = 1 \end{aligned}$$

x	$f(x)$	
-4	$\frac{59}{3} \approx 19.67$	Absolute maximum
1	$-\frac{7}{6} \approx -1.17$	Absolute minimum
-5	$\frac{101}{6} \approx 16.83$	
2	$\frac{5}{3} \approx 1.67$	

15. $f(x) = x^4 - 18x^2 + 1$; $[-4, 4]$

$$\begin{aligned} f'(x) &= 4x^3 - 36x = 0 \\ 4x(x^2 - 9) &= 0 \\ 4x(x+3)(x-3) &= 0 \end{aligned}$$

$$x = 0 \quad \text{or} \quad x = -3 \quad \text{or} \quad x = 3$$

x	$f(x)$	
-4	-31	
-3	-80	Absolute minimum
0	1	Absolute maximum
3	-80	Absolute minimum
4	-31	

17. $f(x) = \frac{1-x}{3+x}$; $[0, 3]$

$$f'(x) = \frac{-4}{(3+x)^2}$$

No critical numbers

x	$f(x)$	
0	$\frac{1}{3}$	Absolute maximum
3	$-\frac{1}{3}$	Absolute minimum

19. $f(x) = \frac{x-1}{x^2+1}$; $[1, 5]$

$$f'(x) = \frac{-x^2 + 2x + 1}{(x^2+1)^2}$$

$f'(x) = 0$ when

$$\begin{aligned} -x^2 + 2x + 1 &= 0 \\ x &= 1 \pm \sqrt{2}, \end{aligned}$$

but $1 - \sqrt{2}$ is not in $[1, 5]$.

x	$f(x)$	
1	0	Absolute minimum
5	$\frac{2}{13} \approx 0.15$	
$1 + \sqrt{2}$	$\frac{\sqrt{2} - 1}{2} \approx 0.21$	Absolute maximum

21. $f(x) = (x^2 - 4)^{1/3}$; $[-2, 3]$

$$f'(x) = \frac{1}{3}(x^2 - 4)^{-2/3}(2x) = \frac{2x}{3(x^2 - 4)^{2/3}}$$

$f'(x) = 0$ when $2x = 0$

$x = 0$

$f'(x)$ is undefined at $x = -2$ and $x = 2$, but $f(x)$ is defined there, so -2 and 2 are also critical numbers.

x	$f(x)$	
-2	0	
0	$(-4)^{1/3} \approx -1.587$	Absolute minimum
2	0	
3	$5^{1/3} \approx 1.710$	Absolute maximum

23. $f(x) = 5x^{2/3} + 2x^{5/3}$; $[-2, 1]$

$$f'(x) = \frac{10}{3}x^{-1/3} + \frac{10}{3}x^{2/3} = \frac{10}{3x^{1/3}} + \frac{10x^{2/3}}{3} = \frac{10x + 10}{3x^{1/3}} = \frac{10(x+1)}{3\sqrt[3]{x}}$$

$f'(x) = 0$ when $10(x + 1) = 0$

$x + 1 = 0$

$x = -1.$

$f'(x)$ is undefined at $x = 0$, but $f(x)$ is defined at $x = 0$, so 0 is also a critical number.

x	$f(x)$	
-2	1.587	
-1	3	
0	0	Absolute minimum
1	7	Absolute maximum

25. $f(x) = x^2 - 8 \ln x$; $[1, 4]$

$$f'(x) = 2x - \frac{8}{x}$$

$f'(x) = 0$ when $2x - \frac{8}{x} = 0$

$$2x = \frac{8}{x}$$

$$2x^2 = 8$$

$$x^2 = 4$$

$$x = -2 \text{ or } x = 2$$

but $x = -2$ is not in the given interval.

Although $f'(x)$ fails to exist at $x = 0$, 0 is not in the specified domain for $f(x)$, so 0 is not a critical number.

x	$f(x)$	
1	1	
2	-1.545	Absolute minimum
4	4.910	Absolute maximum

27. $f(x) = x + e^{-3x}$; $[-1, 3]$

$f'(x) = 1 - 3e^{-3x}$

$f'(x) = 0$ when $1 - 3e^{-3x} = 0$

$$-3e^{-3x} = -1$$

$$e^{-3x} = \frac{1}{3}$$

$$-3x = \ln\frac{1}{3}$$

$$x = \frac{\ln 3}{3}$$

x	$f(x)$	
-1	19.09	Absolute maximum
$\frac{\ln 3}{3}$	0.6995	Absolute minimum
3	3.000	

29. $f(x) = \dfrac{-5x^4 + 2x^3 + 3x^2 + 9}{x^4 - x^3 + x^2 + 7}$; $[-1, 1]$

The indicated domain tells us the x-values to use for the viewing window, but we must experiment to find a suitable range for the y-values. In order to show the absolute extrema on $[-1, 1]$, we find that a suitable window is $[-1, 1]$ by $[0, 1.5]$ with Xscl = 0.1, Yscl = 0.1.

From the graph, we se that on $[-1, 1]$, f has an absolute maximum of 1.356 at about 0.6085 and an absolute minimum of 0.5 at -1.

31. $f(x) = 2x + \dfrac{8}{x^2} + 1$, $x > 0$

$$f'(x) = 2 - \frac{16}{x^3} = \frac{2x^3 - 16}{x^3} = \frac{2(x-2)(x^2 + 2x + 4)}{x^3}$$

Since the specified domain is $(0, \infty)$, a critical number is $x = 2$.

x	$f(x)$
2	7

There is an absolute minimum at $x = 2$; there is no absolute maximum, as can be seen by looking at the graph of f.

33. $f(x) = -3x^4 + 8x^3 + 18x^2 + 2$
$f'(x) = -12x^3 + 24x^2 + 36x$
$= -12x(x^2 - 2x - 3)$
$= -12x(x-3)(x+1)$

Critical numbers are 0, 3, and -1.

x	$f(x)$
-1	9
0	2
3	137

There is an absolute maximum at $x = 3$; there is no absolute minimum, as can be seen by looking at the graph of f.

35. $f(x) = \dfrac{x-1}{x^2+2x+6}$

$$f'(x) = \frac{(x^2+2x+6)(1) - (x-1)(2x+2)}{(x^2+2x+6)^2}$$
$$= \frac{x^2+2x+6-2x^2+2}{(x^2+2x+6)^2}$$
$$= \frac{-x^2+2x+8}{(x^2+2x+6)^2}$$
$$= \frac{-(x^2-2x-8)}{(x^2+2x+6)^2}$$
$$= \frac{-(x-4)(x+2)}{(x^2+2x+6)^2}$$

Critical numbers are 4 and -2.

x	$f(x)$
-2	$-\frac{1}{2}$
4	0.1

There is an absolute maximum at $x = 4$ and an absolute minimum at $x = -2$. This can be verified by looking at the graph of f.

37. $f(x) = \dfrac{\ln x}{x^3}$

$$f'(x) = \frac{x^3 \cdot \frac{1}{x} - 3x^2 \ln x}{x^6}$$
$$= \frac{x^2 - 3x^2 \ln x}{x^6}$$
$$= \frac{x^2(1 - 3\ln x)}{x^6}$$
$$= \frac{1 - 3\ln x}{x^4}$$

$f'(x) = 0$ when $x = e^{1/3}$, and $f'(x)$ does not exist when $x \le 0$. The only critical number is $e^{1/3}$.

x	$f(x)$
$e^{1/3}$	$\frac{1}{3}e^{-1} \approx 0.1226$

There is an absolute maximum of 0.1226 at $x = e^{1/3}$. There is no absolute minimum, as can be seen by looking at the graph of f.

39. Let $P(x)$ be the perimeter of the rectangle with vertices $(0,0)$, $(x,0)$, $(x, f(x))$, and $(0, f(x))$ for $x > 0$ when $f(x) = e^{-2x}$.

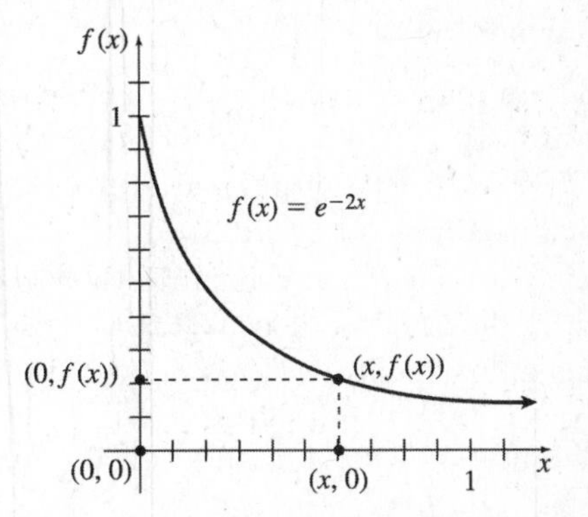

The length of the rectangle is x and the width is given by e^{-2x}. Therefore, an equation for the perimeter is

$$P(x) = x + e^{-2x} + x + e^{-2x} = 2(x + e^{-2x}).$$

$P'(x) = 2 - 4e^{-2x}$
$P'(x) = 0$ when $2 - 4e^{-2x} = 0$
$-4e^{-2x} = -2$
$e^{-2x} = \frac{1}{2}$
$e^{2x} = 2$
$2x = \ln 2$
$x = \frac{\ln 2}{2}$

x	$P(x)$
$\frac{\ln 2}{2}$	$1 + \ln 2 \approx 1.693$

There is an absolute minimum of 1.693 at $x = \frac{\ln 2}{2}$. There is no absolute maximum, as can be seen by looking at the graph of P. Therefore, the correct statement is **a**.

41. **(a)** By looking at the graph, there are relative maxima of 413 in 1997, 341 in 2000, and 134 in 2004. There are relative minima of 290 in 1996, 313 in 1998, and 131 in 2003.

(b) Annual bank burglaries reached an absolute maximum of 413 in 1997 and an absolute minimum of 131 in 2003.

43. $P(x) = -x^3 + 9x^2 + 120x - 400,\ x \geq 5$

$$\begin{aligned} P'(x) &= -3x^2 + 18x + 120 \\ &= -3(x^2 - 6x - 40) \\ &= \ \ -3(x-10)(x+4) = 0 \\ x &= 10 \quad \text{or} \quad x = -4 \end{aligned}$$

-4 is not relevant since $x \geq 5$, so the only critical number is 10.

The graph of $P'(x)$ is a parabola that opens downward, so $P'(x) > 0$ on the interval $[5, 10)$ and $P'(x) < 0$ on the interval $(10, \infty)$. Thus, $P(x)$ is a maximum at $x = 10$.

Since x is measured in hundred thousands, 10 hundred thousand or 1,000,000 tires must be sold to maximize profit.

Also,

$$\begin{aligned} P(10) &= -(10)^3 + 9(10)^2 + 120(10) - 400 \\ &= 700. \end{aligned}$$

The maximum profit is \$700 thousand or \$700,000.

45. $C(x) = x^3 + 37x + 250$

(a) $1 \leq x \leq 10$

$$\begin{aligned} \overline{C}(x) &= \frac{C(x)}{x} = \frac{x^3 + 37x + 250}{x} \\ &= x^2 + 37 + \frac{250}{x} \end{aligned}$$

$$\begin{aligned} \overline{C}'(x) &= 2x - \frac{250}{x^2} \\ &= \frac{2x^3 - 250}{x^2} = 0 \text{ when} \\ 2x^3 &= 250 \\ x^3 &= 125 \\ x &= 5. \end{aligned}$$

Test for relative minimum.

$$\begin{aligned} \overline{C}'(4) &= -7.625 < 0 \\ \overline{C}'(6) &\approx 5.0556 > 0 \\ \overline{C}(5) &= 112 \\ \overline{C}(1) &= 1 + 37 + 250 = 288 \\ \overline{C}(10) &= 100 + 37 + 25 = 162 \end{aligned}$$

The minimum on the interval $1 \leq x \leq 10$ is 112.

(b) $10 \leq x \leq 20$

There are no critical values in this interval. Check the endpoints.

$$\begin{aligned} \overline{C}(10) &= 162 \\ \overline{C}(20) &= 400 + 37 + 12.5 = 449.5 \end{aligned}$$

The minimum on the interval $10 \leq x \leq 20$ is 162.

47. The value $x = 11$ minimizes $\frac{f(x)}{x}$ because this is the point where the line from the origin to the curve is tangent to the curve.
A production level of 11 units results in the minimum cost per unit.

49. The value $x = 100$ maximizes $\frac{f(x)}{x}$ because this is the point where the line from the origin to the curve is tangent to the curve.
A production level of 100 units results in the maximum profit per item produced.

51. $S(x) = -x^3 + 3x^2 + 360x + 5000;\ 6 \leq x \leq 20$

$$\begin{aligned} S'(x) &= -3x^2 + 6x + 360 \\ &= -3(x^2 - 2x - 120) \\ S'(x) &= -3(x-12)(x+10) = 0 \\ x &= 12 \quad \text{or} \quad x = -10 \ (\text{not in the interval}) \end{aligned}$$

x	$f(x)$
6	7052
12	8024
10	7900

12° is the temperature that produces the maximum number of salmon.

53. The function is defined on the interval $[15, 46]$. We look first for critical numbers in the interval. We find

$$R'(T) = -0.00021T^2 + 0.0802T - 1.6572$$

Using our graphing calculator, we find one critical number in the interval at about 21.92

T	$R(T)$
15	81.01
21.92	79.29
46	98.89

The relative humidity is minimized at about 21.92°C.

55. $M(x) = -0.015x^2 + 1.31x - 7.3, 30 \leq x \leq 60$
$M'(x) = -0.03x + 1.31 = 0$
$x \approx 43.7$

x	$M(x)$
30	18.5
43.7	21.30
60	17.3

The absolute maximum of 21.30 mpg occurs at 43.7 mph. The absolute minimum of 17.3 mpg occurs at 60 mph.

57. Total area $= A(x)$

$$= \pi\left(\frac{x}{2\pi}\right)^2 + \left(\frac{12-x}{4}\right)^2$$
$$= \frac{x^2}{4\pi} + \frac{(12-x)^2}{16}$$
$$A'(x) = \frac{x}{2\pi} - \frac{12-x}{8} = 0$$
$$\frac{4x - \pi(12-x)}{8\pi} = 0$$
$$x = \frac{12\pi}{4+\pi} \approx 5.28$$

x	Area
0	9
5.28	5.04
12	11.46

The total area is maximized when all 12 feet of wire are used to form the circle.

59. (a) $I(p) = -p \ln p - (1-p) \ln (1-p)$

$$I'(p) = -p\left(\frac{1}{p}\right) + (\ln p)(-1) - \left[(1-p)\frac{-1}{1-p} + [\ln (1-p)](-1)\right]$$
$$= -1 - \ln p + 1 + \ln (1-p)$$
$$= -\ln p + \ln (1-p)$$

(b) $-\ln p + \ln (1-p) = 0$
$$\ln (1-p) = \ln p$$
$$1 - p = p$$
$$1 = 2p$$
$$\frac{1}{2} = p$$

$$I'(0.25) = 1.0986$$
$$I'(0.75) = -1.099$$

There is a relative maximum of 0.693 at $p = \frac{1}{2}$.

6.2 Applications of Extrema

1. $x + y = 180, P = xy$

(a) $y = 180 - x$

(b) $P = xy = x(180 - x)$

(c) Since $y = 180 - x$ and x and y are nonnegative numbers, $x \geq 0$ and $180 - x \geq 0$ or $x \leq 180$. The domain of P is $[0, 180]$.

(d) $P'(x) = 180 - 2x$
$$180 - 2x = 0$$
$$2(90 - x) = 0$$
$$x = 90$$

(e)

x	P
0	0
90	8100
180	0

(f) From the chart, the maximum value of P is 8100; this occurs when $x = 90$ and $y = 90$.

3. $x + y = 90$

Minimize x^2y.

(a) $y = 90 - x$

(b) Let $P = x^2y = x^2(90 - x)$
$= 90x^2 - x^3$.

(c) Since $y = 90 - x$ and x and y are nonnegative numbers, the domain of P is $[0, 90]$.

(d) $P' = 180x - 3x^2$

$$180x - 3x^2 = 0$$
$$3x(60 - x) = 0$$
$$x = 0 \text{ or } x = 60$$

(e)

x	P
0	0
60	108,000
90	0

(f) The maximum value of x^2y occurs when $x = 60$ and $y = 30$. The maximum value is 108,000.

5. $C(x) = \frac{1}{2}x^3 + 2x^2 - 3x + 35$

The average cost function is

$$\begin{aligned} A(x) = \overline{C}(x) &= \frac{C(x)}{x} \\ &= \frac{\frac{1}{2}x^3 + 2x^2 - 3x + 35}{x} \\ &= \frac{1}{2}x^2 + 2x - 3 + \frac{35}{x} \\ \text{or } &\frac{1}{2}x^2 + 2x - 3 + 35x^{-1}. \end{aligned}$$

Then

$$\begin{aligned} A'(x) &= x + 2 - 35x^{-2} \\ \text{or } &x + 2 - \frac{35}{x^2}. \end{aligned}$$

Graph $y = A'(x)$ on a graphing calculator. A suitable choice for the viewing window is $[0, 10]$ by $[-10, 10]$. (Negative values of x are not meaningful in this application.) Using the calculator, we see that the graph has an x-intercept or "zero" at $x \approx 2.722$. Thus, 2.722 is a critical number.
Now graph $y = A(x)$ and use this graph to confirm that a minimum occurs at $x \approx 2.722$.

Thus, the average cost is smallest at $x \approx 2.722$.

7. $p(x) = 160 - \frac{x}{10}$

(a) Revenue from sale of x thousand candy bars:

$$\begin{aligned} R(x) &= 1000xp \\ &= 1000x\left(160 - \frac{x}{10}\right) \\ &= 160{,}000x - 100x^2 \end{aligned}$$

(b) $R'(x) = 160{,}000 - 200x$

$$\begin{aligned} 160{,}000 - 200x &= 0 \\ 160{,}000 &= 200x \\ 800 &= x \end{aligned}$$

The maximum revenue occurs when 800 thousand bars are sold.

(c) $R(800) = 160{,}000(800) - 100(800)^2$
$= 64{,}000{,}000$

The maximum revenue is 64,000,000 cents.

9. Let $x =$ the width
and $y =$ the length.

(a) The perimeter is

$$\begin{aligned} P &= 2x + y \\ &= 1400, \end{aligned}$$

so

$$y = 1400 - 2x.$$

(b) Area $= xy = x(1400 - 2x)$
$A(x) = 1400x - 2x^2$

(c) $A' = 1400 - 4x$

$$\begin{aligned} 1400 - 4x &= 0 \\ 1400 &= 4x \\ 350 &= x \end{aligned}$$

$A'' = -4$, which implies that $x = 350$ m leads to the maximum area.

(d) If $x = 350$,

$$y = 1400 - 2(350) = 700.$$

The maximum area is $(350)(700) = 245{,}000 \text{ m}^2$.

11. Let $x =$ the width of the rectangle
$y =$ the total length of the rectangle.

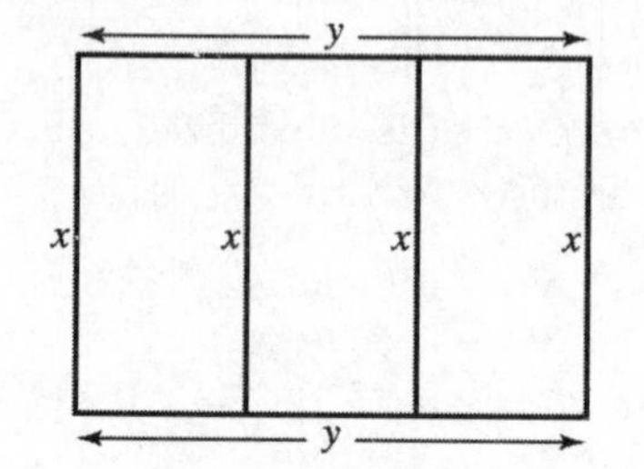

An equation for the fencing is

$$\begin{aligned} 3600 &= 4x + 2y \\ 2y &= 3600 - 4x \\ y &= 1800 - 2x. \end{aligned}$$

Area $= xy = x(1800 - 2x)$
$A(x) = 1800x - 2x^2$

$A' = 1800 - 4x$

$$\begin{aligned} 1800 - 4x &= 0 \\ 1800 &= 4x \\ 450 &= x \end{aligned}$$

$A'' = -4$, which implies that $x = 450$ is the location of a maximum.

If $x = 450$, $y = 1800 - 2(450) = 900$.
The maximum area is

$$(450)(900) = 405{,}000 \text{ m}^2.$$

13. Let $x =$ length at \$1.50 per meter
$y =$ width at \$3 per meter.

$$xy = 25{,}600$$

$$y = \frac{25{,}600}{x}$$

$$\text{Perimeter} = x + 2y = x + \frac{51{,}200}{x}$$

$$\text{Cost} = C(x) = x(1.5) + \frac{51{,}200}{x}(3)$$
$$= 1.5x + \frac{153{,}600}{x}$$

Minimize cost:

$$C'(x) = 1.5 - \frac{153{,}600}{x^2}$$

$$1.5 - \frac{153{,}600}{x^2} = 0$$

$$1.5 = \frac{153{,}600}{x^2}$$
$$1.5x^2 = 153{,}600$$
$$x^2 = 102{,}400$$
$$x = 320$$

$$y = \frac{25{,}600}{320} = 80$$

320 m at \$1.50 per meter will cost \$480. 160 m at \$3 per meter will cost \$480. The total cost will be \$960.

15. Let $x =$ the number of days to wait.

$\frac{12,000}{100} = 120 =$ the number of 100-lb groups collected already.
Then $7.5 - 0.15x =$ the price per 100 lb;
$4x =$ the number of 100-lb groups collected per day;
$120 + 4x =$ total number of 100-lb groups collected.

$$\text{Revenue} = R(x)$$
$$= (7.5 - 0.15x)(120 + 4x)$$
$$= 900 + 12x - 0.6x^2$$

$$R'(x) = 12 - 1.2x = 0$$
$$x = 10$$

$R''(x) = -1.2 < 0$ so $R(x)$ is maximized at $x = 10$.

The scouts should wait 10 days at which time their income will be maximized at

$$R(10) = 900 + 12(10) - 0.6(10)^2 = \$960.$$

17. Let $x =$ the number of refunds.
Then $535 - 5x =$ the cost per passenger
and $85 + x =$ the number of passengers.

(a) $\text{Revenue} = R(x) = (535 - 5x)(85 + x)$
$= 45{,}475 + 110x - 5x^2$
$R'(x) = 110 - 10x = 0$
$x = 11$

$R''(x) = -10 < 0$, so $R(x)$ is maximized when $x = 11$.
Thus, the number of passengers that will maximize revenue is $85 + 11 = 96$.

(b) $R(11) = 45{,}475 + 110(11) - 5(11)^2$
$= 46{,}080$

The maximum revenue is \$46,080.

19. Let $x =$ the length of a side of the top and bottom.
Then $x^2 =$ the area of the top and bottom
and $(3)(2x^2) =$ the cost for the top and bottom.

Let $y =$ depth of box.
Then $xy =$ the area of one side,
$4xy =$ the total area of the sides,
and $(1.50)(4xy) =$ the cost of the sides.

The total cost is

$$C(x) = (3)(2x^2) + (1.50)(4xy) = 6x^2 + 6xy.$$

The volume is

$$V = 16{,}000 = x^2 y.$$

$$y = \frac{16{,}000}{x^2}$$

$$C(x) = 6x^2 + 6x\left(\frac{16{,}000}{x^2}\right) = 6x^2 + \frac{96{,}000}{x}$$

$$C'(x) = 12x - \frac{96{,}000}{x^2} = 0$$
$$x^3 = 8000$$
$$x = 20$$

$C''(x) = 12 + \frac{192{,}000}{x^3} > 0$ at $x = 20$, which implies that $C(x)$ is minimized when $x = 20$.

$$y = \frac{16{,}000}{(20)^2} = 40$$

So the dimensions of the box are x by x by y, or 20 cm by 20 cm by 40 cm.

$$C(20) = 6(20)^2 + \frac{96,000}{20} = 7200$$

The minimum total cost is \$7200.

21. **(a)** $S = 2\pi r^2 + 2\pi rh,\ V = \pi r^2 h$

$$S = 2\pi r^2 + \frac{2V}{r}$$

Treat V as a constant.

$$S' = 4\pi r - \frac{2V}{r^2}$$

$$\begin{aligned} 4\pi r - \frac{2V}{r^2} &= 0 \\ \frac{4\pi r^3 - 2V}{r^2} &= 0 \\ 4\pi r^3 - 2V &= 0 \\ 2\pi r^3 - V &= 0 \\ 2\pi r^3 &= V \\ 2\pi r^3 &= \pi r^2 h \\ 2r &= h \end{aligned}$$

23. Let $\quad x =$ the length of the side of the cutout square.
Then $3 - 2x =$ the width of the box
and $8 - 2x =$ the length of the box.

$$\begin{aligned} V(x) &= x(3-2x)(8-2x) \\ &= 4x^3 - 22x^2 + 24x \end{aligned}$$

The domain of V is $\left(0, \frac{3}{2}\right)$.

Maximize the volume.

$$\begin{aligned} V'(x) &= 12x^2 - 44x + 24 \\ 12x^2 - 44x + 24 &= 0 \\ 4(3x^2 - 11x + 6) &= 0 \\ 4(3x-2)(x-3) &= 0 \\ x = \frac{2}{3} \quad &\text{or} \quad x = 3 \end{aligned}$$

3 is not in the domain of V.

$$V''(x) = 24x - 44$$

$$V''\left(\frac{2}{3}\right) = -28 < 0$$

This implies that V is maximized when $x = \frac{2}{3}$. The box will have maximum volume when $x = \frac{2}{3}$ ft or 8 in.

25. Let $x =$ the width of printed material and $y =$ the length of printed material.

Then, the area of the printed material is

$$xy = 36,$$

$$\text{so} \quad y = \frac{36}{x}.$$

Also, $x + 2 =$ the width of a page
and $\ y + 3 =$ the length of a page.

The area of a page is

$$\begin{aligned} A &= (x+2)(y+3) \\ &= xy + 2y + 3x + 6 \\ &= 36 + 2\left(\frac{36}{x}\right) + 3x + 6 \\ &= 42 + \frac{72}{x} + 3x. \end{aligned}$$

$$\begin{aligned} A' &= -\frac{72}{x^2} + 3 = 0 \\ x^2 &= 24 \\ x &= \sqrt{24} \\ &= 2\sqrt{6} \end{aligned}$$

(We discard $x = -2\sqrt{6}$ once we must have $x > 0$.) $A'' = \frac{216}{x^3} > 0$ when $x = 2\sqrt{6}$, which implies that A is minimized when $x = 2\sqrt{6}$.

$$y = \frac{36}{x} = \frac{36}{2\sqrt{6}} = \frac{18}{\sqrt{6}} = \frac{18\sqrt{6}}{6} = 3\sqrt{6}$$

The width of a page is

$$\begin{aligned} x + 2 &= 2\sqrt{6} + 2 \\ &\approx 6.9 \text{ in.} \end{aligned}$$

The length of a page is

$$\begin{aligned} y + 3 &= 3\sqrt{6} + 3 \\ &\approx 10.3 \text{ in.} \end{aligned}$$

27. Distance on shore: $7 - x$ miles
Cost on shore: \$400 per mile
Distance underwater: $\sqrt{x^2 + 36}$
Cost underwater: \$500 per mile
Find the distance from A, that is, $7 - x$, to minimize cost, $C(x)$.

$$\begin{aligned} C(x) &= (7-x)(400) + (\sqrt{x^2+36})(500) \\ &= 2800 - 400x + 500(x^2+36)^{1/2} \end{aligned}$$

$$\begin{aligned} C'(x) &= -400 + 500\left(\frac{1}{2}\right)(x^2+36)^{-1/2}(2x) \\ &= -400 + \frac{500x}{\sqrt{x^2+36}} \end{aligned}$$

If $C'(x) = 0$,

$$\frac{500x}{\sqrt{x^2+36}} = 400$$
$$\frac{5x}{4} = \sqrt{x^2+36}$$
$$\frac{25}{16}x^2 = x^2 + 36$$
$$\frac{9}{16}x^2 = 36$$
$$x^2 = \frac{36 \cdot 16}{9}$$
$$x = \frac{6 \cdot 4}{3} = 8.$$

(Discard the negative solution.)
$x = 8$ is impossible since Point A is only 7 miles from point C.
Check the endpoints.

x	$C(x)$
0	5800
7	4610

The cost is minimized when $x = 7$.

$7 - x = 7 - 7 = 0$, so the company should angle the cable at Point A.

29. From Example 4, we know that the surface area of the can is given by

$$S = 2\pi r^2 + \frac{2000}{r}.$$

Aluminum costs 3¢/cm^2, so the cost of the aluminum to make the can is

$$0.03\left(2\pi r^2 + \frac{2000}{r}\right) = 0.06\pi r^2 + \frac{60}{r}.$$

The perimeter (or circumference) of the circular top is $2\pi r$. Since there is a 2¢/cm charge to seal the top and bottom, the sealing cost is

$$0.02(2)(2\pi r) = 0.08\pi r.$$

Thus, the total cost is given by the function

$$C(r) = 0.06\pi r^2 + \frac{60}{r} + 0.08\pi r$$
$$= 0.06\pi r^2 + 60r^{-1} + 0.08\pi r.$$

Then

$$C'(r) = 0.12\pi r - 60r^{-2} + 0.08\pi$$
$$= 0.12\pi r - \frac{60}{r^2} + 0.08\pi.$$

Graph

$$y = 0.12\pi x - \frac{60}{x^2} + 0.08\pi$$

on a graphing calculator. Since r must be positive in this application, our window should not include negative values of x. A suitable choice for the viewing window is $[0, 10]$ by $[-10, 10]$. From the graph, we find that $C'(x) = 0$ when $x \approx 5.206$.
Thus, the cost is minimized when the radius is about 5.206 cm.

We can find the corresponding height by using the equation

$$h = \frac{1000}{\pi r^2}$$

from Example 4.
If $r = 5.206$,

$$h = \frac{1000}{\pi(5.206)^2} \approx 11.75.$$

To minimize cost, the can should have radius 5.206 cm and height 11.75 cm.

31. In Exercises 29 and 30, we found that the cost of the aluminum to make the can is $0.06\pi r^2 + \frac{60}{r}$, the cost to seal the top and bottom is $0.08\pi r$, and the cost to seal the vertical seam is $\frac{10}{\pi r^2}$.
Thus, the total cost is now given by the function

$$C(r) = 0.06\pi r^2 + \frac{60}{r} + 0.08\pi r + \frac{10}{\pi r^2}$$
$$\text{or} \quad 0.06\pi r^2 + 60r^{-1} + 0.08\pi r + \frac{10}{\pi}r^{-2}.$$

Then

$$C'(r) = 0.12\pi r - 60r^{-2} + 0.08\pi - \frac{20}{\pi}r^{-3}$$
$$\text{or} \quad 0.12\pi r - \frac{60}{r^2} + 0.08\pi - \frac{20}{\pi r^3}.$$

Graph

$$y = 0.12\pi r - \frac{60}{r^2} + 0.08\pi - \frac{20}{\pi r^3}$$

on a graphing calculator. A suitable choice for the viewing window is $[0, 10]$ by $[-10, 10]$. From the graph, we find that $C'(x) = 0$ when $x \approx 5.242$.
Thus, the cost is minimized when the radius is about 5.242 cm.

To find the corresponding height, use the equation

$$h = \frac{1000}{\pi r^2}$$

from Example 4.
If $r = 5.242$,

$$h = \frac{1000}{\pi(5.242)^2} \approx 11.58.$$

To minimize cost, the can should have radius 5.242 cm and height 11.58 cm.

33. $N(t) = 20\left[\frac{t}{12} - \ln\left(\frac{t}{12}\right)\right] + 30;$
$1 \le t \le 15$

$$\begin{aligned} N'(t) &= 20\left[\frac{1}{12} - \frac{12}{t}\left(\frac{1}{12}\right)\right] \\ &= 20\left(\frac{1}{12} - \frac{1}{t}\right) \\ &= \frac{20(t-12)}{12t} \end{aligned}$$

$N'(t) = 0$ when

$$\begin{aligned} t - 12 &= 0 \\ t &= 12. \end{aligned}$$

$N''(t)$ does not exist at $t = 0$, but 0 is not in the domain of N.
Thus, 12 is the only critical number.

To find the absolute extrema on $[1, 15]$, evaluate N at the critical number and at the endpoints.

t	$N(t)$
1	81.365
12	50
15	50.537

Use this table to answer the questions in (a)-(d).

(a) The number of bacteria will be a minimum at $t = 12$, which represents 12 days.

(b) The minimum number of bacteria is given by $N(12) = 50$, which represents 50 bacteria per ml.

(c) The number of bacteria will be a maximum at $t = 1$, which represents 1 day.

(d) The maximum number of bacteria is given by $N(1) = 81.365$, which represents 81.365 bacteria per ml.

35. $H(S) = f(S) - S$
$f(S) = 12S^{0.25}$
$H(S) = 12S^{0.25} - S$
$H'(S) = 3S^{-0.75} - 1$

$H'(S) = 0$ when

$$\begin{aligned} 3S^{-0.75} - 1 &= 0 \\ S^{-0.75} &= \frac{1}{3} \\ \frac{1}{S^{0.75}} &= \frac{1}{3} \\ S^{0.75} &= 3 \\ S^{3/4} &= 3 \\ S &= 3^{4/3} \\ S &= 4.327. \end{aligned}$$

The number of creatures needed to sustain the population is $S_0 = 4.327$ thousand.

$H''(S) = \frac{-2.25}{S^{1.75}} < 0$ when $S = 4.327$, so $H(S)$ is maximized.

$$\begin{aligned} H(4.327) &= 12(4.327)^{0.25} - 4.327 \\ &\approx 12.98 \end{aligned}$$

The maximum sustainable harvest is 12.98 thousand.

37. (a) $H(S) = f(S) - S$
$= Se^{r(1-S/P)} - S$

$$H'(S) = Se^{r(1-S/P)}\left(\tfrac{-r}{P}\right) + e^{r(1-S/P)} - 1$$

Note that

$$\begin{aligned} f(S) &= Se^{r(1-S/P)} \\ f'(S) &= Se^{r(1-S/P)}\left(-\frac{r}{P}\right) + e^{r(1-S/P)}. \end{aligned}$$

$$\begin{aligned} H'(S) = Se^{r(1-S/P)}\left(-\frac{r}{P}\right) + e^{r(1-S/p)} - 1 &= 0 \\ Se^{r(1-S/P)}\left(-\frac{r}{P}\right) + e^{r(1-S/P)} &= 1 \\ f'(S) &= 1 \end{aligned}$$

(b) $f(S) = Se^{r(1-S/P)}$

$$f'(S) = \left(-\frac{r}{P}\right)Se^{r(1-S/P)} + e^{r(1-S/P)}$$

Set $f'(S_0) = 1$

$$\begin{aligned} e^{r(1-S_0/P)}\left[\frac{-rS_0}{P} + 1\right] &= 1 \\ e^{r(1-S_0/P)} &= \frac{1}{\frac{-rS_0}{P} + 1} \end{aligned}$$

Using $H(S)$ from part (a), we get

$$H(S_0) = S_0e^{r(1-S_0/P)} - S_0$$
$$= S_0(e^{r(1-S_0/P)} - 1)$$
$$= S_0\left(\frac{1}{1-\frac{rS_0}{P}} - 1\right).$$

39. $r = 0.4$, $P = 500$

$$f(S) = Se^{r(1-S/P)}$$
$$f'(S) = -\frac{0.4}{500}Se^{0.4(1-S/500)} + e^{0.4(1-S/500)}$$
$$f'(S_0) = -0.0008S_0e^{0.4(1-S_0/500)} + e^{0.4(1-S_0/500)}$$

Graph

$$Y_1 = -0.0008x_0e^{0.4(1-x/500)} + e^{0.4(1-x/500)}$$

and

$$Y_2 = 1$$

on the same screen. A suitable choice for the viewing window is $[0, 300]$ by $[0.5, 1.5]$ with Xscl = 50, Yscl = 0.5. By zooming or using the "intersect" option, we find that the graphs intersect when $x \approx 237.10$.

The maximum sustainable harvest is 237.10.

41. Let x = distance from P to A.

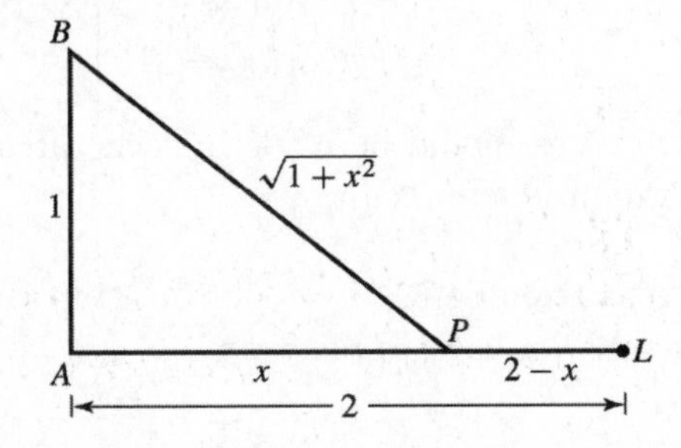

Energy used over land: 1 unit per mile
Energy used over water: $\frac{10}{9}$ units per mile
Distance over land: $(2-x)$ mi
Distance over water: $\sqrt{1+x^2}$ mi
Find the location of P to minimize energy used.

$$E(x) = 1(2-x) + \frac{10}{9}\sqrt{1+x^2}, \text{ where } 0 \le x \le 2.$$

$$E'(x) = -1 + \frac{10}{9}\left(\frac{1}{2}\right)(1+x^2)^{-1/2}(2x)$$

If $E'(x) = 0$,

$$\frac{10}{9}x(1+x^2)^{-1/2} = 1$$
$$\frac{10x}{9(1+x^2)^{1/2}} = 1$$
$$\frac{10}{9}x = (1+x^2)^{1/2}$$
$$\frac{100}{81}x^2 = 1 + x^2$$
$$\frac{19}{81}x^2 = 1$$
$$x^2 = \frac{81}{19}$$
$$x = \frac{9}{\sqrt{19}}$$
$$= \frac{9\sqrt{19}}{19}$$
$$\approx 2.06.$$

This value cannot give the absolute maximum since the total distance from A to L is just 2 miles. Test the endpoints of the domain.

x	$E(x)$
0	$3\frac{1}{9} \approx 3.1111$
2	2.4845

Point P must be at Point L.

43. **(a)** Solve the given equation for effective power for T, time.

$$\frac{kE}{T} = aSv^3 + I$$
$$\frac{kE}{aSv^3 + I} = T$$

Since distance is velocity, v, times time, T, we have

$$D(v) = v\frac{kE}{aSv^3 + I}$$
$$= \frac{kEv}{aSv^3 + I}.$$

(b) $$D'(v) = \frac{(aSv^3 + I)kE - kEv(3aSv^2)}{(aSv^3 + I)^2}$$
$$= \frac{kE(aSv^3 + I - 3aSv^3)}{(aSv^3 + I)^2}$$
$$= \frac{kE(I - 2aSv^3)}{(aSv^3 + I)^2}$$

Find the critical numbers by solving $D'(v) = 0$ for v.

$$I - 2aSv^3 = 0$$
$$2aSv^3 = I$$
$$v^3 = \frac{I}{2aS}$$
$$v = \left(\frac{I}{2aS}\right)^{1/3}$$

45. Let $8 - x =$ the distance the hunter will travel on the river.

Then $\sqrt{9 + x^2} =$ the distance he will travel on land.

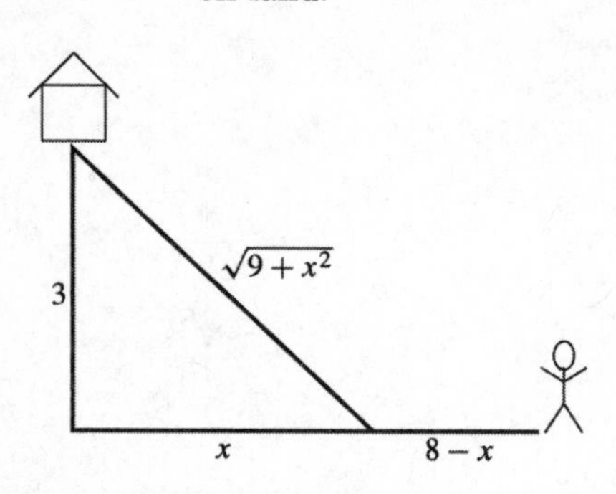

The rate on the river is 5 mph, the rate on land is 2 mph. Using $t = \frac{d}{r}$,

$\frac{8-x}{5} =$ the time on the river,

$\frac{\sqrt{9+x^2}}{2} =$ the time on land.

The total time is

$$T(x) = \frac{8 - x}{5} + \frac{\sqrt{9 + x^2}}{2}$$
$$= \frac{8}{5} - \frac{1}{5}x + \frac{1}{2}(9 + x^2)^{1/2}.$$
$$T' = -\frac{1}{5} + \frac{1}{4} \cdot 2x(9 + x^2)^{-1/2}$$
$$-\frac{1}{5} + \frac{x}{2(9 + x^2)^{1/2}} = 0$$
$$\frac{1}{5} = \frac{x}{2(9 + x^2)^{1/2}}$$
$$2(9 + x^2)^{1/2} = 5x$$
$$4(9 + x^2) = 25x^2$$
$$36 + 4x^2 = 25x^2$$
$$36 = 21x^2$$
$$\frac{6}{\sqrt{21}} = x$$
$$\frac{6\sqrt{21}}{21} = \frac{2\sqrt{21}}{7} = x$$

x	$T(x)$
0	3.1
$\frac{2\sqrt{21}}{7}$	2.98
8	4.27

Since the minimum time is 2.98 hr, the hunter should travel $8 - \frac{2\sqrt{21}}{7} = \frac{56 - 2\sqrt{21}}{7}$ or about 6.7 miles along the river.

6.3 Further Business Applications: Economic Lot Size; Economic Order Quantity; Elasticity of Demand

1. When $q < \sqrt{\frac{2fM}{k}}$, $T'(q) < -\frac{k}{2} + \frac{k}{2} = 0$; and when $q > \sqrt{\frac{2fM}{k}}$, $T'(q) > -\frac{k}{2} + \frac{k}{2} = 0$. Since the function $T(q)$ is decreasing before $q = \sqrt{\frac{2fM}{k}}$ and increasing after $q = \sqrt{\frac{2fM}{k}}$, there must be a relative minimum at $q = \sqrt{\frac{2fM}{k}}$. By the critical point theorem, there is an absolute minimum there.

3. The economic order quantity formula assumes that M, the total units needed per year, is known. Thus, c is the correct answer.

5. Use equation (3) with $k = 9, M = 13{,}950$, and $f = 31$.

$$q = \sqrt{\frac{2fM}{k}}$$
$$= \sqrt{\frac{2(31)(13{,}950)}{9}}$$
$$= \sqrt{96{,}100} = 310$$

310 cases should be made in each batch to minimize production costs.

7. From Exercise 5, $M = 13{,}950$ and $q = 310$. The number of batches per year is

$$\frac{M}{q} = \frac{13{,}950}{310} = 45$$

45 cases should be made in each batch to minimize production costs.

9. Here $k = 1, M = 900$, and $f = 5$. We have

$$q = \sqrt{\frac{2fM}{k}}$$
$$= \sqrt{\frac{2(5)(900)}{1}}$$
$$= \sqrt{9000} \approx 94.9$$

$T(94) \approx 94.872$ and $T(95) \approx 94.868$, so ordering 95 bottles per order minimizes the annual costs.

11. Use $q = \sqrt{\frac{fM}{k}}$ from Exercise 10 with $k = 6$, $M = 5000$, and $f = 1000$.

$$q = \sqrt{\frac{fM}{k}} = \sqrt{\frac{(1000)(5000)}{6}} \approx 912.9$$

with $T(q) = \frac{fM}{q} + kq$ (assume $g = 0$ since the subsequent cost per book is so low that it can be ignored), $T(912) \approx 10{,}954.456$ and $T(913) \approx 10{,}954.451$. So, 913 books should be printed in each print run.

13. Use $q = \sqrt{\frac{2fM}{k_1+2k_2}}$ from Exercise 12 with $k_1 = 1$, $k_2 = 2$, $M = 30{,}000$, and $f = 750$. Also, note that $g = 8$.

$$\begin{aligned} q &= \sqrt{\frac{2fM}{k_1 + 2k_2}} \\ &= \sqrt{\frac{2(750)(30{,}000)}{1 + 2(2)}} \\ &= \sqrt{9{,}000{,}000} = 3000 \end{aligned}$$

The number of production runs each year to minimize her total costs is

$$\frac{M}{q} = \frac{30{,}000}{3000} = 10.$$

15. $q = 50 - \frac{p}{4}$

(a) $\frac{dq}{dp} = -\frac{1}{4}$

$$\begin{aligned} E &= -\frac{p}{q} \cdot \frac{dq}{dp} \\ &= -\frac{p}{50 - \frac{p}{4}}\left(-\frac{1}{4}\right) \\ &= -\frac{p}{\frac{200-p}{4}}\left(-\frac{1}{4}\right) \\ &= \frac{p}{200 - p} \end{aligned}$$

(b) $R = pq$

$$\frac{dR}{dp} = q(1 - E)$$

When R is maximum, $q(1 - E) = 0$.
Since $q = 0$ means no revenue, set $1 - E = 0$.

$$E = 1$$

From (a),

$$\begin{aligned} \frac{p}{200 - p} &= 1 \\ p &= 200 - p \\ p &= 100. \\ q &= 50 - \frac{p}{4} \\ &= 50 - \frac{100}{4} \\ &= 25 \end{aligned}$$

Total revenue is maximized if $q = 25$.

17. (a) $q = 37{,}500 - 5p^2$

$$\begin{aligned} \frac{dq}{dp} &= -10p \\ E &= \frac{-p}{q} \cdot \frac{dq}{dp} \\ &= \frac{-p}{37{,}500 - 5p^2}(-10p) \\ &= \frac{10p^2}{37{,}500 - 5p^2} \\ &= \frac{2p^2}{7500 - p^2} \end{aligned}$$

(b) $R = pq$

$$\frac{dR}{dp} = q(1 - E)$$

When R is maximum, $q(1 - E) = 0$. Since $q = 0$ means no revenue, set $1 - E = 0$.

$$E = 1$$

From (a),

$$\begin{aligned} \frac{2p^2}{7500 - p^2} &= 1 \\ 2p^2 &= 7500 - p^2 \\ 3p^2 &= 7500 \\ p^2 &= 2500 \\ p &= \pm 50. \end{aligned}$$

Since p must be positive, $p = 50$.

$$\begin{aligned} q &= 37{,}500 - 5p^2 \\ &= 37{,}500 - 5(50)^2 \\ &= 37{,}500 - 5(2500) \\ &= 37{,}500 - 12{,}500 \\ &= 25{,}000. \end{aligned}$$

19. $p = 400e^{-0.2q}$

In order to find the derivative $\frac{dq}{dp}$, we first need to solve for q in the equation $p = 400e^{-0.2q}$.

(a)

$$\frac{p}{400} = e^{-0.2q}$$

$$\ln\left(\frac{p}{400}\right) = \ln\left(e^{-0.2q}\right) = -0.2q$$

$$q = \frac{\ln \frac{p}{400}}{-0.2} = -5\ln\left(\frac{p}{400}\right)$$

Now

$$\frac{dq}{dp} = -5\frac{1}{\frac{p}{400}} \cdot \frac{1}{400} = \frac{-5}{p}, \text{ and}$$

$$E = -\frac{p}{q} \cdot \frac{dq}{dp} = -\frac{p}{q} \cdot \frac{-5}{p} = \frac{5}{q}.$$

(b) $R = pq$

$$\frac{dR}{dp} = q(1 - E)$$

When R is maximum, $q(1 - E) = 0$. Since $q = 0$ means no revenue, set $1 - E = 0$.

$$E = 1$$

From part (a),

$$\frac{5}{q} = 1$$

$$5 = q$$

21. $q = 400 - 0.2p^2$

$$\frac{dq}{dp} = 0 - 0.4p$$

$$E = -\frac{p}{q} \cdot \frac{dq}{dp}$$

$$E = -\frac{p}{400 - 0.2p^2}(-0.4p)$$

$$= \frac{0.4p^2}{400 - 0.2p^2}$$

(a) If $p = \$20$,

$$E = \frac{(0.4)(20)^2}{400 - 0.2(20)^2}$$

$$= 0.5.$$

Since $E < 1$, demand is inelastic. This indicates that total revenue increases as price increases.

(b) If $p = \$40$,

$$E = \frac{(0.4)(40)^2}{400 - 0.2(40)^2}$$

$$= 8.$$

Since $E > 1$, demand is elastic. This indicates that total revenue decreases as price increases.

23. **(a)** $q = 55.2 - 0.022p$

$$\frac{dq}{dp} = -0.022$$

$$E = -\frac{p}{q} \cdot \frac{dq}{dp}$$

$$= \frac{-p}{55.2 - 0.022p} \cdot (-0.022)$$

$$= \frac{0.022p}{55.2 - 0.022p}$$

When $p = \$166.10$,

$$E = \frac{3.6542}{55.2 - 3.6542}$$

$$\approx 0.071.$$

(b) Since $E < 1$, the demand for airfare is inelastic at this price.

(c) $R = pq$

$$\frac{dR}{dp} = q(1 - E)$$

When R is a maximum, $q(1 - E) = 0$. Since $q = 0$ means no revenue, set $1 - E = 0$.

$$E = 1$$

From (a),

$$\frac{0.022p}{55.2 - 0.022p} = 1$$

$$0.022p = 55.2 - 0.022p$$

$$0.044p = 55.2$$

$$p \approx 1255$$

Total revenue is maximized if $p \approx \$1255$.

25. $q = m - np$ for $0 \le p \le \frac{m}{n}$

$$\frac{dq}{dp} = -n$$

$$E = -\frac{p}{q} \cdot \frac{dq}{dp}$$

$$E = -\frac{p}{m - np}(-n)$$

$$E = \frac{pn}{m - np} = 1$$

$$pn = m - np$$

$$2np = m$$

$$p = \frac{m}{2n}$$

Thus, $E = 1$ when $p = \frac{m}{2n}$, or at the midpoint of the demand curve on the interval $0 \le p \le \frac{m}{n}$.

27. **(a)**
$$q = Cp^{-k}$$
$$\frac{dq}{dp} = -Ckp^{-k-1}$$
$$E = \frac{-p}{q} \cdot \frac{dq}{dp}$$
$$= \frac{-p}{Cp^{-k}}(-Ckp^{-k-1})$$
$$= \frac{kp^{-k}}{p^{-k}} = k$$

29. The demand function $q(p)$ is positive and increasing, so $\frac{dq}{dp}$ is positive. Since p_0 and q_0 are also positive, the elasticity $E = -\frac{p_0}{q_0} \cdot \frac{dq}{dp}$ is negative.

6.4 Implicit Differentiation

1. $6x^2 + 5y^2 = 36$

$$\frac{d}{dx}(6x^2 + 5y^2) = \frac{d}{dx}(36)$$
$$\frac{d}{dx}(6x^2) + \frac{d}{dx}(5y^2) = \frac{d}{dx}(36)$$
$$12x + 5 \cdot 2y\frac{dy}{dx} = 0$$
$$10y\frac{dy}{dx} = -12x$$
$$\frac{dy}{dx} = -\frac{6x}{5y}$$

3. $8x^2 - 10xy + 3y^2 = 26$

$$\frac{d}{dx}(8x^2 - 10xy + 3y^2) = \frac{d}{dx}(26)$$
$$16x - \frac{d}{dx}(10xy) + \frac{d}{dx}(3y^2) = 0$$
$$16x - 10x\frac{dy}{dx} - y\frac{d}{dx}(10x) + 6y\frac{dy}{dx} = 0$$
$$16x - 10x\frac{dy}{dx} - 10y + 6y\frac{dy}{dx} = 0$$
$$(-10x + 6y)\frac{dy}{dx} = -16x + 10y$$
$$\frac{dy}{dx} = \frac{-16x + 10y}{-10x + 6y}$$
$$\frac{dy}{dx} = \frac{8x - 5y}{5x - 3y}$$

5. $5x^3 = 3y^2 + 4y$

$$\frac{d}{dx}(5x^3) = \frac{d}{dx}(3y^2 + 4y)$$
$$15x^2 = \frac{d}{dx}(3y^2) + \frac{d}{dx}(4y)$$
$$15x^2 = 6y\frac{dy}{dx} + 4\frac{dy}{dx}$$
$$\frac{15x^2}{6y + 4} = \frac{dy}{dx}$$

7. $3x^2 = \dfrac{2 - y}{2 + y}$

$$\frac{d}{dx}(3x^2) = \frac{d}{dx}\left(\frac{2 - y}{2 + y}\right)$$
$$6x = \frac{(2 + y)\frac{d}{dx}(2 - y) - (2 - y)\frac{d}{dx}(2 + y)}{(2 + y)^2}$$
$$6x = \frac{(2 + y)\left(-\frac{dy}{dx}\right) - (2 - y)\frac{dy}{dx}}{(2 + y)^2}$$
$$6x = \frac{-4\frac{dy}{dx}}{(2 + y)^2}$$
$$6x(2 + y)^2 = -4\frac{dy}{dx}$$
$$-\frac{3x(2 + y)^2}{2} = \frac{dy}{dx}$$

9. $2\sqrt{x} + 4\sqrt{y} = 5y$

$$\frac{d}{dx}(2x^{1/2} + 4y^{1/2}) = \frac{d}{dx}(5y)$$
$$x^{-1/2} + 2y^{-1/2}\frac{dy}{dx} = 5\frac{dy}{dx}$$
$$(2y^{-1/2} - 5)\frac{dy}{dx} = -x^{-1/2}$$
$$\frac{dy}{dx} = \frac{x^{-1/2}}{5 - 2y^{-1/2}}\left(\frac{x^{1/2}y^{1/2}}{x^{1/2}y^{1/2}}\right)$$
$$= \frac{y^{1/2}}{x^{1/2}(5y^{1/2} - 2)}$$
$$= \frac{\sqrt{y}}{\sqrt{x}(5\sqrt{y} - 2)}$$

11. $x^4y^3 + 4x^{3/2} = 6y^{3/2} + 5$

$$\frac{d}{dx}(x^4y^3 + 4x^{3/2}) = \frac{d}{dx}(6y^{3/2} + 5)$$
$$\frac{d}{dx}(x^4y^3) + \frac{d}{dx}(4x^{3/2}) = \frac{d}{dx}(6y^{3/2}) + \frac{d}{dx}(5)$$
$$4x^3y^3 + x^4 \cdot 3y^2\frac{dy}{dx} + 6x^{1/2} = 9y^{1/2}\frac{dy}{dx} + 0$$
$$4x^3y^3 + 6x^{1/2} = 9y^{1/2}\frac{dy}{dx} - 3x^4y^2\frac{dy}{dx}$$
$$4x^3y^3 + 6x^{1/2} = (9y^{1/2} - 3x^4y^2)\frac{dy}{dx}$$
$$\frac{4x^3y^3 + 6x^{1/2}}{9y^{1/2} - 3x^4y^2} = \frac{dy}{dx}$$

13. $e^{x^2y} = 5x + 4y + 2$

$$\frac{d}{dx}(e^{x^2y}) = \frac{d}{dx}(5x + 4y + 2)$$
$$e^{x^2y}\frac{d}{dx}(x^2y) = \frac{d}{dx}(5x) + \frac{d}{dx}(4y) + \frac{d}{dx}(2)$$
$$e^{x^2y}\left(2xy + x^2\frac{dy}{dx}\right) = 5 + 4\frac{dy}{dx} + 0$$
$$2xye^{x^2y} + x^2e^{x^2y}\frac{dy}{dx} = 5 + 4\frac{dy}{dx}$$
$$x^2e^{x^2y}\frac{dy}{dx} - 4\frac{dy}{dx} = 5 - 2xye^{x^2y}$$
$$(x^2e^{x^2y} - 4)\frac{dy}{dx} = 5 - 2xye^{x^2y}$$
$$\frac{dy}{dx} = \frac{5 - 2xye^{x^2y}}{x^2e^{x^2y} - 4}$$

15. $x + \ln y = x^2y^3$

$$\frac{d}{dx}(x + \ln y) = \frac{d}{dx}(x^2y^3)$$
$$1 + \frac{1}{y}\frac{dy}{dx} = 2xy^3 + 3x^2y^2\frac{dy}{dx}$$
$$\frac{1}{y}\frac{dy}{dx} - 3x^2y^2\frac{dy}{dx} = 2xy^3 - 1$$
$$\left(\frac{1}{y} - 3x^2y^2\right)\frac{dy}{dx} = 2xy^3 - 1$$
$$\frac{dy}{dx} = \frac{2xy^3 - 1}{\frac{1}{y} - 3x^2y^2}$$
$$= \frac{y(2xy^3 - 1)}{1 - 3x^2y^3}$$

17. $x^2 + y^2 = 25$; tangent at $(-3, 4)$

$$\frac{d}{dx}(x^2 + y^2) = \frac{d}{dx}(25)$$
$$2x + 2y\frac{dy}{dx} = 0$$
$$2y\frac{dy}{dx} = -2x$$
$$\frac{dy}{dx} = -\frac{x}{y}$$
$$m = -\frac{x}{y} = -\frac{-3}{4} = \frac{3}{4}$$
$$y - y_1 = m(x - x_1)$$
$$y - 4 = \frac{3}{4}[x - (-3)]$$
$$4y - 16 = 3x + 9$$
$$4y = 3x + 25$$
$$y = \frac{3}{4}x + \frac{25}{4}$$

19. $x^2y^2 = 1$; tangent at $(-1, 1)$

$$\frac{d}{dx}(x^2y^2) = \frac{d}{dx}(1)$$
$$x^2\frac{d}{dx}(y^2) + y^2\frac{d}{dx}(x^2) = 0$$
$$x^2(2y)\frac{dy}{dx} + y^2(2x) = 0$$
$$2x^2y\frac{dy}{dx} = -2xy^2$$
$$\frac{dy}{dx} = \frac{-2xy^2}{2x^2y} = -\frac{y}{x}$$
$$m = -\frac{y}{x} = -\frac{1}{-1} = 1$$
$$y - 1 = 1[x - (-1)]$$
$$y = x + 1 + 1$$
$$y = x + 2$$

21. $2y^2 - \sqrt{x} = 4$; tangent at $(16, 2)$

$$\frac{d}{dx}(2y^2 - \sqrt{x}) = \frac{d}{dx}(4)$$
$$4y\frac{dy}{dx} - \frac{1}{2}x^{-1/2} = 0$$
$$4y\frac{dy}{dx} = \frac{1}{2x^{1/2}}$$
$$\frac{dy}{dx} = \frac{1}{8yx^{1/2}}$$

$$m = \frac{1}{8yx^{1/2}} = \frac{1}{8(2)(16)^{1/2}} = \frac{1}{8(2)(4)} = \frac{1}{64}$$

$$y - 2 = \frac{1}{64}(x - 16)$$

$$64y - 128 = x - 16$$

$$64y = x + 112$$

$$y = \frac{x}{64} + \frac{7}{4}$$

23. $e^{x^2+y^2} = xe^{5y} - y^2e^{5x/2}$; tangent at $(2, 1)$

$$\frac{d}{dx}(e^{x^2+y^2}) = \frac{d}{dx}(xe^{5y} - y^2e^{5x/2})$$

$$e^{x^2+y^2} \cdot \frac{d}{dx}(x^2 + y^2) = e^{5y} + x\frac{d}{dx}(e^{5y}) - \left[2y\frac{dy}{dx}e^{5x/2} + y^2e^{5x/2}\frac{d}{dx}\left(\frac{5x}{2}\right)\right]$$

$$e^{x^2+y^2}\left(2x + 2y\frac{dy}{dx}\right) = e^{5y} + x \cdot 5e^{5y}\frac{dy}{dx} - 2ye^{5x/2}\frac{dy}{dx} - \frac{5}{2}y^2e^{5x/2}$$

$$(2ye^{x^2+y^2} - 5xe^{5y} + 2ye^{5x/2})\frac{dy}{dx} = -2xe^{x^2+y^2} + e^{5y} - \frac{5}{2}y^2e^{5x/2}$$

$$\frac{dy}{dx} = \frac{-2xe^{x^2+y^2} + e^{5y} - \frac{5}{2}y^2e^{5x/2}}{2ye^{x^2+y^2} - 5xe^{5y} + 2ye^{5x/2}}$$

$$m = \frac{-4e^5 + e^5 - \frac{5}{2}e^5}{2e^5 - 10e^5 + 2e^5} = \frac{-\frac{11}{2}e^5}{-6e^5} = \frac{11}{12}$$

$$y - 1 = \frac{11}{12}(x - 2)$$

$$y = \frac{11}{12}x - \frac{5}{6}$$

25. $\ln(x + y) = x^3y^2 + \ln(x^2 + 2) - 4$; tangent at $(1, 2)$

$$\frac{d}{dx}[\ln(x + y)] = \frac{d}{dx}[x^3y^2 + \ln(x^2 + 2) - 4]$$

$$\frac{1}{x+y} \cdot \frac{d}{dx}(x + y) = 3x^2y^2 + x^3 \cdot 2y\frac{dy}{dx} + \frac{1}{x^2+2} \cdot \frac{d}{dx}(x^2 + 2) - \frac{d}{dx}(4)$$

$$\left(\frac{1}{x+y} - 2x^3y\right)\frac{dy}{dx} = 3x^2y^2 + \frac{2x}{x^2+2} - \frac{1}{x+y}$$

$$\frac{dy}{dx} = \frac{3x^2y^2 + \frac{2x}{x^2+2} - \frac{1}{x+y}}{\frac{1}{x+y} - 2x^3y}$$

$$m = \frac{3 \cdot 1 \cdot 4 + \frac{2 \cdot 1}{3} - \frac{1}{3}}{\frac{1}{3} - 2 \cdot 1 \cdot 2} = \frac{\frac{37}{3}}{\frac{-11}{3}} = -\frac{37}{11}$$

$$y - 2 = -\frac{37}{11}(x - 1)$$

$$y = -\frac{37}{11}x + \frac{59}{11}$$

27. $y^3 + xy - y = 8x^4$; $x = 1$

First, find the y-value of the point.

$$\begin{aligned} y^3 + (1)y - y &= 8(1)^4 \\ y^3 &= 8 \\ y &= 2 \end{aligned}$$

The point is $(1, 2)$.

Find $\frac{dy}{dx}$.

$$\begin{aligned} 3y^2 \frac{dy}{dx} + x\frac{dy}{dx} + y - \frac{dy}{dx} &= 32x^3 \\ (3y^2 + x - 1)\frac{dy}{dx} &= 32x^3 - y \\ \frac{dy}{dx} &= \frac{32x^3 - y}{3y^2 + x - 1} \end{aligned}$$

At $(1, 2)$,

$$\frac{dy}{dx} = \frac{32(1)^3 - 2}{3(2)^2 + 1 - 1} = \frac{30}{12} = \frac{5}{2}.$$

$$\begin{aligned} y - 2 &= \frac{5}{2}(x - 1) \\ y - 2 &= \frac{5}{2}x - \frac{5}{2} \\ y &= \frac{5}{2}x - \frac{1}{2} \end{aligned}$$

29. $y^3 + xy^2 + 1 = x + 2y^2$; $x = 2$

Find the y-value of the point.

$$\begin{aligned} y^3 + 2y^2 + 1 &= 2 + 2y^2 \\ y^3 + 1 &= 2 \\ y^3 &= 1 \\ y &= 1 \end{aligned}$$

The point is $(2, 1)$.

Find $\frac{dy}{dx}$.

$$\begin{aligned} 3y^2 \frac{dy}{dx} + x\,2y\frac{dy}{dx} + y^2 &= 1 + 4y\frac{dy}{dx} \\ 3y^2 \frac{dy}{dx} + 2xy\frac{dy}{dx} - 4y\frac{dy}{dx} &= 1 - y^2 \\ (3y^2 + 2xy - 4y)\frac{dy}{dx} &= 1 - y^2 \\ \frac{dy}{dx} &= \frac{1 - y^2}{3y^2 + 2xy - 4y} \end{aligned}$$

At $(2, 1)$,

$$\frac{dy}{dx} = \frac{1 - 1^2}{3(1)^2 + 2(2)(1) - 4(1)} = 0.$$

$$\begin{aligned} y - 0 &= 0(x - 2) \\ y &= 1 \end{aligned}$$

31. $2y^3(x - 3) + x\sqrt{y} = 3$; $x = 3$

Find the y-value of the point.

$$\begin{aligned} 2y^3(3 - 3) + 3\sqrt{y} &= 3 \\ 3\sqrt{y} &= 3 \\ \sqrt{y} &= 1 \\ y &= 1 \end{aligned}$$

The point is $(3, 1)$

Find $\frac{dy}{dx}$.

$$\begin{aligned} &2y^3(1) + 6y^2(x - 3)\frac{dy}{dx} \\ &\quad + x\left(\frac{1}{2}\right)y^{-1/2}\frac{dy}{dx} + \sqrt{y} = 0 \\ &6y^2(x - 3)\frac{dy}{dx} + \frac{x}{2\sqrt{y}}\frac{dy}{dx} = -2y^3 - \sqrt{y} \\ &\left[6y^2(x - 3) + \frac{x}{2\sqrt{y}}\right]\frac{dy}{dx} = -2y^3 - \sqrt{y} \\ &\frac{dy}{dx} = \frac{-2y^3 - \sqrt{y}}{6y^2(x - 3) + \frac{x}{2\sqrt{y}}} \\ &\quad = \frac{-4y^{7/2} - 2y}{12y^{5/2}(x - 3) + x} \end{aligned}$$

At $(3, 1)$,

$$\frac{dy}{dx} = \frac{-4(1) - 2}{12(1)(3 - 3) + 3} = \frac{-6}{3} = -2.$$

$$\begin{aligned} y - 1 &= -2(x - 3) \\ y - 1 &= -2x + 6 \\ y &= -2x + 7 \end{aligned}$$

33. $x^2 + y^2 = 100$

(a) Lines are tangent at points where $x = 6$. By substituting $x = 6$ in the equation, we find that the points are $(6, 8)$ and $(6, -8)$.

$$\begin{aligned} \frac{d}{dx}(x^2 + y^2) &= \frac{d}{dx}(100) \\ 2x + 2y\frac{dy}{dx} &= 0 \\ 2y\frac{dy}{dx} &= -2x \\ dy &= -\frac{x}{y} \end{aligned}$$

$$m_1 = -\frac{x}{y} = -\frac{6}{8} = -\frac{3}{4}$$

$$m_2 = -\frac{x}{y} = -\frac{6}{-8} = \frac{3}{4}$$

First tangent:

$$y - 8 = -\frac{3}{4}(x-6)$$
$$y = -\frac{3}{4}x + \frac{25}{2}$$

Second tangent:

$$y - (-8) = \frac{3}{4}(x-6)$$
$$y + 8 = \frac{3}{4}x - \frac{18}{4}$$
$$y = \frac{3}{4}x - \frac{25}{2}$$

(b)

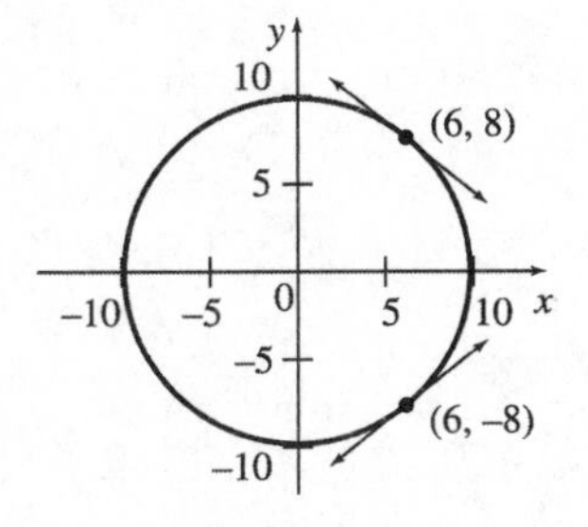

35. $3(x^2+y^2)^2 = 25(x^2-y^2)$; $(2,1)$

Find $\frac{dy}{dx}$.

$$6(x^2+y^2)\frac{d}{dx}(x^2+y^2) = 25\frac{d}{dx}(x^2-y^2)$$
$$6(x^2+y^2)\left(2x+2y\frac{dy}{dx}\right) = 25\left(2x-2y\frac{dy}{dx}\right)$$
$$12x^3+12x^2y\frac{dy}{dx}+12xy^2+12y^3\frac{dy}{dx} = 50x-50y\frac{dy}{dx}$$
$$12x^2y\frac{dy}{dx}+12y^3\frac{dy}{dx}+50y\frac{dy}{dx} = -12x^3-12xy^2+50x$$
$$(12x^2y+12y^3+50y)\frac{dy}{dx} = -12x^3-12xy^2+50x$$
$$\frac{dy}{dx} = \frac{-12x^3-12xy^2+50x}{12x^2y+12y^3+50y}$$

At $(2,1)$,

$$\frac{dy}{dx} = \frac{-12(2)^3-12(2)(1)^2+50(2)}{12(2)^2+12(1)^3+50(1)}$$
$$= \frac{-20}{110}$$
$$= -\frac{2}{11}$$
$$y-1 = -\frac{2}{11}(x-2)$$
$$y-1 = -\frac{2}{11}x+\frac{4}{11}$$
$$y = -\frac{2}{11}x+\frac{15}{11}$$

37. $2(x^2+y^2)^2 = 25xy^2$; $(2,1)$

Find $\frac{dy}{dx}$.

$$4(x^2+y^2)\frac{d}{dx}(x^2+y^2) = 25\frac{d}{dx}(xy^2)$$
$$4(x^2+y^2)\left(2x+2y\frac{dy}{dx}\right) = 25\left(y^2+2xy\frac{dy}{dx}\right)$$
$$8x^3+8x^2y\frac{dy}{dx}+8xy^2+8y^3\frac{dy}{dx} = 25y^2+50xy\frac{dy}{dx}$$
$$8x^2y\frac{dy}{dx}+8y^3\frac{dy}{dx}-50xy\frac{dy}{dx} = -8x^3-8xy^2+25y^2$$
$$(8x^2y+8y^3-50xy)\left(\frac{dy}{dx}\right) = -8x^3-8xy^2+25y^2$$
$$\frac{dy}{dx} = \frac{-8x^3-8xy^2+25y^2}{8x^2y+8y^3-50xy}$$

At $(2,1)$,

$$\frac{dy}{dx} = \frac{-8(2)^3-8(2)(1)^2+25(1)^2}{8(2)^2(1)+8(1)^3-50(2)(1)}$$
$$= \frac{-55}{-60} = \frac{11}{12}$$
$$y-1 = \frac{11}{12}(x-2)$$
$$y-1 = \frac{11}{12}x-\frac{11}{6}$$
$$y = \frac{11}{12}x-\frac{5}{6}$$

39. $y^2 = x^3+ax+b$

$$\frac{d}{dx}(y^2) = \frac{d}{dx}(x^3+ax+b)$$
$$2y\frac{dy}{dx} = 3x^2+a$$
$$\frac{dy}{dx} = \frac{3x^2+a}{2y}$$

41. $\sqrt{u}+\sqrt{2v+1}=5$

$$\frac{dv}{du}\left(\sqrt{u}+\sqrt{2v+1}\right)=\frac{dv}{du}(5)$$

$$\frac{1}{2}u^{-1/2}+\frac{1}{2}(2v+1)^{-1/2}(2)\frac{dv}{du}=0$$

$$(2v+1)^{-1/2}\frac{dv}{du}=-\frac{1}{2}u^{-1/2}$$

$$\frac{dv}{du}=-\frac{(2v+1)^{1/2}}{2u^{1/2}}$$

43. $C^2=x^2+100\sqrt{x}+50$

(a) $2C\dfrac{dC}{dx}=2x+\dfrac{1}{2}(100)x^{-1/2}$

$$\frac{dC}{dx}=\frac{2x+50x^{-1/2}}{2C}$$

$$\frac{dC}{dx}=\frac{x+25x^{-1/2}}{C}\cdot\frac{x^{1/2}}{x^{1/2}}$$

$$\frac{dC}{dx}=\frac{x^{3/2}+25}{Cx^{1/2}}$$

When $x=5$, the approximate increase in cost of an additional unit is

$$\frac{(5)^{3/2}+25}{(5^2+100\sqrt{5}+50)^{1/2}(5)^{1/2}}=\frac{36.18}{(17.28)\sqrt{5}}$$

$$\approx 0.94.$$

(b) $900(x-5)^2+25R^2=22,500$

$$R^2=900-36(x-5)^2$$

$$2R\frac{dR}{dx}=-72(x-5)$$

$$\frac{dR}{dx}=\frac{-36(x-5)}{R}=\frac{180-36x}{R}$$

When $x=5$, the approximate change in revenue for a unit increase in sales is

$$\frac{180-36(5)}{R}=\frac{0}{R}=0.$$

45. $b-a=(b+a)^3$

$$\frac{d}{db}(b-a)=\frac{d}{db}[(b+a)^3]$$

$$1-\frac{da}{db}=3(b+a)^2\frac{d}{db}(b+a)$$

$$1-\frac{da}{db}=3(b+a)^2\left(1+\frac{da}{db}\right)$$

$$1-\frac{da}{db}=3(b+a)^2+3(b+a)^2\frac{da}{db}$$

$$-\frac{da}{db}-3(b+a)^2\frac{da}{db}=3(b+a)^2-1$$

$$[-1-3(b+a)^2]\frac{da}{db}=3(b+a)^2-1$$

$$\frac{da}{db}=\frac{3(b+a)^2-1}{-1-3(b+a)^2}$$

$$\frac{da}{db}=0$$

$$3(b+a)^2-1=0$$

$$(b+a)^2=\frac{1}{3}$$

$$b+a=\frac{1}{\sqrt{3}}$$

Since $b-a=(b+a)^3=\left(\dfrac{1}{\sqrt{3}}\right)^3=\dfrac{1}{3\sqrt{3}}.$

$$b+a=\frac{1}{\sqrt{3}}$$

$$-(b-a)=-\frac{1}{3\sqrt{3}}$$

$$2a=\frac{2}{3\sqrt{3}}$$

$$a=\frac{1}{3\sqrt{3}}$$

47. $s^3-4st+2t^3-5t=0$

$$3s^2\frac{ds}{dt}-\left(4t\frac{ds}{dt}+4s\right)+6t^2-5=0$$

$$3s^2\frac{ds}{dt}-4t\frac{ds}{dt}-4s+6t^2-5=0$$

$$\frac{ds}{dt}(3s^2-4t)=4s-6t^2+5$$

$$\frac{ds}{dt}=\frac{4s-6t^2+5}{3s^2-4t}$$

6.5 Related Rates

1. $y^2 - 8x^3 = -55;\ \dfrac{dx}{dt} = -4, x = 2, y = 3$

$$\begin{aligned} 2y\frac{dy}{dt} - 24x^2\frac{dx}{dt} &= 0 \\ y\frac{dy}{dt} &= 12x^2\frac{dx}{dt} \\ 3\frac{dy}{dt} &= 48(-4) \\ \frac{dy}{dt} &= -64 \end{aligned}$$

3. $2xy - 5x + 3y^3 = -51;\ \dfrac{dx}{dt} = -6, x = 3, y = -2$

$$\begin{aligned} 2x\frac{dy}{dt} + 2y\frac{dx}{dt} - 5\frac{dx}{dt} + 9y^2\frac{dy}{dt} &= 0 \\ (2x+9y^2)\frac{dy}{dt} + (2y-5)\frac{dx}{dt} &= 0 \\ (2x+9y^2)\frac{dy}{dt} &= (5-2y)\frac{dx}{dt} \\ \frac{dy}{dt} &= \frac{5-2y}{2x+9y^2}\cdot\frac{dx}{dt} \\ &= \frac{5-2(-2)}{2(3)+9(-2)^2}\cdot(-6) \\ &= \frac{9}{42}\cdot(-6) = \frac{-54}{42} = -\frac{9}{7} \end{aligned}$$

5. $\dfrac{x^2+y}{x-y} = 9;\ \dfrac{dx}{dt} = 2,\ x = 4,\ y = 2$

$$\frac{(x-y)\left(2x\frac{dx}{dt}+\frac{dy}{dt}\right)-(x^2+y)\left(\frac{dx}{dt}-\frac{dy}{dt}\right)}{(x-y)^2} = 0$$

$$\frac{2x(x-y)\frac{dx}{dt} + (x-y)\frac{dy}{dt} - (x^2+y)\frac{dx}{dt} + (x^2+y)\frac{dy}{dt}}{(x-y)^2} = 0$$

$$[2x(x-y)-(x^2+y)]\frac{dx}{dt} + [(x-y)+(x^2+y)]\frac{dy}{dt} = 0$$

$$\begin{aligned} \frac{dy}{dt} &= \frac{[(x^2+y)-2x(x-y)]\frac{dx}{dt}}{(x-y)+(x^2+y)} \\ \frac{dy}{dt} &= \frac{(-x^2+y+2xy)\frac{dx}{dt}}{x+x^2} \\ &= \frac{[-(4)^2+2+2(4)(2)](2)}{4+4^2} \\ &= \frac{4}{20} = \frac{1}{5} \end{aligned}$$

7. $xe^y = 3 + \ln x;\ \dfrac{dx}{dt} = 6, x = 2, y = 0$

$$\begin{aligned} e^y\frac{dx}{dt} + xe^y\frac{dy}{dt} &= 0 + \frac{1}{x}\frac{dx}{dt} \\ xe^y\frac{dy}{dt} &= \left(\frac{1}{x} - e^y\right)\frac{dx}{dt} \\ \frac{dy}{dt} &= \frac{\left(\frac{1}{x}-e^y\right)\frac{dx}{dt}}{xe^y} \\ &= \frac{(1-xe^y)\frac{dx}{dt}}{x^2e^y} \\ &= \frac{[1-(2)e^0](6)}{2^2e^0} \\ &= \frac{-6}{4} = -\frac{3}{2} \end{aligned}$$

9. $C = 0.2x^2 + 10{,}000;\ x = 80,\ \dfrac{dx}{dt} = 12$

$$\frac{dC}{dt} = 0.2(2x)\frac{dx}{dt} = 0.2(160)(12) = 384$$

The cost is changing at a rate of \$384 per month.

11. $R = 50x - 0.4x^2;\ C = 5x + 15;\ x = 40;\ \frac{dx}{dt} = 10$

(a) $$\begin{aligned} \frac{dR}{dt} &= 50\frac{dx}{dt} - 0.8x\frac{dx}{dt} \\ &= 50(10) - 0.8(40)(10) \\ &= 500 - 320 \\ &= 180 \end{aligned}$$

Revenue is increasing at a rate of \$180 per day.

(b) $\dfrac{dC}{dt} = 5\dfrac{dx}{dt} = 5(10) = 50$

Cost is increasing at a rate of \$50 per day.

(c) Profit = Revenue − Cost

$P = R - C$

$$\frac{dP}{dt} = \frac{dR}{dt} - \frac{dC}{dt} = 180 - 50 = 130$$

Profit is increasing at a rate of \$130 per day.

13. $pq = 8000; p = 3.50,\ \frac{dp}{dt} = 0.15$

$$\begin{aligned} pq &= 8000 \\ p\frac{dq}{dt} + q\frac{dp}{dt} &= 0 \\ \frac{dq}{dt} &= \frac{-q\frac{dp}{dt}}{p} \\ &= \frac{-\left(\frac{8000}{3.50}\right)(0.15)}{3.50} \\ &\approx -98 \end{aligned}$$

Demand is decreasing at a rate of approximately 98 units per unit time.

15. $V = k(R^2 - r^2)$; $k = 555.6$, $R = 0.02$ mm, $\frac{dR}{dt} = 0.003$ mm per minute; r is constant.

$$V = k(R^2 - r^2)$$
$$V = 555.6(R^2 - r^2)$$
$$\frac{dV}{dt} = 555.6\left(2R\frac{dR}{dt} - 0\right)$$
$$= 555.6(2)(0.02)(0.003)$$
$$= 0.067 \text{ mm/min}$$

17. $b = 0.22m^{0.87}$

$$\frac{db}{dt} = 0.22(0.87)m^{-0.13}\frac{dm}{dt}$$
$$= 0.1914m^{-0.13}\frac{dm}{dt}$$
$$\frac{dm}{dt} = \frac{m^{0.13}}{0.1914}\frac{db}{dt}$$
$$= \frac{25^{0.13}}{0.1914}(0.25)$$
$$\approx 1.9849$$

The rate of change of the total weight is about 1.9849 g/day.

19. $r = 140.2m^{0.75}$

(a) $\frac{dr}{dt} = 140.2(0.75)m^{-0.25}\frac{dm}{dt}$

$$= 105.15m^{-0.25}\frac{dm}{dt}$$

(b) $\frac{dr}{dt} = 105.15(250)^{-0.25}(2)$

$$\approx 52.89$$

The rate of change of the average daily metabolic rate is about 52.89 kcal/day^2.

21. $C = \frac{1}{10}(T - 60)^2 + 100$

$$\frac{dC}{dt} = \frac{1}{5}(T - 60)\frac{dT}{dt}$$

If $T = 76°$ and $\frac{dT}{dt} = 8$,

$$\frac{dC}{dt} = \frac{1}{5}(76 - 60)(8) = \frac{1}{5}(16)(8)$$
$$= 25.6.$$

The crime rate is rising at the rate of 25.6 crimes/month.

23. Let $x =$ the distance of the base of the ladder from the base of the building;
$y =$ the distance up the side of the building to the top of the ladder.

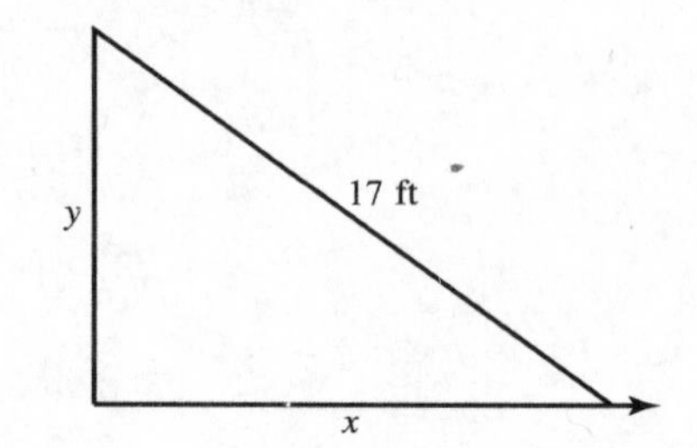

Find $\frac{dy}{dt}$ when $x = 8$ ft and $\frac{dx}{dt} = 9$ ft/min.

Since $y = \sqrt{17^2 - x^2}$, when $x = 8$,

$$y = 15.$$

By the Pythagorean theorem,

$$x^2 + y^2 = 17^2.$$

$$\frac{d}{dt}(x^2 + y^2) = \frac{d}{dt}(17^2)$$
$$2x\frac{dx}{dt} + 2y\frac{dy}{dt} = 0$$
$$2y\frac{dy}{dt} = -2x\frac{dx}{dt}$$
$$\frac{dy}{dt} = \frac{-2x}{2y}\cdot\frac{dx}{dt} = -\frac{x}{y}\cdot\frac{dx}{dt}$$
$$= -\frac{8}{15}(9)$$
$$= -\frac{24}{5}$$

The ladder is sliding down the building at the rate of $\frac{24}{5}$ ft/min.

25. Let $r =$ the radius of the circle formed by the ripple.

Find $\frac{dA}{dt}$ when $r = 4$ ft and $\frac{dr}{dt} = 2$ ft/min.

$$A = \pi r^2$$
$$\frac{dA}{dt} = 2\pi r\frac{dr}{dt}$$
$$= 2\pi(4)(2)$$
$$= 16\pi$$

The area is changing at the rate of 16π ft^2/min.

27. $V = x^3, x = 3$ cm, and $\frac{dV}{dt} = 2$ cm^3/min

$$\frac{dV}{dt} = 3x^2 \frac{dx}{dt}$$
$$\frac{dx}{dt} = \frac{1}{3x^2}\frac{dV}{dt}$$
$$= \frac{1}{3 \cdot 3^2}(2)$$
$$= \frac{2}{27} \text{ cm/min}$$

29. Let $y =$ the length of the man's shadow;
$x =$ the distance of the man from the lamp post;
$h =$ the height of the lamp post.

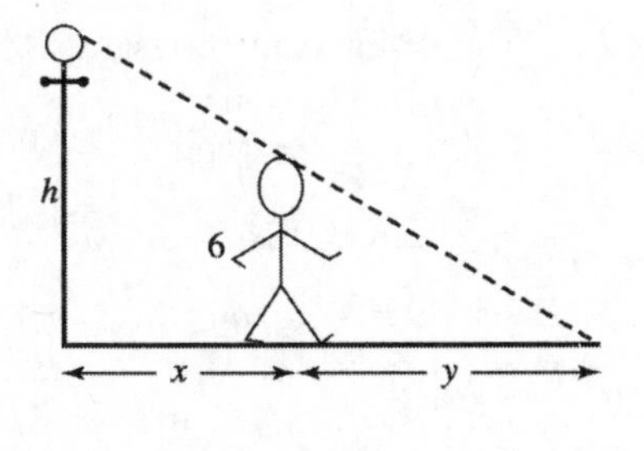

$$\frac{dx}{dt} = 50 \text{ ft/min}$$

Find $\frac{dy}{dt}$ when $x = 25$ ft.

Now $\frac{h}{x+y} = \frac{6}{y}$, by similar triangles.

When $x = 8,\ y = 10$,

$$\frac{h}{18} = \frac{6}{10}$$
$$h = 10.8.$$

$$\frac{10.8}{x+y} = \frac{6}{y},$$
$$10.8y = 6x + 6y$$
$$4.8y = 6x$$
$$y = 1.25x$$

$$\frac{dy}{dt} = 1.25\frac{dx}{dt}$$
$$= 1.25(50)$$
$$\frac{dy}{dt} = 62.5$$

The length of the shadow is increasing at the rate of 62.5 ft/min.

31. Let $x =$ the distance from the docks
$s =$ the length of the rope.

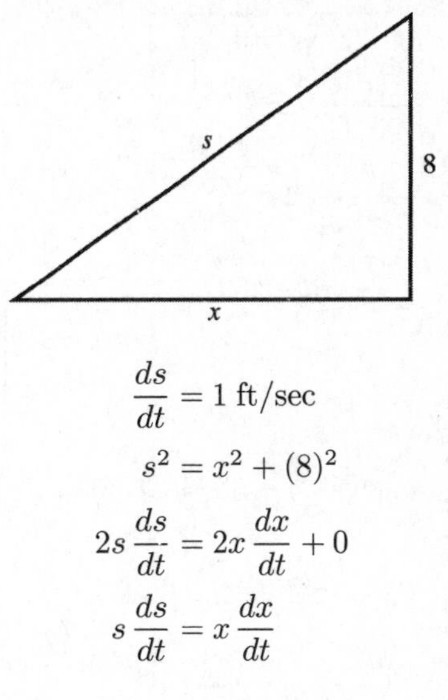

$$\frac{ds}{dt} = 1 \text{ ft/sec}$$
$$s^2 = x^2 + (8)^2$$
$$2s\frac{ds}{dt} = 2x\frac{dx}{dt} + 0$$
$$s\frac{ds}{dt} = x\frac{dx}{dt}$$

If $x = 8$,

$$s = \sqrt{(8)^2 + (8)^2} = \sqrt{128} = 8\sqrt{2}.$$

Then,

$$8\sqrt{2}(1) = 8\frac{dx}{dt}$$
$$\frac{dx}{dt} = \sqrt{2} \approx 1.41$$

The boat is approaching the deck at $\sqrt{2} \approx 1.41$ ft/sec.

6.6 Differentials: Linear Approximation

1. $y = 2x^3 - 5x;\ x = -2,\ \Delta x = 0.1$
$dy = (6x^2 - 5)\,dx$
$\Delta y \approx (6x^2 - 5)\,\Delta x \approx [6(-2)^2 - 5](0.1) \approx 1.9$

3. $y = x^3 - 2x^2 + 3,\ x = 1,\ \Delta x = -0.1$
$dy = (3x^2 - 4x)\,dx$
$\approx (3x^2 - 4x)\,\Delta x$
$= [3(1^2) - 4(1)](-0.1)$
$= 0.1$

5. $y = \sqrt{3x + 2},\ x = 4,\ \Delta x = 0.15$

$$dy = 3\left(\frac{1}{2}(3x+2)^{-1/2}\right)dx$$

$$\Delta y \approx \frac{3}{2\sqrt{3x+2}}\Delta x \approx \frac{3}{2(3.74)}(0.15) \approx 0.060$$

7. $y = \dfrac{2x-5}{x+1}$; $x = 2,\ \Delta x = -0.03$

$$\begin{aligned} dy &= \frac{(x+1)(2)-(2x-5)(1)}{(x+1)^2}\,dx \\ &= \frac{7}{(x+1)^2}\,dx \\ &= \frac{7}{(x+1)^2}\,\Delta x \\ &= \frac{7}{(2+1)^2}(-0.03) \\ &= -0.023 \end{aligned}$$

9. $\sqrt{145}$

We know $\sqrt{144} = 12$, so $f(x) = \sqrt{x}$, $x = 144$, $dx = 1$.

$$\begin{aligned} \frac{dy}{dx} &= \frac{1}{2}x^{-1/2} \\ dy &= \frac{1}{2\sqrt{x}}\,dx \\ dy &= \frac{1}{2\sqrt{144}}(1) = \frac{1}{24} \\ \sqrt{145} \approx f(x) + dy &= 12 + \frac{1}{24} \\ &\approx 12.0417 \end{aligned}$$

By calculator, $\sqrt{145} \approx 12.0416$.
The difference is $|12.0417 - 12.0416| = 0.0001$.

11. $\sqrt{0.99}$

We know $\sqrt{1} = 1$, so $f(x) = \sqrt{x}$, $x = 1$, $dx = -0.01$.

$$\begin{aligned} \frac{dy}{dx} &= \frac{1}{2}x^{-1/2} \\ dy &= \frac{1}{2\sqrt{x}}\,dx \\ dy &= \frac{1}{2\sqrt{1}}(-0.01) = -0.005 \\ \sqrt{0.99} \approx f(x) + dy &= 1 - 0.005 \\ &= 0.995 \end{aligned}$$

By calculator, $\sqrt{0.99} \approx 0.9950$.
The difference is $|0.995 - 0.9950| = 0$.

13. $e^{0.01}$

We know $e^0 = 1$, so $f(x) = e^x$, $x = 0$, $dx = 0.01$.

$$\begin{aligned} \frac{dy}{dx} &= e^x \\ dy &= e^x\,dx \\ dy &= e^0(0.01) = 0.01 \\ e^{0.01} &\approx f(x) + dy = 1 + 0.01 = 1.01 \end{aligned}$$

By calculator, $e^{0.01} \approx 1.0101$.
The difference is $|1.01 - 1.0101| = 0.0001$.

15. $\ln 1.05$

We know $\ln 1 = 0$, so $f(x) = \ln x$, $x = 1$, $dx = 0.05$.

$$\begin{aligned} \frac{dy}{dx} &= \frac{1}{x} \\ dy &= \frac{1}{x}\,dx \\ dy &= \frac{1}{1}(0.05) = 0.05 \end{aligned}$$

$$\ln 1.05 \approx f(x) + dy = 0 + 0.05 = 0.05$$

By calculator, $\ln 1.05 \approx 0.0488$.
The difference is $|0.05 - 0.0488| = 0.0012$.

17. Let $D =$ the demand in thousands of pounds;
$x =$ the price in dollars.
$D(q) = -3q^3 - 2q^2 + 1500$

(a) $q = 2,\ \Delta q = 0.10$

$$\begin{aligned} dD &= (-9q^2 - 4q)\,dq \\ \Delta D &\approx (-9q^2 - 4q)\,\Delta q \\ &\approx [-9(4) - 4(2)](0.10) \\ &\approx -4.4 \text{ thousand pounds} \end{aligned}$$

(b) $q = 6,\ \Delta q = 0.15$

$$\begin{aligned} \Delta D &\approx [-9(36) - 4(6)](0.15) \\ &\approx -52.2 \text{ thousand pounds} \end{aligned}$$

19. $R(x) = 12{,}000\ln(0.01x + 1)$
$x = 100,\ \Delta x = 1$

$$\begin{aligned} dR &= \frac{12{,}000}{0.01x + 1}(0.01)\,dx \\ \Delta R &\approx \frac{120}{0.01x + 1}\,\Delta x \\ &\approx \frac{120}{0.01(100) + 1}(1) \\ &\approx \$60 \end{aligned}$$

21. If a cube is given a coating 0.1 in. thick, each edge increases in length by twice that amount, or 0.2 in. because there is a face at both ends of the edge.

$V = x^3$, $x = 4$, $\Delta x = 0.2$

$$\begin{aligned} dV &= 3x^2\,dx \\ \Delta V &\approx 3x^2\,\Delta x \\ &= 3(4^2)(0.2) \\ &= 9.6 \end{aligned}$$

For 1000 cubes $9.6(1000) = 9600$ in.^{3} of coating should be ordered.

23. (a) $A(x) = y = 0.003631x^3 - 0.03746x^2 + 0.1012x + 0.009$

Let $x = 1,\ dx = 0.2.$

$$\frac{dy}{dx} = 0.010893x^2 - 0.07492x + 0.1012$$

$$dy = (0.010893x^2 - 0.07492x + 0.1012)\,dx$$

$$\begin{aligned}\Delta y &\approx (0.010893x^2 - 0.07492x + 0.1012)\,\Delta x\\ &\approx (0.010893 \cdot 1^2 - 0.07492 \cdot 1 + 0.1012) \cdot 0.2\\ &\approx 0.007435\end{aligned}$$

The alcohol concentration increases by about 0.74 percent.

(b) $\Delta y \approx (0.010893 \cdot 3^2 - 0.07492 \cdot 3 + 0.1012) \cdot 0.2 \approx -0.005105$

The alcohol concentration decreases by about 0.51 percent.

25. $P(x) = \dfrac{25x}{8 + x^2}$

$$dP = \frac{(8 + x^2)(25) - 25x(2x)}{(8 + x^2)^2}\,dx = \frac{(8 + x^2)(25) - 25x(2x)}{(8 + x^2)^2}\,\Delta x$$

(a) $x = 2,\ \Delta x = 0.5$

$$dP = \frac{[(8 + 4)(25) - (25)(2)(4)](0.5)}{(8 + 4)^2} = 0.347 \text{ million}$$

(b) $x = 3,\ \Delta x = 0.25$

$$dP = \frac{[(8 + 9)(25) - 25(3)(6)]0.25}{(8 + 9)^2} \approx -0.022 \text{ million}$$

27. r changes from 14 mm to 16 mm, so $\Delta r = 2$.

$$\begin{aligned}V &= \frac{4}{3}\pi r^3\\ dV &= \frac{4}{3}(3)\pi r^2\,dr\\ \Delta V &\approx 4\pi r^2\,\Delta r = 4\pi(14)^2(2) = 1568\pi \text{ mm}^3\end{aligned}$$

29. r increases from 20 mm to 22 mm, so $\Delta r = 2$.

$$\begin{aligned}A &= \pi r^2\\ dA &= 2\pi r\,dr\\ \Delta A &\approx 2\pi r\,\Delta r = 2\pi(20)(2) = 80\pi \text{ mm}^2\end{aligned}$$

31. $W(t) = -3.5 + 197.5e^{-e^{-0.01394(t-108.4)}}$

(a) $dW = 197.5e^{-e^{-0.01394(t-108.4)}}(-1)e^{-0.01394(t-108.4)}(-0.01394)dt = 2.75315e^{-e^{-0.01394(t-108.4)}}e^{-0.01394(t-108.4)}dt$

We are given $t = 80$ and $dt = 90 - 80 = 10$.

$$dW \approx 9.258$$

The pig will gain about 9.3 kg.

(b) The actual weight gain is calculated as

$$W(90) - W(80) \approx 50.736 - 41.202 = 9.534$$

or about 9.5 kg.

33. $r = 3$ cm, $\Delta r = -0.2$ cm

$$\begin{aligned} V &= \frac{4}{3}\pi r^3 \\ dV &= 4\pi r^2\,dr \\ \Delta V &\approx 4\pi r^2\,\Delta r \\ &= 4\pi(9)(-0.2) \\ &= -7.2\pi \text{ cm}^3 \end{aligned}$$

35. $V = \frac{1}{3}\pi r^2 h;\ h = 13, dh = 0.2$

$$\begin{aligned} V &= \frac{1}{3}\pi\left(\frac{h}{15}\right)^2 h \\ &= \frac{\pi}{775}h^3 \\ dV &= \frac{\pi}{775}\cdot 3h^2 dh \\ &= \frac{\pi}{225}h^2 dh \\ \Delta V &\approx \frac{\pi}{225}h^2\Delta h \\ &\approx \frac{\pi}{225}(13^2)(0.2) \\ &\approx 0.472 \text{ cm}^3 \end{aligned}$$

37. $A = x^2;\ x = 4, dA = 0.01$

$$\begin{aligned} dA &= 2x\,dx \\ \Delta A &\approx 2x\,\Delta x \\ \Delta x &\approx \frac{\Delta A}{2x} \approx \frac{0.01}{2(4)} \approx 0.00125 \text{ cm} \end{aligned}$$

39. $V = \frac{4}{3}\pi r^3;\ r = 5.81,\ \Delta r = \pm 0.003$

$$\begin{aligned} dV &= \frac{4}{3}\pi(3r^2)\,dr \\ \Delta V &\approx \frac{4}{3}\pi(3r^2)\,\Delta r \\ &= 4\pi(5.81)^2(\pm 0.003) \\ &= \pm 0.405\pi \approx \pm 1.273 \text{ in.}^3 \end{aligned}$$

41. $h = 7.284$ in., $r = 1.09 \pm 0.007$ in.

$$\begin{aligned} V &= \frac{1}{3}\pi r^2 h \\ dV &= \frac{2}{3}\pi r h\,dr \\ \Delta V &\approx \frac{2}{3}\pi r h\,\Delta r \\ &= \frac{2}{3}\pi(1.09)(7.284)(0.007) \\ &= \pm 0.116 \text{ in.}^3 \end{aligned}$$

Chapter 6 Review Exercises

1. $f(x) = -x^3 + 6x^2 + 1;\ [-1, 6]$
$f'(x) = -3x^2 + 12x = 0$ when $x = 0, 4$.

$$\begin{aligned} f(-1) &= 8 \\ f(0) &= 1 \\ f(4) &= 33 \\ f(6) &= 1 \end{aligned}$$

Absolute maximum of 33 at 4; absolute minimum of 1 at 0 and 6.

3. $f(x) = x^3 + 2x^2 - 15x + 3;\ [-4, 2]$
$f'(x) = 3x^2 + 4x - 15 = 0$ when
$(3x-5)(x+3) = 0$

$$x = \frac{5}{3} \quad \text{or} \quad x = -3.$$

$$\begin{aligned} f(-4) &= 31 \\ f(-3) &= 39 \\ f\left(\frac{5}{3}\right) &= -\frac{319}{27} \\ f(2) &= -11 \end{aligned}$$

Absolute maximum of 39 at -3; absolute minimum of $-\frac{319}{27}$ at $\frac{5}{3}$

7. (a) $f(x) = \frac{2\ln x}{x^2};\ [1, 4]$

$$\begin{aligned} f'(x) &= \frac{x^2\left(\frac{2}{x}\right) - (2\ln x)(2x)}{x^4} \\ &= \frac{2x - 4x\ln x}{x^4} \\ &= \frac{2 - 4\ln x}{x^3} \end{aligned}$$

$f'(x) = 0$ when

$$\begin{aligned} 2 - 4\ln x &= 0 \\ 2 &= 4\ln x \\ 0.5 &= \ln x \\ e^{0.5} &= x \\ x &\approx 1.6487. \end{aligned}$$

x	$f(x)$
1	0
$e^{0.5}$	0.36788
4	0.17329

Maximum is 0.37; minimum is 0.

(b) $[2, 5]$

Note that the critical number of f is not in the domain, so we only test the endpoints.

x	$f(x)$
2	0.34657
5	0.12876

Maximum is 0.35, minimum is 0.13.

11. $x^2 - 4y^2 = 3x^3y^4$

$$\begin{aligned}\frac{d}{dx}(x^2 - 4y^2) &= \frac{d}{dx}(3x^3y^4)\\ 2x - 8y\frac{dy}{dx} &= 9x^2y^4 + 3x^3 \cdot 4y^3\frac{dy}{dx}\\ (-8y - 3x^3 \cdot 4y^3)\frac{dy}{dx} &= 9x^2y^4 - 2x\\ \frac{dy}{dx} &= \frac{2x - 9x^2y^4}{8y + 12x^3y^3}\end{aligned}$$

13. $2\sqrt{y-1} = 9x^{2/3} + y$

$$\begin{aligned}\frac{d}{dx}[2(y-1)^{1/2}] &= \frac{d}{dx}(9x^{2/3} + y)\\ 2 \cdot \frac{1}{2} \cdot (y-1)^{-1/2}\frac{dy}{dx} &= 6x^{-1/3} + \frac{dy}{dx}\\ [(y-1)^{-1/2} - 1]\frac{dy}{dx} &= 6x^{-1/3}\\ \frac{1 - \sqrt{y-1}}{\sqrt{y-1}} \cdot \frac{dy}{dx} &= \frac{6}{x^{1/3}}\\ \frac{dy}{dx} &= \frac{6\sqrt{y-1}}{x^{1/3}(1 - \sqrt{y-1})}\end{aligned}$$

15. $\dfrac{6+5x}{2-3y} = \dfrac{1}{5x}$

$$\begin{aligned}5x(6+5x) &= 2 - 3y\\ 30x + 25x^2 &= 2 - 3y\\ \frac{d}{dx}(30x + 25x^2) &= \frac{d}{dx}(2 - 3y)\\ 30 + 50x &= -3\frac{dy}{dx}\\ -\frac{30 + 50x}{3} &= \frac{dy}{dx}\end{aligned}$$

17. $\ln(xy+1) = 2xy^3 + 4$

$$\begin{aligned}\frac{d}{dx}[\ln(xy+1)] &= \frac{d}{dx}(2xy^3 + 4)\\ \frac{1}{xy+1} \cdot \frac{d}{dx}(xy+1) &= 2y^3 + 2x \cdot 3y^2\frac{dy}{dx} + \frac{d}{dx}(4)\\ \frac{1}{xy+1}\left(y + x\frac{dy}{dx} + \frac{d}{dx}(1)\right) &= 2y^3 + 6xy^2\frac{dy}{dx}\\ \frac{y}{xy+1} + \frac{x}{xy+1} \cdot \frac{dy}{dx} &= 2y^3 + 6xy^2\frac{dy}{dx}\\ \left(\frac{x}{xy+1} - 6xy^2\right)\frac{dy}{dx} &= 2y^3 - \frac{y}{xy+1}\\ \frac{dy}{dx} &= \frac{2y^3 - \frac{y}{xy+1}}{\frac{x}{xy+1} - 6xy^2}\\ &= \frac{2y^3(xy+1) - y}{x - 6xy^2(xy+1)}\\ &= \frac{2xy^4 + 2y^3 - y}{x - 6x^2y^3 - 6xy^2}\end{aligned}$$

21. $y = 8x^3 - 7x^2$, $\frac{dx}{dt} = 4$, $x = 2$

$$\begin{aligned}\frac{dy}{dt} &= \frac{d}{dt}(8x^3 - 7x^2)\\ &= 24x^2\frac{dx}{dt} - 14x\frac{dx}{dt}\\ &= 24(2)^2(4) - 14(2)(4)\\ &= 272\end{aligned}$$

23. $y = \dfrac{1+\sqrt{x}}{1-\sqrt{x}}$, $\dfrac{dx}{dt} = -4$, $x = 4$

$$\begin{aligned}\frac{dy}{dt} &= \frac{d}{dt}\left[\frac{1+\sqrt{x}}{1-\sqrt{x}}\right]\\ &= \frac{(1-\sqrt{x})\left(\frac{1}{2}x^{-1/2}\frac{dx}{dt}\right) - (1+\sqrt{x})\left(-\frac{1}{2}\right)\left(x^{-1/2}\frac{dx}{dt}\right)}{(1-\sqrt{x})^2}\\ &= \frac{(1-2)\left(\frac{1}{2\cdot 2}\right)(-4) - (1+2)\left(\frac{-1}{2\cdot 2}\right)(-4)}{(1-2)^2}\\ &= \frac{1-3}{1} = -2\end{aligned}$$

25. $y = xe^{3x}$; $\dfrac{dx}{dt} = -2, x = 1$

$$\begin{aligned}\frac{dy}{dt} &= \frac{d}{dt}(xe^{3x})\\ &= \frac{dx}{dt} \cdot e^{3x} + x \cdot \frac{d}{dt}(e^{3x})\\ &= \frac{dx}{dt} \cdot e^{3x} + xe^{3x} \cdot 3\frac{dx}{dt}\\ &= (1+3x)e^{3x}\frac{dx}{dt}\\ &= (1 + 3 \cdot 1)e^{3(1)}(-2) = -8e^3\end{aligned}$$

29. $y = \dfrac{3x-7}{2x+1}$; $x = 2$, $\Delta x = 0.003$

$$dy = \frac{(3)(2x+1)-(2)(3x-7)}{(2x+1)^2}\,dx$$

$$\begin{aligned} dy &= \frac{17}{(2x+1)^2}\,dx \\ &\approx \frac{17}{(2x+1)^2}\,\Delta x \\ &= \frac{17}{(2[2]+1)^2}(0.003) \\ &= 0.00204 \end{aligned}$$

33. Let $x =$ the length and width of a side of the base;
$h =$ the height.

The volume is 32 m^3; the base is square and there is no top. Find the height, length, and width for minimum surface area.

$$\begin{aligned} \text{Volume} &= x^2h \\ x^2h &= 32 \\ h &= \frac{32}{x^2} \end{aligned}$$

$$\begin{aligned} \text{Surface area} &= x^2 + 4xh \\ A &= x^2 + 4x\left(\frac{32}{x^2}\right) \\ &= x^2 + 128x^{-1} \\ A' &= 2x - 128x^{-2} \end{aligned}$$

If $A' = 0$,

$$\begin{aligned} \frac{2x^3-128}{x^2} &= 0 \\ x^3 &= 64 \\ x &= 4. \end{aligned}$$

$$\begin{aligned} A''(x) &= 2 + 2(128)x^{-3} \\ A''(4) &= 6 > 0 \end{aligned}$$

The minimum is at $x = 4$, where

$$h = \frac{32}{4^2} = 2.$$

The dimensions are 2 m by 4 m by 4 m.

35. Volume of cylinder $= \pi r^2 h$
Surface area of cylinder open at one end $= 2\pi rh + \pi r^2$.

$$\begin{aligned} V &= \pi r^2 h = 27\pi \\ h &= \frac{27\pi}{\pi r^2} = \frac{27}{r^2} \\ A &= 2\pi r\left(\frac{27}{r^2}\right) + \pi r^2 \\ &= 54\pi r^{-1} + \pi r^2 \\ A' &= -54\pi r^{-2} + 2\pi r \end{aligned}$$

If $A' = 0$,

$$\begin{aligned} 2\pi r &= \frac{54\pi}{r^2} \\ r^3 &= 27 \\ r &= 3. \end{aligned}$$

If $r = 3$,

$$A'' = 108\pi r^{-3} + 2\pi > 0,$$

so the value at $r = 3$ is a minimum.

For the minimum cost, the radius of the bottom should be 3 inches.

37. Here $k = 0.15$, $M = 20{,}000$, and $f = 12$. We have

$$\begin{aligned} q &= \sqrt{\frac{2fM}{k}} = \sqrt{\frac{2(12)20{,}000}{0.15}} \\ &= \sqrt{3{,}200{,}000} \approx 1789 \end{aligned}$$

Ordering 1789 rolls each time minimizes annual cost.

39. Use equation (3) from Section 6.3 with $k = 1$, $M = 128{,}000$, and $f = 10$.

$$\begin{aligned} q &= \sqrt{\frac{2fM}{k}} = \sqrt{\frac{2(10)(128{,}000)}{1}} \\ &= \sqrt{2{,}560{,}000} = 1600 \end{aligned}$$

The number of lots that should be produced annually is

$$\frac{M}{q} = \frac{128{,}000}{1600} = 80.$$

41. $A = \pi r^2$; $\frac{dr}{dt} = 4$ ft/min, $r = 7$ ft

$$\frac{dA}{dt} = 2\pi r\,\frac{dr}{dt}$$
$$\frac{dA}{dt} = 2\pi(7)(4)$$
$$\frac{dA}{dt} = 56\pi$$

The rate of change of the area is 56π ft^2/min.

43. **(a)**

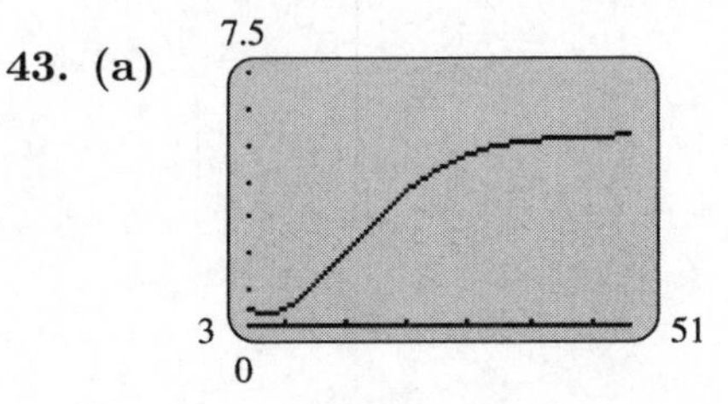

(b) We use a graphing calculator to graph

$$M'(t) = -0.4321173 + 0.1129024t - 0.0061518t^2 + 0.0001260t^3 - 0.0000008925t^4$$

on $[3, 51]$ by $[0, 0.3]$. We find the maximum value of $M'(t)$ on this graph at about 15.41, or on about the 15th day.

45. Let $x =$ the distance from the base of the ladder to the building;
$y =$ the height on the building at the top of the ladder.

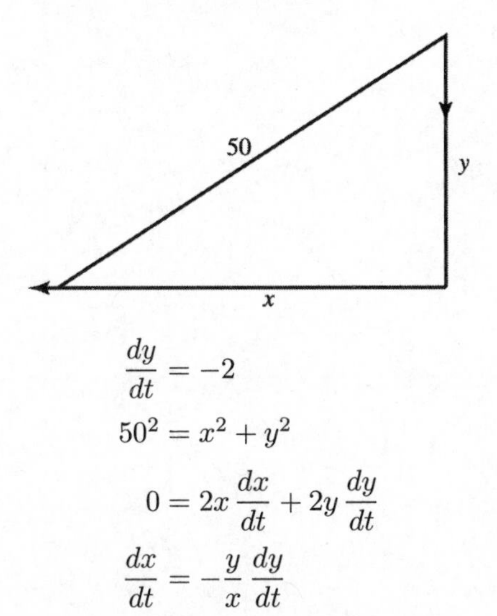

$$\frac{dy}{dt} = -2$$
$$50^2 = x^2 + y^2$$
$$0 = 2x\,\frac{dx}{dt} + 2y\,\frac{dy}{dt}$$
$$\frac{dx}{dt} = -\frac{y}{x}\frac{dy}{dt}$$

When $x = 30$, $y = \sqrt{2500 - (30)^2} = 40$.
So
$$\frac{dx}{dt} = \frac{-40}{30}(-2) = \frac{80}{30} = \frac{8}{3}$$

The base of the ladder is slipping away from the building at a rate of $\frac{8}{3}$ ft/min.

47. Let $x =$ one-half the width of the triangular cross section;
$h =$ the height of the water;
$V =$ the volume of the water.

$$\frac{dV}{dt} = 3.5\ ft^3/\text{min}$$

Find $\frac{dV}{dt}$ when $h = \frac{1}{3}$.

$$V = \left(\begin{array}{c}\text{Area of}\\ \text{triangular}\\ \text{side}\end{array}\right)(\text{length})$$

4 ft, 2 ft, 1 ft

Area of triangular cross section

$$= \frac{1}{2}(\text{base})(\text{altitude})$$
$$= \frac{1}{2}(2x)(h) = xh$$

2 ft, 1 ft, 1 ft, 1 ft, x, x, h

By similar triangles, $\frac{2x}{h} = \frac{2}{1}$, so $x = h$.

$$V = (xh)(4) = h^2 \cdot 4 = 4h^2$$
$$\frac{dV}{dt} = 8h\,\frac{dh}{dt}$$
$$\frac{1}{8h} \cdot \frac{dV}{dt} = \frac{dh}{dt}$$
$$\frac{1}{8\left(\frac{1}{3}\right)}(3.5) = \frac{dh}{dt}$$
$$\frac{dh}{dt} = \frac{21}{16} = 1.3125$$

The depth of water is changing at the rate of 1.3125 ft/min.

49. $A = s^2$; $s = 9.2$, $\Delta s = \pm 0.04$

$$ds = 2s\,ds$$
$$\Delta A \approx 2s\,\Delta s = 2(9.2)(\pm 0.04) = \pm 0.736 \text{ in.}^2$$

51. We need to minimize y. Note that $x > 0$.

$$\frac{dy}{dx} = \frac{x}{8} - \frac{2}{x}$$

Set the derivative equal to 0.

$$\frac{x}{8} - \frac{2}{x} = 0$$
$$\frac{x}{8} = \frac{2}{x}$$
$$x^2 = 16$$
$$x = 4$$

Since $\lim_{x\to 0} y = \infty$, $\lim_{x\to\infty} y = \infty$, and $x = 4$ is the only critical value in $(0, \infty)$, $x = 4$ produces a minimum value.

$$\begin{aligned} y &= \frac{4^2}{16} - 2\ln 4 + \frac{1}{4} + 2\ln 6 \\ &= 1.25 + 2(\ln 6 - \ln 4) \\ &= 1.25 + 2\ln 1.5 \end{aligned}$$

The y coordinate of the Southern most point of the second boat's path is $1.25 + 2\ln 1.5$.

53. Distance on shore: $40 - x$ feet
Speed on shore: 5 feet per second
Distance in water: $\sqrt{x^2 + 40^2}$ feet
Speed in water: 3 feet per second

The total travel time t is $t = t_1 + t_2 = \frac{d_1}{v_1} + \frac{d_2}{v_2}$.

$$\begin{aligned} t(x) &= \frac{40 - x}{5} + \frac{\sqrt{x^2 + 40^2}}{3} \\ &= 8 - \frac{x}{5} + \frac{\sqrt{x^2 + 1600}}{3} \\ t'(x) &= -\frac{1}{5} + \frac{1}{3} \cdot \frac{1}{2}(x^2 + 1600)^{-1/2}(2x) \\ &= -\frac{1}{5} + \frac{x}{3\sqrt{x^2 + 1600}} \end{aligned}$$

Minimize the travel time $t(x)$. If $t'(x) = 0$:

$$\frac{x}{3\sqrt{x^2 + 1600}} = \frac{1}{5}$$
$$5x = 3\sqrt{x^2 + 1600}$$
$$\frac{5x}{3} = \sqrt{x^2 + 1600}$$
$$\frac{25}{9}x^2 = x^2 + 1600$$
$$\frac{16}{9}x^2 = 1600$$
$$x^2 = \frac{1600 \cdot 9}{16}$$
$$x = \frac{40 \cdot 3}{4} = 30$$

(Discard the negative solution.)

To minimize the time, he should walk $40 - x = 40 - 30 = 10$ ft along the shore before paddling toward the desired destination. The minimum travel time is

$$\frac{40 - 30}{5} + \frac{\sqrt{30^2 + 40^2}}{3} \approx 18.67 \text{ seconds.}$$

Chapter 6 Test

[6.1]

Find the locations of any absolute extrema for the functions with graphs as follows.

1. **2.**

Find the locations of all absolute extrema for the functions defined as follows, with the specified domains.

3. $f(x) = x^3 - 12x$; $[0, 4]$

4. $f(x) = \frac{x^2 + 4}{x}$; $[-3, -1]$

5. Why is it important to check the endpoints of the domain when checking for absolute extrema?

[6.2]

6. Find two nonnegative numbers x and y such that $x + y = 50$ and $P = xy$ is maximized.

7. A travel agency offers a tour to the Bahamas for 12 people at $800 each. For each person more than the original 12, up to a total of 30 people, who signs up for the cruise, the fare is reduced by $20. What tour group size produces the greatest revenue for the travel agency?

8. $320 is available for fencing for a rectangular garden. The fencing for the two sides parallel to the back of the house costs $6 per linear foot, while the fencing for the other sides costs $2 per linear foot. Find the dimensions that will maximize the area of the garden.

[6.3]

9. A small beauty supply store sells 600 hair dryers each year. It costs the store $3 a year to store one hair dryer for one year. The fixed cost of placing the order is $36. Find the optimum number of dryers per order.

10. A demand function is given by

$$q = 100 - 2p$$

where q is the number of units produced and p is the price in dollars.

(a) Find E, the elasticity of demand. **(b)** Find the price that maximizes the total revenue.

11. What does elasticity of demand measure? What does it mean for demand to be elastic? What does it mean for demand to be inelastic?

[6.4]

12. Find $\frac{dy}{dx}$ for $x^3 - x^2y + y^2 = 0$.

13. Find an equation for the tangent line to the graph of $2x^2 - y^2 = 2$ at $(3, 4)$.

14. Suppose $3x^2 + 4y^2 + 6 = 0$. Use implicit differentiation to find $\frac{dy}{dx}$. Explain why your result is meaningless.

[6.5]

15. Find $\frac{dy}{dt}$ if $y = 3x^3 - 2x^2$, $\frac{dx}{dt} = -2$, and $x = 3$.

16. Find $\frac{dy}{dt}$ if $2xy - y^2 = x$, $x = -1$, $y = 3$, and $\frac{dx}{dt} = 2$.

17. When solving related rates problems, how should you interpret a negative derivative?

18. A 25-foot ladder is leaning against a vertical wall. If the bottom of the ladder is pulled horizontally away from the wall at 3 feet per second, how fast is the top of the ladder sliding down the wall when the bottom is 15 feet from the wall?

19. A real estate developer estimates that his monthly sales are given by

$$S = 30y - xy - \frac{y^2}{3000},$$

where y is the average cost of a new house in the development and x percent is the current interest rate for home mortgages. If the current rate is 11% and is rising at a rate of $\frac{1}{4}$% per month and the current average price of a new home is \$90,000 and is increasing at a rate of \$500 per month, how fast are his expected sales changing.

Given the revenue and cost functions $R = 30q - 0.2q^2$ and $C = 12q + 20$, where q is the daily production (and sales), answer Problems 20-22 when 50 units are produced and the rate of change of production is 8 units per day.

20. Find the rate of change of revenue with respect to time.

21. Find the rate of change of cost with respect to time.

22. Find the rate of change of profit with respect to time.

[6.6]

23. If $y = \frac{2x-3}{x^2-7}$, find dy.

24. Find the value of dy if $y = \sqrt{5 - x^2}$, $x = 1$, $\Delta x = 0.01$.

25. The radius of a sphere is claimed to be 5 cm with a possible error of 0.01 cm. Use differentials to estimate the possible error in the volume of the sphere.

Chapter 6 Test Answers

1. Absolute maximum at x_3

2. No absolute extrema

3. Absolute maximum of 16 at 4; absolute minimum of -16 at 2

4. Absolute maximum of -4 at -2; absolute minimum of -5 at -1

5. The smallest or largest value in a closed interval may occur at the endpoints.

6. 25 and 25

7. 26 people

8. $13\frac{1}{2}$ ft on each side parallel to the back, 40 ft on each of the other sides

9. 120

10. (a) $E = \frac{2p}{100-2p}$ (b) \$25

11. The instantaneous responsiveness of demand to price; the relative change in demand is greater than the relative change in price; the relative change in demand is less than the relative change in price.

12. $\frac{dy}{dx} = \frac{2xy-3x^2}{2y-x^2}$

13. $3x - 2y = 1$

14. $\frac{-3x}{4y}$; no function exists such that $3x^2+4y^2+6 = 0$.

15. -138

16. 1.25

17. A decrease in rate

18. $-\frac{9}{4}$ ft/sec

19. Decreasing at \$43,000/month

20. Increasing at a rate of \$80/day

21. Increasing at a rate of \$96/day

22. Decreasing at a rate of \$16/day

23. $dy = \frac{-2x^2+6x-14}{(x^2-7)^2}dx$

24. -0.005

25. 3.14 cm^3

Chapter 7

INTEGRATION

7.1 Antiderivatives

1. If $F(x)$ and $G(x)$ are both antiderivatives of $f(x)$, then there is a constant C such that

$$F(x) - G(x) = C.$$

The two functions can differ only by a constant.

5. $$\int 6\,dk = 6\int 1\,dk$$
$$= 6\int k^0\,dy$$
$$= 6\cdot\frac{1}{1}k^{0+1} + C$$
$$= 6k + C$$

7. $$\int (2z+3)\,dz$$
$$= 2\int z\,dz + 3\int z^0\,dz$$
$$= 2\cdot\frac{1}{1+1}z^{1+1} + 3\cdot\frac{1}{0+1}z^{0+1} + C$$
$$= z^2 + 3z + C$$

9. $$\int (6t^2 - 8t + 7)\,dt$$
$$= 6\int t^2\,dt - 8\int t\,dt + 7\int t^0\,dt$$
$$= \frac{6t^3}{3} - \frac{8t^2}{2} + 7t + C$$
$$= 2t^3 - 4t^2 + 7t + C$$

11. $$\int (4z^3 + 3z^2 + 2z - 6)\,dz$$
$$= 4\int z^3\,dz + 3\int z^2\,dz + 2\int z\,dz$$
$$- 6\int z^0\,dz$$
$$= \frac{4z^4}{4} + \frac{3z^3}{3} + \frac{2z^2}{2} - 6z + C$$
$$= z^4 + z^3 + z^2 - 6z + C$$

13. $$\int (5\sqrt{z} + \sqrt{2})\,dz = 5\int z^{1/2}\,dz + \sqrt{2}\int dz$$
$$= \frac{5z^{3/2}}{\frac{3}{2}} + \sqrt{2}z + C$$
$$= 5\left(\frac{2}{3}\right)z^{3/2} + \sqrt{2}z + C$$
$$= \frac{10z^{3/2}}{3} + \sqrt{2}z + C$$

15. $$\int 5x(x^2-8)dx = \int (5x^3 - 40x)dx$$
$$= \frac{5x^4}{4} - \frac{40x^2}{2} + C$$
$$= \frac{5x^4}{4} - 20x^2 + C$$

17. $$\int (4\sqrt{v} - 3v^{3/2})\,dv$$
$$= 4\int v^{1/2}\,dv - 3\int v^{3/2}\,dv$$
$$= \frac{4v^{3/2}}{\frac{3}{2}} - \frac{3v^{5/2}}{\frac{5}{2}} + C$$
$$= \frac{8v^{3/2}}{3} - \frac{6v^{5/2}}{5} + C$$

19. $$\int (10u^{3/2} - 14u^{5/2})\,du$$
$$= 10\int u^{3/2}\,du - 14\int u^{5/2}\,du$$
$$= \frac{10u^{5/2}}{\frac{5}{2}} - \frac{14u^{7/2}}{\frac{7}{2}} + C$$
$$= 10\left(\frac{2}{5}\right)u^{5/2} - 14\left(\frac{2}{7}\right)u^{7/2} + C$$
$$= 4u^{5/2} - 4u^{7/2} + C$$

21. $$\int \left(\frac{7}{z^2}\right)\,dz = \int 7z^{-2}\,dz$$
$$= 7\int z^{-2}dz$$
$$= 7\left(\frac{z^{-2+1}}{-2+1}\right) + C$$
$$= \frac{7z^{-1}}{-1} + C$$
$$= -\frac{7}{z} + C$$

23. $\int \left(\frac{\pi^3}{y^3} - \frac{\sqrt{\pi}}{\sqrt{y}}\right) dy = \int \pi^3 y^{-3}\, dy - \int \sqrt{\pi} y^{-1/2}\, dy$

$= \pi^3 \int y^{-3}\, dy - \sqrt{\pi} \int y^{-1/2}\, dy$

$= \pi^3 \left(\frac{y^{-2}}{-2}\right) - \sqrt{\pi}\left(\frac{y^{1/2}}{\frac{1}{2}}\right) + C$

$= -\frac{\pi^3}{2y^2} - 2\sqrt{\pi y} + C$

25. $\int (-9t^{-2.5} - 2t^{-1})\, dt$

$= -9 \int t^{-2.5}\, dt - 2 \int t^{-1}\, dt$

$= \frac{-9t^{-1.5}}{-1.5} - 2 \int \frac{dt}{t}$

$= 6t^{-1.5} - 2 \ln |t| + C$

27. $\int \frac{1}{3x^2} dx = \int \frac{1}{3} x^{-2} dx$

$= \frac{1}{3} \int x^{-2} dx$

$= \frac{1}{3}\left(\frac{x^{-1}}{-1}\right) + C$

$= -\frac{1}{3} x^{-1} + C$

$= -\frac{1}{3x} + C$

29. $\int 3e^{-0.2x}\, dx = 3 \int e^{-0.2x}\, dx$

$= 3\left(\frac{1}{-0.2}\right) e^{-0.2x} + C$

$= \frac{3(e^{-0.2x})}{-0.2} + C$

$= -15e^{-0.2x} + C$

31. $\int \left(-\frac{3}{x} + 4e^{-0.4x} + e^{0.1}\right) dx$

$= -3 \int \frac{dx}{x} + 4 \int e^{-0.4x}\, dx + e^{0.1} \int dx$

$= -3 \ln |x| + \frac{4e^{-0.4x}}{-0.4} + e^{0.1}x + C$

$= -3 \ln |x| - 10e^{-0.4x} + e^{0.1}x + C$

33. $\int \left(\frac{1 + 2t^3}{4t}\right) dt = \int \left(\frac{1}{4t} + \frac{t^2}{2}\right) dt$

$= \frac{1}{4} \int \frac{1}{t}\, dt + \frac{1}{2} \int t^2\, dt$

$= \frac{1}{4} \ln |t| + \frac{1}{2}\left(\frac{t^3}{3}\right) + C$

$= \frac{1}{4} \ln |t| + \frac{t^3}{6} + C$

35. $\int (e^{2u} + 4u)\, du = \frac{e^{2u}}{2} + \frac{4u^2}{2} + C$

$= \frac{e^{2u}}{2} + 2u^2 + C$

37. $\int (x+1)^2\, dx = \int (x^2 + 2x + 1)\, dx$

$= \frac{x^3}{3} + \frac{2x^2}{2} + x + C$

$= \frac{x^3}{3} + x^2 + x + C$

39. $\int \frac{\sqrt{x} + 1}{\sqrt[3]{x}}\, dx = \int \left(\frac{\sqrt{x}}{\sqrt[3]{x}} + \frac{1}{\sqrt[3]{x}}\right) dx$

$= \int (x^{(1/2 - 1/3)} + x^{-1/3})\, dx$

$= \int x^{1/6}\, dx + \int x^{-1/3}\, dx$

$= \frac{x^{7/6}}{\frac{7}{6}} + \frac{x^{2/3}}{\frac{2}{3}} + C$

$= \frac{6x^{7/6}}{7} + \frac{3x^{2/3}}{2} + C$

41. $\int 10^x dx = \frac{10^x}{\ln 10} + C$

43. Find $f(x)$ such that $f'(x) = x^{2/3}$, and $\left(1, \frac{3}{5}\right)$ is on the curve.

$\int x^{2/3} dx = \frac{x^{5/3}}{\frac{5}{3}} + C$

$f(x) = \frac{3x^{5/3}}{5} + C$

Since $\left(1, \frac{3}{5}\right)$ is on the curve,

$f(1) = \frac{3}{5}.$

$f(1) = \frac{3(1)^{5/3}}{5} + C = \frac{3}{5}$

$\frac{3}{5} + C = \frac{3}{5}$

$C = 0.$

Thus,

$$f(x) = \frac{3x^{5/3}}{5}.$$

45. $C'(x) = 4x - 5$; fixed cost is \$8.

$$C(x) = \int (4x - 5)\, dx$$
$$= \frac{4x^2}{2} - 5x + k$$
$$= 2x^2 - 5x + k$$
$$C(0) = 2(0)^2 - 5(0) + k = k$$

Since $C(0) = 8$, $k = 8$.

Thus,

$$C(x) = 2x^2 - 5x + 8.$$

47. $C'(x) = 0.03e^{0.01x}$; fixed cost is \$8.

$$C(x) = \int 0.03e^{0.01x}\, dx$$
$$= 0.03 \int e^{0.01x}\, dx$$
$$= 0.03\left(\frac{1}{0.01}e^{0.01x}\right) + k$$
$$= 3e^{0.01x} + k$$
$$C(0) = 3e^{0.01(0)} + k = 3(1) + k$$
$$= 3 + k$$

Since $C(0) = 8$, $3 + k = 8$, and $k = 5$.
Thus,

$$C(x) = 3e^{0.01x} + 5.$$

49. $C'(x) = x^{2/3} + 2$; 8 units cost \$58.

$$C(x) = \int (x^{2/3} + 2)\, dx$$
$$= \frac{3x^{5/3}}{5} + 2x + k$$
$$C(8) = \frac{3(8)^{5/3}}{5} + 2(8) + k$$
$$= \frac{3(32)}{5} + 16 + k$$

Since $C(8) = 58$,

$$58 - 16 - \frac{96}{5} = k$$
$$\frac{114}{5} = k.$$

Thus,

$$C(x) = \frac{3x^{5/3}}{5} + 2x + \frac{114}{5}.$$

51. $C'(x) = 5x - \frac{1}{x}$; 10 units cost \$94.20, so

$$C(10) = 94.20.$$

$$C(x) = \int \left(5x - \frac{1}{x}\right) dx = \frac{5x^2}{2} - \ln|x| + k$$
$$C(10) = \frac{5(10)^2}{2} - \ln{(10)} + k$$
$$= 250 - 2.30 + k.$$

Since $C(10) = 94.20$,

$$94.20 = 247.70 + k$$
$$-153.50 = k.$$

Thus, $C(x) = \frac{5x^2}{2} - \ln|x| - 153.50$.

53. $R'(x) = 175 - 0.02x - 0.03x^2$

$$R = \int (175 - 0.02x - 0.03x^2)dx$$
$$= 175x - 0.01x^2 - 0.01x^3 + C.$$

If $x = 0$, then $R = 0$ (no items sold means no revenue), and

$$0 = 175(0) - 0.01(0)^2 - 0.01(0)^3 + C$$
$$0 = C.$$

Thus, $R = 175x - 0.01x^2 - 0.01x^3$

gives the revenue function. Now, recall that $R = xp$, where p is the demand function. Then

$$175x - 0.01x^2 - 0.01x^3 = xp$$
$$175x - 0.01x - 0.01x^2 = p, \text{ the demand function.}$$

55. $R'(x) = 500 - 0.15\sqrt{x}$

$$R = \int (500 - 0.15\sqrt{x})dx$$
$$= 500x - 0.1x^{3/2} + C.$$

If $x = 0$, $R = 0$ (no items sold means no revenue), and

$$0 = 500(0) - 0.1(0)^{3/2} + C$$
$$0 = C.$$

Thus, $R = 500x - 0.1x^{3/2}$ gives the revenue function. Now, recall that $R = xp$, where p is the demand function. Then

$$500x - 0.1x^{3/2} = xp$$
$$500 - 0.1\sqrt{x} = p, \text{ the demand function.}$$

57. $f'(t) = 1.498t + 1.626$

(a) $$\begin{aligned} f(t) &= \int (1.498t + 1.626)dt \\ &= 0.749t^2 + 1.626t + C \end{aligned}$$

In 1992 $(t = 2)$, $f(t) = 8.893$, and

$$\begin{aligned} 8.893 &= 0.749(2)^2 + 1.626(2) + C \\ 8.893 &= 6.248 + C \\ 2.645 &= C \end{aligned}$$

Thus, $f(t) = 0.749t^2 + 1.626t + 2.645$.

(b) In 2006, $t = 16$, and

$$\begin{aligned} f(16) &= 0.749(16)^2 + 1.626(16) + 2.645 \\ &= 191.744 + 26.016 + 2.645 \\ &= 220.405 \end{aligned}$$

The function predicted approximately 220 million subscribers in 2006.

59. (a) $P'(x) = 50x^3 + 30x^2$; profit is -40 when no cheese is sold.

$$\begin{aligned} P(x) &= \int (50x^3 + 30x^2)\, dx \\ &= \frac{25x^4}{2} + 10x^3 + k \end{aligned}$$

$$P(0) = \frac{25(0)^4}{2} + 10(0)^3 + k$$

Since

$$P(0) = -40,$$
$$-40 = k.$$

Thus,

$$P(x) = \frac{25x^4}{2} + 10x^3 - 40.$$

(b) $P(2) = \dfrac{25(2)^4}{2} + 10(2)^3 - 40 = 240$

The profit from selling 200 lbs of Brie cheese is \$240.

61. $$\begin{aligned} \int \frac{g(x)}{x} dx &= \int \frac{a - bx}{x} dx \\ &= \int \left(\frac{a}{x} - b\right) dx \\ &= a \int \frac{dx}{x} - b \int dx \\ &= a \ln |x| - bx + C \end{aligned}$$

Since x represents a positive quantity, the absolute value sign can be dropped.

$$\int \frac{g(x)}{x} dx = a \ln x - bx + C$$

63. $N'(t) = Ae^{kt}$

(a) $N(t) = \dfrac{A}{k}e^{kt} + C$

$A = 50$, $N(t) = 300$ when $t = 0$.

$$N(0) = \frac{50}{k}e^0 + C = 300$$
$$N'(5) = 250$$

Therefore,

$$\begin{aligned} N'(5) = 50e^{5k} &= 250 \\ e^{5k} &= 5 \\ 5k &= \ln 5 \\ k &= \frac{\ln 5}{5}. \end{aligned}$$

$$\begin{aligned} N(0) = \frac{50}{\frac{\ln 5}{5}} &+ C = 300 \\ \frac{250}{\ln 5} &+ C = 300 \\ C &= 300 - \frac{250}{\ln 5} \approx 144.67 \end{aligned}$$

$$\begin{aligned} N(t) &= \frac{50}{\frac{\ln 5}{5}} e^{(\ln 5/5)t} + 144.67 \\ &= 155.3337e^{0.321888t} + 144.67 \end{aligned}$$

(b) $$\begin{aligned} N(12) &= 155.3337e^{0.321888(12)} + 144.67 \\ &\approx 7537 \end{aligned}$$

There are 7537 cells present after 12 days.

65. $a(t) = 5t^2 + 4$

$$\begin{aligned} v(t) &= \int (5t^2 + 4)\, dt \\ &= \frac{5t^3}{3} + 4t + C \end{aligned}$$

$$v(0) = \frac{5(0)^3}{3} + 4(0) + C$$

Since $v(0) = 6$, $C = 6$.

$$v(t) = \frac{5t^3}{3} + 4t + 6$$

67. $a(t) = -32$

$$v(t) = \int -32\,dt = -32t + C_1$$

$v(0) = -32(0) + C_1$

Since $v(0) = 0$, $C_1 = 0$.

$$v(t) = -32t$$

$$s(t) = \int -32t\,dt = \frac{-32t^2}{2} + C_2 = -16t^2 + C_2$$

At $t = 0$, the plane is at 6400 ft.
That is, $s(0) = 6400$.

$$s(0) = -16(0)^2 + C_2$$
$$6400 = 0 + C_2$$
$$C_2 = 6400$$
$$s(t) = -16t^2 + 6400$$

When the object hits the ground, $s(t) = 0$.

$$-16t^2 + 6400 = 0$$
$$-16t^2 = -6400$$
$$t^2 = 400$$
$$t = \pm 20$$

Discard -20 since time must be positive.
The object hits the ground in 20 sec.

69. $a(t) = \frac{15}{2}\sqrt{t} = 3e^{-t}$

$$v(t) = \int \left(\frac{15}{2}\sqrt{t} + 3e^{-t}\right) dt = \int \left(\frac{15}{2}t^{1/2} + 3e^{-t}\right) dt = \frac{15}{2}\left(\frac{t^{3/2}}{\frac{3}{2}}\right) + 3\left(\frac{1}{-1}e^{-t}\right) + C_1 = 5t^{3/2} - 3e^{-t} + C_1$$

$v(0) = 5(0)^{3/2} - 3e^{-0} + C_1 = -3 + C_1$

Since $v(0) = -3$, $C_1 = 0$.

$$v(t) = 5t^{3/2} - 3e^{-t}$$

$$s(t) = \int (5t^{3/2} - 3e^{-t})\,dt = 5\left(\frac{t^{5/2}}{\frac{5}{2}}\right) - 3\left(-\frac{1}{1}e^{-t}\right) + C_2 = 2t^{5/2} + 3e^{-t} + C_2$$

$s(0) = 2(0)^{5/2} + 3e^{-0} + C_2 = 3 + C_2$

Since $s(0) = 4$, $C_2 = 1$.
Thus,

$$s(t) = 2t^{5/2} + 3e^{-t} + 1.$$

71. First find $v(t)$ by integrating $a(t)$:

$$v(t) = \int (-32)dt = -32t + k.$$

When $t = 5, v(t) = 0$:

$$0 = -32(5) + k$$
$$160 = k$$

and

$$v(t) = -32t + 160.$$

Now integrate $v(t)$ to find $h(t)$.

$$h(t) = \int (-32t + 160)dt = -16t^2 + 160t + C$$

Since $h(t) = 412$ when $t = 5$, we can substitute these values into the equation for $h(t)$ to get $C = 12$ and

$$h(t) = -16t^2 + 160t + 12.$$

Therefore, from the equation given in Exercise 70, the initial velocity v_0 is 160 ft/sec and the initial height of the rocket h_0 is 12 ft.

73. **(a)** First find $B(t)$ by integrating $B'(t)$:

$$B(t) = \int 9.2935e^{0.02955t}dt \approx 314.5e^{0.02955t} + k$$

When $t = 0, B(t) = 792.3$:

$$792.3 = 314.5e^{0.02955(0)} + k$$
$$477.8 = k$$

and

$$B(t) = 314.5e^{0.02955t} + 477.8.$$

(b) In 2012, $t = 42$.

$$B(42) = 314.5e^{0.02955(42)} + 477.8 \approx 1565.8$$

About 1,566,000 bachelor's degrees will be conferred in 2012.

7.2 Substitution

3. $\int 4(2x+3)^4\,dx = 2\int 2(2x+3)^4\,dx$

Let $u = 2x+3$, so that $du = 2\,dx$.

$$= 2\int u^4\,du$$
$$= \frac{2\cdot u^5}{5} + C$$
$$= \frac{2(2x+3)^5}{5} + C$$

5. $\int \frac{2\,dm}{(2m+1)^3} = \int 2(2m+1)^{-3}\,dm$

Let $u = 2m+1$, so that $du = 2\,dm$.

$$= \int u^{-3}\,du$$
$$= \frac{u^{-2}}{-2} + C$$
$$= \frac{-(2m+1)^{-2}}{2} + C$$

7. $\int \frac{2x+2}{(x^2+2x-4)^4}\,dx$

$$= \int (2x+2)(x^2+2x-4)^{-4}dx$$

Let $w = x^2+2x-4$, so that $dw = (2x+2)dx$.

$$= \int w^{-4}dw$$
$$= \frac{w^{-3}}{-3} + C$$
$$= -\frac{(x^2+2x-4)^{-3}}{3} + C$$
$$= -\frac{1}{3(x^2+2x-4)^3} + C$$

9. $\int z\sqrt{4z^2-5}\,dz = \int z(4z^2-5)^{1/2}\,dz$

$$= \frac{1}{8}\int 8z(4z^2-5)^{1/2}\,dz$$

Let $u = 4z^2-5$, so that $du = 8z\,dz$.

$$= \frac{1}{8}\int u^{1/2}\,du$$
$$= \frac{1}{8}\cdot\frac{u^{3/2}}{\frac{3}{2}} + C$$
$$= \frac{1}{8}\cdot\left(\frac{2}{3}\right)u^{3/2} + C$$
$$= \frac{(4z^2-5)^{3/2}}{12} + C$$

11. $\int 3x^2 e^{2x^3}\,dx = \frac{1}{2}\int 2\cdot 3x^2 e^{2x^3}\,dx$

Let $u = 2x^3$, so that $du = 6x^2\,dx$.

$$= \frac{1}{2}\int e^u\,du$$
$$= \frac{1}{2}e^u + C$$
$$= \frac{e^{2x^3}}{2} + C$$

13. $\int (1-t)e^{2t-t^2}\,dt$

$$= \frac{1}{2}\int 2(1-t)e^{2t-t^2}\,dt$$

Let $u = 2t-t^2$, so that $du = (2-2t)\,dt$.

$$= \frac{1}{2}\int e^u\,du$$
$$= \frac{e^u}{2} + C$$
$$= \frac{e^{2t-t^2}}{2} + C$$

15. $\int \frac{e^{1/z}}{z^2}\,dz = -\int e^{1/z}\cdot\frac{-1}{z^2}\,dz$

Let $u = \frac{1}{z}$, so that $du = \frac{-1}{z^2}\,dx$.

$$= -\int e^u\,du$$
$$= -e^u + C$$
$$= -e^{1/z} + C$$

17. $\int (x^3+2x)(x^4+4x^2+7)^8dx$

$$= \frac{1}{4}\int (x^4+4x^2+7)^8(4x^3+8x)dx$$

Let $u = x^4+4x^2+7$, so that $du = (4x^3+8x)dx$

$$= \frac{1}{4}\int u^8du = \frac{1}{4}\int\left(\frac{u^9}{9}\right) + C$$
$$= \frac{u^9}{36} + C = \frac{(x^4+4x^2+7)^9}{36} + C$$

19. $\int \frac{2x+1}{(x^2+x)^3}\,dx$

$$= \int (2x+1)(x^2+x)^{-3}\,dx$$

Let $u = x^2+x$, so that $du = (2x+1)\,dx$.

$$= \int u^{-3}\,du = \frac{u^{-2}}{-2} + C$$
$$= \frac{-1}{2u^2} + C = \frac{-1}{2(x^2+x)^2} + C$$

21. $\int p(p+1)^5\,dp$

Let $u = p+1$, so that $du = dp$; also, $p = u-1$.

$$= \int (u-1)u^5\,du$$
$$= \int (u^6 - u^5)\,du$$
$$= \frac{u^7}{7} - \frac{u^6}{6} + C$$
$$= \frac{(p+1)^7}{7} - \frac{(p+1)^6}{6} + C$$

23. $\int \frac{u}{\sqrt{u-1}}\,du$

$$= \int u(u-1)^{-1/2}\,du$$

Let $w = u-1$, so that $dw = du$ and $u = w+1$.

$$= \int (w+1)w^{-1/2}\,dw$$
$$= \int (w^{1/2} + w^{-1/2})\,dw$$
$$= \frac{w^{3/2}}{\frac{3}{2}} + \frac{w^{1/2}}{\frac{1}{2}} + C$$
$$= \frac{2(u-1)^{3/2}}{3} + 2(u-1)^{1/2} + C$$

25. $\int (\sqrt{x^2+12x})(x+6)\,dx$

$$= \int (x^2+12x)^{1/2}(x+6)\,dx$$

Let $x^2 + 12x = u$, so that

$$(2x+12)\,dx = du$$
$$2(x+6)\,dx = du.$$

$$= \frac{1}{2}\int u^{1/2}\,du = \frac{1}{2}\left(\frac{2}{3}\right)u^{3/2} + C$$
$$= \frac{(x^2+12x)^{3/2}}{3} + C$$

27. $\int \frac{t}{t^2+2}\,dt$

Let $t^2 + 2 = u$, so that $2t\,dt = du$.

$$= \frac{1}{2}\int \frac{du}{u}$$
$$= \frac{1}{2}\ln|u| + C$$
$$= \frac{\ln(t^2+2)}{2} + C$$

29. $\int \frac{(1+3\ln x)^2}{x}\,dx$

Let $u = 1 + 3\ln x$, so that $du = \frac{3}{x}\,dx$.

$$= \frac{1}{3}\int \frac{3(1+3\ln x)^2}{x}\,dx$$
$$= \frac{1}{3}\int u^2\,du$$
$$= \frac{1}{3}\cdot\frac{u^3}{3} + C$$
$$= \frac{(1+3\ln x)^3}{9} + C$$

31. $\int \frac{e^{2x}}{e^{2x}+5}\,dx$

Let $u = e^{2x} + 5$, so that $du = 2e^{2x}\,dx$.

$$= \frac{1}{2}\int \frac{du}{u}$$
$$= \frac{1}{2}\ln|u| + C$$
$$= \frac{1}{2}\ln\left|e^{2x}+5\right| + C$$
$$= \frac{1}{2}\ln(e^{2x}+5) + C$$

33. $\int \frac{\log x}{x}\,dx$

Let $u = \log x$, so that $du = \frac{1}{(\ln 10)x}\,dx$.

$$= (\ln 10)\int \frac{\log x}{(\ln 10)x}\,dx$$
$$= (\ln 10)\int u\,du$$
$$= (\ln 10)\left(\frac{u^2}{2}\right) + C$$
$$= \frac{(\ln 10)(\log x)^2}{2} + C$$

35. $\int x8^{3x^2+1}dx$

Let $u = 3x^2 + 1$, so that $du = 6x\,dx$.

$$= \frac{1}{6}\int 6x\cdot 8^{3x^2+1}dx$$
$$= \frac{1}{6}\int 8^u\,du$$
$$= \frac{1}{6}\left(\frac{8^u}{\ln 8}\right) + C$$
$$= \frac{8^{3x^2+1}}{6\ln 8} + C$$

39. **(a)** $R'(x) = 4x(x^2 + 27{,}000)^{-2/3}$

$$R(x) = \int 4x(x^2 + 27{,}000)^{-2/3}\,dx = 2\int 2x(x^2 + 27{,}000)^{-2/3}\,dx$$

Let $u = x^2 + 27{,}000$, so that $du = 2x\,dx$.

$$R = 2\int u^{-2/3}\,du = 2 \cdot 3u^{1/3} + C = 6(x^2 + 27{,}000)^{1/3} + C$$
$$R(125) = 6(125^2 + 27{,}000)^{1/3} + C$$

Since $R(125) = 29.591$,

$$6(125^2 + 27{,}000)^{1/3} + C = 29.591$$
$$C = -180$$

Thus,

$$R(x) = 6(x^2 + 27{,}000)^{1/3} - 180.$$

(b)

$$R(x) = 6(x^2 + 27,000)^{1/3} - 180 \geq 40$$
$$6(x^2 + 27{,}000)^{1/3} \geq 220$$
$$(x^2 + 27{,}000)^{1/3} \geq 36.6667$$
$$x^2 + 27{,}000 \geq 49{,}296.43$$
$$x^2 \geq 22{,}296.43$$
$$x \geq 149.4$$

For a revenue of at least \$40,000, 150 players must be sold.

41. $C'(x) = \dfrac{60x}{5x^2 + e}$

(a) Let $u = 5x^2 + e$, so that $du = 10x\,dx$.

$$C(x) = \int C'(x)\,dx = \int \frac{60x}{5x^2 + e}\,dx = 6\int \frac{du}{u} = 6\ln|u| + C = 6\ln|5x^2 + e| + C$$

Since $C(0) = 10, C = 4$.

Therefore,

$$C(x) = 6\ln|5x^2 + e| + 4 = 6\ln(5x^2 + e) + 4.$$

(b) $C(5) = 6\ln(5 \cdot 5^2 + e) + 4 \approx 33.099$

Since this represents \$33,099 dollars which is greater than \$20,000, a new source of investment income should be sought.

43. $f'(t) = 4.0674 \cdot 10^{-4} t(t - 1970)^{0.4}$

(a) Let $u = t - 1970$. To get the t outside the parentheses in terms of u, solve $u = t - 1970$ for t to get $t = u + 1970$. Then $dt = du$ and we can substitute as follows.

$$\begin{aligned} f(t) = \int f'(t)dt &= \int 4.0674 \cdot 10^{-4} t(t-1970)^{0.4} dt \\ &= \int 4.0674 \cdot 10^{-4}(u+1970)(u)^{0.4} du \\ &= 4.0674 \cdot 10^{-4} \int (u+1970)(u)^{0.4} du \\ &= 4.0674 \cdot 10^{-4} \int (u^{1.4} + 1970u^{0.4}) du \\ &= 4.0674 \cdot 10^{-4} \left(\frac{u^{2.4}}{2.4} + \frac{1970u^{1.4}}{1.4} \right) + C \\ &= 4.0674 \cdot 10^{-4} \left[\frac{(t-1970)^{2.4}}{2.4} + \frac{1970(t-1970)^{1.4}}{1.4} \right] + C \end{aligned}$$

Since $f(1970) = 61.298, C = 61.298$.

Therefore,

$$f(t) = 4.0674 \cdot 10^{-4} \left[\frac{(t-1970)^{2.4}}{2.4} + \frac{1970(t-1970)^{1.4}}{1.4} \right] + 61.298.$$

(b) $f(2015) = 4.0674 \cdot 10^{-4} \left[\frac{(2015-1970)^{2.4}}{2.4} + \frac{1970(2015-1970)^{1.4}}{1.4} \right] + 61.298 \approx 180.9.$

In the year 2015, there will be about 181,000 local transit vehicles.

7.3 Area and the Definite Integral

3. $f(x) = 2x + 5$, $x_1 = 0$, $x_2 = 2$, $x_3 = 4$, $x_4 = 6$, and $\Delta x = 2$

(a) $\sum_{i=1}^{4} f(x_i)\Delta x$

$$\begin{aligned} &= f(x_1)\Delta x + f(x_2)\Delta x + f(x_3)\Delta x + f(x_4)\Delta x \\ &= f(0)(2) + f(2)(2) + f(4)(2) + f(6)(2) \\ &= [2(0)+5](2) + [2(2)+5](2) + [2(4)+5](2) + [2(6)+5](2) \\ &= 10 + 9(2) + 13(2) + 17(2) \\ &= 88 \end{aligned}$$

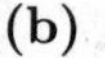

(b)

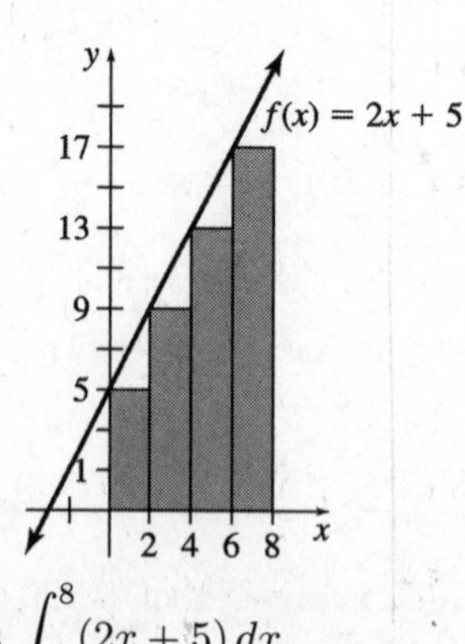

The sum of these rectangles approximates $\int_0^8 (2x + 5)\, dx$.

7. $f(x) = 2x + 5$ from $x = 2$ to $x = 4$

For $n = 4$ rectangles:

$$\Delta x = \frac{4-2}{4} = 0.5$$

(a) Using the left endpoints:

i	x_i	$f(x_i)$
1	2	9
2	2.5	10
3	3	11
4	3.5	12

$$A = \sum_{1}^{4} f(x_i)\Delta x = 9\,(0.5) + 10\,(0.5) + 11\,(0.5) + 12\,(0.5) = 21$$

(b) Using the right endpoints:

i	x_i	$f(x_i)$
1	2.5	10
2	3	11
3	3.5	12
4	4	13

$$A = 10(0.5) + 11(0.5) + 12(0.5) + 13(0.5) = 23$$

(c) Average $= \dfrac{21+23}{2} = \dfrac{44}{2} = 22$

(d) Using the midpoints:

i	x_i	$f(x_i)$
1	2.25	9.5
2	2.75	10.5
3	3.25	11.5
4	3.75	12.5

LOL wut?

$$A = \sum_{1}^{4} f(x_i)\Delta x = 9.5\,(0.5) + 10.5\,(0.5) + 11.5\,(0.5) + 12.5\,(0.5) = 22$$

9. $f(x) = -x^2 + 4$ from $x = -2$ to $x = 2$

For $n = 4$ rectangles:

$$\Delta x = \frac{2-(-2)}{4} = 1$$

(a) Using the left endpoints:

i	x_i	$f(x_i)$
1	-2	$-(-2)^2 + 4 = 0$
2	-1	$-(-1)^2 + 4 = 3$
3	0	$-(0)^2 + 4 = 4$
4	1	$-(1)^2 + 4 = 3$

$$A = \sum_{i=1}^{4} f(x_i)\Delta x = (0)(1) + (3)(1) + (4)(1) + (3)(1) = 10$$

(b) Using the right endpoints:

i	x_i	$f(x_i)$
1	-1	3
2	0	4
3	1	3
4	2	0

$$\text{Area} = 1(3) + 1(4) + 1(3) + 1(0) = 10$$

(c) Average $= \dfrac{10+10}{2} = 10$

(d) Using the midpoints:

i	x_i	$f(x_i)$
1	$-\frac{3}{2}$	$\frac{7}{4}$
2	$-\frac{1}{2}$	$\frac{15}{4}$
3	$\frac{1}{2}$	$\frac{15}{4}$
4	$\frac{3}{2}$	$\frac{7}{4}$

$$A = \sum_{i=1}^{4} f(x_i)\Delta x = \frac{7}{4}(1) + \frac{15}{4}(1) + \frac{15}{4}(1) + \frac{7}{4}(1) = 11$$

11. $f(x) = e^x + 1$ from $x = -2$ to $x = 2$

For $n = 4$ rectangles:

$$\Delta x = \frac{2-(-2)}{4} = 1$$

(a) Using the left endpoints:

i	x_i	$f(x_i)$
1	-2	$e^{-2}+1$
2	-1	$e^{-1}+1$
3	0	$e^0+1=2$
4	1	e^1+1

$$A = \sum_{i=1}^{4} f(x_i)\Delta x = \sum_{i=1}^{4} f(x_i)(1) = \sum_{i=1}^{4} f(x_i) = (e^{-2}+1) + (e^{-1}+1) + 2 + e^1 + 1 \approx 8.2215 \approx 8.22$$

(b) Using the right endpoints:

i	x_i	$f(x_i)$
1	-1	$e^{-1}+1$
2	0	2
3	1	$e+1$
4	2	e^2+1

$$\text{Area} = 1(e^{-1}+1) + 1(2) + 1(e+1) + 1(e^2+1) \approx 15.4752 \approx 15.48$$

(c) Average $= \dfrac{8.2215 + 15.4752}{2} = 11.84835 \approx 11.85$

(d) Using the midpoints:

i	x_i	$f(x_i)$
1	$-\frac{3}{2}$	$e^{-3/2}+1$
2	$-\frac{1}{2}$	$e^{-1/2}+1$
3	$\frac{1}{2}$	$e^{1/2}+1$
4	$\frac{3}{2}$	$e^{3/2}+1$

$$A=\sum_{i=1}^{4} f(x_i)\Delta x=(e^{-3/2}+1)(1)+(e^{-1/2}+1)(1)+(e^{1/2}+1)(1)+(e^{3/2}+1)(1)\approx 10.9601\approx 10.96$$

13. $f(x)=\frac{2}{x}$ from $x=1$ to $x=9$

For $n=4$ rectangles:

$$\Delta x=\frac{9-1}{4}=2$$

(a) Using the left endpoints:

i	x_i	$f(x_i)$
1	1	$\frac{2}{1}=2$
2	3	$\frac{2}{3}$
3	5	$\frac{2}{5}=0.4$
4	7	$\frac{2}{7}$

$$A=\sum_{i=1}^{4} f(x_i)\Delta x=(2)(2)+\frac{2}{3}(2)+(0.4)(2)+\left(\frac{2}{7}\right)(2)\approx 6.7048\approx 6.70$$

(b) Using the right endpoints:

i	x_i	$f(x_i)$
1	3	$\frac{2}{3}$
2	5	$\frac{2}{5}$
3	7	$\frac{2}{7}$
4	9	$\frac{2}{9}$

$$\text{Area}=2\left(\frac{2}{3}\right)+2\left(\frac{2}{5}\right)+2\left(\frac{2}{7}\right)+2\left(\frac{2}{9}\right)=\frac{4}{3}+\frac{4}{5}+\frac{4}{7}+\frac{4}{9}\approx 3.1492\approx 3.15$$

(c) $\text{Average}=\dfrac{6.7+3.15}{2}=4.93$

(d) Using the midpoints:

i	x_i	$f(x_i)$
1	2	1
2	4	$\frac{1}{2}$
3	6	$\frac{1}{3}$
4	8	$\frac{1}{4}$

$$A = \sum_{i=1}^{4} f(x_i)\Delta x = 1(2) + \frac{1}{2}(2) + \frac{1}{3}(2) + \frac{1}{4}(2) \approx 4.1667 \approx 4.17$$

15. $\int_0^5 (5 - x)\, dx$

Graph $y = 5 - x$.

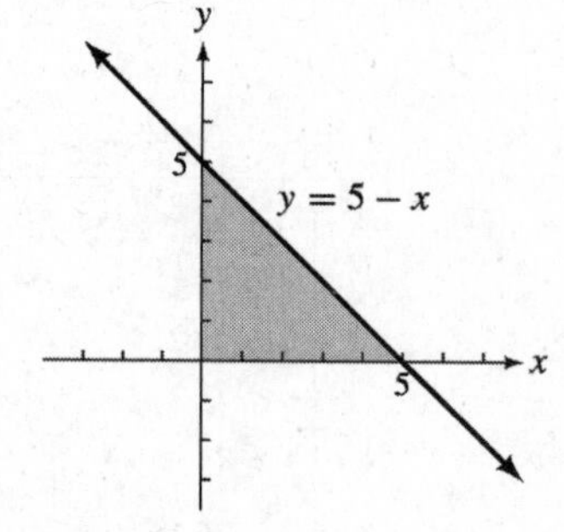

$\int_0^5 (5 - x)\, dx$ is the area of a triangle with base $= 5 - 0 = 5$ and altitude $= 5$.

$$\text{Area} = \frac{1}{2}(\text{altitude})(\text{base}) = \frac{1}{2}(5)(5) = 12.5$$

17. (a) $\int_0^2 f(x)dx$ is the area of a rectangle with width $x = 2$ and length $y = 4$. The rectangle has area $2 \cdot 4 = 8$.

$\int_2^6 f(x)dx$ is the area of one-fourth of a circle that has radius 4. The area is $\frac{1}{4}\pi r^2 = \frac{1}{4}\pi(4)^2 = 4\pi$.

Therefore, $\int_2^6 f(x)dx = 8 + 4\pi$.

(b) $\int_0^2 f(x)dx$ is the area of one-fourth of a circle that has radius 2. The area is $\frac{1}{4}\pi r^2 = \frac{1}{4}\pi(2)^2 = \pi$.

$\int_2^6 f(x)dx$ is the area of a triangle with base 4 and height 2. The triangle has area $\frac{1}{2} \cdot 4 \cdot 2 = 4$.

Therefore, $\int_0^6 f(x)dx = 4 + \pi$.

19. $\int_{-4}^{0} \sqrt{16 - x^2}\, dx$

Graph $y = \sqrt{16 - x^2}$.

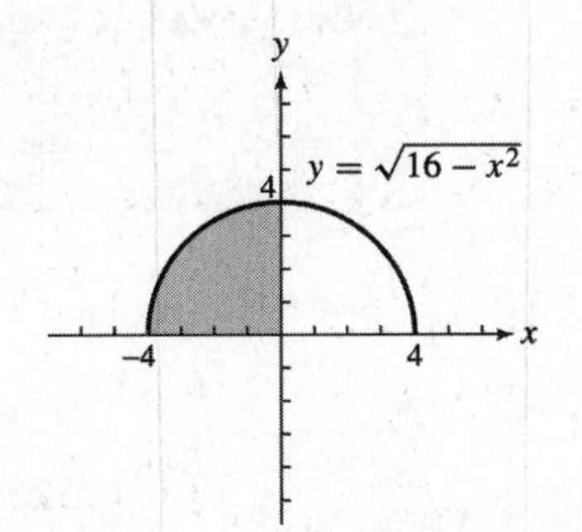

$\int_{-4}^{0} \sqrt{16 - x^2}\, dx$ is the area of the portion of the circle in the second quadrant, which is one-fourth of a circle. The circle has radius 4.

$$\text{Area} = \frac{1}{4}\pi r^2 = \frac{1}{4}\pi(4)^2 = 4\pi$$

21. $\int_{2}^{5} (1 + 2x)\, dx$

Graph $y = 1 + 2x$.

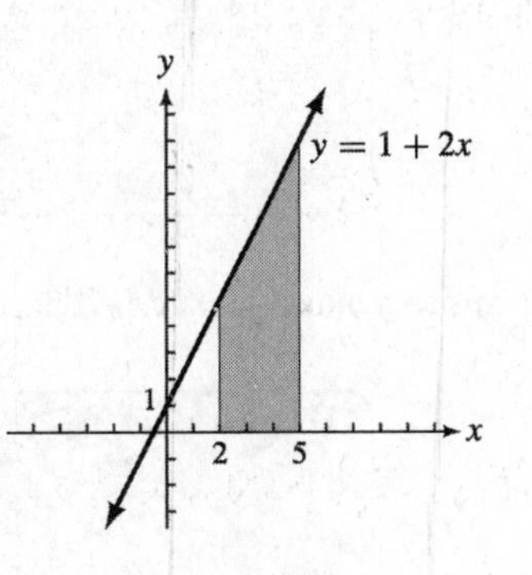

$\int_{2}^{5} (1 + 2x)\, dx$ is the area of the trapezoid with $B = 11$, $b = 5$, and $h = 3$. The formula for the area is

$$A = \frac{1}{2}(B + b)h,$$

so we have

$$A = \frac{1}{2}(11 + 5)(3) = 24.$$

23. **(a)** With $n = 10$, $\Delta x = \frac{1-0}{10} = 0.1$, and $x_1 = 0 + 0.1 = 0.1$, use the command seq(X^3,X, 0.1, 0.1, 0.1) →L1. The resulting screen is:

```
seq(X^3,X,.1,1,.
1)→L1
{.001 .008 .027...
```

(b) Since $\sum_{i=1}^{n} f(x_i)\Delta x = \Delta x\left(\sum_{i=1}^{n} f(x_i)\right)$, use the command 0.1*sum(L1) to approximate $\int_0^1 x^3 dx$. The resulting screen is:

```
.1*sum(L1)
                .3025
```

$$\int_0^1 x^3 dx \approx 0.3025$$

(c) With $n = 100$, $\Delta x = \frac{1-0}{100} = 0.01$, and $x_1 = 0 + 0.01 = 0.01$, use the command seq(X^3,X, 0.01, 0.1, 0.01) →L1. The resulting screen is:

```
seq(X^3,X,.01,1,
.01)→L1
{1E-6 8E-6 2.7E...
```

Use the command 0.01*sum(L1) to approximate $\int_0^1 x^3 dx$. The resulting screen is:

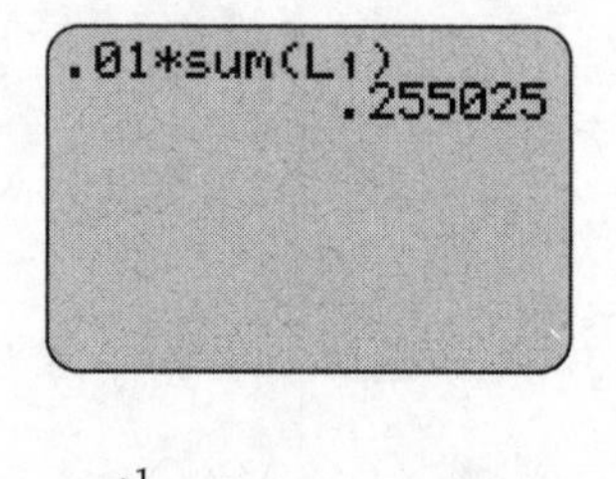

$$\int_0^1 x^3 dx \approx 0.255025$$

(d) With $n = 500$, $\Delta x = \frac{1-0}{500} = 0.002$, and $x_1 = 0+0.002 = 0.002$, use the command seq (X^3,X, 0.002, 1, 0.002) →L1. The resulting screen is:

```
seq(X^3,X,.002,1
,.002)→L1
{8E-9 6.4E-8 2....
```

Use the command 0.002*sum(L1) to approximate $\int_0^1 x^3 dx$. The resulting screen is:

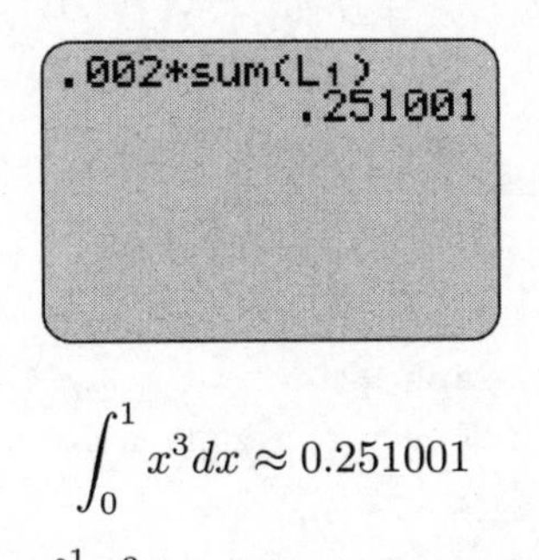

$$\int_0^1 x^3 dx \approx 0.251001$$

(e) As n gets larger the approximation for $\int_0^1 x^3 dx$ seems to be approaching 0.25 or $\frac{1}{4}$. We estimate $\int_0^1 x^3 dx = \frac{1}{4}$.

For Exercises 25–33, readings on the graphs and answers may vary.

25. Left endpoints:

Read values of the function on the graph every 5 years from 1980 to 2000. These values give us the heights of 5 rectangles. The width of each rectangle is $\Delta x = 5$. We estimate the area under the curve as

$$A = \sum_{i=1}^{5} f(x_i)\Delta x = 702.7(5) + 818(5) + 902.9(5) + 962.1(5) + 1084.1(5) = 22{,}349$$

Right endpoints:

Read values of the function from the graph every 5 years from 1985 to 2005. These values give the heights of 5 rectangles. The width of each rectangle is $\Delta x = 5$. We estimate the area under the curve as

$$A = \sum_{i=1}^{5} f(x_i)\Delta x = 818(5) + 902.9(5) + 962.1(5) + 1084.1(5) + 1128.3(5) = 24{,}477$$

Average:

$$\frac{22{,}349 + 24{,}477}{2} = \frac{46{,}826}{2} = 23{,}413$$

The area under the curve represents the total U.S. coal consumption. We estimate this consumption as about 23,413 million short tons.

27. (a) Left endpoints:

Read values of the function from the graph for every 14 days from 18 Feb. through 30 Apr. The values give the heights of 6 rectangles. The width of each rectangle is $\Delta x = 14$. We estimate the area under the curve as

$$A = \sum_{i=1}^{6} f(x_i)\Delta x = 0(14) + 15(14) + 33(14) + 40(14) + 16(14) + 5(14) = 1526.$$

Right endpoints:

Read values of the function from the graph for every 14 days from 4 Mar. through 13 May. Now we estimate the area under the curve as

$$A = \sum_{i=1}^{6} f(x_i)\Delta x = 15(14) + 33(14) + 40(14) + 16(14) + 5(14) + 1(14) = 1540.$$

Average:

$$\frac{1526 + 1540}{2} = 1533$$

There were about 1533 cases of the disease.

(b) Left endpoints:

Read values of the function from the graph for every 14 days from 18 Feb. through 30 Apr. The values give the heights of 6 rectangles. The width of each rectangle is $\Delta x = 14$. We estimate the area under the curve as

$$A = \sum_{i=1}^{6} f(x_i)\Delta x = 0(14) + 10(14) + 15(14) + 10(14) + 3(14) + 1(14) = 546.$$

Right endpoints:

Read values of the function from the graph for every 14 days from 4 Mar. through 13 May. Now we estimate the area under the curve as

$$A = \sum_{i=1}^{6} f(x_i)\Delta x = 10(14) + 15(14) + 10(14) + 3(14) + 1(14) + 1(14) = 560.$$

Average:

$$\frac{546 + 560}{2} = 553$$

There would have been about 553 cases of the disease.

29. Read the value of the function for every 5 sec from $x = 2.5$ to $x = 12.5$. These are the midpoints of rectangles with width $\Delta x = 5$. Then read the function for $x = 17$, which is the midpoint of a rectangle with width $\Delta x = 4$.

$$\begin{aligned} \sum_{i=1}^{4} f(x_i)\Delta x &\approx 36(5) + 63(5) + 84(5) + 95(4) \approx 1295 \\ \frac{1295}{3600}(5280) &\approx 1900 \end{aligned}$$

The Porsche 928 traveled about 1900 ft.

31. Left endpoints:

Read values of the function from the table for every number of seconds from 2.0 to 19.3. These values give the heights of 10 rectangles. The width of each rectangle varies. We estimate the area under the curve as

$$\begin{aligned} &\sum_{i=1}^{10} f(x_i)\Delta x \\ &= 30(2.9 - 2.0) + 40(4.1 - 2.9) + 50(5.3 - 4.1) + 60(6.9 - 5.3) + 70(8.7 - 6.9) + 80(10.7 - 8.7) \\ &\quad + 90(13.2 - 10.7) + 100(16.1 - 13.2) + 110(19.3 - 16.1) + 120(23.4 - 19.3) \\ &= 1876 \end{aligned}$$

$$\frac{5280}{3600}(1876) \approx 2751$$

Right endpoints:

Read values of the function from the table for every number of seconds from 2.0 to 23.4. These values give the heights of 11 rectangles. The width of each rectangle varies. We estimate the area under the curve as

$$\sum_{i=1}^{11} f(x_i)\Delta x$$

$$\begin{aligned} &= 30(2.0-0)+40(2.9-2.0)+50(4.1-2.9)+60(5.3-4.1)+70(6.9-5.3)+80(8.7-6.9) \\ &\quad +90(10.7-8.7)+100(13.2-10.7)+110(16.1-13.2)+120(19.3-16.1)+130(23.4-19.3) \\ &= 2150 \end{aligned}$$

$$\frac{5280}{3600}(2150) \approx 3153$$

Average:

$$\frac{2751+3153}{2} = \frac{5904}{2}$$
$$= 2952$$

The distance traveled by the Mercedes-Benz S550 is about 2952 ft.

33. (a) Read values of the function on the plain glass graph every 2 hr from 6 to 6. These are at midpoints of the widths $\Delta x = 2$ and represent the heights of the rectangles.

$$f(x_i)\Delta x = 132(2)+215(2)+150(2)+44(2)+34(2)+26(2)+12(2) \approx 1226$$

The total heat gain was about 1230 BTUs per square foot.

(b) Read values on the ShadeScreen graph every 2 hr from 6 to 6.

$$\sum f(x_i)\Delta x = 38(2)+25(2)+16(2)+12(2)+10(2)+10(2)+5(2) \approx 232$$

The total heat gain was about 230 BTUs per square foot.

35. (a) The area of a trapezoid is

$$A = \frac{1}{2}h(b_1+b_2) = \frac{1}{2}(6)(1+2) = 9.$$

Car A has traveled 9 ft.

(b) Car A is furthest ahead of car B at 2 sec. Notice that from $t=0$ to $t=2$, $v(t)$ is larger for car A than for car B. For $t>2$, $v(t)$ is larger for car B than for car A.

(c) As seen in part (a), car A drove 9 ft after 2 sec. The distance of car B can be calculated as follows:

$$\frac{2-0}{4} = \frac{1}{2} = \text{width}$$

$$\text{Distance} = \frac{1}{2}\cdot v(0.25)+\frac{1}{2}v(0.75)+\frac{1}{2}v(1.25)+\frac{1}{2}v(1.75) = \frac{1}{2}(0.2)+\frac{1}{2}(1)+\frac{1}{2}(2.6)+\frac{1}{2}(5) = 4.4$$

$$9-4.4 = 4.6$$

The furthest car A can get ahead of car B is about 4.6 ft.

(d) At $t=3$, car A travels $\frac{1}{2}(6)(2+3) = 15$ ft and car B travels approximately 13 ft.
At $t=3.5$, car A travels $\frac{1}{2}(6)(2.5+3.5) = 18$ ft and car B travels approximately 18.25 ft. Therefore, car B catches up with car A between 3 and 3.5 sec.

37. Using the left endpoints:

$$\text{Distance} = v_0(1) + v_1(1) + v_2(1) = 10 + 6.5 + 6 = 22.5 \text{ ft}$$

Using the right endpoints:

$$\text{Distance} = v_1(1) + v_2(1) + v_3(1) = 6.5 + 6 + 5.5 = 18 \text{ ft}$$

39. **(a)** Read values from the graph for every hour from 1 A.M. through 11 P.M. The values give the heights of 23 rectangles. The width of each rectangle is $\Delta x = 1$. We estimate the area under the curve as

$$\begin{aligned} A &= \sum_{i=1}^{23} f(x_i)\Delta x \\ &= 500(1) + 550(1) + 800(1) + 1600(1) + 4000(1) + 7000(1) + 7000(1) + 5900(1) + 4500(1) \\ &\quad + 3500(1) + 3100(1) + 3100(1) + 3500(1) + 3800(1) + 4100(1) + 4800(1) + 4750(1) \\ &\quad + 4000(1) + 2500(1) + 2250(1) + 1800(1) + 1500(1) + 1050(1) \\ &= 75{,}600 \end{aligned}$$

(b) Read values from the graph for every hour from 1 A.M. through 11 P.M. The values give the heights of 23 rectangles. The width of each rectangle is $\Delta x = 1$. We estimate the area under the curve as

$$\begin{aligned} A &= \sum_{i=1}^{23} f(x_i)\Delta x \\ &= 500(1) + 400(1) + 400(1) + 700(1) + 1500(1) + 3000(1) + 4100(1) + 3900(1) + 3200(1) \\ &\quad + 3600(1) + 4000(1) + 4000(1) + 4300(1) + 5200(1) + 6000(1) + 6500(1) + 6400(1) \\ &\quad + 6000(1) + 4700(1) + 3100(1) + 2600(1) + 1900(1) + 1300(1) \\ &= 77{,}300 \end{aligned}$$

7.4 The Fundamental Theorem of Calculus

1. $\displaystyle\int_{-2}^{4} (-3)dp = -3\int_{-2}^{4} dp = -3 \cdot p\Big|_{-2}^{4} = -3[4 - (-2)] = -18$

3.
$$\begin{aligned} \int_{-1}^{2} (5t - 3)dt &= 5\int_{-1}^{2} t\,dt - 3\int_{-1}^{2} dt \\ &= \frac{5}{2}t^2\Big|_{-1}^{2} - 3t\Big|_{-1}^{2} \\ &= \frac{5}{2}[2^2 - (-1)^2] - 3[2 - (-1)] \\ &= \frac{5}{2}(4 - 1) - 3(2 + 1) \\ &= \frac{15}{2} - 9 \\ &= \frac{15}{2} - \frac{18}{2} \\ &= -\frac{3}{2} \end{aligned}$$

5. $\int_0^2 (5x^2 - 4x + 2)\,dx$

$$= 5\int_0^2 x^2\,dx - 4\int_0^2 x\,dx + 2\int_0^2 dx$$

$$= \frac{5x^3}{3}\Big|_0^2 - 2x^2\Big|_0^2 + 2x\Big|_0^2$$

$$= \frac{5}{3}(2^3 - 0^3) - 2(2^2 - 0^2) + 2(2 - 0)$$

$$= \frac{5}{3}(8) - 2(4) + 2(2)$$

$$= \frac{40 - 24 + 12}{3} = \frac{28}{3}$$

7. $\int_0^2 3\sqrt{4u+1}\,du$

Let $4u + 1 = x$, so that $4\,du = dx$.

When $u = 0$, $x = 4(0) + 1 = 1$.
When $u = 2$, $x = 4(2) + 1 = 9$.

$$\int_0^2 3\sqrt{4u+1}\,du$$

$$= \frac{3}{4}\int_0^2 \sqrt{4u+1}\,(4\,du)$$

$$= \frac{3}{4}\int_1^9 x^{1/2}\,dx$$

$$= \frac{3}{4}\cdot\frac{x^{3/2}}{3/2}\Big|_1^9$$

$$= \frac{3}{4}\cdot\frac{2}{3}(9^{3/2} - 1^{3/2})$$

$$= \frac{1}{2}(27 - 1) = \frac{26}{2}$$

$$= 13$$

9. $\int_0^4 2(t^{1/2} - t)\,dt = 2\int_0^4 t^{1/2}dt - 2\int_0^4 t\,dt$

$$= 2\cdot\frac{t^{3/2}}{\frac{3}{2}}\Big|_0^4 - 2\cdot\frac{t^2}{2}\Big|_0^4$$

$$= \frac{4}{3}(4^{3/2} - 0^{3/2}) - (4^2 - 0^2)$$

$$= \frac{32}{3} - 16$$

$$= -\frac{16}{3}$$

11. $\int_1^4 (5y\sqrt{y} + 3\sqrt{y})\,dy$

$$= 5\int_1^4 y^{3/2}\,dy + 3\int_1^4 y^{1/2}\,dy$$

$$= 5\left(\frac{y^{5/2}}{\frac{5}{2}}\right)\Big|_1^4 + 3\left(\frac{y^{3/2}}{\frac{3}{2}}\right)\Big|_1^4$$

$$= 2y^{5/2}\Big|_1^4 + 2y^{3/2}\Big|_1^4$$

$$= 2(4^{5/2} - 1) + 2(4^{3/2} - 1)$$
$$= 2(32 - 1) + 2(8 - 1)$$
$$= 62 + 14$$
$$= 76$$

13. $\int_4^6 \frac{2}{(2x-7)^2}\,dx$

Let $u = 2x - 7$, so that $du = 2\,dx$.

When $x = 6$, $u = 2\cdot 6 - 7 = 5$.
When $x = 4$, $u = 2\cdot 4 - 7 = 1$.

$$\int_4^6 \frac{2}{(2x-7)^2}\,dx = \int_1^5 u^{-2}du$$

$$= \frac{u^{-1}}{-1}\Big|_1^5$$

$$= -u^{-1}\big|_1^5$$

$$= -\left(\frac{1}{5} - 1\right)$$

$$= -\left(-\frac{4}{5}\right)$$

$$= \frac{4}{5}$$

15. $\int_1^5 (6n^{-2} - n^{-3})dn = 6\int_1^5 n^{-2}dn - \int_1^5 n^{-3}dn$

$$= 6\cdot\frac{n^{-1}}{-1}\Big|_1^5 - \frac{n^{-2}}{-2}\Big|_1^5$$

$$= \frac{-6}{n}\Big|_1^5 + \frac{1}{2n^2}\Big|_1^5$$

$$= \frac{-6}{5} - \left(\frac{-6}{1}\right) + \left[\frac{1}{2(25)} - \frac{1}{2(1)}\right]$$

$$= \frac{-6}{5} + \frac{6}{1} + \frac{1}{50} - \frac{1}{2}$$

$$= \frac{108}{25}$$

17. $\int_{-3}^{-2}\left(2e^{-0.1y}+\frac{3}{y}\right)dy$

$$= 2\int_{-3}^{-2} e^{-0.1y}dy + \int_{-3}^{-2}\frac{3}{y}\,dy$$
$$= 2\cdot\frac{e^{-0.1y}}{-0.1}\Big|_{-3}^{-2} + 3\ln|y|\Big|_{-3}^{-2}$$
$$= -20e^{-0.1y}\Big|_{-3}^{-2} + 3\ln|y|\Big|_{-3}^{-2}$$
$$= 20e^{0.3} - 20e^{0.2} + 3\ln 2 - 3\ln 3$$
$$\approx 1.353$$

19. $\int_1^2\left(e^{4u}-\frac{1}{(u+1)^2}\right)du$

$$= \int_1^2 e^{4u}du - \int_1^2\frac{1}{(u+1)^2}du$$
$$= \frac{e^{4u}}{4}\Big|_1^2 - \frac{-1}{u+1}\Big|_1^2$$
$$= \frac{e^8}{4} - \frac{e^4}{4} + \frac{1}{2+1} - \frac{1}{1+1}$$
$$= \frac{e^8}{4} - \frac{e^4}{4} - \frac{1}{6}$$
$$\approx 731.4$$

21. $\int_{-1}^{0} y(2y^2-3)^5\,dy$

Let $u = 2y^2 - 3$, so that

$$du = 4y\,dy \text{ and } \frac{1}{4}\,du = y\,dy.$$

When $y=-1,\ u = 2(-1)^2 - 3 = -1.$
When $y = 0,\ u = 2(0)^2 - 3 = -3.$

$$\frac{1}{4}\int_{-1}^{-3} u^5\,du = \frac{1}{4}\cdot\frac{u^6}{6}\Big|_{-1}^{-3}$$
$$= \frac{1}{24}u^6\Big|_{-1}^{-3}$$
$$= \frac{1}{24}(-3)^6 - \frac{1}{24}(-1)^6$$
$$= \frac{729}{24} - \frac{1}{24}$$
$$= \frac{728}{24}$$
$$= \frac{91}{3}$$

23. $\int_1^{64}\frac{\sqrt{z}-2}{\sqrt[3]{z}}\,dz$

$$= \int_1^{64}\left(\frac{z^{1/2}}{z^{1/3}} - 2z^{-1/3}\right)dz$$
$$= \int_1^{64} z^{1/6}\,dz - 2\int_1^{64} z^{-1/3}\,dz$$
$$= \frac{z^{7/6}}{\frac{7}{6}}\Big|_1^{64} - 2\frac{z^{2/3}}{\frac{2}{3}}$$
$$= \frac{6z^{7/6}}{7}\Big|_1^{64} - 3z^{2/3}\Big|_1^{64}$$
$$= \frac{6(64)^{7/6}}{7} - \frac{6(1)^{7/6}}{7} - 3(64^{2/3} - 1^{2/3})$$
$$= \frac{6(128)}{7} - \frac{6}{7} - 3(16-1)$$
$$= \frac{768-6-315}{7}$$
$$= \frac{447}{7} \approx 63.86$$

25. $\int_1^2\frac{\ln\,x}{x}\,dx$

Let $u = \ln\,x$, so that

$$du = \frac{1}{x}\,dx.$$

When $x = 1,\ u = \ln\,1 = 0.$
When $x = 2,\ u = \ln\,2.$

$$\int_0^{\ln\,2} u\,du = \frac{u^2}{2}\Big|_0^{\ln\,2}$$
$$= \frac{(\ln\,2)^2}{2} - 0$$
$$= \frac{(\ln\,2)^2}{2}$$
$$\approx 0.2402$$

27. $\int_0^8 x^{1/3}\sqrt{x^{4/3}+9}\,dx$

Let $u = x^{4/3} + 9$, so that

$$du = \frac{4}{3}x^{1/3}\,dx \text{ and } \frac{3}{4}\,du = x^{1/3}\,dx.$$

When $x = 0,\ u = 0^{4/3} + 9 = 9.$
When $x = 8,\ u = 8^{4/3} + 9 = 25.$

$$\frac{3}{4}\int_9^{25}\sqrt{u}\,du = \frac{3}{4}\int_9^{25} u^{1/2}\,du$$
$$= \frac{3}{4}\cdot\frac{u^{3/2}}{\frac{3}{2}}\bigg|_9^{25}$$
$$= \frac{1}{2}u^{3/2}\bigg|_9^{25}$$
$$= \frac{1}{2}(25)^{3/2} - \frac{1}{2}(9)^{3/2}$$
$$= \frac{125}{2} - \frac{27}{2}$$
$$= 49$$

29. $\displaystyle\int_0^1 \frac{e^{2t}}{(3+e^{2t})^2}\,dt$

Let $u = 3 + e^{2t}$, so that $du = 2e^{2t}dt$.

When $x = 1, u = 3 + e^{2\cdot 1} = 3 + e^2$.
When $x = 0, u = 3 + e^{2\cdot 0} = 4$.

$$\int_0^1 \frac{e^{2t}}{(3+e^{2t})^2}\,dt = \frac{1}{2}\int_4^{3+e^2} u^{-2}du$$
$$= \frac{1}{2}\cdot\frac{u^{-1}}{-1}\bigg|_4^{3+e^2}$$
$$= \frac{-1}{2u}\bigg|_4^{3+e^2}$$
$$= \frac{1}{8} - \frac{1}{2(3+e^2)}$$
$$\approx 0.07687$$

31. $f(x) = 2x - 14; [6, 10]$

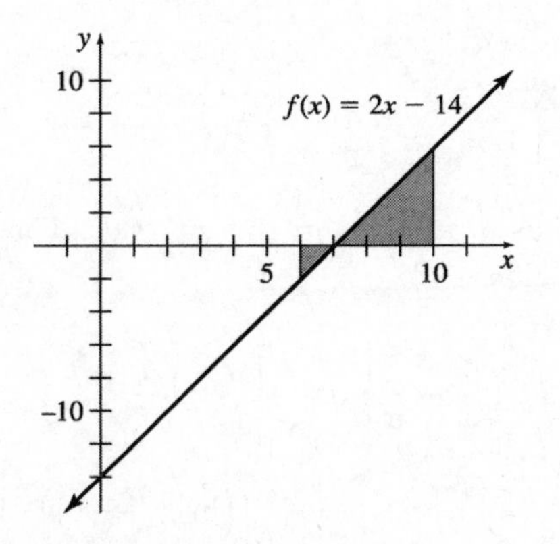

The graph crosses the x-axis at

$$0 = 2x - 14$$
$$2x = 14$$
$$x = 7.$$

This location is in the interval. The area of the region is

$$\left|\int_6^7 (2x-14)dx\right| + \int_7^{10}(2x-14)dx$$
$$= \left|(x^2-14x)\big|_6^7\right| + (x^2-14x)\big|_7^{10}$$
$$= |(7^2-98)-(6^2-84)|$$
$$+ (10^2-140)-(7^2-98)$$
$$= |-1| + (-40) - (-49)$$
$$= 10.$$

33. $f(x) = 2 - 2x^2;\ [0, 5]$

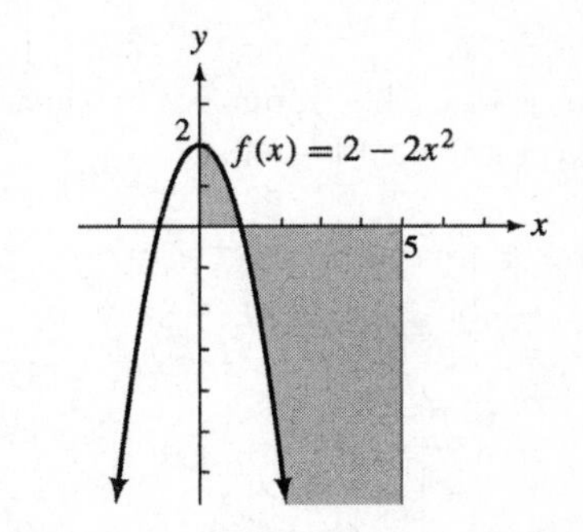

Find the points where the graph crosses the x-axis by solving $2 - 2x^2 = 0$.

$$2 - 2x^2 = 0$$
$$2x^2 = 2$$
$$x^2 = 1$$
$$x = \pm 1.$$

The only solution in the interval $[0, 5]$ is 1. The total area is

$$\int_0^1 (2-2x^2)\,dx + \left|\int_1^5 (2-2x^2)\,dx\right|$$
$$= \left(2x - \frac{2x^3}{3}\right)\bigg|_0^1 + \left|\left(2x - \frac{2x^3}{3}\right)\bigg|_1^5\right|$$
$$= 2 - \frac{2}{3} + \left|10 - \frac{2(5^3)}{3} - 2 + \frac{2}{3}\right|$$
$$= \frac{4}{3} + \left|\frac{-224}{3}\right|$$
$$= \frac{228}{3}$$
$$= 76.$$

35. $f(x) = x^3;\ [-1, 3]$

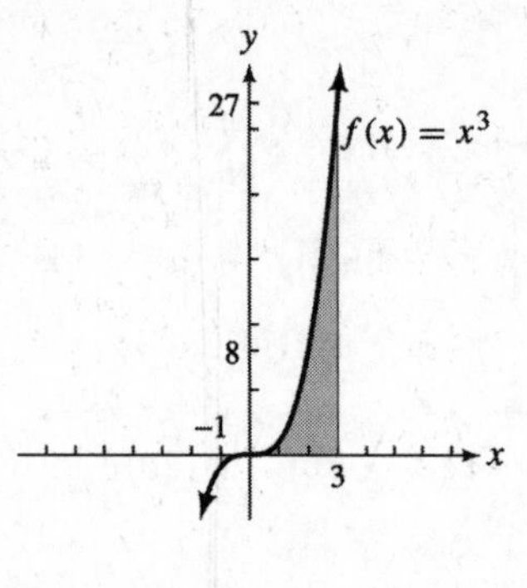

The solution

$$x^3 = 0$$
$$x = 0$$

indicates that the graph crosses the x-axis at 0 in the given interval $[-1, 3]$.

The total area is

$$\left|\int_{-1}^{0} x^3\,dx\right| + \int_{0}^{3} x^3\,dx$$
$$= \left|\frac{x^4}{4}\Big|_{-1}^{0}\right| + \left|\frac{x^4}{4}\right|_{0}^{3}$$
$$= \left|\left(0 - \frac{1}{4}\right)\right| + \left(\frac{3^4}{4} - 0\right)$$
$$= \frac{1}{4} + \frac{81}{4} = \frac{82}{4}$$
$$= \frac{41}{2}.$$

37. $f(x) = e^x - 1;\ [-1, 2]$

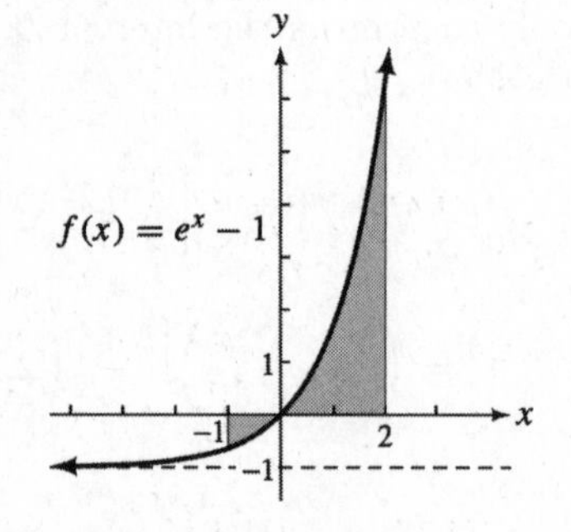

Solve

$$e^x - 1 = 0.$$
$$e^x = 1$$
$$x \ln e = \ln 1$$
$$x = 0$$

The graph crosses the x-axis at 0 in the given interval $[-1, 2]$.

The total area is

$$\left|\int_{-1}^{0} (e^x - 1)\,dx\right| + \int_{0}^{2} (e^x - 1)\,dx$$
$$= \left|(e^x - x)\Big|_{-1}^{0}\right| + (e^x - x)\Big|_{0}^{2}$$
$$= |(1 - 0) - (e^{-1} + 1)|$$
$$+ (e^2 - 2) - (1 - 0)$$
$$= \left|1 - e^{-1} - 1\right| + e^2 - 2 - 1$$
$$= \frac{1}{e} + e^2 - 3$$
$$\approx 4.757.$$

39. $f(x) = \frac{1}{x} - \frac{1}{e};\ [1, e^2]$

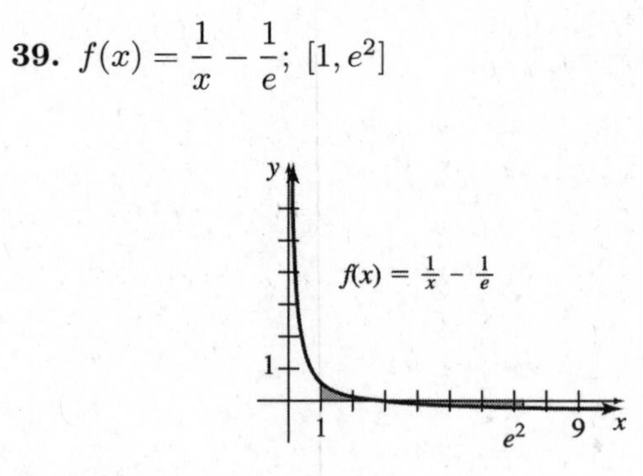

The graph crosses the x-axis at

$$0 = \frac{1}{x} - \frac{1}{e}$$
$$\frac{1}{x} = \frac{1}{e}$$
$$x = e.$$

This location is in the interval. The area of the region is

$$\int_{1}^{e} \left(\frac{1}{x} - \frac{1}{e}\right) dx + \left|\int_{e}^{e^2} \left(\frac{1}{x} - \frac{1}{e}\right) dx\right|$$
$$= \left|\ln|x| - \frac{x}{e}\right|_{1}^{e} + \left|\left(\ln|x| - \frac{x}{e}\right)\Big|_{e}^{e^2}\right|$$
$$= 0 - \left(-\frac{1}{e}\right) + |(2 - e) - 0|$$
$$= \frac{1}{e} + |2 - e|$$
$$= e - 2 + \frac{1}{e}.$$

41. $y = 4 - x^2$; $[0, 3]$

From the graph, we see that the total area is

$$\int_0^2 (4 - x^2)\,dx + \left|\int_2^3 (4 - x^2)\,dx\right|$$
$$= \left(4x - \frac{x^3}{3}\right)\Big|_0^2 + \left|\left(4x - \frac{x^3}{3}\right)\Big|_2^3\right|$$
$$= \left[\left(8 - \frac{8}{3}\right) - 0\right]$$
$$+ \left|\left[(12 - 9) - \left(8 - \frac{8}{3}\right)\right]\right|$$
$$= \frac{16}{3} + \left|3 - \frac{16}{3}\right|$$
$$= \frac{16}{3} + \frac{7}{3}$$
$$= \frac{23}{3}$$

43. $y = e^x - e$; $[0, 2]$

From the graph, we see that the total area is

$$\left|\int_0^1 (e^x - e)\,dx\right| + \int_1^2 (e^x - e)\,dx$$
$$= \left|(e^x - xe)\Big|_0^1\right| + (e^x - xe)\Big|_1^2$$
$$= |(e^1 - e) - (e^0 + 0)|$$
$$+ (e^2 - 2e) - (e^1 - e)$$
$$= |-1| + e^2 - 2e$$
$$= 1 + e^2 - 2e$$
$$\approx 2.952.$$

45. $\int_a^c f(x)\,dx = \int_a^b f(x)\,dx + \int_b^c f(x)\,dx$

47. $\int_0^{16} f(x)\,dx = \int_0^2 f(x)\,dx + \int_2^5 f(x)\,dx$

$$+ \int_5^8 f(x)\,dx + \int_8^{16} f(x)\,dx$$
$$= \frac{1}{2}\cdot 2(1+3) + \frac{\pi(3^2)}{4} - \frac{\pi(3^2)}{4} - \frac{1}{2}(3)(8)$$
$$= 4 + \frac{9}{4}\pi - \frac{9}{4}\pi - 12$$
$$= -8$$

49. Prove: $\int_a^b f(x)\,dx = \int_a^c f(x)\,dx + \int_c^b f(x)\,dx.$

Let $F(x)$ be an antiderivative of $f(x)$.

$$\int_a^c f(x)\,dx + \int_c^b f(x)\,dx$$
$$= F(x)\Big|_a^c + F(x)\Big|_c^b$$
$$= [F(c) - F(a)] + [F(b) - F(c)]$$
$$= F(c) - F(a) + F(b) - F(c)$$
$$= F(b) - F(a)$$
$$= \int_a^b f(x)\,dx$$

51. $\int_{-1}^4 f(x)\,dx$

$$= \int_{-1}^0 (2x+3)\,dx \int_0^4 \left(-\frac{x}{4} - 3\right)\,dx$$
$$= (x^2 + 3x)\Big|_{-1}^0 + \left(-\frac{x^2}{8} - 3x\right)\Big|_0^4$$
$$= -(1 - 3) + (-2 - 12)$$
$$= 2 - 14$$
$$= -12$$

53. (a) $g(t) = t^4$ and $c = 1$, use substitution.

$$f(x) = \int_c^x g(t)dt$$
$$= \int_1^x t^4 dt$$
$$= \frac{t^5}{5}\Big|_1^x$$
$$= \frac{x^5}{5} - \frac{(1)^5}{5}$$
$$= \frac{x^5}{5} - \frac{1}{5}$$

(b) $f'(x) = \frac{d}{dx}(f(x))$

$$= \frac{d}{dx}\left(\frac{x^5}{5} - \frac{1}{5}\right)$$
$$= \frac{1}{5}\cdot\frac{d}{dx}(x^5) - \frac{d}{dx}\left(\frac{1}{5}\right)$$
$$= \frac{1}{5}\cdot 5x^4 - 0$$
$$= x^4$$

Since $g(t) = t^4$, then $g(x) = x^4$ and we see $f'(x) = g(x)$.

(c) Let $g(t) = e^{t^2}$ and $c = 0$, then $f(x) = \int_0^x e^{t^2}\,dt$.

$f(1) = \int_0^1 e^{t^2}\,dt$ and $f(1.01) = \int_0^{1.01} e^{t^2}\,dt$.

Use the fnInt command in the Math menu of your calculator to find $\int_0^1 e^{x^2}\,dx$ and $\int_0^{1.01} e^{x^2}\,dx$. The resulting screens are:

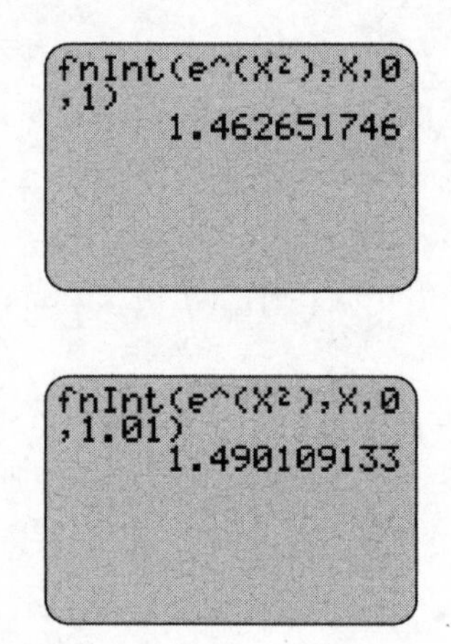

$f(1) \approx 1.46265$
$f(1.01) \approx 1.49011$

Use $\dfrac{f(1+h) - f(1)}{h}$ to approximate $f'(1)$ with $h = 0.01$

$$\begin{aligned}\frac{f(1+h) - f(1)}{h} &= \frac{f(1.01) - f(1)}{0.01}\\ &\approx \frac{1.49011 - 1.46265}{0.01}\\ &= 2.746\end{aligned}$$

So $f'(1) \approx 2.746$, and $g(1) = e^{1^2} = e \approx 2.718$.

55. $P'(t) = (3t+3)(t^2+2t+2)^{1/3}$

(a) $\int_0^3 3(t+1)(t^2+2t+2)^{1/3}\,dt$

Let $u = t^2 + 2t + 2$, so that

$du = (2t+2)\,dt$ and $\frac{1}{2}\,du = (t+1)\,dt$.

When $t = 0$, $u = 0^2 + 2 \cdot 0 + 2 = 2$.

When $t = 3$, $u = 3^2 + 2 \cdot 3 + 2 = 17$.

$$\begin{aligned}\frac{3}{2}\int_2^{17} u^{1/3}\,du &= \frac{3}{2} \cdot \frac{u^{4/3}}{\frac{4}{3}}\Bigg|_2^{17}\\ &= \frac{9}{8}u^{4/3}\Bigg|_2^{17}\\ &= \frac{9}{8}(17)^{4/3} - \frac{9}{8}(2)^{4/3}\\ &\approx 46.341\end{aligned}$$

Total profits for the first 3 yr were

$$\frac{9000}{8}(17^{4/3} - 2^{4/3}) \approx \$46,341.$$

(b) $\int_3^4 3(t+1)(t^2+2t+2)^{1/3}\,dt$

Let $u = t^2 + 2t + 2$, so that

$du = (2t+2)\,dt = 2(t+1)\,dt$ and

$\frac{3}{2}\,du = 3(t+1)\,dt$.

When $t = 3$, $u = 3^2 + 2 \cdot 3 + 2 = 17$.
When $t = 4$, $u = 4^2 + 2 \cdot 4 + 2 = 26$.

$$\begin{aligned}\frac{3}{2}\int_{17}^{26} u^{1/3}\,du &= \frac{9}{8}u^{4/3}\Bigg|_{17}^{26}\\ &= \frac{9}{8}(26)^{4/3} - \frac{9}{8}(17)^{4/3}\\ &\approx 37.477\end{aligned}$$

Profit in the fourth year was

$$\frac{9000}{8}(26^{4/3} - 17^{4/3}) \approx \$37,477.$$

(c) $\lim_{t\to\infty} P'(t)$

$$\begin{aligned}&= \lim_{t\to\infty} (3t+3)(t^2+2t+2)^{1/3}\\ &= \infty\end{aligned}$$

The annual profit is slowly increasing without bound.

57. $P'(t) = 140t^{5/2}$

$$\begin{aligned}\int_0^4 140t^{5/2}\,dt &= 140 \cdot \frac{t^{7/2}}{\frac{7}{2}}\Bigg|_0^4\\ &= 40t^{7/2}\Bigg|_0^4\\ &= 5120\end{aligned}$$

Since 5120 is above the total level of acceptable pollution (4850), the factory cannot operate for 4 years without killing all the fish in the lake.

59. Growth rate is $0.6 + \frac{4}{(t+1)^3}$ ft/yr.

(a) Total growth in the second year is

$$\int_1^2 \left[0.6 + \frac{4}{(t+1)^3}\right] dt$$
$$= \left[0.6t + \frac{4}{-2(t+1)^2}\right]\Bigg|_1^2$$
$$= \left[0.6(2) - \frac{2}{(2+1)^2}\right] - \left[0.6(1) - \frac{2}{(1+1)^2}\right]$$
$$= \frac{44}{45} - \frac{1}{10}$$
$$\approx 0.8778 \text{ ft.}$$

(b) Total growth in the third year is

$$\int_2^3 \left[0.6 + \frac{4}{(t+1)^3}\right] dt$$
$$= \left[0.6t + \frac{4}{-2(t+1)^2}\right]\Bigg|_2^3$$
$$= \left[0.6(3) - \frac{2}{(3+1)^2}\right] - \left[0.6(2) - \frac{2}{(2+1)^2}\right]$$
$$= \frac{67}{40} - \frac{44}{45}$$
$$\approx 0.6972 \text{ ft.}$$

61. $R'(t) = \frac{5}{t+1} + \frac{2}{\sqrt{t+1}}$

(a) Total reaction from $t = 1$ to $t = 12$ is

$$\int_1^{12} \left(\frac{5}{t+1} + \frac{2}{\sqrt{t+1}}\right) dt$$
$$= \left[5\ln(t+1) + 4\sqrt{t+1}\,\right]\Big|_1^{12}$$
$$= (5\ln 13 + 4\sqrt{13}) - (5\ln 2 + 4\sqrt{2})$$
$$\approx 18.12.$$

(b) Total reaction from $t = 12$ to $t = 24$ is

$$\int_{12}^{24} \left(\frac{5}{t+1} + \frac{2}{\sqrt{t+1}}\right) dt$$
$$= \left[5\ln(t+1) + 4\sqrt{t+1}\,\right]\Big|_{12}^{24}$$
$$= (5\ln 25 + 4\sqrt{25}) - (5\ln 13 + 4\sqrt{13})$$
$$\approx 8.847.$$

63. **(b)** $\int_0^{60} n(x)\,dx$

(c) $\int_5^{10} \sqrt{5x+1}\,dx$

Let $u = 5x + 1$. Then $du = 5\,dx$.
When $x = 5$, $u = 26$; when $x = 10$, $u = 51$.

$$\frac{1}{5}\int_{26}^{51} u^{1/2}\,du$$
$$= \frac{1}{5} \cdot \frac{u^{3/2}}{\frac{3}{2}}\Bigg|_{26}^{51}$$
$$= \frac{2}{15}u^{3/2}\Big|_{26}^{51}$$
$$= \frac{2}{15}(51^{3/2} - 26^{3/2})$$
$$\approx 30.89 \text{ million}$$

65. $v = k(R^2 - r^2)$

(a)
$$Q(R) = \int_0^R 2\pi v r\,dr$$
$$= \int_0^R 2\pi k(R^2 - r^2)r\,dr$$
$$= 2\pi k \int_0^R (R^2 r - r^2)\,dr$$
$$= 2\pi k \left(\frac{R^2 r^2}{2} - \frac{r^4}{4}\right)\Bigg|_0^R$$
$$= 2\pi k \left(\frac{R^4}{2} - \frac{R^4}{4}\right)$$
$$= 2\pi k \left(\frac{R^4}{4}\right)$$
$$= \frac{\pi k R^4}{2}$$

(b)
$$Q(0.4) = \frac{\pi k (0.4)^4}{2}$$
$$= 0.04k \text{ mm/min}$$

67. $E(t) = 753t^{-0.1321}$

(a) Since t is the age of the beagle in years, to convert the formula to days, let $T = 365t$, or $t = \frac{T}{365}$.

$$E(T) = 753\left(\frac{T}{365}\right)^{-0.1321} \approx 1642T^{-0.1321}$$

Now, replace T with t.

$$E(t) = 1642t^{-0.1321}$$

(b) The beagle's age in days after one year is 365 days and after 3 years she is 1095 days old.

$$\int_{365}^{1095} 1642t^{-0.1321}\,dt$$

$$= 1642\frac{1}{0.8679}t^{0.8679}\Big|_{365}^{1095}$$

$$\approx 1892(1{,}095^{0.8679} - 365^{0.8679})$$

$$\approx 505{,}155$$

The beagle's total energy requirements are about 505,000 kJ/W$^{0.67}$.

69. (a) $f(x) = 40.1 + 2.03x - 0.741x^2$

$$\int_0^9 (40.1 + 2.03x - 0.741x^2)\,dx$$

$$= \left(40.1x + \frac{2.03}{2}x^2 - \frac{0.741}{3}x^3\right)\Big|_0^9$$

$$= (40.1x + 1.015x^2 - 0.247x^3)\Big|_0^9$$

$$\approx 263$$

This integral shows that the total population aged 0 to 90 was about 263 million.

(b) $$\int_{3.5}^{5.5} (40.1 + 2.03x - 0.741x^2)\,dx$$

$$= (40.1x + 1.015x^2 - 0.247x^3\Big|_{3.5}^{5.5}$$

$$\approx 210.1591 - 142.1936$$

$$= 67.9655$$

The number of baby boomers is about 68 million.

71. $c'(t) = ke^{rt}$

(a) $c'(t) = 1.2e^{0.04t}$

(b) The amount of oil that the company will sell in the next ten years is given by the integral $\int_0^{10} 1.2e^{0.04t}dt$.

(c) $$\int_0^{10} 1.2e^{0.04t}dx = \frac{1.2e^{0.04t}}{0.04}\Big|_0^{10}$$

$$= 30e^{0.04t}\Big|_0^{10}$$

$$= 30e^{0.4} - 30$$

$$\approx 14.75$$

This represents about 14.75 billion barrels of oil.

(d) $$\int_0^T 1.2e^{0.04t}\,dt = 30e^{0.04t}\Big|_0^T$$

$$= 30e^{0.04T} - 30$$

Solve

$$20 = 30e^{0.04T} - 30.$$

$$50 = 30e^{0.04T}$$

$$\frac{5}{3} = e^{0.04T}$$

$$\ln\frac{5}{3} = 0.04T\ \ln e$$

$$T = \frac{\ln\frac{5}{3}}{0.04}$$

$$\approx 12.8$$

The oil will last about 12.8 years.

(e) $$\int_0^T 1.2e^{0.02t}\,dt = 60e^{0.02t}\Big|_0^T$$

$$= 60e^{0.02T} - 60$$

Solve

$$20 = 60e^{0.02T} - 60.$$

$$80 = 60e^{0.02T}$$

$$\frac{4}{3} = e^{0.02T}$$

$$\ln\frac{4}{3} = 0.02T\ \ln e$$

$$T = \frac{\ln\frac{4}{3}}{0.02} \approx 14.4$$

The oil will last about 14.4 years.

7.5 The Area Between Two Curves

1. $x = -2, x = 1, y = 2x^2 + 5, y = 0$

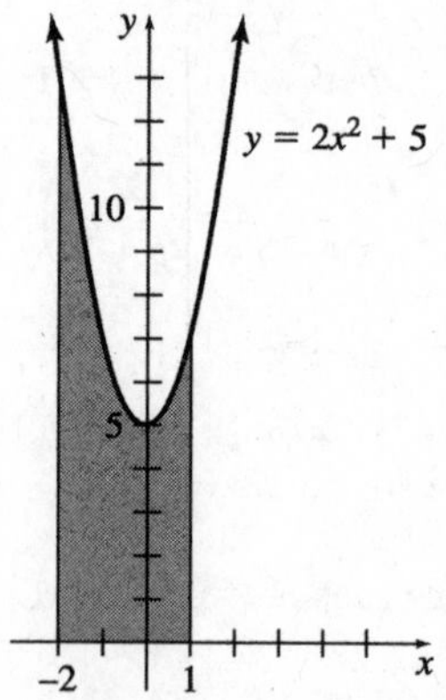

$$\int_{-2}^{1}[(2x^2+5) - 0] = \left(\frac{2x^3}{3} + 5x\right)\Big|_{-2}^{1}$$

$$= \left(\frac{2}{3} + 5\right) - \left(-\frac{16}{3} - 10\right)$$

$$= 21$$

3. $x = -3, x = 1, y = x^3 + 1, y = 0$

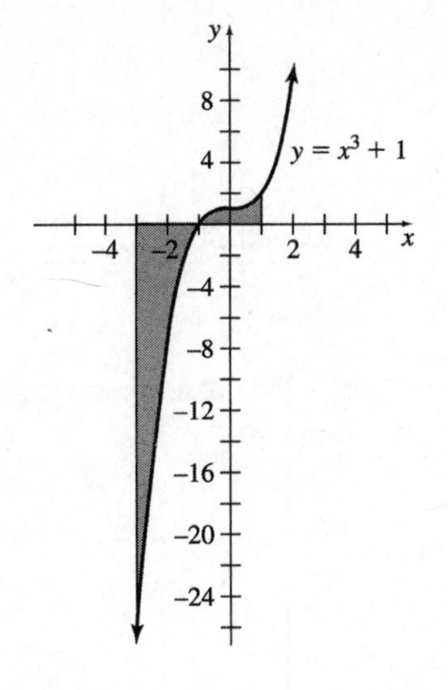

To find the points of intersection of the graphs, substitute for y.

$$x^3 + 1 = 0$$
$$x^3 = -1$$
$$x = -1$$

The region is composed of two separate regions because $y = x^3 + 1$ intersects $y = 0$ at $x = -1$.
Let $f(x) = x^3 + 1$, $g(x) = 0$.
In the interval $[-3, -1]$, $g(x) \geq f(x)$.
In the interval $[-1, 1]$, $f(x) \geq g(x)$.

$$\int_{-3}^{-1} [0 - (x^3 + 1)]dx + \int_{-1}^{1} [(x^3 + 1) - 0]dx$$
$$= \left(\frac{-x^4}{4} - x\right)\Big|_{-3}^{-1} + \left(\frac{x^4}{4} + x\right)\Big|_{-1}^{1}$$
$$= \left(-\frac{1}{4} + 1\right) - \left(-\frac{81}{4} + 3\right) + \left(\frac{1}{4} + 1\right) - \left(\frac{1}{4} - 1\right)$$
$$= 20$$

5. $x = -2,\ x = 1,\ y = 2x,\ y = x^2 - 3$

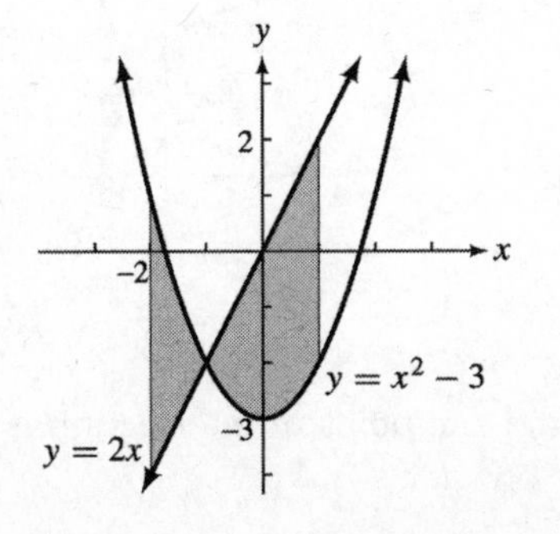

Find the points of intersection of the graphs of $y = 2x$ and $y = x^2 - 3$ by substituting for y.

$$2x = x^2 - 3$$
$$0 = x^2 - 2x - 3$$
$$0 = (x - 3)(x + 1)$$

The only intersection in $[-2, 1]$ is at $x = -1$.
In the interval $[-2, -1]$, $(x^2 - 3) \geq 2x$.
In the interval $[-1, 1]$, $2x \geq (x^2 - 3)$.

$$\int_{-2}^{-1} [(x^2 - 3) - (2x)]\,dx + \int_{-1}^{1} [(2x) - (x^2 - 3)]\,dx$$
$$= \int_{-2}^{-1} (x^2 - 3 - 2x)\,dx$$
$$+ \int_{-1}^{1} (2x - x^2 + 3)\,dx$$
$$= \left(\frac{x^3}{3} - 3x - x^2\right)\Big|_{-2}^{-1}$$
$$+ \left(x^2 - \frac{x^3}{3} + 3x\right)\Big|_{-1}^{1}$$
$$= -\frac{1}{3} + 3 - 1 - \left(-\frac{8}{3} + 6 - 4\right) + 1 - \frac{1}{3} + 3$$
$$- \left(1 + \frac{1}{3} - 3\right)$$
$$= \frac{5}{3} + 6 = \frac{23}{3}$$

7. $y = x^2 - 30$
$y = 10 - 3x$

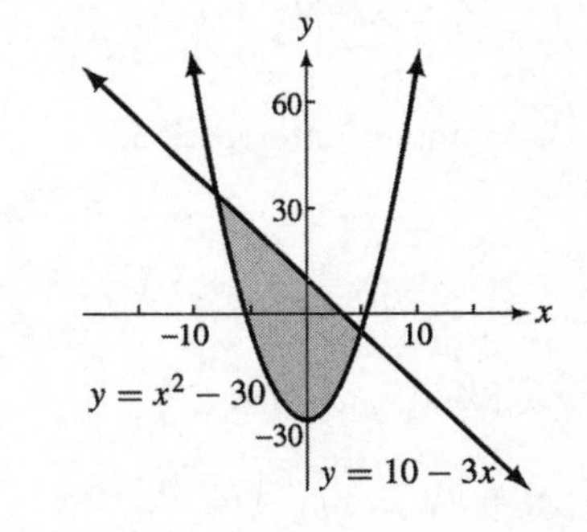

Find the points of intersection.

$$x^2 - 30 = 10 - 3x$$
$$x^2 + 3x - 40 = 0$$
$$(x + 8)(x - 5) = 0$$
$$x = -8 \quad \text{or} \quad x = 5$$

Let $f(x) = 10 - 3x$ and $g(x) = x^2 - 30$.

The area between the curves is given by

$$\int_{-8}^{5} [f(x) - g(x)]\,dx$$

$$= \int_{-8}^{5} [(10 - 3x) - (x^2 - 30)]\,dx$$

$$= \int_{-8}^{5} (-x^2 - 3x + 40)\,dx$$

$$= \left(\frac{-x^3}{3} - \frac{3x^2}{2} + 40x\right)\Big|_{-8}^{5}$$

$$= \frac{-5^3}{3} - \frac{3(5)^2}{2} + 40(5)$$

$$- \left[\frac{-(-8)^3}{3} - \frac{3(-8)^2}{2} + 40(-8)\right]$$

$$= \frac{-125}{3} - \frac{75}{2} + 200 - \frac{512}{3} + \frac{192}{2} + 320$$

$$\approx 366.1667.$$

9. $y = x^2$, $y = 2x$

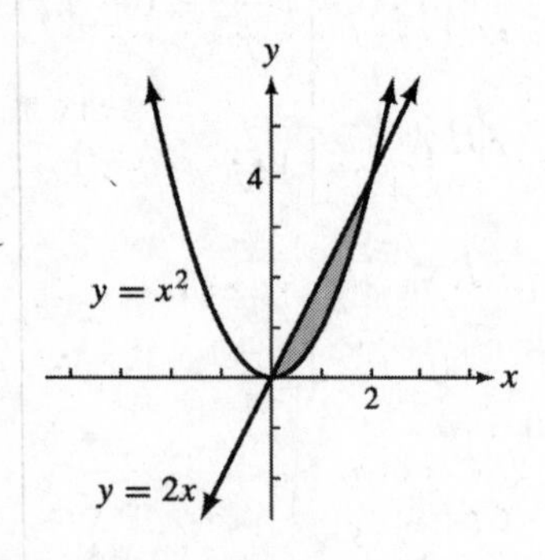

Find the points of intersection.

$$\begin{aligned} x^2 &= 2x \\ x^2 - 2x &= 0 \\ x(x-2) &= 0 \\ x = 0 \quad &\text{or} \quad x = 2 \end{aligned}$$

Let $f(x) = 2x$ and $g(x) = x^2$.
The area between the curves is given by

$$\int_0^2 [f(x) - g(x)]\,dx = \int_0^2 (2x - x^2)\,dx$$

$$= \left(\frac{2x^2}{2} - \frac{x^3}{3}\right)\Big|_0^2$$

$$= 4 - \frac{8}{3} = \frac{4}{3}.$$

11. $x = 1, x = 6, y = \frac{1}{x}, y = \frac{1}{2}$

To find the points of intersection of the graphs, substitute for y.

$$\frac{1}{x} = \frac{1}{2}$$

$$x = 2$$

The region is composed of two separate regions because $y = \frac{1}{x}$ intersects $y = \frac{1}{2}$ at $x = 2$.

Let $f(x) = \frac{1}{x}, g(x) = \frac{1}{2}$.

In the interval $[1, 2]$, $f(x) \geq g(x)$.
In the interval $[2, 6]$, $g(x) \geq f(x)$.

$$\int_1^2 \left(\frac{1}{x} - \frac{1}{2}\right) dx + \int_2^6 \left(\frac{1}{2} - \frac{1}{x}\right) dx$$

$$= \left(\ln|x| - \frac{x}{2}\right)\Big|_1^2 + \left(\frac{x}{2} - \ln|x|\right)\Big|_2^6$$

$$= (\ln 2 - 1) - \left(0 - \frac{1}{2}\right) + (3 - \ln 6) - (1 - \ln 2)$$

$$= 2\ln 2 - \ln 6 + \frac{3}{2}$$

$$\approx 1.095$$

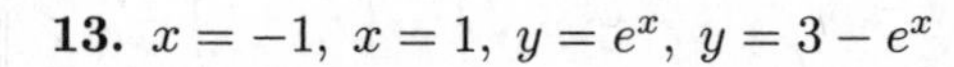

13. $x = -1,\ x = 1,\ y = e^x,\ y = 3 - e^x$

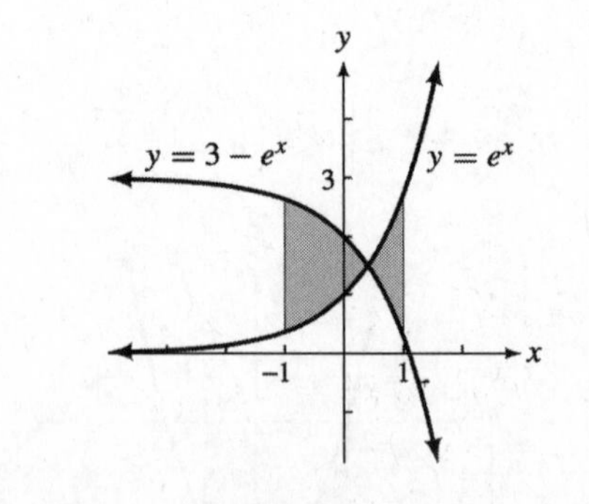

To find the point of intersection, set $e^x = 3 - e^x$ and solve for x.

$$\begin{aligned} e^x &= 3 - e^x \\ 2e^x &= 3 \\ e^x &= \frac{3}{2} \\ \ln e^x &= \ln \frac{3}{2} \\ x \ln e &= \ln \frac{3}{2} \\ x &= \ln \frac{3}{2} \end{aligned}$$

The area of the region between the curves from $x = -1$ to $x = 1$ is

$$\int_{-1}^{\ln 3/2} [(3 - e^x) - e^x]\, dx + \int_{\ln 3/2}^{1} [e^x - (3 - e^x)]\, dx$$

$$= \int_{-1}^{\ln 3/2} (3 - 2e^x)\, dx + \int_{\ln 3/2}^{1} (2e^x - 3)\, dx$$

$$= (3x - 2e^x)\Big|_{-1}^{\ln 3/2} + (2e^x - 3x)\Big|_{\ln 3/2}^{1}$$

$$= \left[\left(3 \ln \frac{3}{2} - 2e^{\ln 3/2}\right) - [3(-1) - 2e^{-1}]\right] + \left[2e^1 - 3(1) - \left(2e^{\ln 3/2} - 3 \ln \frac{3}{2}\right)\right]$$

$$= \left[\left(3 \ln \frac{3}{2} - 3\right) - \left(-3 - \frac{2}{e}\right)\right] + \left[2e - 3 - \left(3 - 3 \ln \frac{3}{2}\right)\right]$$

$$= 6 \ln \frac{3}{2} + \frac{2}{e} + 2e - 6 \approx 2.605.$$

15. $x = -1, x = 2, y = 2e^{2x}, y = e^{2x} + 1$

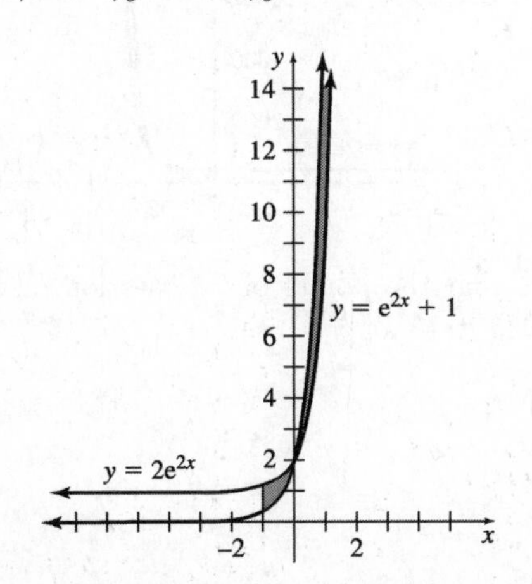

To find the points of intersection of the graphs, substitute for y.

$$\begin{aligned} 2e^{2x} &= e^{2x} + 1 \\ e^{2x} &= 1 \\ 2x &= 0 \\ x &= 0 \end{aligned}$$

The region is composed of two separate regions because $y = 2e^{2x}$ intersects $y = e^{2x} + 1$ at $x = 0$.
Let $f(x) = 2e^{2x}, g(x) = e^{2x} + 1$.
In the interval $[-1, 0]$, $g(x) \geq f(x)$.
In the interval $[0, 2]$, $f(x) \geq g(x)$.

$$\int_{-1}^{0} (e^{2x} + 1 - 2e^{2x})dx + \int_{0}^{2} [2e^{2x} - (e^{2x} + 1)]dx$$

$$= \left(-\frac{e^{2x}}{2} + x\right)\Big|_{-1}^{0} + \left(\frac{e^{2x}}{2} - x\right)\Big|_{0}^{2}$$

$$= \left(-\frac{1}{2} + 0\right) - \left(-\frac{e^{-2}}{2} - 1\right) + \left(\frac{e^4}{2} - 2\right) - \left(\frac{1}{2} - 0\right)$$

$$= \frac{e^{-2} + e^4}{2} - 2$$

$$\approx 25.37$$

17. $y = x^3 - x^2 + x + 1,\ y = 2x^2 - x + 1$

Find the points of intersection.

$$\begin{aligned} x^3 - x^2 + x + 1 &= 2x^2 - x + 1 \\ x^3 - 3x^2 + 2x &= 0 \\ x(x^2 - 3x + 2) &= 0 \\ x(x - 2)(x - 1) &= 0 \end{aligned}$$

The points of intersection are at $x = 0$, $x = 1$, and $x = 2$.

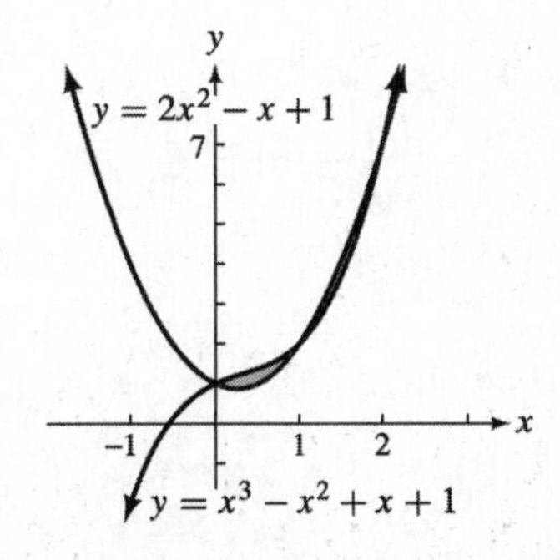

Area between the curves is

$$\int_0^1 [(x^3 - x^2 + x + 1) - (2x^2 - x + 1)]\,dx$$

$$+ \int_1^2 [(2x^2 - x + 1) - (x^3 - x^2 + x + 1)]\,dx$$

$$= \int_0^1 (x^3 - 3x^2 + 2x)\,dx + \int_0^1 (-x^3 + 3x^2 - 2x)\,dx$$

$$= \left(\frac{x^4}{4} - x^3 + x^2\right)\Bigg|_0^1 + \left(\frac{-x^4}{4} + x^3 - x^2\right)\Bigg|_1^2$$

$$= \left[\left(\frac{1}{4} - 1 + 1\right) - (0)\right]$$

$$+ \left[(-4 + 8 - 4) - \left(-\frac{1}{4} + 1 - 1\right)\right]$$

$$= \frac{1}{4} + \frac{1}{4}$$

$$= \frac{1}{2}.$$

19. $y = x^4 + \ln(x + 10)$,
$y = x^3 + \ln(x + 10)$

Find the points of intersection.

$$x^4 + \ln(x + 10) = x^3 + \ln(x + 10)$$
$$x^4 - x^3 = 0$$
$$x^3(x - 1) = 0$$
$$x = 0 \quad \text{or} \quad x = 1$$

The points of intersection are at $x = 0$ and $x = 1$. The area between the curves is

$$\int_0^1 [(x^3 + \ln(x + 10)) - (x^4 + \ln(x + 10))]\,dx$$

$$= \int_0^1 (x^3 - x^4)\,dx$$

$$= \left(\frac{x^4}{4} - \frac{x^5}{5}\right)\Bigg|_0^1$$

$$= \left(\frac{1}{4} - \frac{1}{5}\right) - (0) = \frac{1}{20}.$$

21. $y = x^{4/3}$, $y = 2x^{1/3}$

Find the points of intersection.

$$x^{4/3} = 2x^{1/3}$$
$$x^{4/3} - 2x^{1/3} = 0$$
$$x^{1/3}(x - 2) = 0$$
$$x = 0 \quad \text{or} \quad x = 2$$

The points of intersection are at $x = 0$ and $x = 2$.

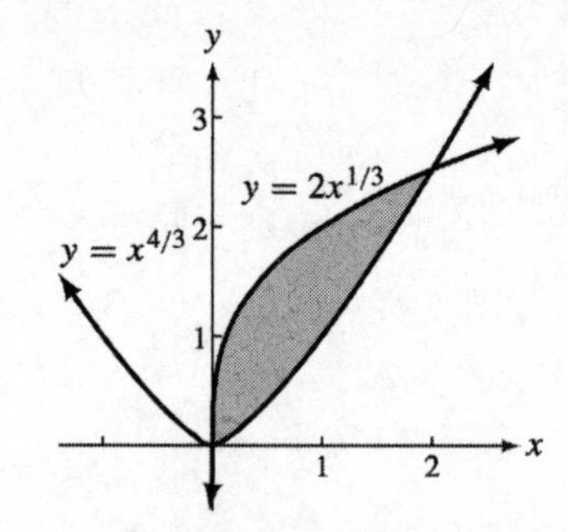

The area between the curves is

$$\int_0^2 (2x^{1/3} - x^{4/3})\,dx = 2\frac{x^{4/3}}{\frac{4}{3}} - \frac{x^{7/3}}{\frac{7}{3}}\Bigg|_0^2$$

$$= \frac{3}{2}x^{4/3} - \frac{3}{7}x^{7/3}\Bigg|_0^2$$

$$= \left[\frac{3}{2}(2)^{4/3} - \frac{3}{7}(2)^{7/3}\right] - 0$$

$$= \frac{3(2^{4/3})}{2} - \frac{3(2^{7/3})}{7}$$

$$\approx 1.62.$$

23. $x = 0, x = 3, y = 2e^{3x}, y = e^{3x} + e^6$

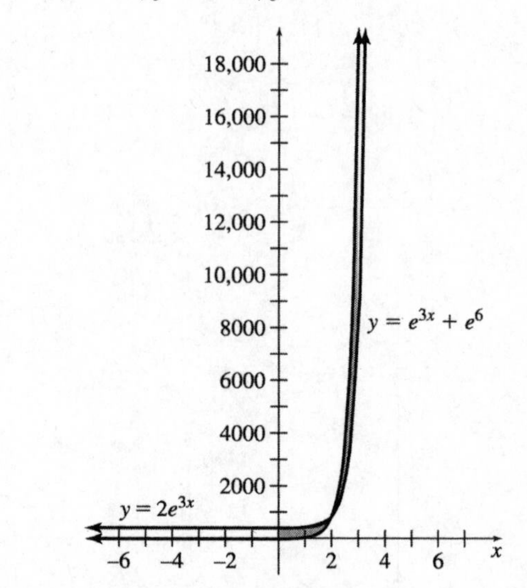

To find the points of intersection of the graphs, substitute for y.

$$2e^{3x} = e^{3x} + e^6$$
$$e^{3x} = e^6$$
$$3x = 6$$
$$x = 2$$

The region is composed of two separate regions because $y = 2e^{3x}$ intersects $y = e^{3x} + e^6$ at $x = 2$.

Let $f(x) = 2e^{3x}$, $g(x) = e^{3x} + e^6$.
In the interval $[0, 2]$, $g(x) \geq f(x)$.
In the interval $[2, 3]$, $f(x) \geq g(x)$.

$$\int_0^2 (e^{3x} + e^6 - 2e^{3x})dx + \int_2^3 [2e^{3x} - (e^{3x} + e^6)]dx$$

$$= \left(-\frac{e^{3x}}{3} + e^6 x\right)\Big|_0^2 + \left(\frac{e^{3x}}{3} - e^6 x\right)\Big|_2^3$$

$$= \left(-\frac{e^6}{3} + 2e^6\right) - \left(-\frac{1}{3} + 0\right) + \left(\frac{e^9}{3} - 3e^6\right) - \left(\frac{e^6}{3} - 2e^6\right)$$

$$= \frac{e^9 + e^6 + 1}{3}$$

$$\approx 2836$$

25. Graph $y_1 = e^x$ and $y_2 = -x^2 - 2x$ on your graphing calculator. Use the intersect command to find the two intersection points. The resulting screens are:

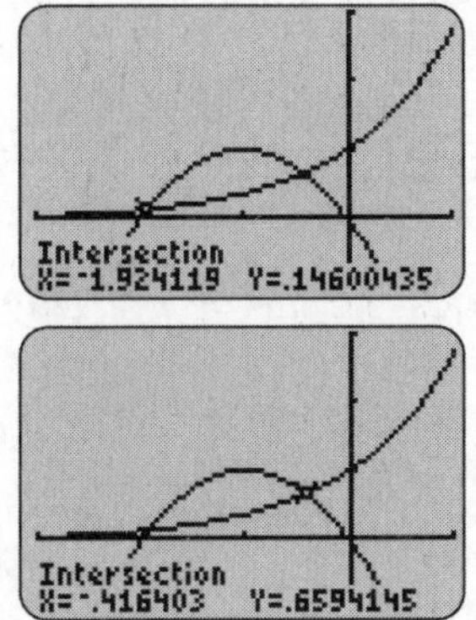

These screens show that $e^x = -x^2 - 2x$ when $x \approx -1.9241$ and $x \approx -0.4164$.
In the interval $[-1.9241, -0.4164]$,

$$e^x < -x^2 - 2x.$$

The area between the curves is given by

$$\int_{-1.9241}^{-0.4164} [(-x^2 - 2x) - e^x]dx.$$

Use the fnInt command to approximate this definite integral.
The resulting screen is:

The last screen shows that the area is approximately 0.6650.

27. **(a)** It is profitable to use the machine until $S'(x) = C'(x)$.

$$150 - x^2 = x^2 + \frac{11}{4}x$$

$$2x^2 + \frac{11}{4}x - 150 = 0$$

$$8x^2 + 11x - 600 = 0$$

$$x = \frac{-11 \pm \sqrt{121 - 4(8)(-600)}}{16}$$

$$= \frac{-11 \pm 139}{16}$$

$x = 8$ or $x = -9.375$

It will be profitable to use this machine for 8 years. Reject the negative solution.

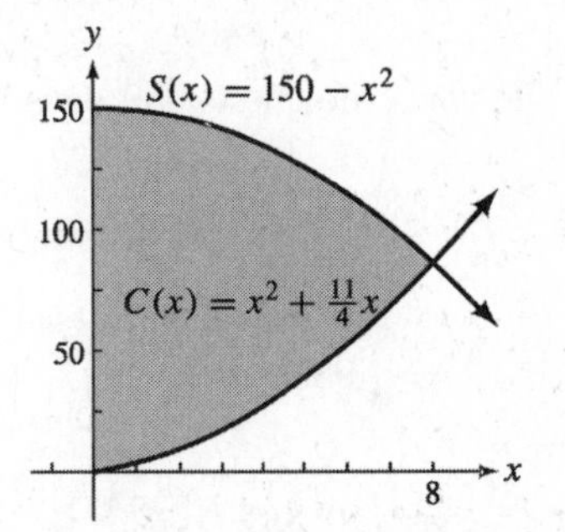

(b) Since $150 - x^2 > x^2 + \frac{11}{4}x$, in the interval $[0, 8]$, the net total savings in the first year are

$$\int_0^1 \left[(150 - x^2) - \left(x^2 + \frac{11}{4}x\right)\right] dx$$

$$= \int_0^1 \left(-2x^2 - \frac{11}{4}x + 150\right) dx$$

$$= \left(\frac{-2x^3}{3} - \frac{11x^2}{8} + 150x\right)\Big|_0^1$$

$$= -\frac{2}{3} - \frac{11}{8} + 150$$

$$\approx \$148.$$

(c) The net total savings over the entire period of use are

$$\int_0^8 \left[(150 - x^2) - \left(x^2 + \frac{11}{4}x\right)\right] dx$$

$$= \left(\frac{-2x^3}{3} - \frac{11x^2}{8} + 150x\right)\Big|_0^8$$

$$= \frac{-2(8^3)}{3} - \frac{11(8^2)}{8} + 150(8)$$

$$= \frac{-1024}{3} - \frac{704}{8} + 1200$$

$$\approx \$771.$$

29. **(a)** $E'(x) = e^{0.1x}$ and $I'(x) = 98.8 - e^{0.1x}$

To find the point of intersection, where profit will be maximized, set the functions equal to each other and solve for x.

$$\begin{aligned} e^{0.1x} &= 98.8 - e^{0.1x} \\ 2e^{0.1x} &= 98.8 \\ e^{0.1x} &= 49.4 \\ 0.1x &= \ln 49.4 \\ x &= \frac{\ln 49.4}{0.1} \\ x &\approx 39 \end{aligned}$$

The optimum number of days for the job to last is 39.

(b) The total income for 39 days is

$$\begin{aligned} &\int_0^{39} (98.8 - e^{0.1x})\, dx \\ &= \left(98.8x - \frac{e^{0.1x}}{0.1}\right)\Bigg|_0^{39} \\ &= \left(98.8x - 10e^{0.1x}\right)\Big|_0^{39} \\ &= [98.8(39) - 10e^{3.9}] - (0 - 10) \\ &= \$3369.18. \end{aligned}$$

(c) The total expenditure for 39 days is

$$\begin{aligned} \int_0^{39} e^{0.1x}\, dx &= \frac{e^{0.1x}}{0.1}\Bigg|_0^{39} \\ &= 10e^{0.1x}\Big|_0^{39} \\ &= 10e^{3.9} - 10 \\ &= \$484.02. \end{aligned}$$

(d) Profit = Income − Expense
= 3369.18 − 484.02
= \$2885.16

31. $S(q) = q^{5/2} + 2q^{3/2} + 50$; $q = 16$ is the equilibrium quantity.

Producers' surplus $= \int_0^{q_0} [p_0 - S(q)]\, dq$, where p_0 is the equilibrium price and q_0 is equilibrium supply.

$$\begin{aligned} p_0 &= S(16) = (16)^{5/2} + 2(16)^{3/2} + 50 \\ &= 1202 \end{aligned}$$

Therefore, the producers' surplus is

$$\begin{aligned} &\int_0^{16} [1202 - (q^{5/2} + 2q^{3/2} + 50)]\, dq \\ &= \int_0^{16} (1152 - q^{5/2} - 2q^{3/2})\, dq \\ &= \left(1152q - \frac{2}{7}q^{7/2} - \frac{4}{5}q^{5/2}\right)\Bigg|_0^{16} \\ &= 1152(16) - \frac{2}{7}(16)^{7/2} - \frac{4}{5}(16)^{5/2} \\ &= 18{,}432 - \frac{32{,}768}{7} - \frac{4096}{5} \\ &= 12{,}931.66. \end{aligned}$$

The producers' surplus is 12,931.66.

33. $D(q) = \frac{200}{(3q+1)^2}$; $q = 3$ is the equilibrium quantity.

$$\text{Consumers' surplus} = \int_0^{q_0} |D(q) - p_0|\, dq$$

$$p_0 = D(3) = 2$$

Therefore, the consumers' surplus is

$$\int_0^3 \left[\frac{200}{(3q+1)^2} - 2\right] dq = \int_0^3 \frac{200}{(3q+1)^2}\, dq - \int_0^3 2\, dq.$$

Let $u = 3q + 1$, so that

$$du = 3\, dq \text{ and } \frac{1}{3}\, du = dq.$$

$$\begin{aligned} \int_0^3 \frac{200}{(3q+1)^2}\, dq - \int_0^3 2\, dq &= \frac{1}{3}\int_1^{10} \frac{200}{u^2}\, du - \int_0^3 2\, dq \\ &= \frac{200}{3}\int_1^{10} u^{-2} du - \int_0^3 2\, dq \\ &= \frac{200}{3} \cdot \frac{u^{-1}}{-1}\Bigg|_1^{10} - 2q\Bigg|_0^3 \\ &= -\frac{200}{3u}\Bigg|_1^{10} - 6 \\ &= -\frac{200}{30} + \frac{200}{3} - 6 \\ &= 54 \end{aligned}$$

35. $S(q) = q^2 + 10q$
$D(q) = 900 - 20q - q^2$

(a) The graphs of the supply and demand functions are parabolas with vertices at $(-5, -25)$ and $(-10, 1900)$, respectively.

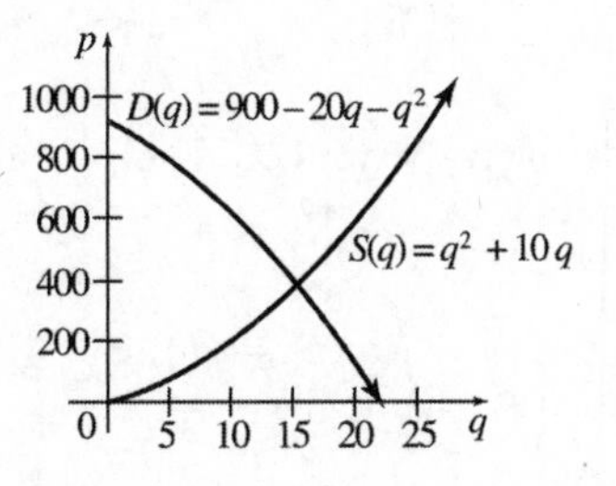

(b) The graphs intersect at the point where the y-coordinates are equal.

$$\begin{aligned} q^2 + 10q &= 900 - 20q - q^2 \\ 2q^2 + 30q - 900 &= 0 \\ q^2 + 15q - 450 &= 0 \\ (q+30)(q-15) &= 0 \\ q = -30 \quad &\text{or} \quad q = 15 \end{aligned}$$

Disregard the negative solution.
The supply and demand functions are in equilibrium when $q = 15$.

$$S(15) = 15^2 + 10(15) = 375$$

The point is $(15, 375)$.

(c) Find the consumers' surplus.

$$\int_0^{q_0} [D(q) - p_0)]\,dq$$

$$p_0 = D(15) = 375$$

$$\begin{aligned} &\int_0^{15} [(900 - 20q - q^2) - 375]\,dq \\ &= \int_0^{15} (525 - 20q - q^2)\,dq \\ &= \left(525q - 10q^2 - \frac{1}{3}q^3\right)\Big|_0^{15} \\ &= \left[525(15) - 10(15)^2 - \frac{1}{3}(15)^3\right] - 0 \\ &= 4500 \end{aligned}$$

The consumer's surplus is \$4500.

(d) Find the producers' surplus.

$$\int_0^{q_0} [p_0 - S(q)]\,dq$$

$$p_0 = S(15) = 375$$

$$\begin{aligned} &\int_0^{15} [375 - (q^2 + 10q)]dq \\ &= \int_0^{15} (375 - q^2 - 10q)\,dq \\ &= \left(375q - \frac{1}{3}q^3 - 5q^2\right)\Big|_0^{15} \\ &= \left[375(15) - \frac{1}{3}(15)^3 - 5(15)^2\right] - 0 \\ &= 3375 \end{aligned}$$

The producer's surplus is \$3375.

37. (a) $S(q) = q^2 + 10q$; $S(q) = 264$ is the price the government set.

$$\begin{aligned} 264 &= q^2 + 10q \\ 0 &= q^2 + 10q - 264 \\ 0 &= (q - 12)(q + 22) \\ q = 12 \quad &\text{or} \quad q = -22 \end{aligned}$$

Only 12 is a meaningful solution here. Thus, 12 units of oil will be produced.

(b) The consumers' surplus is given by

$$\begin{aligned} &\int_0^{12} (900 - 20q - q^2 - 264)dq \\ &= \int_0^{12} (636 - 20q - q^2)dq \\ &= \left(636q - 10q^2 - \frac{1}{3}q^3\right)\Big|_0^{12} \\ &= 636(12) - 10(12)^2 - \frac{1}{3}(12)^3 - 0 \\ &= 5616 \end{aligned}$$

Here the consumer' surplus is \$5616. In this case, the consumers' surplus is $5616 - 4500 = \$1116$ larger.

(c) The producers' surplus is given by

$$\begin{aligned} &\int_0^{12} [264 - (q^2 + 10q)]dq \\ &= \int_0^{12} (264 - q^2 - 10q)dq \\ &= \left(264q - \frac{1}{3}q^3 - 5q^2\right)\Big|_0^{12} \\ &= 264(12) - \frac{1}{3}(12)^3 - 5(12)^2 - 0 \\ &= 1872 \end{aligned}$$

Here the producers' surplus is \$1872. In this case, the producers' surplus is $3375 - 1872 = \$1503$ smaller.

(d) For the equilibrium price, the total consumers' and producers' surplus is

$$4500 + 3375 = \$7875$$

For the government price, the total consumers' and producers' surplus is

$$5616 + 1872 = \$7488.$$

The difference is

$$7875 - 7488 = \$387.$$

39. **(a)** The pollution level in the lake is changing at the rate $f(t) - g(t)$ at any time t. We find the amount of pollution by integrating.

$$\int_0^{12} [f(t) - g(t)]dt$$

$$= \int_0^{12} [10(1 - e^{-0.5t}) - 0.4t]dt$$

$$= \left(10t - 10 \cdot \frac{1}{-0.5} e^{-0.5t} - 0.4 \cdot \frac{1}{2}t^2\right)\Bigg|_0^{12}$$

$$= (20e^{-0.5t} + 10t - 0.2t^2)\Big|_0^{12}$$

$$= [20e^{-0.5(12)} + 10(12) - 0.2(12)^2]$$
$$- [20e^{-0.5(0)} + 10(0) - 0.2(0)^2]$$
$$= (20e^{-6} + 91.2) - (20)$$
$$= 20e^{-6} + 71.2 \approx 71.25$$

After 12 hours, there are about 71.25 gallons.

(b) The graphs of the functions intersect at about 25.00. So the rate that pollution enters the lake equals the rate the pollution is removed at about 25 hours.

(c) $\int_0^{25} [f(t) - g(t)]dt$

$$= (20e^{-0.5t} + 10t - 0.2t^2)\Big|_0^{25}$$
$$= [20e^{-0.5(25)} + 10(25) - 0.2(25)^2] - 20$$
$$= 20e^{-12.5} + 105$$
$$\approx 105$$

After 25 hours, there are about 105 gallons.

(d) For $t > 25, g(t) > f(t)$, and pollution is being removed at the rate $g(t) - f(t)$. So, we want to solve for c, where

$$\int_0^c [f(t) - g(t)]dt = 0.$$

(Altternatively, we could solve for c in

$$\int_{25}^c [g(t) - f(t)dt = 105.$$

One way to do this with a graphing calculator is to graph the function

$$y = \int_0^x [f(t) - g(t)]dt$$

and determine the values of x for which $y = 0$. The first window shows how the function can be defined.

```
Plot1 Plot2 Plot3
\Y1=fnInt(10(1-e
^(-0.5X))-0.4X,X
,0,X)
\Y2=
\Y3=
\Y4=
\Y5=
```

A suitable window for the graph is $[0, 50]$ by $[0, 110]$.

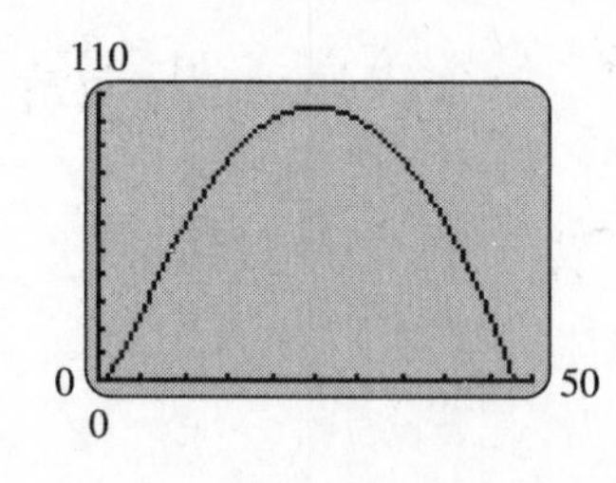

Use the calculator's features to approximate where the graph intersects the x-axis. These are at 0 and about 47.91. Therefore, the pollution will be removed from the lake after about 47.91 hours.

41. $I(x) = 0.9x^2 + 0.1x$

(a) $I(0.1) = 0.9(0.1)^2 + 0.1(0.1)$
$= 0.019$

The lower 10% of income producers earn 1.9% of total income of the population.

(b) $I(0.4) = 0.9(0.4)^2 + 0.1(0.4) = 0.184$

The lower 40% of income producers earn 18.4% of total income of the population.

(c) The graph of $I(x) = x$ is a straight line through the points $(0, 0)$ and $(1, 1)$. The graph of $I(x) = 0.9x^2 + 0.1x$ is a parabola with vertex $\left(-\frac{1}{18}, -\frac{1}{360}\right)$. Restrict the domain to $0 \le x \le 1$.

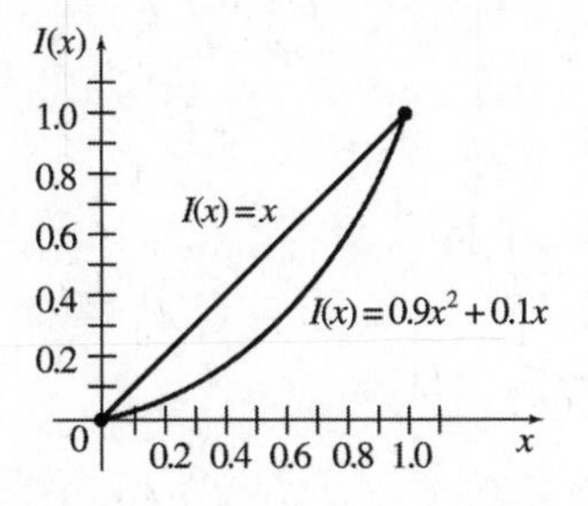

(d) To find the points of intersection, solve

$$x = 0.9x^2 + 0.1x.$$
$$0.9x^2 - 0.9x = 0$$
$$0.9x(x - 1) = 0$$
$$x = 0 \quad \text{or} \quad x = 1$$

The area between the curves is given by

$$\int_0^1 [x - (0.9x^2 + 0.1x)]\, dx$$

$$= \int_0^1 (0.9x - 0.9x^2)\, dx$$

$$= \left(\frac{0.9x^2}{2} - \frac{0.9x^3}{3}\right)\Bigg|_0^1$$

$$= \frac{0.9}{2} - \frac{0.9}{3} = 0.15.$$

7.6 Numerical Integration

1. $\displaystyle\int_0^2 (3x^2 + 2)\, dx$

$n = 4, b = 2, a = 0, f(x) = 3x^2 + 2$

i	x_i	$f(x_i)$
0	0	2
1	$\frac{1}{2}$	2.75
2	1	5
3	$\frac{3}{2}$	8.75
4	2	14

(a) Trapezoidal rule:

$$\int_0^2 (3x^2 + 2)\, dx$$
$$\approx \frac{2 - 0}{4}\left[\frac{1}{2}(2) + 2.75 + 5 + 8.75 + \frac{1}{2}(14)\right]$$
$$= 0.5(24.5)$$
$$= 12.25$$

(b) Simpson's rule:

$$\int_0^2 (3x^2 + 2)\, dx$$
$$\approx \frac{2 - 0}{3(4)}[2 + 4(2.75) + 2(5) + 4(8.75) + 14]$$
$$= \frac{2}{12}(72)$$
$$= 12$$

(c) Exact value:

$$\int_0^2 (3x^2 + 2)\, dx = (x^3 + 2x)\Big|_0^2$$
$$= (8 + 4) - 0$$
$$= 12$$

3. $\displaystyle\int_{-1}^3 \frac{3}{5 - x}\, dx$

$n = 4, b = 3, a = -1,\ f(x) = \dfrac{3}{5 - x}$

i	x_i	$f(x_i)$
0	-1	0.5
1	0	0.6
2	1	0.75
3	2	1
4	3	1.5

(a) Trapezoidal rule:

$$\int_{-1}^{3} \frac{3}{5-x}\,dx$$
$$\approx \frac{3-(-1)}{4}\left[\frac{1}{2}(0.5)+0.6+0.75+1+\frac{1}{2}(1.5)\right]$$
$$= 1(3.35)$$
$$= 3.35$$

(b) Simpson's rule:

$$\int_{-1}^{3} \frac{3}{5-x}\,dx$$
$$\approx \frac{3-(-1)}{3(4)}[0.5+4(0.6)+2(0.75)+4(1)+1.5]$$
$$= \frac{1}{3}\left(\frac{99}{10}\right)$$
$$= \frac{33}{10} \approx 3.3$$

(c) Exact value:

$$\int_{-1}^{3} \frac{3}{5-x}\,dx = -3\ln|5-x|\,\Big|_{-1}^{3}$$
$$= -3(\ln|2| - \ln|6|)$$
$$= 3\ln 3 \approx 3.296$$

5. $\int_{-1}^{2} (2x^3+1)\,dx$

$n=4,\ b=2,\ a=-1,\ f(x)=2x^3+1$

i	x_i	$f(x)$
0	-1	-1
1	$-\frac{1}{4}$	$\frac{31}{32}$
2	$\frac{1}{2}$	$\frac{5}{4}$
3	$\frac{5}{4}$	$\frac{157}{32}$
4	2	17

(a) Trapezoidal rule:

$$\int_{-1}^{2} (2x^3+1)\,dx$$
$$\approx \frac{2-(-1)}{4}\left[\frac{1}{2}(-1)+\frac{31}{32}+\frac{5}{4}+\frac{157}{32}+\frac{1}{2}(17)\right]$$
$$= 0.75(15.125)$$
$$\approx 11.34$$

(b) Simpson's rule:

$$\int_{-1}^{2} (2x^3+1)\,dx$$
$$\approx \frac{2-(-1)}{3(4)}\left[-1+4\left(\frac{31}{32}\right)+2\left(\frac{5}{4}\right)+4\left(\frac{157}{32}\right)+17\right]$$
$$= \frac{1}{4}(42)$$
$$= 10.5$$

(c) Exact value:

$$\int_{-1}^{2} (2x^3+1)\,dx$$
$$= \left(\frac{x^4}{2}+x\right)\Big|_{-1}^{2}$$
$$= (8+2) - \left(\frac{1}{2}-1\right)$$
$$= \frac{21}{2}$$
$$= 10.5$$

7. $\int_{1}^{5} \frac{1}{x^2}\,dx$

$n=4, b=5, a=1, f(x)=\frac{1}{x^2}$

i	x_i	$f(x_i)$
0	1	1
1	2	0.25
2	3	0.1111
3	4	0.0625
4	5	0.04

(a) Trapezoidal rule:

$$\int_{1}^{5} \frac{1}{x^2}\,dx$$
$$\approx \frac{5-1}{4}\left[\frac{1}{2}(1)+0.25+0.1111+0.0625+\frac{1}{2}(0.04)\right]$$
$$\approx 0.9436$$

(b) Simpson's rule:

$$\int_{1}^{5} \frac{1}{x^2}\,dx$$
$$\approx \frac{5-1}{12}[1+4(0.25)+2(0.1111)+4(0.0625)+0.04)]$$
$$\approx 0.8374$$

(c) Exact value:

$$\int_1^5 x^{-2}\,dx = -x^{-1}\Big|_1^5 = -\frac{1}{5}+1 = \frac{4}{5} = 0.8$$

9. $\displaystyle\int_0^1 4xe^{-x^2}\,dx$

$n = 4,\ b = 1,\ a = 0,\ f(x) = 4xe^{-x^2}$

i	x_i	$f(x_i)$
0	0	0
1	$\frac{1}{4}$	$e^{-1/16}$
2	$\frac{1}{2}$	$2e^{-1/4}$
3	$\frac{3}{4}$	$3e^{-9/16}$
4	1	$4e^{-1}$

(a) Trapezoidal rule:

$$\int_0^1 4xe^{-x^2}\,dx \approx \frac{1-0}{4}\left[\frac{1}{2}(0)+e^{-1/16}+2e^{-1/4} + 3e^{-9/16}+\frac{1}{2}(4e^{-1})\right]$$
$$= \frac{1}{4}(e^{-1/16}+2e^{-1/4}+3e^{-9/16}+2e^{-1}) \approx 1.236$$

(b) Simpson's rule:

$$\int_0^1 4xe^{-x^2}\,dx \approx \frac{1-0}{3(4)}[0+4(e^{-1/16})+2(2e^{-1/4}) + 4(3e^{-9/16})+4e^{-1}]$$
$$= \frac{1}{12}(4e^{-1/16}+4e^{-1/4}+12e^{-9/16}+4e^{-1}) \approx 1.265$$

(c) Exact value:

$$\int_0^1 4xe^{-x^2}\,dx = -2e^{-x^2}\Big|_0^1 = (-2e^{-1})-(-2) = 2-2e^{-1} \approx 1.264$$

11. $y = \sqrt{4-x^2}$

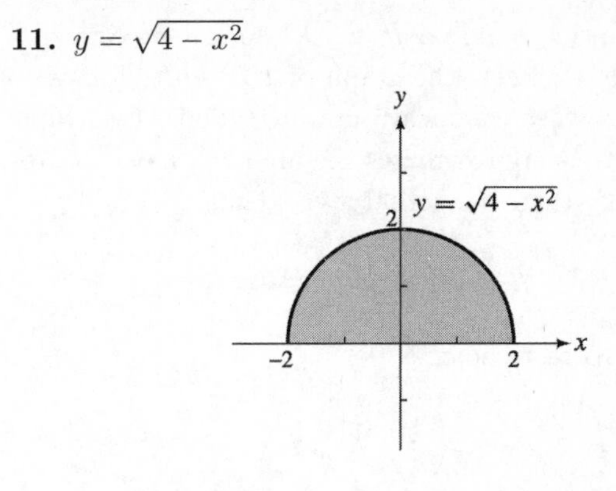

$n = 8,\ b = 2,\ a = -2,\ f(x) = \sqrt{4-x^2}$

i	x_i	y
0	−2.0	0
1	−1.5	1.32289
2	−1.0	1.73205
3	−0.5	1.93649
4	0	2
5	0.5	1.93649
6	1.0	1.73205
7	1.5	1.32289
8	2.0	0

(a) Trapezoidal rule:

$$\int_{-2}^2 \sqrt{4-x^2}\,dx \approx \frac{2-(-2)}{8} \cdot \left[\frac{1}{2}(0)+1.32289+1.73205+\cdots+\frac{1}{2}(0)\right] \approx 5.991$$

(b) Simpson's rule:

$$\int_{-2}^2 \sqrt{4-x^2}\,dx \approx \frac{2-(-2)}{3(8)} \cdot[0+4(1.32289)+2(1.73205)+4(1.93649)+2(2) + 4(1.93649)+2(1.73205)+4(1.32289)+0] \approx 6.167$$

(c) Area of semicircle $= \dfrac{1}{2}\pi r^2 = \dfrac{1}{2}\pi(2)^2 \approx 6.283$

Simpson's rule is more accurate.

13. Since $f(x) > 0$ and $f''(x) > 0$ for all x between a and b, we know the graph of $f(x)$ on the interval from a to b is concave upward. Thus, the trapezoid that approximates the area will have an area greater than the actual area. Thus,

$$T > \int_a^b f(x)\,dx.$$

The correct choice is (b).

15. (a) $\displaystyle\int_0^1 x^4\,dx = \left(\frac{1}{5}\right)x^5\Big|_0^1$

$$= \frac{1}{5}$$

$$= 0.2$$

(b) $n = 4,\ b = 1,\ a = 0,\ f(x) = x^4$

$$\int_0^1 x^4\,dx \approx \frac{1-0}{4}\left[\frac{1}{2}(0) + \frac{1}{256} + \frac{1}{16} + \frac{81}{256} + \frac{1}{2}(1)\right]$$

$$= \frac{1}{4}\left(\frac{226}{256}\right)$$

$$\approx 0.220703$$

$n = 8,\ b = 1,\ a = 0,\ f(x) = x^4$

$$\int_0^1 x^4\,dx \approx \frac{1-0}{8}\left[\frac{1}{2}(0) + \frac{1}{4096} + \frac{1}{256} + \frac{81}{4096} + \frac{1}{16} + \frac{625}{4096} + \frac{81}{256} + \frac{2401}{4096} + \frac{1}{2}(1)\right]$$

$$= \frac{1}{8}\left(\frac{6724}{4096}\right)$$

$$\approx 0.20520$$

$n = 16,\ b = 1,\ a = 0,\ f(x) = x^4$

$$\int_0^1 x^4\,dx \approx \frac{1-0}{16}\left[\frac{1}{2}(0) + \frac{1}{65{,}536} + \frac{1}{4096} + \frac{81}{65{,}536} + \frac{1}{256} + \frac{625}{65{,}536} + \frac{81}{4096} + \frac{2401}{65{,}536} + \frac{1}{16} + \frac{6561}{65{,}536} + \frac{625}{4096} + \frac{14{,}641}{65{,}536} + \frac{81}{256} + \frac{28{,}561}{65{,}536} + \frac{2401}{4096} + \frac{50{,}625}{65{,}536} + \frac{1}{2}(1)\right]$$

$$\approx \frac{1}{16}\left(\frac{211{,}080}{65{,}536}\right)$$

$$\approx 0.201302$$

$n = 32,\ b = 1,\ a = 0,\ f(x) = x^4$

$$\int_0^1 x^4\,dx$$

$$\approx \frac{1-0}{32}\left[\frac{1}{2}(0) + \frac{1}{1{,}048{,}576} + \frac{1}{65{,}536} + \frac{81}{1{,}048{,}576} + \frac{1}{4096} + \frac{625}{1{,}048{,}576} + \frac{81}{65{,}536} + \frac{2401}{1{,}048{,}576} + \frac{1}{256} + \frac{6561}{1{,}048{,}576} + \frac{625}{65{,}536} + \frac{14{,}641}{1{,}048{,}576} + \frac{81}{4096} + \frac{28{,}561}{1{,}048{,}576} + \frac{2401}{65{,}536} + \frac{50{,}625}{1{,}048{,}576} + \frac{1}{16} + \frac{83{,}521}{1{,}048{,}576} + \frac{6561}{65{,}536} + \frac{130{,}321}{1{,}048{,}576} + \frac{625}{4096} + \frac{194{,}481}{1{,}048{,}576} + \frac{14{,}641}{65{,}536} + \frac{279{,}841}{1{,}048{,}576} + \frac{81}{256} + \frac{390{,}625}{1{,}048{,}576} + \frac{28{,}561}{65{,}536} + \frac{531{,}441}{1{,}048{,}576} + \frac{2401}{4096} + \frac{707{,}281}{1{,}048{,}576} + \frac{50{,}625}{65{,}536} + \frac{923{,}521}{1{,}048{,}576} + \frac{1}{2}(1)\right]$$

$$\approx \frac{1}{32}\left(\frac{6{,}721{,}808}{1{,}048{,}576}\right) \approx 0.200325$$

To find error for each value of n, subtract as indicated.

$n = 4$: $(0.220703 - 0.2) = 0.020703$
$n = 8$: $(0.205200 - 0.2) = 0.005200$
$n = 16$: $(0.201302 - 0.2) = 0.001302$
$n = 32$: $(0.200325 - 0.2) = 0.000325$

(c) $p = 1$

$$4^1(0.020703) = 4(0.020703) = 0.082812$$
$$8^1(0.005200) = 8(0.005200) = 0.0416$$

Since these are not the same, try $p = 2$.

$p = 2$:

$$4^2(0.020703) = 16(0.020703) = 0.331248$$
$$8^2(0.005200) = 64(0.005200) = 0.3328$$
$$16^2(0.001302) = 256(0.001302) = 0.333312$$
$$32^2(0.000325) = 1024(0.000325) = 0.3328$$

Since these values are all approximately the same, the correct choice is $p = 2$.

17. (a) $\displaystyle\int_0^1 x^4\,dx = \frac{1}{5}x^5\Big|_0^1$

$$= \frac{1}{5}$$

$$= 0.2$$

(b) $n = 4,\ b = 1,\ a = 0,\ f(x) = x^4$

$$\int_0^1 x^4\,dx \approx \frac{1-0}{3(4)}\left[0 + 4\left(\frac{1}{256}\right) + 2\left(\frac{1}{16}\right) + 4\left(\frac{81}{256}\right) + 1\right]$$

$$= \frac{1}{12}\left(\frac{77}{32}\right)$$

$$\approx 0.2005208$$

$n = 8,\ b = 1,\ a = 0,\ f(x) = x^4$

$$\int_0^1 x^4\,dx \approx \frac{1-0}{3(8)}\left[0 + 4\left(\frac{1}{4096}\right) + 2\left(\frac{1}{256}\right) + 4\left(\frac{81}{4096}\right) + 2\left(\frac{1}{16}\right) + 4\left(\frac{625}{4096}\right) + 2\left(\frac{81}{256}\right) + 4\left(\frac{2401}{4096}\right) + 1\right]$$

$$= \frac{1}{24}\left(\frac{4916}{1024}\right)$$

$$\approx 0.2000326$$

$n = 16,\ b = 1,\ a = 0,\ f(x) = x^4$

$$\int_0^1 x^4\,dx$$

$$\approx \frac{1-0}{3(16)}\left[0 + 4\left(\frac{1}{65{,}536}\right) + 2\left(\frac{1}{4096}\right) + 4\left(\frac{81}{65{,}536}\right) + 2\left(\frac{1}{256}\right) + 4\left(\frac{625}{65{,}536}\right) + 2\left(\frac{81}{4096}\right) + 4\left(\frac{2401}{65{,}536}\right) + 2\left(\frac{1}{16}\right) + 4\left(\frac{6561}{65{,}536}\right) + 2\left(\frac{625}{4096}\right) + 4\left(\frac{14{,}641}{65{,}536}\right) + 2\left(\frac{81}{256}\right) + 4\left(\frac{28{,}561}{65{,}536}\right) + 2\left(\frac{2401}{4096}\right) + 4\left(\frac{50{,}625}{65{,}536} + 1\right)\right]$$

$$= \frac{1}{48}\left(\frac{157{,}288}{16{,}384}\right) \approx 0.2000020$$

$n = 32,\ b = 1,\ a = 0,\ f(x) = x^4$

$$\int_0^1 x^4\,dx$$

$$\approx \frac{1-0}{3(32)}\left[0 + 4\left(\frac{1}{1{,}048{,}576}\right) + 2\left(\frac{1}{65{,}536}\right) + 4\left(\frac{81}{1{,}048{,}576}\right) + 2\left(\frac{1}{4096}\right) + 4\left(\frac{625}{1{,}048{,}576}\right) + 2\left(\frac{625}{65{,}536}\right) + 4\left(\frac{14{,}641}{1{,}048{,}576}\right) + 2\left(\frac{81}{4096}\right) + 4\left(\frac{28{,}561}{1{,}048{,}576}\right) + 2\left(\frac{2401}{65{,}536}\right) + 4\left(\frac{50{,}625}{1{,}048{,}576}\right) + 2\left(\frac{1}{16}\right) + 4\left(\frac{83{,}521}{1{,}048{,}576}\right) + 2\left(\frac{6561}{65{,}536}\right) + 4\left(\frac{130{,}321}{1{,}048{,}576}\right) + 2\left(\frac{625}{4096}\right) + 4\left(\frac{194{,}481}{1{,}048{,}576}\right) + 2\left(\frac{14{,}641}{65{,}536}\right) + 4\left(\frac{279{,}841}{1{,}048{,}576}\right) + 2\left(\frac{81}{256}\right) + 4\left(\frac{390{,}625}{1{,}048{,}576}\right) + 2\left(\frac{28{,}561}{65{,}536}\right) + 4\left(\frac{531{,}441}{1{,}048{,}576}\right) + 2\left(\frac{2401}{4096}\right) + 4\left(\frac{707{,}281}{1{,}048{,}576}\right) + 2\left(\frac{50{,}625}{65{,}536}\right) + 4\left(\frac{923{,}521}{1{,}048{,}576}\right) + 1\right]$$

$$= \frac{1}{96}\left(\frac{50{,}033{,}168}{262{,}144}\right) \approx 0.2000001$$

To find error for each value of n, subtract as indicated.

$n = 4$: $(0.2005208 - 0.2) = 0.0005208$
$n = 8$: $(0.2000326 - 0.2) = 0.0000326$
$n = 16$: $(0.2000020 - 0.2) = 0.0000020$
$n = 32$: $(0.2000001 - 0.2) = 0.0000001$

(c) $p = 1$:

$$4^1(0.0005208) = 4(0.0005208) = 0.0020832$$
$$8^1(0.0000326) = 8(0.0000326) = 0.0002608$$

Try $p = 2$:

$$4^2(0.0005208) = 16(0.0005208) = 0.0083328$$
$$8^2(0.0000326) = 64(0.0000326) = 0.0020864$$

Try $p = 3$:

$$4^3(0.0005208) = 64(0.0005208) = 0.0333312$$
$$8^3(0.0000326) = 512(0.0000326) = 0.0166912$$

Try $p = 4$:

$$4^4(0.0005208) = 256(0.0005208) = 0.1333248$$
$$8^4(0.0000326) = 4096(0.0000326) = 0.1335296$$
$$16^4(0.0000020) = 65536(0.0000020) = 0.131072$$
$$32^4(0.0000001) = 1048576(0.0000001) = 0.1048576$$

These are the closest values we can get; thus, $p = 4$.

19. Midpoint rule:

$n = 4,\ b = 5,\ a = 1, f(x) = \frac{1}{x^2}, \Delta x = 1$

i	x_i	$f(x_i)$
1	$\frac{3}{2}$	$\frac{4}{9}$
2	$\frac{5}{2}$	$\frac{4}{25}$
3	$\frac{7}{2}$	$\frac{4}{49}$
4	$\frac{9}{2}$	$\frac{4}{81}$

$$\int_1^5 \frac{1}{x^2}dx \approx \sum_{i=1}^{4} f(x_i)\Delta x$$
$$= \frac{4}{9}(1) + \frac{4}{25}(1) + \frac{4}{49}(1) + \frac{4}{81}(1)$$
$$\approx 0.7355$$

Simpson's rule:

$m = 8, b = 5, a = 1, f(x) = \frac{1}{x^2}$

i	x_i	$f(x_i)$
0	1	1
1	$\frac{3}{2}$	$\frac{4}{9}$
2	2	$\frac{1}{4}$
3	$\frac{5}{2}$	$\frac{4}{25}$
4	3	$\frac{1}{9}$
5	$\frac{7}{2}$	$\frac{4}{49}$
6	4	$\frac{1}{16}$
7	$\frac{9}{2}$	$\frac{4}{81}$
8	5	$\frac{1}{25}$

$$\int_1^5 \frac{1}{x^2}dx$$
$$\approx \frac{5-1}{3(8)}\left[1 + 4\left(\frac{4}{9}\right) + 2\left(\frac{1}{4}\right) + 4\left(\frac{4}{25}\right) + 2\left(\frac{1}{9}\right) + 4\left(\frac{4}{49}\right) + 2\left(\frac{1}{16}\right) + 4\left(\frac{4}{81}\right) + \frac{1}{25}\right]$$
$$\approx \frac{1}{6}(4.82906)$$
$$\approx 0.8048$$

From #7 part a, $T \approx 0.9436$, when $n = 4$. To verify the formula evaluate $\frac{2M+T}{3}$.

$$\frac{2M+T}{3} \approx \frac{2(0.7355) + 0.9436}{3}$$
$$\approx 0.8048$$

21. (a)

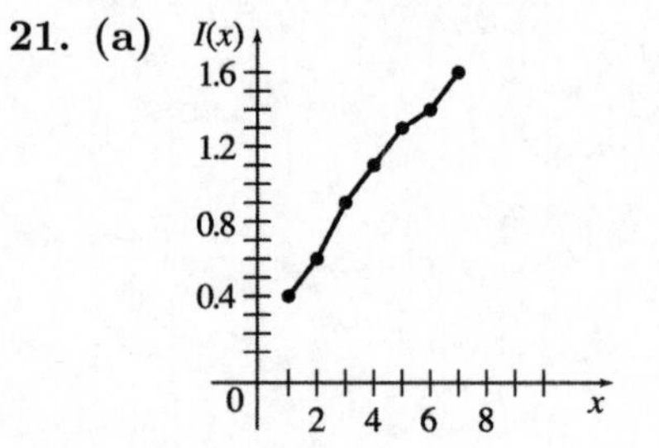

(b) $A = \frac{7-1}{6}\left[\frac{1}{2}(0.4) + 0.6 + 0.9 + 1.1 + 1.3 + 1.4 + \frac{1}{2}(1.6)\right]$

$= 6.3$

(c) $A = \frac{7-1}{3(6)}[0.4 + 4(0.6) + 2(0.9) + 4(1.1) + 2(1.3) + 4(1.4) + 1.6]$

≈ 6.27

23. $y = e^{-t^2} + \dfrac{1}{t+1}$

The total reaction is

$$\int_1^9 \left(e^{-t^2} + \frac{1}{t+1}\right) dt.$$

$n = 8,\ b = 9,\ a = 1,\ f(t) = e^{-t^2} + \frac{1}{t+1}$

i	x_i	$f(x_i)$
0	1	0.8679
1	2	0.3516
2	3	0.2501
3	4	0.2000
4	5	0.1667
5	6	0.1429
6	7	0.1250
7	8	0.1111
8	9	0.1000

(a) Trapezoidal rule:

$$\int_1^9 \left(e^{-t^2} + \frac{1}{t+1}\right) dt$$

$$\approx \frac{9-1}{8}\left[\frac{1}{2}(0.8679) + 0.3516 + 0.2501 + \cdots + \frac{1}{2}(0.1000)\right]$$

$$\approx 1.831$$

(b) Simpson's rule:

$$\int_1^9 \left(e^{-t^2} + \frac{1}{t+1}\right) dt$$

$$\approx \frac{9-1}{3(8)}[0.8679 + 4(0.3516) + 2(0.2501) + 4(0.2000) + 2(0.1667) + 4(0.1429) + 2(0.1250) + 4(0.1111) + 0.1000]$$

$$= \frac{1}{3}(5.2739)$$

$$\approx 1.758$$

25. Note that heights may differ depending on the readings of the graph. Thus, answers may vary. $n = 10,\ b = 20,\ a = 0$

i	x_i	$f(x_i)$
0	0	0
1	2	5
2	4	3
3	6	2
4	8	1.5
5	10	1.2
6	12	1
7	14	0.5
8	16	0.3
9	18	0.2
10	20	0.2

Area under curve for Formulation A

$$= \frac{20-0}{10}\left[\frac{1}{2}(0) + 5 + 3 + 2 + 1.5 + 1.2 + 1 + 0.5 + 0.3 + 0.2 + \frac{1}{2}(0.2)\right]$$

$$= 2(14.8)$$

$$\approx 30 \text{ mcgh/ml}$$

This represents the total amount of drug available to the patient for each ml of blood.

27. As in Exercise 25, readings on the graph may vary, so answers may vary. The area both under the curve for Formulation A and above the minimum effective concentration line in on the interval $\left[\frac{1}{2}, 6\right]$.

Area under curve for Formulation A on $\left[\frac{1}{2}, 1\right]$, with $n = 1$

$$= \frac{1-\frac{1}{2}}{1}\left[\frac{1}{2}(2+6)\right]$$

$$= \frac{1}{2}(4) = 2$$

Area under curve for Formulation A on $[1, 6]$, with $n = 5$

$$= \frac{6-1}{5}\left[\frac{1}{2}(6) + 5 + 4 + 3 + 2.4 + \frac{1}{2}(2)\right]$$
$$= 18.4$$

Area under minimum effective concentration line $[\frac{1}{2}, 6]$

$$= 5.5(2) = 11.0$$

Area under the curve for Formulation A and above minimum effective concentration line

$$= 2 + 18.4 - 11.0$$

$$\approx 9 \text{ mcgh/ml}$$

This represents the total effective amount of drug available to the patient for each ml of blood.

29. $y = b_0 w^{b_1} e^{-b_2 w}$

(a) If $t = 7w$ then $w = \frac{t}{7}$.

$$y = b_0 \left(\frac{t}{7}\right)^{b_1} e^{-b_2 t/7}$$

(b) Replacing the constants with the given values, we have

$$y = 5.955 \left(\frac{t}{7}\right)^{0.233} e^{-0.027t/7} dt$$

In 25 weeks, there are 175 days.

$$\int_0^{175} 5.955 \left(\frac{t}{7}\right)^{0.233} e^{-0.027t/7} dt$$

$n = 10, b = 175, a = 0,$

$$f(t) = 5.955 \left(\frac{t}{7}\right)^{0.233} e^{-0.027t/7}$$

i	t_i	$f(t_i)$
0	0	0
1	17.5	6.89
2	35	7.57
3	52.5	7.78
4	70	7.77
5	87.5	7.65
6	105	7.46
7	122.5	7.23
8	140	6.97
9	157.5	6.70
10	175	6.42

Trapezoidal rule:

$$\int_0^{175} 5.955 \left(\frac{t}{7}\right)^{0.233} e^{-0.027t/7} dt$$

$$\approx \frac{175-0}{10}\left[\frac{1}{2}(0) + 6.89 + 7.57 + 7.78 + 7.77 + 7.65 + 7.46 + 7.23 + 6.97 + 6.70 + \frac{1}{2}(6.42)\right]$$

$$= 17.5(69.23)$$
$$= 1211.525$$

The total milk consumed is about 1212 kg.

Simpson's rule:

$$\int_0^{175} 5.955 \left(\frac{t}{7}\right)^{0.233} e^{-0.027t/7} dt$$

$$\approx \frac{175-0}{3(10)}[0 + 4(6.89) + 2(7.57) + 4(7.78) + 2(7.77) + 4(7.65) + 2(7.46) + 4(7.23) + 2(6.97) + 4(6.70) + 6.42]$$

The total milk consumed is about 1231 kg.

(c) Replacing the constants with the given values, we have

$$y = 8.409 \left(\frac{t}{7}\right)^{0.143} e^{-0.037t/7}.$$

In 25 weeks, there are 175 days.

$$\int_0^{175} 8.409 \left(\frac{t}{7}\right)^{0.143} e^{-0.037t/7} dt$$

$n = 10,\ b = 175,\ a = 0,$

$$f(t) = 8.409 \left(\frac{t}{7}\right)^{0.143} e^{-0.037t/7}$$

i	t_i	$f(t_i)$
0	0	0
1	17.5	8.74
2	35	8.80
3	52.5	8.50
4	70	8.07
5	87.5	7.60
6	105	7.11
7	122.5	6.63
8	140	6.16
9	157.5	5.71
10	175	5.28

Trapezoidal rule:

$$\int_0^{175} 8.409\left(\frac{t}{7}\right)^{0.143} e^{-0.037t/7}dt$$

$$\approx \frac{175-0}{10}\left[\frac{1}{2}(0)+8.74+8.80+8.50 +8.07+7.60+7.11+6.63 +6.16+5.71+\frac{1}{2}(5.28)\right]$$

$$= 17.5(69.96)$$
$$= 1224.30$$

The total milk consumed is about 1224 kg.

Simpson's rule:

$$\int_0^{175} 8.409\left(\frac{t}{7}\right)^{0.143} e^{-0.037t/7}dt$$

$$\approx \frac{175-0}{3(10)}[0+4(8.74)+2(8.80)+4(8.50) +2(8.07)+4(7.60)+2(7.11)+4(6.63) +2(6.16)+4(5.71)+5.28]$$

$$= \frac{35}{6}(214.28)$$
$$= 1249.97$$

The total milk consumed is about 1250 kg.

31. **(a)**

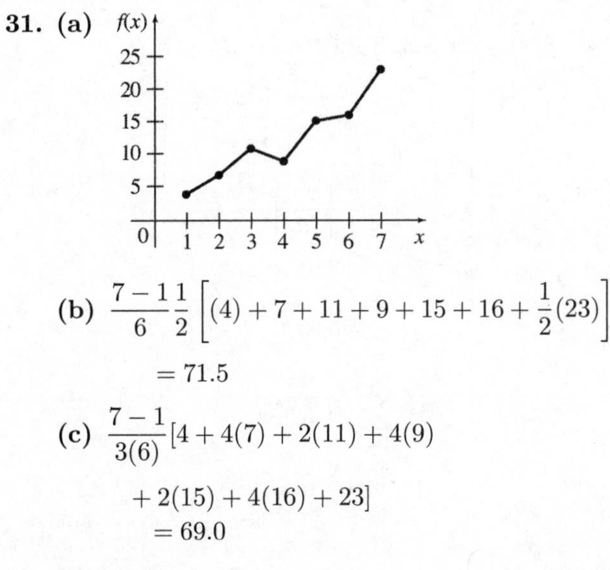

(b) $\frac{7-1}{6}\frac{1}{2}\left[(4)+7+11+9+15+16+\frac{1}{2}(23)\right]$

$= 71.5$

(c) $\frac{7-1}{3(6)}[4+4(7)+2(11)+4(9) +2(15)+4(16)+23]$

$= 69.0$

33. We need to evaluate

$$\int_{12}^{36}(105e^{0.01\sqrt{x}}+32)\,dx.$$

Using a calculator program for Simpson's rule with $n = 20$, we obtain 3413.18 as the value of this integral. This indicates that the total revenue between the twelfth and thirty-sixth months is about 3413.

35. Use a calculator program for Simpson's rule with $n = 20$ to evaluate each of the integrals in this exercise.

(a) $\int_{-1}^{1}\left(\frac{1}{\sqrt{2\pi}}e^{-x^2/2}\right)dx \approx 0.6827$

The probability that a normal random variable is within 1 standard deviation of the mean is about 0.6827.

(b) $\int_{-2}^{2}\left(\frac{1}{\sqrt{2\pi}}e^{-x^2/2}\right)dx \approx 0.9545$

The probability that a normal random variable is within 2 standard deviations of the mean is about 0.9545.

(c) $\int_{-3}^{3}\left(\frac{1}{\sqrt{2\pi}}e^{-x^2/2}\right)dx \approx 0.9973$

The probability that a normal random variable is within 3 standard deviations of the mean is about 0.9973.

Chapter 7 Review Exercises

5. $\int(2x+3)\,dx = \frac{2x^2}{2}+3x+C$

$= x^2+3x+C$

7. $\int(x^2-3x+2)\,dx$

$= \frac{x^3}{3}-\frac{3x^2}{2}+2x+C$

9. $\int 3\sqrt{x}\,dx = 3\int x^{1/2}\,dx$

$= \frac{3x^{3/2}}{\frac{3}{2}}+C$

$= 2x^{3/2}+C$

11. $\int(x^{1/2}+3x^{-2/3})\,dx$

$= \frac{x^{3/2}}{\frac{3}{2}}+\frac{3x^{1/3}}{\frac{1}{3}}+C$

$= \frac{2x^{3/2}}{3}+9x^{1/3}+C$

13. $\int\frac{-4}{x^3}\,dx = \int -4x^{-3}\,dx$

$= \frac{-4x^{-2}}{-2}+C$

$= 2x^{-2}+C$

15. $\int -3e^{2x}\,dx = \frac{-3e^{2x}}{2} + C$

17. $\int xe^{3x^2}\,dx = \frac{1}{6}\int 6xe^{3x^2}\,dx$

Let $u = 3x^2$, so that $du = 6x\,dx$.

$$= \frac{1}{6}\int e^u\,du$$
$$= \frac{1}{6}e^u + C$$
$$= \frac{e^{3x^2}}{6} + C$$

19. $\int \frac{3x}{x^2-1}\,dx = 3\left(\frac{1}{2}\right)\int \frac{2x\,dx}{x^2-1}$

Let $u = x^2 - 1$, so that $du = 2x\,dx$.

$$= \frac{3}{2}\int \frac{du}{u}$$
$$= \frac{3}{2}\ln|u| + C$$
$$= \frac{3\ln|x^2-1|}{2} + C$$

21. $\int \frac{x^2\,dx}{(x^3+5)^4} = \frac{1}{3}\int \frac{3x^2\,dx}{(x^3+5)^4}$

Let $u = x^3 + 5$, so that

$du = 3x^2\,dx.$

$$= \frac{1}{3}\int \frac{du}{u^4}$$
$$= \frac{1}{3}\int u^{-4}\,du$$
$$= \frac{1}{3}\left(\frac{u^{-3}}{-3}\right) + C$$
$$= \frac{-(x^3+5)^{-3}}{9} + C$$

23. $\int \frac{x^3}{e^{3x^4}}\,dx = \int x^3e^{-3x^4}$

$$= -\frac{1}{12}\int -12x^3e^{-3x^4}\,dx$$

Let $u = -3x^4$, so that $du = -12x^3\,dx$.

$$= -\frac{1}{12}\int e^u\,du$$
$$= -\frac{1}{12}e^u + C$$
$$= \frac{-e^{-3x^4}}{12} + C$$

25. $\int \frac{(3\ln x + 2)^4}{x}\,dx$

Let $u = 3\ln x + 2$ so that

$$du = \frac{3}{x}\,dx.$$

$$\int \frac{(3\ln x+2)^4}{x}\,dx = \frac{1}{3}\int \frac{3(3\ln x+2)^4}{x}\,dx$$
$$= \frac{1}{3}\int u^4\,du$$
$$= \frac{1}{3}\cdot\frac{u^5}{5} + C$$
$$= \frac{(3\ln x+2)^5}{15} + C$$

27. $f(x) = 3x + 1,\ x_1 = -1,\ x_2 = 0,\ x_3 = 1,$ $x_4 = 2,\ x_5 = 3$

$f(x_1) = -2,\ f(x_2) = 1,\ f(x_3) = 4,$
$f(x_4) = 7,\ f(x_5) = 10$

$$\sum_{i=1}^{5} f(x_i)$$
$$= f(1) + f(2) + f(3) + f(4) + f(5)$$
$$= -2 + 1 + 4 + 7 + 10$$
$$= 20$$

29. $f(x) = 2x + 3$, from $x = 0$ to $x = 4$

$$\Delta x = \frac{4-0}{4} = 1$$

i	x_i	$f(x_i)$
1	0	3
2	1	5
3	2	7
4	3	9

$$A = \sum_{i=1}^{4} f(x_i)\Delta x$$
$$= 3(1) + 5(1) + 7(1) + 9(1)$$
$$= 24$$

31. **(a)** Since $s(t)$ represents the odometer reading, the distance traveled between $t = 0$ and $t = T$ will be $s(T) - s(0)$.

(b) $\int_0^T v(t)\,dt = s(T) - s(0)$ is equivalent to the Fundamental Theorem of Calculus with $a = 0$, and $b = T$ because $s(t)$ is an antiderivative of $v(t)$.

33. $\int_1^2 (3x^2+5)\,dx = \left(\frac{3x^3}{3}+5x\right)\Big|_1^2$

$= (2^3+10)-(1+5)$
$= 18-6$
$= 12$

35. $\int_1^5 (3x^{-1}+x^{-3})dx = \left(3\ln|x|+\frac{x^{-2}}{-2}\right)\Big|_1^5$

$= \left(3\ln 5-\frac{1}{50}\right)-\left(3\ln 1-\frac{1}{2}\right)$

$= 3\ln 5+\frac{12}{25}\approx 5.308$

37. $\int_0^1 x\sqrt{5x^2+4}\,dx$

Let $u=5x^2+4$, so that

$du = 10x\,dx$ and $\frac{1}{10}\,du = x\,dx.$

When $x=0,\ u=5(0^2)+4=4.$
When $x=1,\ u=5(1^2)+4=9.$

$= \frac{1}{10}\int_4^9 \sqrt{u}\,du = \frac{1}{10}\int_4^9 u^{1/2}\,du$

$= \frac{1}{10}\cdot\frac{u^{3/2}}{3/2}\Big|_4^9 = \frac{1}{15}u^{3/2}\Big|_4^9$

$= \frac{1}{15}(9)^{3/2}-\frac{1}{15}(4)^{3/2}$

$= \frac{27}{15}-\frac{8}{15}$

$= \frac{19}{15}$

39. $\int_0^2 3e^{-2x}\,dx = \frac{-3e^{-2x}}{2}\Big|_0^2$

$= \frac{-3e^{-4}}{2}+\frac{3}{2}$

$= \frac{3(1-e^{-4})}{2}\approx 1.473$

41. $\int_0^{1/2} x\sqrt{1-16x^4}\,dx$

Let $u=4x^2$. Then $du=8x\,dx$.
When $x=0,\ u=0$, and when $x=\frac{1}{2},\ u=1.$

Thus,

$$\int_0^{1/2} x\sqrt{1-16x^4}\,dx = \frac{1}{8}\int_0^1 \sqrt{1-u^2}\,du.$$

Note that this integral represents the area of right upper quarter of a circle centered at the origin with a radius of 1.

Area of circle $=\pi r^2=\pi(1^2)=\pi$

$\int_0^1 \sqrt{1-u^2}\,du = \frac{\pi}{4}$

$\frac{1}{8}\int_0^1 \sqrt{1-u^2}\,du = \frac{1}{8}\cdot\frac{\pi}{4}=\frac{\pi}{32}$

43. $\int_1^{\sqrt{7}} 2x\sqrt{36-(x^2-1)^2}\,dx$

Let $u=x^2-1$. Then $du=2x\,dx$.
When $x=\sqrt{7}, u=(\sqrt{7})^2-1=6.$
When $x=1, u=(\sqrt{1})^2-1=0.$

Thus,

$$\int_1^{\sqrt{7}} 2x\sqrt{36-(x^2-1)^2}\,dx = \int_0^6 \sqrt{36-u^2}\,du.$$

Note that this integral represents the area of a right upper quarter of a circle centered at the origin with a radius of 6.

Area of circle $=\pi r^2=\pi(6)^2=36\pi$

$\int_0^6 \sqrt{36-u^2}\,du = \frac{36\pi}{4}=9\pi$

45. $f(x)=(3x+2)^6; [-2,0]$

Area $= \int_{-2}^0 (3x+2)^6\,dx$

$= \frac{(3x+2)^7}{21}\Big|_{-2}^0$

$= \frac{2^7}{21}-\frac{(-4)^7}{21}$

$= \frac{5504}{7}$

47. $f(x)=1+e^{-x};\ [0,4]$

$\int_0^4 (1+e^{-x})\,dx = (x-e^{-x})\Big|_0^4$

$= (4-e^{-4})-(0-e^0)$
$= 5-e^{-4}$
≈ 4.982

49. $f(x) = x^2 - 4x$; $g(x) = x - 6$

Find the points of intersection.

$$x^2 - 4x = x - 6$$
$$x^2 - 5x + 6 = 0$$
$$(x - 3)(x - 2) = 0$$
$$x = 2 \quad \text{or} \quad x = 3$$

Since $g(x) \geq f(x)$ in the interval $[2, 3]$, the area between the graphs is

$$\int_2^3 [g(x) - f(x)]\,dx$$
$$= \int_2^3 [(x - 6) - (x^2 - 4x)]\,dx$$
$$= \int_2^3 (-x^2 + 5x - 6)\,dx$$
$$= \left(\frac{-x^3}{3} + \frac{5x^2}{2} - 6x\right)\bigg|_2^3$$
$$= \frac{-27}{3} + \frac{5(9)}{2} - 6(3) - \frac{-8}{3}$$
$$- \frac{5(4)}{2} + 6(2)$$
$$= -\frac{19}{3} + \frac{25}{2} - 6 = \frac{1}{6}.$$

51. $f(x) = 5 - x^2$, $g(x) = x^2 - 3$, $x = 0$, $x = 4$

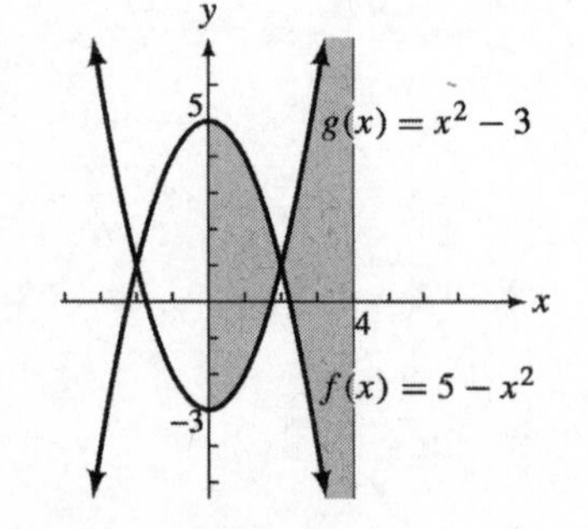

Find the points of intersection.

$$5 - x^2 = x^2 - 3$$
$$8 = 2x^2$$
$$4 = x^2$$
$$\pm 2 = x$$

The curves intersect at $x = 2$ and $x = -2$.

Thus, the area is

$$\int_0^2 [(5 - x^2) - (x^2 - 3)]\,dx$$
$$+ \int_2^4 [(x^2 - 3) - (5 - x^2)]\,dx$$
$$= \int_0^2 (-2x^2 + 8)\,dx + \int_2^4 (2x^2 - 8)\,dx$$
$$= \left(\frac{-2x^3}{3} + 8x\right)\bigg|_0^2 + \left(\frac{2x^3}{3} - 8x\right)\bigg|_2^4$$
$$= \frac{-16}{3} + 16 + \left(\frac{128}{3} - 32\right) - \left(\frac{16}{3} - 16\right)$$
$$= \frac{32}{3} + \frac{128}{3} - 32 - \frac{16}{3} + 16$$
$$= 32.$$

53. $\displaystyle\int_2^{10} \frac{x\,dx}{x - 1}$

Trapezoidal Rule:

$n = 4,\ b = 10,\ a = 2,\ f(x) = \frac{x}{x-1}$

i	x_i	$f(x_i)$
0	2	2
1	4	$\frac{4}{3}$
2	6	$\frac{6}{5}$
3	8	$\frac{8}{7}$
4	10	$\frac{10}{9}$

$$\int_2^{10} \frac{x}{x - 1}\,dx$$
$$\approx \frac{10 - 2}{4}\left[\frac{1}{2}(2) + \frac{4}{3} + \frac{6}{5} + \frac{8}{7} + \frac{1}{2}\left(\frac{10}{9}\right)\right]$$
$$\approx 10.46$$

Exact Value:

Let $u = x - 1$, so that $du = dx$ and $x = u + 1$.

Then

$$\int_2^{10} \frac{x}{x-1}\,dx = \int_1^9 \frac{u+1}{u}\,du$$
$$= \int_1^9 \left(1+\frac{1}{u}\right)du$$
$$= \int_1^9 du + \int_1^9 \frac{1}{u}\,du$$
$$= u\Big|_1^9 + \ln|u|\Big|_1^9$$
$$= (9-1) + (\ln 9 - \ln 1)$$
$$= 8 + \ln 9 \approx 10.20.$$

55. $\displaystyle\int_1^3 \frac{\ln x}{x}\,dx$

Simpson's rule:

$n = 4,\ b = 3,\ a = 1,\ f(x) = \frac{\ln x}{x}$

i	x_i	$f(x_i)$
0	1	0
1	1.5	0.27031
2	2	0.34657
3	2.5	0.36652
4	3	0.3662

$$\int_1^3 \frac{\ln x}{x}\,dx$$
$$\approx \frac{3-1}{3(4)}[0 + 4(0.27031) + 2(0.34657)$$
$$+ 4(0.36652) + 0.3662]$$
$$\approx 0.6011$$

This answer is close to the value of 0.6035 obtained from the exact integral in Exercise 52.

57. $\displaystyle\int_0^1 e^x\sqrt{e^x+4}\,dx$

Simpson's rule:

$n = 4,\ b = 1,\ a = 0,\ f(x) = e^x\sqrt{e^x+4}$

i	x_i	$f(x_i)$
0	0	2.236
1	0.25	2.952
2	0.5	3.919
3	0.75	5.236
4	1	7.046

$$\int_0^1 e^x\sqrt{e^x+4}\,dx$$
$$= \frac{1-0}{3(4)}[2.236 + 4(2.952) + 2(3.919)$$
$$+ 4(5.236) + 7.046$$
$$\approx 4.156$$

This answer is close to the answer of 4.155 obtained from the exact integral in Exercise 54.

59. $\displaystyle\int_{-2}^2 [x(x-1)(x+1)(x-2)(x+2)]^2 dx$

(a) Trapezoidal Rule:

$n = 4, b = -2, a = 2,$

$$f(x) = [x(x-1)(x+1)(x-2)(x+2)]^2$$

i	x_i	$f(x_i)$
0	−2	0
1	−1	0
2	0	0
3	1	0
4	2	0

$$\int_{-2}^2 [x(x-1)(x+1)(x-2)(x+2)]^2 dx$$
$$\approx \frac{2-(-2)}{4}\left[\frac{1}{2}(0) + 0 + 0 + 0 + \frac{1}{2}(0)\right]$$
$$= 0$$

(b) Simpson's Rule

$n = 4, b = 2, a = 2,$

$$f(x) = [x(x-1)(x+1)(x-2)(x+2)]^2$$

i	x_i	$f(x_i)$
0	−2	0
1	−1	0
2	0	0
3	1	0
4	2	0

$$\int_{-2}^2 [x(x-1)(x+1)(x-2)(x+2)]^2 dx$$
$$\approx \frac{2-(-2)}{3(4)}[0 + 4(0) + 2(0) + 4(0) + 0]$$
$$= 0$$

61. $C'(x) = 3\sqrt{2x-1}$; 13 units cost \$270.

$$C(x) = \int 3(2x-1)^{1/2}\,dx = \frac{3}{2}\int 2(2x-1)^{1/2}\,dx$$

Let $u = 2x - 1$, so that

$$du = 2\,dx. = \frac{3}{2}\int u^{1/2}\,du = \frac{3}{2}\left(\frac{u^{3/2}}{3/2}\right) + C = (2x-1)^{3/2} + C$$

$$C(13) = [2(13) - 1]^{3/2} + C$$

Since $C(13) = 270$,

$$\begin{aligned} 270 &= 25^{3/2} + C \\ 270 &= 125 + C \\ C &= 145. \end{aligned}$$

Thus,

$$C(x) = (2x-1)^{3/2} + 145.$$

63. Read values for the rate of investment income accumulation for every 2 years from year 1 to year 9. These are the heights of rectangles with width $\Delta x = 2$.

Total accumulated income $= 11,000(2) + 9000(2) + 12,000(2) + 10,000(2) + 6000(2) \approx \$96,000$

65. $S'(x) = 3\sqrt{2x+1} + 3$

$$S(x) = \int_0^4 (3\sqrt{2x+1} + 3)\,dx = [(2x+1)^{3/2} + 3x]\Big|_0^4 = (27+12) - (1+0) = 38$$

Total sales $=$ \$38,000.

67. $S(q) = q^2 + 5q + 100$
$D(q) = 350 - q^2$
$S(q) = D(q)$ at the equilibrium point.

$$\begin{aligned} q^2 + 5q + 100 &= 350 - q^2 \\ 2q^2 + 5q - 250 &= 0 \\ (-2q + 25)(q - 10) &= 0 \\ q = -\frac{25}{2} \quad &\text{or} \quad q = 10 \end{aligned}$$

Since the number of units produced would not be negative, the equilibrium point occurs when $q = 10$.

Equilibrium supply
$= (10)^2 + 5(10) + 100 = 250$

Equilibrium demand
$= 350 - (10)^2 = 250$

(a) Producers' surplus $= \int_0^{10} [250 - (q^2 + 5q + 100)]\,dx = \int_0^{10} (-q^2 - 5q + 150)\,dx$

$$= \left(\frac{-q^3}{3} - \frac{5q^2}{2} + 150q\right)\Bigg|_0^{10} = \frac{-1000}{3} - \frac{500}{2} + 1500$$

$$= \frac{\$2750}{3} \approx \$916.67$$

(b) Consumers' surplus $= \int_0^{10} [(350 - q^2) - 250]\,dx = \int_0^{10} (100 - q^2)\,dx = \left(100q - \frac{q^3}{3}\right)\Bigg|_0^{10}$

$$= 1000 - \frac{1000}{3} = \frac{\$2000}{3} \approx \$666.67$$

69. **(a)** Total amount $= \frac{1}{2}(2.394) + 2.366 + 2.355 + 2.282 + 2.147 + 2.131 + 2.118 + 2.097 + 2.073 + 1.983 + \frac{1}{2}(1.869)$

$$\approx 21.684$$

This calculation yields a total of 21.684 billion barrels.

71. $f(t) = 100 - t\sqrt{0.4t^2 + 1}$

The total number of additional spiders in the first ten months is

$$\int_0^{10} (100 - t\sqrt{0.4t^2 + 1})dt,$$

where t is the time in months.

$$= \int_0^{10} 100dt - \int_0^{10} t\sqrt{0.4t^2 + 1}\, dt.$$

Let $u = 0.4t^2 + 1$, so that

$$du = 0.8t\, dt \text{ and } \tfrac{1}{0.8}\, du = t\, dt.$$

When $t = 10, u = 41.$
When $t = 0, u = 1.$

$$= \int_0^{10} 100dt - \frac{1}{0.8}\int_1^{41} u^{1/2}\, du$$

$$= 100t\Big|_0^{10} - \frac{5}{4} \cdot \frac{u^{3/2}}{\frac{3}{2}}\Big|_1^{41}$$

$$= 1000 - \frac{5}{6}u^{3/2}\Big|_1^{41}$$

$$\approx 782$$

The total number of additional spiders in the first 10 months is about 782.

73. **(a)** The total area is the area of the triangle on $[0, 12]$ with height 0.024 plus the area of the rectangle on $[12, 17.6]$ with height 0.024.

$$A = \frac{1}{2}(12 - 0)(0.024) + (17.6 - 12)(0.024) = 0.144 + 0.1344 = 0.2784$$

(b) On $[0, 12]$ we defined the function $f(x)$ with slope $\frac{0.024-0}{12-0} = 0.002$ and y-intercept 0.

$$f(x) = 0.002x$$

On $[12, 17.6]$, define $g(x)$ as the constant value.

$$g(x) = 0.024.$$

The area is the sum of the integrals of these two functions.

$$A = \int_0^{12} 0.002xdx + \int_{12}^{17.6} 0.024dx = 0.001x^2\Big|_0^{12} + 0.024x\Big|_{12}^{17.6}$$

$$= 0.001(12^2 - 0^2) + 0.024(17.6 - 12) = 0.144 + 0.1344 = 0.2784$$

75. **(a)** Total amount $= \frac{1}{2}(271{,}553) + 278{,}325 + 274{,}690 + 290{,}525 + 289{,}890$

$$+ 309{,}569 + 317{,}567 + 335{,}869 + 331{,}055 + \frac{1}{2}(331{,}208)$$

$$\approx 2{,}728{,}871$$

This calculation yields a total of about \$2,728,871.

77. For each month, subtract the average temperature from 65° (if it falls below 65°F), then multiply this number times the number of days in the month. The sum is the total number of heating degree days. Readings may vary, but the sum is approximately 4800 degree-days. (The actual value is 4868 degree-days.)

Chapter 7 Test

[7.1]

1. Explain what is meant by an antiderivative of a function $f(x)$.

Find each indefinite integral.

2. $\int (3x^3 - 5x^2 + x + 1)\,dx$

3. $\int \left(\frac{5}{x} + e^{0.5x}\right) dx$

4. $\int \frac{2x^3 - 3x^2}{\sqrt{x}}\,dx$

5. Find the cost function $C(x)$ if the marginal cost function is given by $C'(x) = 200 + 2x^{-1/4}$ and 16 units cost \$4000.

6. A ball is thrown upward at time $t = 0$ with initial velocity of 64 feet per second from a height of 100 feet. Assume that $a(t) = -32$ feet per second per second. Find $v(t)$ and $s(t)$.

[7.2]

Use substitution to find each indefinite integral.

7. $\int 6x^2 (3x^3 - 5)^8\,dx$

8. $\int \frac{6x + 5}{3x^2 + 5x}\,dx$

9. $\int \sqrt[3]{2x^2 - 8x}\,(x - 2)\,dx$

10. $\int 4x^3 e^{-x^4}\,dx$

11. A city's population is predicted to grow at a rate of

$$P'(x) = \frac{400e^{10t}}{1 + e^{10t}}$$

people per year where t is the time in years from the present. Find the total population 3 years from now if $P(0) = 100{,}000$.

[7.3]

12. Evaluate $\sum_{i=1}^{4} \frac{2}{i^2}$.

13. Approximate the area under the graph of $f(x) = x^2 + x$ and above the x-axis from $x = 0$ to $x = 2$ using four rectangles of equal width. Let the height of each rectangle be the function value at the left endpoint.

14. Approximate the value of $\int_1^5 x^2\,dx$ by summing the areas of rectangles. Use four rectangles of equal width. Use the left endpoints, then the right endpoints; then give the average of these answers.

[7.4]

Evaluate the following definite integrals.

15. $\int_1^5 (4x^3 - 5x)\, dx$

16. $\int_1^5 \left(\frac{3}{x^2} + \frac{2}{x}\right) dx$

17. $\int_1^2 \sqrt{3r - 2}\, dr$

18. $\int_0^1 4xe^{x^2+1}\, dx$

19. $\int_{-2}^1 3x\left(x^2 - 4\right)^5 dx$

20. Find the area of the region between the x-axis and $f(x) = e^{x/2}$ on the interval $[0, 2]$.

21. The rate at which a substance grows is given by $R(x) = 500e^{0.5x}$, where x is the time in days. What is the total accumulated growth after 4 days?

[7.5]

22. Find the area of the region enclosed by $f(x) = -x + 4$, $g(x) = -x^2 + 6x - 6$, $x = 2$, and $x = 4$.

23. Find the area of the region enclosed by $f(x) = 5x$ and $g(x) = x^3 - 4x$.

24. A company has determined that the use of a new process would produce a savings rate (in thousands of dollars) of

$$S(x) = 2x + 7,$$

where x is the number of years the process is used. However, the use of this process also creates additional costs (in thousands of dollars) according to the rate-of-cost function

$$C(x) = x^2 + 2x + 3.$$

(a) For how many years does the new process save the company money?

(b) Find the net savings in thousands of dollars over this period.

25. Suppose that the supply function of a commodity is $p = 0.05q + 5$ and the demand function is $p = 12 - 0.02q$.

(a) Find the producers' surplus.

(b) Find the consumers' surplus.

[7.6]

26. Use $n = 4$ to approximate the value of the given integral by the following methods: **(a)** the trapezoidal rule and **(b)** Simpson's rule. **(c)** Find the exact value by integration.

$$\int_0^2 x\sqrt{x^2 + 1}\, dx$$

27. Use $n = 4$ to approximate the value of the given integral by the following methods: **(a)** the trapezoidal rule and **(b)** Simpson's rule.

$$\int_0^2 \frac{1}{\sqrt{1+x^3}}\,dx$$

28. Use Simpson's rule with $n = 6$ to approximate the value of $\int_0^3 \frac{1}{x^2+1}\,dx$.

29. Find the area between the curve $y = e^{-x^2}$ and the x-axis from $x = 0$ to $x = 3$, using the trapezoidal rule with $n = 6$.

30. Find the area between the curve $y = \frac{x}{x^2+1}$ and the x-axis from $x = 1$ to $x = 4$, using Simpson's rule with $n = 6$.

Chapter 7 Test Answers

1. An antiderivative of a function $f(x)$ is a function $F(x)$ such that $F'(x) = f(x)$.

2. $\frac{3}{4}x^4 - \frac{5}{3}x^3 + \frac{x^2}{2} + x + C$

3. $5\ln|x| + 2e^{0.5x} + C$

4. $\frac{4}{7}x^{7/2} - \frac{6}{5}x^{5/2} + C$

5. $C(x) = 200x + \frac{8}{3}x^{3/4} + 778.67$

6. $v(t) = -32t + 64$; $s(t) = -16t^2 + 64t + 100$

7. $\frac{2}{27}\left(3x^3 - 5\right)^9 + C$

8. $\ln\left|3x^2 + 5x\right| + C$

9. $\frac{3}{16}\left(2x^2 - 8x\right)^{4/3} + C$

10. $-e^{-x^4} + C$

11. 101,172

12. $2\frac{61}{72}$ or 2.85

13. 3.25

14. Left endpoints: 30; right endpoints: 54; average: 42

15. 564

16. 5.62

17. $\frac{14}{9}$ or 1.56

18. $2e^2 - 2e$ or 9.34

19. $\frac{729}{4}$ or 182.25

20. $2e - 2$ or 3.44

21. 6389

22. $\frac{10}{3}$ or 3.33

23. 40.5

24. **(a)** 2 years **(b)** $5.33 thousand

25. **(a)** $250 **(b)** $100

26. **(a)** 3.457 **(b)** 3.392 **(c)** 3.393

27. **(a)** 1.397 **(b)** 1.405

28. 1.25

29. 0.89

30. 1.07

Chapter 8

FURTHER TECHNIQUES AND APPLICATIONS OF INTEGRATION

8.1 Integration by Parts

1. $\int xe^x \, dx$

Let $dv = e^x \, dx$ and $u = x$.

Then $v = \int e^x \, dx$ and $du = dx$.

$v = e^x$

Use the formula

$$\int u \, dv = uv - \int v \, du.$$

$$\int xe^x \, dx = xe^x - \int e^x \, dx$$
$$= xe^x - e^x + C$$

3. $\int (4x - 12)e^{-8x} dx$

Let $dv = e^{-8x} \, dx$ and $u = 4x - 12$

Then $v = \int e^{-8x} \, dx$ and $du = 4 \, dx$.

$v = \frac{e^{-8x}}{-8}$

$$\int (4x - 12)e^{-8x} dx$$

$$= (4x - 12)\left(\frac{e^{-8x}}{-8}\right) - \int \left(\frac{e^{-8x}}{-8}\right) \cdot 4 \, dx$$

$$= -\frac{4x}{8}e^{-8x} + \frac{12}{8}e^{-8x} - \left(-\frac{4}{8} \cdot \frac{e^{-8x}}{-8}\right) + C$$

$$= -\frac{x}{2}e^{-8x} + \frac{3}{2}e^{-8x} - \frac{1}{16}e^{-8x} + C$$

$$= \left(-\frac{x}{2} + \frac{23}{16}\right) e^{-8x} + C$$

5. $\int_0^1 \frac{2x + 1}{e^x} \, dx$

$$= \int_0^1 (2x + 1)e^{-x} \, dx$$

Let $dv = e^{-x} \, dx$ and $u = 2x + 1$.

Then $v = \int e^{-x} \, dx$ and $du = 2 \, dx$.

$v = -e^{-x}$

$$\int \frac{2x + 1}{e^x} \, dx$$

$$= -(2x + 1)e^{-x} + \int 2e^{-x} \, dx$$

$$= -(2x + 1)e^{-x} - 2e^{-x}$$

$$\int_0^1 \frac{2x + 1}{e^x} \, dx$$

$$= [-(2x + 1)e^{-x} - 2e^{-x}] \Big|_0^1$$

$$= [-(3)e^{-1} - 2e^{-1}] - (-1 - 2)$$
$$= -5e^{-1} + 3$$
$$\approx 1.161$$

7. $\int \ln 3x \, dx$

Let $dv = dx$ and $u = \ln 3x$.

Then $v = x$ and $du = \frac{1}{x} \, dx$.

$$\int \ln 3x \, dx = x \ln 3x - \int dx$$
$$= x \ln 3x - x$$

$$\int \ln 3x \, dx = (x \ln 3x - x) \Big|_1^9$$
$$= (9 \ln 27 - 9) - (\ln 3 - 1)$$
$$= 9 \ln 3^3 - 9 - \ln 3 + 1$$
$$= 27 \ln 3 - \ln 3 - 8$$
$$= 26 \ln 3 - 8 \approx 20.56$$

9. $\int x \ln dx$

Let $dv = x\,dx$ and $u = \ln x$.

Then $v = \frac{x^2}{2}$ and $du = \frac{1}{x}\,dx$.

$$\int x \ln dx = \frac{x^2}{2} \ln x - \int \frac{x}{2}\,dx$$
$$= \frac{x^2 \ln x}{2} - \frac{x^2}{4} + C$$

11. The area is $\int_2^4 (x-2)e^x\,dx$.

Let $dv = e^x\,dx$ and $u = x - 2$.
Then $v = e^x$ and $du = dx$.

$$\int (x-2)e^x\,dx = (x-2)e^x - \int e^x\,dx$$
$$\int_1^4 (x-2)e^x\,dx = [(x-2)e^x - e^x]\Big|_2^4$$
$$= (2e^4 - e^4) - (0 - e^2)$$
$$= e^4 + e^2 \approx 61.99$$

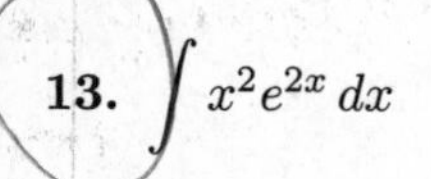

13. $\int x^2 e^{2x}\,dx$

Let $u = x^2$ and $dv = e^{2x}\,dx$.
Use column integration.

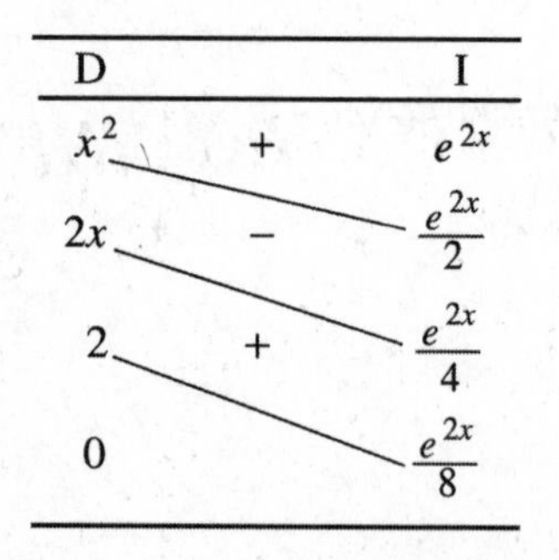

D		I
x^2	+	e^{2x}
$2x$	−	$\frac{e^{2x}}{2}$
2	+	$\frac{e^{2x}}{4}$
0		$\frac{e^{2x}}{8}$

$$\int x^2 e^{2x}\,dx = x^2\left(\frac{e^{2x}}{2}\right) - 2x\left(\frac{e^{2x}}{4}\right) + \frac{2e^{2x}}{8} + C$$
$$= \frac{x^2 e^{2x}}{2} - \frac{xe^{2x}}{2} + \frac{e^{2x}}{4} + C$$

15. $\int_0^5 x\sqrt[3]{x^2+2}\,dx$

$$= \int_0^5 x(x^2+2)^{1/2}\,dx$$
$$= \frac{1}{2}\int_0^5 2x(x^2+2)^{1/3}\,dx$$

Let $u = x^2 + 2$. Then $du = 2x\,dx$.
If $x = 5,\ u = 27$. If $x = 0,\ u = 2$.

$$= \frac{1}{2}\int_2^{27} u^{1/3}\,du$$
$$= \frac{1}{2}\left(\frac{u^{4/3}}{1}\right)\left(\frac{3}{4}\right)\Big|_2^{27}$$
$$= \frac{3}{8}(27)^{4/3} - \frac{3}{8}(2)^{4/3}$$
$$= \frac{243}{8} - \frac{3(2^{4/3})}{8}$$
$$= \frac{243}{8} - \frac{3\sqrt[3]{2}}{4} \approx 29.43$$

17. $\int (8x+10)\ln(5x)dx$

Let $dv = (8x+10)\,dx$ and $u = \ln(5x)$.

Then $v = 4x^2 + 10x$ and $du = \frac{1}{x}\,dx$.

$$\int (8x+10)\ln(5x)dx$$
$$= (4x^2+10x)\ln(5x) - \int (4x^2+10x)\left(\frac{1}{x}\right)dx$$
$$= (4x^2+10x)\ln(5x) - \int (4x+10)dx$$
$$= (4x^2+10x)\ln(5x) - 2x^2 - 10x + C$$

19. $\int x^2\sqrt{x+4}\,dx$

Let $u = x^2$ and $dv = (x+4)^{1/2}$. Use column integration.

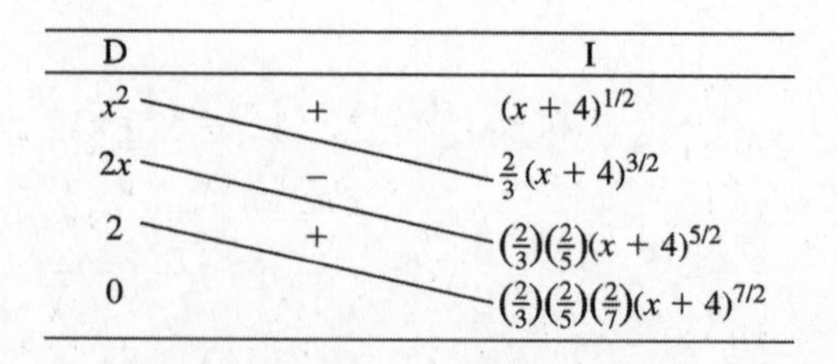

D		I
x^2	+	$(x+4)^{1/2}$
$2x$	−	$\frac{2}{3}(x+4)^{3/2}$
2	+	$\left(\frac{2}{3}\right)\left(\frac{2}{5}\right)(x+4)^{5/2}$
0		$\left(\frac{2}{3}\right)\left(\frac{2}{5}\right)\left(\frac{2}{7}\right)(x+4)^{7/2}$

$$\int x^2\sqrt{x+4}\,dx$$

$$= x^2(x+4)^{3/2}\left(\frac{2}{3}\right) - 2x(x+4)^{5/2}\left(\frac{2}{3}\right)\left(\frac{2}{5}\right)$$

$$+ 2(x+4)^{7/2}\left(\frac{2}{3}\right)\left(\frac{2}{5}\right)\left(\frac{2}{7}\right) + C$$

$$= \frac{2}{3}x^2(x+4)^{3/2} - \frac{8}{15}x(x+4)^{5/2}$$

$$+ \frac{16}{105}(x+4)^{7/2} + C$$

21. $\int_0^1 \frac{x^3\,dx}{\sqrt{3+x^2}} = \int_0^1 x^3(3x+x^2)^{-1/2}\,dx$

Let $dv = x(3+x^2)^{-1/2}\,dx$ and $u = x^2$.

Then $v = \frac{2(3+x^2)^{1/2}}{2}$

$v = (3+x^2)^{1/2}$ and $du = 2x\,dx$.

$$\int \frac{x^3\,dx}{\sqrt{3+x^2}}$$

$$= x^2(3+x^2)^{1/2} - \int 2x(3+x^2)^{1/2}\,dx$$

$$= x^2(3+x^2)^{1/2} - \frac{2}{3}(3+x^2)^{3/2}$$

$$\int_0^1 \frac{x^3\,dx}{\sqrt{3+x^2}}$$

$$= [x^2(3+x^2)^{1/2} - \frac{2}{3}(3+x^2)^{3/2}]\Big|_0^1$$

$$= 4^{1/2} - \frac{2}{3}(4^{3/2}) - 0 + \frac{2}{3}(3^{3/2})$$

$$= 2 - \frac{2}{3}(8) + \frac{2}{3}(3^{3/2})$$

$$= -\frac{10}{3} + 2\sqrt{3}$$

$$\approx 0.1308$$

23. $\int \frac{16}{\sqrt{x^2+16}}\,dx$

Use entry 5 from the table of integrals with $a = 4$.

$$\int \frac{16}{\sqrt{x^2+16}}\,dx = 16\int \frac{1}{\sqrt{x^2+4^2}}\,dx$$

$$= 16\ln\left|x+\sqrt{x^2+16}\right| + C$$

25. $\int \frac{3}{x\sqrt{121-x^2}}\,dx$

$$= 3\int \frac{dx}{x\sqrt{11^2-x^2}}$$

If $a = 11$, this integral matches entry 9 in the table.

$$= 3\left(-\frac{1}{11}\ln\left|\frac{11+\sqrt{121-x^2}}{x}\right|\right) + C$$

$$= -\frac{3}{11}\ln\left|\frac{11+\sqrt{121-x^2}}{x}\right| + C$$

27. $\int \frac{-6}{x(4x+6)^2}\,dx$

Use entry 14 from the table of integrals with $a = 4$ and $b = 6$.

$$\int \frac{-6}{x(4x+6)^2}\,dx$$

$$= -6\int \frac{1}{x(4x+6)^2}\,dx$$

$$= -6\left[\frac{1}{6(4x+6)} + \frac{1}{6^2}\ln\left|\frac{x}{4x+6}\right|\right] + C$$

$$= \frac{-1}{(4x+6)} - \frac{1}{6}\ln\left|\frac{x}{4x+6}\right| + C$$

31. First find the indefinite integral using integration by parts.

$$\int u\,dv = uv - \int v\,du$$

Now substitute the given values.

$$\int_0^1 u\,dv = uv\Big|_0^1 - \int_0^1 v\,du$$

$$= [u(1)v(1) - u(0)v(0)] - 4$$

$$= (3)(-4) - (2)(1) - 4 = -18$$

33. The area between the x-axis and the nonnegative function $h(x) = s(x)\frac{dr}{dx}$ on the interval [0, 2] is

$$\int_0^2 s(x)\frac{dr}{dx}\,dx = \int_0^2 s(x)\,dr.$$

The area between the x-axis and the nonnegative function $d(x) = r(x)\frac{ds}{dx}$ on the interval [0, 2] is

$$\int_0^2 r(x)\frac{ds}{dx}\,dx = \int_0^2 r(x)\,ds.$$

Rewrite the integration by parts formula in terms of r and s.

$$\int r\,ds = rs - \int s\,dr$$

Therefore, substituting the given values, we have

$$r(x)\,ds = \int_0^2 - \int_0^2 s(x)\,dr$$
$$10 = [r(2)s(2) - r(0)s(0)] - 5$$
$$15 = r(2)s(2) - 0 \cdot s(0)$$
$$15 = r(2)s(2).$$

35. $\int x^n \cdot \ln |x| \, dx, n \neq -1$

Let $u = \ln |x|$ and $dv = x^n \, dx$.

Use column integration.

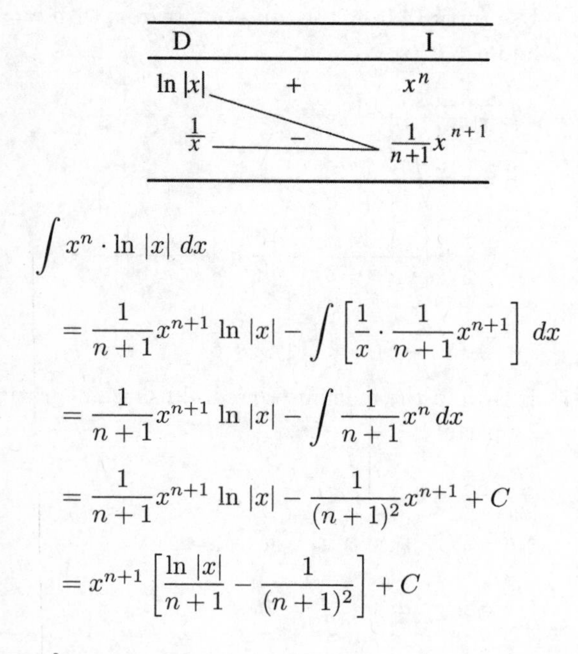

$$\int x^n \cdot \ln |x| \, dx$$
$$= \frac{1}{n+1}x^{n+1} \ln |x| - \int \left[\frac{1}{x} \cdot \frac{1}{n+1}x^{n+1}\right] dx$$
$$= \frac{1}{n+1}x^{n+1} \ln |x| - \int \frac{1}{n+1}x^n \, dx$$
$$= \frac{1}{n+1}x^{n+1} \ln |x| - \frac{1}{(n+1)^2}x^{n+1} + C$$
$$= x^{n+1}\left[\frac{\ln |x|}{n+1} - \frac{1}{(n+1)^2}\right] + C$$

37. $\int x\sqrt{x+1}\,dx$

(a) Let $u = x$ and $dv = \sqrt{x+1}\,dx$.
Use column integration.

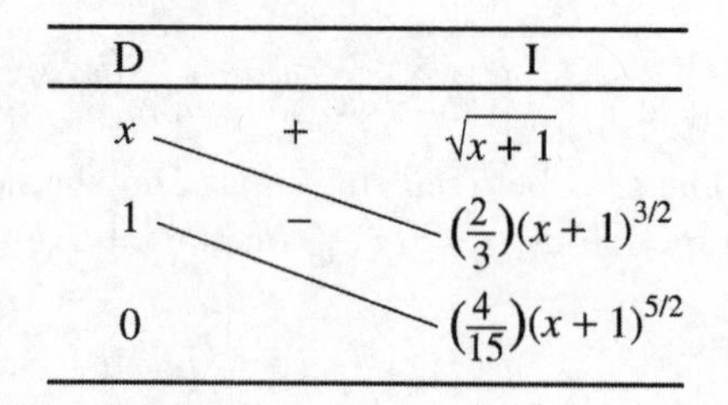

$$\int x\sqrt{x+1}\,dx$$
$$= \left(\frac{2}{3}\right) x(x+1)^{3/2} - \left(\frac{4}{15}\right)(x+1)^{5/2} + C$$

(b) Let $u = x + 1$; then $u - 1 = x$ and $du = dx$.

$$\int x\sqrt{x+1}\,dx$$
$$= \int (u-1)u^{1/2}\,du = \int (u^{3/2} - u^{1/2})\,du$$
$$= \frac{2}{5}u^{5/2} - \frac{2}{3}u^{3/2} + C$$
$$= \frac{2}{5}(x+1)^{5/2} - \frac{2}{3}(x+1)^{3/2} + C$$

(c) Both results factor as $\frac{2}{15}(x+1)^{3/2}(3x-2)+C$, so they are equivalent.

39. $R = \int_0^{12} (x+1)\ln(x+1)dx$

Let $u = \ln(x+1)$ and $dv = (x+1)dx$.

Then $du = \frac{1}{x+1}dx$ and $v = \frac{1}{2}(x+1)^2$.

$$\int (x+1)\ln(x+1)dx$$
$$= \frac{1}{2}(x+1)^2\ln(x+1) - \int \left[\frac{1}{2}(x+1)^2 \cdot \frac{1}{x+1}\right] dx$$
$$= \frac{1}{2}(x+1)^2\ln(x+1) - \int \frac{1}{2}(x+1)dx$$
$$= \frac{1}{2}(x+1)^2\ln(x+1) - \frac{1}{4}(x+1)^2 + C$$

$$\int_0^{12} (x+1)\ln(x+1)dx$$
$$= \left[\frac{1}{2}(x+1)^2\ln(x+1) - \frac{1}{4}(x+1)^2\right]\Bigg|_0^{12}$$
$$= \frac{169}{2}\ln 13 - 42 \approx \$174.74$$

41. The total accumulated growth of the microbe population during the first 2 days is given by

$$\int_0^2 27xe^{3x}\,dx.$$

Let $dv = e^{3x}\,dx$ and $u = 27x$.

Then $v = \frac{e^{3x}}{3}$ and $du = 27\,dx$.

$$\int 27xe^{3x}dx = 27x \cdot \frac{e^{3x}}{3} - \int \frac{e^{3x}}{3} \cdot 27\,dx$$
$$= 9xe^{3x} - 3e^{3x}$$
$$\int_0^2 27xe^{3x}dx = \left(9xe^{3x} - 3e^{3x}\right)\Big|_0^2$$
$$= (18e^6 - 3e^6) - (0 - 3)$$
$$= 15e^6 + 3 \approx 6054$$

43. $\int_0^6 (-10.28 + 175.9te^{-t/1.3})dt$

$$= -10.28t + 175.9\int te^{-t/1.3}dt$$

Evaluate this integral using integration by parts.
Let $u = t$ and $dv = e^{-t/1.3}dt$.
Then $du = dt$ and $v = -1.3e^{-t/1.3}$.

$$\int te^{-t/1.3}dt$$
$$= (t)(-1.3e^{-t/1.3}) - \int(-1.3e^{-t/1.3})dt$$
$$= -1.3te^{-t/1.3} - 1.69e^{-t/1.3} + C$$

Substitute this expression in the earlier expression.

$$-10.28t + 175.9(-1.3te^{-t/1.3} - 1.69e^{-t/1.3})\Big|_0^6$$
$$= -10.28t - 228.67te^{-t/1.3} - 297.271e^{-t/1.3}\Big|_0^6$$
$$= (-61.68 - 1669.291e^{-6/1.3}) - (-297.271)$$
$$\approx 219.07$$

The total thermic energy is about 219 kJ.

8.2 Volume and Average Value

1. $f(x) = x,\ y = 0,\ x = 0,\ x = 3$

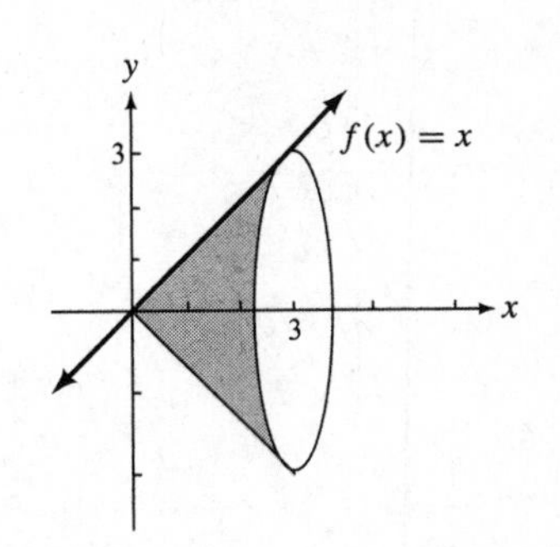

$$V = \pi\int_0^3 x^2\,dx = \frac{\pi x^3}{2}\Big|_0^3 = \frac{\pi(27)}{3} - 0 = 9\pi$$

3. $f(x) = 2x + 1,\ y = 0,\ x = 0,\ x = 4$

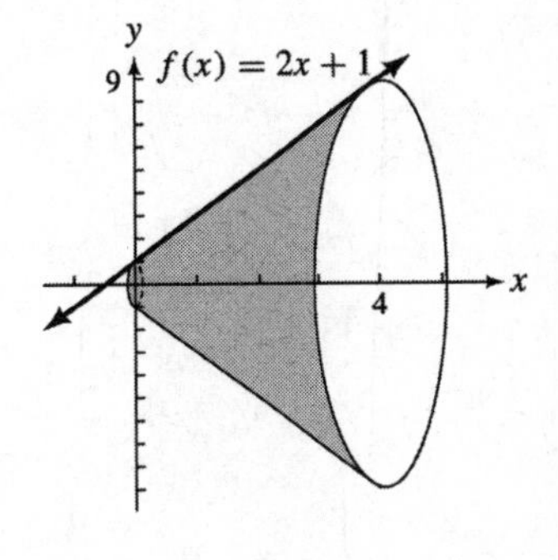

$$V = \pi\int_0^4 (2x+1)^2\,dx$$

Let $u = 2x + 1$. Then $du = 2\,dx$.
If $x = 4$, $u = 9$. If $x = 0$, $u = 1$.

$$V = \frac{1}{2}\pi\int_0^4 2(2x+1)^2\,dx$$
$$= \frac{1}{2}\pi\int_1^9 u^2\,du$$
$$= \frac{\pi}{2}\left(\frac{u^3}{3}\right)\Big|_1^9$$
$$= \frac{\pi}{2}\left(\frac{729}{3} - \frac{1}{3}\right)$$
$$= \frac{728\pi}{6}$$
$$= \frac{364\pi}{3}$$

5. $f(x) = \frac{1}{3}x + 2,\ y = 0,\ x = 1,\ x = 3$

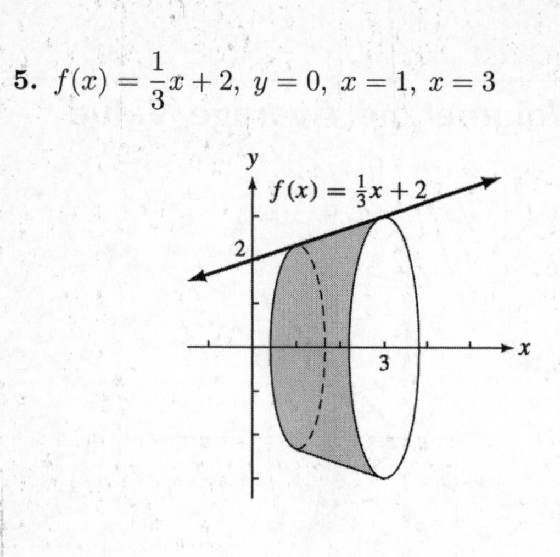

$$\begin{aligned}
V &= \pi \int_1^3 \left(\frac{1}{3}x + 2\right)^2 dx \\
&= 3\pi \int_1^3 \frac{1}{3}\left(\frac{1}{3}x + 2\right)^2 dx \\
&= 3\pi \frac{\left(\frac{1}{3}x + 2\right)^3}{3}\Bigg|_1^3 \\
&= \pi \left(\frac{1}{3}x + 2\right)^3 \Bigg|_1^3 \\
&= 27\pi - \frac{343\pi}{27} \\
&= \frac{386\pi}{27}
\end{aligned}$$

7. $f(x) = \sqrt{x},\ y = 0,\ x = 1,\ x = 4$

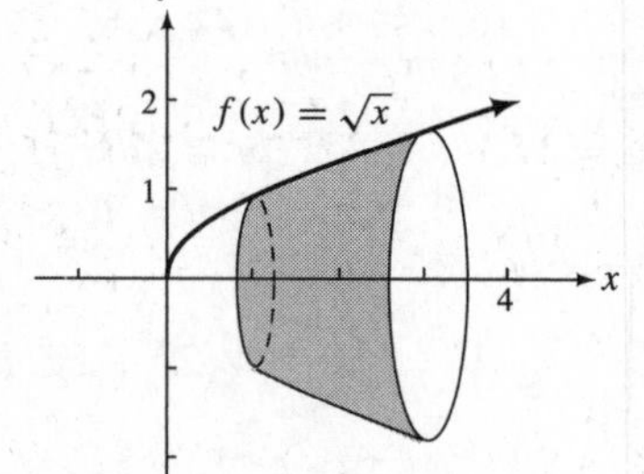

$$\begin{aligned}
V &= \pi \int_1^4 (\sqrt{x})^2\, dx = \pi \int_1^4 x\, dx \\
&= \frac{\pi x^2}{2}\Bigg|_1^4 \\
&= 8\pi - \frac{\pi}{2} \\
&= \frac{15\pi}{2}
\end{aligned}$$

9. $f(x) = \sqrt{2x + 1},\ y = 0,\ x = 1,\ x = 4$

$$\begin{aligned}
V &= \pi \int_1^4 (\sqrt{2x + 1})^2\, dx \\
&= \pi \int_1^4 (2x + 1)\, dx \\
&= \pi \left(\frac{2x^2}{2} + x\right)\Bigg|_1^4 \\
&= \pi[(16 + 4) - 2] \\
&= 18\pi
\end{aligned}$$

11. $f(x) = e^x;\ y = 0,\ x = 0,\ x = 2$

$$\begin{aligned}
V = \pi \int_0^2 e^{2x}\, dx &= \frac{\pi e^{2x}}{2}\Bigg|_0^2 \\
&= \frac{\pi e^4}{2} - \frac{\pi}{2} \\
&= \frac{\pi}{2}(e^4 - 1) \\
&\approx 84.19
\end{aligned}$$

13. $f(x) = \frac{2}{\sqrt{x}}, y = 0, x = 1, x = 3$

$$\begin{aligned}
V &= \pi \int_1^3 \left(\frac{2}{\sqrt{x}}\right)^2 dx \\
&= \pi \int_1^3 \frac{4}{x}\, dx \\
&= 4\pi \ln|x|\Bigg|_1^3 \\
&= 4\pi(\ln 3 - \ln 1) \\
&= 4\pi \ln 3 \approx 13.81
\end{aligned}$$

15. $f(x) = x^2,\ y = 0,\ x = 1,\ x = 5$

$$V = \pi \int_1^5 x^4\, dx = \frac{\pi x^5}{5}\Bigg|_1^5 = 625\pi - \frac{\pi}{5} = \frac{3124\pi}{5}$$

17. $f(x) = 1 - x^2,\ y = 0$

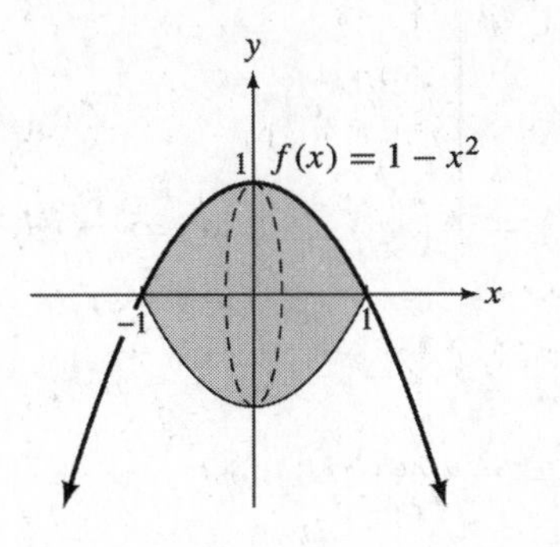

Since $f(x) = 1 - x^2$ intersects $y = 0$ where

$$\begin{aligned} 1 - x^2 &= 0 \\ x &= \pm 1, \\ a = -1 \text{ and } b &= 1. \end{aligned}$$

$$\begin{aligned} V &= \pi \int_{-1}^{1} (1 - x^2)^2\,dx \\ &= \pi \int_{-1}^{1} (1 - 2x^2 + x^4)\,dx \\ &= \pi \left(x - \frac{2x^3}{3} + \frac{x^5}{5} \right) \Big|_{-1}^{1} \\ &= \pi \left(1 - \frac{2}{3} + \frac{1}{5} \right) - \pi \left(-1 + \frac{2}{3} - \frac{1}{5} \right) \\ &= 2\pi - \frac{4\pi}{3} + \frac{2\pi}{5} \\ &= \frac{16\pi}{15} \end{aligned}$$

19. $f(x) = \sqrt{1 - x^2}$
$r = \sqrt{1} = 1$

$$\begin{aligned} V &= \pi \int_{-1}^{1} (\sqrt{1 - x^2})^2\,dx \\ &= \pi \int_{-1}^{1} (1 - x^2)\,dx \\ &= \pi \left(x - \frac{x^3}{3} \right) \Big|_{-1}^{1} \\ &= \pi \left(1 - \frac{1}{3} \right) - \pi \left(-1 + \frac{1}{3} \right) \\ &= 2\pi - \frac{2}{3}\pi \\ &= \frac{4\pi}{3} \end{aligned}$$

21. $f(x) = \sqrt{r^2 - x^2}$

$$\begin{aligned} V &= \pi \int_{-r}^{r} (\sqrt{r^2 - x^2})^2\,dx \\ &= \pi \int_{-r}^{r} (r^2 - x^2)\,dx \\ &= \pi \left(r^2 x - \frac{x^3}{3} \right) \Big|_{-r}^{r} \\ &= \pi \left(r^3 - \frac{r^3}{3} \right) - \pi \left(-r^3 + \frac{r^3}{3} \right) \\ &= 2r^3\pi - \left(\frac{2r^3\pi}{3} \right) \\ &= \frac{4r^3\pi}{3} \end{aligned}$$

23. $f(x) = r,\ x = 0,\ x = h$

Graph $f(x) = r$; then show the solid of revolution formed by rotating about the x-axis the region bounded by $f(x)$, $x = 0$, $x = h$.

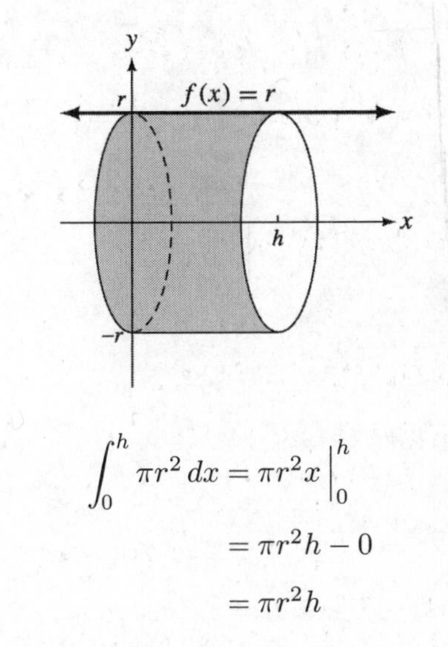

$$\begin{aligned} \int_0^h \pi r^2\,dx &= \pi r^2 x \Big|_0^h \\ &= \pi r^2 h - 0 \\ &= \pi r^2 h \end{aligned}$$

25. $f(x) = x^2 - 4;\ [0, 5]$

Average value

$$\begin{aligned} &= \frac{1}{5 - 0} \int_0^5 (x^2 - 4)dx \\ &= \frac{1}{5} \left(\frac{x^3}{3} - 4x \right) \Big|_0^5 \\ &= \frac{1}{5} \left[\left(\frac{125}{3} - 20 \right) - 0 \right] \\ &= \frac{13}{3} \approx 4.333 \end{aligned}$$

27. $f(x) = \sqrt{x+1}$; $[3, 8]$

Average value

$$= \frac{1}{8-3}\int_3^8 \sqrt{x+1}\,dx$$

$$= \frac{1}{5}\int_3^8 (x+1)^{1/2}\,dx$$

$$= \frac{1}{5}\cdot\frac{2}{3}(x+1)^{3/2}\Big|_3^8$$

$$= \frac{2}{15}(9^{3/2} - 4^{3/2})$$

$$= \frac{2}{15}(27-8) = \frac{38}{15} \approx 2.533$$

29. $f(x) = e^{x/7}$; $[0, 7]$

Average value

$$= \frac{1}{7-0}\int_0^7 e^{x/7}dx$$

$$= \frac{1}{7}\cdot 7e^{x/7}\Big|_0^7$$

$$= e^{x/7}\Big|_0^7 = e^1 - e^0$$

$e - 1 \approx 1.718$

31. $f(x) = x^2e^{2x}$; $[0, 2]$

$$\text{Average value} = \frac{1}{2-0}\int_0^2 x^2e^{2x}\,dx$$

Let $u = x^2$ and $dv = e^{2x}\,dx$.
Use column integration.

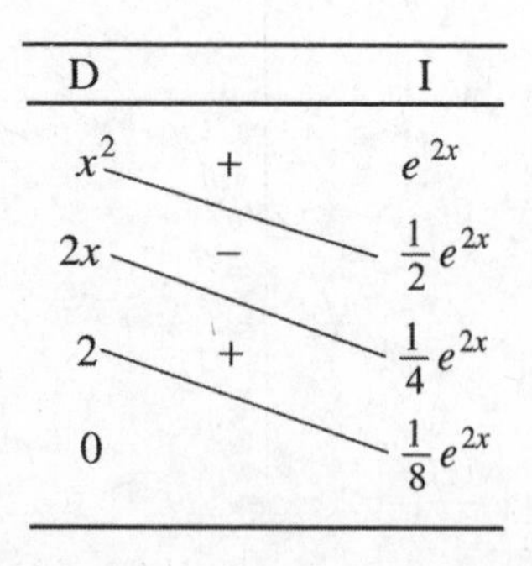

D		I
x^2	+	e^{2x}
$2x$	−	$\frac{1}{2}e^{2x}$
2	+	$\frac{1}{4}e^{2x}$
0		$\frac{1}{8}e^{2x}$

$$\frac{1}{2-0}\int_0^2 x^2e^{2x}\,dx$$

$$= \frac{1}{2}\left[(x^2)\left(\frac{1}{2}\right)e^{2x} - (2x)\left(\frac{1}{4}\right)e^{2x} + 2\left(\frac{1}{8}\right)e^{2x}\right]\Big|_0^2$$

$$= \frac{1}{2}\left(2e^4 - e^4 + \frac{1}{4}e^4 - \frac{1}{4}\right)$$

$$= \frac{5e^4-1}{8} \approx 34.00$$

33. $f(x) = e^{-x^2}, y = 0, x = -1, x = 1$

$$V = \pi\int_{-1}^1 (e^{-x^2})^2dx$$

$$= \pi\int_{-2}^2 e^{-2x^2}\,dx$$

Using an integration feature on a graphing calculator to evaluate the integral, we obtain $3.758249634 \approx 3.758$.

35. Use the formula for average value with $a = 0$ and $b = 6$.

$$\frac{1}{6-0}\int_0^6 (37 + 6e^{-0.03t})dt$$

$$= \frac{1}{6}\left(37t + \frac{6}{-0.03}e^{-0.03t}\right)\Big|_0^6$$

$$= \frac{1}{6}(37t - 200e^{-0.03t})\Big|_0^6$$

$$= \frac{1}{6}[(222 - 200e^{-0.18}) - (0 - 200)]$$

$$= \frac{1}{6}(422 - 200e^{-0.18})$$

$$\approx 42.49$$

The average price is \$42.49.

37. Use the formula for average value with $a = 0$ and $b = 6$. The average price is

$$\frac{1}{30-0}\int_0^{30} (600 - 20\sqrt{30t})dt$$

$$= \frac{1}{30}\left(600t - 20\sqrt{30}\cdot\frac{2}{3}t^{3/2}\right)\Big|_0^{30}$$

$$= \frac{1}{30}\left(600t - \frac{40\sqrt{30}}{3}t^{3/2}\right)\Big|_0^{30}$$

$$= \frac{1}{30}(18{,}000 - 12{,}000)$$

$$= 200 \text{ cases}$$

39. $R(t) = te^{-0.1t}$

"During the nth hour" corresponds to the interval $(n-1, n)$.
The average intensity during nth hour is

$$\frac{1}{n-(n-1)}\int_{n-1}^{n} te^{-0.1t}\,dt = \int_{n-1}^{n} te^{-0.1t}\,dt$$

Let $u = t$ and $dv = e^{-0.1t}\,dt$.

D		I
t	$+$	$e^{-0.1t}$
1	$-$	$-10e^{-0.1t}$
0		$100e^{-0.1t}$

$$\int_{n-1}^{n} te^{-0.1t}\,dt$$

$$= (-10te^{-0.1t} - 100e^{-0.1t})\Big|_{n-1}^{n}$$

(a) Second hour, $n = 2$

Average intensity

$$= -10e^{-0.2}(12) + 10e^{-0.1}(11)$$
$$= 110e^{-0.1} - 120e^{-0.2}$$
$$\approx 1.284$$

(b) Twelfth hour, $n = 12$

Average intensity

$$= -10e^{-1.2}(12+10) + 10e^{-1.1}(11+10)$$
$$= 210e^{-1.1} - 220e^{-1.2}$$
$$\approx 3.640$$

(c) Twenty-fourth hour, $n = 24$

Average intensity

$$= -10e^{-2.4}(24+10) + 10e^{-2.3}(23+10)$$
$$= 330e^{-2.3} - 340e^{-2.4}$$
$$\approx 2.241$$

41. For each part below, use

Average value

$$= \frac{1}{b-a}\int_a^b 45\ln(t+1)dt$$

$$= \frac{45}{b-a}\int_a^b \ln(t+1)dt.$$

Solve the integral using integration by parts.

Let $\quad u = \ln(t+1) \quad$ and $\quad dv = dt.$

Then $\quad du = \frac{1}{t+1}\,dt \quad$ and $\quad v = t.$

$$\int \ln(t+1)\,dt$$
$$= t\ln(t+1) - \int \frac{t}{t+1}\,dt$$
$$= t\ln(t+1) - \int\left(1 - \frac{1}{t+1}\right)dt$$
$$= t\ln(t+1) - t + \ln(t+1) + C$$
$$= (t+1)\ln(t+1) - t + C$$

Therefore

Average value

$$= \frac{1}{b-a}\int_a^b 45\ln(t+1)dt$$

$$= \frac{45}{b-a}[(t+1)\ln(t+1) - t]\Big|_a^b.$$

(a) The average number of items produced daily after 5 days is

$$\frac{45}{5-0}[(t+1)\ln(t+1) - t]\Big|_0^5$$
$$= 9[(6\ln 6 - 5) - (\ln 1 - 0)]$$
$$= 9(6\ln 6 - 5)$$
$$\approx 51.76.$$

(b) The average number of items produced daily after 9 days is

$$\frac{45}{9-0}[(t+1)\ln(t+1) - t]\Big|_0^9$$
$$= 5(10\ln 10 - 9)$$
$$\approx 70.13.$$

(c) The average number of items produced daily after 30 days is

$$\frac{45}{30-0}[(t+1)\ln(t+1) - t]\Big|_0^{30}$$
$$= \frac{3}{2}(31\ln 31 - 30)$$
$$\approx 114.7.$$

43. From Exercise 22, the volume of an ellipsoid with horizontal axis of length $2a$ and vertical axis of length 2b is

$$V = \frac{4ab^2\pi}{3}.$$

For the Earth, $a = 6{,}356{,}752.3142$ and $b = 6{,}378{,}137$.

$$V = \frac{4(6{,}356{,}752.3142)(6{,}378{,}137)^2\pi}{3}$$
$$\approx 1.083 \times 10^{21}$$

The volume of the Earth is about 1.083×10^{21} cubic meters (m^3).

8.3 Continuous Money Flow

1. $f(x) = 1000$

(a) $$P = \int_0^{10} 1000e^{-0.08x}\,dx = \frac{1000}{-0.08}e^{-0.08x}\Big|_0^{10} = -12{,}500(e^{-0.8} - e^0) = -12{,}500(e^{-0.8} - 1) \approx 6883.387949$$

(We will use this value for P in part (b). Store it in your calculator without rounding.)

The present value is \$6883.39.

(b) $$A = e^{0.08(10)}\int_0^{10} 1000e^{-0.08x}\,dx = e^{0.8}P \approx 15{,}319.26161$$

The accumulated value is \$15,319.26.

3. $f(x) = 500$

(a) $$P = \int_0^{10} 500e^{-0.08x}\,dx = \frac{500}{-0.08}e^{-0.08x}\Big|_0^{10} = -6250(e^{-0.8} - e^0) \approx 3441.693974$$

The present value is \$3441.69.

(b) $$A = e^{0.08(10)}\int_0^{10} 500e^{-0.08x}\,dx = e^{0.8}P \approx 7659.630803$$

The accumulated value is \$7659.63.

5. $f(x) = 400e^{0.03x}$

(a) $$P = \int_0^{10} 400e^{0.03x}e^{-0.08x}\,dx = 400\int_0^{10} e^{-0.05x}\,dx = \frac{400}{-0.05}e^{-0.05x}\Big|_0^{10} = -8000(e^{-0.5} - e^0) \approx 3147.754722$$

The present value is \$3147.75.

(b) $$A = e^{0.08(10)}\int_0^{10} 400e^{0.03x}e^{-0.08x}\,dx = e^{0.8}P \approx 7005.456967$$

The accumulated value is \$7005.46.

7. $f(x) = 5000e^{-0.01x}$

(a) $$P = \int_0^{10} 5000e^{-0.01x}e^{-0.08x}\,dx = 5000\int_0^{10} e^{-0.09x}\,dx = \frac{5000}{-0.09}e^{-0.09x}\Big|_0^{10} = -\frac{5000}{0.09}(e^{-0.9} - e^0) \approx 32{,}968.35224$$

The present value is \$32,968.35.

(b) $$A = e^{0.08(10)}\int_0^{10} 5000e^{-0.01x}e^{-0.08x}\,dx = e^{0.8}P \approx 73{,}372.41725$$

The accumulated value is \$73,372.42.

9. $f(x) = 25x$

(a) $$P = \int_0^{10} 25xe^{-0.08x}\,dx = 25\int_0^{10} xe^{-0.08x}\,dx$$

Find the antiderivative using integration by parts.

Let $u = x$ and $dv = e^{-0.08x}\,dx$.
Then $du = dx$ and $v = \frac{1}{-0.08}e^{-0.08x} = -12.5e^{-0.08x}$.

$$\int xe^{-0.08x}dx$$
$$= x(-12.5e^{-0.08x}) - \int(-12.5e^{-0.08x})dx$$
$$= -12.5xe^{-0.08x} + 12.5\int e^{-0.08x}dx$$
$$= -12.5xe^{-0.08x} + \frac{12.5}{-0.08}e^{-0.08x} + C$$
$$= -(12.5x + 156.25)e^{-0.08x} + C$$

Therefore;

$$P = 25[-(12.5x + 156.25)e^{-0.08x}]\Big|_0^{10}$$
$$= [-25(12.5x + 156.25)e^{-0.08x}]\Big|_0^{10}$$
$$= (-7031.25e^{-0.8}) - (-3906.25e^0)$$
$$\approx 746.9057211.$$

The present value is \$746.91.

(b) $A = e^{0.08(10)}\int_0^{10} 25xe^{-0.08x}dx$

$$= e^{0.8}P$$
$$\approx 1662.269252$$

The accumulated value is \$1662.27.

11. $f(x) = 0.01x + 100$

(a) $P = \int_0^{10}(0.01x + 100)e^{-0.08x}dx$

$$= \int_0^{10} 0.01xe^{-0.08x}dx + \int_0^{10} 100e^{-0.08x}dx$$
$$= 0.01\int_0^{10} xe^{-0.08x}dx + 100\int_0^{10} e^{-0.08x}dx$$

From Exercise 9, we know that

$$\int xe^{-0.08x}dx = -(12.5x + 156.25)e^{-0.08x} + C$$

From Exercise 1, we know that

$$\int e^{-0.08x}dx = -12.5e^{-0.08x} + C$$

Substitute the given expressions and simplify.

$$P = \{0.01[-(12.5x + 156.25)e^{-0.08x}] + 100(-12.5e^{-0.08x})\}\Big|_0^{10}$$
$$= [-(0.125x + 1251.5625)e^{-0.08x}]\Big|_0^{10}$$
$$= (-1252.8125e^{-0.8}) - (-1251.5625e^0)$$
$$\approx 688.6375571$$

The present value is \$688.64.

(b) $A = e^{0.08(10)}\int_0^{10}(0.01x + 100)e^{-0.08x}dx$

$$= e^{0.8}P$$
$$\approx 1532.591068$$

The accumulated value is \$1532.59.

13. $f(x) = 1000x - 100x^2$

(a) $P = \int_0^{10}(1000x - 100x^2)e^{-0.08x}dx$

$$= 1000\int_0^{10} xe^{-0.08x}dx - 100\int_0^{10} x^2e^{-0.08x}dx$$

From Exercise 9, we know that

$$\int xe^{-0.08x}dx = -(12.5x + 156.25)e^{-0.08x} + C$$

Evaluate the antiderivative $\int x^2e^{-0.08x}dx$ using column integration. (Note that $\frac{1}{-0.08} = -12.5$.)

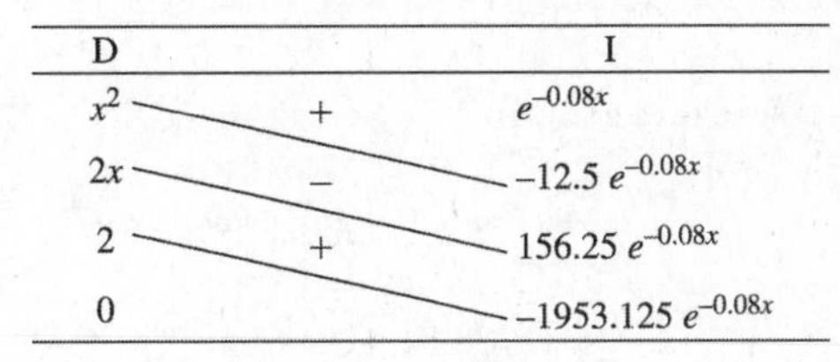

D		I
x^2	+	$e^{-0.08x}$
$2x$	−	$-12.5\ e^{-0.08x}$
2	+	$156.25\ e^{-0.08x}$
0		$-1953.125\ e^{-0.08x}$

Thus,

$$\int_0^{10} x^2e^{-0.08x}dx$$
$$= (x^2)(-12.5e^{-0.08x}) - (2x)(156.25e^{-0.08x}) + (2)(-1953.125e^{-0.08x}) + C$$
$$= -(12.5x^2 + 312.5x + 3906.25)e^{-0.08x} + C.$$

Therefore:

$$P = \{1000[-(12.5x + 156.25)e^{-0.08x}] - 100[-(12.5x^2 + 312.5x + 3906.25)e^{-0.08x}]\}\Big|_0^{10}$$

Collect like terms and simplify.

$$P = [(1250x^2 + 18{,}750x + 234{,}375)e^{-0.08x}]\Big|_0^{10}$$
$$= (546{,}875e^{-0.8}) - (234{,}375e^0)$$
$$\approx 11{,}351.77725$$

The principal value is \$11,351.78.

(b) $A = e^{0.08(10)}\int_0^{10}(1000x - 100x^2)e^{-0.08x}dx$

$$= e^{0.8}P$$
$$\approx 25{,}263.84488$$

The accumulated value is \$25,263.84.

15. $A = e^{0.14(3)} \int_0^3 20{,}000e^{-0.14x}\,dx$

$= e^{0.42}\left(\frac{20{,}000}{-0.14}e^{-0.14x}\right)\Big|_0^3$

$= e^{0.42}\left(\frac{20{,}000}{-0.14}e^{-0.42} + \frac{20{,}000}{0.14}\right)$

$\approx \$74{,}565.94$

17. (a) Present value

$= \int_0^8 5000e^{-0.01x}e^{-0.08x}\,dx$

$= \int_0^8 5000e^{-0.09x}\,dx$

$= \left(\frac{5000}{-0.09}e^{-0.09x}\right)\Big|_0^8$

$= \frac{5000e^{-0.72}}{-0.09} + \frac{5000}{0.09}$

$\approx \$28{,}513.76$

(b) Final amount

$= e^{0.08(8)} \int_0^8 5000e^{-0.01x}e^{-0.08x}\,dx$

$\approx e^{0.64}(28{,}513.76)$

$\approx \$54{,}075.81$

19. $P = \int_0^5 (1500 - 60x^2)e^{-0.10x}\,dx$

$= \int_0^5 1500e^{-0.10x}\,dx - \int_0^5 60x^2e^{-0.10x}\,dx$

$= 1500 \int_0^5 e^{-0.10x}\,dx$

$- 60 \int_0^5 x^2e^{-0.10x}\,dx$

Find the second integral by column integration.

D		I
x^2	+	$e^{-0.10x}$
$2x$	−	$\frac{e^{-0.10x}}{-0.10}$
2	+	$\frac{e^{-0.10x}}{0.01}$
0		$\frac{e^{-0.10x}}{-0.001}$

$\int x^2e^{-0.10x}\,dx$

$= \frac{x^2e^{-0.10x}}{-0.10} - \frac{2xe^{-0.10x}}{0.01} + \frac{2e^{-0.10x}}{-0.001} + C$

Now add the first integral to this result.

$1500 \int_0^5 e^{-0.10x}\,dx - 60 \int_0^5 x^2e^{-0.10x}\,dx$

$= \frac{1500}{-0.10}e^{-0.10x}\Big|_0^5$

$- 60\left(-\frac{x^2e^{-0.10x}}{0.10} - \frac{2xe^{-0.10x}}{0.01}\right.$

$\left.- \frac{2e^{-0.10x}}{0.001}\right)\Big|_0^5$

$= -15{,}000e^{-0.5} + 15{,}000$

$+ 60\left(\frac{25e^{-0.5}}{0.1} + \frac{10e^{-0.5}}{0.01} + \frac{2e^{-0.5}}{0.001}\right.$

$\left.- 0 - 0 - \frac{2}{0.001}\right)$

$= -15{,}000e^{-0.5} + 15{,}000 + 15{,}000e^{-0.5}$
$+ 60{,}000e^{-0.5} + 120{,}000e^{-0.5} - 120{,}000$
$= 180{,}000e^{-0.5} - 105{,}000$
$\approx \$4175.52$

8.4 Improper Integrals

1. $\int_3^\infty \frac{1}{x^2}\,dx = \lim_{b\to\infty} \int_3^b x^{-2}\,dx = \lim_{b\to\infty} \int (-x^{-1})\Big|_3^b$

$= \lim_{b\to\infty}\left(-\frac{1}{b} + \frac{1}{3}\right)$

$= \lim_{b\to\infty}\left(-\frac{1}{b}\right) + \lim_{b\to\infty}\frac{1}{3}$

As $b\to\infty$, $-\frac{1}{b}\to 0$. The integral is convergent.

$\int_3^\infty \frac{1}{x^2}\,dx = 0 + \frac{1}{3} = \frac{1}{3}$

3. $\int_4^\infty \frac{2}{\sqrt{x}}\,dx = \lim_{b\to\infty} \int_4^b 2x^{-1/2}dx$

$= \lim_{b\to\infty} 4x^{1/2}\Big|_4^b$

$= \lim_{b\to\infty} (4\sqrt{b} - 4\sqrt{4})$

$= \lim_{b\to\infty} 4\sqrt{b} - 8$

As $b\to\infty$, $4\sqrt{b}\to\infty$. The integral diverges.

5. $\int_{-\infty}^{-1} \frac{2}{x^3}\,dx = \int_{-\infty}^{-1} 2x^{-3}\,dx = \lim_{a\to-\infty} \int_a^{-1} 2x^{-3}\,dx$

$= \lim_{a\to-\infty} \left(\frac{2x^{-2}}{-2}\right)\Big|_a^{-1} = \lim_{a\to-\infty} \left(-1+\frac{1}{a^2}\right)$

As $a \to -\infty$, $\frac{1}{a^2} \to 0$. The integral is convergent.

$$\int_{-\infty}^{-1} \frac{2}{x^3}\,dx = -1+0 = -1$$

7. $\int_1^{\infty} \frac{1}{x^{1.0001}}\,dx$

$= \int_1^{\infty} x^{-1.0001}\,dx$

$= \lim_{b\to\infty} \int_1^b x^{-1.0001}\,dx$

$= \lim_{b\to\infty} \left(\frac{x^{-0.0001}}{-0.0001}\right)\Big|_1^b$

$= \lim_{b\to\infty} \left(-\frac{1}{(0.0001)b^{0.0001}} + \frac{1}{0.0001}\right)$

As $b \to \infty$, $-\frac{1}{0.0001b^{0.0001}} \to 0$.

The integral is convergent.

$$\int_1^{\infty} \frac{1}{x^{1.0001}}\,dx = 0 + \frac{1}{0.0001} = 10{,}000$$

9. $\int_{-\infty}^{-10} x^{-2}\,dx = \lim_{a\to-\infty} \int_a^{-10} x^{-2}\,dx$

$= \lim_{a\to-\infty} (-x^{-1})\Big|_a^{-10}$

$= \lim_{a\to-\infty} \left(\frac{1}{10}+\frac{1}{a}\right)$

$= \frac{1}{10}+0$

$= \frac{1}{10}$

The integral is convergent.

11. $\int_{-\infty}^{-1} x^{-8/3}\,dx = \lim_{a\to-\infty} \int_a^{-1} x^{-8/3}\,dx$

$= \lim_{a\to-\infty} \left(-\frac{3}{5}x^{-5/3}\right)\Big|_a^{-1}$

$= \lim_{a\to-\infty} \left(\frac{3}{5}+\frac{3}{5a^{5/3}}\right)$

$= \frac{3}{5}+0$

$= \frac{3}{5}$

The integral is convergent.

13. $\int_0^{\infty} 8e^{-8x}\,dx = \lim_{b\to\infty} \int_0^b 8e^{-8x}\,dx$

$= \lim_{b\to\infty} \left(\frac{8e^{-8x}}{-8}\right)\Big|_0^b$

$= \lim_{b\to\infty} (-e^{-8b}+1)$

$= \lim_{b\to\infty} \left(-\frac{1}{e^{8b}}+1\right)$

$= 0+1 = 1$

The integral is convergent.

15. $\int_{-\infty}^{0} 1000e^x\,dx = \lim_{a\to-\infty} \int_a^0 1000e^x\,dx$

$= \lim_{a\to-\infty} (1000e^x)\Big|_a^0$

$= \lim_{a\to-\infty} (1000-1000e^a)$

As a approaches $-\infty$, e^a is in the denominator of a fraction.

As $a \to \infty$, $-1000e^a \to 0$. The integral is convergent.

$$\int_{-\infty}^{0} 1000e^x\,dx = 1000-0 = 1000$$

17. $\int_{-\infty}^{-1} \ln|x|\,dx = \lim_{a\to-\infty} \int_a^{-1} \ln|x|\,dx$

Let $u = \ln|x|$ and $dv = dx$.

Then $du = \frac{1}{x}\,dx$ and $v = x$.

$\int \ln|x|\,dx = x\ln|x| - \int \frac{x}{x}\,dx$

$= x\ln|x| - x + C$

$\int_{-\infty}^{-1} \ln|x|\,dx = \lim_{a\to-\infty} (x\ln|x|-x)\Big|_a^{-1}$

$= \lim_{a\to-\infty} (-\ln 1 + 1 - a\ln|a| + a)$

$= \lim_{a\to-\infty} (1+a-a\ln|a|)$

The integral is divergent, since as $a \to -\infty$.

$$(a - a\ln|a|) = -a(-1+\ln|a|) \to \infty.$$

19. $\int_0^{\infty} \frac{dx}{(x+1)^2} = \lim_{b\to\infty} \int_0^b \frac{dx}{(x+1)^2}$ *Use substitution*

$= \lim_{b\to\infty} -(x+1)^{-1}\Big|_0^b$

$= \lim_{b\to\infty} \left(\frac{-1}{b+1}+1\right)$

As $b \to \infty$, $-\frac{1}{b+1} \to 0$. The integral is convergent.

$$\int_0^\infty \frac{dx}{(x+1)^2} = 0 + 1 = 1$$

21. $\displaystyle\int_{-\infty}^{-1} \frac{2x-1}{x^2-x}\,dx$

$$= \lim_{a\to-\infty} \int_a^{-1} \frac{2x-1}{x^2-x}\,dx \quad \textit{Use substitution}$$

$$= \lim_{a\to-\infty} \ln\left|x^2-x\right|\Big|_a^{-1}$$

$$= \lim_{a\to-\infty} \left(\ln 2 - \ln\left|a^2-a\right|\right)$$

As $a \to -\infty$, $\ln\left|a^2-a\right| \to \infty$. The integral is divergent.

23. $\displaystyle\int_2^\infty \frac{1}{x\ln x}\,dx$

$$= \lim_{b\to\infty} \int_2^b \frac{1}{x\ln x}\,dx \quad \textit{Use substitution}$$

$$= \lim_{b\to\infty} \left[\ln(\ln x)\Big|_2^b\right]$$

$$= \lim_{b\to\infty} \left[\ln(\ln b) - \ln(\ln 2)\right]$$

As $b \to \infty$, $\ln(\ln b) \to \infty$. The integral is divergent.

25. $\displaystyle\int_0^\infty xe^{4x}\,dx = \lim_{b\to\infty}\int_0^b xe^{4x}\,dx$

Let $dv = e^{4x}\,dx$ and $u = x$.

Then $v = \frac{1}{4}e^{4x}$ and $du = dx$.

$$\int xe^{4x}\,dx = \frac{x}{4}e^{4x} - \int \frac{1}{4}e^{4x}\,dx$$

$$= \frac{x}{4}e^{4x} - \frac{1}{16}e^{4x} + C$$

$$= \frac{1}{16}(4x-1)e^{4x} + C$$

$$\int_0^\infty xe^{4x}\,dx$$

$$= \lim_{b\to\infty}\left[\frac{1}{16}(4x-1)e^{4x}\right]\Bigg|_0^b$$

$$= \lim_{b\to\infty}\left[\frac{1}{16}(4b-1)e^{4b} - \frac{1}{16}(-1)(1)\right]$$

$$= \lim_{b\to\infty}\left[\frac{1}{16}(4b-1)e^{4b} + \frac{1}{16}\right]$$

As $b \to \infty$, $\frac{1}{16}(4b-1)e^{4b} \to \infty$. The integral is divergent.

27. $\displaystyle\int_{-\infty}^\infty x^3e^{-x^4}\,dx = \int_{-\infty}^0 x^3e^{-x^4}\,dx + \int_0^\infty x^3e^{-x^4}\,dx$

We evaluate each of the two improper integrals on the right.

$$\int_{-\infty}^0 x^3e^{-x^4}\,dx = \lim_{b\to-\infty}\int_b^0 x^3e^{-x^4}\,dx \quad \textit{Use substitution}$$

$$= \lim_{b\to-\infty}\left[-\frac{1}{4}e^{-x^4}\Big|_b^0\right]$$

$$= \lim_{b\to-\infty}\left[-\frac{1}{4} + \frac{1}{4e^{b^4}}\right]$$

As $b \to -\infty$, $\frac{1}{4e^{b^4}} \to 0$. The integral is convergent.

$$\int_{-\infty}^0 x^3e^{-x^4}\,dx = -\frac{1}{4} + 0 = -\frac{1}{4}$$

$$\int_0^\infty x^3e^{-x^4}\,dx = \lim_{b\to\infty}\int_0^b x^3e^{-x^4}\,dx \quad \textit{Use substitution}$$

$$= \lim_{b\to\infty}\left[-\frac{1}{4}e^{-x^4}\Big|_0^b\right]$$

$$= \lim_{b\to\infty}\left[-\frac{1}{4e^{b^4}} + \frac{1}{4}\right]$$

As $b \to \infty$, $-\frac{1}{4e^{b^4}} \to 0$. The integral is convergent.

$$\int_0^\infty x^3e^{-x^4}\,dx = 0 + \frac{1}{4} = \frac{1}{4}$$

Since each of the improper integrals converges, the original improper integral converges.

$$\int_{-\infty}^\infty x^3e^{-x^4}\,dx = -\frac{1}{4} + \frac{1}{4} = 0$$

29. $\displaystyle\int_{-\infty}^\infty \frac{x}{x^2+1}\,dx = \int_{-\infty}^0 \frac{x}{x^2+1}\,dx + \int_0^\infty \frac{x}{x^2+1}\,dx$

We evaluate the first improper integral on the right.

$$\int_{-\infty}^0 \frac{x}{x^2+1}\,dx = \lim_{b\to-\infty}\int_b^0 \frac{x}{x^2+1}\,dx \quad \textit{Use substitution}$$

$$= \lim_{b\to-\infty}\left[\frac{1}{2}\ln(x^2+1)\Big|_b^0\right]$$

$$= \lim_{b\to-\infty}\left[0 - \frac{1}{2}\ln(b^2+1)\right]$$

As $b \to -\infty$, $\ln(b^2+1) \to \infty$. The integral is divergent. Since one of the two improper integrals on the right diverges, the original improper integral diverges.

31. $f(x) = \dfrac{1}{x-1}$ for $(-\infty, 0]$

$$\int_{-\infty}^{0} \frac{1}{x-1}\,dx = \lim_{a\to-\infty} \int_{a}^{0} \frac{dx}{x-1}$$

$$= \lim_{a\to-\infty} \left(\ln|x-1|\right)\Big|_{a}^{0}$$

$$= \lim_{a\to-\infty} \left(\ln|-1| - \ln|a-1|\right)$$

But $\lim_{a\to-\infty} (\ln|a-1|) = \infty$.

The integral is divergent, so the area cannot be found.

33. $f(x) = \dfrac{1}{(x-1)^2}$ for $(-\infty, 0]$

$$\int_{-\infty}^{0} \frac{1}{(x-1)^2}$$

$$= \lim_{a\to-\infty} \int_{a}^{0} \frac{1}{(x-1)^2} \quad \textit{Use substitution}$$

$$= \lim_{a\to-\infty} -(x-1)^{-1}\Big|_{a}^{0}$$

$$= \lim_{a\to-\infty} \left(-\frac{1}{-1} + \frac{1}{a-1}\right)$$

As $a\to-\infty$, $\frac{1}{a-1}\to 0$. The integral is convergent.

$$= 1 + 0 = 1$$

Therefore, the area is 1.

35. $\displaystyle\int_{-\infty}^{\infty} xe^{-x^2}\,dx$

Let $u = -x^2$, so that $du = -2x\,dx$.

$$= \lim_{a\to-\infty} \left(-\frac{1}{2}\int_{a}^{0} -2xe^{-x^2}\,dx\right)$$

$$+ \lim_{b\to\infty} \left(-\frac{1}{2}\int_{0}^{b} -2xe^{-x^2}\,dx\right)$$

$$= \lim_{a\to-\infty} \left(-\frac{1}{2}e^{-x^2}\right)\Big|_{a}^{0}$$

$$+ \lim_{b\to\infty} \left(-\frac{1}{2}e^{-x^2}\right)\Big|_{0}^{b}$$

$$= \lim_{a\to-\infty} \left(-\frac{1}{2} + \frac{1}{2e^{-a^2}}\right)$$

$$+ \lim_{b\to\infty} \left(-\frac{1}{2e^{b^2}} + \frac{1}{2}\right)$$

$$= -\frac{1}{2} + \frac{1}{2} = 0$$

37. $\displaystyle\int_{1}^{\infty} \frac{1}{x^p}\,dx$

Case 1a $p < 1$:

$$\int_{1}^{\infty} \frac{1}{x^p}\,dx$$

$$= \int_{1}^{\infty} x^{-p}\,dx$$

$$= \lim_{a\to\infty} \int_{1}^{a} x^{-p}\,dx$$

$$= \lim_{a\to\infty} \left[\frac{x^{-p+1}}{(-p+1)}\Big|_{1}^{a}\right]$$

$$= \lim_{a\to\infty} \left[\frac{1}{(-p+1)}(a^{-p+1} - 1)\right]$$

$$= \lim_{a\to\infty} \left[\frac{1}{(-p+1)}a^{1-p} - \frac{1}{(-p+1)}\right]$$

Since $p < 1$, $1 - p$ is positive and, as $a\to\infty$, $a^{1-p}\to\infty$. The integral diverges.

Case 1b $p = 1$:

$$\int_{1}^{\infty} \frac{1}{x^p}\,dx = \int_{1}^{\infty} \frac{1}{x}\,dx$$

$$= \lim_{a\to\infty} \int_{1}^{a} \frac{1}{x}\,dx$$

$$= \lim_{a\to\infty} \left(\ln|x|\Big|_{1}^{a}\right)$$

$$= \lim_{a\to\infty} (\ln|a| - \ln 1)$$

$$= \lim_{a\to\infty} \ln|a|$$

As $a\to\infty$, $\ln|a|\to\infty$. The integral diverges.

Therefore, $\displaystyle\int_{1}^{\infty} \frac{1}{x^p}$ diverges when $p \le 1$.

Case 2 $p > 1$:

$$\int_{1}^{\infty} \frac{1}{x^p}\,dx = \lim_{a\to\infty} \int_{1}^{a} x^{-p}\,dx$$

$$= \lim_{a\to\infty} \left(\frac{x^{-p+1}}{-p+1}\Big|_{1}^{a}\right)$$

$$= \lim_{a\to\infty} \left[\frac{a^{-p+1}}{(-p+1)} - \frac{1}{(-p+1)}\right]$$

Since $p > 1$, $-p + 1 < 0$; thus as $a\to\infty$, $\frac{a^{-p+1}}{(-p+1)}\to 0$.

Hence,

$$\lim_{a\to\infty}\left[\frac{a^{-p+1}}{(-p+1)}-\frac{1}{(-p+1)}\right]=0-\frac{1}{(-p+1)}$$
$$=\frac{-1}{-p+1}$$
$$=\frac{1}{p-1}.$$

The integral converges.

39. **(a)** Use the $fnInt$ feature on a graphing utility to obtain

$$\int_1^{20}\frac{1}{\sqrt{1+x^2}}\,dx\approx 2.808;$$
$$\int_1^{50}\frac{1}{\sqrt{1+x^2}}\,dx\approx 3.724;$$
$$\int_1^{100}\frac{1}{\sqrt{1+x^2}}\,dx\approx 4.417;$$
$$\int_1^{1000}\frac{1}{\sqrt{1+x^2}}\,dx\approx 6.720;$$
$$\int_1^{10,000}\frac{1}{\sqrt{1+x^2}}\,dx\approx 9.022.$$

(b) Since the values of the integrals in part a do not appear to be approaching some fixed finite number but get bigger, the integral $\int_1^\infty \frac{1}{\sqrt{1+x^2}}\,dx$ appears to be divergent.

(c) Use the $fnInt$ feature on a graphing utility to obtain

$$\int_1^{20}\frac{1}{\sqrt{1+x^4}}\,dx\approx 0.8770;$$
$$\int_1^{50}\frac{1}{\sqrt{1+x^4}}\,dx\approx 0.9070;$$
$$\int_1^{100}\frac{1}{\sqrt{1+x^4}}\,dx\approx 0.9170;$$
$$\int_1^{1000}\frac{1}{\sqrt{1+x^4}}\,dx\approx 0.9260;$$
$$\int_1^{10,000}\frac{1}{\sqrt{1+x^4}}\,dx\approx 0.9269.$$

(d) Since the values of the integrals in part c appear to be approaching some fixed finite number, the integral $\int_1^\infty \frac{1}{\sqrt{1+x^4}}\,dx$ appears to be convergent.

(e) For large x, we may consider $1+x^2\approx x^2$ and $1+x^4\approx x^4$.
Thus,

$$\frac{1}{\sqrt{1+x^4}}\approx\frac{1}{\sqrt{x^2}}=\frac{1}{x}\text{ and}$$
$$\frac{1}{\sqrt{1+x^4}}\approx\frac{1}{\sqrt{x^4}}=\frac{1}{x^2}.$$

In Example 1(a) on page 455, we showed that $\int_1^\infty \frac{1}{x}dx$ diverges. Thus, we might guess that $\int_1^\infty \frac{1}{\sqrt{1+x^2}}\,dx$ diverges as well. In Exercise 1, we saw that $\int_2^\infty \frac{1}{x^2}dx$ converges. Thus, we might guess that $\int_1^\infty \frac{1}{\sqrt{1+x^4}}\,dx$ converges as well.

41. **(a)** Use the $fnInt$ feature on a graphing utility to obtain

$$\int_0^{10} e^{-0.00001x}dx\approx 9.9995;$$
$$\int_0^{50} e^{-0.00001x}dx\approx 49.9875;$$
$$\int_0^{100} e^{-0.00001x}dx\approx 99.9500;$$
$$\int_0^{1000} e^{-0.00001x}dx\approx 995.0166.$$

(b) Since the values of the integrals in part a do not appear to be approaching some fixed finite number, the integral $\int_0^\infty e^{-0.00001x}dx$ appears to be divergent.

(c) $\int_0^\infty e^{-0.00001x}dx$

$$=\lim_{b\to\infty}\int_0^b e^{-0.00001x}dx$$
$$=\lim_{b\to\infty}\left[\frac{e^{-0.00001x}}{-0.00001}\bigg|_0^b\right]$$
$$=\lim_{b\to\infty}\left[-\frac{1}{0.00001e^{0.00001b}}+\frac{1}{0.00001}\right]$$
$$=0+100{,}000=100{,}000$$

43. $\displaystyle\int_0^\infty 1{,}000{,}000e^{-0.05t}dt$

$$=\lim_{b\to\infty}\int_0^b 1{,}000{,}000e^{-0.05t}dt$$
$$=\lim_{b\to\infty}\left(\frac{1{,}000{,}000}{-0.05}e^{-0.05t}\right)\bigg|_0^b$$
$$=-20{,}000{,}000\left[\lim_{b\to\infty}(e^{-0.05b})-e^0\right]$$

As $b \to \infty$, $e^{-0.05b} = \frac{1}{e^{0.05b}} \to 0$. The integral converges.

$$\int_0^{\infty} 1{,}000{,}000e^{-0.05t}dt$$
$$= -20{,}000{,}000(0-1)$$
$$= 20{,}000{,}000$$

The capital value is \$20,000,000.

45. $$\int_0^{\infty} 1200e^{0.03t}e^{-0.07t}dt$$

$$= \lim_{b\to\infty} \int_0^b 1200e^{-0.04t}dt$$

$$= \lim_{b\to\infty} \left(\frac{1200}{-0.04}e^{-0.04t}\right)\Bigg|_0^b$$

$$= -30{,}000\left[\lim_{b\to\infty}(e^{-0.04b}) - e^0\right]$$

As $b \to \infty$, $e^{-0.04b} = \frac{1}{e^{0.04b}} \to 0$. The integral converges.

$$\int_0^{\infty} 1200e^{0.03t}e^{-0.07t}dt = -30{,}000(0-1)$$
$$= 30{,}000$$

The capital value is \$30,000.

47. $$\int_0^{\infty} 3000e^{-0.1t}\,dt = \lim_{b\to\infty}\int_0^b 3000e^{-0.1t}\,dt$$
$$= \lim_{b\to\infty} \frac{3000e^{-0.1b}}{-0.1}\Bigg|_0^b$$
$$= \lim_{b\to\infty}\left(\frac{3000e^{-0.1b}}{-0.1} + \frac{3000}{0.1}\right)$$
$$= 0 + 30{,}000$$
$$= \$30{,}000$$

49. $$S = N\int_0^{\infty} \frac{a(1-e^{-kt})}{k}e^{-bt}dt$$
$$= \frac{Na}{k}\lim_{c\to\infty}\int_0^c (1-e^{-kt})(e^{-bt})dt$$
$$= \frac{Na}{k}\lim_{c\to\infty}\int_0^c (e^{-bt} - e^{-(b+k)t})dt$$
$$= \frac{Na}{k}\lim_{c\to\infty}\left[-\frac{1}{b}e^{-bt} + \frac{1}{b+k}e^{-(b+k)t}\right]\Bigg|_0^c$$
$$= \frac{Na}{k}\lim_{c\to\infty}\left[\left(-\frac{1}{b}e^{-bc} + \frac{1}{b+k}e^{-(b+k)c}\right) - \left(-\frac{1}{b}e^0 + \frac{1}{b+k}e^0\right)\right]$$
$$= \frac{Na}{k}\left(0+0+\frac{1}{b} - \frac{1}{b+k}\right)$$
$$= \frac{Na}{k}\cdot\frac{(b+k)-b}{b(b+k)}$$
$$= \frac{Nak}{kb(b+k)}$$
$$= \frac{Na}{b(b+k)}$$

51. $$\int_0^{\infty} 50e^{-0.06t}\,dt = 50\lim_{b\to\infty}\int_0^b e^{-0.06t}\,dt$$
$$= 50\lim_{b\to\infty}\frac{e^{-0.06t}}{-0.06}\Bigg|_0^b$$
$$= \frac{50}{-0.06}\lim_{b\to\infty}(e^{-0.06b} - e^0)$$
$$= -\frac{50}{0.06}(0-1) = \frac{50}{0.06}$$
$$\approx 833.3$$

Chapter 8 Review Exercises

5. $$\int \frac{3x}{\sqrt{x-2}}\,dx = \int 3x(x-2)^{-1/2}\,dx$$

Let $u = 3x$ and $dv = (x-2)^{-1/2}\,dx$.
Then $du = 3\,dx$ and $v = 2(x-2)^{1/2}$.

$$\int \frac{3x}{\sqrt{x-2}}\,dx$$
$$= 6x(x-2)^{1/2} - 6\int (x-2)^{1/2}\,dx$$
$$= 6x(x-2)^{1/2} - \frac{6(x-2)^{3/2}}{\frac{3}{2}} + C$$
$$= 6x(x-2)^{1/2} - 4(x-2)^{3/2} + C$$

7. $\int (3x+6)e^{-3x}\,dx$

Let $u = 3x+6$ and $dv = e^{-3x}dx$.
Then $du = 3\,dx$ and $v = \frac{1}{-3}e^{-3x}$.

$$\int (3x+6)e^{-3x}dx$$
$$= (3x+6)\left(-\frac{1}{3}e^{-3x}\right) - \int\left(-\frac{1}{3}e^{-3x}\right)3\,dx$$
$$= -(x+2)e^{-3x} + \int e^{-3x}dx$$
$$= -(x+2)e^{-3x} - \frac{1}{3}e^{-3x} + C$$

9. $\int (x-1)\ln|x|\,dx$

Let $u = \ln|x|$ and $dv = (x-1)\,dx$.

Then $du = \frac{1}{x}\,dx$ and $v = \frac{x^2}{2} - x$.

$$\int (x-1)\ln|x|\,dx$$
$$= \left(\frac{x^2}{2} - x\right)\ln|x| - \int\left(\frac{x}{2} - 1\right)dx$$
$$= \left(\frac{x^2}{2} - x\right)\ln|x| - \frac{x^2}{4} + x + C$$

11. $\int \frac{x}{\sqrt{16+8x^2}}\,dx$

Use substitution.
Let $u = 16+8x^2$. Then $du = 16x\,dx$.

$$\int \frac{x}{\sqrt{16+8x^2}}\,dx = \frac{1}{16}\int\frac{16x}{\sqrt{16+8x^2}}\,dx$$
$$= \frac{1}{16}\int\frac{1}{\sqrt{u}}\,du$$
$$= \frac{1}{16}\int u^{-1/2}du$$
$$= \frac{1}{16}(2)u^{1/2} + C$$
$$= \frac{1}{8}(16+8x^2)^{1/2} + C$$
$$= \frac{1}{8}\sqrt{16+8x^2} + C$$

13. $\int_0^1 x^2e^{x/2}\,dx$

Let $u = x^2$ and $dv = e^{x/2}\,dx$.
Use column integration.

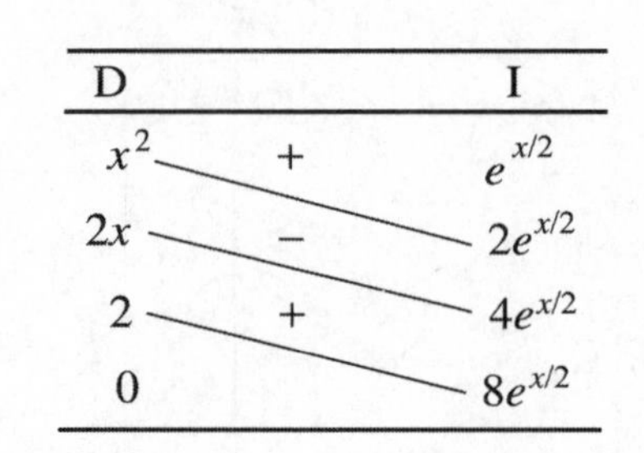

D		I
x^2	+	$e^{x/2}$
$2x$	−	$2e^{x/2}$
2	+	$4e^{x/2}$
0		$8e^{x/2}$

$$\int_0^1 x^2e^{x/2}\,dx = (2x^2e^{x/2} - 8xe^{x/2} + 16e^{x/2})\Big|_0^1$$
$$= 2e^{1/2} - 8e^{1/2} + 16e^{1/2} - 16$$
$$= 10e^{1/2} - 16 \approx 0.4872$$

15. $A = \int_1^3 x^3(x^2-1)^{1/3}\,dx$

Let $u = x^2$ and $dv = x(x^2-1)^{1/3}\,dx$.
Then $du = 2x\,dx$ and $v = \frac{3}{8}(x^2-1)^{4/3}$.

$$\int x^3(x^2-1)^{1/3}\,dx$$
$$= \frac{3x^2}{8}(x^2-1)^{4/3} - \frac{3}{4}\int x(x^2-1)^{4/3}\,dx$$
$$= \frac{3x^2}{8}(x^2-1)^{4/3} - \frac{3}{4}\left[\frac{1}{2}\cdot\frac{3}{7}(x^2-1)^{7/3}\right]$$
$$= \frac{3x^2}{8}(x^2-1)^{4/3} - \frac{9}{56}(x^2-1)^{7/3} + C$$
$$A = \left[\frac{3x^2}{8}(x^2-1)^{4/3} - \frac{9}{56}(x^2-1)^{7/3}\right]\Bigg|_0^3$$
$$= \frac{3}{8}(144) - \frac{9}{56}(128) = 54 - \frac{144}{7} = \frac{234}{7} \approx 33.43$$

17. $f(x) = \sqrt{x-4};\ y = 0;\ x = 13$

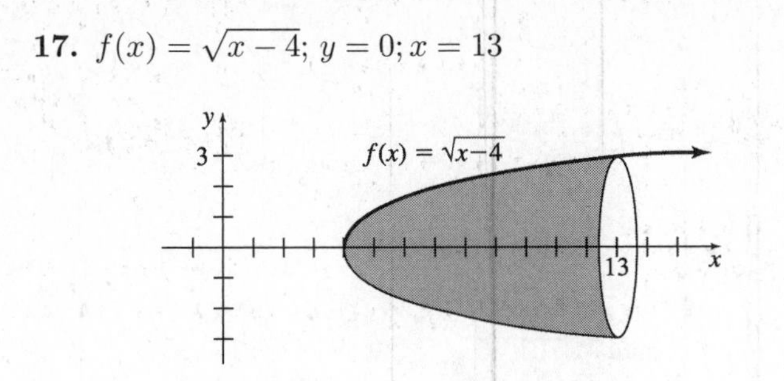

Since $f(x) = \sqrt{x-4}$ intersects $y = 0$ at $x = 4$, the integral has lower bound $a = 4$.

$$V = \pi\int_4^{13}(\sqrt{x-4})^2dx$$
$$= \pi\int_4^{13}(x-4)dx$$
$$= \pi\left(\frac{x^2}{2} - 4x\right)\Bigg|_4^{13}$$
$$= \pi\left[\left(\frac{169}{2} - 52\right) - (8-16)\right]$$
$$= \pi\left(\frac{65}{2} + 8\right) = \frac{81}{2}\pi \approx 127.2$$

19. $f(x) = \dfrac{1}{\sqrt{x-1}},\ y = 0,\ x = 2,\ x = 4$

$$V = \pi\int_2^4\left(\frac{1}{\sqrt{x-1}}\right)^2dx$$
$$= \pi\int_2^4\frac{dx}{x-1}$$
$$= \pi\left(\ln|x-1|\right)\Big|_2^4$$
$$= \pi\ln 3 \approx 3.451$$

21. $f(x) = \dfrac{x^2}{4},\ y = 0,\ x = 4$

Since $f(x) = \frac{x^2}{4}$ intersects $y = 0$ at $x = 0$, the integral has a lower bound, $a = 0$.

$$V = \pi\int_0^4\left(\frac{x^2}{4}\right)^2dx = \pi\int_0^4\frac{x^4}{16}$$
$$= \frac{\pi}{16}\left(\frac{x^5}{5}\right)\Bigg|_0^4$$
$$= \frac{\pi}{16}\left(\frac{1024}{5}\right)$$
$$= \frac{64\pi}{5} \approx 40.21$$

25. $$\text{Average value} = \frac{1}{2-0}\int_0^2 7x^2(x^3+1)^6dx$$
$$= \frac{7}{2}\int_0^2 x^2(x^3+1)^6dx$$

Let $u = x^3 + 1$. Then $du = 3x^2dx$.

$$\int x^2(x^3+1)^6dx = \frac{1}{3}\int 3x^2(x^3+1)^6dx$$
$$= \frac{1}{3}\int u^6du$$
$$= \frac{1}{3}\cdot\frac{1}{7}u^7 + C$$
$$= \frac{1}{21}(x^3+1)^7 + C$$

$$\frac{7}{2}\int_0^2 x^2(x^3+1)^6dx = \frac{7}{2}\cdot\frac{1}{21}(x^3+1)^7\Big|_0^2$$
$$= \frac{1}{6}(9^7 - 1^7)$$
$$= \frac{1}{6}(4{,}782{,}969 - 1)$$
$$= \frac{2{,}391{,}484}{3}$$

27. $$\int_{-\infty}^{-5}x^{-2}dx = \lim_{a\to-\infty}\int_a^{-5}x^{-2}dx$$
$$= \lim_{a\to-\infty}\left(\frac{x^{-1}}{-1}\right)\Bigg|_a^{-5}$$
$$= \lim_{a\to-\infty}\left(-\frac{1}{x}\right)\Bigg|_a^{-5}$$
$$= \frac{1}{5} + \lim_{a\to-\infty}\left(\frac{1}{a}\right)$$

As $a \to -\infty$, $\frac{1}{a} \to 0$. The integral converges.

$$\int_{-\infty}^{-5}x^{-2}dx = \frac{1}{5} + 0 = \frac{1}{5}$$

29. $$\int_1^\infty 6e^{-x}\,dx = \lim_{b\to\infty}\int_1^b 6e^{-x}\,dx$$
$$= \lim_{b\to\infty} -6e^{-x}\Big|_1^b$$
$$= \lim_{b\to\infty}(-6e^{-b} + 6e^{-1})$$
$$= \lim_{b\to\infty}\left(\frac{-6}{e^b} + \frac{6}{e}\right)$$

As $b \to \infty$, $e^b \to \infty$, so $\frac{-6}{e^b} \to 0$. The integral converges.

$$\int_1^\infty 6e^{-x}\,dx = 0 + \frac{6}{e} = \frac{6}{e} \approx 2.207$$

31. $\int_4^\infty \ln(5x)dx = \lim_{b\to\infty} \int_4^b \ln(5x)dx$

Let $u = \ln(5x)$ and $dv = dx$.

Then $du = \frac{1}{x}dx$ and $v = x$.

$$\int \ln(5x)dx = x\ln(5x) - \int x \cdot \frac{1}{x}\,dx$$
$$= x\ln(5x) - \int dx$$
$$= x\ln(5x) - x + C$$

$$\lim_{b\to\infty}\int_4^b \ln(5x)dx = \lim_{b\to\infty}[x\ln(5x) - x]\Big|_4^b$$
$$= \lim_{b\to\infty}[b\ln(5b) - b] - (4\ln 20 - 4)$$

As $b\to\infty$, $b\ln(5b) - b \to \infty$. The integral diverges.

33. $f(x) = 3e^{-x}$ for $[0,\infty)$

$$A = \int_0^\infty 3e^{-x}\,dx$$
$$= \lim_{b\to\infty}\int_0^b 3e^{-x}\,dx$$
$$= \lim_{b\to\infty}(-3e^{-x})\Big|_0^b$$
$$= \lim_{b\to\infty}\left(\frac{-3}{e^b} + 3\right)$$

As $b\to\infty$, $\frac{-3}{e^b}\to 0$.

$A = 0 + 3 = 3$

35. $R' = x(x-50)^{1/2}$

$$R = \int_{50}^{75} x(x-50)^{1/2}\,dx$$

Let $u = x$ and $dv = (x-50)^{1/2}$.

Then $du = dx$ and $v = \frac{2}{3}(x-50)^{3/2}$.

$$\int x(x-50)^{1/2}\,dx$$
$$= \frac{2}{3}x(x-50)^{3/2} - \frac{2}{3}\int (x-50)^{3/2}\,dx$$
$$= \frac{2}{3}x(x-50)^{3/2} - \frac{2}{3}\cdot\frac{2}{5}(x-50)^{5/2}$$

$$R = \left[\frac{2}{3}x(x-50)^{3/2} - \frac{4}{15}(x-50)^{5/2}\right]\Big|_{50}^{75}$$
$$= \frac{2}{3}(75)(25^{3/2}) - \frac{4}{15}(25^{5/2})$$
$$= 6250 - \frac{2500}{3}$$
$$= \frac{16{,}250}{3} \approx 5416.67$$

37. $f(x) = 25{,}000$; 12 yr; 10%

$$P = \int_0^{12} 25{,}000e^{-0.10x}\,dx$$
$$= 25{,}000\left(\frac{e^{-0.10x}}{-0.10}\right)\Big|_0^{12}$$
$$\approx 250{,}000(-0.3012 + 1)$$
$$\approx \$174{,}701.45$$

39. $f(x) = 15x$; 18 mo; 8%

$$P = \int_0^{1.5} 15xe^{-0.08x}dx$$
$$= 15\int_0^{1.5} xe^{-0.08x}dx$$

Find the antiderivative using integration by parts.

Let $u = x$ and $dv = e^{-0.08x}dx$.

Then $du = dx$ and $v = \frac{1}{-0.08}e^{-0.08x}$
$= -12.5e^{-0.08x}$.

$$\int xe^{-0.08x}dx = -12.5xe^{-0.08x} - \int(-12.5e^{-0.08x})dx$$
$$= -12.5xe^{-0.08x} - 156.25e^{-0.08x} + C$$

$$P = 15\int_0^{1.5} xe^{-0.08x}dx$$
$$= 15(-12.5xe^{-0.08x} - 156.25e^{-0.08x})\Big|_0^{1.5}$$
$$= 15[(-18.75e^{-0.12} - 156.25e^{-0.12}) - (0 - 156.25)]$$
$$= 15(-175e^{-0.12} + 156.25)$$
$$\approx 15.58385362$$

The present value is \$15.58.

41. $f(x) = 500e^{-0.04x}$; 8 yr; 10% per yr

$$A = e^{0.1(8)}\int_0^8 500e^{-0.04x}\cdot e^{-0.1x}dx$$
$$= e^{0.8}\int_0^8 500e^{-0.14x}dx$$
$$= e^{0.8}\left(\frac{500}{-0.14}e^{-0.14x}\right)\Big|_0^8$$
$$= e^{0.8}\left[\frac{500}{-0.14}(e^{-1.12} - 1)\right]$$
$$\approx 5354.971041$$

The accumulated value is \$5354.97.

43. $f(x) = 1000 + 200x$; 10 yr; 9% per yr

$$e^{(0.09)(10)} \int_0^{10} (1000 + 200x)e^{-0.09x}\,dx$$

$$= e^{0.9}\left[\frac{1000}{-0.09}e^{0.09x} + \frac{200}{(0.09)^2}(-0.09x - 1)e^{-0.09x}\right]\Bigg|_0^{10}$$

$$= e^{0.9}\left[\frac{1000}{-0.09}(e^{-0.9} - 1) + \frac{200}{(0.09)^2}(-1.9e^{-0.9} + 1)\right]$$

$$\approx \$30{,}035.17$$

45. $e^{0.105(10)} \int_0^{10} 10{,}000e^{-0.105x}\,dx$

$$= e^{1.05}\left(\frac{10{,}000e^{-0.105x}}{-0.105}\right)\Bigg|_0^{10}$$

$$= \frac{10{,}000e^{1.05}}{-0.105}(e^{-1.05} - 1)$$

$$\approx -272{,}157.25(-0.65006)$$

$$\approx \$176{,}919.15$$

47. $\int_0^5 0.5xe^{-x}\,dx$

$$= 0.5\int_0^5 xe^{-x}\,dx$$

Let $u = x$ and $dv = e^{-x}\,dx$.

Then $du = dx$ and $v = \frac{e^{-x}}{-1}$.

$$\int xe^{-x}\,dx = \frac{xe^{-x}}{-1} + \int e^{-x}\,dx$$

$$= -xe^{-x} + \frac{e^{-x}}{-1}$$

$$0.5\int_0^5 xe^{-x}\,dx = 0.5(-xe^{-x} - e^{-x})\Big|_0^5$$

$$= 0.5(-5e^{-5} - e^{-5} + e^0)$$

$$\approx 0.4798$$

The total reaction over the first 5 hr is 0.4798.

49. $\int_0^{320} 1.87t^{1.49}e^{-0.189(\ln t)^2}\,dt$

$n = 8, b = 320, a = 1, f(t) = 1.87t^{1.49}e^{-0.189(\ln t)^2}$

i	t_i	$f(t_i)$
0	1	1.8700
1	41	34.9086
2	81	33.9149
3	121	30.7147
4	161	27.5809
5	201	24.8344
6	241	22.4794
7	281	20.4622
8	321	18.7255

(a) Trapezoidal rule:

$$\int_1^{321} 1.87t^{1.49}e^{-0.189(\ln t)^2}\,dt$$

$$\approx \frac{321-1}{8}\left[\frac{1}{2}(1.87)+34.9086+33.9149+30.7147\right.$$
$$+27.5809+24.8344+22.4794+20.4622$$
$$\left.+\frac{1}{2}(18.7255)\right]$$

$$\approx 8208$$

The total amount of milk produced is about 8208 kg.

(b) Simpson's rule:

$$\int_1^{321} 1.87t^{1.49}e^{-0.189(\ln t)^2}\,dt$$

$$\approx \frac{321-1}{3(8)}[1.87 + 4(34.9086) + 2(33.9149)$$
$$+ 4(30.7147) + 2(27.5809) + 4(24.8344)$$
$$+ 2(22.4794) + 4(20.4622) + 18.7255]$$
$$\approx 8430$$

The total amount of milk produced is about 8430 kg.

(c) Using a graphing calculator's fnInt feature:

$$\int_1^{321} 1.87t^{1.49}e^{-0.189(\ln t)^2}\,dt \approx 8558.$$

The total amount of milk produced is about 8558 kg.

Chapter 8 Test

[8.1]

Use integration by parts to find the following integrals.

1. $\int x \ln|5x|\, dx$

2. $\int x\sqrt{2x-1}\, dx$

3. $\int (2x-1)\, e^x\, dx$

4. $\int \frac{x+3}{(3x-1)^4}\, dx$

5. $\int_4^{11} x\sqrt[3]{x-3}\, dx$

6. Given that the marginal profit in dollars earned from the sale of x computers is

$$P'(x) = xe^{0.001x} - 100,$$

find the total profit from the sale of the first 500 computers.

[8.2]

Find the volume of the solid of revolution formed by rotating each of the following bounded regions about the x-axis.

7. $f(x) = 3x - 1,\ y = 0,\ x = 4$

8. $f(x) = \frac{1}{\sqrt{3x+1}},\ y = 0,\ x = 0,\ x = 5$

9. $f(x) = e^x,\ y = 0,\ x = -1,\ x = 2$

10. $f(x) = x^2,\ y = 0,\ x = 2$

11. Find the average value of $f(x) = x^3 - x^2$ on the interval $[-1, 1]$.

12. The rate of depreciation t years after purchase of a certain machine is

$$D = 10{,}000(t-7),\ 0 \le t \le 6.$$

What is the average depreciation over the first 3 years?

[8.3]

13. The rate of flow of an investment is $2000 per year for 10 years. Find the present value if the annual interest rate is 11% compounded continuously.

14. The rate of flow of an investment is given by

$$f(x) = 200e^{-0.1x}$$

for 10 years. Find the present value if the annual interest rate is 8% compounded continuously.

15. The rate of flow of an investment is given by the function $f(x) = 1000$. Find the final amount at the end of 15 years at an interest rate of 11% compounded continuously.

16. An investment scheme is expected to produce a continuous flow of money, starting at $2000 and increasing exponentially at 4% per year for 8 years. Find the present value at an interest rate of 12% compounded continuously.

[8.4]

Find the value of each integral that converges.

17. $\int_2^{\infty} \frac{14}{(x+1)^2}\,dx$

18. $\int_{-\infty}^{2} \frac{4}{x+1}\,dx$

19. $\int_1^{\infty} e^{-3x}\,dx$

20. $\int_1^{\infty} \frac{1}{x}\,dx$

21. $\int_1^{\infty} x^{-1/3}\,dx$

22. $\int_{100}^{\infty} x^{-3/2}\,dx$

Find the area between the graph of the function and the x-axis over the given interval, if possible.

23. $f(x) = 8e^{-x}$ on $[1, \infty)$

24. $f(x) = \frac{1}{(2x-1)^2}$ on $[2, \infty)$

25. Find the capital value of an asset that generates income at an annual rate of $4000 if the interest rate is 8% compounded continuously.

Chapter 8 Test Answers

1. $\frac{1}{2}x^2 \ln|5x| - \frac{1}{4}x^2 + C$

2. $\frac{x}{3}(2x-1)^{3/2} - \frac{1}{15}(2x-1)^{5/2} + C$

3. $(2x-3)e^x + C$

4. $-\frac{1}{9}\left[\frac{x+3}{(3x-1)^3}\right] - \frac{1}{54(3x-1)^2} + C$ or $-\frac{9x+17}{54(3x-1)^3} + C$

5. $\frac{2469}{28}$ or 88.18

6. \$125,639.36

7. $\frac{1331}{9}\pi$ or 464.6

8. $\frac{4}{3}\pi\ln(2)$ or 2.903

9. $\frac{\pi}{2}(e^4 - e^{-2})$ or 85.55

10. $\frac{32\pi}{5}$ or 20.11

11. $-\frac{1}{3}$

12. 55,000 per year

13. \$12,129.62

14. \$927.45

15. \$38,245.27

16. \$11,817.69

17. $\frac{14}{3}$

18. Divergent

19. $\frac{1}{3e^3}$ or 0.017

20. Divergent

21. Divergent

22. $\frac{1}{5}$

23. $\frac{8}{e}$ or 2.943

24. $\frac{1}{6}$

25. \$50,000

Chapter 9

MULTIVARIABLE CALCULUS

9.1 Functions of Several Variables

1. $f(x, y) = 2x - 3y + 5$

(a) $f(2, -1) = 2(2) - 3(-1) + 5 = 12$

(b) $f(-4, 1) = 2(-4) - 3(1) + 5 = -6$

(c) $f(-2, -3) = 2(-2) - 3(-3) + 5 = 10$

(d) $f(0, 8) = 2(0) - 3(8) + 5 = -19$

3. $h(x, y) = \sqrt{x^2 + 2y^2}$

(a) $h(5, 3) = \sqrt{25 + 2(9)} = \sqrt{43}$

(b) $h(2, 4) = \sqrt{4 + 32} = 6$

(c) $h(-1, -3) = \sqrt{1 + 18} = \sqrt{19}$

(d) $h(-3, -1) = \sqrt{9 + 2} = \sqrt{11}$

5. $x + y + z = 9$

If $x = 0$ and $y = 0$, $z = 9$.
If $x = 0$ and $z = 0$, $y = 9$.
If $y = 0$ and $z = 0$, $x = 9$.

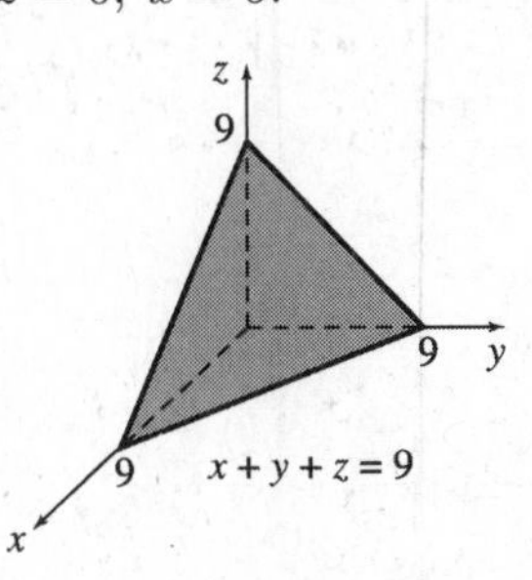

7. $2x + 3y + 4z = 12$

If $x = 0$ and $y = 0$, $z = 3$.
If $x = 0$ and $z = 0$, $y = 4$.
If $y = 0$ and $z = 0$, $x = 6$.

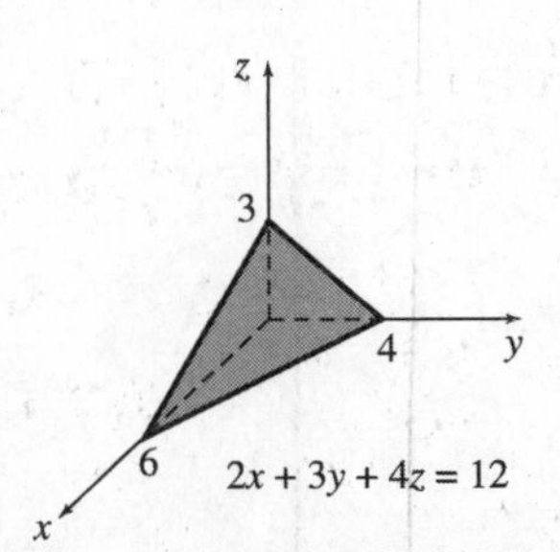

9. $x + y = 4$

If $x = 0$, $y = 4$.
If $y = 0$, $x = 4$.

There is no z-intercept.

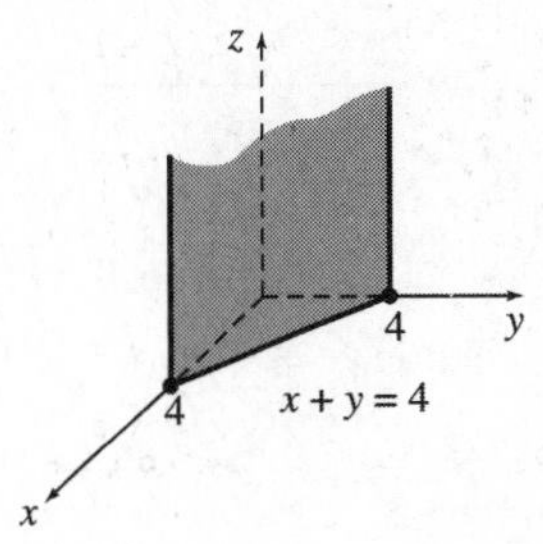

11. $x = 5$

The point $(5, 0, 0)$ is on the graph.
There are no $y-$ or z-intercepts.
The plane is parallel to the yz-plane.

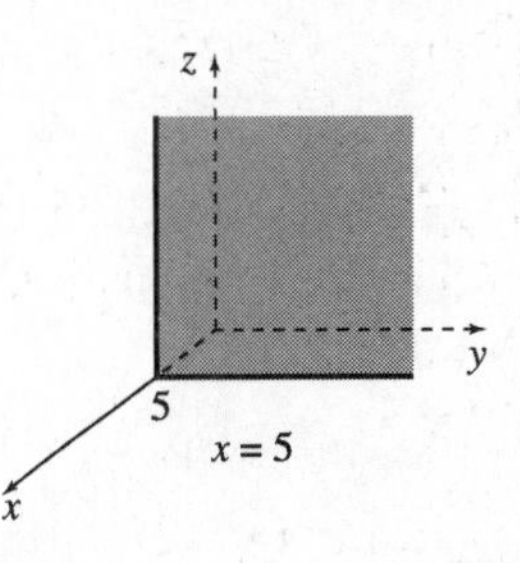

13. $3x + 2y + z = 24$

For $z = 0$, $3x + 2y = 24$. Graph the line $3x + 2y = 24$ in the xy-plane.
For $z = 2$, $3x + 2y = 22$. Graph the line $3x + 2y = 22$ in the plane $z = 2$.
For $z = 4$, $3x + 2y = 20$. Graph the line $3x + 2y = 20$ in the plane $z = 4$.

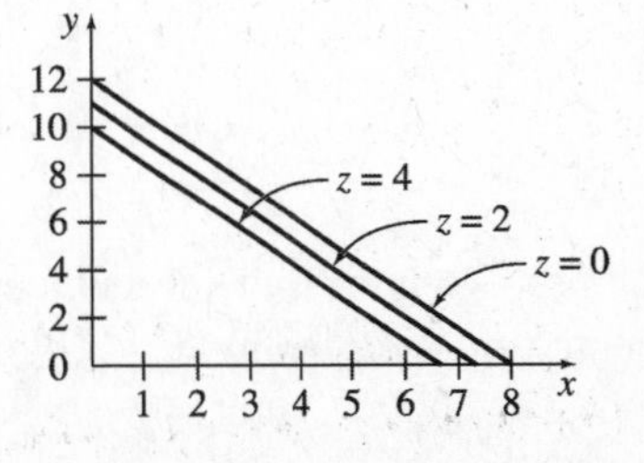

15. $y^2 - x = -z$

For $z = 0$, $x = y^2$. Graph $x = y^2$ in the xy-plane.
For $z = 2$, $x = y^2 + 2$. Graph $x = y^2 + 2$ in the plane $z = 2$.
For $z = 4$, $x = y^2 + 4$. Graph $x = y^2 + 4$ in the plane $z = 4$.

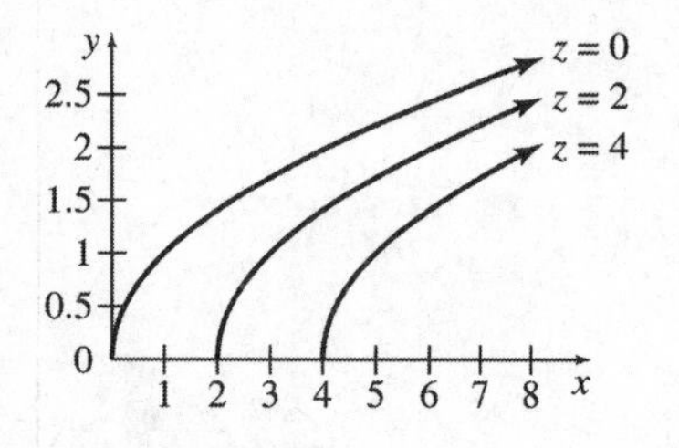

21. $z = x^2 + y^2$

The xz-trace is

$$z = x^2 + 0 = x^2.$$

The yz-trace is

$$z = 0 + y^2 = y^2.$$

Both are parabolas with vertices at the origin that open upward.
The xy-trace is

$$0 = x^2 + y^2.$$

This is a point, the origin.
The equation is represented by a paraboloid, as shown in (c).

23. $x^2 - y^2 = z$

The xz-trace is

$$x^2 = z,$$

which is a parabola with vertex at the origin that opens upward.
The yz-trace is

$$-y^2 = z,$$

which is a parabola with vertex at the origin that opens downward.
The xy-trace is

$$\begin{aligned} x^2 - y^2 &= 0 \\ x^2 &= y^2 \\ x = y \quad &\text{or} \quad x = -y, \end{aligned}$$

which are two lines that intersect at the origin.
The equation is represented by a hyperbolic paraboloid, as shown in (e).

25. $\dfrac{x^2}{16} + \dfrac{y^2}{25} + \dfrac{z^2}{4} = 1$

xz-trace:

$\dfrac{x^2}{16} + \dfrac{z^2}{4} = 1$, an ellipse

yz-trace:

$\dfrac{y^2}{25} + \dfrac{z^2}{4} = 1$, an ellipse

xy-trace:

$\dfrac{x^2}{16} + \dfrac{y^2}{25} = 1$, an ellipse

The graph is an ellipsoid, as shown in (b).

27. $f(x, y) = 4x^2 - 2y^2$

(a) $$\begin{aligned} &\frac{f(x+h, y) - f(x, y)}{h} \\ &= \frac{[4(x+h)^2 - 2y^2] - [4x^2 - 2y^2]}{h} \\ &= \frac{4x^2 + 8xh + 4h^2 - 2y^2 - 4x^2 + 2y^2}{h} \\ &= \frac{h(8x + 4h)}{h} = 8x + 4h \end{aligned}$$

(b) $$\begin{aligned} &\frac{f(x, y+h) - f(x, y)}{h} \\ &= \frac{[4x^2 - 2(y+h)^2] - [4x^2 - 2y^2]}{h} \\ &= \frac{4x^2 - 2y^2 - 4yh - 2h^2 - 4x^2 + 2y^2}{h} \\ &= \frac{h(-4y - 2h)}{h} = -4y - 2h \end{aligned}$$

(c) $$\lim_{h \to 0} \frac{f(x+h, y) - f(x, y)}{h} = \lim_{h \to 0} (8x + 4h) = 8x + 4(0) = 8x$$

(d) $$\lim_{h \to 0} \frac{f(x, y+h) - f(x, y)}{h} = \lim_{h \to 0} (-4y - 2h) = -4y - 2(0) = -4y$$

29. $f(x, y) = xye^{x^2+y^2}$

(a) $\lim\limits_{h \to 0} \dfrac{f(1+h, 1) - f(1, 1)}{h}$

$$= \lim_{h \to 0} \frac{(1+h)(1)e^{1+2h+h^2+1} - (1)(1)e^{1+1}}{h}$$

$$= \lim_{h \to 0} \frac{(1+h)e^{2+2h+h^2} - e^2}{h}$$

$$= e^2 \lim_{h \to 0} \frac{(1+h)e^{2h+h^2} - 1}{h}$$

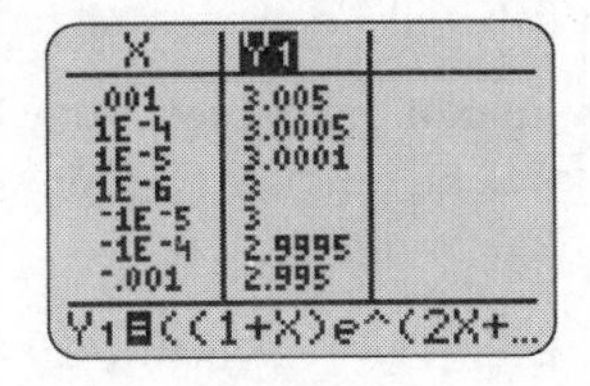

The graphing calculator indicates that $\lim\limits_{h \to 0} \frac{(1+h)e^{2h+h^2} - 1}{h} = 3$, thus $\lim\limits_{h \to 0} \frac{f(1+h,1) - f(1,1)}{h} = 3e^2$.
The slope of the tangent line in the direction of x at $(1, 1)$ is $3e^2$.

(b) $\lim\limits_{h \to 0} \dfrac{f(1, 1+h) - f(1, 1)}{h}$

$$= \lim_{h \to 0} \frac{(1)(1+h)e^{1+1+2h+h^2} - (1)(1)e^{1+1}}{h}$$

$$= \lim_{h \to 0} \frac{(1+h)e^{2+2h+h^2} - e^2}{h}$$

$$= e^2 \lim_{h \to 0} \frac{(1+h)e^{2h+h^2} - 1}{h}$$

So, this limit reduces to the exact same limit as in part a. Therefore, since

$$\lim_{h \to 0} \frac{(1+h)e^{2h+h^2} - 1}{h} = 3,$$

then

$$\lim_{h \to 0} \frac{f(1, 1+h) - f(1, 1)}{h} = 3e^2.$$

The slope of the tangent line in the direction of y at $(1, 1)$ is $3e^2$.

31. $P(x, y) = 100 \left[\dfrac{3}{5}x^{-2/5} + \dfrac{2}{5}y^{-2/5}\right]^{-5}$

(a) $P(32, 1)$

$$= 100 \left[\frac{3}{5}(32)^{-2/5} + \frac{2}{5}(1)^{-2/5}\right]^{-5}$$

$$= 100 \left[\frac{3}{5}\left(\frac{1}{4}\right) + \frac{2}{5}(1)\right]^{-5}$$

$$= 100 \left(\frac{11}{20}\right)^{-5}$$

$$= 100 \left(\frac{20}{11}\right)^{5}$$

$$\approx 1986.95$$

The production is approximately 1987 cameras.

(b) $P(1, 32)$

$$= 100 \left[\frac{3}{5}(1)^{-2/5} + \frac{2}{5}(32)^{-2/5}\right]^{-5}$$

$$= 100 \left[\frac{3}{5}(1) + \frac{2}{5}\left(\frac{1}{4}\right)\right]^{-5}$$

$$= 100 \left(\frac{7}{10}\right)^{-5}$$

$$= 100 \left(\frac{10}{7}\right)^{5}$$

$$\approx 595$$

The production is approximately 595 cameras.

(c) 32 work hours means that $x = 32$. 243 units of capital means that $y = 243$.

$P(32, 243)$

$$= 100 \left[\frac{3}{5}(32)^{-2/5} + \frac{2}{5}(243)^{-2/5}\right]^{-5}$$

$$= 100 \left[\frac{3}{5}\left(\frac{1}{4}\right) + \frac{2}{5}\left(\frac{1}{9}\right)\right]^{-5}$$

$$= 100 \left(\frac{7}{36}\right)^{-5}$$

$$= 100 \left(\frac{36}{7}\right)^{5}$$

$$\approx 359{,}767.81$$

The production is approximately 359,768 cameras.

33. $M = f(40, 0.06, 0.28)$

$$= \frac{(1+0.06)^{40}(1-0.28)+0.28}{[1+(1-0.28)(0.06)]^{40}}$$

$$= \frac{(1.06)^{40}(0.72)+0.28}{[1+(0.72)(0.06)]^{40}}$$

$$\approx 1.416$$

The multiplier is 1.42. Since $M > 1$, the IRA account grows faster.

35. $z = x^{0.6}y^{0.4}$ where $z = 500$

$$500 = x^{3/5}y^{2/5}$$

$$\frac{500}{x^{3/5}} = y^{2/5}$$

$$\left(\frac{500}{x^{3/5}}\right)^{5/2} = (y^{2/5})^{5/2}$$

$$y = \frac{(500)^{5/2}}{x^{3/5}}$$

$$y \approx \frac{5{,}590{,}170}{x^{3/2}}$$

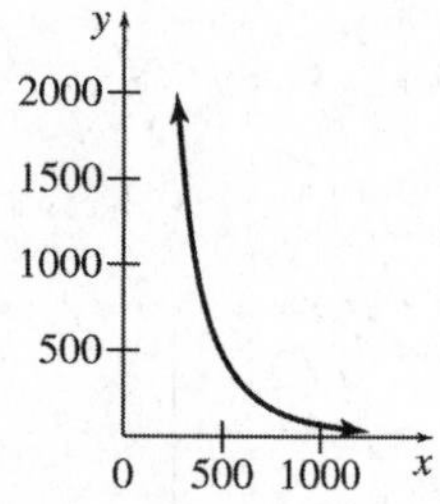

37. The cost function, C, is the sum of the products of the unit costs times the quantities x, y, and z. Therefore,

$$C(x, y, z) = 250x + 150y + 75z.$$

39. $F = \dfrac{v^2}{g\ell}$

(a) $2.56 = \dfrac{v^2}{(9.81)(0.09)}$

$$v^2 = (2.56)(9.81)(0.09)$$
$$v^2 \approx 2.260$$
$$v \approx 1.5$$

For a ferret, this change occurs at 1.5 m/sec.

$$2.56 = \frac{v^2}{(9.81)(1.2)}$$

$$v^2 = (2.56)(9.81)(1.2)$$
$$v^2 \approx 30.136$$
$$v \approx 5.5$$

For a rhinoceros, this change occurs at 5.5 m/sec.

(b) $0.025 = \dfrac{v^2}{(9.81)(4)}$

$$v^2 = (0.025)(9.81)(4)$$
$$v^2 = 0.981$$
$$v \approx 0.99$$

The sauropods were traveling at 1 m/sec.

41. $P(W, R, A) = 48 - 2.43W - 1.81R - 1.22A$

(a) $P(5, 15, 0) = 48 - 2.43(5) - 1.81(15) - 1.22(0) = 8.7$

8.7% of fish will be intolerant to pollution.

(b) The maximum percentage will occur when the variable factors are a minimum, or when $W = 0, R = 0$, and $A = 0$.

$$P(0, 0, 0) = 48 - 2.43(0) - 1.81(0) - 1.22(0) = 48$$

48% of fish will be intolerant to pollution.

(c) Any combination of values of W, R, and A that result in $P = 0$ is a scenario that will drive the percentage of fish intolerant to pollution to zero.

If $R = 0$ and $A = 0$:

$$P(W, 0, 0) = 48 - 2.43W - 1.81(0) - 1.22(0) = 48 - 2.43W.$$

$$48 - 2.43W = 0$$

$$W = \frac{48}{2.43} \approx 19.75$$

So $W = 19.75, R = 0, A = 0$ is one scenario.

If $W = 10$ and $R = 10$:

$$P(10, 10, A) = 48 - 2.43(10) - 1.81(10) - 1.22A = 5.6 - 1.22A$$

$$5.6 - 1.22A = 0$$

$$A = \frac{5.6}{1.22} \approx 4.59$$

So $W = 10, R = 10, A = 4.59$ is another scenario.

(d) Since the coefficient of W is greater than the coefficients of R and A, a change in W will affect the value of P more than an equal change in R or A. Thus, the percentage of wetland (W) has the greatest influence on P.

43. $A(L,T,U,C) = 53.02 + 0.383L + 0.0015T + 0.0028U - 0.0003C$

(a) $$A(266,\ 107{,}484,\ 31{,}697,\ 24{,}870) = 53.02 + 0.383(266) + 0.0015(107{,}484) + 0.0028(31{,}697) - 0.0003(24{,}870) \approx 397$$

The estimated number of accidents is 397.

45. $\ln(T) = 5.49 - 3.00\ln(F) + 0.18\ln(C)$

(a) $$e^{\ln(T)} = e^{5.49 - 3.00\ln(F) + 0.18\ln(C)}$$

$$T = e^{5.49}e^{-3.00\ln(F)}e^{0.18\ln(C)} = \frac{e^{5.49}e^{\ln(C^{0.18})}}{e^{\ln(F^3)}}$$

$$T \approx \frac{242.257C^{0.18}}{F^3}$$

(b) Replace F with 2 and C with 40 in the preceding formula.

$$T \approx \frac{242.257(40)^{0.18}}{(2)^3} \approx 58.82$$

T is about 58.8%. In other words, a tethered sow spends nearly 59% of the time doing repetitive behavior when she is fed 2 kg of food per day and a neighboring sow spends 40% of the time doing repetitive behavior.

47. Let the area be given by $g(L, W, H)$.
Then,

$$g(L, W, H) = 2LW + 2WH + 2LH \text{ ft}^2.$$

9.2 Partial Derivatives

1. $z = f(x, y) = 6x^2 - 4xy + 9y^2$

(a) $\dfrac{\partial z}{\partial x} = 12x - 4y$

(b) $\dfrac{\partial z}{\partial y} = -4x + 18y$

(c) $\dfrac{\partial f}{\partial x}(2, 3) = 12(2) - 4(3) = 12$

(d) $f_y(1, -2) = -4(1) + 18(-2) = -40$

3. $f(x, y) = -4xy + 6y^3 + 5$

$f_x(x, y) = -4y$
$f_y(x, y) = -4x + 18y^2$
$f_x(2, -1) = -4(-1) = 4$
$f_y(-4, 3) = -4(-4) + 18(3)^{2=16+18(9)} = 178$

5. $f(x, y) = 5x^2y^3$
$f_x(x, y) = 10xy^3$
$f_y(x, y) = 15x^2y^2$
$f_x(2, -1) = 10(2)(-1)^3 = -20$
$f_y(-4, 3) = 15(-4)^2(3)^2 = 2160$

7. $f(x,y) = e^{x+y}$

$f_x(x,y) = e^{x+y}$

$f_y(x,y) = e^{x+y}$

$$f_x(2,-1) = e^{2-1} = e^1 = e$$

$$f_y(-4,3) = e^{-4+3} = e^{-1} = \frac{1}{e}$$

9. $f(x,y) = -6e^{4x-3y}$

$f_x(x,y) = -24e^{4x-3y}$

$f_y(x,y) = 18e^{4x-3y}$

$f_x(2,-1) = -24e^{4(2)-3(-1)} = -24e^{11}$

$f_y(-4,3) = 18e^{4(-4)-3(3)} = 18e^{-25}$

11. $f(x,y) = \dfrac{x^2+y^3}{x^3-y^2}$

$$f_x(x,y) = \frac{2x(x^3-y^2) - 3x^2(x^2+y^3)}{(x^3-y^2)^2} = \frac{2x^4 - 2xy^2 - 3x^4 - 3x^2y^3}{(x^3-y^2)^2} = \frac{-x^4 - 2xy^2 - 3x^2y^3}{(x^3-y^2)^2}$$

$$f_y(x,y) = \frac{3y^2(x^3-y^2) - (-2y)(x^2+y^3)}{(x^3-y^2)^2} = \frac{3x^3y^2 - 3y^4 + 2x^2y + 2y^4}{(x^3-y^2)^2} = \frac{3x^3y^2 - y^4 + 2x^2y}{(x^3-y^2)^2}$$

$$f_x(2,-1) = \frac{-2^4 - 2(2)(-1)^2 - 3(2^2)(-1)^3}{[2^3-(-1)^2]^2} = -\frac{8}{49}$$

$$f_y(-4,3) = \frac{3(-4)^3(3)^2 - 3^4 + 2(-4)^2(3)}{[(-4)^3-3^2]^2} = -\frac{1713}{5329}$$

13. $f(x,y) = \ln|1+5x^3y^2|$

$$f_x(x,y) = \frac{1}{1+5x^3y^2} \cdot 15x^2y^2 = \frac{15x^2y^2}{1+5x^3y^2}$$

$$f_y(x,y) = \frac{1}{1+5x^3y^2} \cdot 10x^3y = \frac{10x^3y}{1+5x^3y^2}$$

$$f_x(2,-1) = \frac{15(2)^2(-1)^2}{1+5(2)^3(-1)^2} = \frac{60}{41}$$

$$f_y(-4,3) = \frac{10(-4)^3(3)}{1+5(-4)^3(3)^2} = \frac{1920}{2879}$$

15. $f(x,y) = xe^{x^2y}$

$$f_x(x,y) = e^{x^2y} \cdot 1 + x(2xy)(e^{x^2y}) = e^{x^2y}(1+2x^2y)$$

$f_y(x,y) = x^3e^{x^2y}$

$f_x(2,-1) = e^{-4}(1-8) = -7e^{-4}$

$f_y(-4,3) = -64e^{48}$

17. $f(x,y) = \sqrt{x^4+3xy+y^4+10}$

$$f_x(x,y) = \frac{4x^3+3y}{2\sqrt{x^4+3xy+y^4+10}}$$

$$f_y(x,y) = \frac{3x+4y^3}{2\sqrt{x^4+3xy+y^4+10}}$$

$$f_x(2,-1) = \frac{4(2)^3+3(-1)}{2\sqrt{2^4+3(2)(-1)+(-1)^4+10}} = \frac{29}{2\sqrt{21}}$$

$$f_y(-4,3) = \frac{3(-4)+4(3)^3}{2\sqrt{(-4)^4+3(-4)(3)+3^4+10}} = \frac{48}{\sqrt{311}}$$

19. $f(x,y) = \dfrac{3x^2y}{e^{xy}+2}$

$$f_x(x,y) = \frac{6xy(e^{xy}+2) - ye^{xy}(3x^2y)}{(e^{xy}+2)^2} = \frac{6xy(e^{xy}+2) - 3x^2y^2e^{xy}}{(e^{xy}+2)^2}$$

$$f_y(x,y) = \frac{3x^2(e^{xy}+2) - xe^{xy}(3x^2y)}{(e^{xy}+2)^2} = \frac{3x^2(e^{xy}+2) - 3x^3ye^{xy}}{(e^{xy}+2)^2}$$

$$f_x(2,-1) = \frac{6(2)(-1)(e^{2(-1)}+2) - 3(2)^2(-1)^2e^{2(-1)}}{(e^{2(-1)}+2)^2} = \frac{-12e^{-2} - 24 - 12e^{-2}}{(e^{-2}+2)^2} = \frac{-24(e^{-2}+1)}{(e^{-2}+2)^2}$$

$$f_y(-4,3) = \frac{3(-4)^2(e^{(-4)(3)}+2) - 3(-4)^3(3)e^{(-4)(3)}}{(e^{(-4)(3)}+2)^2} = \frac{48e^{-12} + 96 + 576e^{-12}}{(e^{-12}+2)^2} = \frac{624e^{-12}+96}{(e^{-12}+2)^2}$$

21. $f(x,y) = 4x^2y^2 - 16x^2 + 4y$

$$f_x(x,y) = 8xy^2 - 32x$$
$$f_y(x,y) = 8x^2y + 4$$
$$f_{xx}(x,y) = 8y^2 - 32$$
$$f_{yy}(x,y) = 8x^2$$
$$f_{xy}(x,y) = f_{yx}(x,y) = 16xy$$

23. $R(x,y) = 4x^2 - 5xy^3 + 12y^2x^2$

$$R_x(x,y) = 8x - 5y^3 + 24y^2x$$
$$R_y(x,y) = -15xy^2 + 24yx^2$$
$$R_{xx}(x,y) = 8 + 24y^2$$
$$R_{yy}(x,y) = -30xy + 24x^2$$
$$R_{xy}(x,y) = -15y^2 + 48xy = R_{yx}(x,y)$$

25. $r(x,y) = \dfrac{6y}{x+y}$

$$r_x(x,y) = \frac{(x+y)(0) - 6y(1)}{(x+y)^2} = -6y(x+y)^{-2}$$
$$r_y(x,y) = \frac{(x+y)(6) - 6y(1)}{(x+y)^2} = 6x(x+y)^{-2}$$
$$r_{xx}(x,y) = -6y(-2)(x+y)^{-3}(1) = \frac{12y}{(x+y)^3}$$
$$r_{yy}(x,y) = 6x(-2)(x+y)^{-3}(1) = -\frac{12x}{(x+y)^3}$$
$$\begin{aligned} r_{xy}(x,y) &= r_{yx}(x,y) \\ &= -6y(-2)(x+y)^{-3}(1) + (x+y)^{-2}(-6) \\ &= \frac{12y - 6(x+y)}{(x+y)^3} = \frac{6y - 6x}{(x+y)^3} \end{aligned}$$

27. $z = 9ye^x$

$$z_x = 9ye^x$$
$$z_y = 9e^x$$
$$z_{xx} = 9ye^x$$
$$z_{yy} = 0$$
$$z_{xy} = z_{yx} = 9e^x$$

29. $r = \ln|x+y|$

$$r_x = \frac{1}{x+y}$$
$$r_y = \frac{1}{x+y}$$
$$r_{xx} = \frac{-1}{(x+y)^2}$$
$$r_{yy} = \frac{-1}{(x+y)^2}$$
$$r_{xy} = r_{yx} = \frac{-1}{(x+y)^2}$$

31. $z = x \ln|xy|$

$$z_x = \ln|xy| + 1$$
$$z_y = \frac{x}{y}$$
$$z_{xx} = \frac{1}{x}$$
$$z_{yy} = -xy^{-2} = \frac{-x}{y^2}$$
$$z_{xy} = z_{yz} = \frac{1}{y}$$

33. $f(x,y) = 6x^2 + 6y^2 + 6xy + 36x - 5$

First, $f_x = 12x + 6y + 36$ and $f_y = 12y + 6x$. We must solve the system

$$12x + 6y + 36 = 0$$
$$12y + 6x = 0.$$

Multiply both sides of the first equation by -2 and add.

$$\begin{array}{rrr} -24x & -\,12y & -\,72 = 0 \\ 6x & +\,12y & = 0 \\ \hline -18x & & -\,72 = 0 \\ & & x = -4 \end{array}$$

Substitute into either equation to get $y = 2$. The solution is $x = -4, y = 2$.

35. $f(x,y) = 9xy - x^3 - y^3 - 6$

First, $f_x = 9y - 3x^2$ and $f_y = 9x - 3y^2$. We must solve the system

$$9y - 3x^2 = 0$$
$$9x - 3y^2 = 0.$$

From the first equation, $y = \frac{1}{3}x^2$. Substitute into the second equation to get

$$9x - 3\left(\frac{1}{3}x^2\right)^2 = 0$$
$$9x - 3\left(\frac{1}{9}x^4\right) = 0$$
$$9x - \frac{1}{3}x^4 = 0.$$

Multiply by 3 to get

$$27x - x^4 = 0.$$

Now factor.

$$x(27 - x^3) = 0$$

Set each factor equal to 0.

$$x = 0 \quad \text{or} \quad 27 - x^3 = 0$$
$$x = 3$$

Substitute into $y = \frac{x^2}{3}$.

$$y = 0 \quad \text{or} \quad y = 3$$

The solutions are $x = 0,\ y = 0$ and $x = 3,\ y = 3$.

37. $f(x, y, z) = x^4 + 2yz^2 + z^4$
$f_x(x, y, z) = 4x^3$
$f_y(x, y, z) = 2z^2$
$f_z(x, y, z) = 4yz + 4z^3$
$f_{yz}(x, y, z) = 4z$

39. $f(x, y, z) = \dfrac{6x - 5y}{4z + 5}$

$$f_x(x, y, z) = \frac{6}{4z + 5}$$

$$f_y(x, y, z) = \frac{-5}{4z + 5}$$

$$f_z(x, y, z) = \frac{-4(6x - 5y)}{(4z + 5)^2}$$

$$f_{yz}(x, y, z) = \frac{20}{(4z + 5)^2}$$

41. $f(x, y, z) = \ln \left|x^2 - 5xz^2 + y^4\right|$

$$f_x(x, y, z) = \frac{2x - 5z^2}{x^2 - 5xz^2 + y^4}$$

$$f_y(x, y, z) = \frac{4y^3}{x^2 - 5xz^2 + y^4}$$

$$f_z(x, y, z) = \frac{-10xz}{x^2 - 5xz^2 + y^4}$$

$$f_{yz}(x, y, z) = \frac{4y^3(10zx)}{(x^2 - 5xz^2 + y^4)^2} = \frac{40xy^3z}{(x^2 - 5xz^2 + y^4)^2}$$

43. $f(x, y) = \left(x + \dfrac{y}{2}\right)^{x+y/2}$

(a) $f_x(1, 2) = \displaystyle\lim_{h \to 0} \frac{f(1 + h, 2) - f(1, 2)}{h}$

We will use a small value for h. Let $h = 0.00001$.

$$f_x(1, 2) \approx \frac{f(1.00001, 2) - f(1, 2)}{0.00001}$$
$$\approx \frac{\left(1.00001 + \frac{2}{2}\right)^{1.00001+2/2} - \left(1 + \frac{2}{2}\right)^{1+2/2}}{0.00001}$$
$$\approx \frac{2.00001^{2.00001} - 2^2}{0.00001}$$
$$\approx 6.773$$

(b) $f_y(1, 2) = \displaystyle\lim_{h \to 0} \frac{f(1, 2 + h) - f(1, 2)}{h}$

Again, let $h = 0.00001$.

$$f_y(1, 2) \approx \frac{f(1, 200001) - f(1, 2)}{0.00001}$$
$$\approx \frac{\left(1 + \frac{2.00001}{2}\right)^{1+2.00001/2} - \left(1 + \frac{2}{2}\right)^{1+2/2}}{0.00001}$$
$$\approx \frac{2.000005^{2.000005} - 2^2}{0.00001}$$
$$\approx 3.386$$

45. $M(x, y) = 45x^2 + 40y^2 - 20xy + 50$

(a) $M_y(x, y) = 80y - 20x$
$M_y(4, 2) = 80(2) - 20(4) = 80$

(b) $M_x(x, y) = 90x - 20y$
$M_x(3, 6) = 90(3) - 20(6) = 150$

(c) $\dfrac{\partial M}{\partial x}(2, 5) = 90(2) - 20(5) = 80$

(d) $\dfrac{\partial M}{\partial y}(6, 7) = 80(7) - 20(6) = 440$

47. $f(p, i) = 99p - 0.5pi - 0.0025p^2$

(a) $f(19{,}400, 8)$
$= 99(19{,}400) - 0.5(19{,}400)(8) - 0.0025(19{,}400)^2$
$= \$902{,}100$

The weekly sales are \$902,100.

(b) $f_p(p, i) = 99 - 0.5i - 0.005p$, which represents the rate of change in weekly sales revenue per unit change in price when the interest rate remains constant.

$f_i(p, i) = -0.5p$, which represents the rate of change in weekly sales revenue per unit change in interest rate when the list price remains constant.

(c) $p = 19{,}400$ remains constant and i changes by 1 unit from 8 to 9.

$$f_i(p, i) = f_i(19{,}400, 8)$$
$$= -0.5(19{,}400)$$
$$= -9700$$

Therefore, sales revenue declines by \$9700.

49. $f(x,y) = \left(\frac{1}{4}x^{-1/4} + \frac{3}{4}y^{-1/4}\right)^{-4}$

(a) $$f(16,81) = \left[\frac{1}{4}(16)^{-1/4} + \frac{3}{4}(81)^{-1/4}\right]^{-4} = \left(\frac{1}{4}\cdot\frac{1}{2} + \frac{3}{4}\cdot\frac{1}{3}\right)^{-4} = \left(\frac{3}{8}\right)^{-4} \approx 50.56790123$$

50.57 hundred units are produced.

(b) $$f_x(x,y) = -4\left(\frac{1}{4}x^{-1/4} + \frac{3}{4}y^{-1/4}\right)^{-5}\left[\frac{1}{4}\left(-\frac{1}{4}\right)x^{-5/4}\right] = \frac{1}{4}x^{-5/4}\left(\frac{1}{4}x^{-1/4} + \frac{3}{4}y^{-1/4}\right)^{-5}$$

$$f_x(16,81) = \frac{1}{4}(16)^{-5/4}\left[\frac{1}{4}(16)^{-1/4} + \frac{3}{4}(81)^{-1/4}\right]^{-5} = \frac{1}{4}\left(\frac{1}{32}\right)\left(\frac{3}{8}\right)^{-5} = \frac{256}{243} \approx 1.053497942$$

$f_x(16,81) = 1.053$ hundred units and is the rate at which production is changing when labor changes by one unit (from 16 to 17) and capital remains constant.

$$f_y(x,y) = -4\left(\frac{1}{4}x^{-1/4} + \frac{3}{4}y^{-1/4}\right)^{-5}\left[\frac{3}{4}\left(-\frac{1}{4}\right)y^{-5/4}\right] = \frac{3}{4}y^{-5/4}\left(\frac{1}{4}x^{-1/4} + \frac{3}{4}y^{-1/4}\right)^{-5}$$

$$f_y(16,81) = \frac{3}{4}(81)^{-5/4}\left[\frac{1}{4}(16)^{-1/4} + \frac{3}{4}(81)^{-1/4}\right]^{-5} = \frac{3}{4}\left(\frac{1}{243}\right)\left(\frac{3}{8}\right)^{-5} = \frac{8192}{19{,}683} \approx 0.4161967180$$

$f_y(16,81) = 0.4162$ hundred units and is the rate at which production is changing when capital changes by one unit (from 81 to 82) and labor remains constant.

(c) If labor is increased by one unit, production would increase at the rate

$$f_x(x,y) = \frac{1}{4}x^{-5/4}\left(\frac{1}{4}x^{-1/4} + \frac{3}{4}y^{-1/4}\right)^{-5}$$

hundred units.

51. $z = x^{0.4}y^{0.6}$

The marginal productivity of labor is

$$\frac{\partial z}{\partial x} = 0.4x^{-0.6}y^{0.6} + x^{0.4}\cdot 0 = 0.4x^{-0.6}y^{0.6}.$$

The marginal productivity of capital is

$$\frac{\partial z}{\partial y} = x^{0.4}(0.6y^{-0.4}) + y^{0.6}\cdot 0 = 0.6x^{0.4}y^{-0.4}.$$

53. $f(w,v) = 25.92w^{0.68} + \frac{3.62w^{0.75}}{v}$

(a) $$f(300,10) = 25.92(300)^{0.68} + \frac{3.62(300)^{0.75}}{10} \approx 1279.46$$

The value is about 1279 kcal/hr.

(b) $$f_w(w,v) = 25.92(0.68)w^{-0.32} + \frac{3.62(0.75)w^{-0.25}}{v} = \frac{17.6256}{w^{0.32}} + \frac{2.715}{w^{0.25}v}$$

$$f_w(300,10) = \frac{17.6256}{(300)^{0.32}} + \frac{2.715}{(300)^{0.25}(10)} \approx 2.906$$

The value is about 2.906 kcal/hr/g. This means the instantaneous rate of change of energy usage for a 300 kg animal traveling at 10 kilometers per hour to walk or run 1 kilometer is about 2.906 kcal/hr/g.

55. $A(W,H) = 0.202W^{0.425}H^{0.725}$

(a) $$\frac{\partial A}{\partial W}(72,1.8) = 0.08585(72)^{-0.575}(1.8)^{0.725} = 0.01124$$

(b) $$\frac{\partial A}{\partial H}(70,1.6) = 0.14645(70)^{0.425}(1.6)^{-0.275} \approx 0.7829$$

57. $f(n, c) = \frac{1}{8}n^2 - \frac{1}{5}c + \frac{1937}{8}$

(a) $f(4, 1200) = \frac{1}{8}(4)^2 - \frac{1}{5}(1200) + \frac{1937}{8}$
$= 2 - 240 + \frac{1937}{8} = 4.125$

The client could expect to lose 4.125 lb.

(b) $\frac{\partial f}{\partial n} = \frac{1}{8}(2n) - \frac{1}{5}(0) + 0 = \frac{1}{4}n,$

which represents the rate of change of weight loss per unit change in number of workouts.

(c) $f_n(3, 1100) = \frac{1}{4}(3) = \frac{3}{4}$ lb

represents an additional weight loss by adding the fourth workout.

59. $R(x, t) = x^2(a - x)t^2e^{-t} = (ax^2 - x^3)t^2e^{-t}$

(a) $\frac{\partial R}{\partial x} = (2ax - 3x^2)t^2e^{-t}$

(b) $\frac{\partial R}{\partial t} = x^2(a - x) \cdot [t^2 \cdot (-e^{-t}) + e^{-t} \cdot 2t]$
$= x^2(a - x)(-t^2 + 2t)e^{-t}$

(c) $\frac{\partial^2 R}{\partial x^2} = (2a - 6x)t^2e^{-t}$

(d) $\frac{\partial^2 R}{\partial x \partial t} = (2ax - 3x^2)(-t^2 + 2t)e^{-t}$

(e) $\frac{\partial R}{\partial x}$ gives the rate of change of the reaction per unit of change in the amount of drug administered.
$\frac{\partial R}{\partial t}$ gives the rate of change of the reaction for a 1-hour change in the time after the drug is administered.

61. $W(V, T)$

$= 91.4 - \frac{(10.45 + 6.69\sqrt{V} - 0.447V)(91.4 - T)}{22}$

(a) $W(20, 10)$

$= 91.4 - \frac{(10.45 + 6.69\sqrt{20} - 0.447(20))(91.4 - 10)}{22}$

≈ -24.9

The wind chill is -24.9°F when the wind speed is 20 mph and the temperature is 10°F.

(b) Solve

$-25 = 91.4 - \frac{(10.45 + 6.69\sqrt{V} - 0.447V)(91.4 - 5)}{22}$

for V.

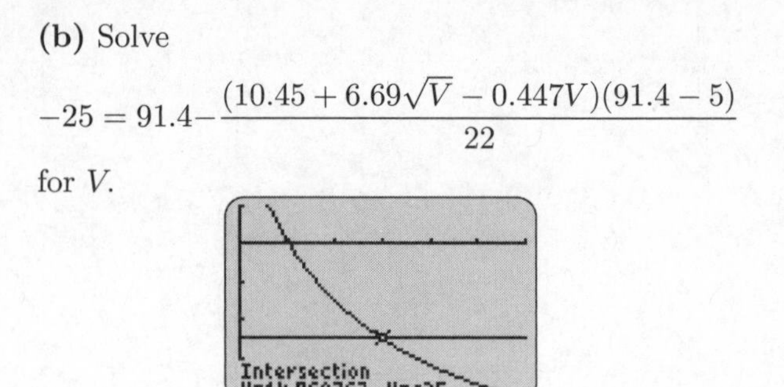

The wind speed is approximately 15 mph.

(c) $W_V = -\frac{1}{22}\left(\frac{6.69}{2\sqrt{V}} - 0.447\right)(91.4 - T)$

$W_T = -\frac{1}{22}(10.45 + 6.69\sqrt{V} - 0.447V)(-1)$
$= \frac{1}{22}(10.45 + 6.69\sqrt{V} - 0.447V)$

$W_V(20, 10) = \frac{1}{22}\left(\frac{6.69}{2\sqrt{20}} - 0.447\right)(91.4 - 10)$
≈ -1.114

When the temperature is held fixed at 10°F, the wind chill decreases approximately 1.1 degrees when the wind velocity increases by 1 mph.

$W_T(20, 10) = \frac{1}{22}[10.45 + 6.69\sqrt{20} - 0.447(20)]$
≈ 1.429

When the wind velocity is held fixed at 20 mph, the wind chill increases approximately 1.429°F when the temperature increases from 10°F to 11°F.

(d) A sample table is

T/V	5	10	15	20
30	27	16	9	4
20	16	3	−5	−11
10	6	−9	−18	−25
0	−5	−21	−32	−39

63. The rate of change in lung capacity with respect to age can be found by comparing the change in two lung capacity measurements to the difference in the respective ages when the height is held constant. So for a woman 58 inches tall, at age 20 the measured lung capacity is 1900 ml, and at age 25 the measured lung capacity is 1850 ml. So the rate of change in lung capacity with respect to age is

$$\frac{1900 - 1850}{20 - 25} = \frac{50}{-5} = -10 \text{ ml per year.}$$

The rate of change in lung capacity with respect to height can be found by comparing the change in two lung capacity measurements to the difference in the respective heights when the age is held constant. So for a 20-year old woman the measured lung capacity for a woman 58 inches tall is 1900 ml and the measured lung capacity for a woman 60 inches tall is 2100 ml. So the rate of change in lung capacity with respect to height is

$$\frac{1900-2100}{58-60} = \frac{-200}{-2} = 100 \text{ ml per in.}$$

The two rates of change remain constant throughout the table.

65. $F = \dfrac{mgR^2}{r^2} = mgR^2r^{-2}$

(a) $F_m = \dfrac{gR^2}{r^2}$ is the approximate rate of change in gravitational force per unit change in mass while distance is held constant.

$F_r = \dfrac{-2mgR^2}{r^3}$ is the approximate rate of change in gravitational force per unit change in distance while mass is held constant.

(b) $F_m = \frac{gR^2}{r^2}$, where all quantities are positive. Therefore, $F_m > 0$.

$F_r = \frac{-2mgR^2}{r^3}$, where m, g, R^2, and r^3 are positive.
Therefore, $F_r < 0$.
These results are reasonable since gravitational force increases when mass increases (m is in the numerator) and gravitational force decreases when distance increases (r is in the denominator).

67. $T = (s,w) = 105 + 265\log_2\left(\dfrac{2s}{w}\right)$

(a) $$T(3,0.5) = 105 + 265\log_2\left[\frac{2(3)}{0.5}\right]$$
$$= 105 + 265\log_2 12$$
$$\approx 1055$$

(b) $$T(s,w) = 105 + 265\frac{\ln\left(\frac{2s}{w}\right)}{\ln 2}$$
$$= 105 + \frac{265}{\ln 2}[\ln(2s) - \ln(w)]$$
$$T_s(s,w) = \frac{265}{\ln 2}\left(\frac{1}{s}\right)$$
$$T_w(s,w) = -\frac{265}{\ln 2}\left(\frac{1}{w}\right)$$
$$T_s(3,0.5) = \frac{265}{3\ln 2} \approx 127.4 \text{ msec/ft}$$

If the distance the object is being moved increases from 3 feet to 4 feet, while keeping w fixed at 0.5 foot, the time to move the object increases by approximately 127.4 msec.

$$T_w(3,0.5) = -\frac{265}{0.5\ln 2} \approx -764.5 \text{ msec/ft}$$

If the width of the target area is increased by 1 foot, while keeping the distance fixed at 3 feet, the movement time decreases by approximately 764.5 msec.

9.3 Maxima and Minima

1. $f(x,y) = xy + y - 2x$

$f_x(x,y) = y-2,\ f_y(x,y) = x+1$

If $f_x(x,y) = 0,\ y = 2$.
If $f_y(x,y) = 0,\ x = -1$.

Therefore, $(-1,2)$ is the critical point.

$$f_{xx}(x,y) = 0$$
$$f_{yy}(x,y) = 0$$
$$f_{xy}(x,y) = 1$$

For $(-1,2)$,

$$D = 0\cdot 0 - 1^2 = -1 < 0.$$

A saddle point is at $(-1,2)$.

3. $f(x,y) = 3x^2 - 4xy + 2y^2 + 6x - 10$
$f_x(x,y) = 6x - 4y + 6$
$f_y(x,y) = -4x + 4y$

Solve the system $f_x(x,y) = 0, f_y(x,y) = 0$.

$$\begin{array}{r} 6x - 4y + 6 = 0 \\ -4x + 4y \quad\ \ = 0 \\ \hline 2x \quad\ \ + 6 = 0 \\ x = -3 \end{array}$$

$$-4(-3) + 4y = 0$$
$$y = -3$$

Therefore, $(-3,-3)$ is a critical point.

$$f_{xx}(x,y) = 6$$
$$f_{yy}(x,y) = 4$$
$$f_{xy}(x,y) = -4$$
$$D = 6\cdot 4 - (-4)^2 = 8 > 0$$

Since $f_{xx}(x,y) = 6 > 0$, there is a relative minimum at $(-3,-3)$.

5. $f(x, y) = x^2 - xy + y^2 + 2x + 2y + 6$

$f_x(x, y) = 2x - y + 2,\ f_y(x, y) = -x + 2y + 2$

Solve the system $f_x(x, y) = 0,\ f_y(x, y) = 0$.

$$\begin{aligned} 2x - \ \ y + 2 &= 0 \\ -x + 2y + 2 &= 0 \end{aligned}$$

$$\begin{aligned} 2x - \ \ y + 2 &= 0 \\ -2x + 4y + 4 &= 0 \\ \hline 3y + 6 &= 0 \\ y &= -2 \end{aligned}$$

$$\begin{aligned} -x + 2(-2) + 2 &= 0 \\ x &= -2 \end{aligned}$$

$(-2, -2)$ is the critical point.

$$\begin{aligned} f_{xx}(x, y) &= 2 \\ f_{yy}(x, y) &= 2 \\ f_{xy}(x, y) &= -1 \end{aligned}$$

For $(-2, -2)$,

$$D = (2)(2) - (-1)^2 = 3 > 0.$$

Since $f_{xx}(x, y) > 0$, a relative minimum is at $(-2, -2)$.

7. $f(x, y) = x^2 + 3xy + 3y^2 - 6x + 3y$

$f_x(x, y) = 2x + 3y - 6,\ f_y(x, y) = 3x + 6y + 3$

Solve the system $f_x(x, y) = 0,\ f_y(x, y) = 0$.

$$\begin{aligned} 2x + 3y - 6 &= 0 \\ 3x + 6y + 3 &= 0 \end{aligned}$$

$$\begin{aligned} -4x - 6y + 12 &= 0 \\ 3x + 6y + \ \ 3 &= 0 \\ \hline -x + 15 &= 0 \\ x &= 15 \end{aligned}$$

$$\begin{aligned} 3(15) + 6y + 3 &= 0 \\ 6y &= -48 \\ y &= -8 \end{aligned}$$

$(15, -8)$ is the critical point.

$$\begin{aligned} f_{xx}(x, y) &= 2 \\ f_{yy}(x, y) &= 6 \\ f_{xy}(x, y) &= 3 \end{aligned}$$

For $(15, -8)$,

$$D = 2 \cdot 6 - 9 = 3 > 0.$$

Since $f_{xx}(x, y) > 0$, a relative minimum is at $(15, -8)$.

9. $f(x, y) = 4xy - 10x^2 - 4y^2 + 8x + 8y + 9$

$f_x(x, y) = 4y - 20x + 8,\ f_y(x, y) = 4x - 8y + 8$

$$\begin{aligned} 4y - 20x + 8 &= 0 \\ 4x - \ \ 8y + 8 &= 0 \end{aligned}$$

$$\begin{aligned} 4y - 20x + \ \ 8 &= 0 \\ -4y + \ \ 2x + \ \ 4 &= 0 \\ \hline -18x + 12 &= 0 \\ x &= \frac{2}{3} \end{aligned}$$

$$4y - 20\left(\frac{2}{3}\right) + 8 = 0$$

The critical point is $\left(\frac{2}{3}, \frac{4}{3}\right)$.

$$\begin{aligned} f_{xx}(x, y) &= -20 \\ f_{yy}(x, y) &= -8 \\ f_{xy}(x, y) &= 4 \end{aligned}$$

For $\left(\frac{2}{3}, \frac{4}{3}\right)$,

$$D = (-20)(-8) - 16 = 144 > 0.$$

Since $f_{xx}(x, y) < 0$, a relative maximum is at $\left(\frac{2}{3}, \frac{4}{3}\right)$.

11. $f(x, y) = x^2 + xy - 2x - 2y + 2$

$f_x(x, y) = 2x + y - 2,\ f_y(x, y) = x - 2$

$$\begin{aligned} 2x + y - 2 &= 0 \\ x \qquad - 2 &= 0 \\ x &= 2 \\ 2(2) + y - 2 &= 0 \\ y &= -2 \end{aligned}$$

The critical point is $(2, -2)$.

$$\begin{aligned} f_{xx}(x, y) &= 2 \\ f_{yy}(x, y) &= 0 \\ f_{xy}(x, y) &= 1 \end{aligned}$$

For $(2, -2)$,

$$D = 2 \cdot 0 - 1^2 = -1 < 0.$$

A saddle point is at $(2, -2)$.

13. $f(x,y) = 3x^2 + 2y^3 - 18xy + 42$
$f_x(x,y) = 6x - 18y$
$f_y(x,y) = 6y^2 - 18x$

If $f_x(x,y) = 0, 6x-18y = 0$, or $x = 3y$. Substitute $3y$ for x in $f_y(x,y) = 0$ and solve for y.

$$6y^2 - 18(3y) = 0$$
$$6y(y-9) = 0$$

$$y = 0 \quad \text{or} \quad y = 9$$
Then $$x = 0 \quad \text{or} \quad x = 27.$$

Therefore, $(0,0)$ and $(27,9)$ are critical points.

$$f_{xx}(x,y) = 6$$
$$f_{yy}(x,y) = 12y$$
$$f_{xy}(x,y) = -18$$

For $(0,0)$,

$$D = 6 \cdot 12(0) - (-18)^2 = -324 < 0.$$

There is a saddle point at $(0,0)$.

For $(27,9)$,

$$D = 6 \cdot 12(9) - (-18)^2 = 324 > 0.$$

Since $f_{xx}(x,y) = 6 > 0$, there is a relative minimum at $(27,9)$.

15. $f(x,y) = x^2 + 4y^3 - 6xy - 1$

$f_x(x,y) = 2x - 6y,\ f_y(x,y) = 12y^2 - 6x$

Solve $f_x(x,y) = 0$ for x.

$$2x + 6y = 0$$
$$x = 3y$$

Substitute for x in $12y^2 - 6x = 0$.

$$12y^2 - 6(3y) = 0$$
$$6y(2y-3) = 0$$
$$y = 0 \quad \text{or} \quad y = \frac{3}{2}$$
Then $$x = 0 \quad \text{or} \quad x = \frac{9}{2}.$$

The critical points are $(0,0)$ and $\left(\frac{9}{2}, \frac{3}{2}\right)$.

$$f_{xx}(x,y) = 2$$
$$f_{yy}(x,y) = 24y$$
$$f_{xy}(x,y) = -6$$

For $(0,0)$,

$$D = 2 \cdot 24(0) - (-6)^2 = -36 < 0.$$

A saddle point is at $(0,0)$.

For $\left(\frac{9}{2}, \frac{3}{2}\right)$,

$$D = 2 \cdot 24\left(\frac{3}{2}\right) - (-6)^2$$
$$= 36 > 0.$$

Since $f_{xx}(x,y) > 0$, a relative minimum is at $\left(\frac{9}{2}, \frac{3}{2}\right)$.

17. $f(x,y) = e^{x(y+1)}$
$f_x(x,y) = (y+1)e^{x(y+1)}$
$f_y(x,y) = xe^{x(y+1)}$

If $$f_x(x,y) = 0$$
$$(y+1)e^{x(y+1)} = 0$$
$$y + 1 = 0$$
$$y = -1.$$

If $$f_y(x,y) = 0$$
$$xe^{x(y+1)} = 0$$
$$x = 0.$$

Therefore, $(0,-1)$ is a critical point.

$$f_{xx}(x,y) = (y+1)^2 e^{x(y+1)}$$
$$f_{yy}(x,y) = x^2 e^{x(y+1)}$$
$$f_{xy}(x,y) = (y+1)e^{x(y+1)} \cdot x + e^{x(y+1)} \cdot 1$$
$$= (xy + x + 1)e^{x(y+1)}$$

For $(0,-1)$,

$$f_{xx}(0,-1) = (0)^2 e^0 = 0$$
$$f_{yy}(0,-1) = (0)^2 e^0 = 0$$
$$f_{xy}(0,-1) = (0+0+1)e^0 = 1$$
$$D = 0 \cdot 0 - 1^2 = -1 < 0$$

There is a saddle point at $(0,-1)$.

21. $z = -3xy + x^3 - y^3 + \frac{1}{8}$

$f_x(x,y) = -3y + 3x^2,\ f_y(x,y) = -3x - 3y^2$

Solve the system $f_x = 0,\ f_y = 0$.

$$-3y + 3x^2 = 0$$
$$-3x - 3y^2 = 0$$

$$-y + x^2 = 0$$
$$-x - y^2 = 0$$

Solve the first equation for y, substitute into the second, and solve for x.

$$\begin{aligned} y &= x^2 \\ -x - x^4 &= 0 \\ x(1+x^3) &= 0 \\ x = 0 \quad &\text{or} \quad x = -1 \\ \text{Then} \qquad y = 0 \quad &\text{or} \quad y = 1. \end{aligned}$$

The critical points are $(0,0)$ and $(-1,1)$.

$$\begin{aligned} f_{xx}(x,y) &= 6x \\ f_{yy}(x,y) &= -6y \\ f_{xy}(x,y) &= -3 \end{aligned}$$

For $(0,0)$,

$$D = 0 \cdot 0 - (-3)^2 = -9 < 0.$$

A saddle point is at $(0,0)$.
For $(-1,1)$,

$$D = -6(-6) - (-3)^2 = 27 > 0.$$

$$f_{xx}(x,y) = 6(-1) = -6 < 0.$$

$$\begin{aligned} f(-1,1) &= -3(-1)(1) + (-1)^3 - 1^3 + \frac{1}{8} \\ &= \frac{9}{8} \end{aligned}$$

A relative maximum of $\frac{9}{8}$ is at $(-1,1)$.
The equation matches graph (a).

23. $z = y^4 - 2y^2 + x^2 - \dfrac{17}{16}$

$f_x(x,y) = 2x,\ f_y(x,y) = 4y^3 - 4y$

Solve the system $f_x = 0,\ f_y = 0$.

$$\begin{aligned} 2x &= 0 \quad (1) \\ 4y^3 - 4y &= 0 \quad (2) \\ 4y(y^2-1) &= 0 \\ 4y(y+1)(y-1) &= 0 \end{aligned}$$

Equation (1) gives $x = 0$ and equation (2) gives $y = 0,\ y = -1$, or $y = 1$.
The critical points are $(0,0)$, $(0,-1)$, and $(0,1)$.

$$\begin{aligned} f_{xx}(x,y) &= 2, \\ f_{yy}(x,y) &= 12y^2 - 4, \\ f_{xy}(x,y) &= 0 \end{aligned}$$

For $(0,0)$,

$$D = 2(12 \cdot 0^2 - 4) - 0 = -8 < 0.$$

A saddle point is at $(0,0)$.
For $(0,-1)$,

$$D = 2[12(-1)^2 - 4] - 0 = 16 > 0.$$

$$f_{xx}(x,y) = 2 > 0$$

$$\begin{aligned} f(0,-1) &= (-1)^4 - 2(-1)^2 + 0^2 - \frac{17}{16} \\ &= -2\tfrac{1}{16} \end{aligned}$$

A relative minimum of $-2\frac{1}{16}$ is at $(0,-1)$.
For $(0,1)$,

$$D = 2(12 \cdot 1^2 - 4) - 0 = 16 > 0$$

$$f_{xx}(x,y) = 2 > 0$$

$$\begin{aligned} f(0,1) &= 1^4 - 2 \cdot 1^2 + 0^2 - \frac{17}{16} \\ &= -\frac{33}{16} \end{aligned}$$

A relative minimum of $-\frac{33}{16}$ is at $(0,-1)$.
The equation matches graph (b).

25. $z = -x^4 + y^4 + 2x^2 - 2y^2 + \dfrac{1}{16}$

$f_x(x,y) = -4x^3 + 4x,\ f_y(x,y) = 4y^3 - 4y$

Solve $f_x(x,y) = 0,\ f_y(x,y) = 0$.

$$\begin{aligned} -4x^3 + 4x &= 0 \quad (1) \\ 4y^3 - 4y &= 0 \quad (2) \end{aligned}$$

$$\begin{aligned} -4x(x^2-1) &= 0 \quad (1) \\ -4x(x+1)(x-1) &= 0 \end{aligned}$$

$$\begin{aligned} 4y(y^2-1) &= 0 \quad (2) \\ 4y(y+1)(y-1) &= 0 \end{aligned}$$

Equation (1) gives $x = 0,\ -1$, or 1.
Equation (2) gives $y = 0,\ -1$, or 1.
Critical points are $(0,0)$, $(0,-1)$, $(0,1)$, $(-1,0)$, $(-1,-1)$, $(-1,1)$, $(1,0)$, $(1,-1)$, $(1,1)$.

$$\begin{aligned} f_{xx}(x,y) &= -12x^2 + 4, \\ f_{yy}(x,y) &= 12y^2 - 4 \\ f_{xy}(x,y) &= 0 \end{aligned}$$

For $(0,0)$,

$$D = 4(-4) - 0 = -16 < 0.$$

For $(0, -1)$,

$$D = 4(8) - 0 = 32 > 0,$$

and $f_{xx}(x, y) = 4 > 0$.

$$f(0, -1) = -\frac{15}{16}$$

For $(0, 1)$,

$$D = 4(8) - 0 = 32 > 0,$$

and $f_{xx}(x, y) = 4 > 0$.

$$f(0, 1) = -\frac{15}{16}$$

For $(-1, 0)$,

$$D = -8(-4) - 0 = 32 > 0,$$

and $f_{xx}(x, y) = -8 < 0$.

$$f(-1, 0) = \frac{17}{16}$$

For $(-1, -1)$,

$$D = -8(8) - 0 = -64 < 0.$$

For $(-1, 1)$,

$$D = -8(8) - 0 = -64 < 0.$$

For $(1, 0)$,

$$D = -8(-4) = 32 > 0,$$

and $f_{xx}(x, y) = -8 < 0$.

$$f(1, 0) = 1\tfrac{1}{16}$$

For $(1, -1)$,

$$D = -8(8) - 0 = -64 < 0.$$

For $(1, 1)$,

$$D = -8(8) - 0 = -64 < 0.$$

Saddle points are at $(0, 0)$, $(-1, -1)$, $(-1, 1)$, $(1, -1)$, and $(1, 1)$.
Relative maximum of $\frac{17}{16}$ is at $(-1, 0)$ and $(1, 0)$.
Relative minimum of $-\frac{15}{16}$ is at $(0, -1)$ and $(0, 1)$.
The equation matches graph (e).

27. $f(x, y) = 1 - x^4 - y^4$

$f_x(x, y) = -4x^3$, $f_y(x, y) = -4y^3$

The system

$f_x(x, y) = -4x^3 = 0, f_y(x, y) = -4y^3 = 0$

gives the critical point $(0, 0)$.

$$f_{xx}(x, y) = -12x^2$$
$$f_{yy}(x, y) = -12y^3$$
$$f_{xy}(x, y) = 0$$

For $(0, 0)$,

$$D = 0 \cdot 0 - 0^2 = 0.$$

Therefore, the test gives no information. Examine a graph of the function drawn by using level curves.
If $f(x, y) = 1$, then $x^4 + y^4 = 0$. The level curve is the point $(0, 0, 1)$.
If $f(x, y) = 0$, then $x^4 + y^4 = 1$. The level curve is the circle with center $(0, 0, 0)$ and radius 1.
If $f(x, y) = -15$, then $x^4 + y^4 = 16$. The level curve is the curve with center $(0, 0, -15)$ and radius 2.
The xz-trace is

$$z = 1 - x^4.$$

This curve has a maximum at $(0, 0, 1)$ and opens downward.
The yz-trace is

$$z = 1 - y^4.$$

This curve also has a maximum at $(0, 0, 1)$ and opens downward.

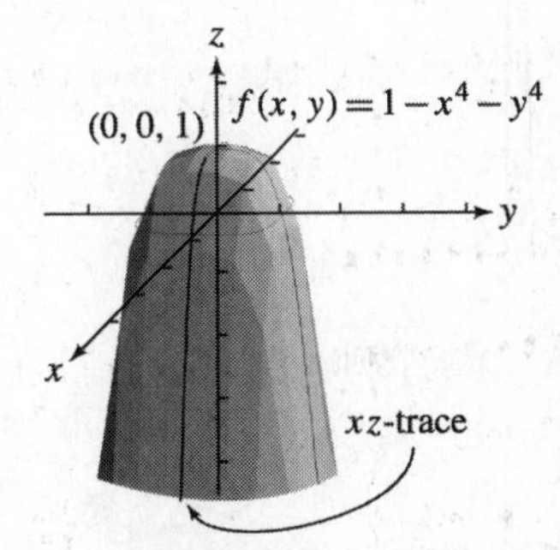

If $f(x, y) > 1$, then $x^4 + y^4 < 0$, which is impossible, so the function does not exist. Thus, the function has a relative maximum of 1 at $(0, 0)$.

31. $f(x, y) = x^2(y+1)^2 + k(x+1)^2y^2$

(a) $f_x(x, y) = 2x + 2ky^2(x+1)$
$f_y(x, y) = 2x^2(y+1) + 2k(x+1)^2y$
$f_x(0, 0) = 2(0) + 2k(0)^2(0+1) = 0$
$f_y(0, 0) = 2(0)^2(0+1) + 2k(0+1)^2(0) = 0$

Thus, $(0, 0)$ is a critical point for all values of k.

(b) $f_{xx}(x, y) = 2 + 2ky^2$
$f_{yy}(x, y) = 2x^2 + 2k(x+1)^2$
$f_{xy}(x, y) = 4ky(x+1)$
$f_{xx}(0, 0) = 2 + 2k(0)^2 = 2$
$f_{yy}(0, 0) = 2(0)^2 = 2k(0+1)^2 = 2k$
$f_{xy}(0, 0) = 4k(0)(0+1) = 0$
$D = 2 \cdot 2k - 0^2 = 4k$

$(0, 0)$ is a relative minimum when $4k > 0$, hence when $k > 0$. When $k = 0$, $D = 0$ so the test for relative extrema gives no information. But if $k = 0$, $f(xy) = x^2(y+1)^2$, which is always greater than or equal to $f(0, 0) = 0$. So $(0, 0)$ is a relative minimum for $k \geq 0$.

33. $P(x, y) = 1500 + 36x - 1.5x^2 + 120y - 2y^2$
$P_x(x, y) = 36 - 3x$
$P_y(x, y) = 120 - 4y$

If $P_x = 0, x = 12$. If $P_y = 0, y = 30$. Therefore, $(12, 30)$ is a critical point.

$P_{xx}(x, y) = -3$
$P_{yy}(x, y) = -4$
$P_{xy}(x, y) = 0$
$D = (-3) \cdot (-4) - 0^2 = 12 > 0$

Since $P_{xx} = -3 < 0$, there is a relative maximum at $(12, 30)$.

$$P(12, 30) = 1500 + 36(12) - 1.5(12)^2 + 120(30) - 2(30)^2$$
$$= 3516 \text{ (hundred dollars)}$$

The maximum profit is \$351,600 when the cost of a unit of labor is 12 and the cost of a unit of goods is 30.

35. $C(x, y) = 2x^2 + 2y^2 - 3xy + 4x - 94y + 4200$

$C_x(x, y) = 4x - 3y + 4$
$C_y(x, y) = 4y - 3x - 94$

Solve the system $C_x(x, y) = 0, C_y(x, y) = 0$.

$$\begin{aligned} 4x - 3y + 4 &= 0 \\ -3x + 4y - 94 &= 0 \end{aligned}$$

$$\begin{aligned} 12x - 9y + 12 &= 0 \\ -12x + 16y - 376 &= 0 \\ \hline 7y - 364 &= 0 \\ y &= 52 \end{aligned}$$

$$\begin{aligned} 4x - 3(52) + 4 &= 0 \\ 4x &= 152 \\ x &= 38 \end{aligned}$$

Therefore, $(38, 52)$ is a critical point.

$C_{xx} = 4$
$C_{yy} = 4$
$C_{xy} = -3$
$D = (4)(4) - (-3)^2 = 7 > 0$

Since $C_{xx} = 4 > 0$, there is a relative minimum at $(38, 52)$.

$$\begin{aligned} &C(38, 52) \\ &= 2(38)^2 + 2(52)^2 - 3(38)(52) \\ &\quad + 4(38) - 94(52) + 4200 \\ &= 1832 \end{aligned}$$

38 units of electrical tape and 52 units of packing tape should be produced to yield a minimum cost of \$1832.

37. $P(x, y) = 36xy - x^3 - 8y^3$
$P_x(x, y) = 36y - 3x^2$
$P_y(x, y) = 36x - 24y^2$

$$\begin{aligned} P_x(x, y) &= 0 \\ 36y - 3x^2 &= 0 \\ 36y &= 3x^2 \\ y &= \frac{1}{12}x^2 \end{aligned}$$

$$\begin{aligned} P_x(x, y) &= 0 \\ 36x - 24y^2 &= 0 \\ 36x &= 24y^2 \\ x &= \frac{2}{3}y^2 \end{aligned}$$

Use substitution to solve the system of equations

$$y = \frac{1}{12}x^2$$
$$x = \frac{2}{3}y^2.$$

$$y = \frac{1}{12}\left(\frac{2}{3}y^2\right)^2$$
$$y = \frac{1}{12}\left(\frac{4}{9}\right)y^4$$
$$y = \frac{1}{27}y^4$$
$$\frac{1}{27}y^4 - y = 0$$
$$\left(\frac{1}{27}y^3 - 1\right)y = 0$$
$$\frac{1}{27}y^3 - 1 = 0 \quad \text{or} \quad y = 0$$
$$\frac{1}{27}y^3 = 1 \quad \text{or} \quad y = 0$$
$$y^3 = 27 \quad \text{or} \quad y = 0$$
$$y = 3 \quad \text{or} \quad y = 0$$

If $y = 3, x = \frac{2}{3}(3)^2 = 6.$

If $y = 0, x = \frac{2}{3}(0)^2 = 0.$

The critical points are $(6, 3)$ and $(0, 0)$.

$$P_{xx}(x,y) = -6x$$
$$P_{yy}(x,y) = -48y$$
$$P_{xy}(x,y) = 36$$
$$P_{xx}(6,3) = -36$$
$$P_{yy}(6,3) = -144$$
$$P_{xy}(6,3) = 36$$

$$D = (-36)(-144) - (36)^2 = 3888$$

Here $D > 0$ and $P_{xx} < 0$, so there is a relative maximum at $(6, 3)$.

$$P_{xx}(0,0) = 0$$
$$P_{yy}(0,0) = 0$$
$$P_{xy}(0,0) = 36$$

$$D = 0 \cdot 0 - 36^2 = -1296$$

Since $D < 0$, there is a saddle point at $(0, 0)$.

$$\begin{aligned} P(6,3) &= 36(6)(3) - (6)^3 - 8(3)^3 \\ &= 648 - 216 - 216 \\ &= 216 \end{aligned}$$

So 6 tons of steel and 3 tons of aluminum produce a maximum profit of $216,000.

39. $P(\alpha, r, s) = \alpha(3r^2(1-r) + r^3) + (1-\alpha)(3s^2(1-s) + s^3)$

(a)
$$\begin{aligned} P(0.9, 0.5, 0.6) &= 0.9[3(0.5)^2(1-0.5) + (0.5)^3] \\ &\quad + (1-0.9)[3(0.6)^2(1-0.6) + (0.6)^3] \\ &= 0.5148 \\ P(0.1, 0.8, 0.4) &= 0.1[3(0.8)^2(1-0.8) + (0.8)^3] \\ &\quad + (1-0.1)[3(0.4)^2(1-0.4) + (0.4)^3] \\ &= 0.4064 \end{aligned}$$

The jury is less likely to make the correct decision in the second situation.

(b) If $r = s = 1$ then $P(\alpha, 1, 1) = 1$, so the jury always makes a correct decision. These values do not depend on α, but in a real-life situation α is likely to influence r and s.

(c) When P reaches a maximum, P_α, P_r, and P_s equal 0.

$$\begin{aligned} P_\alpha(\alpha, r, s) &= 3r^2(1-r) + r^3 \\ &\quad - (3s^2(1-s) + s^3) \\ &= 3r^2 - 2r^3 - (3s^2 - 2s^3) \end{aligned}$$

$$\begin{aligned} P_r(\alpha, r, s) &= \alpha(6r(1-r) - 3r^2 + 3r^2) \\ &= 6\alpha r(1-r) \\ P_s(\alpha, r, s) &= (1-\alpha)(6s(1-s) - 3s^2 + 3s^2) \\ &= 6s(1-\alpha)(1-s) \\ P_\alpha(\alpha, r, s) &= 0 \quad \text{when} \quad r = s. \end{aligned}$$

Since $P_r(\alpha, r, s) = 6\alpha r(1-r)$, and $P_s(\alpha, r, s) = 6(1-\alpha)s(1-s)$, then P_α, P_r, and P_s are simultaneously 0 at the points $(\alpha, 1, 1)$ and $(\alpha, 0, 0)$. So $(\alpha, 1, 1)$ and $(\alpha, 0, 0)$ are critical points.
$P(\alpha, 0, 0) = 0$ while $P(\alpha, 1, 1) = 1$
Since $P(\alpha, r, s)$ represents a probability, $0 \le P(\alpha, r, s) \le 1$. Thus, $P(\alpha, 1, 1) = 1$ is a maximum value of the function.

41. $E(t, T) = 436.16 - 10.57t - 5.46T - 0.02t^2 + 0.02T^2 + 0.08tT$

(a) $E(0, 0) = 436.16.$

The value of E before cooking is 436.16 kJ/mol.

(b)
$$\begin{aligned} E(10, 180) &= 436.16 - 10.57(10) - 5.46(180) \\ &\quad - 0.02(10)^2 + 0.02(180)^2 \\ &\quad + 0.08(10)(180) \\ &= 137.66 \end{aligned}$$

After cooking for 10 minutes at 180°C, the total change in color is 137.66 kJ/mol.

(c) $E_t = -10.57 - 0.04t + 0.08T$
$E_T = -5.46 + 0.04T + 0.08t$

Solve the system $E_t = 0, E_T = 0$.

$$\begin{aligned} -0.04t + 0.08T - 10.57 &= 0 \\ 0.08t + 0.04T - 5.46 &= 0 \end{aligned}$$

$$\begin{aligned} -0.04t + 0.08T - 10.57 &= 0 \\ -0.16t - 0.08T + 10.92 &= 0 \\ \hline -0.20t \qquad + 0.35 &= 0 \\ t &= 1.75 \end{aligned}$$

$$\begin{aligned} -0.04(1.75) + 0.08T - 10.57 &= 0 \\ 0.08T - 10.64 &= 0 \\ T &= 133 \end{aligned}$$

$(1.75, 133)$ is a critical point.

$$\begin{aligned} E_{tt} &= -0.04 \\ E_{TT} &= 0.04 \\ E_{tT} &= 0.08 \\ D &= (-0.04)(0.04) - (0.08)^2 = -0.008 < 0 \end{aligned}$$

$(1.75, 133)$ is a saddle point.

9.4 Lagrange Multipliers

1. Maximize $f(x, y) = 4xy$,
subject to $x + y = 16$.

1. $g(x, y) = x + y - 16$

2. $F(x, y, \lambda) = 4xy - \lambda(x + y - 16)$.

3. $F_x(x, y, \lambda) = 4y - \lambda$
$F_y(x, y, \lambda) = 4x - \lambda$
$F_\lambda(x, y, \lambda) = -(x + y - 16)$

4.
$$\begin{aligned} 4y - \lambda &= 0 \quad (1) \\ 4x - \lambda &= 0 \quad (2) \\ x + y - 16 &= 0 \quad (3) \end{aligned}$$

5. Equations (1) and (2) give $\lambda = 4y$ and $\lambda = 4x$. Thus,

$$\begin{aligned} 4y &= 4x \\ y &= x. \end{aligned}$$

Substituting into equation (3),

$$\begin{aligned} x + (x) - 16 &= 0 \\ x &= 8. \end{aligned}$$

So $y = 8.$

Maximum is $f(8, 8) = 4(8)(8) = 256$.

3. Maximize $f(x, y) = xy^2$,
subject to $x + 2y = 15$.

1. $g(x, y) = x + 2y - 15$

2. $F(x, y, \lambda) = xy^2 - \lambda(x + 2y - 15)$

3. $F_x(x, y, \lambda) = y^2 - \lambda$
$F_y(x, y, \lambda) = 2xy - 2\lambda$
$F_\lambda(x, y, \lambda) = -(x + 2y - 15)$

4.
$$\begin{aligned} y^2 - \lambda &= 0 \quad (1) \\ 2xy - 2\lambda &= 0 \quad (2) \\ x + 2y - 15 &= 0 \quad (3) \end{aligned}$$

5. Equations (1) and (2) give $\lambda = y^2$ and $\lambda = xy$. Thus,

$$\begin{aligned} y^2 &= xy \\ y(y - x) &= 0 \\ y = 0 \quad &\text{or} \quad y = x \end{aligned}$$

Substituting $y = 0$ into equation (3),

$$\begin{aligned} x + 2(0) - 15 &= 0 \\ x &= 15. \end{aligned}$$

Substituting $y = x$ into equation (3)

$$\begin{aligned} x + 2(x) - 15 &= 0 \\ x &= 5. \end{aligned}$$

So $y = x = 5.$

Thus,

$$\begin{aligned} f(15, 0) &= 15(0)^2 = 0, \text{ and} \\ f(5, 5) &= 5(5)^2 = 125. \end{aligned}$$

Since $f(5, 5) > f(15, 0), f(5, 5) = 125$ is a maximum.

5. Minimize $f(x, y) = x^2 + 2y^2 - xy$,
subject to $x + y = 8$.

1. $g(x, y) = x + y - 8$

2. $F(x, y, \lambda)$
$= x^2 + 2y^2 - xy - \lambda(x + y - 8)$

3. $F_x(x, y, \lambda) = 2x - y - \lambda$
$F_y(x, y, \lambda) = 4y - x - \lambda$
$F_\lambda(x, y, \lambda) = -(x + y - 8)$

4. $2x - y - \lambda = 0$
$4y - x - \lambda = 0$
$x + y - 8 = 0$

5. Subtracting the second equation from the first equation to eliminate λ gives the new system of equations

$$x + y = 8$$
$$3x - 5y = 0.$$

Solve this system.

$$\begin{aligned} 5x + 5y &= 40 \\ 3x - 5y &= \ \ 0 \\ \hline 8x &= 40 \\ x &= 5 \end{aligned}$$

But $x + y = 8$, so $y = 3$.

Thus, $\quad f(5, 3) = 25 + 18 - 15 = 28$

is a minimum.

7. Maximize $f(x, y) = x^2 - 10y^2$, subject to $x - y = 18$.

1. $g(x, y) = x - y - 18$

2. $F(x, y, \lambda)$
$= x^2 - 10y^2 - \lambda(x - y - 18)$

3. $F_x(x, y, \lambda) = 2x - \lambda$
$F_y(x, y, \lambda) = -20y - \lambda$
$F_\lambda(x, y, \lambda) = -(x - y - 18)$

4.
$$\begin{aligned} 2x - \lambda &= 0 \\ -20y + \lambda &= 0 \\ x - y - 18 &= 0 \end{aligned}$$

5. Adding the first two equations to eliminate λ gives

$$\begin{aligned} 2x - 20y &= 0 \\ x &= 10y. \end{aligned}$$

Substituting $x = 10y$ in the third equation gives

$$\begin{aligned} 10y - y &= 18 \\ y &= 2 \end{aligned}$$

Thus, $\quad x = 20.$

$$\begin{aligned} f(20, 2) &= 20^2 - 10(2)^2 \\ &= 400 - 40 = 360. \end{aligned}$$

$f(20, 2) = 360$ is a relative maximum.

9. Maximize $f(x, y, z) = xyz^2$, subject to $x + y + z = 6$.

1. $g(x, y, z) = x + y + z - 6$

2. $F(x, y, \lambda)$
$= xyz^2 - \lambda(x + y + z - 6)$

3. $F_x(x, y, z, \lambda) = yz^2 - \lambda$
$F_y(x, y, z, \lambda) = xz^2 - \lambda$
$F_z(x, y, z, \lambda) = 2zxy - \lambda$
$F_\lambda(x, y, z, \lambda) = -(x + y + z - 6)$

4. Setting F_x, F_y, F_z and F_λ equal to zero yields

$$\begin{aligned} yz^2 - \lambda &= 0 \quad (1) \\ xz^2 - \lambda &= 0 \quad (2) \\ 2xyz - \lambda &= 0 \quad (3) \\ x + y + z - 6 &= 0. \quad (4) \end{aligned}$$

5. $\lambda = yz^2$, $\lambda = xz^2$, and $\lambda = 2xyz$

$$\begin{aligned} yz^2 &= xz^2 \\ z^2(y - x) &= 0 \\ x = y \quad &\text{or} \quad z = 0 \\ yz^2 &= 2xyz \end{aligned}$$

$$\begin{aligned} 2xyz - yz^2 &= 0 \\ yz(2x - z) &= 0 \\ y = 0 \quad \text{or} \quad z = 0 \quad &\text{or} \quad z = 2x \end{aligned}$$

In a similar way, the third equation

$$xz^2 = 2xyz$$

implies that $x = 0$ or $z = 0$ or $z = 2y$.

By the nature of the function to be maximized, $f(x, y, z) = xyz^2$, a nonzero maximum can come only from those points with nonzero coordinates. Therefore, assume $y = x$ and $z = 2y = 2x$. If $y = x$ and $z = 2x$ are substituted into equation (4), then

$$\begin{aligned} x + x + 2x - 6 &= 0 \\ x &= \frac{3}{2}. \end{aligned}$$

Thus, $y = \frac{3}{2}$ and $z = 3$, and

$$\begin{aligned} f\left(\frac{3}{2}, \frac{3}{2}, 3\right) &= \frac{3}{2} \cdot \frac{3}{2} \cdot 9 \\ &= \frac{81}{4} > 0. \end{aligned}$$

So, $f\left(\frac{3}{2}, \frac{3}{2}, 3\right) = \frac{81}{4} = 20.25$ is a maximum.

11. The problem can be restated as

Maximize $f(x, y) = 3xy^2$,
subject to $x + y = 24, x > 0, y > 0$.

1. $g(x, y) = x + y - 24$

2. $F(x, y, \lambda) = 3xy^2 - \lambda(x + y - 24)$

3. $F_x(x, y, \lambda) = 3y^2 - \lambda$
$F_y(x, y, \lambda) = 6xy - \lambda$
$F_\lambda(x, y, \lambda) = -(x + y - 24)$

4.
$$\begin{aligned} 3y^2 - \lambda &= 0 \quad (1) \\ 6xy - \lambda &= 0 \quad (2) \\ x + y - 24 &= 0 \quad (3) \end{aligned}$$

5. Equations (1) and (2) give $\lambda = 3y^2$ and $\lambda = 6xy$. Thus,

$$\begin{aligned} 3y^2 &= 6xy \\ 3y^2 - 6xy &= 0 \\ 3y(y - 2x) &= 0 \\ y = 0 \quad &\text{or} \quad y = 2x. \end{aligned}$$

Substituting $y = 0$ into equation (3),

$$\begin{aligned} x + (0) - 24 &= 0 \\ x &= 24. \end{aligned}$$

Substituting $y = 2x$ into equation (3),

$$\begin{aligned} x + (2x) - 24 &= 0 \\ 3x - 24 &= 0 \\ x &= 8. \end{aligned}$$

So $\qquad y = 2x = 16.$

Thus,

$$\begin{aligned} f(24, 0) &= 3(24)(0)^2 = 0, \text{ and} \\ f(8, 16) &= 3(8)(16)^2 = 6144. \end{aligned}$$

Since $f(8, 16) > f(24, 0), x = 8$ and $y = 16$ will maximize $f(x, y) = 3xy^2$.

13. Let x, y, and z be three numbers such that

$$x + y + z = 90$$
and
$$f(x, y, z) = xyz.$$

1. $g(x, y, z) = x + y + z - 90$

2. $F(x, y, z)$
$= xyz - \lambda(x + y + z - 90)$

3. $F_x(x, y, z, \lambda) = yz - \lambda$
$F_y(x, y, z, \lambda) = xz - \lambda$
$F_\lambda(x, y, z, \lambda) = xy - \lambda$
$F_\lambda(x, y, z, \lambda) = -(x + y + z - 90)$

4.
$$\begin{aligned} yz - \lambda &= 0 \quad (1) \\ xz - \lambda &= 0 \quad (2) \\ xy - \lambda &= 0 \quad (3) \\ x + y + z - 90 &= 0 \quad (4) \end{aligned}$$

5. $\lambda = yz$, $\lambda = xz$, and $\lambda = xy$

$$\begin{aligned} yz &= xz \\ yz - xz &= 0 \\ (y - x)z &= 0 \\ y - x = 0 \quad &\text{or} \quad z = 0 \\ xz - xy &= 0 \\ x(z - y) &= 0 \\ x = 0 \quad &\text{or} \quad z - y = 0 \end{aligned}$$

Since $x = 0$ or $z = 0$ would not maximize $f(x, y, z) = xyz$, then $y - x = 0$ and $z - y = 0$ imply that $y = x = z$.
Substituting into equation (4) gives

$$\begin{aligned} x + x + x - 90 &= 0 \\ x &= 30. \end{aligned}$$

$x = y = z = 30$ will maximize $f(x, y, z) = xyz$.
The numbers are 30, 30, and 30.

17. Consider the constraint and solve for x in terms of y.

$$\begin{aligned} x + 2y &= 15 \\ x &= 15 - 2y \end{aligned}$$

Then

$$\begin{aligned} f(x, y) &= xy^2 \\ &= (15 - 2y)y^2 \\ &= -2y^3 + 15y^2 \end{aligned}$$

So, $f(x, y) = -2y^3 + 15y^2 = f(y)$. Notice that f is unbounded; more specifically,

$$\lim_{y \to \infty} f(y) = -\infty$$
and
$$\lim_{y \to -\infty} f(y) = \infty.$$

Therefore f, subject to the given constraint, has neither an absolute maximum nor an absolute minimum.

21. Let x be the length of the fence opposite the building and y be the length of each end. The area is then xy and the total cost is $25x+15(2y)$. Restate the problem as follows.

Maximize xy,
subject to $25x+30y=2400$.

1. $g(x,y)=25x+30y-2400$

2. $F(x,y,\lambda)=xy-\lambda(25x+30y-2400)$

3. $F_x(x,y,\lambda)=y-25\lambda$
$F_y(x,y,\lambda)=x-30\lambda$
$F_\lambda(x,y,\lambda)=-(25x+30y-2400)$

4.
$$y-25\lambda=0 \quad (1)$$
$$x-30\lambda=0 \quad (2)$$
$$25x+30y-2400=0 \quad (3)$$

5. Equations (1) and (2) give $\lambda=\frac{y}{25}$ and $\lambda=\frac{x}{30}$. Thus,

$$\frac{y}{25}=\frac{x}{30}$$
$$y=\frac{5}{6}x.$$

Substituting $y=\frac{5}{6}x$ into equation (3) gives

$$25x+30\left(\frac{5}{6}x\right)-2400=0$$
$$50x-2400=0$$
$$x=48.$$

So
$$y=\frac{5}{6}x=40.$$

The dimensions are 48 ft (opposite the building) by 40 ft (the ends).

23. Maximize $P(x,y)=-x^2-y^2+4x+8y$,
subject to $x+y=6$.

1. $g(x,y)=x+y-6$

2. $F(x,y,\lambda)$
$=-x^2-y^2+4x+8y-\lambda(x+y-6)$

3. $F_x(x,y,\lambda)=-2x+4-\lambda$
$F_y(x,y,\lambda)=-2y+8-\lambda$
$F_\lambda(x,y,\lambda)=-(x+y-6)$

4. $-2x+4-\lambda=0$
$-2y+8-\lambda=0$
$x+y-6=0$

5. $\lambda=-(2x-4)$ and $\lambda=-(2y-8)$
$2x-4=2y-8$
$y=x+2$

Substituting into $x+y-6=0$, we have

$x+(x+2)-6=0$ so $x=2$ and $y=4$.

25. Maximize $f(x,y)=12x^{3/4}y^{1/4}$,
subject to $100x+180y=25,200$.

1. $g(x,y)=100x+180y-25,200$

2. $F(x,y,\lambda)$
$=12x^{3/4}y^{1/4}$
$-\lambda(100x+180y-25{,}200)$

3.
$$F_x(x,y,\lambda)=\frac{3}{4}(12x^{-1/4}y^{1/4})-100\lambda$$
$$=\frac{9y^{1/4}}{x^{1/4}}-100\lambda$$
$$F_y(x,y,\lambda)=\frac{1}{4}(12x^{3/4}y^{-3/4})-180\lambda$$
$$=\frac{3x^{3/4}}{y^{3/4}}-180\lambda$$
$$F_\lambda(x,y,\lambda)=-(100x+180y-25{,}200)$$

4.
$$\frac{9y^{1/4}}{x^{1/4}}-100\lambda=0$$
$$\frac{3x^{3/4}}{y^{3/4}}-180\lambda=0$$
$$100x+180y-25,200=0$$

5. $\lambda=\dfrac{9y^{1/4}}{100x^{1/4}}$ and $\lambda=\dfrac{3x^{3/4}}{180y^{3/4}}$
$$=\frac{x^{3/4}}{60y^{3/4}}$$

$$\frac{9y^{1/4}}{100x^{1/4}}=\frac{x^{3/4}}{60y^{3/4}}$$
$$100x=540y$$
$$x=\frac{27y}{5}$$

Substitute into

$$100x+180y-25,200=0.$$
$$100\left(\frac{27y}{5}\right)+180y=25{,}200$$
$$540y+180y=25{,}200$$
$$720y=25{,}200$$
$$y=35$$
$$x=\frac{27(35)}{5}=189$$

Production will be maximized with 189 units of labor and 35 units of capital.

27. If x and y are the dimensions of the field, we must maximize $f(x, y) = xy$ subject to $x + 2y = 600$.

1. $g(x, y) = x + 2y - 600$

2. $F(x, y, \lambda)$
$= xy - \lambda(x + 2y - 600)$

3. $F_x(x, y, \lambda) = y - \lambda$
$F_y(x, y, \lambda) = x - 2\lambda$
$F_\lambda(x, y, \lambda) = -(x + 2y - 600)$

4.
$$x - \lambda = 0$$
$$x - 2\lambda = 0$$
$$x + 2y - 600 = 0$$

5. $\lambda = y$ and $\lambda = \dfrac{x}{2}$

$$y = \frac{x}{2}$$

Substituting into $x + 2y - 600 = 0$, we have

$x + 2\left(\frac{x}{2}\right) - 600 = 0$, so $x = 300$ and $y = 150$.

The largest area is $(300)(150) = 45{,}000 \text{ m}^2$.

29. Let x be the radius of the can and y be the height.

Minimize surface area $f(x, y) = 2\pi xy + 2\pi x^2$, subject to the constraint that $\pi x^2 y = 25$.

1. $g(x, y) = \pi x^2 y - 25$

2. $F(x, y, \lambda)$
$= 2\pi xy + 2\pi x^2 - \lambda(\pi x^2 y - 25)$

3. $F_x(x, y, \lambda) = 2\pi y + 4\pi x - 2\lambda \pi xy$
$F_y(x, y, \lambda) = 2\pi x - \lambda \pi x^2$
$F_\lambda(x, y, \lambda) = -(\pi x^2 y - 25)$

4. $2\pi y + 4\pi x - 2\lambda \pi xy = 0$
$$2\pi x - \lambda \pi x^2 = 0$$
$$\pi x^2 y - 25 = 0$$

5. $\lambda = \dfrac{2x + y}{xy}$ and $\lambda = \dfrac{2}{x}$

$$\frac{2x + y}{xy} = \frac{2}{x}$$
$$2x^2 + xy = 2xy$$
$$2x^2 - xy = 0$$
$$x = 0 \quad \text{or} \quad y = 2x$$

$x = 0$ is impossible.

Substituting $y = 2x$ into $\pi x^2 y - 25 = 0$, we have

$\pi x^2(2x) - 25 = 0$, so $x = \sqrt[3]{\frac{25}{2\pi}} \approx 1.585$ inches

and $y = 2\sqrt[3]{\frac{25}{2\pi}} \approx 3.169$ inches.

The can with minimum surface area will have a radius of approximately 1.585 inches and a height of approximately 3.169 inches.

31. If the box is x by x by y, we must minimize surface area $f(x, y) = 2x^2 + 4xy$, subject to $x^2 y = 185$.

1. $g(x, y) = x^2 y - 185$

2. $F(x, y, \lambda)$
$= 2x^2 + 4xy - \lambda(x^2 y - 185)$

3. $F_x(x, y, \lambda) = 4x + 4y - 2\lambda xy$
$F_y(x, y, \lambda) = 4x - \lambda x^2$
$F_\lambda(x, y, \lambda) = -(x^2 y - 185)$

4. $4x + 4y - 2\lambda xy = 0$
$$4x - \lambda x^2 = 0$$
$$x^2 y - 185 = 0$$

5. $\lambda = \dfrac{2x + 2y}{xy}$ and $\lambda = \dfrac{4}{x}$

$$\frac{2x + 2y}{xy} = \frac{4}{x}$$
$$2x^2 + 2xy = 4xy$$
$$2x^2 - 2xy = 0$$
$$2x(x - y) = 0$$
$$x = 0 \quad \text{or} \quad y = x$$

$x = 0$ is impossible.

Substituting $y = x$ into $x^2 y - 185 = 0$, we have

$$y = x = \sqrt[3]{185} \approx 5.698.$$

The dimensions are 5.698 inches by 5.698 inches by 5.698 inches.

33. Let the dimensions of the bottom be x by y, and let the height be z. We must minimize $f(x, y, z) = xy + 2xz + 2yz$ subject to $xyz = 32$.

1. $g(x, y, z) = xyz - 32$

2. $F(x, y, z, \lambda)$
$= xy + 2xz + 2yz - \lambda(xyz - 32)$

3. $F_x(x, y, z, \lambda) = y + 2z - \lambda yz$
$F_y(x, y, z, \lambda) = x + 2z - \lambda xz$
$F_z(x, y, z, \lambda) = 2x + 2y - \lambda xy$
$F_\lambda(x, y, z, \lambda) = -(xyz - 32)$

4.
$$y + 2z - \lambda yz = 0$$
$$x + 2z - \lambda xz = 0$$
$$2x + 2y - \lambda xy = 0$$
$$xyz - 32 = 0$$

5.
$$\lambda = \frac{y+2z}{yz}$$
$$\lambda = \frac{x+2z}{xz}$$
$$\lambda = \frac{2x+2y}{xy}$$
$$xyz = 32$$

$$\frac{y+2z}{yz} = \frac{x+2z}{xz}$$
$$xyz + 2xz^2 = xyz + 2yz^2$$
$$2z^2(x-y) = 0$$
$$z^2 = 0 \quad \text{or} \quad x - y = 0$$
$$z = 0 \text{ (impossible) or } x = y$$

$$\frac{x+2z}{xz} = \frac{2x+2y}{xy}$$
$$x^2y + 2xyz = 2x^2z + 2xyz$$
$$x^2(y-2z) = 0$$
$$x^2 = 0 \quad \text{or} \quad y - 2z = 0$$
$$x = 0 \text{ (impossible)} \quad \text{or} \quad y = 2z$$

Since $x = y$ and $y = 2z$ and since $xyz = 32$, we have

$$(2z)(2z)z = 32$$
$$z^3 = 8$$
$$z = 2.$$

If $z = 2$, $y = 4$ and $x = 4$.
The dimensions are 4 feet by 4 feet for the base and 2 feet for the height.

35. (a)
$$P(r,s,t) = rs(1-t) + (1-r)st + r(1-s)t + rst$$
$$g(r,s,t) = r+s+t-\alpha$$
$$F(r,s,t,\lambda) = rs(1-t) + (1-r)st + r(1-s)t + rst - \lambda(r+s+t-\alpha)$$

(b)

3.
$$F_r(r,s,t,\lambda) = s(1-t) - st + (1-s)t + st - \lambda = s + t - 2st - \lambda$$
$$F_s(r,s,t,\lambda) = r(1-t) + (1-r)t - rt + rt - \lambda = r + t - 2rt - \lambda$$
$$F_t(r,s,t,\lambda) = -rs + (1-r)s + r(1-s) + rs - \lambda = r + s - 2rs - \lambda$$
$$F_\lambda(r,s,t,\lambda) = -(r+s+t-\alpha)$$

4.
$$s + t - 2st - \lambda = 0$$
$$r + t - 2rt - \lambda = 0$$
$$r + s - 2rs - \lambda = 0$$
$$r + s + t - \alpha = 0$$

5.
$$-\lambda = 2st - s - t$$
$$-\lambda = 2rt - r - t$$
$$-\lambda = 2rs - r - s$$
$$r + s + t = \alpha$$

$$2st - s - t = 2rt - r - t$$
$$s(2t-1) = r(2t-1)$$
$$(s-r)(2t-1) = 0$$
$$s = r \quad \text{or} \quad t = \frac{1}{2}$$

$$2rt - r - t = 2rs - r - s$$
$$t(2r-1) = s(2r-1)$$
$$(t-s)(2r-1) = 0$$
$$t = s \quad \text{or} \quad r = \frac{1}{2}$$

$$2st - s - t = 2rs - r - s$$
$$t(2s-1) = r(2s-1)$$
$$(t-r)(2s-1) = 0$$
$$t = r \quad \text{or} \quad s = \frac{1}{2}$$

Since r, s and t are probabilities, $0 \le r, s, t \le 1$. Also, $r + s + t = \alpha = 0.75$. If $t = \frac{1}{2}$, then either $t = s = \frac{1}{2}$ or $r = \frac{1}{2}$, both of which are impossible (the third value would have to be -0.25 to get a sum of 0.75).
Thus, $r = s = t = 0.25$.

(c) Now we have $r + s + t = \alpha = 3$. If $t = \frac{1}{2}$, then either $t = s = \frac{1}{2}$ or $r = \frac{1}{2}$, both of which are impossible (the third value would have to be 2 to get a sum of 3).
Thus, $r = s = t = 1.0$.

9.5 Total Differentials and Approximations

1. Let $z = f(x,y) = \sqrt{x^2+y^2}$.

Then

$$dz = f_x(x,y)\,dx + f_y(x,y)\,dy$$
$$= \frac{1}{2}(x^2+y^2)^{-1/2}(2x)\,dx + \frac{1}{2}(x^2+y^2)^{-1/2}(2y)\,dy$$
$$= \frac{x\,dx + y\,dy}{\sqrt{x^2+y^2}}.$$

To approximate $\sqrt{8.05^2+5.97^2}$, we let $x = 8$, $dx = 0.05$, $y = 6$ and $dy = -0.03$.

$$\begin{aligned} dz &= \frac{8(0.05)+6(-0.03)}{\sqrt{8^2+6^2}} \\ &= \frac{4}{5}(0.05)+\frac{3}{5}(-0.03) \\ &= 0.04-0.018 = 0.022 \end{aligned}$$

$$\begin{aligned} f(8.05, 5.97) &= f(8,6)+\Delta z \\ &\approx f(8,6)+dz \\ &= \sqrt{8^2+6^2}+0.222 \\ &= 10.022 \end{aligned}$$

Thus, $\sqrt{8.05^2+5.97^2} \approx 10.022$.

Using a calculator, $\sqrt{8.05^2+5.97^2} \approx 10.0221$.

The absolute value of the difference of the two results is $|10.022 - 10.0221| = 0.0001$.

3. Let $z = f(x,y) = (x^2+y^2)^{1/3}$.

Then

$$\begin{aligned} dz &= f_x(x,y)\,dx + f_y(x,y)\,dy \\ dz &= \frac{1}{3}(x^2+y^2)^{-2/3}(2x)\,dx \\ &\quad + \frac{1}{3}(x^2+y^2)^{-2/3}(2y)\,dy \\ &= \frac{2x}{3(x^2+y^2)^{2/3}}\,dx + \frac{2y}{3(x^2+y^2)^{2/3}}\,dy \end{aligned}$$

To approximate $(1.92^2+2.1^2)^{1/3}$, we let $x = 2$, $dx = -0.08$, $y = 2$, and $dy = 0.1$.

$$\begin{aligned} dz &= \frac{2(2)}{3[(2)^2+(2)^2]^{2/3}}(-0.08) \\ &\quad + \frac{2(2)}{3[(2)^2+(2)^2]^{2/3}}(0.1) \\ &= \frac{4}{12}(-0.08)+\frac{4}{12}(0.1) \\ &= 0.00\overline{6} \end{aligned}$$

$$\begin{aligned} f(1.92, 2.1) &= f(2,2)+\Delta z \\ &\approx f(2,2)+dz \\ &= 2+0.00\overline{6} \\ f(1.92, 2.1) &\approx 2.0067 \end{aligned}$$

Using a calculator, $(1.92^2+2.1^2)^{1/3} \approx 2.0080$. The absolute value of the difference of the two results is $|2.0067 - 2.0080| = 0.0013$.

5. Let $z = f(x,y) = xe^y$.

Then

$$\begin{aligned} dz &= f_x(x,y)\,dx + f_y(x,y)\,dy \\ &= e^y\,dx + xe^y\,dy. \end{aligned}$$

To approximate $1.03e^{0.04}$, we let $x = 1$, $dx = 0.03$, $y = 0$, and $dy = 0.04$.

$$\begin{aligned} dz &= e^0(0.03)+1\cdot e^0(0.04) \\ &= 0.07 \end{aligned}$$

$$\begin{aligned} f(1.03, 0.04) &= f(1,0)+\Delta z \\ &\approx f(1,0)+dz \\ &= 1\cdot e^0+0.07 \\ &= 1.07 \end{aligned}$$

Thus, $1.03e^{0.04} \approx 1.07$.
Using a calculator, $1.03e^{0.04} \approx 1.0720$.
The absolute value of the difference of the two results is $|1.07 - 1.0720| = 0.0020$.

7. Let $z = f(x,y) = x \ln y$.

Then

$$\begin{aligned} dz &= f_x(x,y)\,dx + f_y(x,y)\,dy \\ &= \ln y\,dx + \frac{x}{y}\,dy \end{aligned}$$

To approximate $0.99 \ln 0.98$, we let $x = 1$, $dx = -0.01$, $y = 1$, and $dy = -0.02$.

$$\begin{aligned} dz &= \ln(1)\cdot(-0.01)+\frac{1}{1}(-0.02) \\ &= -0.02 \end{aligned}$$

$$\begin{aligned} f(0.99, 0.98) &= f(1,1)+\Delta z \\ &\approx f(1,1)+dz \\ &= 1\cdot\ln(1)-0.02 \\ &\approx -0.02 \end{aligned}$$

Thus, $0.99 \ln 0.98 \approx -0.02$.

Using a calculator, $0.99 \ln 0.98 \approx -0.0200$. The absolute value of the difference of the two results is $|-0.02-(-0.0200)| = 0$.

9. $z = f(x,y) = 2x^2+4xy+y^2$
$x = 5, dx = 0.03, y = -1, dy = -0.02$

$$\begin{aligned} f_x(x,y) &= 4x+4y \\ f_y(x,y) &= 4x+2y \\ dz &= (4x+4y)dx+(4x+2y)dy \\ &= [4(5)+4(-1)](0.03) \\ &\quad + [4(5)+2(-1)](-0.02) \\ &= 0.48-0.36 = 0.12 \end{aligned}$$

11. $z = \dfrac{y^2 + 3x}{y^2 - x}$, $x = 4$, $y = -4$,

$dx = 0.01$, $dy = 0.03$

$$dz = \frac{(y^2 - x)\cdot 3 - (y^2 + 3x)\cdot(-1)}{(y^2 - x)^2}\,dx + \frac{(y^2 - x)\cdot 2y - (y^2 + 3x)\cdot 2y}{(y^2 - x)^2}\,dx$$

$$= \frac{4y^2}{(y^2 - x)^2}\,dx - \frac{8xy}{(y^2 - x)^2}\,dy$$

$$= \frac{4(-4)^2}{[(-4)^2 - 4]^2}(0.01) - \frac{8(4)(-4)}{[(-4)^2 - 4]^2}(0.03)$$

$$\approx 0.0311$$

13. $w = \dfrac{5x^2 + y^2}{z + 1}$

$x = -2$, $y = 1$, $z = 1$

$dx = 0.02$, $dy = -0.03$, $dz = 0.02$

$$f_x(x, y) = \frac{(z + 1)10x - (5x^2 + y^2)(0)}{(z + 1)^2} = \frac{10x}{z + 1}$$

$$f_y(x, y) = \frac{(z + 1)(2y) - (5x^2 + y^2)(0)}{(z + 1)^2} = \frac{2y}{z + 1}$$

$$f_z(x, y) = \frac{(z + 1)(0) - (5x^2 + y^2)(1)}{(z + 1)^2} = \frac{-5x^2 - y^2}{(z + 1)^2}$$

$$dw = \frac{10x}{z + 1}\,dx + \frac{2y}{z + 1}\,dy + \frac{-5x^2 - y^2}{(z + 1)^2}\,dz$$

Substitute the given values.

$$dw = \frac{-20}{2}(0.02) + \frac{2}{2}(-0.03) + \frac{[-5(4) - 1](0.02)}{(2)^2}$$

$$= -0.2 - 0.03 - \frac{21}{4}(0.02) = -0.335$$

15. The volume of the can is

$$V = \pi r^2 h,$$

with $r = 2.5$ cm, $h = 14$ cm, $dr = 0.08$, $dh = 0.16$.

$$dV = 2\pi r h\,dr + \pi r^2\,dh$$
$$= 2\pi(2.5)(14)(0.08) + \pi(2.5)^2(0.16)$$
$$\approx 20.73$$

Approximately 20.73 cm^3 of aluminum are needed.

17. The volume of the box is

$$V = LWH$$

with $L = 10$, $W = 9$, and $H = 18$.
Since 0.1 inch is applied to each side and each dimension has a side at each end,

$$dL = dW = dH = 2(0.1) = 0.2$$
$$dV = WH\,dL + LH\,dW + LW\,dH.$$

Substitute.

$$dV = (9)(18)(0.2) + (10)(18)(0.2) + (10)(9)(0.2)$$
$$= 86.4$$

Approximately 86.4 in.^{3} are needed.

19. $z = x^{0.65}y^{0.35}$
$x = 50$, $y = 29$,
$dx = 52 - 50 = 2$
$dy = 27 - 29 = -2$

$$f_x(x, y) = y^{0.35}(0.65)(x^{-0.35}) = 0.65\left(\frac{y}{x}\right)^{0.35}$$

$$f_y(x, y) = (x^{0.65})(0.35)(y^{-0.65}) = 0.35\left(\frac{x}{y}\right)^{0.65}$$

$$dz = 0.65\left(\frac{y}{x}\right)^{0.35} dx + 0.35\left(\frac{x}{y}\right)^{0.65} dy$$

Substitute.

$$dz = 0.65\left(\frac{29}{50}\right)^{0.35}(2) + 0.35\left(\frac{50}{29}\right)^{0.65}(-2)$$
$$= 0.07694 \text{ unit}$$

21. The volume of the bone is

$$V = \pi r^2 h,$$

with $h = 7$, $r = 1.4$, $dr = 0.09$, $dh = 2(0.09) = 0.18$

$$dV = 2\pi r h\,dr + \pi r^2\,dh$$
$$= 2\pi(1.4)(7)(0.09) + \pi(1.4)^2(0.18)$$
$$= 6.65$$

6.65 cm^3 of preservative are used.

23. $C = \dfrac{b}{a-v} = b(a-v)^{-1}$

$$\begin{aligned} a &= 160, \\ b &= 200,\ v = 125 \\ da &= 145 - 160 = -15 \\ db &= 190 - 200 = -10 \\ dv &= 130 - 125 = 5 \end{aligned}$$

$$\begin{aligned} dC &= -b(a-v)^{-2}\,da \\ &\quad + \frac{1}{a-v}\,db + b(a-v)^{-2}\,dv \\ &= \frac{-b}{(a-v)^2}\,da + \frac{1}{a-v}\,db + \frac{b}{(a-v)^2}\,dv \\ &= \frac{-200}{(160-125)^2}(-15) + \frac{1}{160-125}(-10) \\ &\quad + \frac{200}{(160-125)^2}(5) \\ &\approx 2.98 \text{ liters} \end{aligned}$$

25. $C(t,g) = 0.6(0.96)^{(210t/1500)-1} + \dfrac{gt}{126t-900}[1-(0.96)^{(210t/1500)-1}]$

(a) $C(180,8) = 0.6(0.96)^{(210(180)/1500)-1} + \dfrac{(8)(180)}{126(180)-900} \cdot [1-(0.96)^{(210(180)/1500)-1}] \approx 0.2649$

(b) $C_t(t,g)$

$$\begin{aligned} &= 0.6(\ln 0.96)\left(\frac{210}{1500}\right)(0.96)^{(210t/1500)-1} \\ &\quad + \frac{g(126t-900)-126(gt)}{(126t-900)^2} \\ &\quad \cdot [1-(0.96)^{(210t/1500)-1}] \\ &\quad - \frac{gt}{126t-900}(\ln 0.96)\left(\frac{210}{1500}\right)(0.96)^{(210t/1500)-1} \end{aligned}$$

$C_g(t,g)$

$$= \frac{t}{126t-900}[1-(0.96)^{(210t/1500)-1}]$$

$$\begin{aligned} &C(180-10, 8+1) \\ &\quad \approx C(180,8) + C_t(180,8)\cdot(-10) \\ &\qquad + C_g(180,8)\cdot(1) \\ &\quad \approx 0.2649 + (-0.00115)(-10) + 0.00519(1) \\ &\quad \approx 0.2816 \end{aligned}$$

$C(170,9) \approx 0.2817$

The approximation is very good.

27. $P(A,B,D) = \dfrac{1}{1+e^{3.68-0.016A-0.77B-0.12D}}$

(a) Since bird pecking is present, $B = 1$.

$$\begin{aligned} P(150,1,20) &= \frac{1}{1+e^{3.68-0.016(150)-0.77(1)-0.12(20)}} \\ &= \frac{1}{1+e^{-1.89}} \approx 0.8688 \end{aligned}$$

The probability is about 87%.

(b) Since bird pecking is not present, $B = 0$.

$$\begin{aligned} P(150,0,20) &= \frac{1}{1+e^{3.68-0.016(150)-0.77(0)-0.12(20)}} \\ &= \frac{1}{1+e^{-1.12}} \approx 0.7540 \end{aligned}$$

The probability is about 75%.

(c) Let $B = 0$. To simplify the notation, let $X = 3.68 - 0.016A - 0.12D$. Then

$$\begin{aligned} P(A,0,D) &= \frac{1}{1+e^{3.68-0.016A-0.12D}} \\ &= \frac{1}{1+e^X}. \end{aligned}$$

Some other values that we'll need are

$$\begin{aligned} dA &= 160 - 150 = 10 \\ dD &= 25 - 20 = 4 \\ X(150,20) &= 3.68 - 0.016(150) - 0.12(20) \\ &= -1.12 \end{aligned}$$

$$\begin{aligned} X_A &= \frac{\partial X}{\partial A} = -0.016 \\ X_D &= \frac{\partial X}{\partial D} = -0.12. \end{aligned}$$

$$\begin{aligned} P_A(A,0,D) &= \frac{X_A e^X}{(1+e^X)^2} = \frac{0.016e^X}{(1+e^X)^2} \\ P_D(A,0,D) &= \frac{X_D e^X}{(1+e^X)^2} = \frac{0.12e^X}{(1+e^X)^2} \end{aligned}$$

$$\begin{aligned} dP &= P_A(A,0,D)\,dA + P_D(A,0,D)\,dD \\ &= \frac{0.016e^X}{(1+e^X)^2}\,dA + \frac{0.12e^X}{(1+e^X)^2}\,dD \end{aligned}$$

Substituting the given and calculated values,

$$\begin{aligned} dP &= \frac{0.016e^{-1.12}}{(1+e^{-1.12})^2}(10) + \frac{0.12e^{-1.12}}{(1+e^{-1.12})^2}(5) \\ &= (0.016\cdot 10 + 0.12\cdot 5)\frac{e^{-1.12}}{(1+e^{-1.12})^2} \\ &\approx 0.76 \cdot 0.1855 \approx 0.14. \end{aligned}$$

Therefore,

$$\begin{aligned}P(160,0,25) &= P(150,0,20)+\Delta P\\ &\approx P(150,0,20)+dP\\ &= 0.75+0.14=0.89.\end{aligned}$$

The probability is about 89%.

Using a calculator, $P(160,0,25)\approx 0.8676$, or about 87%.

29. The area is $A=\frac{1}{2}bh$ with $b=15.8$ cm, $h=37.5$ cm, $db=1.1$ cm, and $dh=0.8$ cm.

$$\begin{aligned}dA &= \frac{1}{2}b\,dh+\frac{1}{2}h\,db\\ &= \frac{1}{2}(15.8)(0.8)+\frac{1}{2}(37.5)(1.1)\\ &= 26.945\end{aligned}$$

The maximum possible error is 26.945 cm^2.

31. Let $z=f(L,W,H)=LWH$

Then

$$\begin{aligned}dz &= f_L(L,W,H)\,dL+f_W(L,W,H)\,dW\\ &\quad + f_H(L,W,H)\,dH\\ &= WHdL+LHdW+LWdH.\end{aligned}$$

A maximum 1% error in each measurement means that the maximum values of dL, dW, and dH are given by $dL=0.01L$, $dW=0.01W$, and $dH=0.01H$.
Therefore,

$$\begin{aligned}dz &= WH(0.01L)+LH(0.01W)+LW(0.01H)\\ &= 0.01LWH+0.01LWH+0.01LWH\\ &= 0.03LWH.\end{aligned}$$

Thus, an estimate of the maximum error in calculating the volume is 3%.

9.6 Double Integrals

1. $$\begin{aligned}\int_0^5 (x^4y+y)\,dx &= \left(\frac{x^5y}{5}+xy\right)\Big|_0^5\\ &= (625y+5y)-0=630y\end{aligned}$$

3. $$\begin{aligned}\int_1^7 \sqrt{x+6y}\,dy &= \frac{1}{6}\cdot\frac{2}{3}(x+6y)^{3/2}\Big|_1^7\\ &= \frac{1}{9}[(x+42)^{3/2}-(x+6)^{3/2}]\end{aligned}$$

5. $$\begin{aligned}&\int_4^5 x\sqrt{x^2+3y}\,dy\\ &= \int_4^5 x(x^2+3y)^{1/2}\,dy\\ &= \frac{2x}{9}(x^2+3y)^{3/2}\Big|_4^5\\ &= \frac{2x}{9}[(x^2+15)^{3/2}-(x^2+12)^{3/2}]\end{aligned}$$

7. $$\begin{aligned}\int_4^9 \frac{3+5y}{\sqrt{x}}\,dx &= (3+5y)\int_4^9 x^{-1/2}\,dx\\ &= (3+5y)2x^{1/2}\Big|_4^9\\ &= (3+5y)2[\sqrt{9}-\sqrt{4}]\\ &= 6+10y\end{aligned}$$

9. $$\begin{aligned}\int_2^6 e^{2x+3y}\,dx &= \frac{1}{2}e^{2x+3y}\Big|_2^6\\ &= \frac{1}{2}(e^{12+3y}-e^{4+3y})\end{aligned}$$

11. $$\int_0^3 ye^{4x+y^2}\,dy$$

Let $u=4x+y^2$; then $du=2y\,dy$.
If $y=0$ then $u=4x$.
If $y=3$ then $u=4x+9$.

$$\begin{aligned}\int_{4x}^{4x+9} e^u\cdot\frac{1}{2}du &= \frac{1}{2}e^u\Big|_{4x}^{4x+9}\\ &= \frac{1}{2}(e^{4x+9}-e^{4x})\end{aligned}$$

13. $$\int_1^2\int_0^5 (x^4y+y)dx\,dy$$

From Exercise 1

$$\int_0^5 (x^4y+y)dx=630y.$$

Therefore,

$$\begin{aligned}\int_1^2\left[\int_0^5 (x^4y+y)dx\right]dy &= \int_1^2 630y\,dy\\ &= 315y^2\Big|_1^2\\ &= 315(4-1)=945.\end{aligned}$$

15. $\int_0^1 \left[\int_3^6 x\sqrt{x^2+3y}\,dx\right] dy$

From Exercise 6,

$$\int_3^6 x\sqrt{x^2+3y}\,dx$$

$$= \frac{1}{3}[(36+3y)^{3/2} - (9+3y)^3].$$

$$\int_0^1 \left[\int_3^6 x\sqrt{x^2+3y}\,dx\right] dy$$

$$= \int_0^1 \frac{1}{3}[(36+3y)^{3/2} - (9+3y)^{3/2}]\,dy$$

Let $u = 36+3y$. Then $du = 3\,dy$.
When $y = 0$, $u = 36$.
When $y = 1, u = 39$.
Let $z = 9+3y$. Then $dz = 3\,dy$.
When $y = 0$, $z = 9$.
When $y = 1$, $z = 12$.

$$\frac{1}{9}\left[\int_{36}^{39} u^{3/2}\,du - \int_9^{12} z^{3/2}\,dz\right]$$

$$= \frac{1}{9}\cdot\frac{2}{5}[(39)^{5/2} - (36)^{5/2} - (12)^{5/2} + (9)^{5/2}]$$

$$= \frac{2}{45}[(39)^{5/2} - (12)^{5/2} - 6^5 + 3^5]$$

$$= \frac{2}{45}(39^{5/2} - 12^{5/2} - 7533)$$

17. $\int_1^2 \left[\int_4^9 \frac{3+5y}{\sqrt{x}}\,dx\right] dy$

From Exercise 7,

$$\int_4^9 \frac{3+5y}{\sqrt{x}}\,dx = 6+10y.$$

$$\int_1^2 \left[\int_4^9 \frac{3+5y}{\sqrt{x}}\,dx\right] dy$$

$$= \int_1^2 (6+10y)\,dy$$

$$= 6y\Big|_1^2 + 5y^2\Big|_1^2$$

$$= 6(2-1) + 5(4-1)$$
$$= 6+15$$
$$= 21$$

19. $\int_1^3 \int_1^3 \frac{dy\,dx}{xy} = \int_1^3 \left[\int_1^3 \frac{1}{xy}\,dy\right] dx$

$$= \int_1^3 \left(\frac{1}{x}\ln|y|\right)\Big|_1^3 dx$$

$$= \int_1^3 \frac{\ln 3}{x}\,dx$$

$$= (\ln 3)\ln|x|\Big|_1^3$$

$$= (\ln 3)(\ln 3 - 0) = (\ln 3)^2$$

21. $\int_2^4 \int_3^5 \left(\frac{x}{y} + \frac{y}{3}\right) dx\,dy$

$$= \int_2^4 \left(\frac{x^2}{2y} + \frac{yx}{3}\right)\Big|_3^5 dy$$

$$= \int_2^4 \left[\frac{25}{2y} + \frac{5y}{3} - \left(\frac{9}{2y} + \frac{3y}{3}\right)\right] dy$$

$$= \int_2^4 \left(\frac{16}{2y} + \frac{2y}{3}\right) dy$$

$$= \left(8\ln|y| + \frac{y^2}{3}\right)\Big|_2^4$$

$$= 8(\ln 4 - \ln 2) + \frac{16}{3} - \frac{4}{3}$$

$$= 8\ln\frac{4}{2} + \frac{12}{3}$$

$$= 8\ln 2 + 4$$

23. $\iint_R (3x^2+4y)dx\,dy;\ 0 \le x \le 3, 1 \le y \le 4$

$$\iint_R (3x^2+4y)dx\,dy = \int_1^4 \int_0^3 (3x^2+4y)dx\,dy$$

$$= \int_1^4 (x^3+4xy)\Big|_0^3 dy$$

$$= \int_1^4 (27+12y)dy$$

$$= (27y+6y^2)\Big|_1^4$$

$$= (108+96) - (27+6) = 171$$

25. $\iint_R \sqrt{x+y}\,dy\,dx;\ 1 \le x \le 3,\ 0 \le y \le 1$

$$\iint_R \sqrt{x+y}\,dy\,dx = \int_1^3 \int_0^1 (x+y)^{1/2}\,dy\,dx$$
$$= \int_1^3 \left[\frac{2}{3}(x+y)^{3/2}\right]\Big|_0^1 dx$$
$$= \int_1^3 \frac{2}{3}[(x+1)^{3/2} - x^{3/2}]\,dx$$
$$= \frac{2}{3}\cdot\frac{2}{5}[(x+1)^{5/2} - x^{5/2}]\Big|_1^3$$
$$= \frac{4}{15}(4^{5/2} - 3^{5/2} - 2^{5/2} + 1^{5/2})$$
$$= \frac{4}{15}(32 - 3^{5/2} - 2^{5/2} + 1)$$
$$= \frac{4}{15}(33 - 3^{5/2} - 2^{5/2})$$

27. $\iint_R \frac{3}{(x+y)^2}\,dy\,dx; 2 \le x \le 4, 1 \le y \le 6$

$$\iint_R \frac{3}{(x+y)^2}\,dy\,dx = 3\int_2^4 \int_1^6 (x+y)^{-2}\,dy\,dx$$
$$= -3\int_2^4 (x+y)^{-1}\Big|_1^6 dx$$
$$= -3\int_2^4 \left(\frac{1}{x+6} - \frac{1}{x+1}\right) dx$$
$$= -3(\ln|x+6| - \ln|x+1|)\Big|_2^4$$
$$= -3\left(\ln\left|\frac{x+6}{x+1}\right|\right)\Big|_2^4$$
$$= -3\left(\ln 2 - \ln\frac{8}{3}\right) = -3\ln\frac{2}{\frac{8}{3}}$$
$$= -3\ln\frac{3}{4} \text{ or } 3\ln\frac{4}{3}$$

29. $\iint_R ye^{(x+y^2)}\,dx\,dy;\ 2 \le x \le 3,\ 0 \le y \le 2$

$$\iint_R ye^{(x+y^2)}\,dx\,dy$$
$$= \int_0^2 \int_2^3 ye^{x+y^2}\,dx\,dy$$
$$= \int_0^2 ye^{x+y^2}\Big|_2^3 dy$$
$$= \int_0^2 (ye^{3+y^2} - ye^{2+y^2})\,dy$$
$$= e^3\int_0^2 ye^{y^2}\,dy - e^2\int_0^2 ye^{y^2}\,dy$$
$$= \frac{e^3}{2}(e^{y^2})\Big|_0^2 - \frac{e^2}{2}(e^{y^2})\Big|_0^2$$
$$= \frac{e^3}{2}(e^4 - e^0) - \frac{e^2}{2}(e^4 - e^0)$$
$$= \frac{1}{2}(e^7 - e^6 - e^3 + e^2)$$

31. $z = 8x + 4y + 3;\ -1 \le x \le 1, 0 \le y \le 3$

$$V = \int_{-1}^1 \int_0^3 (8x + 4y + 3)\,dy\,dx$$
$$= \int_{-1}^1 (8xy + 2y^2 + 3y)\Big|_0^3 dx$$
$$= \int_{-1}^1 (24x + 18 + 9 - 0)\,dx$$
$$= \int_{-1}^1 (24x + 27)\,dx$$
$$= (12x^2 + 27x)\Big|_{-1}^1$$
$$= (12 + 27) - (12 - 27) = 54$$

33. $z = x^2; 0 \le x \le 2, 0 \le y \le 5$

$$V = \int_0^2 \int_0^5 x^2\,dy\,dx$$
$$= \int_0^2 x^2y\Big|_0^5 dx$$
$$= \int_0^2 5x^2 dx$$
$$= \frac{5}{3}x^3\Big|_0^2$$
$$= \frac{40}{3}$$

35. $z = x\sqrt{x^2+y};\ 0 \le x \le 1,\ 0 \le y \le 1$

$$V = \int_0^1 \int_0^1 x\sqrt{x^2+y}\,dx\,dy$$

Let $u = x^2 + y$. Then $du = 2x\,dx$.
When $x = 0,\ u = y$.
When $x = 1,\ u = 1 + y$.

$$= \int_0^1 \left[\int_y^{1+y} u^{1/2}\,du\right] dy$$
$$= \int_0^1 \frac{1}{2}\left(\frac{2}{3}u^{3/2}\right)\Big|_y^{1+y} dy$$
$$= \int_0^1 \frac{1}{3}[(1+y)^{3/2} - y^{3/2}]\,dy$$
$$= \frac{1}{3}\cdot\frac{2}{5}[(1+y)^{5/2} - y^{5/2}]\Big|_0^1$$
$$= \frac{2}{15}(2^{5/2} - 1 - 1)$$
$$= \frac{2}{15}(2^{5/2} - 2)$$

37. $z = \dfrac{xy}{(x^2+y^2)^2};\ 1 \le x \le 2,\ 1 \le y \le 4$

$$V = \int_1^2 \int_1^4 \frac{xy}{(x^2+y^2)^2}\,dy\,dx$$
$$= \int_1^2 \left[\int_1^4 xy(x^2+y^2)^{-2}\,dy\right] dx$$
$$= \int_1^2 \left[\int_1^4 \frac{1}{2}x(x^2+y^2)^{-2}(2y)\,dy\right] dx$$
$$= \int_1^2 \left[-\frac{1}{2}x(x^2+y^2)^{-1}\right]\Big|_1^4 dx$$
$$= \int_1^2 \left[-\frac{1}{2}x(x^2+16)^{-1} + \frac{1}{2}x(x^2+1)^{-1}\right] dx$$
$$= -\frac{1}{2}\int_1^2 \frac{1}{2}(x^2+16)^{-1}(2x)\,dx$$
$$+ \frac{1}{2}\int_1^2 \frac{1}{2}(x^2+1)^{-1}(2x)\,dx$$
$$= -\frac{1}{2}\cdot\frac{1}{2}\ln|x^2+16|\Big|_1^2 + \frac{1}{2}\cdot\frac{1}{2}\ln|x^2+1|\Big|_1^2$$
$$= -\frac{1}{4}\cdot\ln 20 + \frac{1}{4}\ln 17 + \frac{1}{4}\ln 5 - \frac{1}{4}\ln 2$$
$$= \frac{1}{4}(-\ln 20 + \ln 17 + \ln 5 - \ln 2)$$
$$= \frac{1}{4}\ln\frac{(17)(5)}{(20)(2)}$$
$$= \frac{1}{4}\ln\frac{17}{8}$$

39. $\displaystyle\iint_R xe^{xy}\,dx\,dy;\ 0 \le x \le 2;\ 0 \le y \le 1$

$$\iint_R xe^{xy}\,dx\,dy$$
$$= \int_0^2 \int_0^1 xe^{xy}\,dy\,dx$$
$$= \int_0^2 \frac{x}{x}e^{xy}\Big|_0^1 dx$$
$$= \int_0^2 (e^x - e^0)\,dx$$
$$= (e^x - x)\Big|_0^2$$
$$= e^2 - 2 - e^0 + 0$$
$$= e^2 - 3$$

41. $\displaystyle\int_2^4 \int_2^{x^2} (x^2+y^2)\,dy\,dx$

$$= \int_2^4 \left(x^2y + \frac{y^3}{3}\right)\Big|_2^{x^2} dx$$
$$= \int_2^4 \left(x^4 + \frac{x^6}{3} - 2x^2 - \frac{8}{3}\right) dx$$
$$= \left(\frac{x^5}{5} + \frac{x^7}{21} - \frac{2}{3}x^3 - \frac{8}{3}x\right)\Big|_2^4$$
$$= \frac{1024}{5} + \frac{16{,}384}{21} - \frac{2}{3}(64) - \frac{8}{3}(4)$$
$$- \left(\frac{32}{5} + \frac{128}{21} - \frac{16}{3} - \frac{16}{3}\right)$$
$$= \frac{1024}{5} - \frac{32}{5} + \frac{16{,}384 - 128}{21}$$
$$- \frac{128}{3} - \frac{32}{3} - \left(\frac{-32}{3}\right)$$
$$= \frac{992}{5} + \frac{16{,}256}{21} - \frac{128}{3}$$
$$= \frac{20{,}832}{105} + \frac{81{,}280}{105} - \frac{4480}{105}$$
$$= \frac{97{,}632}{105}$$

43. $\int_0^4 \int_0^x \sqrt{xy}\, dy\, dx$

$$= \int_0^4 \int_0^x (xy)^{1/2}\, dy\, dx$$

$$= \int_0^4 \left[\frac{2(xy)^{3/2}}{3x}\right]\Bigg|_0^x dx$$

$$= \frac{2}{3}\int_0^4 \left[\frac{(\sqrt{x^2})^3}{x} - \frac{0}{x}\right] dx$$

$$= \frac{2}{3}\int_0^4 x^2\, dx = \frac{2}{3}\cdot\frac{x^3}{3}\bigg|_0^4 = \frac{2}{9}(64)$$

$$= \frac{128}{9}$$

45. $\int_2^6 \int_{2y}^{4y} \frac{1}{x}\, dx\, dy$

$$= \int_2^6 (\ln|x|)\bigg|_{2y}^{4y} dy$$

$$= \int_2^6 (\ln|4y| - \ln|2y|)\, dy$$

$$= \int_2^6 \ln\left|\frac{4y}{2y}\right| dy$$

$$= \int_2^6 \ln 2\, dy$$

$$= (\ln 2)y\bigg|_2^6$$

$$= (\ln 2)(6-2) = 4\ln 2$$

Note: We can write $4\ln 2$ as $\ln 2^4$, or $\ln 16$.

47. $\int_0^4 \int_1^{e^x} \frac{x}{y}\, dy\, dx$

$$= \int_0^4 (x\ \ln\ |y|)\bigg|_1^{e^x} dx$$

$$= \int_0^4 (x\ \ln\ e^x - x\ \ln\ 1)\, dx$$

$$= \int_0^4 x^2\, dx = \frac{x^3}{3}\bigg|_0^4 = \frac{64}{3}$$

49. $\int_0^{\ln 2} \int_{e^y}^2 \frac{1}{\ln x}\, dx\, dy$

Changing the order of integration,

$$\int_0^{\ln 2} \int_{e^y}^2 \frac{1}{\ln x}\, dx\, dy$$

$$= \int_1^2 \int_0^{\ln x} \frac{1}{\ln x}\, dy\, dx$$

$$= \int_1^2 \left[\frac{1}{\ln x}\, y\bigg|_0^{\ln x}\right] dx$$

$$= \int_1^2 (1-0)\, dx$$

$$= x\bigg|_1^2$$

$$= 2 - 1 = 1$$

51. $\iint_R (5x+8y)\, dy\, dx;\ 1 \le x \le 3, 0 \le y \le x-1$

$$\iint_R (5x+8y)\, dy\, dx$$

$$= \int_1^3 \int_0^{x-1} (5x+8y)\, dy\, dx$$

$$= \int_1^3 (5xy+4y^2)\bigg|_0^{x-1} dx$$

$$= \int_1^3 [5x(x-1) + 4(x-1)^2 - 0]\, dx$$

$$= \int_1^3 (9x^2 - 13x + 4)\, dx$$

$$= \left(3x^3 - \frac{13}{2}x^2 + 4x\right)\bigg|_1^3$$

$$= \left(81 - \frac{117}{2} + 12\right) - \left(3 - \frac{13}{2} + 4\right)$$

$$= 34$$

53. $\iint_R (4 - 4x^2)\,dy\,dx;\ 0 \le x \le 1,$

$0 \le y \le 2 - 2x$

$$\iint_R (4 - 4x^2)\,dy\,dx$$

$$= \int_0^2 \int_0^{2-2x} 4(1 - x^2)\,dy\,dx$$

$$= \int_0^1 [4(1 - x^2)y]\Big|_0^{2(1-x)}\,dx$$

$$= \int_0^1 4(1 - x^2)(2)(1 - x)\,dx$$

$$= 8\int_0^1 (1 - x - x^2 + x^3)\,dx$$

$$= 8\left(x - \frac{x^2}{2} - \frac{x^3}{3} + \frac{x^4}{4}\right)\Big|_0^1$$

$$= 8\left(1 - \frac{1}{2} - \frac{1}{3} + \frac{1}{4}\right)$$

$$= 8\left(\frac{1}{2} - \frac{1}{12}\right)$$

$$= 8 \cdot \frac{5}{12} = \frac{10}{3}$$

55. $\iint_R e^{x/y^2}\,dx\,dy;\ 1 \le y \le 2,\ 0 \le x \le y^2$

$$\iint_R e^{x/y^2}\,dx\,dy$$

$$= \int_1^2 \int_0^{y^2} e^{x/y^2}\,dx\,dy$$

$$= \int_1^2 [y^2 e^{x/y^2}]\Big|_0^{y^2}\,dy$$

$$= \int_1^2 (y^2 e^{y^2/y^2} - y^2 e^0)\,dy$$

$$= \int_1^2 (ey^2 - y^2)\,dy$$

$$= (e - 1)\,\frac{y^3}{3}\Big|_1^2$$

$$= (e - 1)\left(\frac{8}{3} - \frac{1}{3}\right)$$

$$= \frac{7(e - 1)}{3}$$

57. $\iint_R x^3y\,dy\,dx;$ R bounded by $y = x^2,\ y = 2x$

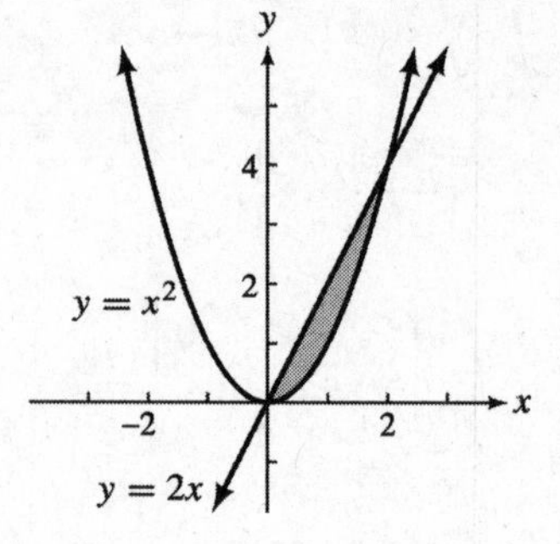

The points of intersection can be determined by solving the following system for x.

$$y = x^2$$
$$y = 2x$$

$$x^2 = 2x$$
$$x(x - 2) = 0$$
$$x = 0 \quad \text{or} \quad x = 2$$

Therefore,

$$\iint_R x^3y\,dx\,dy$$

$$= \int_0^2 \int_{x^2}^{2x} x^3y\,dy\,dx$$

$$= \int_0^2 \left(x^3\,\frac{y^2}{2}\right)\Big|_{x^2}^{2x}\,dx$$

$$= \int_0^2 \left[x^3\,\frac{(4x^2)}{2} - x^3\,\frac{(x^4)}{2}\right]dx$$

$$= \int_0^2 \left(2x^5 - \frac{x^7}{2}\right)dx$$

$$= \left(\frac{1}{3}x^6 - \frac{1}{16}x^8\right)\Big|_0^2$$

$$= \frac{1}{3} \cdot 2^6 - \frac{1}{16} \cdot 2^8$$

$$= \frac{64}{3} - 16$$

$$= \frac{16}{3}.$$

59. $\displaystyle\iint_R \frac{dy\,dx}{y}$; R bounded by $y = x$, $y = \frac{1}{x}$, $x = 2$.

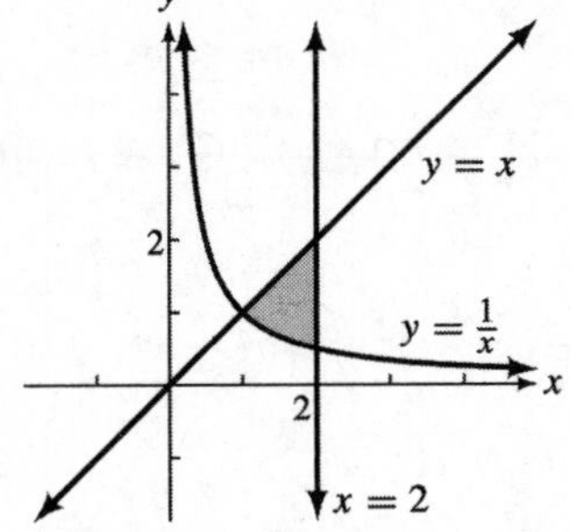

The graphs of $y = x$ and $y = \frac{1}{x}$ intersect at $(1, 1)$.

$$\int_1^2 \int_{1/x}^x \frac{dy}{y}\,dx = \int_1^2 \ln y \Big|_{1/x}^x \,dx = \int_1^2 \left(\ln x - \ln \frac{1}{x}\right) dx = \int_1^2 2 \ln x\,dx$$

$$= 2(x \ln x - x)\Big|_1^2 = 2[(2 \ln 2 - 2) - (\ln 1 - 1)] = 4 \ln 2 - 2$$

63. $f(x, y) = x^2 + y^2$; $0 \le x \le 2$, $0 \le y \le 3$

The area of region R is

$$A = (2 - 0)(3 - 0) = 6.$$

The average value of

$$f(x, y) = x^2 + y^2$$

over R is

$$\frac{1}{A}\iint_R (x^2 + y^2)\,dy\,dx = \frac{1}{6}\int_0^2 \int_0^3 (x^2 + y^2)\,dy\,dx = \frac{1}{6}\int_0^2 \left(x^2 y + \frac{y^3}{3}\right)\Big|_0^3 dx = \frac{1}{6}\int_0^2 (3x^2 + 9)\,dx$$

$$= \frac{1}{6}(x^3 + 9x)\Big|_0^2 = \frac{1}{6}(8 + 18 - 0)_0^2 = \frac{1}{6} \cdot 26 = \frac{13}{3}.$$

65. $f(x, y) = e^{2x+y}$; $1 \le x \le 2$, $2 \le y \le 3$

The area of region R is

$$A = (2 - 1)(3 - 2) = 1.$$

The average value of f over R is

$$\frac{1}{A}\iint_R e^{2x+y}\,dy\,dx = \iint_R e^{2x+y}\,dy\,dx = \int_1^2 \int_2^3 e^{2x+y}\,dy\,dx = \int_1^2 \left(e^{2x+y}\right)\Big|_2^3 dx$$

$$= \int_1^2 \left(e^{2x+3} - e^{2x+2}\right) dx = \frac{1}{2}\left(e^{2x+3} - e^{2x+2}\right)\Big|_1^2$$

$$= \frac{1}{2}(e^{4+3} - e^{2+3} - e^{4+2} + e^4) = \frac{e^7 - e^6 - e^5 + e^4}{2}.$$

67. $C(x,y) = \frac{1}{9}x^2 + 2x + y^2 + 5y + 100$

The area of region R is

$A(80-40)(70-30) = 1600.$

The average cost is

$$\frac{1}{A}\iint\limits_R C(x,y)\,dy\,dx = \frac{1}{1600}\int_{30}^{70}\int_{40}^{80}\left(\frac{1}{9}x^2 + 2x + y^2 + 5y + 100\right)dy\,dx$$

$$= \frac{1}{1600}\int_{30}^{70}\left(\frac{1}{27}x^3 + x^2 + xy^2 + 5xy + 100x\right)\Big|_{40}^{80} dy$$

$$= \int_{30}^{70}\left[\left(\frac{320}{27} + 4 + \frac{1}{20}y^2 + \frac{1}{4}y + 5\right) - \left(\frac{40}{27} + 1 + \frac{1}{40}y^2 + \frac{1}{8}y + \frac{5}{2}\right)\right] dy$$

$$= \int_{30}^{70}\left(\frac{1}{40}y^2 + \frac{1}{8}y + \frac{857}{54}\right) dy = \left(\frac{1}{120}y^3 + \frac{1}{16}y^2 + \frac{857}{54}y\right)\Big|_{30}^{70}$$

$$= \left(\frac{8575}{3} + \frac{1225}{4} + \frac{29{,}995}{27}\right) - \left(225 + \frac{225}{4} + \frac{4285}{9}\right) = \frac{94{,}990}{27} \approx 3518.$$

The average cost is about \$3518.

69. $P(x,y) = -(x-100)^2 - (y-50)^2 + 2000$

Area $= (150-100)(80-40) = (50)(40) = 2000$

The average weekly profit is

$$\frac{1}{2000}\iint\limits_R [-(x-100)^2 - (y-50)^2 + 2000]\,dy\,dx$$

$$= \frac{1}{2000}\int_{100}^{150}\int_{40}^{80} [-(x-100)^2 - (y-50)^2 + 2000]\,dy\,dx$$

$$= \frac{1}{2000}\int_{100}^{150}\left[-(x-100)^2 y - \frac{(y-50)^3}{3} + 2000y\right]\Big|_{40}^{80} dx$$

$$= \frac{1}{2000}\int_{100}^{150}\left[-(x-100)^2(80-40) - \frac{(80-50)^3}{3} + \frac{(40-50)^3}{3} + 2000(80-40)\right] dx$$

$$= \frac{1}{2000}\int_{100}^{150}\left[-40(x-100)^2 - \frac{30^3}{3} + \frac{(-10)^3}{3} + 2000(40)\right] dx$$

$$= \frac{1}{2000}\int_{100}^{150}\left[-40(x-100)^2 - \frac{28{,}000}{3} + 80{,}000\right] dx$$

$$= \frac{1}{2000}\cdot\left[\frac{-40(x-100)^3}{3} - \frac{28{,}000}{3}x + 80{,}000x\right]\Big|_{100}^{150}$$

$$= \frac{1}{2000}\left[-\frac{40}{3}(150-100)^3 + \frac{40}{3}(100-100)^3 - \frac{28{,}000}{3}(150-100) + 80{,}000(150-100)\right]$$

$$= \frac{1}{2000}\left[-\frac{40}{3}(50)^3 + \frac{40}{3}\cdot 0 - \frac{28{,}000}{3}(50) + 80{,}000(50)\right]$$

$$= \frac{1}{2000}\left(\frac{-5{,}000{,}000 - 1{,}400{,}000 + 12{,}000{,}000}{3}\right)$$

$$= \frac{1}{2000}\left(\frac{5{,}600{,}000}{3}\right) = \$933.33.$$

71. $T(x, y) = x^4 + 16y^4 - 32xy + 40$

Area $= (4 - 0)(2 - 0) = 8$

The average time is

$$\frac{1}{8}\iint_R (x^4 + 16y^4 - 32xy + 40)dy\,dx$$
$$= \frac{1}{8}\int_0^4 \int_0^2 (x^4 + 16y^4 - 32xy + 40)dy\,dx$$
$$= \frac{1}{8}\int_0^4 \left(x^4 y + \frac{16y^5}{5} - \frac{32xy^2}{2} + 40y\right)\bigg|_0^2 dx$$
$$= \frac{1}{8}\int_0^4 \left[x^4(2-0) + \frac{16(2-0)^5}{5} - \frac{32x(2-0)^2}{2} + 40(2-0)\right] dx$$
$$= \frac{1}{8}\int_0^4 \left(2x^4 + \frac{512}{5} - 64x + 80\right) dx$$
$$= \frac{1}{8}\left(\frac{2x^5}{5} + \frac{512}{5} - \frac{64x^2}{2} + 80\right)\bigg|_0^4$$
$$= \frac{1}{8}\left[\frac{2(4-0)^5}{5} + \frac{512}{5}(4-0) - \frac{64(4-0)^2}{2} + 80(4-0)\right]$$
$$= \frac{1}{8}\left(\frac{2048}{5} + \frac{2048}{5} - 512 + 320\right) = 78.4 \text{ hours}$$

Chapter 9 Review Exercises

5. $f(x, y) = 2x^2y^2 - 7x + 4y$
$f(-1, 2) = 2(1)(4) - 7(-1) + 4(2) = 23$
$f(6, -3) = 2(36)(9) - 7(6) + 4(-3) = 594$

7. $f(x, y) = \dfrac{\sqrt{x^2 + y^2}}{x - y}$

$$f(-1, 2) = \frac{\sqrt{1+4}}{-1-2} = -\frac{\sqrt{5}}{3}$$

$$f(6, -3) = \frac{\sqrt{36+9}}{6+3} = \frac{\sqrt{45}}{9} = \frac{\sqrt{5}}{3}$$

9. $x + 2y + 6z = 6$

x-intercept: $y = 0,\ z = 0$
$x = 6$

y-intercept: $x = 0,\ z = 0$
$2y = 6$
$y = 3$

z-intercept: $x = 0,\ y = 0$
$6z = 6$
$z = 1$

11. $4x + 3z = 12$

No y-intercept
x-intercept: $y = 0,\ z = 0$
$4x = 12$
$x = 3$

z-intercept: $x = 0,\ y = 0$
$3z = 12$
$z = 4$

13. $y = 4$

No x-intercept, no z-intercept
The graph is a plane parallel to the xz-plane.

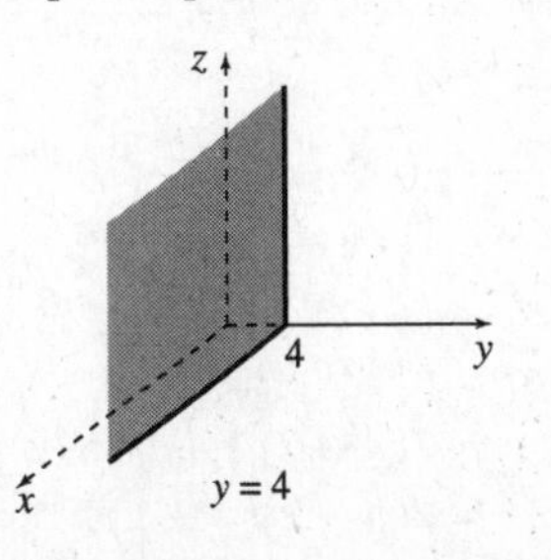

15. $z = f(x,y) = \dfrac{x+y^2}{x-y^2}$

(a) $$\frac{\partial z}{\partial y} = \frac{(x-y^2)\cdot 2y - (x+y^2)(-2y)}{(x-y^2)^2} = \frac{4xy}{(x-y^2)^2}$$

(b) $$\frac{\partial z}{\partial x} = \frac{(x-y^2)\cdot 1 - (x+y^2)\cdot 1}{(x-y^2)^2} = \frac{-2y^2}{(x-y^2)^2} = -2y^2(x-y^2)^{-2}$$

$$\left(\frac{\partial z}{\partial x}\right)(0,2) = \frac{-8}{(-4)^2} = -\frac{1}{2}$$

(c) $$f_{xx}(x,y) = 4y^2(x-y^2)^{-3} = \frac{4y^2}{(x-y^2)^3}$$

$$f_{xx}(-1,0) = \frac{0}{1} = 0$$

17. $$f(x,y) = 5x^4y^3 - 6x^5y$$
$$f_x(x,y) = 20x^3y^3 - 30x^4y$$
$$f_y(x,y) = 15x^4y^2 - 6x^5$$

19. $f(x,y) = \dfrac{2x+5y^2}{3x^2+y^2}$

$$f_x(x,y) = \frac{(3x^2+y^2)\cdot 2 - (2x+5y^2)\cdot 6x}{(3x^2+y^2)^2} = \frac{2y^2-6x^2-30xy^2}{(3x^2+y^2)^2}$$

$$f_y(x,y) = \frac{(3x^2+y^2)\cdot 10y - (2x+5y^2)\cdot 2y}{(3x^2+y^2)^2} = \frac{30x^2y-4xy}{(3x^2+y^2)^2}$$

21. $f(x,y) = (y-2)^2e^{x+2y}$

$$f_x(x,y) = (y-2)^2e^{x+2y}$$
$$f_y(x,y) = e^{x+2y}\cdot 2(y-2) + (y-2)^2\cdot 2e^{x+2y} = 2(y-2)[1+(y-2)]e^{x+2y} = 2(y-2)(y-1)e^{x+2y}$$

23. $f(x,y) = \ln|2-x^2y^3|$

$$f_x(x,y) = \frac{1}{2-x^2y^3}\cdot(-2xy^3) = \frac{-2xy^3}{2-x^2y^3}$$

$$f_y(x,y) = \frac{1}{2-x^2y^3}\cdot(-3x^2y^2) = \frac{-3x^2y^2}{2-x^2y^3}$$

25. $$f(x,y) = -3x^2y^3 + x^3y$$
$$f_x(x,y) = -6xy^3 + 3x^2y$$
$$f_{xx}(x,y) = -6y^3 + 6xy$$
$$f_{xy}(x,y) = -18xy^2 + 3x^2$$

27. $f(x,y) = \dfrac{3x+y}{x-1}$

$$f_x(x,y) = \frac{(x-1)\cdot 3 - (3x+y)\cdot 1}{(x-1)^2} = \frac{-3-y}{(x-1)^2} = (-3-y)(x-1)^{-2}$$

$$f_{xx}(x,y) = -2(-3-y)(x-1)^{-3} = \frac{2(3+y)}{(x-1)^3}$$

$$f_{xy}(x,y) = \frac{-1}{(x-1)^2}$$

29. $$f(x,y) = ye^{x^2}$$
$$f_x(x,y) = 2xye^{x^2}$$
$$f_{xx}(x,y) = 2xy\cdot 2xe^{x^2} + e^{x^2}\cdot 2y = 2ye^{x^2}(2x^2+1)$$
$$f_{xy}(x,y) = 2xe^{x^2}$$

31. $f(x,y) = \ln|1+3xy^2|$

$$f_x(x,y) = \frac{1}{1+3xy^2}\cdot 3y^2 = \frac{3y^2}{1+3xy^2} = 3y^2(1+3xy^2)^{-1}$$

$$f_{xx}(x,y) = 3y^2\cdot(-3y^2)(1+3xy^2)^{-2} = \frac{-9y^4}{(1+3xy^2)^2}$$

$$f_{xy}(x,y) = \frac{(1+3xy^2)\cdot 6y - 3y^2(6xy)}{(1+3xy^2)^2} = \frac{6y}{(1+3xy^2)^2}$$

33. $z = x^2 + y^2 + 9x - 8y + 1$

$z_x(x, y) = 2x + 9,\ z_y(x, y) = 2y - 8$

$$2x + 9 = 0$$
$$x = -\frac{9}{2}$$
$$2y - 8 = 0$$
$$y = 4$$

$z_{xx}(x, y) = 2,\ z_{yy}(x, y) = 2,\ z_{xy}(x, y) = 0$

$D = 2(2) - (0)^2 = 4 > 0$ and $z_{xx}(x, y) > 0$.

Relative minimum at $\left(-\frac{9}{2}, 4\right)$

35. $z = x^3 - 8y^2 + 6xy + 4$

$z_x(x, y) = 3x^2 + 6y,\ z_y(x, y) = -16y + 6x$

$$3x^2 + 6y = 0$$
$$x^2 + 2y = 0$$
$$y = -\frac{x^2}{2}$$
$$-16y + 6x = 0$$
$$-8y + 3x = 0$$

Substituting, we have

$$-8\left(-\frac{x^2}{2}\right) + 3x = 0$$
$$4x^2 + 3x = 0$$
$$x(4x + 3) = 0$$
$$x = 0 \quad \text{or} \quad x = -\frac{3}{4}$$
$$y = 0 \quad \text{or} \quad y = -\frac{9}{32}.$$

$z_{xx}(x, y) = 6x,\ z_{yy}(x, y) = -16,\ z_{xy}(x, y) = 6$

$D = 6x(-16) - (6)^2 = -96x - 36$

At $(0, 0)$, $D = -36 < 0$.
Saddle point at $(0, 0)$.

At $\left(-\frac{3}{4}, -\frac{9}{32}\right)$, $D = 36 > 0$ and $z_{xx}(x, y) = -\frac{9}{2} < 0$.

Relative maximum at $\left(-\frac{3}{4}, -\frac{9}{32}\right)$

37. $f(x, y) = 2x^2 + 4xy + 4y^2 - 3x + 5y - 15$

$$f_x(x, y) = 4x + 4y - 3$$
$$f_y(x, y) = 4x + 8y + 5$$

Solve the system $f_x(x, y) = 0, f_y(x, y) = 0$.

$$4x + 4y - 3 = 0$$
$$4x + 8y + 5 = 0$$

$$\begin{array}{r} -4x - 4y + 3 = 0 \\ 4x + 8y + 5 = 0 \\ \hline 4y + 8 = 0 \\ y = -2 \end{array}$$

$$4x + 4(-2) - 3 = 0$$
$$4x = 11$$
$$x = \frac{11}{4}$$

Therefore, $\left(\frac{11}{4}, -2\right)$ is a critical point.

$$f_{xx}(x, y) = 4$$
$$f_{yy}(x, y) = 8$$
$$f_{xy}(x, y) = 4$$
$$D = 4 \cdot 8 - 4^2 = 16 > 0$$

Since $f_{xx} = 4 > 0$, there is a relative minimum at $\left(\frac{11}{4}, -2\right)$.

39. $f(x, y) = 7x^2 + y^2 - 3x + 6y - 5xy$

$f_x(x, y) = 14x - 3 - 5y,\ f_y(x, y) = 2y + 6 - 5x$

$$\begin{array}{r} 14x - 5y - 3 = 0 \\ -5x + 2y + 6 = 0 \\ \hline \end{array}$$

$$\begin{array}{r} 28x - 10y - 6 = 0 \\ -25x + 10y + 30 = 0 \\ \hline 3x \qquad + 24 = 0 \\ x = -8 \end{array}$$

$$-5(-8) + 2y + 6 = 0$$
$$2y = -46$$
$$y = -23$$

$f_{xx}(x, y) = 14,\ f_{yy}(x, y) = 2,\ f_{xy}(x, y) = -5$

$D = 14(2) - (-5)^2 = 3 > 0$ and $f_{xx}(x, y) > 0$.

Relative minimum at $(-8, -23)$

41. $f(x, y) = x^2y;\ x + y = 4$

1. $g(x) = x + y - 4$

2. $F(x, y, \lambda) = x^2y - \lambda(x + y - 4)$

3. $F_x(x, y, \lambda) = 2xy - \lambda$
$F_y(x, y, \lambda) = x^2 - \lambda$
$F_\lambda(x, y, \lambda) = -(x + y - 4)$

4.
$$\begin{aligned} 2xy - \lambda &= 0 \quad (1) \\ x^2 - \lambda &= 0 \quad (2) \\ x + y - 4 &= 0 \quad (3) \end{aligned}$$

5.
$$\begin{aligned} \lambda &= 2xy \\ \lambda &= x^2 \\ 2xy &= x^2 \end{aligned}$$

$$\begin{aligned} 2xy - x^2 &= 0 \\ x(2y - x) &= 0 \\ x = 0 \quad \text{or} \quad 2y &= x \end{aligned}$$

Substituting into equation (3) gives

$$y = 4 \quad \text{or} \quad y = \frac{4}{3}.$$

If $y = \frac{4}{3}, x = \frac{8}{3}$.

The critical points are $(0, 4)$ and $\left(\frac{8}{3}, \frac{4}{3}\right)$.

$$f(0, 4) = 0$$

$$f\left(\frac{8}{3}, \frac{4}{3}\right) = \frac{64}{9} \cdot \frac{4}{3} = \frac{256}{27}$$

Therefore, f has a minimum of 0 at $(0, 4)$ and a maximum of $\frac{256}{27}$ at $\left(\frac{8}{3}, \frac{4}{3}\right)$.

43. Let x and y be the numbers such that $x + y = 80$ and $f(x, y) = x^2y$.

1. $g(x) = x + y - 80$

2. $F(x, y, \lambda) = x^2y - \lambda(x + y - 80)$

3. $F_x(x, y, \lambda) = 2xy - \lambda$
$F_y(x, y, \lambda) = x^2 - \lambda$
$F_\lambda(x, y, \lambda) = -(x + y - 80)$

4.
$$\begin{aligned} 2xy - \lambda &= 0 \quad (1) \\ x^2 - \lambda &= 0 \quad (2) \\ x + y - 80 &= 0 \quad (3) \end{aligned}$$

5.
$$\begin{aligned} \lambda &= 2xy \\ \lambda &= x^2 \\ 2xy &= x^2 \end{aligned}$$

$$\begin{aligned} 2xy - x^2 &= 0 \\ x(2y - x) &= 0 \\ x = 0 \quad \text{or} \quad x &= 2y \end{aligned}$$

Substituting into equation (3) gives

$$\begin{aligned} y = 80 \quad \text{or} \quad 2y + y - 80 &= 0 \\ 3y &= 80 \\ y &= \frac{80}{3} \\ \text{and} \quad x &= \frac{160}{3}. \end{aligned}$$

$$f(0, 80) = 0 \cdot 80^2 = 0$$

$$\begin{aligned} f\left(\frac{160}{3}, \frac{80}{3}\right) &= \frac{(160)^2}{9} \frac{(80)}{3} \\ &= \frac{2{,}048{,}000}{27} > f(0, 80) \end{aligned}$$

f has a maximum at $\left(\frac{160}{3}, \frac{80}{3}\right)$.

Therefore, if $x = \frac{160}{3}$ and $y = \frac{80}{3}$, then x^2y is maximized.

45. No, a maximum does not exist without the requirement that x and y are positive. Consider maximizing x^2y with $x + y = 80$. If $x < 0$, then $y = 80 - x > 80$. The values of y and x^2y will increase as the value of $|x|$ increases, so there is no maximum. A similar situation occurs by xy^2 when $x + y = 50$, by taking $y < 0$ and considering what happens to the values of x and xy^2 as $|y|$ increases.

47. $z = (x, y) = \dfrac{x + 5y}{x - 2y}$

$x = 1, y = -2, dx = -0.04, dy = 0.02$

$$\begin{aligned} f_x(x, y) &= \frac{(x - 2y)(1) - (x + 5y)(1)}{(x - 2y)^2} \\ &= \frac{-7y}{(x - 2y)^2} \end{aligned}$$

$$\begin{aligned} f_y(x, y) &= \frac{(x - 2y)(5) - (x + 5y)(-2)}{(x - 2y)^2} \\ &= \frac{7x}{(x - 2y)^2} \end{aligned}$$

$$\begin{aligned} dz &= \frac{-7y}{(x - 2y)^2}\, dx + \frac{7x}{(x - 2y)^2}\, dy \\ &= \frac{-7(-2)}{[1 - 2(-2)]^2}(-0.04) \\ &\quad + \frac{7(1)}{[1 - 2(-2)]^2}(0.02) \\ &= -0.0224 + 0.0056 = -0.0168 \end{aligned}$$

49. Let $z = f(x, y) = \sqrt{x}\, e^y$.
Then

$$dz = f_x(x,y)\,dx + f_y(x,y)\,dy$$
$$= \frac{1}{2}x^{-1/2}e^y\,dx + x^{1/2}e^y\,dy$$
$$= \frac{e^y}{2\sqrt{x}}\,dx + \sqrt{x}e^y\,dy.$$

To approximate $\sqrt{4.06}e^{0.04}$, let $x = 4$, $dx = 0.06$, $y = 0$, and $dy = 0.04$.

Therefore,

$$dz = \frac{e^0}{2\sqrt{4}}(0.06) + \sqrt{4}e^0(0.04)$$
$$dz = \frac{1}{4}(0.06) + 2(0.04)$$
$$dz = 0.095.$$

$$f(4.06, 0.04) = f(4,0) + \Delta z$$
$$\approx f(4,0) + dz$$
$$= \sqrt{4}e^0 + 0.095$$
$$f(4.06, 0.04) \approx 2.095$$

Using a calculator, $\sqrt{4.06}e^{0.04} \approx 2.0972$. The absolute value of the difference of the two results is $|2.095 - 2.0972| = 0.0022$.

51. $$\int_1^5 e^{3x+5y}\,dx = \frac{1}{3}e^{3x+5y}\Big|_1^5 = \frac{1}{3}(e^{15+5y} - e^{3+5y})$$

53. $$\int_1^3 y^2(7x+11y^3)^{-1/2}\,dy$$
$$= \frac{2}{33}(7x+11y^3)^{1/2}\Big|_1^3$$
$$= \frac{2}{33}[(7x+297)^{1/2} - (7x+11)^{1/2}]$$

55. $$\int_0^3\int_0^5 (2x+6y+y^2)\,dy\,dx$$
$$= \int_0^3 \left(2xy + 3y^2 + \frac{1}{3}y^3\right)\Big|_0^5\,dx$$
$$= \int_0^3 \left[\left(10x + 75 + \frac{125}{3}\right) - 0\right]dx$$
$$= \int_0^3 \left(10x + \frac{350}{3}\right)dx$$
$$= \left(5x^2 + \frac{350}{3}x\right)\Big|_0^3$$
$$= (45 + 350) - 0 = 395$$

57. (See Exercise 51.)

$$\int_1^2 \left[\int_3^5 (e^{2x-7y})\,dx\right] dy$$
$$= \int_1^2 \frac{1}{2}(e^{10-7y} - e^{6-7y})\,dy$$
$$= -\frac{1}{14}(e^{10-7y} - e^{6-7y})\Big|_1^2$$
$$= -\frac{1}{14}(e^{-4} - e^{-8} - e^3 + e^{-1})$$
$$= \frac{e^3 + e^{-8} - e^{-4} - e^{-1}}{14}$$

59. $$\int_1^2\int_1^2 \frac{dx\,dy}{x} = \int_1^2 \ln x\Big|_1^2\,dy$$
$$= \int_1^2 \ln 2\,dy$$
$$= y\ln 2\ \Big|_1^2$$
$$= 2\ln 2 - \ln 2$$
$$= \ln 2$$

61. $\iint_R \sqrt{2x+y}\,dx\,dy;\ 1 \le x \le 3,\ 2 \le y \le 5$

$$\iint_R \sqrt{2x+y}\,dx\,dy$$
$$= \int_1^3\int_2^5 (2x+y)^{1/2}\,dy\,dx$$
$$= \int_1^3 \frac{2}{3}(2x+y)^{3/2}\Big|_2^5\,dx$$
$$= \int_1^3 \frac{2}{3}[(2x+5)^{3/2} - (2x+2)^{3/2}]\,dx$$
$$= \frac{2}{15}[(2x+5)^{5/2} - (2x+2)^{5/2}]\Big|_1^3$$
$$= \frac{2}{15}(11^{5/2} - 8^{5/2} - 7^{5/2} + 4^{5/2})$$
$$= \frac{2}{15}(11^{5/2} - 8^{5/2} - 7^{5/2} + 32)$$

63. $\iint\limits_R ye^{y^2+x}\,dx\,dy; 0 \le x \le 1, 0 \le y \le 1$

$$\iint\limits_R ye^{y^2+x}\,dx\,dy$$
$$= \int_0^1 \int_0^1 ye^{y^2+x}\,dy\,dx$$
$$= \int_0^1 \frac{1}{2}e^{y^2+x}\Big|_0^1\,dx$$
$$= \int_0^1 \frac{1}{2}[e^{1+x} - e^x]\,dx$$
$$= \frac{1}{2}[e^{1+x} - e^x]\Big|_0^1$$
$$= \frac{1}{2}[e^2 - e - e + 1]$$
$$= \frac{e^2 - 2e + 1}{2}$$

65. $z = x^2 + y^2; 3 \le x \le 5, 2 \le y \le 4$

$$V = \iint\limits_R (x^2 + y^2)\,dy\,dx$$
$$= \int_3^5 \int_2^4 (x^2 + y^2)\,dy\,dx$$
$$= \int_3^5 \left[x^2y + \frac{y^3}{3}\right]\Big|_2^4\,dx$$
$$= \int_3^5 \left[4x^2 + \frac{64}{3} - 2x^2 - \frac{8}{3}\right]dx$$
$$= \int_3^5 \left(2x^2 + \frac{56}{3}\right)dx = \left(\frac{2x^3}{3} + \frac{56x}{3}\right)\Big|_3^5$$
$$= \frac{250}{3} + \frac{280}{3} - 18 - 56$$
$$= \frac{308}{3}$$

67. $\int_1^2 \int_2^{2x^2} y\,dy\,dx$

$$= \int_1^2 \frac{1}{2}y^2\Big|_2^{2x^2}\,dx$$
$$= \int_1^2 (2x^4 - 2)\,dx$$
$$= \left(\frac{2}{5}x^5 - 2x\right)\Big|_1^2$$
$$= \left(\frac{64}{5} - 4\right) - \left(\frac{2}{5} - 2\right)$$
$$= \frac{52}{5}$$

69. $\int_0^1 \int_y^{\sqrt{y}} x\,dx\,dy = \int_0^1 \frac{x^2}{2}\Big|_y^{\sqrt{y}}\,dy$

$$= \int_0^1 \frac{1}{2}(y - y^2)\,dy$$
$$= \frac{1}{2}\left(\frac{y^2}{2} - \frac{y^3}{3}\right)\Big|_0^1$$
$$= \frac{1}{2}\left(\frac{1}{2} - \frac{1}{3}\right) = \frac{1}{12}$$

71. $\int_0^8 \int_{x/2}^4 \sqrt{y^2+4}\,dy\,dx$

Change the order of integration.

$$\int_0^8 \int_{x/2}^4 \sqrt{y^2+4}\,dy\,dx$$
$$= \int_0^4 \int_0^{2y} \sqrt{y^2+4}\,dx\,dy$$
$$= \int_0^4 (y^2+4)^{1/2}\,x\Big|_0^{2y}\,dy$$
$$= \int_0^4 [(y^2+4)^{1/2}(2y)] - (y^2+4)(0)]\,dy$$
$$= \int_0^4 (y^2+4)^{1/2}(2y)\,dy$$
$$= \frac{(y^2+4)^{3/2}}{\frac{3}{2}}\Big|_0^4$$
$$= \frac{2}{3}(4^2+4)^{3/2} - \frac{2}{3}(0^2+4)^{3/2}$$
$$= \frac{2}{3}(20^{3/2} - 4^{3/2})$$
$$= \frac{2}{3}(20\sqrt{20} - 8)$$
$$= \frac{2}{3}(40\sqrt{5} - 8)$$
$$= \frac{16}{3}(5\sqrt{5} - 1)$$

73. $\iint_R (2 - x^2 - y^2)\,dy\,dx; 0 \le x \le 1, x^2 \le y \le x$

$$\int_0^1 \int_{x^2}^x (2 - x^2 - y^2)\,dy\,dx$$

$$= \int_0^1 \left(2y - x^2y - \frac{y^3}{3}\right)\Big|_{x^2}^x \,dx$$

$$= \int_0^1 \left(2x - x^3 - \frac{x^3}{3} - 2x^2 + x^4 + \frac{x^6}{3}\right) dx$$

$$= \int_0^1 \left(2x - 2x^2 - \frac{4x^3}{3} + x^4 + \frac{x^6}{3}\right) dx$$

$$= \left(x^2 - \frac{2x^3}{3} - \frac{x^4}{3} + \frac{x^5}{5} + \frac{x^7}{21}\right)\Big|_0^1$$

$$= 1 - \frac{2}{3} - \frac{1}{3} + \frac{1}{5} + \frac{1}{21} = \frac{26}{105}$$

75. $c(x, y) = 2x + y^2 + 4xy + 25$

(a) $c_x = 2 + 4y$
$c_x(640, 6) = 2 + 4(6)$
$= 26$

For an additional 1 MB of memory, the approximate change in cost is $26.

(b) $c_y = 2y + 4x$
$c_y(640, 6) = 2(6) + 4(640)$
$= 2572$

For an additional hour of labor, the approximate change in cost is $2572.

77. **(a)** Minimize $c(x, y)$

$$= x^2 + 5y^2 + 4xy - 70x - 164y + 1800$$

$c_x = 2x + 4y - 70$
$c_y = 10y + 4x - 164$

$$\begin{array}{r} 2x + 4y - 70 = 0 \\ \underline{4x + 10y - 164 = 0} \end{array}$$

$$\begin{array}{r} -4x - 8y + 140 = 0 \\ \underline{4x + 10y - 164 = 0} \\ 2y - 24 = 0 \\ y = 12 \end{array}$$

$$\begin{array}{r} 4x + 10(12) - 164 = 0 \\ 4x = 44 \\ x = 11 \end{array}$$

Extremum at $(11, 12)$

$c_{xx} = 2,\ c_{yy} = 10,\ c_{xy} = 4$

For $(11, 12)$,

$D = (2)(10) - 16 = 4 > 0$ and

$c_{xx}(11, 12) = 2 > 0.$

There is a relative minimum at $(11, 12)$.

(b) $c(11, 12)$

$$\begin{aligned} &= (11)^2 + 5(12)^2 + 4(11)(12) \\ &\quad - 70(11) - 164(12) + 1800 \\ &= 121 + 720 + 528 - 770 \\ &\quad - 1968 + 1800 \\ &= \$431 \end{aligned}$$

79. $V = \frac{1}{3}\pi r^2 h$

$r = 2$ cm, $h = 8$ cm
$dr = 0.21$ cm, $dh = 0.21$ cm

$$\begin{aligned} dV &= \frac{\pi}{3}(2rh\,dr + r^2\,dh) \\ &= \frac{\pi}{3}[2(2)(8)(0.21) + 4(0.21)] \\ &= \frac{\pi}{3}(6.72 + 0.84) \\ &= \frac{\pi}{3}(7.56) \\ &\approx 7.92\,\text{cm}^3 \end{aligned}$$

81. $V = \frac{1}{3}\pi r^2 h$

$r = 2.9\,\text{cm}$, $h = 11.4\,\text{cm}$
$dr = dh = 0.2\,\text{cm}$

$$\begin{aligned} dV &= \frac{\pi}{3}(2rh\,dr + r^2\,dh) \\ &= \frac{\pi}{3}[2(2.9)(11.4)(0.2) + (2.9)^2(0.2)] \\ &= \frac{\pi}{3}[13.224 + 1.682) \\ &= \frac{\pi}{3}(14.906) \\ &\approx 15.6\,\text{cm}^3 \end{aligned}$$

83. Assume that blood vessels are cylindrical.

$$\begin{aligned} V &= \pi r^2 h \\ r &= 0.7,\ h = 2.7 \\ dr &= dh = \pm.1 \\ dV &= 2\pi rh\,dr + \pi r^2\,dh \\ &= 2\pi(0.7)(2.7)(\pm 0.1) + \pi(0.7)^2(\pm 0.1) \\ &\approx \pm 1.341 \end{aligned}$$

The possible error is 1.341 cm^3.

85. $L(w,t) = (0.00082t + 0.0955)e^{(\ln w + 10.49)/2.842}$

(a) $L(450,4) = [0.00082(4) + 0.0955]e^{(\ln(450)+10.49)/2.842}$
≈ 33.982

The length is about 33.98 cm.

(b) $L_w(w,t) = (0.00082t + 0.0955)e^{(\ln w + 10.49)/2.842} \cdot \dfrac{1}{2.842w}$

$L_w(450,7) \approx 0.02723$

The approximate change in the length of a trout if its weight increases from 450 to 451 g while age is held constant at 7 yr is 0.027 cm.

$$L_t(w,t) = 0.00082e^{(\ln w + 10.49)/2.842}$$
$$L_t(450,7) \approx 0.2821$$

The approximate change in the length of a trout if its age increases from 7 to 8 yr while weight is held constant at 450 g is 0.28 cm.

87. $f(a,b) = \dfrac{1}{4}b\sqrt{4a^2 - b^2}$

(a) $f(3,2) = \dfrac{1}{4}(2)\sqrt{4(3)^2 - 2^2}$
$= \dfrac{1}{2}\sqrt{32}$
$= 2\sqrt{2}$
≈ 2.828

The area of the bottom of the planter is approximately 2.828 ft^2.

(b) $A = \dfrac{1}{4}b\sqrt{4a^2 - b^2}$

$$dA = \frac{1}{4}b \cdot \frac{1}{2}(4a^2 - b^2)^{-1/2}(8a)da + \left[\frac{1}{4}b \cdot \frac{1}{2}(4a^2 - b^2)^{-1/2}(-2b) + \frac{1}{4}(4a^2 - b^2)^{1/2}\right] db$$

$$dA = \frac{ab}{\sqrt{4a^2 - b^2}} da + \frac{1}{4}\left(\frac{-b^2}{\sqrt{4a^2 - b^2}} + \sqrt{4a^2 - b^2}\right) db$$

If $a = 3$, $b = 2$, $da = 0$, and $db = 0.5$,

$$dA = \frac{1}{4}\left(\frac{-2^2}{\sqrt{4(3)^2 - 2^2}} + \sqrt{4(3)^2 - 2^2}\right)(0.5)$$

$dA \approx 0.6187.$

The approximate effect on the area is an increase of 0.6187 ft^2.

89. Maximize $f(x,y) = xy$, subject to $2x + y = 400$.

1. $g(x,y) = 2x + y - 400$

2. $F(x,y,\lambda) = xy - \lambda(2x + y - 400)$

3. $F_x = y - 2\lambda$
$F_y = x - \lambda$
$F_\lambda = -(2x + y - 400)$

4.
$$y - 2\lambda = 0$$
$$x - \lambda = 0$$
$$2x + y - 400 = 0$$

5. $\lambda = \dfrac{y}{2}$, $\lambda = x$

$$\frac{y}{2} = x$$
$$y = 2x$$

Substituting into $2x + y - 400$, we have

$$2x + 2x - 400 = 0,$$

so $x = 100$, $y = 200$.

Dimensions are 100 feet by 200 feet for maximum area of 20,000 ft^2.

Chapter 9 Test

[9.1]

1. Find $f(-2, 1)$ for $f(x, y) = 2x^2 - 4xy + 7y$.

2. Find $g(-1, 3)$ for $g(x, y) = \sqrt{2x^2 - xy^2}$.

3. Complete the ordered triples $(0, 0, \quad)$, $(0, \quad, 0)$ and $(\quad, 0, 0)$ for the plane $2x - 4y + 8z = 8$.

4. Graph the first octant portion of the plane $4x + 2y + 8z = 8$.

5. Let $f(x, y) = x^2 + 2y^2$. Find $\dfrac{f(x+h, y) - f(x, y)}{h}$.

[9.2]

6. Let $z = f(x, y) = 3x^3 - 5x^2y + 4y^2$. Find each of the following.

(a) $\dfrac{\partial z}{\partial x}$ **(b)** $\dfrac{\partial z}{\partial y}(1, 1)$ **(c)** f_{xx}

7. Let $f(x, y) = \dfrac{x}{2x + y^2}$. Find each of the following.

(a) $f_x(1, -1)$ **(b)** $f_y(2, 1)$

8. Let $f(x, y) = \sqrt{2x^2 - y^2}$. Find each of the following.

(a) f_x **(b)** f_y **(c)** f_{xy}

9. Let $f(x, y) = xe^{y^2}$. Find each of the following.

(a) f_x **(b)** f_y **(c)** f_{yx}

10. Let $f(x, y) = \ln\left(2x^2y^2 + 1\right)$. Find each of the following.

(a) f_{xx} **(b)** f_{xy} **(c)** f_{yy}

11. The production function for a certain country is

$$z = 2x^5y^4,$$

where x represents the amount of labor and y the amount of capital. Find the marginal productivity of

(a) labor; **(b)** capital.

[9.3]

Find all points where the functions defined below have any relative extrema. Find any saddle points.

12. $z = 4x^2 + 2y^2 - 8x$

13. $z = 2x^2 - 4xy + y^3$

14. $z = 1 - 2y - x^2 - 2xy - 2y^2$

15. A company manufactures two calculator models. The total revenue from x thousand solar calculators and y thousand battery-operated calculators is

$$R(x, y) = 4000 - 5y^2 - 8x^2 - 2xy + 42y + 102x.$$

Find x and y so that revenue is maximized.

[9.4]

16. Use Lagrange multipliers to find extrema of $f(x, y) = x^2 - 6xy$, subject to the constraint $x + y = 7$.

17. Find two numbers whose sum is 30 such that x^2y is maximized.

18. A closed box with square ends must have a volume of 64 cubic inches. Find the dimensions of such a box that has minimum surface area.

[9.5]

19. Use the total differential to approximate the quantity $\sqrt{3.05^2 + 4.02^2}$. Then use a calculator to approximate the quantity and give the absolute value of the difference in the two results to four decimal places.

20. **(a)** Find dz for $z = e^{x+y} \ln xy$.

(b) Evaluate dz when $x = 1,\ y = 1,\ dx = 0.02$ and $dy = 0.01$.

21. A sphere of radius 3 feet is to receive an insulating coating $\frac{1}{2}$ inch thick. Use differentials to approximate the volume of the coating needed.

[9.6]

22. Evaluate $\displaystyle\int_0^4 \int_1^4 \sqrt{x + 3y}\, dx\, dy$.

23. Find the volume under the surface $z = 2xy$ and above the rectangle with boundaries $0 \le x \le 1,\ 0 \le y \le 4$.

24. Evaluate $\displaystyle\int_0^2 \int_{y/2}^3 (x + y)\ dx\, dy$.

25. Use the region R with boundaries $0 \le y \le 2$ and $0 \le x \le y$ to evaluate

$$\iint_R (6 - x - y)\ dx\, dy.$$

Chapter 9 Test Answers

1. 23

2. $\sqrt{11}$

3. $(0,0,1)$, $(0,-2,0)$, $(4,0,0)$

4.

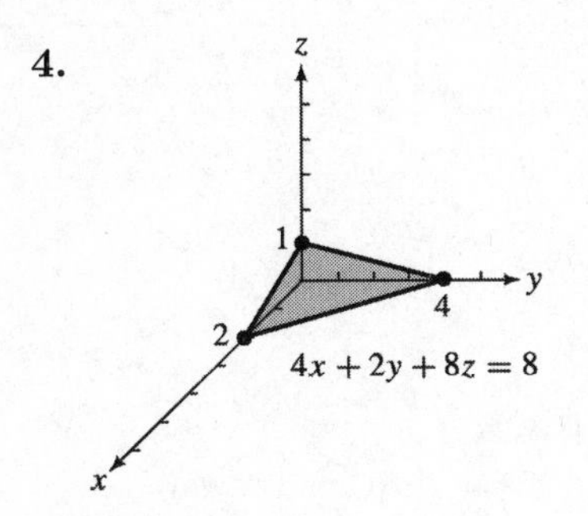

5. $2x + h$

6. **(a)** $9x^2 - 10xy$ **(b)** 3 **(c)** $18x - 10y$

7. **(a)** $\frac{1}{9}$ **(b)** $\frac{-4}{25}$

8. **(a)** $\frac{2x}{\sqrt{2x^2-y^2}}$ **(b)** $\frac{-y}{\sqrt{2x^2-y^2}}$ **(c)** $\frac{2xy}{(2x^2-y^2)^{3/2}}$

9. **(a)** e^{y^2} **(b)** $2xye^{y^2}$ **(c)** $2ye^{y^2}$

10. **(a)** $\frac{-8x^2y^4+4y^2}{(2x^2y^2+1)^2}$ **(b)** $\frac{8xy}{(2x^2y^2+1)^2}$ **(c)** $\frac{4x^2-8x^4y^2}{(2x^2y^2+1)^2}$

11. **(a)** $10x^4y^4$ **(b)** $8x^5y^3$

12. Relative minimum of -4 at $(1,0)$

13. Relative minimum of $-\frac{32}{27}$ at $\left(\frac{4}{3}, \frac{4}{3}\right)$, saddle point at $(0,0)$

14. Relative maximum of 2 at $(1,-1)$

15. 6 thousand solar calculators and 3 thousand battery-operated calculators

16. Minimum of -63 at $(3,4)$

17. $x = 20,\ y = 10$

18. 4 inches by 4 inches by 4 inches

19. 5.046; 5.0461; 0.0001

20. **(a)** $\left(e^{x+y}\ln(xy) + \frac{e^{x+y}}{x}\right) dx + \left(e^{x+y}\ln xy + \frac{e^{x+y}}{y}\right) dy$ **(b)** $0.03e^2 \approx 0.222$

21. 4.71 cu ft

22. $\frac{4}{45}\left(993 - 13^{5/2}\right)$

23. 8

24. $\frac{40}{3}$

25. 8

Chapter 10

DIFFERENTIAL EQUATIONS

10.1 Solutions of Elementary and Separable Differential Equations

1. $\frac{dy}{dx} = -4x + 6x^2$

$$y = \int (-4x + 6x^2)\,dx$$
$$= -2x^2 + 2x^3 + C$$

3. $4x^3 - 2\frac{dy}{dx} = 0$

Solve for $\frac{dy}{dx}$.

$$\frac{dy}{dx} = 2x^3$$
$$y = 2\int x^3\,dx$$
$$= 2\left(\frac{x^4}{4}\right) + C$$
$$= \frac{x^4}{2} + C$$

5. $y\frac{dy}{dx} = x^2$

Separate the variables and take antiderivatives.

$$\int y\,dy = \int x^2\,dx$$
$$\frac{y^2}{2} = \frac{x^3}{3} + K$$
$$y^2 = \frac{2}{3}x^3 + 2K$$
$$y^2 = \frac{2}{3}x^3 + C$$

7. $\frac{dy}{dx} = 2xy$

$$\int \frac{dy}{y} = \int 2x\,dx$$
$$\ln|y| = \frac{2x^2}{2} + C$$
$$\ln|y| = x^2 + C$$
$$e^{\ln|y|} = e^{x^2+C}$$
$$y = \pm e^{x^2+C}$$
$$y = \pm e^{x^2} \cdot e^C$$
$$y = ke^{x^2}$$

9. $\frac{dy}{dx} = 3x^2y - 2xy$

$$\frac{dy}{dx} = y(3x^2 - 2x)$$
$$\int \frac{dy}{y} = \int (3x^2 - 2x)\,dx$$
$$\ln|y| = \frac{3x^3}{3} - \frac{2x^2}{2} + C$$
$$e^{\ln|y|} = e^{x^3-x^2+C}$$
$$y = \pm(e^{x^3-x^2})e^C$$
$$y = ke^{x^3-x^2}$$

11. $\frac{dy}{dx} = \frac{y}{x},\ x > 0$

$$\int \frac{dy}{y} = \int \frac{dx}{x}$$
$$\ln|y| = \ln x + C$$
$$e^{\ln|y|} = e^{\ln x + C}$$
$$y = \pm e^{\ln x} \cdot e^C$$
$$y = Me^{\ln x}$$
$$y = Mx$$

13. $\frac{dy}{dx} = y - 6$

$$\int \frac{dy}{y-6} = \int dx$$
$$\ln|y-6| = x + C$$
$$e^{\ln|y-6|} = e^{x+C}$$
$$y - 6 = \pm e^x \cdot e^C$$
$$y - 6 = Me^x$$
$$y = Me^x + 6$$

15. $\frac{dy}{dx} = y^2e^{2x}$

$$\int y^{-2}\,dy = \int e^{2x}\,dx$$
$$-y^{-1} = \frac{1}{2}e^2 + C$$
$$-\frac{1}{y} = \frac{1}{2}e^2 + C$$
$$y = \frac{-1}{\frac{1}{2}e^2 + C}$$

17. $\dfrac{dy}{dx} + 3x^2 = 2x$

$$\frac{dy}{dx} = 2x - 3x^2$$

$$y = \frac{2x^2}{2} - \frac{3x^3}{3} + C$$

$$y = x^2 - x^3 + C$$

Since $y = 5$ when $x = 0$,

$$5 = 0 - 0 + C$$
$$C = 5.$$

Thus,

$$y = x^2 - x^3 + 5.$$

19. $2\dfrac{dy}{dx} = 4xe^{-x}$

$$\frac{dy}{dx} = 2xe^{-x}$$

Use the table of integrals or integrate by parts.

$$y = 2(-x-1)e^{-x} + C$$

Since $y = 42$ when $x = 0$,

$$42 = 2(0-1)(1) + C$$
$$42 = -2 + C$$
$$C = 44.$$

Thus,

$$y = -2xe^{-x} - 2e^{-x} + 44.$$

21. $\dfrac{dy}{dx} = \dfrac{x^3}{y}$; $y = 5$ when $x = 0$.

$$\int y\,dy = \int x^3\,dx$$

$$\frac{y^2}{2} = \frac{x^4}{4} + C$$

$$y^2 = \frac{1}{2}x^4 + 2C$$

$$y^2 = \frac{1}{2}x^4 + k$$

Since $y = 5$ when $x = 0$,

$$25 = 0 + k$$
$$k = 25.$$

So $y^2 = \dfrac{1}{2}x^4 + 25.$

23. $(2x+3)y = \dfrac{dy}{dx}$; $y = 1$ when $x = 0$.

$$\int (2x+3)\,dx = \int \frac{dy}{y}$$

$$\frac{2x^2}{2} + 3x + C = \ln|y|$$

$$e^{x^2+3x+C} = e^{\ln|y|}$$

$$y = (e^{x^2+3x})(\pm e^C)$$

$$y = ke^{x^2+3x}$$

Since $y = 1$ when $x = 0$.

$$1 = ke^{0+0}$$
$$k = 1.$$

So $y = e^{x^2+3x}$.

25. $\dfrac{dy}{dx} = 4x^3 - 3x^2 + x$; $y = 0$ when $x = 1$.

$$y = \int (4x^3 - 3x^2 + x)\,dx$$

$$= x^4 - x^3 + \frac{x^2}{2} + C$$

Substitute.

$$0 = 1 - 1 + \frac{1}{2} + C$$

$$-\frac{1}{2} = C$$

$$y = x^4 - x^3 + \frac{x^2}{2} - \frac{1}{2}$$

27. $\dfrac{dy}{dx} = \dfrac{y^2}{x}$; $y = 3$ when $x = e$.

$$\int y^{-2}\,dy = \int \frac{dx}{x}$$

$$-y^{-1} = \ln|x| + C$$

$$-\frac{1}{y} = \ln|x| + C$$

$$y = \frac{-1}{\ln|x| + C}$$

Since $y = 3$ when $x = e$,

$$3 = \frac{-1}{\ln e + C}$$

$$3 = \frac{-1}{1 + C}$$

$$3 + 3C = -1$$
$$3C = -4$$
$$C = -\frac{4}{3}.$$

So $$y = \frac{-1}{\ln|x| - \frac{4}{3}} = \frac{-3}{3\ln|x| - 4}.$$

29. $\frac{dy}{dx} = (y-1)^2 e^{x-1}$; $y = 2$ when $x = 1$.

$$\frac{dy}{(y-1)^2} = e^{x-1}\,dx$$

$$\int (y-1)^{-2}\,dy = \int e^{x-1}\,dx$$

$$\frac{(y-1)^{-1}}{-1} = e^{x-1} + C$$

$$-\frac{1}{y-1} = e^{x-1} + C$$

$$-(y-1) = \frac{1}{e^{x-1}+C}$$

$$-y+1 = \frac{1}{e^{x-1}+C}$$

$$1 - \frac{1}{e^{x-1}+C} = y$$

$$y = \frac{e^{x-1}+C}{e^{x-1}+C} - \frac{1}{e^{x-1}+C}$$

$$y = \frac{e^{x-1}+C-1}{e^{x-1}+C}$$

$y = 2$, when $x = 1$.

$$2 = \frac{e^0+C-1}{e^0+C}$$

$$2 = \frac{C}{1+C}$$

$$2 + 2C = C$$

$$C = -2$$

$$y = \frac{e^{x-1}-3}{e^{x-1}-2}.$$

31. $\frac{dy}{dx} = \frac{k}{N}(N-y)y$

(a) $\frac{N\,dy}{(N-y)y} = k\,dx$

Since $\frac{1}{y} + \frac{1}{N-y} = \frac{N}{(N-y)y}$,

$$\int \frac{dy}{y} + \int \frac{dy}{N-y} = k\,dx$$

$$\ln\left|\frac{y}{N-y}\right| = kx + C$$

$$\frac{y}{N-y} = Ce^{kx}.$$

For $0 < y < N$, $Ce^{kx} > 0$.
For $0 < N < y$, $Ce^{kx} < 0$.
Solve for y.

$$y = \frac{Ce^{kx}N}{1+Ce^{kx}} = \frac{N}{1+C^{-1}e^{-kx}}$$

Let $b = C^{-1} > 0$ for $0 < y < N$.

$$y = \frac{N}{1+be^{-kx}}$$

Let $-b = C^{-1} < 0$ for $0 < N < y$.

$$y = \frac{N}{1-be^{-kx}}$$

(b) For $0 < y < N$; $t = 0$, $y = y_0$.

$$y_0 = \frac{N}{1+be^0} = \frac{N}{1+b}$$

Solve for b.

$$b = \frac{N-y_0}{y_0}$$

(c) For $0 < N < y$; $t = 0$, $y = y_0$.

$$y_0 = \frac{N}{1-be^0} = \frac{N}{1-b}$$

Solve for b.

$$b = \frac{y_0-N}{y_0}$$

33. **(a)** $0 < y_0 < N$ implies that $y_0 > 0$, $N > 0$, and $N - y_0 > 0$.

Therefore,

$$b = \frac{N-y_0}{y_0} > 0.$$

Also, $e^{-kx} > 0$ for all x, which implies that $1 + be^{-kx} > 1$.

(1) $y(x) = \frac{N}{1+be^{-kx}} < N$ since $1 + be^{-kx} > 1$.

(2) $y(x) = \frac{N}{1+be^{-kx}} > 0$ since $N > 0$ and $1 + be^{-kx} > 0$.

Combining statements (1) and (2), we have

$$0 < \frac{N}{1+be^{-kx}} = y(x)$$

$$= \frac{N}{1+be^{-kx}} < N$$

or $0 < y(x) < N$ for all x.

(b) $\lim\limits_{x\to\infty} \frac{N}{1+be^{-kx}} = \frac{N}{1+b(0)} = N$

$$\lim_{x\to-\infty} \frac{N}{1+be^{-kx}} = 0$$

Note that as $x \to -\infty$, $1+be^{-kx}$ becomes infinitely large.
Therefore, the horizontal asymptotes are $y = N$ and $y = 0$.

(c) $y'(x) = \dfrac{(1+be^{-kx})(0) - N(-kbe^{-kx})}{(1+be^{-kx})^2}$

$= \dfrac{Nkbe^{-kx}}{(1+be^{-kx})^2} > 0$ for all x.

Therefore, $y(x)$ is an increasing function.

(d) To find $y''(x)$, apply the quotient rule to find the derivative of $y'(x)$. The numerator of $y''(x)$ is

$$\begin{aligned} y''(x) &= (1+be^{-kx})^2(-Nk^2be^{-kx}) \\ &\quad - Nkbe^{-kx}[-2kbe^{-kx}(1+be^{-kx})] \\ &= -Nkbe^{-kx}(1+be^{-kx}) \\ &\quad \cdot [k(1+be^{-kx}) - 2kbe^{-kx}] \\ &= -Nkbe^{-kx}(1+be^{-kx})(k - kbe^{-kx}), \end{aligned}$$

and the denominator is

$$[(1+be^{-kt})^2]^2 = (1+be^{-kx})^4.$$

Thus,

$$y''(x) = \frac{-Nkbe^{-kx}(k - kbe^{-kx})}{(1+be^{-kx})^3}.$$

$y''(x) = 0$ when

$$\begin{aligned} k - kbe^{-kx} &= 0 \\ be^{-kx} &= 1 \\ e^{-kx} &= \frac{1}{b} \\ -kx &= \ln\left(\frac{1}{b}\right) \\ x &= -\frac{\ln\left(\frac{1}{b}\right)}{k} \\ &= \frac{\ln\left(\frac{1}{b}\right)^{-1}}{k} = \frac{\ln b}{k}. \end{aligned}$$

When $x = \frac{\ln b}{k}$,

$$\begin{aligned} y &= \frac{N}{1+be^{-k\left(\frac{\ln b}{k}\right)}} = \frac{N}{1+be^{(-\ln b)}} \\ &= \frac{N}{1+be^{\ln(1/b)}} = \frac{N}{1+b\left(\frac{1}{b}\right)} = \frac{N}{2}. \end{aligned}$$

Therefore, $\left(\frac{\ln b}{k}, \frac{N}{2}\right)$ is a point of inflection.

(e) To locate the maximum of $\frac{dy}{dx}$, we must consider, from part (d),

$$\frac{d}{dx}\left(\frac{dy}{dx}\right) = \frac{-Nkbe^{-kx}(k - kbe^{-kx})}{(1+be^{-kx})^3}.$$

Since $y''(x) > 0$ for $x < \dfrac{\ln b}{k}$ and

$y''(x) < 0$ for $x > \dfrac{\ln b}{k}$,

we know that $x = \frac{\ln b}{k}$ locates a relative maximum of $\frac{dy}{dx}$.

35. $\dfrac{dy}{dx} = \dfrac{100}{32 - 4x}$

$$y = 100\left(-\frac{1}{4}\right)\ln|32 - 4x| + C$$

$$y = -25\ln|32 - 4x| + C$$

Now, $y = 1000$ when $x = 0$.

$$\begin{aligned} 1000 &= -25\ln|32| + C \\ C &= 1000 + 25\ln 32 \\ C &\approx 1086.64 \end{aligned}$$

Thus,

$$y = -25\ln|32 - 4x| + 1086.64.$$

(a) Let $x = 3$.

$$\begin{aligned} y &= -25\ln|32 - 12| + 1086.64 \\ &\approx \$1011.75 \end{aligned}$$

(b) Let $x = 5$.

$$\begin{aligned} y &= -25\ln|32 - 20| + 1086.64 \\ &\approx \$1024.52 \end{aligned}$$

(c) Advertising expenditures can never reach \$8000. If $x = 8$, the denominator becomes zero.

37. $\dfrac{dy}{dt} = -0.05y$

See Example 4.

$$\begin{aligned} \int \frac{dy}{y} &= \int -0.05\,dt \\ \ln|y| &= -0.05t + C \\ e^{\ln|y|} &= e^{-0.05t + C} \\ e^{\ln|y|} &= e^{-0.05t} \cdot e^C \\ |y| &= e^{-0.05t} \cdot e^C \\ y &= Me^{-0.05t} \end{aligned}$$

Let $y = 1$ when $t = 0$.

Solve for M:

$$\begin{aligned} 1 &= Me^0 \\ M &= 1. \end{aligned}$$

So $y = e^{-0.05t}$.

If $y = 0.50$,

$$\begin{aligned} 0.50 &= e^{-0.05t} \\ \ln 0.5 &= -0.05t \ln e \\ t &= \frac{-\ln 0.5}{0.05} \\ &\approx 13.9 \end{aligned}$$

It will take about 13.9 years for \$1 to lose half its value.

39. $E = -\frac{p}{q} \cdot \frac{dq}{dp}$ with $p > 0$ and $q > 0$

If $E = 2$,

$$\begin{aligned} 2 &= -\frac{p}{q} \cdot \frac{dq}{dp} \\ \frac{2}{p}\, dp &= -\frac{1}{q}\, dq \\ \int \frac{2}{p}\, dp &= -\int \frac{1}{q}\, dq \\ 2 \ln p &= -\ln q + K \\ \ln p^2 + \ln q &= K \\ \ln (p^2 q) &= K \\ p^2 q &= e^K \\ p^2 q &= C \\ q &= \frac{C}{p^2}. \end{aligned}$$

41.
$$\begin{aligned} \frac{dA}{dt} &= Ai \\ \frac{dA}{A} &= i\, dt \\ \int \frac{dA}{A} &= \int i\, dt \\ \ln A &= it + C \\ e^{\ln A} &= e^{it+C} \\ A &= Me^{it} \end{aligned}$$

When $t = 0$, $A = 5000$. Therefore, $M = 5000$. Find i so that $A = 20{,}000$ when $t = 24$.

$$\begin{aligned} 20{,}000 &= 5000e^{24i} \\ 4 &= e^{24i} \\ \ln 4 &= 24i \\ i &= \frac{\ln 4}{24} \\ &= \frac{2 \ln 2}{24} \\ &= \frac{\ln 2}{12} \end{aligned}$$

The answer is d.

43. (a) $\frac{dI}{dW} = 0.088(2.4 - I)$

Separate the variables and take antiderivatives.

$$\begin{aligned} \int \frac{dI}{2.4 - I} &= \int 0.088 dW \\ -\ln |2.4 - I| &= 0.088W + k \end{aligned}$$

Solve for I.

$$\begin{aligned} \ln |2.4 - I| &= -0.088W - k \\ |2.4 - I| &= e^{-0.088W - k} = e^{-k} e^{-0.088W} \\ I - 2.4 &= Ce^{-0.088W}, \text{ where } C = \pm e^{-k}. \\ I &= 2.4 + Ce^{-0.088W} \end{aligned}$$

Since $I(0) = 1$, then

$$\begin{aligned} 1 &= 2.4 + Ce^0 \\ C &= 1 - 2.4 = -1.4. \end{aligned}$$

Therefore, $I = 2.4 - 1.4e^{-0.088W}$.

(b) Note that as W gets larger and larger $e^{-0.088W}$ approaches 0, so

$$\begin{aligned} \lim_{W \to \infty} I &= \lim_{W \to \infty} (2.4 - 1.4e^{-0.088W}) \\ &= 2.4 - 1.4(0) = 2.4, \end{aligned}$$

so I approaches 2.4.

45. (a) $\frac{dw}{dt} = k(C - 17.5w)$

C being constant implies that the calorie intake per day is constant.

(b) pounds/day $= k$ (calories/day)

$$\frac{\text{pounds/day}}{\text{calories/day}} = k$$

The units of k are pounds/calorie.

(c) Since 3500 calories is equivalent to 1 pound, $k = \frac{1}{3500}$ and

$$\frac{dw}{dt} = \frac{1}{3500}(C - 17.5w).$$

(d) $\frac{dw}{dt} = \frac{1}{3500}(C - 17.5w)$; $w = w_0$ when $t = 0$.

$$\frac{3500}{C - 17.5w}\,dw = dt$$
$$\frac{3500}{-17.5}\int \frac{-17.5}{C - 17.5w}\,dw = \int dt$$
$$-200 \ln |C - 17.5w| = t + k$$
$$\ln |C - 17.5w| = -0.005t - 0.005k$$
$$|C - 17.5w| = e^{-0.005t - 0.005k}$$
$$|C - 17.5w| = e^{-0.005t} \cdot e^{-0.005k}$$
$$C - 17.5w = e^{-0.005M}e^{-0.005t}$$
$$-17.5w = -C + e^{-0.005M}e^{-0.005t}$$
$$w = \frac{C}{17.5} - \frac{e^{-0.005M}}{17.5}e^{-0.005t}$$

(e) Since $w = w_0$ when $t = 0$,

$$w_0 = \frac{C}{17.5} - \frac{e^{-0.005M}}{17.5}(1)$$
$$w_0 - \frac{C}{17.5} = -\frac{e^{-0.005M}}{17.5}$$
$$\frac{e^{-0.005M}}{17.5} = \frac{C}{17.5} - w_0.$$

Therefore,

$$w = \frac{C}{17.5} - \left(\frac{C}{17.5} - w_0\right)e^{-0.005t}$$
$$w = \frac{C}{17.5} + \left(w_0 - \frac{C}{17.5}\right)e^{-0.005t}.$$

47. (a)

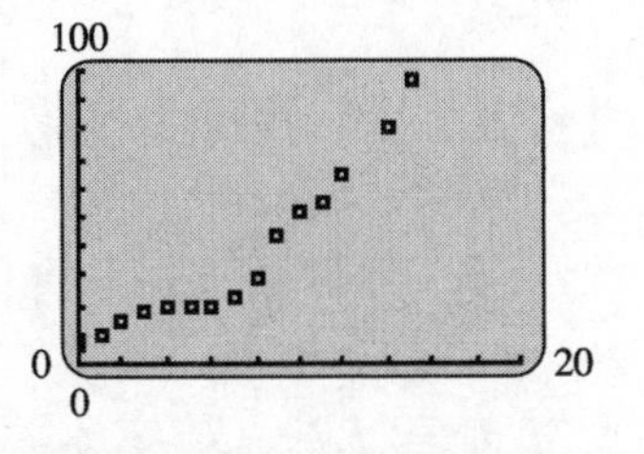

(b) The logistic regression equation is

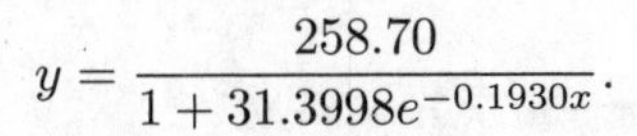

$$y = \frac{258.70}{1 + 31.3998e^{-0.1930x}}.$$

(c) The logistic equation fits the data well.

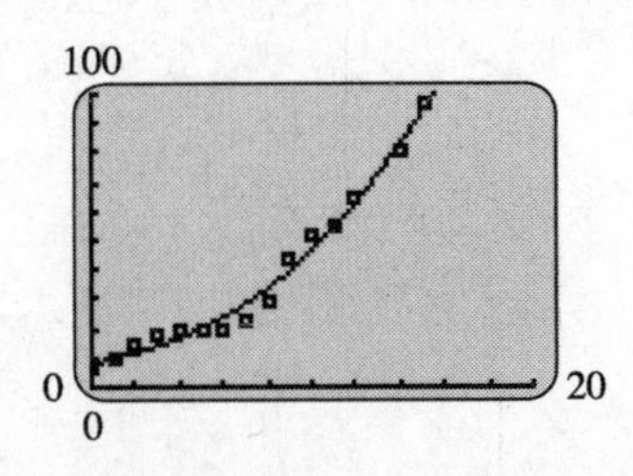

(d) As x gets very large, the value of the function approaches 259 (rounded to the nearest whole number).

49. $\frac{dy}{dt} = ky$

First separate the variables and integrate.

$$\frac{dy}{y} = k\,dt$$
$$\int \frac{dy}{y} = \int k\,dt$$
$$\ln |y| = kt + C.$$

Solve for y.

$$|y| = e^{kt + C_1} = e^{C_1}e^{kt}$$
$$y = Ce^{kt}, \text{ where } C = \pm e^{C_1}.$$
$$y(0) = 35.6, \text{ so } 35.6 = Ce^0 = C, \text{ and}$$
$$y = 35.6e^{kt}.$$

Since $y(50) = 102.6$, then $102.6 = 35.6e^{50k}$.

Solve for k.

$$e^{50k} = \frac{102.6}{35.6}$$
$$50k = \ln\left(\frac{102.6}{35.6}\right)$$
$$k = \frac{\ln\left(\frac{102.6}{35.6}\right)}{50} \approx 0.02117, \text{ so}$$
$$y = 35.6e^{0.02117t}.$$

51. (a)

$$\frac{dy}{dt} = ky$$
$$\int \frac{dy}{y} = \int k\,dt$$
$$\ln |y| = kt + C$$
$$e^{\ln |y|} = e^{kt + C}$$
$$y = \pm(e^{kt})(e^C)$$
$$y = Me^{kt}$$

If $y = 1$ when $t = 0$ and $y = 5$ when $t = 2$, we have the system of equations

$$1 = Me^{k(0)}$$
$$5 = Me^{2k}.$$

$$1 = M(1)$$
$$M = 1$$

Substitute.

$$\begin{aligned} 5 &= (1)e^{2k} \\ e^{2k} &= 5 \\ 2k \ln e &= \ln 5 \\ k &= \frac{\ln 5}{2} \\ &\approx 0.8 \end{aligned}$$

(b) If $k = 0.8$ and $M = 1$,

$$y = e^{0.8t}.$$

When $t = 3$,

$$\begin{aligned} y &= e^{0.8(3)} \\ &= e^{2.4} \\ &\approx 11. \end{aligned}$$

(c) When $t = 5$,

$$\begin{aligned} y &= e^{0.8(5)} \\ &= e^{4} \\ &\approx 55. \end{aligned}$$

(d) When $t = 10$,

$$\begin{aligned} y &= e^{0.8(10)} \\ &= e^{8} \\ &\approx 3000. \end{aligned}$$

53. $\dfrac{dy}{dx} = 7.5e^{-0.3y}$, $y = 0$ when $x = 0$.

$$\begin{aligned} e^{0.3y}\,dy &= 7.5\,dx \\ \int e^{0.3y}\,dy &= \int 7.5\,dx \\ \frac{e^{0.3y}}{0.3} &= 7.5x + C \\ e^{0.3y} &= 2.25x + C \\ 1 &= 0 + C = C \\ e^{0.3y} &= 2.25x + 1 \\ 0.3y &= \ln(2.25x + 1) \\ y &= \frac{\ln(2.25x + 1)}{0.3} \end{aligned}$$

When $x = 8$,

$$\begin{aligned} y &= \frac{\ln[2.25(8) + 1]}{0.3} \\ &\approx 10 \text{ items.} \end{aligned}$$

55. Let $t = 0$ be the time it started snowing.
If h is the height of the snow and if the rate of snowfall is constant, $\frac{dh}{dt} = k_1$, where k_1 is a constant.

$$\frac{dh}{dt} = k_1 \text{ and } h = 0 \text{ when } t = 0.$$

$$\begin{aligned} dh &= k_1\,dt \\ \int dh &= \int k_1\,dt \\ h &= k_1 t + C_1 \end{aligned}$$

Since $h = 0$ and $t = 0$, $0 = k_1(0) + C_1$. Thus, $C_1 = 0$ and $h = k_1 t$.
Since the snowplow removes a constant volume of snow per hour and the volume is proportional to the height of the snow, the rate of travel of the snowplow is inversely proportional to the height of the snow.

$\dfrac{dx}{dt} = \dfrac{k^2}{h}$, where k_2 is a constant.

When $t = T$, $x = 0$.
When $t = T + 1$, $x = 2$.
When $t = T + 2$, $x = 3$.

Since $\dfrac{dy}{dt} = \dfrac{k^2}{h}$ and $h = k_1 t$,

$$\begin{aligned} \frac{dy}{dt} &= \frac{k_2}{k_1 t} \\ \frac{dx}{dt} &= \frac{k_2}{k_1} \cdot \frac{1}{t}. \end{aligned}$$

Let $k_3 = \dfrac{k_2}{k_1}$. Then

$$\begin{aligned} \frac{dx}{dt} &= k_3 \frac{1}{t} \\ dx &= k_3 \frac{1}{t}\,dt \\ \int dx &= \int k_3 \frac{1}{t}\,dt \\ x &= k_3 \ln t + C_2. \end{aligned}$$

Since $x = 0$, when $t = T$,

$$\begin{aligned} 0 &= k_3 \ln T + C_2 \\ C_2 &= -k_3 \ln T. \end{aligned}$$

Thus,

$$\begin{aligned} x &= k_3 \ln t - k_3 \ln T \\ x &= k_3(\ln t - \ln T) \\ x &= k_3 \ln\left(\frac{t}{T}\right). \end{aligned}$$

Since $x = 2$, when $t = T + 1$,

$$2 = k_3 \ln\left(\frac{T+1}{T}\right). \quad (1)$$

Since $x = 3$ when $t = T + 2$,

$$3 = k_3 \ln\left(\frac{T+2}{T}\right). \quad (2)$$

We want to solve for T, so we divide equation (1) by equation (2).

$$\frac{2}{3} = \frac{k_3 \ln\left(\frac{T+1}{T}\right)}{k_3 \ln\left(\frac{T+2}{T}\right)}$$

$$\frac{2}{3} = \frac{\ln(T+1) - \ln T}{\ln(T+2) - \ln T}$$

$$2\ln(T+2) - 2\ln T = 3\ln(T+1) - 3\ln T$$

$$\ln(T+2)^2 - \ln T^2 - \ln(T+1)^3 + \ln T^3 = 0$$

$$\ln\frac{(T+2)^2T^3}{T^2(T+1)^3} = 0$$

$$\frac{T(T+2)^2}{(T+1)^3} = 1$$

$$T(T^2 + 4T + 4) = T^3 + 3T^2 + 3T + 1$$

$$T^3 + 4T^2 + 4T = T^3 + 3T^2 + 3T + 1$$

$$T^2 + T - 1 = 0$$

$$T = \frac{-1 \pm \sqrt{1+4}}{2}$$

$T = \frac{-1-\sqrt{5}}{2}$ is negative and is not a possible solution.

Thus, $T = \frac{-1+\sqrt{5}}{2} \approx 0.618$ hr.

0.618 hr $\approx$ 37 min and 5 sec

Now, 37 min and 5 sec before 8:00 A.M. is 7:22:55 A.M.

Thus, it started snowing at 7:22:55 A.M.

10.2 Linear First-Order Differential Equations

1. $\frac{dy}{dx} + 3y = 6$

$$I(x) = e^{3\int dx} = e^{3x}$$

Multiply each term by e^{3x}.

$$e^{3x}\frac{dy}{dx} + 3e^{3x}y = 6e^{3x}$$

$$D_x(e^{3x}y) = 6e^{3x}$$

Integrate both sides.

$$e^{3x}y = \int 6e^{3x}\,dx$$

$$= 2e^{3x} + C$$

$$y = 2 + Ce^{-3x}$$

3. $\frac{dy}{dx} + 2xy = 4x$

$$I(x) = e^{\int 2x\,dx} = e^{x^2}$$

$$e^{x^2}\frac{dy}{dx} + 2xe^{x^2}y = 4xe^{x^2}$$

$$D_x(e^{x^2}y) = 4xe^{x^2}$$

$$e^{x^2}y = \int 4xe^{x^2}\,dx$$

$$= 2e^{x^2} + C$$

$$y = 2 + Ce^{-x^2}$$

5. $x\frac{dy}{dx} - y - x = 0;\ x > 0$

$$\frac{dy}{dx} - \frac{1}{x}y = 1$$

$$I(x) = e^{-\int 1/x\,dx}$$

$$= e^{-\ln x} = \frac{1}{x}$$

$$\frac{1}{x}\frac{dy}{dx} - \frac{1}{x^2}y = \frac{1}{x}$$

$$D_x\left(\frac{1}{x}y\right) = \frac{1}{x}$$

$$\frac{y}{x} = \int \frac{1}{x}\,dx$$

$$\frac{y}{x} = \ln x + C$$

$$y = x\ln x + Cx$$

7. $2\frac{dy}{dx} - 2xy - x = 0$

$$\frac{dy}{dx} - xy = \frac{x}{2}$$

$$I(x) = e^{-\int x\,dx} = e^{-x^2/2}$$

$$e^{-x^2/2}\frac{dy}{dx} - xe^{-x^2/2}y = \frac{x}{2}e^{-x^2/2}$$

$$D_x(e^{-x^2/2}y) = \frac{x}{2}e^{-x^2/2}$$

$$e^{-x^2/2}y = \int \frac{x}{2}e^{-x^2/2}\,dx$$

$$= \frac{-1}{2}e^{-x^2/2} + C$$

$$y = -\frac{1}{2} + Ce^{x^2/2}$$

9. $x\dfrac{dy}{dx} + 2y = x^2 + 6x;\ x > 0$

$$\frac{dy}{dx} + \frac{2}{x}y = x + 6$$

$$I(x) = e^{\int 2/x\,dx} = e^{2\ln x} = x^2$$

$$\begin{aligned} x^2\frac{dy}{dx} + 2xy &= x^3 + 6x^2 \\ D_x\left(x^2y\right) &= x^3 + 6x^2 \\ x^2y &= \int (x^3 + 6x^2)\,dx \\ &= \frac{x^4}{4} + 2x^3 + C \\ y &= \frac{x^2}{4} + 2x + \frac{C}{x^2} \end{aligned}$$

11. $y - x\dfrac{dy}{dx} = x^3;\ x > 0$

$$\frac{dy}{dx} - \frac{y}{x} = -x^2$$

$$I(x) = e^{-\int 1/x\,dx} = e^{-\ln x} = x^{-1}$$

$$\begin{aligned} \frac{1}{x}\frac{dy}{dx} - \frac{y}{x^2} &= -x \\ D_x\left(\frac{1}{x}y\right) &= -x \\ \frac{y}{x} &= \int -x\,dx \\ &= \frac{-x^2}{2} + C \\ y &= \frac{-x^3}{2} + Cx \end{aligned}$$

13. $\dfrac{dy}{dx} + y = 4e^x;\ y = 50$ when $x = 0$.

$$\begin{aligned} I(x) &= e^{\int dx} = e^x \\ e^x\frac{dy}{dx} + ye^x &= 4e^{2x} \\ D_x\left(e^xy\right) &= 4e^{2x} \\ e^xy &= \int 4e^{2x}\,dx \\ &= 2e^{2x} + C \\ y &= 2e^x + Ce^{-x} \end{aligned}$$

Since $y = 50$ when $x = 0$,

$$\begin{aligned} 50 &= 2e^0 + Ce^0 \\ 50 &= 2 + C \\ C &= 48. \end{aligned}$$

Therefore,

$$y = 2e^x + 48e^{-x}.$$

15. $\dfrac{dy}{dx} - 2xy - 4x = 0;\ y = 20$ when $x = 1$.

$$\begin{aligned} \frac{dy}{dx} - 2xy - 4x &= 0 \\ \frac{dy}{dx} - 2xy &= 4x \end{aligned}$$

$$\begin{aligned} I(x) &= e^{-\int 2x\,dx} = e^{-x^2} \\ e^{-x^2}\frac{dy}{dx} - 2xe^{-x^2}y &= 4xe^{-x^2} \\ D_x\left(e^{-x^2}y\right) &= 4xe^{-x^2} \\ e^{-x^2}y &= \int 4xe^{-x^2}\,dx \\ e^{-x^2}y &= -2e^{-x^2} + C \\ y &= -2 + Ce^{x^2} \end{aligned}$$

Since $y = 20$ when $x = 1$,

$$\begin{aligned} 20 &= -2 + Ce^1 \\ 22 &= Ce \\ C &= \frac{22}{e}. \end{aligned}$$

Therefore,

$$\begin{aligned} y &= -2 + \frac{22}{e}\left(e^{x^2}\right) \\ &= -2 + 22e^{(x^2-1)}. \end{aligned}$$

17. $x\dfrac{dy}{dx} + 5y = x^2;\ y = 12$ when $x = 2$.

$$x\frac{dy}{dx} + 5y = x^2$$

$$\frac{dy}{dx} + \frac{5}{x}y = x$$

$$I(x) = e^{\int 5/x\,dx} = e^{5\ln x} = x^5$$

$$\begin{aligned} x^5\frac{dy}{dx} + 5x^4y &= x^6 \\ D_x\left(x^5y\right) &= x^6 \\ x^5y &= \int x^6\,dx \\ x^5y &= \frac{x^7}{7} + C \\ y &= \frac{x^2}{7} + \frac{C}{x^5} \end{aligned}$$

Since $y = 12$, when $x = 2$,

$$12 = \frac{4}{7} + \frac{C}{32}$$
$$\frac{80}{7} = \frac{C}{32}$$
$$C = \frac{2560}{7}.$$

Therefore,

$$y = \frac{x^2}{7} + \frac{2560}{7x^5}.$$

19. $x\frac{dy}{dx} + (1+x)y = 3;\ y = 50 \text{ when } x = 4$

$$\frac{dy}{dx} + \left(\frac{1+x}{x}\right)y = \frac{3}{x}$$

$$\begin{aligned} I(x) &= e^{\int (1+x)\,dx/x} \\ &= e^{\int (1/x)\,dx + dx} \\ &= e^{(\ln x)+x} \\ &= e^{\ln x} \cdot e^x \\ &= xe^x \end{aligned}$$

$$\begin{aligned} xe^x\frac{dy}{dx} + (1+x)e^x y &= 3e^x \\ D_x(xe^x y) &= 3e^x \\ xe^x y &= \int 3e^x\,dx \\ xe^x y &= 3e^x + C \\ y &= \frac{3}{x} + \frac{C}{xe^x} \end{aligned}$$

Since $y = 50$ when $x = 4$,

$$50 = \frac{3}{4} + \frac{C}{4e^4}$$
$$\frac{197}{4} = \frac{C}{4e^4}$$
$$C = 197e^4.$$

Therefore,

$$\begin{aligned} y &= \frac{3}{x} + \frac{197e^4}{xe^x} \\ &= \frac{3}{x} + \frac{197}{x}e^{4-x} \\ &= \frac{3 + 197e^{4-x}}{x}. \end{aligned}$$

21. $\frac{dy}{dx} = cy - py^2$

(a) Let $y = \frac{1}{z}$ and $\frac{dy}{dx} = -\frac{z'}{z^2}$.

$$\begin{aligned} -\frac{z'}{z^2} &= c\left(\frac{1}{z}\right) - p\left(\frac{1}{z^2}\right) \\ z' &= -cz + p \\ z' + cz &= p \end{aligned}$$

$$I(x) = e^{\int c\,dx} = e^{cx}$$

$$\begin{aligned} D_x(e^{cx} \cdot z) &= \int pe^{cx}\,dx \\ e^{cx} \cdot z &= \frac{p}{c}e^{cx} + K \\ z &= \frac{p}{c} + Ke^{-cx} \\ &= \frac{p + Kce^{-cx}}{c} \end{aligned}$$

Therefore,

$$y = \frac{c}{p + Kce^{-cx}}.$$

(b) Let $z(0) = \frac{1}{y_0}$.

$$\begin{aligned} \frac{1}{y_0} &= \frac{p + Kce^0}{c} = \frac{p + Kc}{c} \\ \frac{c}{y_0} &= p + Kc \\ Kc &= \frac{c}{y_0} - p = \frac{c - py_0}{y_0} \\ K &= \frac{c - py_0}{cy_0} \\ y &= \frac{c}{p + \left(\frac{c - py_0}{cy_0}\right)ce^{-cx}} \end{aligned}$$

From part (a)

$$= \frac{cy_0}{py_0 + (c - py_0)e^{-cx}}$$

(c)
$$\begin{aligned} \lim_{x\to\infty} y &= \lim_{x\to\infty}\left(\frac{cy_0}{py_0 + (c - py_0)e^{-cx}}\right) \\ &= \frac{cy_0}{py_0 - 0} \\ &= \frac{c}{p} \end{aligned}$$

23. (a) $\frac{dC}{dt} = -kC + D(t)$

$$\frac{dC}{dt} + kC = D(t)$$

$$I = e^{\int k\,dt} = e^{kt}$$

$$e^{kt}\frac{dC}{dt} + e^{kt}kC = e^{kt}D(t)$$
$$D_t(e^{kt}C) = e^{kt}D(t)$$
$$e^{kt}C = \int e^{kt}D(t)d(t)$$
$$= \int_0^t e^{kt}D(y)dy$$
$$C(t) = e^{-kt}\int_0^t e^{ky}D(y)dy + C_2$$

If $C(0) = 0$,

$$C(0) = e^0\int_0^0 e^{ky}D(y)dy + C_2$$
$$0 = 0 + C_2$$

Therefore,

$$C(t) = e^{-kt}\int_0^t e^{ky}D(y)dy.$$

(b) Let $D(y) = D$, a constant.

$$C(t) = e^{-kt}\int_0^t e^{ky}D\,dy$$
$$= De^{-kt}\int_0^t e^{ky}dy$$
$$= De^{-kt}\left(\frac{1}{k}\right)(e^{kt} - e^{k(0)})$$
$$C(t) = \frac{D(1 - e^{-kt})}{k}$$

25. $\frac{dy}{dt} = 0.02y + e^t$; $y = 10{,}000$ when $t = 0$.

$$\frac{dy}{dt} - 0.02y = e^t$$

$$I(t) = e^{\int -0.02\,dt} = e^{-0.02t}$$

$$e^{-0.02t}\frac{dy}{dt} - 0.02e^{-0.02t}y = e^{-0.02t}\cdot e^t$$
$$D_t\left(e^{-0.02t}y\right) = e^{0.98t}$$
$$e^{-0.02t}y = \int e^{0.98t}\,dt$$
$$= \frac{e^{0.98t}}{0.98} + C$$
$$y = \frac{e^t}{0.98} + Ce^{0.02t}$$
$$10{,}000 = \frac{1}{0.98} + C$$
$$C \approx 9999$$
$$y \approx \frac{e^t}{0.98} + 9999e^{0.02t}$$
$$= 1.02e^t + 9999e^{0.02t}$$

27. $\frac{dy}{dt} = 0.02y - t$; $y = 10{,}000$ when $t = 0$.

$$\frac{dy}{dt} - 0.02y = -t$$

$$I(t) = e^{\int -0.02} = e^{-0.02t}$$

$$e^{-0.02t}\frac{dy}{dt} - 0.02e^{-0.02t}y = -te^{-0.02t}$$
$$D_t\left(e^{-0.02t}y\right) = -te^{-0.02t}$$
$$e^{-0.02t}y = \int -te^{-0.02t}\,dt$$

Integration by parts:

Let $u = -t$ $\qquad dv = e^{-0.02t}\,dt$

$du = -dt$ $\qquad v = \frac{e^{-0.02t}}{-0.02}$

$$e^{-0.02t} = \frac{te^{-0.02t}}{0.02} - \int \frac{e^{-0.02t}}{0.02}\,dt$$
$$e^{-0.02t}y = \frac{te^{-0.02t}}{0.02} + \frac{e^{-0.02t}}{0.0004} + C$$
$$y = 50t + 2500 + Ce^{0.02t}$$
$$10{,}000 = 2500 + C$$
$$C = 7500$$
$$y = 50t + 2500 + 7500e^{0.02t}$$

29.
$$\frac{dT}{dt} = -k(T - T_0)$$
$$\frac{dT}{dt} = -kT + kT_0$$
$$\frac{dT}{dt} + kT = kT_0$$
$$I(t) = e^{\int k\,dt} = e^{kt}$$

Multiply both sides by e^{kt}.

$$e^{kt}\frac{dT}{dt} + ke^{kt}T = kT_0e^{kt}$$
$$D_t\left(Te^{kt}\right) = kT_0e^{kt}$$
$$Te^{kt} = \int kT_0e^{kt}\,dt$$
$$Te^{kt} = T_0e^{kt} + C$$
$$T = T_0 + Ce^{-kt}$$
$$T = Ce^{-kt} + T_0$$

31. If $T = ce^{-kt} + T_0$,

$$\lim_{t\to\infty} T = \lim_{t\to\infty}(ce^{-kt} + T_0)$$
$$= c(0) + T_0 \text{ since } k > 0$$
$$= T_0.$$

Thus, the temperature approaches T_0 according to Newton's law of cooling. We would expect the temperature of the object to approach the temperature of the surrounding medium.

33. Refer to Exercise 27.

$$T = Ce^{-kt} + T_0$$

(a) In this problem, $T_0 = 10$, $C = 88.6$, and $k = 0.24$.
Therefore,

$$T = 88.6e^{-0.24t} + 10.$$

(b)

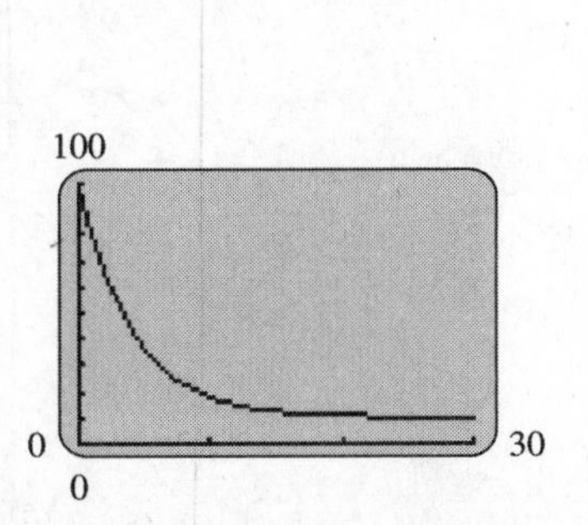

(c) The graph shows the most rapid decrease in the first few hours which is just after death.

(d) If $t = 4$,

$$T = 88.6e^{-0.24(4)} + 10$$
$$T \approx 43.9.$$

The temperature of the body will be 43.9°F after 4 hours.

(e)
$$40 = 88.6e^{-0.24t} + 10$$
$$88.6e^{-0.24t} = 30$$
$$e^{-0.24t} = \frac{30}{88.6}$$
$$-0.24t = \ln\left(\frac{30}{88.6}\right)$$
$$t = \frac{\ln\left(\frac{30}{88.6}\right)}{-0.24}$$
$$t \approx 4.5$$

The body will reach a temperature of 40°F in 4.5 hours.

10.3 Euler's Method

Note: In each step of the calculations shown in this section, all digits should be kept in your calculator as you proceed through Euler's method. Do not round intermediate results.

1. $\frac{dy}{dx} = x^2 + y^2$; $y(0) = 2$, $h = 0.1$. Find $y(0.5)$.

$g(x, y) = x^2 + y^2$
$x_0 = 0$; $y_0 = 2$
$g(x_0, y_0) = 0 + 4 = 4$

$$x_1 = 0.1;\ y_1 = 2 + 4(0.1) = 2.4$$
$$g(x_1, y_1) = (0.1)^2 + (2.4)^2 = 5.77$$

$$x_2 = 0.2;\ y_2 = 2.4 + 5.77(0.1) = 2.977$$
$$g(x_2, y_2) = (0.2)^2 + (2.977)^2 \approx 8.903$$

$$x_3 = 0.3;\ y_3 = 2.977 + 8.903(0.1) \approx 3.867$$
$$g(x_3, y_3) = (0.3)^2 + (3.867)^2 \approx 15.046$$

$$x_4 = 0.4;\ y_4 = 3.867 + 15.046(0.1) \approx 5.372$$
$$g(x_4, y_4) = (0.4)^2 + (5.372)^2 \approx 29.016$$

$x_5 = 0.5;\ y_5 = 5.372 + 29.016(0.1)$
≈ 8.273

These results are tabulated as follows.

x_i	y_i
0	2
0.1	2.4
0.2	2.977
0.3	3.867
0.4	5.372
0.5	8.273

$y(0.5) \approx 8.273.$

Use Euler's method as outlined as in the solution for Exercise 1 in the following exercises. The results are tabulated.

3. $\dfrac{dy}{dx} = 1 + y;\ y(0) = 2, h = 0.1;$ find $y(0.6)$.

x_i	y_i
0	2
0.1	2.3
0.2	2.63
0.3	2.993
0.4	3.3923
0.5	3.8315
0.6	4.31465

$y(0.6) \approx 4.315$

5. $\dfrac{dy}{dx} = x + \sqrt{y};\ y(0) = 1, h = 0.1;$ find $y(0.4)$.

x_i	y_i
0	1
0.1	1.1
0.2	1.215
0.3	1.345
0.4	1.491

$y(0.4) \approx 1.491$

7. $\dfrac{dy}{dx} = 2x\sqrt{1+y^2};\ y(1) = 0, h = 0.1;$ find $y(1.5)$.

x_i	y_i
1	2
1.1	2.447
1.2	3.029
1.3	3.794
1.4	4.815
1.5	6.191

$y(1.5) \approx 6.191.$

9. $\dfrac{dy}{dx} = -4 + x;\ y(0) = 1, h = 0.1,$ find $y(0.4)$.

x_i	y_i
0	1
0.1	0.6
0.2	0.21
0.3	−0.17
0.4	−0.540

$y(0.4) \approx -0.540$
Exact solution:

$$\frac{dy}{dx} = -4 + x$$

$$y = -4x + \frac{x^2}{2} + C$$

At $y(0) = 1$,

$$1 = -4(0) + \frac{0}{2} + C$$

$$C = 1$$

Therefore,

$$y = -4x + \frac{x^2}{2} + 1$$

$$y(0.4) = -4(0.4) + \frac{(0.4)^2}{2} + 1$$

$$= -0.520.$$

11. $\dfrac{dy}{dx} = x^3;\ y(0) = 4,\ h = 0.1,$ find $y(0.4)$.

x_i	y_i
0	4
0.1	4
0.2	4.0001
0.3	4.0009
0.4	4.0036
0.5	4.010

$y(0.5) \approx 4.010$

Exact solution:

$$\frac{dy}{dx} = x^3$$

$$y = \frac{x^4}{4} + C$$

At $y(0) = 4$,

$$4 = \frac{0}{4} + C$$

$$C = 4.$$

Therefore,

$$y = \frac{x^4}{4} + 4$$
$$y(0.5) = \frac{(0.5)^4}{4} + 4 \approx 4.016.$$

13. $\frac{dy}{dx} = 2xy;\ y(1) = 1,\ h = 0.1$, find $y(1.6)$.

$$g(x, y) = 2xy$$

x_i	y_i
1	1
1.1	1.2
1.2	1.464
1.3	1.815
1.4	2.287
1.5	2.927
1.6	3.806

$y(1.6) \approx 3.806$

Exact solution:

$$\frac{dy}{y} = 2x\,dx$$
$$\int \frac{dy}{y} = \int 2x\,dx$$
$$\ln|y| = x^2 + C$$
$$|y| = e^{x^2+C}$$
$$y = ke^{x^2}$$

At $y(1) = 1$,

$$1 = ke^1 = ke$$
$$k = \frac{1}{e}.$$

Therefore,

$$y = \frac{1}{e}(e^{x^2})$$
$$= e^{x^2-1}$$
$$y(1.6) = e^{(1.6)^2-1}$$
$$= 4.759.$$

15. $\frac{dy}{dx} = ye^x;\ y(0) = 2, h = 0.1$, find $y(0.4)$

x_i	y_i
0	2
0.1	2.2
0.2	2.443
0.3	2.742
0.4	3.112

So, $y(0.4) \approx 3.112$.

Exact solution:

$$\frac{dy}{y} = e^x\,dx$$
$$\int \frac{dy}{y} = \int e^x + c$$
$$\ln|y| = e^x + c$$
$$|y| = e^{e^x+c} = e^c e^{e^x}$$
$$y = ke^{e^x}, \text{ where } k = \pm e^c.$$

At $y(0) = 2$, $2 = ke^{e^0} = ke$, so $k = \frac{2}{e}$.

Therefore,

$$y = \frac{2}{e}e^{e^x} = 2e^{e^x-1},$$

so

$$y(0.4) = 2e^{e^{0.4}-1} \approx 3.271.$$

17. $\frac{dy}{dx} + y = 2e^x;\ y(0) = 100, h = 0.1$. Find $y(0.3)$.

x_i	y_i
0	100
0.1	90.2
0.2	81.401
0.3	73.505

$y(0.3) \approx 73.505$.

Exact solution:

$I(x) = e^{\int dx} = e^x$

$e^x \frac{dy}{dx} + e^x y = 2e^x e^x$

$D_x(e^x y) = 2e^{2x}$

$$e^x y = \int 2e^{2x}\,dx + C$$
$$= e^{2x} + C$$
$$y = e^x + Ce^{-x}$$
$$100 = 1 + C$$
$$C = 99$$
$$y = e^x + 99e^{-x}$$
$$y(0.3) = e^{0.3} + 99e^{-0.3} \approx 74.691$$

19. $\frac{dy}{dx} = \sqrt[3]{x},\ y(0) = 0$

Using the program for Euler's method in the Graphing Calculator Manual, the following values are obtained:

x_i	y_i	$y(x_i)$	$y_i - y(x_i)$
0	0	0	0
0.2	0	0.08772053	−0.08772053
0.4	0.11696071	0.22104189	−0.10408118
0.6	0.26432197	0.37954470	−0.11522273
0.8	0.43300850	0.55699066	−0.12398216
1.0	0.61867206	0.75000000	−0.13132794

21. $\frac{dy}{dx} = 4 - y,\ y(0) = 0$

Using the program for Euler's method in the Graphing Calculator Manual, the following values are obtained:

x_i	y_i	$y(x_i)$	$y_i - y(x_i)$
0	0	0	0
0.2	0.8	0.725077	0.07492
0.4	1.44	1.3187198	0.12128
0.6	1.952	1.8047535	0.14725
0.8	2.3616	2.2026841	0.15892
1.0	2.68928	2.5284822	0.16080

23. $\frac{dy}{dx} = \sqrt[3]{x};\ y(0) = 0$

$y = \frac{3}{4}x^{4/3}$ See Exercise 19.

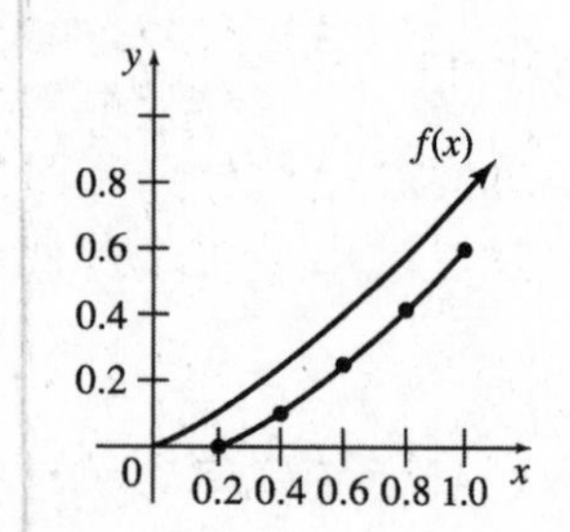

25. $\frac{dy}{dx} = 4 - y;\ y(0) = 0$

$y = 4(1 - e^{-x})$ See Exercise 21.

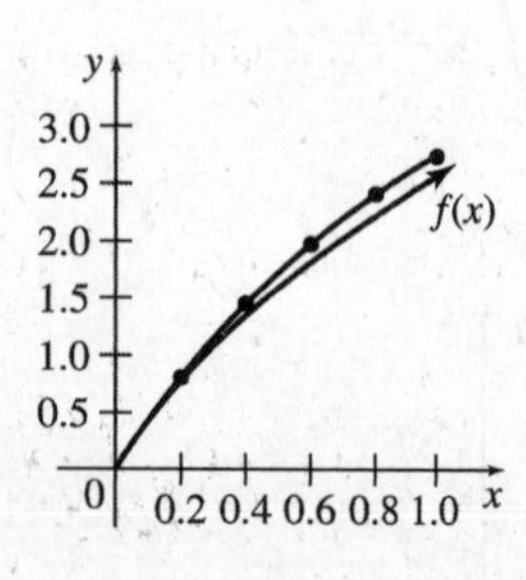

27. $\frac{dy}{dx} = y^2;\ y(0) = 1$

(a)

x_i	y_i
0	1
0.2	1.2
0.4	1.488
0.6	1.9308288
0.8	2.676448771
1.0	4.109124376

Thus, $y(1.0) \approx 4.109$.

(b) $\frac{dy}{dx} = y^2;\ y = 1$ when $x = 0$

$$\frac{dy}{dx} = y^2$$

$$\frac{1}{y^2}\,dy = dx$$

$$\int \frac{1}{y^2}\,dy = \int dx$$

$$-\frac{1}{y} = x + C$$

When $x = 0,\ y = 1$.

$$-\frac{1}{1} = 0 + C$$

$$C = -1$$

$$-\frac{1}{y} = x - 1$$

$$-1 = (x - 1)y$$

$$y = \frac{-1}{x - 1}$$

$$y = \frac{1}{1 - x}$$

As x approaches 1 from the left, y approaches ∞.

29. Let $y =$ the number of algae (in thousands) at time.

$$y \le 500;\ y(0) = 5$$

(a) $\frac{dy}{dt} = 0.01y(500 - y)$

$$= 5y - 0.01y^2$$

(b) Find $y(2)$; $h = 0.5$.

x_i	y_i
0	5
0.5	17.375
1	59.303
1.5	189.976
2	484.462

Therefore, $y(2) \approx 484$, so about 484,000 algae are present when $t = 2$.

31. $\frac{dy}{dt} = 0.05y - 0.1y^{1/2}$; $y(0) = 60$, $h = 1$; find $y(6)$.

x_i	y_i
0	60
1	62.22541
2	64.54785
3	66.97182
4	69.50205
5	72.14347
6	74.90127

Therefore, $y(6) \approx 75$, so about 75 insects are present after 6 weeks.

33. $\frac{dW}{dt} = -0.01189W + 0.92389W^{0.016}$

$W(0) = 3.65$; $h = 1$; find $W(5)$.
Using the program for Euler's method in *The Graphing Calculator Manual,* the following values are obtained:

t_i	W_i
0	3.6500000
1	4.5498301
2	5.4422927
3	6.3268604
4	7.2032009
5	8.0710987

The weight of a goat at 5 weeks is about 8.07 kg.

35. $\frac{dN}{dt} = 0.2(500 - N)N^{1/2}$; $N(0) = 2$, $h = 0.5$;

find $N(3)$.

t_i	N_i
0	2
0.5	72.42784
1	436.3112
1.5	569.3448
2	403.8816
2.5	597.0489
3	359.9139

Therefore, $N(3) \approx 360$, so about 360 people have heard the rumor.

10.4 Applications of Differential Equations

1. $\frac{dA}{dt} = rA + D$; $r = 0.06$, $D = \$5000$

$$\frac{dA}{dt} = 0.06A + 3000$$

$$\int \frac{dA}{0.06A + 3000} = \int dt$$

$$\ln |0.06A + 3000| = 0.06t + C$$

$$|0.06A + 3000| = e^{0.06t+C}$$

$$0.06A + 3000 = Me^{0.06t}$$

$$A = \frac{M}{0.06}e^{0.06t} - \frac{3000}{0.06}$$

$$A = 16.66667Me^{0.06t} - 50{,}000$$

$t = 0$, $A = 5000$

$$5000 = 16.66667Me^{0} - 50{,}000$$

$$M = 3300$$

$$A = 16.66667(3300)e^{0.06t} - 50{,}000$$

$$= 55{,}000e^{0.06t} - 50{,}000$$

When $t = 10$ yr,

$$A = 55{,}000e^{0.06(10)} - 50{,}000$$

$$= \$50{,}216.53.$$

3. $\frac{dA}{dt} = rA + D$; $r = 0.01$; $D = \$50{,}000$;
$t = 0$, $A = 0$

$$\frac{dA}{dt} = 0.1A + 50{,}000$$

$$\frac{dA}{0.1A + 50{,}000} = dt$$

$$\frac{1}{0.1}\ln |0.1A + 50{,}000| = t + C$$

$$\ln |0.1A + 50{,}000| = 0.1t + C$$

$$0.1A + 50{,}000 = e^{0.1t+C}$$

$$= Me^{0.1t}$$

$$A = \frac{M}{0.1}e^{0.1t} - \frac{50{,}000}{0.1}$$

When $t = 0$, $A = 0$.

$$0 = \frac{M}{0.1}(1) - \frac{50{,}000}{0.1}$$

$$M = 50{,}000$$

$$A = \frac{50{,}000}{0.1}e^{0.1t} - \frac{50{,}000}{0.1}$$

$$= 500{,}000(e^{0.1t} - 1)$$

Find t when $A = \$500{,}000$.

$$500{,}000 = 500{,}000(e^{0.1t} - 1)$$
$$e^{0.1t} = 2$$
$$t = \frac{1}{0.1}\ln 2$$
$$\approx 6.9 \text{ yr}$$

It will take about 6.9 years to accumulate \$500,000.

5. **(a)** $\frac{dA}{dt} = rA + D$; $r = 0.06$; $D = -\$1200$;
$A(0) = 8000$

$$\frac{dA}{dt} = 0.06A - 1200$$

(b) To solve for A, first separate the variables.

$$\frac{dA}{0.06A - 1200} = dt$$

Integrate.

$$\int \frac{dA}{0.06A - 1200} = \int dt$$
$$\frac{1}{0.06}\ln|0.06A - 1200| = t + k$$

Solve for A. $\ln|0.06A - 1200| = 0.06t + C$, where $C = 0.06k$.

$$|0.06A - 1200| = e^{0.06t+C} = e^C e^{0.06t}$$
$$0.06A - 1200 = Me^{0.06t}, \text{ where } M = \pm e^C.$$
$$0.06A = 1200 + Me^{0.06t}, \text{ so}$$
$$A = 20{,}000 + C_1e^{0.06t}, \text{ where } C_1 = \frac{M}{0.06}.$$

Solve for C_1. Since $A(0) = 8000$, then

$$8000 = 20{,}000 + C_1e^0, \text{ so}$$
$$C_1 = -12{,}000. \text{ Therefore,}$$
$$A = 20{,}000 - 12{,}000e^{0.06t}.$$

Since $A(2) = 20{,}000 - 12{,}000e^{0.06(2)} \approx 6470.04$, the account will have \$6470.04 left after two years.

(c) If $A = 0$, then $20{,}000 - 12{,}000e^{0.06t} = 0$.

$$12{,}000e^{0.06t} = 20{,}000$$
$$e^{0.06t} = \frac{5}{3}$$
$$0.06t = \ln\left(\frac{5}{3}\right), \text{ so}$$
$$t = \frac{1}{0.06}\ln\left(\frac{5}{3}\right) \approx 8.51.$$

Therefore, the account will be completely depleted in 8.51 years.

7. $\frac{dx}{dt} = -4x + 4xy$, $\frac{dy}{dt} = -3y + 2xy$

$$\frac{dy}{dx} = \frac{\frac{dy}{dt}}{\frac{dx}{dt}} = \frac{-3y + 2xy}{-4x + 4xy} = \frac{y(-3 + 2x)}{x(-4 + 4y)}$$

$$\frac{-4 + 4y}{y}\,dy = \frac{-3 + 2x}{x}\,dx$$
$$\int\left(-\frac{4}{y} + 4\right)dy = \int\left(-\frac{3}{x} + 2\right)dx$$
$$-4\ln y + 4y = -3\ln x + 2x + C$$

(a) When $x = 5, y = 1$, so

$$4 = -3\ln 5 + 10 + C$$
$$-6 + 3\ln 5 = C$$
$$3\ln x - 4\ln y - 2x + 4y = 3\ln 5 - 6$$

(b)
$$-4x + 4xy = 0$$
$$x(-4 + 4y) = 0$$
$$-3y + 2xy = 0$$
$$y(-3 + 2x) = 0$$

The solution $x = 0,\ y = 0$ may be feasible. If not, $x = \frac{3}{2}$ and $y = 1$.

9. Let N = population size, y = number infected.

(a) $N - y = N - \frac{N}{1 + (N-1)e^{-aNt}}$ *Example 4*

$$= \frac{N + N(N-1)e^{-aNt} - N}{1 + (N-1)e^{-aNt}}$$
$$= \frac{N(N-1)e^{-aNt}}{1 + (N-1)e^{-aNt}}$$
$$= \frac{N(N-1)}{e^{aNt} + (N-1)}$$
$$= \frac{N(N-1)}{N - 1 + e^{aNt}}$$

(b) $N = 100,\ a = 0.01$

$$y = \frac{100}{1 + 99e^{-t}} \text{ and } 100 - y = \frac{9900}{99 + e^t}$$

(c) $$y = \frac{100}{1 + 99e^{-t}} = 100(1 + 99e^{-t})^{-1}$$

$$\frac{dy}{dt} = -100(1 + 99e^{-t})^{-2}(-99e^{-t})$$

$$= 9900 \cdot \frac{e^{-t}}{(1 + 99e^{-t})^2}$$

$$\frac{d^2y}{dt} = 9900 \cdot \frac{(1+99e^{-t})^2 \cdot (-e^{-t}) - e^{-t} \cdot 2(1+99e^{-t}) \cdot (-99e^{-t})}{(1 + 99e^{-t})^4}$$

$$= \frac{9900(-e^{-t})(1 - 99e^{-t})}{(1 + 99e^{-t})^3}$$

Since $\frac{-9900e^{-t}}{(1 + 99e^{-t})^3} < 0$ for all t, $\frac{d^2y}{dt}$ changes sign.

Thus, $1 - 99e^{-t}$ changes sign:

$$1 - 99e^{-t} > 0$$
$$\frac{1}{99} > \frac{1}{e^t}$$
$$e^t > 99$$
$$t > \ln 99.$$

Therefore, y is concave upward if $t > \ln 99$ and concave downward if $x < \ln 99$.

If $x = \ln 99$,

$$y = \frac{100}{1 + 99\left(\frac{1}{99}\right)} = 50.$$

Therefore,

$(\ln 99,\ 50) \approx (4.6 \text{ days, } 50 \text{ people})$ is a point of inflection.

The second derivative of $100 - y$ merely has the opposite sign as $\frac{d^2y}{dt}$ above, so $100 - y$ has the same point of inflection as y.

(d) At the point of inflection, the number of infected people is the same as the number of uninfected people.

(e) $$\lim_{t\to\infty} y = \lim_{t\to\infty} \frac{100}{1 + 99e^{-t}}$$
$$= \frac{100}{1 + 99(0)}$$
$$= 100 \text{ people}$$

Therefore,

$$\lim_{t\to\infty} (100 - y) = 100 - \lim_{t\to\infty} y$$
$$= 100 - 100$$
$$= 0 \text{ people.}$$

In general,

$$\lim_{t\to\infty} y = N \text{ and } \lim_{t\to\infty} (N - y) = 0.$$

11. (a) $$\frac{dy}{dt} = kye^{-at}$$

$a = 0.02$, so

$$\frac{dy}{dt} = kye^{-0.02t}.$$

$y = 50$ when $t = 0$;

$y = 300$ when $t = 10$.

$$\frac{1}{y}\,dy = ke^{-0.02t}\,dt$$

$$\ln y = \frac{k}{-0.02}e^{-0.02t} + C$$

To find k and C, solve the system

$$\ln 50 = \frac{k}{-0.02} + C \qquad (1)$$

$$\ln 300 = \frac{k}{-0.02}e^{-0.2} + C. \qquad (2)$$

Subtract equation (1) from equation (2).

$$\ln 300 - \ln 50 = \frac{k}{-0.02}(e^{-0.2} - 1)$$

$$\ln\left(\frac{300}{50}\right) = \ln 6$$

$$= \frac{k}{0.02}(1 - e^{-0.2})$$

$$k = \frac{0.02 \ln 6}{1 - e^{-0.2}} \approx 0.1977$$

Therefore,

$$C = \ln 50 + \frac{k}{0.02} \approx \ln 50 + \frac{0.1977}{0.02}$$
$$\approx 13.7970.$$

So

$$\ln y = \frac{0.1977}{-0.02}e^{-0.02t} + 13.7970$$
$$\approx -9.885e^{-0.02t} + 13.7970.$$

Thus,

$$y \approx e^{-9.885e^{-0.02t}+13.7970}$$
$$= e^{13.7970}e^{-9.885e^{-0.02t}}$$
$$\approx 982{,}000e^{-9.89e^{-0.02t}}.$$

(b) Find t when

$$y = \frac{1}{2}(10{,}000) = 5000.$$
$$\ln 5000 = -9.89e^{-0.02t} + 13.7970$$
$$e^{-0.02t} = \frac{\ln 5000 - 13.7970}{-9.89}$$
$$t = \frac{1}{-0.02}\ln\left[\frac{\ln 5000 - 13.7970}{-9.89}\right]$$
$$\approx 31 \text{ days}$$

13. From equations (10) and (12) in Example 4 in the textbook, the solution to

$$\frac{dy}{dt} = a(N-y)y$$

is

$$y = \frac{N}{1+be^{-aNt}},$$

where y = number of people who have heard the information and

$$b = \frac{N-y_0}{y_0}.$$

If $y_0 = 3$, $N = 500$, and $y = 50$ when $t = 2$, find y.

$$b = \frac{500-3}{3} = \frac{497}{3}, \text{ so}$$
$$y = \frac{500}{1+\frac{497}{3}e^{-500at}}$$
$$50 = \frac{500}{1+\frac{497}{3}e^{-1000a}}$$
$$1+\frac{497}{3}e^{-1000a} = 10$$
$$e^{-1000a} = \frac{27}{497}$$
$$a = -\frac{1}{1000}\ln\left(\frac{27}{497}\right) = \frac{\ln\frac{497}{27}}{1000}$$
$$500a \approx 1.456$$

The number of people who have heard the rumor in t days is given by

$$y = \frac{500}{1+\frac{497}{3}e^{-1.456t}}$$
$$= \frac{1500}{3+497e^{-1.456t}}.$$

When $t = 5$,

$$y = \frac{1500}{3+497e^{-1.456(5)}}$$
$$\approx 449,$$

so in 5 days, about 449 people have heard the rumor.

15. $y_0 = 5$, $N = 100$, and when $t = 1$, $y = 20$.

$$b = \frac{100-5}{5} = 19, \text{ so}$$
$$y = \frac{100}{1+19e^{-100at}}$$
$$20 = \frac{100}{1+19e^{-100a}}$$
$$1+19e^{-100a} = 5$$
$$e^{-100a} = \frac{4}{19}$$
$$a = -\frac{1}{100}\ln\frac{4}{19} = \frac{\ln\left(\frac{19}{4}\right)}{100}$$
$$100a \approx 1.558$$

The number of people who have heard the news in t days is given by

$$y = \frac{100}{1+19e^{-1.558t}}.$$

When $t = 3$,

$$y = \frac{100}{1+19e^{-1.558(3)}}$$
$$\approx 85,$$

so after 3 days, about 85 people have heard the news.

17. Let $y =$ amount (in pounds) of salt at time t. At $t = 0$, $y = 20$ lb and $V = 100$ gallons.

(a) Rate of Salt In

$$= (2 \text{ gal/min}) \cdot (2 \text{ lb/gal})$$
$$= 4 \text{ lb/min}.$$

Rate of Salt Out

$$= (2 \text{ gal/min}) \cdot \left(\tfrac{y}{V} \text{ lb/gal}\right)$$
$$= \frac{2y}{V} \text{ (lb/min)}$$

rate of liquid in
$= \text{rate of liquid out}$

$= 2 \text{ gal/min}$

So $\frac{dV}{dt} = 0$ and $V = 100$ at all times t.
Therefore,

$$\frac{dy}{dt} = 4 - \frac{2y}{100} = \frac{200 - y}{50} \text{ lb/min}$$
$$\frac{dy}{200 - y} = \frac{dt}{50}.$$

$$-\ln(200 - y) = \frac{t}{50} + C$$

$$-\ln 180 = C$$

$$-\ln(200 - y) = \frac{t}{50} - \ln 180$$
$$\ln(200 - y) = \ln 180 - .02t$$
$$200 - y = e^{\ln 180 - .02t}$$
$$= 180e^{-.02t}$$
$$y = 200 - 180e^{-.02t}$$

(b) One hour later, $t = 60$ min,

$$y = 200 - 180e^{-0.02(60)} \approx 146 \text{ lb}.$$

(c) $\lim_{t\to\infty} (200 - 180e^{-0.02t})$
$= 200 - 180(0) = 200$ lb
As $t \to \infty$, y increases, approaching 200 lb.

19. **(a)** Change rate of salt in to (1 gal/min) (1 lb/gal) = 2 lb/min and rate of liquid in to 1 gal/min, so $\frac{dV}{dt} = 1 - 2 = -1$ gal/min.

$V = -t + C$ and $V = 100$ when $t = 0$, so $V = 100 - t$.
Therefore,

$$\frac{dy}{dt} = 2 - \frac{2y}{100 - t}$$
$$\frac{dy}{dt} + \frac{2y}{100 - t} = 2.$$

$$I(x) = e^{\int 2/(100-t)\,dt}$$
$$= e^{-2\ln(100-t)}$$
$$= \frac{1}{(100 - t)^2}$$
$$\frac{1}{(100 - t)^2} y = \int \frac{2}{(100 - t)^2} dt$$
$$= \int 2(100 - t)^{-2} dt$$
$$= 2(100 - t)^{-1} + C$$

$$y = 2(100 - t) + C(100 - t)^2$$
$$20 = 2(100) + C(100)^2$$
$$C = -0.018$$
$$y = 2(100 - t) - 0.018(100 - t)^2$$

(b) When $t = 60$ min,

$$y = 2(100 - 60) - 0.018(100 - 60)^2$$
$$\approx 51 \text{ lb}.$$

(c) The graph of

$$y = 2(100 - t) - 0.018(100 - t)^2$$

is a parabola opening downward with vertex at $t = \frac{400}{9}$ and t-intercepts of $-\frac{100}{9}$ and 100 since $y = 20 + 1.6t - 0.018t^2$.

Thus, y increases from $t = 0$ to $t = \frac{400}{9}$ and decreases from $t = \frac{400}{9}$ to $t = 100$.

21. Let $y =$ amount (in grams) of chemical at time t. At $t = 0$, $y = 5$ g and $V = 100$ liters.

(a) rate of chemical in $= 0$
rate of chemical out

$$= (1 \text{ liter/min}) \left(\tfrac{y}{V} \text{ g/liter}\right)$$
$$= \frac{y}{V} \text{ g/min}$$

rate of liquid In

$= 2$ liters/min;

rate of liquid out

$= 1$ liter/min; so

$$\frac{dV}{dt} = 1 \text{ liter/min}.$$

Therefore,

$V = t + C$ and $100 = C$, so $V = t + 100$.

$$\frac{dy}{dt} = \frac{-y}{t + 100}$$
$$\frac{dy}{y} = \frac{-dt}{t + 100}$$

$$\ln y = -\ln(t+100) + C$$
$$\ln 5 = -\ln 100 + C$$
$$\ln 5 + \ln 100 = C$$
$$C = \ln(500)$$
$$\ln y = -\ln(t+100) + \ln(500)$$
$$= \ln\left(\frac{500}{t+100}\right)$$
$$y = \frac{500}{t+100}$$

(b) When $t = 30$,

$$y = \frac{500}{130} \approx 3.8 \text{ g.}$$

23. Let y = amount (in pounds) of soap at time t. At $t = 0$, $y = 4$ pounds and $V = 200$ gallons.

rate of soap in $= 0$;
rate of soap out

$$= (8 \text{ gal/min})\left(\tfrac{y}{V} \text{ lb/gal}\right)$$
$$= \frac{8y}{V} \text{ lb/min}$$

rate of liquid in
$=$ rate of liquid out
$= 8$ gal/min,

so $\frac{dV}{dt} = 0$, and $V = 200$ for all t.

$$\frac{dy}{dt} = -\frac{8y}{200}$$
$$\frac{1}{y}\,dy = -\frac{1}{25}\,dt$$
$$\ln y = -\frac{t}{25} + C$$
$$\ln 4 = C$$
$$\ln y = -\frac{t}{25} + \ln 4$$
$$y = 4e^{-0.04t}$$

If $y = 1$,

$$1 = 4e^{-0.04t}$$
$$e^{-0.04t} = \frac{1}{4}$$
$$t = -\frac{1}{0.04}\ln\left(\frac{1}{4}\right)$$
$$= 25 \ln 4 \approx 34.7 \text{ min.}$$

Chapter 10 Review Exercises

5. $y\frac{dy}{dx} = 2x + y$

Since you cannot separate the variables, that is, rewrite the equation in the form $g(y)dy = f(x)dx$ where g is a function of y alone and f is a function of x alone, then the equation is not separable. Since you cannot re-write the equation in the form $\frac{dy}{dx} + P(x)y = Q(x)$, then the equation is not a line first-order differential equation. Therefore, the equation is neither linear nor separable.

7. $\sqrt{x}\frac{dy}{dx} = \frac{1+\ln x}{y}$

Since you can rewrite the equation in the form $y\,dy = \frac{1+\ln x^x}{\sqrt{x}}\,dx$, then the equation is separable, but since it cannot be rewritten in the form $\frac{dy}{dx} + P(x)y = Q(x)$, then the equation is not linear.

9. $\frac{dy}{dx} + x = xy$

Since you can rewrite the equation in the form $\frac{dy}{dx} + (-x)y = -x$, then the equation is linear. Since the equation can be rewritten in the form $\frac{dy}{y-1} = x\,dx$, then it is also separable. Therefore, it is both linear and separable.

11. $x\frac{dy}{dx} + y = e^x(1+y)$

Since the equation can be rewritten in the form

$$\frac{dy}{dx} + y\left(\frac{1-e^x}{x}\right) = \frac{e^x}{x},$$

then the equation is linear. Since the equation cannot be rewritten in the form $g(y)dy = f(x)dx$, then the equation is not separable.

13. $\frac{dy}{dx} = 3x^2 + 6x$

$$dy = (3x^2 + 6x)\,dx$$
$$y = x^3 + 3x^2 + C$$

15. $\frac{dy}{dx} = 4e^{2x}$

$$dy = 4e^{2x}\,dx$$
$$y = 2e^{2x} + C$$

17. $\dfrac{dy}{dx} = \dfrac{3x+1}{y}$

$$y\,dy = (3x+1)\,dx$$
$$\frac{y^2}{2} = \frac{3x^2}{2} + x + C_1$$
$$y^2 = 3x^2 + 2x + C$$

19. $\dfrac{dy}{dx} = \dfrac{2y+1}{x}$

$$\frac{dy}{2y+1} = \frac{dx}{x}$$
$$\frac{1}{2}\left(\frac{2\,dy}{2y+1}\right) = \frac{dx}{x}$$
$$\frac{1}{2}\ln|2y+1| = \ln|x| + C_1$$
$$\ln|2y+1|^{1/2} = \ln|x| + \ln k$$
$$\textit{Let } \ln k = C_1$$
$$\ln|2y+1|^{1/2} = \ln k\,|x|$$
$$|2y+1|^{1/2} = k\,|x|$$
$$2y+1 = k^2x^2$$
$$2y+1 = Cx^2$$
$$2y = Cx^2 - 1$$
$$y = \frac{Cx^2-1}{2}$$

21. $\dfrac{dy}{dx} + 6y = 18$

$$I(x) = e^{\int 6\,dx} = e^{6x}$$
$$e^{6x}\frac{dy}{dx} + 6e^{6x}y = 18e^{6x}$$
$$D_x\left(e^{6x}y\right) = 18e^{6x}$$
$$e^{6x}y = \frac{18}{6}e^{6x} + C$$
$$y = 3 + Ce^{-6x}$$

23. $3\dfrac{dy}{dx} + xy - 2x = 0$

$$\frac{dy}{dx} + \frac{x}{3}y = \frac{2x}{3}$$
$$I(x) = e^{\int x/3\,dx} = e^{x^2/6}$$
$$e^{x^2/6}\frac{dy}{dx} + \frac{x}{3}e^{x^2/6}y = \frac{2x}{3}e^{x^2/6}$$
$$D_x\left(e^{x^2/6}y\right) = \frac{2x}{3}e^{x^2/6}$$
$$e^{x^2/6}y = \int \frac{2x}{3}e^{x^2/6}\,dx$$
$$= 2e^{x^2/6} + C$$
$$y = 2 + Ce^{-x^2/6}$$

25. $\dfrac{dy}{dx} = x^2 - 6x$; $y = 3$ when $x = 0$.

$$dy = (x^2 - 6x)\,dx$$
$$y = \frac{x^3}{3} - 3x^2 + C$$

When $x = 0,\ y = 3.$

$$3 = 0 - 0 + C$$
$$C = 3$$
$$y = \frac{x^3}{3} - 3x^2 + 3$$

27. $\dfrac{dy}{dx} = 5(e^{-x} - 1)$; $y = 17$ when $x = 0$.

$$dy = 5(e^{-x} - 1)\,dx$$
$$y = 5(-e^{-x} - x) + C$$
$$= -5e^{-x} - 5x + C$$
$$17 = -5e^0 - 0 + C$$
$$17 = -5 + C$$
$$22 = C$$
$$y = -5e^{-x} - 5x + 22$$

29. $(3-2x)y = \dfrac{dy}{dx}$; $y = 5$ when $x = 0$.

$$(3-2x)dx = \frac{dy}{y}$$
$$3x - x^2 + C = \ln|y|$$
$$e^{3x-x^2+C} = y$$
$$e^{3x-x^2}\cdot e^C = y$$
$$Me^{3x-x^2} = y$$
$$Me^0 = 5$$
$$M = 5$$
$$y = 5e^{3x-x^2}$$

31. $\dfrac{dy}{dx} = \dfrac{1-2x}{y+3}$; $y = 16$ when $x = 0$.

$$(y+3)\,dy = (1-2x)\,dx$$
$$\frac{y^2}{2} + 3y = x - x^2 + C$$
$$\frac{16^2}{2} + 3(16) = 0 + C$$
$$176 = C$$
$$\frac{y^2}{2} + 3y = x - x^2 + 176$$
$$y^2 + 6y = 2x - 2x^2 + 352$$

33. $\dfrac{dy}{dx} = 4x^3 + 2;\ y = 3$ when $x = 1$.

$dy = (4x^3 + 2)\,dx$

$$\int dy = \int (4x^3 + 2)\,dx$$

$$y = x^4 + 2x + C$$

Since $y = 3$ when $x = 1$,

$$\begin{aligned} 3 &= (1)^4 + 2(1) + C \\ 3 &= 3 + C \\ C &= 0 \end{aligned}$$

The particular solution is $y = x^4 + 2x$.

35. $\sqrt{x}\dfrac{dy}{dx} = xy;\ y = 4$ when $x = 1$.

$$\frac{1}{y}\,dy = \frac{x}{\sqrt{x}}\,dx$$

$$\int \frac{1}{y}\,dy = \int x^{1/2}\,dx$$

$$\begin{aligned} \ln|y| &= \frac{x^{3/2}}{\frac{3}{2}} + C_1 \\ \ln|y| &= \frac{2}{3}x^{3/2} + C_1 \\ |y| &= e^{2/3x^{3/2}} + C_1 \\ |y| &= e^{C_1}e^{2/3x^{3/2}} \\ y &= \pm e^{C_1}e^{2/3x^{3/2}} \\ y &= Ce^{2/3x^{3/2}} \end{aligned}$$

Since $y = 4$ when $x = 1$,

$$\begin{aligned} 4 &= Ce^{2/3(1)^{3/2}} \\ C &= \frac{4}{e^{2/3}} \\ C &\approx 2.054 \\ y &= 2.054e^{2/3x^{3/2}}. \end{aligned}$$

39. $\dfrac{dy}{dx} = e^x + y;\ y(0) = 1, h = 0.2$; find $y(0.6)$.

$g(x, y) = e^x + y$

$x_0 = 0;\ y_0 = 1$

$g(x_0, y_0) = e^0 + 1 = 2$

$$\begin{aligned} x_1 = 0.2;\ y_1 &= y_0 + g(x_0, y_0)h \\ &= 1 + 2(0.2) = 1.4 \end{aligned}$$

$g(x_1, y_1) = e^{.2} + 1.4 \approx 2.6214$

$$\begin{aligned} x_2 = 0.4;\ y_2 &= y_1 + g(x_1, y_1)h \\ &= 1.4 + 2.6214(0.2) \\ &\approx 1.9243 \end{aligned}$$

$g(x_2, y_2) = e^{0.4} + 1.9243 \approx 3.4161$

$$\begin{aligned} x_3 = 0.6;\ y_3 &= y_2 + g(x_2, y_2)h \\ &= 1.9243 + 3.4161(0.2) \\ &\approx 2.6075 \end{aligned}$$

x_i	y_i
0	1
0.2	1.4
0.4	1.9243
0.6	2.608

So, $y(0.6) \approx 2.608$.

41. $\dfrac{dy}{dx} = 3 + \sqrt{y},\ y(0) = 0,\ h = 0.2$, find $y(1)$.

x_i	y_i
0	0
0.2	0.6
0.4	1.354919
0.6	2.187722
0.8	3.083541
1	4.034741

Therefore, $y(1) \approx 4.035$.

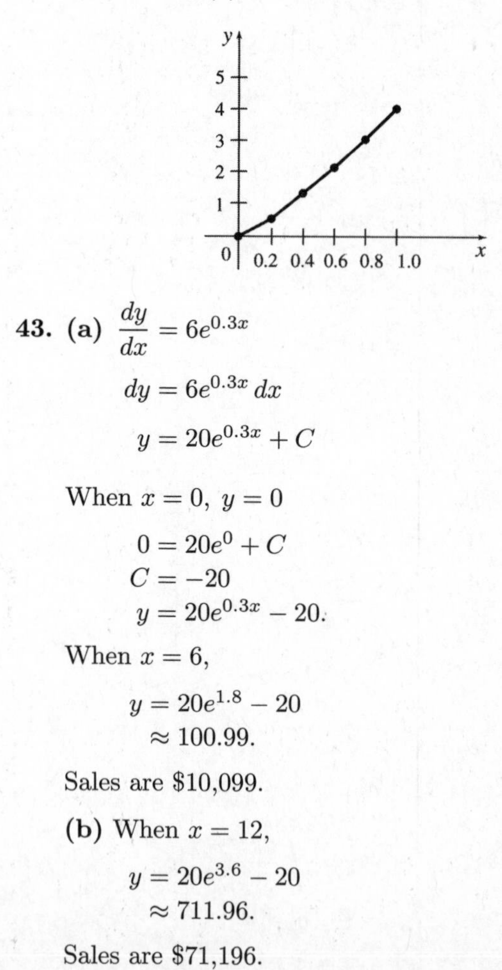

43. (a) $\dfrac{dy}{dx} = 6e^{0.3x}$

$$dy = 6e^{0.3x}\,dx$$

$$y = 20e^{0.3x} + C$$

When $x = 0,\ y = 0$

$$\begin{aligned} 0 &= 20e^0 + C \\ C &= -20 \\ y &= 20e^{0.3x} - 20. \end{aligned}$$

When $x = 6$,

$$\begin{aligned} y &= 20e^{1.8} - 20 \\ &\approx 100.99. \end{aligned}$$

Sales are \$10,099.

(b) When $x = 12$,

$$\begin{aligned} y &= 20e^{3.6} - 20 \\ &\approx 711.96. \end{aligned}$$

Sales are \$71,196.

45. $\frac{dA}{dt} = rA - D$; $t = 0$, $A = \$300{,}000$; $r = 0.05$; $D = \$20{,}000$

$$\frac{dA}{dt} = 0.05A - 20{,}000$$

$$\int \frac{dA}{0.05A - 20{,}000} = \int dt$$

$$\frac{1}{0.05} \ln |0.05A - 20{,}000| = t + C_2$$

$$\ln |0.05A - 20{,}000| = 0.05t + C_1$$

$$0.05A - 20{,}000 = Ce^{0.05t}$$

$$A = 20Ce^{0.05t} + 400{,}000$$

Since $t = 0$ when $A = 300{,}000$,

$$300{,}000 = 20C(1) + 400{,}000$$

$$C = -5000.$$

$$A = 20(-5000)e^{0.05t} + 400{,}000$$

$$A = 400{,}000 - 100{,}000e^{0.05t}.$$

Find t for $A = 0$.

$$0 = 400{,}000 - 100{,}000e^{0.05t}$$

$$e^{0.05t} = \frac{400{,}000}{100{,}000}$$

$$= 4$$

$$\ln e^{0.05t} = \ln 4$$

$$t = 20 \ln 4$$

$$= 27.73$$

It will take about 27.7 yr.

47. $\frac{dy}{dt} = \frac{-10}{1+5t}$; $y = 50$ when $t = 0$.

$$y = -2 \ln (1 + 5t) + C$$

$$50 = -2 \ln 1 + C$$

$$= C$$

$$y = 50 - 2 \ln (1 + 5t)$$

(a) If $t = 24$,

$$y = 50 - 2 \ln [1 + 5(24)]$$

$$\approx 40 \text{ insects.}$$

(b) If $y = 0$,

$$50 = 2 \ln (1 + 5t)$$

$$1 + 5t = e^{25}$$

$$t = \frac{e^{25} - 1}{5}$$

$$\approx 1.44 \times 10^{10} \text{ hours}$$

$$\approx 6 \times 10^8 \text{ days}$$

$$\approx 1.6 \text{ million years.}$$

49.

$$\frac{dx}{dt} = 0.2x - 0.5xy$$

$$\frac{dy}{dt} = -0.3y + 0.4xy$$

$$\frac{dy}{dt} = \frac{\frac{dy}{dt}}{\frac{dx}{dt}}$$

$$= \frac{-0.3y + 0.4xy}{0.2x - 0.5xy}$$

$$= \frac{y(-0.3 + 0.4x)}{x(0.2 - 0.5y)}$$

$$\frac{0.2 - 0.5y}{y} dy = \frac{-0.3 + 0.4x}{x} dx$$

$$\left(\frac{0.2}{y} - 0.5\right) dy = \left(\frac{-0.3}{x} + 0.4\right) dx$$

$$0.2 \ln y - 0.5y = -0.3 \ln x + 0.4x + C$$

$$0.3 \ln x + 0.2 \ln y - 0.4x - 0.5y = C$$

Both growth rates are 0 if

$0.2x - 0.5xy = 0$ and $-0.3y + 0.4xy = 0$.

If $x \neq 0$ and $y \neq 0$, we have

$0.2 - 0.5y = 0$ and $-0.3 + 0.4x = 0$, so

$x = \frac{3}{4}$ units and $y = \frac{2}{5}$ units.

51. Let $y =$ the amount in parts per million (ppm) of smoke at time t.
When $t = 0$, $y = 20$ ppm, $V = 15{,}000$ ft^3 cu ft.

rate of smoke in $= 5$ ppm,
rate of smoke out
$= (1200 \text{ ft}^3/\text{min})(\frac{y}{V} \text{ ppm/ft}^3)$

Rate of air in $=$ rate of air out, so $V = 15{,}000$ ft^3 for all t.

$$\frac{dy}{dt} = 5 - \frac{1200y}{15{,}000} = 5 - \frac{2y}{25}$$

$$= \frac{125 - 2y}{25}$$

$$\frac{1}{125 - 2y} dy = \frac{dt}{25}$$

$$-\frac{1}{2} \ln (125 - 2y) = \frac{t}{25} + C$$

$$-\frac{1}{2} \ln (125 - 2(20)) = C$$

$$C = -\frac{1}{2} \ln 85$$

If $y = 10$,

$$-\frac{1}{2}\ln[125 - 2(10)] = \frac{t}{25} - \frac{1}{2}\ln 85$$

$$\ln 105 = \ln 85 - \frac{2t}{25}$$

$$t = \frac{25}{2}[\ln 85 - \ln 105],$$

which is negative.
It is impossible to reduce y to 10 ppm.

53. $y = \frac{N}{1 + be^{-kx}}$; $y = y_i$ when $x = x_i$,

$i = 1, 2, 3$.

x_1, x_2, x_3 are equally spaced:

$x_3 = 2x_2 - x_1$, so $x_1 + x_3 = 2x_2$, or $x_2 = \frac{x_1 + x_3}{2}$.

Show $N = \frac{\frac{1}{y_1} + \frac{1}{y_3} - \frac{2}{y_2}}{\frac{1}{y_1 y_3} - \frac{1}{y_2^2}}$.

Let

$$\begin{aligned} A &= \frac{1}{y_1} + \frac{1}{y_3} - \frac{2}{y_2} \\ &= \frac{1 + be^{-kx_1}}{N} + \frac{1 + be^{-kx_3}}{N} \\ &\quad - \frac{2(1 + be^{-kx_2})}{N} \\ &= \frac{1}{N}(1 + be^{-kx_1} + 1 + be^{-kx_3} - 2 - 2be^{-kx_2}) \\ &= \frac{b}{N}[e^{-kx_1} + e^{-kx_3} - 2e^{-kx_2}] \end{aligned}$$

Let

$$\begin{aligned} B &= \frac{1}{y_1 y_3} - \frac{1}{{y_2}^2} \\ &= \frac{(1 + be^{-kx_1})(1 + be^{-kx_3})}{N^2} \\ &\quad - \frac{(1 + be^{-kx_2})^2}{N^2} \\ &= \frac{1}{N^2}[1 + be^{-kx_1} + be^{-kx_3} + b^2 e^{-k(x_1 + x_3)} \\ &\quad - 1 - 2be^{-kx_2} - b^2 e^{-2kx_2}] \\ &= \frac{b}{N^2}[e^{-kx_1} + e^{-kx_3} + be^{-k(2x_2)} \\ &\quad - 2e^{-kx_2} - be^{-2kx_2}] \\ &= \frac{b}{N^2}[e^{-kx_1} + e^{-kx_3} - 2e^{-kx_2}] \end{aligned}$$

Clearly, $\frac{A}{B} = N$.

Hence,

$$N = \frac{\frac{1}{y_1} + \frac{1}{y_3} - \frac{2}{y_2}}{\frac{1}{y_1 y_3} - \frac{1}{y_2^2}}.$$

55. From Exercise 53,

$$N = \frac{\frac{1}{40} + \frac{1}{204} - \frac{2}{106}}{\frac{1}{40(204)} - \frac{1}{(106)^2}} \approx 329.$$

(a) $y_0 = 40$,

$$\begin{aligned} b &= \frac{N - y_0}{y_0} = \frac{329 - 40}{40} \\ &\approx 7.23. \end{aligned}$$

If 1920 corresponds to $x = 5$ decades, then

$$\begin{aligned} 106 &= \frac{329}{1 + 7.23e^{-5k}} \\ 1 + 7.23e^{-5k} &= \frac{329}{106} \\ e^{-5k} &= \frac{1}{7.23}\left(\frac{329}{106} - 1\right) \\ \text{so} \qquad k &= -\frac{1}{5}\ln\left[\frac{1}{7.23}\left(\frac{329}{106} - 1\right)\right] \\ &\approx 0.247. \end{aligned}$$

(b) $y = \frac{329}{1 + 7.23e^{-0.247x}}$

In 2000, $x = 13$. If $x = 13$,

$$y = \frac{329}{1 + 7.23e^{-3.211}} \approx 255 \text{ million}$$

The predicated population is 255 million which is less than the table value of 281.4 million.

(c) In 2030, $x = 16$, so

$$\begin{aligned} y &= \frac{329}{1 + 7.23e^{-0.247(16)}} \\ &\approx 289 \text{ million}. \end{aligned}$$

In 2050, $x = 18$, so

$$\begin{aligned} y &= \frac{329}{1 + 7.23e^{-0.247(18)}} \\ &\approx 303 \text{ million}. \end{aligned}$$

57. (a) $\frac{dx}{dt} = 1 - kx$

Separate the variables and integrate.

$$\int \frac{dx}{1-kx} = \int dt$$

$$-\frac{1}{k}\ln|1-kx| = t + C_1$$

Solve for x.

$$\ln|1-kx| = -kt - kC_1$$
$$|1-kx| = e^{-kC_1}e^{-kt}$$

$1 - kx = Me^{-kt}$, where $M = \pm e^{-kC_1}$.

$$x = -\frac{1}{k}(Me^{-kt} - 1) = \frac{1}{k} + Ce^{-kt},$$

where $C = -\frac{M}{k}$.

(b) Write the linear first-order differential equation in the linear form

$$\frac{dx}{dt} + kx = 1.$$

The integrating factor is $I(t) = e^{\int k\,dt} = e^{kt}$. Multiply both sides of the differential equation by $I(t)$.

$$\frac{dx}{dt}e^{kt} + kxe^{kt} = e^{kt}$$

Replace the left side of this equation by

$$D_t(xe^{kt}) = \frac{dx}{dt}e^{kt} + xke^{kt}.$$
$$D_t(xe^{kt}) = e^{kt}$$

Integrate both sides with respect to t.

$$xe^{kt} = \int e^{kt}dt = \frac{1}{k}e^{kt} + C$$

Solve for x.

$$x = \frac{1}{k} + Ce^{-kt}.$$

(c) Since $k > 0$, then as t gets larger and larger, Ce^{-kt} approaches 0, so $\lim_{t\to\infty}\left(\frac{1}{k} + Ce^{-kt}\right) = \frac{1}{k}$.

59.

$$\frac{dT}{dt} = k(T - T_F)$$
$$\frac{dT}{dt} = k(T - 300°)$$
$$T(0) = 40°$$
$$T(1) = 150°$$
$$\frac{dT}{T-300} = k\,dt$$
$$\ln|T-300| = kt + C_1$$
$$|T-300| = e^{kt+C_1}$$
$$T - 300 = Ce^{kt}$$
$$T = Ce^{kt} + 300$$

Since $T(0) = 40°$,

$$40 = Ce^0 + 300$$
$$C = -260.$$

Therefore,

$$T = -260e^{kt} + 300.$$

At $T(1) = 150°$,

$$150 = -260e^k + 300$$
$$-150 = -260e^k$$
$$\frac{15}{26} = e^k$$
$$\ln\left(\frac{15}{26}\right) = k\ln e$$
$$k = \ln\left(\frac{15}{26}\right)$$
$$k = -0.55.$$

Therefore,

$$T = -260e^{-.055t} + 300.$$

At $t = 2$,

$$T = -260e^{-0.55(2)} + 300$$
$$= 213°.$$

61. **(a)**

$$\frac{dv}{dt} = G^2 - K^2v^2$$

$$dv = (G^2 - K^2v^2)\,dt$$

$$\frac{1}{G^2 - K^2v^2}\,dv = dt$$

$$\int \frac{1}{G^2 - K^2v^2}\,dv = \int dt$$

$$\frac{1}{K^2}\int \frac{1}{\left(\frac{G}{K}\right)^2 - v^2} = \int dt$$

Use entry 7 from the table of integrals.

$$\frac{1}{K^2}\cdot\frac{1}{2\frac{G}{K}}\ln\left|\frac{\frac{G}{K}+v}{\frac{G}{K}-v}\right| = t + C_1$$

$$\frac{1}{2GK}\ln\left|\frac{G+Kv}{G-Kv}\right| = t + C_1$$

$$\ln\left|\frac{G+Kv}{G-Kv}\right| = 2GKt + C_2$$

Since $v < \frac{G}{K}$, $Kv < G$ and $\frac{G+Kv}{G-Kv}$ is positive.

$$\ln\left(\frac{G+Kv}{G-Kv}\right) = 2GK(t) + C_2$$

When $t = 0$, $v = 0$, so

$$\ln\left(\frac{G+0}{G-0}\right) = 2GK(0) + C_2$$

$$\ln 1 = C_2$$

$$C_2 = 0.$$

Thus,

$$\ln\left(\frac{G+Kv}{G-Kv}\right) = 2GKt$$

$$\frac{G+Kv}{G-Kv} = e^{2GKt}$$

$$G + Kv = Ge^{2GKt} - Kve^{2GKt}$$

$$Kv + Kve^{2GKt} = Ge^{2GKt} - G$$

$$vK(e^{2GKt}+1) = G(e^{2GKt}-1)$$

$$v = \frac{G(e^{2GKt}-1)}{K(e^{2GKt}+1)}$$

$$v = \frac{G}{K}\cdot\frac{e^{2GKt}-1}{e^{2GKt}+1}$$

(b) $$\lim_{t\to\infty} v = \lim_{t\to\infty}\left(\frac{G}{K}\cdot\frac{e^{2GKt}-1}{e^{2GKt}+1}\right)$$

$$\lim_{t\to\infty} v = \frac{G}{K}\lim_{t\to\infty}\frac{e^{2GKt}-1}{e^{2GKt}+1}$$

$$\lim_{t\to\infty} v = \frac{G}{K}\cdot 1 = \frac{G}{K}$$

A falling object in the presence of air resistance has a limiting velocity, $\frac{G}{k}$.

(c) $$\lim_{t\to\infty} v = \frac{G}{k}$$

$$88 = \frac{G}{k}$$

$$k = \frac{G}{88}$$

$$Gk = \frac{G^2}{88}$$

$$Gk = \frac{32}{88}$$

$$2\,Gk = \frac{64}{88}$$

$$2\,Gk \approx 0.727$$

$$v = \frac{G}{k}\,\frac{e^{2Gkt}-1}{e^{2Gkt}+1}$$

$$v = 88\,\frac{e^{0.727t}-1}{e^{0.727t}+1}$$

Chapter 10 Test

[10.1]

Find the general solutions for the following differential equations.

1. $\frac{dy}{dx} = 3x^2 + 4x - 5$ **2.** $\frac{dy}{dx} = 5e^{2x}$ **3.** $\frac{dy}{dx} = \frac{2x}{x^2+5}$

Find particular solutions for the following differential equations.

4. $\frac{dy}{dx} = 4x^3 + 2x^2 + 1$; $y = 4$ when $x = 1$

5. $\frac{dy}{dx} = \frac{1}{3x+1}$; $y = 4$ when $x = 3$

6. $\frac{dy}{dx} = 4\left(e^{-x} - 1\right)$; $y = 5$ when $x = 0$

7. After use of an insecticide, the rate of decline of an insect population is given by

$$\frac{dy}{dt} = \frac{-8}{1+4t},$$

where t is the number of hours after the insecticide is applied. If there were 40 insects initially, how many are left after 20 hours?

Find general solutions for the following differential equations. (Some solutions may give y implicitly.)

8. $\frac{dy}{dx} = \frac{2x+1}{y-1}$ **9.** $\frac{dy}{dx} = \frac{3y}{x+1}$ **10.** $\frac{dy}{dx} = \frac{e^x - x}{y+1}$

Find particular solutions of the following differential equations.

11. $\sqrt{y}\frac{dy}{dx} = xy$; $y = 4$ when $x = 2$

12. $\frac{dy}{dx} = e^y \cdot x^3$; $y = 0$ when $x = 0$

[10.2]

Find the general solution for each differential equation.

13. $y' + 4y = 12$

14. $y' - xy = 2x$

Find the particular solution for each differential equation.

15. $y' + 3y = 10$; $y = 5$ when $x = 0$

16. $x^2\frac{dy}{dx} + 2x^3y + 2x^3 = 0$; $y = 4$ when $x = 0$

17. $e^xy' + e^xy = x + 1$; $y = \frac{2}{e}$ when $x = 1$

[10.3]

Use Euler's method to approximate the indicated function value for $y = f(x)$ to three decimal places using $h = 0.1$.

18. $y' = 2x - y^{-1}$; $f(0) = 1$; find $f(0.3)$.

19. $y' = xy$; $f(0) = 1$; find $f(0.4)$.

20. Let $y = f(x)$ and $y' = \frac{x^2}{2} + 3$ with $f(0) = 0$. Use Euler's method with $h = 0.1$ to approximate $f(0.3)$ to three decimal places. Then solve the differential equation and find $f(0.3)$ to three decimal places. Also, find $y_3 - f(x_3)$.

21. In Euler's method, large errors may occur. What can be done to reduce the error?

[10.4]

22. Explain why differential equations are useful in problem solving.

23. At her birth, Sally's grandparents deposit \$5000 into a savings account earning 8% interest compounded continuously. If they make continuous deposits at a rate of \$2000 per year through her twenty-first birthday, how much money will accumulate in the account?

24. A deposit of \$20,000 is made into a savings account earning 5% interest compounded continuously. If continuous withdrawals are made at a rate of \$2000 per year, how much money will be left in the account after 2 years? Approximately how long will it take to deplete the account?

25. Find an equation relating x to y given the following equations, which describe the interaction of two competing species and their growth rates.

$$\frac{dx}{dt} = 6x - 4xy$$

$$\frac{dy}{dt} = -8y + 12xy$$

Find the values of x and y for which both growth rates are 0.

26. In a population of size N, the rate at which an influenza epidemic spreads is given by

$$\frac{dy}{dt} = a(N - y) \cdot y,$$

where y is the number of people infected and a is a constant.

(a) If 40 people in a community of 10,000 people are infected at the beginning of an epidemic, and 100 people are infected 5 days later, write an equation for the number of people infected after t days.

(b) How many people are infected after 10 days?

27. A tank presently contains 100 gallons of a solution of dissolved salt and water, which is kept uniform by stirring. While pure water is allowed to flow into the tank at a rate of 5 gallons per minute, the mixture flows out of the tank at a rate of 3 gallons per minute. How much salt will remain in the tank after t minutes if 10 pounds of salt are in the mixture originally?

28. A tank filled with a salt solution has solution flowing in and out. The number of pounds y of salt in the tank is given by

$$y = 25\left(1 - e^{-0.02t}\right)$$

where t is the time in hours. Will there ever be 25 pounds of salt in the solution? Explain.

Chapter 10 Test Answers

1. $y = x^3 + 2x^2 - 5x + C$

2. $y = \frac{5}{2}e^{2x} + C$

3. $y = \ln\left(x^2 + 5\right) + C$

4. $y = x^4 + \frac{2}{3}x^3 + x + \frac{4}{3}$

5. $y = \frac{1}{3}\ln|3x + 1| + 4 - \frac{1}{3}\ln 10$

6. $y = -4e^{-x} - 4x + 9$

7. 31

8. $\frac{y^2}{2} - y = x^2 + x + C$

9. $y = C\,(x + 1)^3$

10. $\frac{y^2}{2} + y = e^x - \frac{x^2}{2} + C$

11. $y^{1/2} = \frac{x^2}{4} + 1$

12. $e^{-y} = 1 - \frac{x^4}{4}$ or $y = -\ln\left|1 - \frac{x^4}{4}\right|$

13. $y = 3 + Ce^{-4x}$

14. $y = Ce^{x^2/2} - 2$

15. $y = \frac{10}{3} + \frac{5}{3}e^{-3x}$

16. $y = 5e^{-x^2} - 1$

17. $y = \frac{x^2}{2}e^{-x} + xe^{-x} + \frac{1}{2}e^{-x}$

18. 0.725

19. 1.061

20. 0.903; $y = \frac{1}{6}x^3 + 3x$; 0.905; -0.002

21. The error can be reduced by using more subintervals of smaller width.

22. Many problems involve rates of change which in turn lead to differential equations.

23. \$135,966.68

24. \$17,896.58; 13.9 yr

25. $3\ln y - 2y = -4\ln x + 6x + C$; $x = \frac{2}{3}$, $y = \frac{3}{2}$

26. **(a)** $y = \frac{10{,}000}{1+249e^{-0.184t}}$ **(b)** 247

27. $y = 10 \cdot \left(\frac{50}{t+50}\right)^{3/2}$

28. As $t \to \infty$, $e^{-0.02t} \to 0$, so $\lim\limits_{t\to\infty} y = \lim\limits_{t\to\infty} 25\left(1 - e^{-0.02t}\right) = 25$. The limiting value is 25. Theoretically, there will never be 25 lb of salt in the solution. However, as time increases the amount of salt will approach 25 lb.

Chapter 11

PROBABILITY AND CALCULUS

11.1 Continuous Probability Models

1. $f(x) = \frac{1}{9}x - \frac{1}{18}$; [2, 5]

Show that condition 1 holds.

Since $2 \le x \le 5$,

$$\frac{2}{9} \le \frac{1}{9}x \le \frac{5}{9}$$

$$\frac{1}{6} \le \frac{1}{9}x - \frac{1}{18} \le \frac{1}{2}.$$

Hence, $f(x) \ge 0$ on [2, 5].

Show that condition 2 holds.

$$\int_2^5 \left(\frac{1}{9}x - \frac{1}{18}\right) dx = \frac{1}{9}\int_2^5 \left(x - \frac{1}{2}\right) dx$$

$$= \frac{1}{9}\left(\frac{x^2}{2} - \frac{1}{2}x\right)\Bigg|_2^5$$

$$= \frac{1}{9}\left(\frac{25}{2} - \frac{5}{2} - \frac{4}{2} + 1\right)$$

$$= \frac{1}{9}(8+1)$$

$$= 1$$

Yes, $f(x)$ is a probability density function.

3. $f(x) = \frac{1}{21}x^2$; [1, 4]

Since $x^2 \ge 0$, $f(x) \ge 0$ on [1, 4].

$$\frac{1}{21}\int_1^4 x^2\, dx = \frac{1}{21}\left(\frac{x^3}{3}\right)\Bigg|_1^4$$

$$= \frac{1}{21}\left(\frac{64}{3} - \frac{1}{3}\right) = 1$$

Yes, $f(x)$ is a probability density function.

5. $f(x) = 4x^3$; [0, 3]

$$4\int_0^3 x^3\, dx = 4\left(\frac{x^4}{4}\right)\Bigg|_0^3$$

$$= 4\left(\frac{81}{4} - 0\right)$$

$$= 81 \ne 1$$

No, $f(x)$ is not a probability density function.

7. $f(x) = \frac{x^2}{16}$; [−2, 2]

$$\frac{1}{16}\int_{-2}^2 x^2\, dx = \frac{1}{16}\left(\frac{x^3}{3}\right)\Bigg|_{-2}^2$$

$$= \frac{1}{16}\left(\frac{8}{3} + \frac{8}{3}\right)$$

$$= \frac{1}{3} \ne 1$$

No, $f(x)$ is not a probability density function.

9. $f(x) = \frac{5}{3}x^2 - \frac{5}{90}$; $[-1, 1]$

Let $x = 0$. Then $f(x) = f(0) = -\frac{5}{90} < 0$.

So $f(x) < 0$ for at least one x-value in $[-1, 1]$.
No, $f(x)$ is not a probability density function.

11. $f(x) = kx^{1/2}$; [1, 4]

$$\int_1^4 kx^{1/2}\, dx = \frac{2}{3}kx^{3/2}\Bigg|_1^4$$

$$= \frac{2}{3}k(8-1)$$

$$= \frac{14}{3}k$$

If $\frac{14}{3}k = 1$,

$$k = \frac{3}{14}.$$

Notice that $f(x) = \frac{3}{4}x^{1/2} \ge 0$ for all x in [1, 4].

13. $f(x) = kx^2$; [0, 5]

$$\int_0^5 kx^2\, dx = k\frac{x^3}{3}\Bigg|_0^5$$

$$= k\left(\frac{125}{3} - 0\right)$$

$$= k\left(\frac{125}{3}\right)$$

If $k\left(\frac{124}{3}\right) = 1$,

$$k = \frac{3}{125}.$$

Notice that $f(x) = \frac{3}{125}x^2 \ge 0$ for all x in [0, 5].

15. $f(x) = kx;\ [0,\ 3]$

$$\int_0^3 kx\,dx = k\frac{x^2}{2}\bigg|_0^3$$
$$= k\left(\frac{9}{2} - 0\right)$$
$$= \frac{9}{2}k$$

If $\frac{9}{2}k = 1$,

$$k = \frac{2}{9}.$$

Notice that $f(x) = \frac{2}{9}x \geq 0$ for all x in [0, 3].

17. $f(x) = kx;\ [1,\ 5]$

$$\int_1^5 kx\,dx = k\frac{x^2}{2}\bigg|_1^5$$
$$= k\left(\frac{25}{2} - \frac{1}{2}\right)$$
$$= 12k$$

If $12k = 1$,

$$k = \frac{1}{12}.$$

Notice that $f(x) = \frac{1}{12}x \geq 0$ for all x in $[1, 5]$.

19. For the probability density function $f(x) = \frac{1}{9}x - \frac{1}{18}$ on $[2, 5]$, the cumulative distribution function is

$$F(x) = \int_a^x f(t)\,dt$$
$$= \int_a^x \left(\frac{1}{9}t - \frac{1}{18}\right)\,dt$$
$$= \left(\frac{1}{18}t^2 - \frac{1}{18}t\right)\bigg|_2^x$$
$$= \frac{1}{18}[(x^2 - x) - (4 - 2)]$$
$$= \frac{1}{18}(x^2 - x - 2), 2 \leq x \leq 5.$$

21. For the probability density function $f(x) = \frac{x^2}{21}$ on $[1, 4]$, the cumulative distribution function is

$$F(x) = \int_1^x \frac{t^2}{21}\,dt$$
$$= \frac{t^3}{63}\bigg|_1^x$$
$$= \frac{1}{63}(x^3 - 1), 1 \leq x \leq 4.$$

23. The value of k was found to be $\frac{3}{14}$. For the probability density function $f(x) = \frac{3}{14}x^{1/2}$ on $[1, 4]$, the cumulative distribution function is

$$F(x) = \int_1^x \frac{3}{14}t^{1/2}\,dt$$
$$= \frac{3}{14}\cdot\frac{2}{3}t^{3/2}\bigg|_1^x$$
$$= \frac{1}{7}(x^{3/2} - 1), 1 \leq x \leq 4.$$

25. The total area under the graph of a probability density function always equals 1.

29. $f(x) = \frac{1}{2}(1+x)^{-3/2};\ [0,\ \infty)$

$$\frac{1}{2}\int_0^\infty (1+x)^{-3/2}\,dx$$
$$= \lim_{a\to\infty}\frac{1}{2}\int_0^a (1+x)^{-3/2}\,dx$$
$$= \lim_{a\to\infty}\frac{1}{2}(1+x)^{-1/2}\left(\frac{-2}{1}\right)\bigg|_0^a$$
$$= \lim_{a\to\infty}[-(1+a)^{-1/2} + 1]$$
$$= \lim_{a\to\infty}\left(\frac{-1}{\sqrt{1+a}} + 1\right)$$
$$= 0 + 1 = 1$$

Since $x \geq 0$, $f(x) \geq 0$.
$f(x)$ is a probability density function.

(a) $P(0 \leq X \leq 2)$

$$= \frac{1}{2}\int_0^2 (1+x)^{-3/2}\,dx$$
$$= -(1+x)^{-1/2}\bigg|_0^2$$
$$= -3^{-1/2} + 1$$
$$\approx 0.4226$$

(b) $P(1 \leq X \leq 3)$

$$= \frac{1}{2}\int_1^3 (1+x)^{-3/2}\,dx$$
$$= -(1+x)^{-1/2}\bigg|_1^3$$
$$= -4^{-1/2} + 2^{-1/2}$$
$$\approx 0.2071$$

(c) $P(X \geq 5)$

$$= \frac{1}{2}\int_5^{\infty} (1+x)^{-3/2}\, dx$$
$$= \lim_{a\to\infty} \frac{1}{2}\int_5^{a} (1+x)^{-3/2}\, dx$$
$$= \lim_{a\to\infty} [-(1+x)^{-1/2}]\Big|_5^a$$
$$= \lim_{a\to\infty} [-(1+a)^{-1/2} + 6^{-1/2}]$$
$$= \lim_{a\to\infty} \left(\frac{-1}{\sqrt{1+a}} + 6^{-1/2}\right)$$
$$\approx 0 + 0.4082$$
$$= 0.4082$$

31. $f(x) = \frac{1}{2}e^{-x/2}$; $[0, \infty)$

$$\frac{1}{2}\int_0^{\infty} e^{-x/2}\, dx$$
$$= \lim_{a\to\infty} \frac{1}{2}\int_0^{a} e^{-x/2}\, dx$$
$$= \lim_{a\to\infty} \frac{1}{2}\left(\frac{-2}{1}e^{-x/2}\right)\Big|_0^a$$
$$= \lim_{a\to\infty} -e^{-x/2}\Big|_0^a$$
$$= \lim_{a\to\infty} \left(\frac{-1}{e^{a/2}} + 1\right)$$
$$= 0 + 1$$
$$= 1$$

$f(x) > 0$ for all x.
$f(x)$ is a probability density function.

(a) $P(0 \leq X \leq 1) = \frac{1}{2}\int_0^1 e^{-x/2}\, dx$

$$= -e^{-x/2}\Big|_0^1$$
$$= \frac{-1}{e^{1/2}} + 1$$
$$\approx 0.3935$$

(b) $P(1 \leq X \leq 3) = \frac{1}{2}\int_1^3 e^{-x/2}\, dx$

$$= -e^{-x/2}\Big|_1^3$$
$$= \frac{-1}{e^{3/2}} + \frac{1}{e^{1/2}}$$
$$\approx 0.3834$$

(c) $P(X \geq 2) = \frac{1}{2}\int_2^{\infty} e^{-x/2}\, dx$

$$= \lim_{a\to\infty} \frac{1}{2}\int_2^{a} e^{-x/2}\, dx$$
$$= \lim_{a\to\infty} (-e^{-x/2})\Big|_2^a$$
$$= \lim_{a\to\infty} \left(\frac{-1}{e^{a/2}} + \frac{1}{e}\right)$$
$$\approx 0.3679$$

33. $f(x) = \begin{cases} \dfrac{x^3}{12} & \text{if } 0 \leq x \leq 2 \\ \dfrac{16}{3x^3} & \text{if } x > 2 \end{cases}$

First, note that $f(x) > 0$ for $x > 0$. Next,

$$\int_0^{\infty} f(x)dx = \int_0^2 \frac{x^3}{12}dx + \lim_{a\to\infty} \int_2^a \frac{16}{3x^3}dx$$
$$= \left(\frac{x^4}{48}\right)\Big|_0^2 + \lim_{a\to\infty} \left(-\frac{8}{3x^2}\right)\Big|_2^a$$
$$= \left(\frac{1}{3} - 0\right) + \left[\lim_{a\to\infty} \left(-\frac{8}{3a^2}\right) - \left(-\frac{8}{12}\right)\right]$$
$$= \frac{1}{3} + \frac{2}{3}$$
$$= 1.$$

Therefore, $f(x)$ is a probability density function.

(a) $P(0 \leq X \leq 2) = \int_0^2 f(x)dx$

$$= \left(\frac{x^4}{48}\right)\Big|_0^2$$
$$= \frac{1}{3}$$

(b) $P(X \geq 2) = P(X > 2)$

$$= \int_2^{\infty} \frac{16}{3x^3}\, dx$$
$$= \lim_{a\to\infty} \int_2^a \frac{16}{3x^3}\, dx$$
$$= \lim_{a\to\infty} \left(-\frac{8}{3x^2}\right)\Big|_2^a$$
$$= \lim_{a\to\infty} \left(-\frac{8}{3a^2}\right) - \left(-\frac{8}{3\cdot 2^2}\right)$$
$$= 0 - \left(-\frac{2}{3}\right)$$
$$= \frac{2}{3}$$

(c) $P(1 \le X \le 3) = \int_1^2 \frac{x^3}{12}\,dx + \int_2^3 \frac{16}{3x^3}\,dx$

$$= \left(\frac{x^4}{48}\right)\Big|_1^2 + \left(-\frac{8}{3x^2}\right)\Big|_2^3$$

$$= \left(\frac{1}{3} - \frac{1}{48}\right) + \left(-\frac{8}{27} + \frac{2}{3}\right)$$

$$= \frac{295}{432}$$

35. $f(x) = \frac{1}{2}e^{-x/2}$; $[0, \infty)$

(a) $P(0 \le X \le 12) = \frac{1}{2}\int_0^{12} e^{-x/2}\,dx$

$$= -e^{-x/2}\Big|_0^{12}$$

$$= \frac{-1}{e^6} + 1$$

$$\approx 0.9975$$

(b) $P(12 \le X \le 20) = \frac{1}{2}\int_{12}^{20} e^{-x/2}\,dx$

$$= -e^{-x/2}\Big|_{12}^{20}$$

$$= \frac{-1}{e^{10}} + \frac{1}{e^6}$$

$$\approx 0.0024$$

(c) $F(x) = \int_0^x \frac{1}{2}e^{-1/2}\,dt$

$$= \frac{1}{2}(-2)e^{-t/2}\Big|_0^x$$

$$= -(e^{-x/2} - 1)$$

$$= 1 - e^{-x/2}, x \ge 0.$$

(d) $F(6) = 1 - e^{-6/2}$

$$= 1 - e^{-3}$$

$$\approx 0.9502$$

The probability is 0.9502.

37. If $f(x)$ is proportional to $(10+x)^{-2}$, then, for some value of k, $f(x) = k(10+x)^{-2}$ on $[0, 40]$. Find k. We know the total probability must equal 1.

$$\int_0^{10} k(10+x)^{-2}\,dx = -k(10+x)^{-1}\Big|_0^{40}$$

$$= -k(50^{-1} - 10^{-1})^{-1}$$

$$= -k\left(\frac{1}{50} - \frac{1}{10}\right)$$

$$= \frac{2}{25}x$$

If $\frac{2}{25}k = 1$, then $k = \frac{25}{2}$. Therefore

$$f(x) = \frac{25}{2}(10+x)^{-2}, 0 \le x \le 40$$

So the probability density function is

$$F(x) = \int_0^x \frac{25}{2}(10+t)^{-2}\,dt$$

$$= -\frac{25}{2}(10+t)^{-1}\Big|_0^x$$

$$= -\frac{25}{2}[(10+x)^{-1} - 10^{-1}]$$

$$= -\frac{25}{2}\left(\frac{1}{10+x} - \frac{1}{10}\right)$$

$$= \frac{25}{2}\left(\frac{1}{10} - \frac{1}{10+x}\right)$$

$$F(6) = \frac{25}{2}\left(\frac{1}{10} - \frac{1}{106}\right)$$

$$\approx 0.47$$

The correct answer choice is **c**.

39. $f(x) = \frac{1}{2\sqrt{x}}$; $[1, 4]$

(a) $P(3 \le X \le 4) = \int_3^4 \left(\frac{1}{2\sqrt{x}}\right)dx$

$$= \frac{1}{2}\int_3^4 x^{-1/2}\,dx$$

$$= \frac{1}{2}(2)x^{1/2}\Big|_3^4$$

$$= 2 - 3^{1/2} \approx 0.2679$$

(b) $P(1 \le X \le 2) = \int_1^2 \left(\frac{1}{2\sqrt{x}}\right)dx$

$$= \frac{1}{2}(2)x^{1/2}\Big|_1^2$$

$$= 2^{1/2} - 1 = 0.4142$$

(c) $P(2 \le X \le 3) = \int_2^3 \left(\frac{1}{2\sqrt{x}}\right)dx$

$$= \frac{1}{2}(2)x^{1/2}\Big|_2^3$$

$$= 3^{1/2} - 2^{1/2} = 0.3178$$

41. $f(x) = 1.185 \cdot 10^{-9} x^{4.5222} - 0.049846x$

(a) $P(0 \leq X \leq 150) = \int_0^{150} 1.185 \cdot 10^{-9} x^{4.5222} e^{-0.049846x}\, dx \approx 0.8131$

(b) $P(100 \leq x \leq 200) = \int_{100}^{200} 1.185 \cdot 10^{-9} x^{4.5222} e^{-0.049846x}\, dx \approx 0.4901$

43. (a)

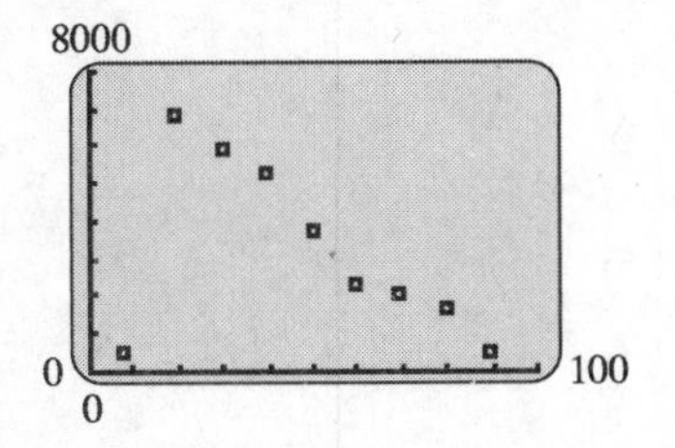

Of the types of functions available using the regression feature of a graphing utility, a polynomial function best matches the data.

(b) The function

$N(x) = -0.00272454x^4 + 0.614038x^3 - 48.0160x^2 + 1418.53x - 7202.78$

provided by the calculator models this data well, as illustrated by the following graph.

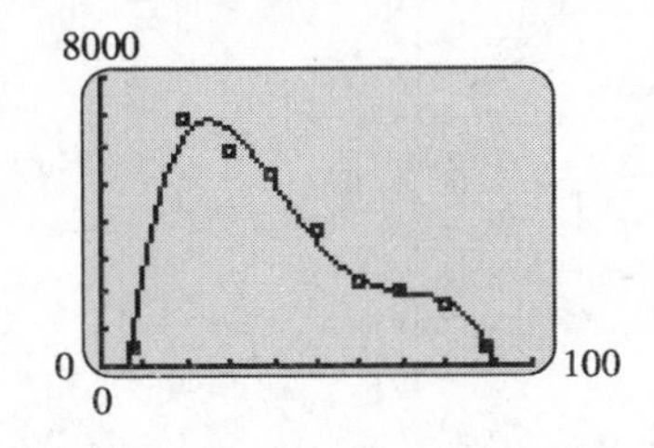

(c) Using the integration feature on our calculator, we find that

$$\int_{6.4}^{91.5} N(x)\, dx \approx 304{,}337.$$

So,

$$S(x) = \frac{1}{304{,}337} N(x) = \frac{1}{304{,}337}(-0.00272454x^4 + 0.614038x^3 - 48.0160x^2 + 1418.53x - 7202.78)$$

will be a probability density function for $[6.4, 91.5]$, because

$$\int_{6.4}^{91.5} S(x)\, dx = 1, \text{ and } S(x) \geq 0$$

for all x in $[6.4, 91.5]$.

(d) Again, using the integration feature on our calculator,

$$P(X < 25) = \int_{6.4}^{25} S(x)\, dx \approx 0.2917$$

$$P(45 < X < 65) = \int_{45}^{65} S(x)\, dx \approx 0.1919$$

$$P(X > 75) = \int_{75}^{91.5} S(x)\, dx \approx 0.0728.$$

From the table, the actual probabilities are

$$P(X < 25) = \frac{489 + 6795}{28{,}854} \approx 0.2524$$
$$P(45 < X < 65) = \frac{3722 + 2306}{28{,}854} \approx 0.2089$$
$$P(X > 75) = \frac{1738 + 561}{28{,}854} \approx 0.0797.$$

45. $f(x) = \dfrac{5.5 - x}{15}$; $[0, 5]$

(a) $P(3 \le X \le 5) = \displaystyle\int_3^5 \frac{5.5 - x}{15}\,dx = 0.2 = \left(\frac{5.5}{15}x - \frac{1}{15}\cdot\frac{x^2}{2}\right)\Big|_3^5 = \left(\frac{5.5}{15}\cdot 5 - \frac{1}{15}\cdot\frac{5^2}{2}\right) - \left(\frac{5.5}{15}\cdot 3 - \frac{1}{15}\cdot\frac{3^2}{2}\right)$

(b) $P(0 \le X \le 2) = \displaystyle\int_0^2 \frac{5.5 - x}{15}\,dx = \left(\frac{5.5}{15}x - \frac{1}{15}\cdot\frac{x^2}{2}\right)\Big|_0^2 = -\left(\frac{5.5}{15}\cdot 2 - \frac{1}{15}\cdot\frac{2^2}{2}\right) - \left(\frac{5.5}{15}\cdot 0 - \frac{1}{15}\cdot\frac{0^2}{2}\right) = 0.6$

(c) $P(1 \le X \le 4) = \displaystyle\int_1^4 \frac{5.5 - x}{15}\,dx = \left(\frac{5.5}{15}x - \frac{1}{15}\cdot\frac{x^2}{2}\right)\Big|_1^4 = \left(\frac{5.5}{15}\cdot 4 - \frac{1}{15}\cdot\frac{4^2}{2}\right) - \left(\frac{5.5}{15}\cdot 1 - \frac{1}{15}\cdot\frac{1^2}{2}\right) = 0.6$

47. $f(t) = \dfrac{1}{3650.1}e^{-t/3650.1}$

(a) $P(365 < T < 1095) = \displaystyle\int_{365}^{1095} \frac{1}{3650.1}e^{-t/3650.1}dt = (-e^{-t/3650.1})\Big|_{365}^{1095} = -e^{-1095/3650.1} + e^{-365/3650.1} \approx 0.16$

(b) $P(T > 7300) = \displaystyle\int_{7300}^{\infty} \frac{1}{3650.1}e^{-t/3650.1}dT = \lim_{b\to\infty}\int_{7300}^{b} \frac{1}{3650.1}e^{-t/3650.1}dt = \lim_{b\to\infty}(-e^{-t/3650.1})\Big|_{7300}^{b}$

$= \displaystyle\lim_{b\to\infty}(-e^{-b/3650.1} + e^{-7300/3650.1}) = 0 + e^{-7300/3650.1} \approx 0.14$

49. $f(x) = 0.06049e^{-0.03211}$; $[16, 84]$

(a) $P(16 \le X \le 25) = \displaystyle\int_{16}^{25} f(x)dx = \int_{16}^{25} 0.06049e^{-0.03211x}dx = \frac{0.06049}{-0.03211}(e^{-0.03211x})\Big|_{16}^{25}$

$\approx -1.88384(e^{-0.03211x})\Big|_{16}^{25} = -1.88384(e^{-0.03211\cdot 25} - e^{-0.03211\cdot 16})$

≈ 0.2829

(b) $P(35 \le X \le 84) = \displaystyle\int_{35}^{84} 0.06049e^{-0.03211x}dx \approx -1.88384(e^{-0.03211x})\Big|_{35}^{84} \approx 0.4853$

(c) $P(21 \le X \le 30) = \displaystyle\int_{21}^{30} 0.06049e^{-0.03211x}dx \approx -1.88384(e^{-0.03211x})\Big|_{21}^{30} \approx 0.2409$

(d) $F(x) = \displaystyle\int_{16}^{x} 0.06049e^{-0.03211t}dt = 0.06049\cdot\frac{1}{-0.03211t}e^{-0.03211t}\Big|_{16}^{x}$

$= -1.8838(e^{-0.03211x} - e^{-0.03211\cdot 16}) = 1.8838(0.5982 - e^{-0.03211x}), 16 \le x \le 84$

(e) $F(21) = 1.8838(0.5982 - e^{-0.03211\cdot 21}) = 1.8838(0.0887) = 0.1671$

The probability is 0.1671.

11.2 Expected Value and Variance of Continuous Random Variables

1. $f(x) = \frac{1}{4}$; [3, 7]

$$E(X) = \mu = \int_3^7 \frac{1}{4}x\,dx = \frac{1}{4}\left(\frac{x^2}{2}\right)\Bigg|_3^7$$
$$= \frac{49}{8} - \frac{9}{8}$$
$$= 5$$

$$\text{Var}(X) = \int_3^7 (x-5)^2\left(\frac{1}{4}\right)dx$$
$$= \frac{1}{4}\cdot\frac{(x-5)^3}{3}\Bigg|_3^7$$
$$= \frac{8}{12} + \frac{8}{12}$$
$$= \frac{4}{3} \approx 1.33$$
$$\sigma \approx \sqrt{\text{Var}(X)}$$
$$= \sqrt{\frac{4}{3}}$$
$$\approx 1.15$$

3. $f(x) = \frac{x}{8} - \frac{1}{4}$; [2, 6]

$$\mu = \int_2^6 x\left(\frac{x}{8} - \frac{1}{4}\right)dx$$
$$= \int_2^6 \left(\frac{x^2}{8} - \frac{x}{4}\right)dx$$
$$= \left(\frac{x^3}{24} - \frac{x^2}{8}\right)\Bigg|_2^6$$
$$= \left(\frac{216}{24} - \frac{36}{8}\right) - \left(\frac{8}{24} - \frac{4}{8}\right)$$
$$= \frac{208}{24} - 4$$
$$= \frac{26}{3} - 4$$
$$= \frac{14}{3} \approx 4.67$$

Use the alternative formula to find

$$\text{Var}(X) = \int_2^6 x^2\left(\frac{x}{8} - \frac{1}{4}\right)dx - \left(\frac{14}{3}\right)^2$$
$$= \int_2^6 \left(\frac{x^3}{8} - \frac{x^2}{4}\right)dx - \frac{196}{9}$$
$$= \left(\frac{x^4}{32} - \frac{x^3}{12}\right)\Bigg|_2^6 - \frac{196}{9}$$
$$= \left(\frac{1296}{32} - \frac{216}{12}\right) - \left(\frac{16}{32} - \frac{8}{12}\right) - \frac{196}{9}$$
$$\approx 0.89.$$
$$\sigma = \sqrt{\text{Var}(X)} \approx \sqrt{0.89} \approx 0.94$$

5. $f(x) = 1 - \frac{1}{\sqrt{x}}$; [1, 4]

$$\mu = \int_1^4 x(1 - x^{-1/2})dx = \int_1^4 (x - x^{1/2})dx$$
$$= \left(\frac{x^2}{2} - \frac{2x^{3/2}}{3}\right)\Bigg|_1^4 = \frac{16}{2} - \frac{16}{3} - \frac{1}{2} + \frac{2}{3}$$
$$= \frac{17}{6} \approx 2.83$$

$$\text{Var}(X) = \int_1^4 x^2(1 - x^{-1/2})dx - \left(\frac{17}{6}\right)^2$$
$$= \int_1^4 (x^2 - x^{3/2})dx - \frac{289}{36}$$
$$= \left(\frac{x^3}{3} - \frac{2x^{5/2}}{5}\right)\Bigg|_1^4 - \frac{289}{36}$$
$$= \frac{64}{3} - \frac{64}{5} - \frac{1}{3} + \frac{2}{5} - \frac{289}{36}$$
$$\approx 0.57$$
$$\sigma \approx \sqrt{\text{Var}(X)} \approx 0.76$$

7. $f(x) = 4x^{-5}$; $[1, \infty)$

$$\mu = \int_1^\infty x(4x^{-5})dx$$
$$= \lim_{a\to\infty} \int_1^a 4x^{-4}\,dx$$
$$= \lim_{a\to\infty} \left(\frac{4x^{-3}}{-3}\right)\Bigg|_1^a$$
$$= \lim_{a\to\infty} \left(\frac{-4}{3a^3} + \frac{4}{3}\right)$$
$$= \frac{4}{3} \approx 1.33$$

$$\text{Var}(X) = \int_1^\infty x^2(4x^{-5})dx - \left(\frac{4}{3}\right)^2$$
$$= \lim_{a\to\infty} \int_1^a 4x^{-3}\,dx - \frac{16}{9}$$
$$= \lim_{a\to\infty} \left(\frac{4x^{-2}}{-2}\right)\Bigg|_1^a - \frac{16}{9}$$
$$= \lim_{a\to\infty} \left(\frac{-2}{a^2} + 2\right) - \frac{16}{9}$$
$$= 2 - \frac{16}{9} = \frac{2}{9} \approx 0.22$$
$$\sigma = \sqrt{\text{Var}(X)} = \sqrt{\frac{2}{9}} \approx 0.47$$

11. $f(x) = \dfrac{\sqrt{x}}{18}$; $[0,\ 9]$

(a) $E(X) = \mu = \displaystyle\int_0^9 \frac{x\sqrt{x}}{18}\,dx$
$$= \int_0^9 \frac{x^{3/2}}{18}\,dx$$
$$= \frac{2x^{5/2}}{90}\Bigg|_0^9 = \frac{x^{5/2}}{45}\Bigg|_0^9$$
$$= \frac{243}{45} = \frac{27}{5} = 5.40$$

(b) $\text{Var}(X) = \displaystyle\int_0^9 \frac{x^2\sqrt{x}}{18}\,dx - \left(\frac{27}{5}\right)^2$
$$= \int_0^9 \frac{x^{5/2}}{18}\,dx - \left(\frac{27}{5}\right)^2$$
$$= \frac{x^{7/2}}{63}\Bigg|_0^9 - \left(\frac{27}{5}\right)^2$$
$$= \frac{2187}{63} - \left(\frac{27}{5}\right)^2 \approx 5.55$$

(c) $\sigma = \sqrt{\text{Var}(X)} \approx 2.36$

(d) $P(5.40 < X \le 9)$
$$= \int_{5.4}^9 \frac{x^{1/2}}{18}\,dx$$
$$= \frac{x^{3/2}}{27}\Bigg|_{5.4}^9$$
$$= \frac{27}{27} - \frac{(5.4)^{1.5}}{27}$$
$$\approx 0.54$$

(e) $P(5.40 - 2.36 \le X \le 5.40 + 2.36)$
$$= \int_{3.04}^{7.76} \frac{x^{1/2}}{18}\,dx$$
$$= \frac{x^{3/2}}{27}\Bigg|_{3.04}^{7.76}$$
$$= \frac{7.76^{3/2}}{27} - \frac{3.04^{3/2}}{27}$$
$$\approx 0.60$$

13. $f(x) = \dfrac{1}{2}x$; $[0,\ 2]$

(a) $E(X) = \mu = \displaystyle\int_0^2 \frac{1}{2}x^2\,dx = \frac{x^3}{6}\Bigg|_0^2$
$$= \frac{8}{6} = \frac{4}{3} \approx 1.33$$

(b) $\text{Var}(X) = \displaystyle\int_0^1 \frac{1}{2}x^3\,dx - \frac{16}{9}$
$$= \frac{x^4}{8}\Bigg|_0^2 - \frac{16}{9}$$
$$= 2 - \frac{16}{9} = \frac{2}{9} \approx 0.22$$

(c) $\sigma = \sqrt{\text{Var}(X)} = \sqrt{\frac{2}{9}} \approx 0.47$

(d) $P\left(\frac{4}{3} < X \le 2\right) = \displaystyle\int_{4/3}^2 \frac{x}{2}\,dx$
$$= \frac{x^2}{4}\Bigg|_{4/3}^2$$
$$= 1 - \frac{16}{36} \approx 0.56$$

(e) $P\left(\frac{4}{3} - 0.47 \le X \le \frac{4}{3} + 0.47\right)$
$$= \int_{0.86}^{1.8} \frac{x}{2}\,dx = \frac{x^2}{4}\Bigg|_{0.86}^{1.8}$$
$$= \frac{1.8^2}{4} - \frac{0.86^2}{4} \approx 0.63$$

15. $f(x) = \dfrac{1}{4}$; $[3,\ 7]$

(a) $m = \text{median}$: $\displaystyle\int_3^m \frac{1}{4}\,dx = \frac{1}{2}$
$$\frac{1}{4}x\Bigg|_3^m = \frac{1}{2}$$
$$\frac{m}{4} - \frac{3}{4} = \frac{1}{2}$$
$$m - 3 = 2$$
$$m = 5$$

(b) $E(X) = \mu = 5$ (from Exercise 1)

$$P(X = 5) = \int_5^5 \frac{1}{4}\,dx = 0$$

17. $f(x) = \frac{x}{8} - \frac{1}{4}$; $[2,\ 6]$

(a) $m =$ median:

$$\int_2^m \left(\frac{x}{8} - \frac{1}{4}\right) dx = \frac{1}{2}$$

$$\left(\frac{x^2}{16} - \frac{x}{4}\right)\Big|_2^m = \frac{1}{2}$$

$$\frac{m^2}{16} - \frac{m}{4} - \frac{1}{4} + \frac{1}{2} = \frac{1}{2}$$

$$m^2 - 4m - 4 + 8 = 8$$

$$m^2 - 4m - 4 = 0$$

$$m = \frac{4 \pm \sqrt{16 + 16(1)}}{2}$$

Reject $\frac{4-\sqrt{32}}{2}$ since it is not in $[2,\ 6]$.

$$m = \frac{4 + \sqrt{32}}{2} \approx 4.828$$

(b) $E(X) = \mu = \frac{14}{3}$ (from Exercise 3)

$$P\left(\tfrac{14}{3} \le X \le 4.828\right)$$

$$= \int_{4.667}^{4.828} \left(\frac{x}{8} - \frac{1}{4}\right) dx$$

$$= \left(\frac{x^2}{16} - \frac{x}{4}\right)\Big|_{4.667}^{4.828}$$

$$= \frac{4.828^2}{16} - \frac{4.828}{4} - \frac{4.667^2}{16} + \frac{4.667}{4}$$

$$\approx 0.0553$$

19. $f(x) = 4x^{-5}$; $[1,\ \infty)$

(a) $m =$ median:

$$\int_1^m 4x^{-5}\,dx = \frac{1}{2}$$

$$\frac{4x^{-4}}{-4}\Big|_1^m = \frac{1}{2}$$

$$-m^{-4} + 1 = \frac{1}{2}$$

$$1 - \frac{1}{m^4} = \frac{1}{2}$$

$$2m^4 - 2 = m^4$$

$$m^4 = 2$$

$$m = \sqrt[4]{2} \approx 1.189$$

(b) $E(X) = \mu = \frac{4}{3}$ (from Exercise 7)

$$P(1.19 \le X \le \tfrac{4}{3}) \approx \int_{1.189}^{1.333} 4x^{-5}\,dx$$

$$\approx -x^{-4}\Big|_{1.189}^{1.333}$$

$$\approx -\frac{1}{1.333^4} + \frac{1}{1.189^4}$$

$$\approx 0.1836$$

21. $f(x) = \begin{cases} \frac{x^3}{12} & \text{if } 0 \le x \le 2 \\ \frac{16}{3x^3} & \text{if } x > 2 \end{cases}$

Expected value:

$$E(X) = \mu = \int_0^\infty x f(x) dx$$

$$= \int_0^2 x\left(\frac{x^3}{12}\right) dx + \lim_{a\to\infty} \int_2^a x\left(\frac{16}{3x^3}\right) dx$$

$$= \int_0^2 \frac{x^4}{12}\,dx + \lim_{a\to\infty} \int_2^a \frac{16}{3x^2}\,dx$$

$$= \left(\frac{x^5}{60}\right)\Big|_0^2 + \lim_{a\to\infty}\left(-\frac{16}{3x}\right)\Big|_2^a$$

$$= \left(\frac{8}{15} - 0\right) + \left[\lim_{a\to\infty}\left(-\frac{16}{3a}\right) - \left(-\frac{16}{6}\right)\right]$$

$$= \frac{16}{5}$$

Variance:

$$\text{Var}(X) = \int_0^\infty x^2 f(x) dx - \mu^2$$

$$= \int_0^2 x^2\left(\frac{x^3}{12}\right) dx + \int_2^\infty \left(\frac{16}{3x^3}\right) dx - \left(\frac{16}{5}\right)^2$$

Examine the second integral.

$$\int_2^\infty x^2\left(\frac{16}{3x^3}\right) dx = \lim_{a\to\infty} \int_2^a x^2\left(\frac{16}{3x^3}\right) dx$$

$$= \lim_{a\to\infty} \int_2^a \frac{16}{3x} dx$$

$$= \lim_{a\to\infty} \frac{16}{3} \ln|a| - \frac{16}{3} \ln|2|$$

Since the limit diverges, neither the variance nor the standard deviation exists.

23. $f(x) = \begin{cases} \dfrac{|x|}{10} & \text{for } -2 \le x \le 4 \\ 0 & \text{otherwise} \end{cases}$

First, note that

$$|x| = \begin{cases} -x \text{ for } & -2 \le x \le 0 \\ x \text{ for } & 0 \le x \le 4 \end{cases}$$

The expected value is

$$E(X) = \mu = \int_{-2}^{4} x \cdot \frac{|x|}{10}\,dx = \int_{-2}^{0} x \cdot \frac{-x}{10}\,dx + \int_{0}^{4} x \cdot \frac{x}{10}\,dx = \int_{-2}^{0} -\frac{x^2}{10}\,dx + \int_{0}^{4} \frac{x^2}{10}\,dx$$

$$= -\left.\frac{x^3}{30}\right|_{-2}^{0} + \left.\frac{x^3}{30}\right|_{0}^{4} = -\left(0 - \frac{-8}{30}\right) + \left(\frac{64}{30} - 0\right) = \frac{56}{30} = \frac{28}{15}$$

The correct answer choice is **d**.

25. $f(x) = \frac{1}{11}\left(1 + \frac{3}{\sqrt{x}}\right)$; [4, 9]

(a) From Exercise 6, $\mu \approx 6.41$ yr.

(b) $\sigma \approx 1.45$ yr.

(c) $P(X > 6.41) = \int_{6.41}^{9} \frac{1}{11}(1 + 3x^{-1/2})dx = \left.\frac{1}{11}(x + 6x^{1/2})\right|_{6.41}^{9} = \frac{1}{11}[9 + 18 - 6.41 - 6(6.41)^{1/2}] \approx 0.49$

27. Using the hint, we have

$$\text{loss not paid} = \begin{cases} x \text{ for } 0.6 < x < 2 \\ 2 \text{ for } x > 2 \end{cases}$$

Therefore, the mean of the manufacturer's annual losses not paid will be

$$\mu = \int_{0.6}^{0} x \cdot f(x)dx + \int_{2}^{\infty} 2 \cdot f(x)\,dx = \int_{0.6}^{2} x\frac{2.5(0.6)^{2.5}}{x^{3.5}}\,dx + \int_{2}^{\infty} 2\,\frac{2.5(0.6)^{2.5}}{x^{3.5}}\,dx$$

$$= 2.5(0.6)^{2.5}\int_{0.6}^{2} \frac{1}{x^{2.5}}\,dx + 5(0.6)^{2.5}\int_{2}^{\infty} \frac{1}{x^{3.5}}\,dx = 2.5(0.6)^{2.5}\left(\frac{1}{-1.5}\right)\left.\frac{1}{x^{1.5}}\right|_{0.6}^{2} + 5(0.6)^{2.5}\left(\frac{1}{-2.5}\right)\left.\frac{1}{x^{2.5}}\right|_{2}^{\infty}$$

$$= -\frac{5}{3}(0.6)^{2.5}\left(\frac{1}{2^{1.5}} - \frac{1}{0.6^{1.5}}\right) - 2(0.6)^{2.5}\left(0 - \frac{1}{2^{2.5}}\right) \approx 0.8357 + 0.0986 \approx 0.93$$

The correct answer choice is **c**.

29. Since the probability density function is proportional to $(1+x)^{-4}$, we have $f(x) = k(1+x)^{-4}$, $0 < x < \infty$. To determine k, solve the equation $\int_0^\infty kf(x)dx = 1$.

$$\int_{0}^{\infty} k(1+x)^{-4}dx = 1$$

$$k\left(-\frac{1}{3}\right)(1+x)^{-3}\Big|_{0}^{\infty} = 1$$

$$-\frac{k}{3}(0-1) = 1$$

$$\frac{k}{3} = 1$$

$$k = 3$$

Thus, $f(x) = 3(1+x)^{-4}, 0 < x < \infty$.

The expected monthly claims are

$$\int_0^\infty x \cdot 3(1+x)^{-4}dx = 3\int_0^\infty \frac{x}{(1+x)^4}\,dx$$

The antiderivative can be found using the substitution $u = 1 - x$.

$$\int \frac{x}{(1+x)^4}\,dx = \int \frac{u-1}{u^4}\,du$$
$$= \int \left(\frac{1}{u^3} - \frac{1}{u^4}\right) du$$
$$= -\frac{1}{2u^2} + \frac{1}{3u^3}$$

Resubstitute $u = 1 + x$.

$$3\int_0^\infty \frac{x}{(1+x)^4}\,dx = 3\left(-\frac{1}{2(1+x)^2} + \frac{1}{3(1+x)^3}\right)\Bigg|_0^\infty$$
$$= 3\left[0 - \left(-\frac{1}{2} + \frac{1}{3}\right)\right]$$
$$= 3\left(\frac{1}{6}\right)$$
$$= \frac{1}{2}$$

The correct answer choice is **c**.

31. $f(x) = \dfrac{1}{(\ln 20)x}$; $[1,\ 20]$

(a) $\mu = \displaystyle\int_1^{20} x \cdot \frac{1}{(\ln 20)x}\,dx$

$$= \int_1^{20} \frac{1}{\ln 20}\,dx$$
$$= \frac{x}{\ln 20}\Bigg|_1^{20}$$
$$= \frac{19}{\ln 20} \approx 6.342 \text{ seconds}$$

(b) $\operatorname{Var}(X) = \displaystyle\int_1^{20} x^2 \cdot \frac{1}{(\ln 20)x}\,dx - \mu^2$

$$= \int_1^{20} \frac{x}{\ln 20}\,dx - \mu^2$$
$$= \frac{x^2}{2\ln 20}\Bigg|_1^{20} - (6.34)^2$$
$$= \frac{399}{2\ln 20} - (6.34)^2$$
$$\approx 26.40$$
$$\sigma \approx \sqrt{26.40}$$
$$\approx 5.138 \text{ sec}$$

(c) $P(6.34 - 5.14 < X < 6.34 + 5.14)$

$$= P(1.2 < X < 11.48)$$
$$= \int_{1.2}^{11.48} \frac{1}{(\ln 20)x}\,dx$$
$$= \frac{\ln x}{\ln 20}\Bigg|_{1.2}^{11.48}$$
$$= \frac{1}{\ln 20}(\ln 11.48 - \ln 1.2)$$
$$\approx 0.7538$$

(d) The median clotting time is the value of m such that $\int_a^m f(x)dx = \frac{1}{2}$.

$$\int_1^m \frac{1}{(\ln 20)x}\,dx = \frac{1}{2}$$
$$\frac{1}{\ln 20}\ln x\Bigg|_1^m = \frac{1}{2}$$
$$\frac{1}{\ln 20}(\ln m - 0) = \frac{1}{2}$$
$$\ln m = \frac{\ln 20}{2}$$
$$m = e^{\ln 20/2} \approx 4.472$$

33. $f(x) = \dfrac{1}{2\sqrt{x}}$; $[1,\ 4]$

(a) $\mu = \displaystyle\int_1^4 x \cdot \frac{1}{2\sqrt{x}}\,dx$

$$= \int_1^4 \frac{x^{1/2}}{2}\,dx$$
$$= \frac{x^{3/2}}{3}\Bigg|_1^4$$
$$= \frac{1}{3}(8-1)$$
$$= \frac{7}{3} \approx 2.333 \text{ cm}$$

(b) $\operatorname{Var}(X) = \displaystyle\int_1^4 x^2 \cdot \frac{1}{2\sqrt{x}}\,dx - \left(\frac{7}{3}\right)^2$

$$= \int_1^4 \frac{x^{3/2}}{2}\,dx - \frac{49}{9}$$
$$= \frac{x^{5/2}}{5}\Bigg|_1^4 - \frac{49}{9}$$
$$= \frac{1}{5}(32-1) - \frac{49}{9}$$
$$\approx 0.7556$$
$$\sigma = \sqrt{\operatorname{Var}(X)}$$
$$\approx 0.8692 \text{ cm}$$

(c) $P(X > 2.33 + 2(0.87))$
$= P(x > 4.07)$
$= 0$

The probability is 0 since two standard deviations falls out of the given interval [1, 4].

(d) The median petal length is the value of m such that $\int_a^m f(x)dx = \frac{1}{2}$.

$$\int_1^m \frac{1}{2\sqrt{x}}\,dx = \frac{1}{2}$$
$$\sqrt{x}\Big|_1^m = \frac{1}{2}$$
$$\sqrt{m} - 1 = \frac{1}{2}$$
$$\sqrt{m} = \frac{3}{2}$$
$$m = \frac{9}{4} = 2.25$$

The median petal length is 2.25 cm.

35. $f(x) = 1.185 \cdot 10^{-9} x^{4.5222} e^{-0.049846x}$

$$E(X) = \int_1^{1000} x\,f(x)\,dx$$

Using the integration function on our calculator.

$E(X) \approx 110.80$

The expected size is about 111.

37. $f(x) = \dfrac{5.5 - x}{15}$; [0, 5]

(a)
$$\mu = \int_0^5 x\left(\frac{5.5 - x}{15}\right)dx$$
$$= \int_0^5 \left(\frac{5.5}{15}x - \frac{1}{15}x^2\right)dx$$
$$= \frac{5.5}{30}x^2 - \frac{1}{45}x^3\Big|_0^5$$
$$= \left(\frac{5.5}{30}\cdot 25 - \frac{1}{45}\cdot 125\right) - 0$$
$$\approx 1.806$$

(b)
$$\text{Var}(X) = \int_0^5 x^2\left(\frac{5.5 - x}{15}\right)dx - \mu^2$$
$$= \int_0^5 \left(\frac{5.5}{15}x^2 - \frac{1}{15}x^3\right)dx - \mu^2$$
$$= \left(\frac{5.5}{45}x^3 - \frac{1}{60}x^4\right)\Big|_0^5 - \mu^2$$
$$= \frac{5.5}{45}\cdot 125 - \frac{1}{60}\cdot 625 - 0 - \mu^2$$
$$\approx 1.60108$$
$$\sigma = \sqrt{\text{Var}(X)}$$
$$\approx 1.265$$

(c) $P(X \le \mu - \sigma)$
$$= P(X \le 1.806 - 1.265)$$
$$= P(X \le 0.541)$$
$$= \int_0^{0.541} \frac{5.5 - x}{15}\,dx$$
$$= \left(\frac{5.5}{15}x - \frac{1}{30}x^2\right)\Big|_0^{0.541}$$
$$= \left(\frac{5.5}{15}(0.541) - \frac{1}{30}(0.541)^2 - 0\right)$$
$$\approx 0.1886$$

39. $f(x) = \dfrac{0.1906}{x^{0.5012}}$; or [16, 44]

(a) Expected value:

$$E(X) = \mu = \int_{16}^{44} x\,\frac{0.1906}{x^{0.5012}}\,dx$$
$$= \int_{16}^{44} 0.1906x^{0.4988}\,dx$$
$$= \frac{0.1906}{1.4988}x^{1.4988}\Big|_{16}^{44}$$
$$= \frac{0.1906}{1.4988}(44^{1.4988} - 16^{1.4988})$$
$$\approx 28.8358$$
$$\approx 28.84 \text{ years}$$

(b) Standard deviation:

$$\begin{aligned}\mathrm{Var}(X) &= \int_{16}^{44} x^2 \frac{0.1906}{x^{0.5012}}\,dx - 28.8358^2 = \int_{16}^{44} 0.1906x^{1.4988}dx - 28.8358^2 \\ &= \frac{0.1906}{2.4988}x^{2.4988}\Big|_{16}^{44} - 28.8358^2 = \frac{0.1906}{2.4988}(44^{2.4988} - 16^{2.4988}) - 28.8358^2 \\ &\approx 65.75\end{aligned}$$

$$\sigma = \sqrt{\mathrm{Var}(X)} = \sqrt{65.75} \approx 8.109$$

(c)
$$\begin{aligned}P(16 \le X \le 28.8 - 8.1) &= \int_{16}^{20.7} \frac{0.1906}{x^{0.5012}}\,dx \\ &= \frac{0.1906}{0.4988}x^{0.4988}\Big|_{16}^{20.7} \\ &= \frac{0.1906}{0.4988}(20.7^{0.4988} - 16^{0.4988}) \\ &\approx 0.2088\end{aligned}$$

(d) To find the median age, find the value of m such that $\int_a^m f(x)dx = \frac{1}{2}$.

$$\begin{aligned}\int_{16}^{m} \frac{0.1906}{x^{0.5012}}\,dx &= \frac{1}{2} \\ \frac{0.1906}{0.4988}x^{0.4988}\Big|_{16}^{m} &= \frac{1}{2} \\ \frac{0.1906}{0.4988}(m^{0.4988} - 16^{0.4988}) &= \frac{1}{2} \\ m^{0.4988} - 16^{0.4988} &= \frac{0.4988}{2 \cdot 0.1906} \\ m^{0.4988} &= \frac{0.4988}{0.3812} + 16^{0.4988} \\ 0.4988 \ln m &= \ln\left(\frac{0.4988}{0.3812} + 16^{0.4988}\right) \\ \ln m &= \frac{1}{0.4988}\ln\left(\frac{0.4988}{0.3812} + 16^{0.4988}\right) \\ m &= e^{\frac{1}{0.4988}\ln\left(\frac{0.4988}{0.3812}+16^{0.4988}\right)} \approx 28.27\end{aligned}$$

The median age is 28.27 years.

11.3 Special Probability Density Functions

1. $f(x) = \frac{5}{7}$ for x in $[3,\ 4.4]$

This is a uniform distribution:
$a = 3,\ b = 4.4$.

(a) $\mu = \frac{1}{2}(4.4 + 3) = \frac{1}{2}(7.4)$
$= 3.7$ cm

(b) $\sigma = \frac{1}{\sqrt{12}}(4.4 - 3)$
$= \frac{1}{\sqrt{12}}(1.4)$
≈ 0.4041 cm

(c) $P(3.7 < X < 3.7 + 0.4041)$
$= P(3.7 < X < 4.1041$

$$= \int_{3.7}^{4.1041} \frac{5}{7}\, dx$$

$$= \frac{5}{7}x\Big|_{3.7}^{4.1041}$$

≈ 0.2886

3. $f(t) = 4e^{-4t}$ for t in $[0,\ \infty)$

This is an exponential distribution: $a = 4$.

(a) $\mu = \frac{1}{4} = 0.25$ year

(b) $\sigma = \frac{1}{4} = 0.25$ year

(c) $P(0.25 < T < 0.25 + 0.25)$
$= P(0.25 < T < 0.5)$

$$= \int_{0.25}^{0.5} 4e^{-4t}\, dt$$

$$= -e^{-4t}\Big|_{0.25}^{0.5}$$

$$= -\frac{1}{e^{-2}} + \frac{1}{e^{-1}}$$

≈ 0.2325

5. $f(t) = \frac{e^{-t/3}}{3}$ for t in $[0,\ \infty)$

This is an exponential distribution: $a = \frac{1}{3}$.

(a) $\mu = \frac{1}{\frac{1}{3}} = 3$ days

(b) $\sigma = \frac{1}{\frac{1}{3}} = 3$ days

(c) $P(3 < T < 3 + 3) = P(3 < T < 6)$

$$= \int_3^6 \frac{e^{-t/3}}{3}\, dt$$

$$= -e^{-t/3}\Big|_3^6$$

$$= -\frac{1}{e^{-2}} + \frac{1}{e^{-1}}$$

≈ 0.2325

In Exercises 7–13, use the table in the Appendix for areas under the normal curve.

7. $z = 3.50$

Area to the left of $z = 3.50$ is 0.9998. Given mean $\mu = z - 0$, so area to left of μ is 0.5.
Area between μ and z is

$$0.9998 - 0.5 = 0.4998.$$

Therefore, this area represents 49.98% of total area under normal curve.

9. Between $z = 1.28$ and $z = 2.05$

Area to left of $z = 2.05$ is 0.9798 and area to left of $z = 1.28$ is 0.8997.

$$0.9798 - 0.8997 = 0.0801$$

Percent of total area $= 8.01\%$

11. Since $10\% = 0.10$, the z-score that corresponds to the area of 0.10 to the left of z is -1.28.

13. 18% of the total area to the right of z means $1 - 0.18$ of the total area is to the left of z.

$$1 - 0.18 = 0.82$$

The closest z-score that corresponds to the area of 0.82 is 0.92.

19. Let m be the median of the exponential distribution $f(x) = ae^{-ax}$ for $[0, \infty)$.

$$\int_0^m ae^{-ax}\,dx = 0.5$$

$$-e^{-ax}\Big|_0^m = 0.5$$

$$-e^{-am} + 1 = 0.5$$

$$0.5 = e^{-am}$$

$$-am = \ln 0.5$$

$$m = -\frac{\ln 0.5}{a}$$

$$\text{or} \quad -am = \ln \frac{1}{2}$$

$$-am = -\ln 2$$

$$m = \frac{\ln 2}{a}$$

21. The area that is to the left of x is

$$A = \int_{-\infty}^{x} \frac{1}{\sigma\sqrt{2\pi}} e^{-\frac{(t-\mu)^2}{2\sigma^2}}\,dt.$$

Let $u = \frac{t-\mu}{\sigma}$. Then $du = \frac{1}{\sigma}\,dt$ and $dt = \sigma du$.
If $t = x$,

$$u = \frac{x-\mu}{\sigma} = z.$$

As $t \to -\infty$, $\mu \to -\infty$.
Therefore,

$$A = \int_{-\infty}^{z} \frac{1}{\sigma\sqrt{2\pi}} e^{(-1/2)u^2}\sigma\,du$$

$$= \frac{\sigma}{\sigma}\int_{-\infty}^{z} \frac{1}{\sqrt{2\pi}} e^{-u^2/2}\,du$$

$$= \int_{-\infty}^{z} \frac{1}{\sqrt{2\pi}} e^{-u^2/2}\,du.$$

This is the area to the left of z for the standard normal curve.

In Exercises 23–25, use Simpson's rule with $n = 100$ or the integration feature on a graphing calculator to approximate the integrals. Answers may vary slightly from those given here depending on the method that is used.

23. **(a)** $\int_0^{50} 0.5e^{-0.5x}\,dx \approx 1.00000$

(b) $\int_0^{50} 0.5xe^{-0.5x}\,dx \approx 1.99999$

(c) $\int_0^{50} 0.5x^2e^{-0.5x}\,dx = 8.00003$

25. $\int_{-\infty}^{\infty} \frac{1}{\sqrt{2\pi}} e^{-x^2/2}\,dx$

$$\approx \int_{-4}^{4} \frac{1}{\sqrt{2\pi}} e^{-x^2/2}\,dx$$

(a) $\mu = 1.75 \times 10^{-14} \approx 0$

(b) $\sigma = 0.999433 \approx 1$

27. The probability density function for the uniform distribution is $f(x) = \frac{1}{b-a}$ for x in $[a, b]$.

The cumulative distribution function for f is

$$F(x) = P(X \le x)$$

$$= \int_a^x f(t)\,dt$$

$$= \int_a^x \frac{1}{b-a}\,dt$$

$$= \frac{1}{b-a} t\Big|_a^x$$

$$= \frac{1}{b-a}(x-a)$$

$$= \frac{x-a}{b-a}, a \le x \le b.$$

29. For a uniform distribution,

$$f(x) = \frac{1}{b-a} \text{ for } [a, b].$$

Thus, we have

$$f(x) = \frac{1}{85-10} = \frac{1}{75}$$

for $[10, 85]$.

(a) $\mu = \frac{1}{2}(10 + 85) = \frac{1}{2}(95)$

$= 47.5$ thousands

Therefore, the agent sells \$47,500 in insurance.

(b) $P(50 < X < 85) = \int_{50}^{85} \frac{1}{75}\,dx$

$$= \frac{x}{75}\Big|_{50}^{85}$$

$$= \frac{85}{75} - \frac{50}{75}$$

$$= \frac{35}{75} = 0.4667$$

31. **(a)** Since we have an exponential distribution with $\mu = 4.25$,

$$\mu = \frac{1}{a} = 4.25$$
$$a = 0.235.$$

Therefore, $f(x) = 0.235e^{-0.235x}$ on $[0, \infty)$.

(b) $P(X > 10)$

$$= \int_{10}^{\infty} 0.235e^{-0.235x} dx$$
$$= \lim_{a\to\infty} \int_{10}^{a} 0.235e^{-0.235x} dx$$
$$= \lim_{a\to\infty} (-e^{-0.235x})\Big|_{10}^{a}$$
$$= \lim_{a\to\infty} \left(-\frac{1}{e^{0.235a}} + \frac{1}{e^{2.35}}\right)$$
$$= \frac{1}{e^{2.35}} = 0.09537$$

33. **(a)** $\mu = 2.5$, $\sigma = 0.2$, $x = 2.7$

$$z = \frac{2.7 - 2.5}{0.2} = 1$$

Area to the right of $z = 1$ is

$$1 - 0.8413 = 0.1587.$$

Probability $= 0.1587$

(b) Within 1.2 standard deviations of the mean is the area between $z = -1.2$ and $z = 1.2$.
Area to left of $z = 1.2 = 0.8849$
Area to the left of $z = -1.2 = 0.1151$

$$0.8849 - 0.1151 = 0.7698$$

Probability $= 0.7698$

35. If X has a uniform distribution on $[0, 1000]$, then its density function is $f(x) = \frac{1}{1000}$ for x in $[0, 1000]$. The expected payment with no deductible is

$$E(X) = \int_0^{1000} x \cdot \frac{1}{1000} dx$$
$$= \frac{1}{1000} \cdot \frac{1}{2}x^2 \Big|_0^{1000}$$
$$= \frac{1}{2000} \cdot (1000^2 - 0)$$
$$= 500.$$

Now, let the deductible be D. According to the hint,

$$\text{payment} = \begin{cases} 0 & \text{for } x \le D \\ x - D & \text{for } x > D. \end{cases}$$

The expected payment with the deductible is therefore

$$E(X) = \int_0^D 0 \cdot \frac{1}{1000} dx + \int_D^{1000} (x - D) \cdot \frac{1}{1000} dx$$
$$= 0 + \frac{1}{1000} \cdot \left(\frac{1}{2}x^2 - Dx\right)\Big|_D^{1000}$$
$$= \frac{1}{1000}\left[\left(\frac{1}{2}1000^2 - 1000D\right) - \left(\frac{1}{2}D^2 - D^2\right)\right]$$
$$= 500 - D + \frac{1}{2000}D^2.$$

For this amount to be 25% of the amount with no deductible, we must have

$$500 - D + \frac{1}{2000}D^2 = 0.25 \cdot 500$$
$$\frac{1}{2000}D^2 - D + 500 = 125$$
$$\frac{1}{2000}D^2 - D + 375 = 0$$
$$D^2 - 2000D + 750{,}000 = 0$$
$$(D - 1500)(D - 500) = 0$$
$$D = 1500 \quad \text{or} \quad D = 500$$

We reject $D = 1500$ since it is not in $[0, 1000]$. Therefore, $D = 500$. The correct answer choice is **c**.

37. Let the random variable X be the lifetime of the printer in years. Then it has exponential distribution $f(x) = ae^{-ax}$ for $x \ge 0$.

If the mean is 2 years, then $\frac{1}{a} = 2$, or $a = \frac{1}{2}$ and the function is $f(x) = \frac{1}{2}e^{-x/2}$ for $x \ge 0$.

We wish to find $P(0 \le X \le 1)$ and $P(1 \le X \le 2)$.

$$P(0 \le X \le 1) = \int_0^1 \frac{1}{2}e^{-x/2} dx$$
$$= -e^{-x/2}\Big|_0^1$$
$$= -e^{-1/2} + 1$$

$$P(1 \le X \le 2) = \int_1^2 \frac{1}{2}e^{-x/2} dx$$
$$= -e^{-x/2}\Big|_1^2$$
$$= -e^{-1} + e^{-1/2}$$

If 100 printers are sold, then

$(1 - e^{-1/2} + 1)(100)$ will fail in the first year,

and

$(e^{-1/2} - e^{-1/2})(100)$ will fail in the second year.

The manufacturer pays a full refund on those failing the first year and one-half refund on those failing during the second year.

$$\begin{aligned}\text{Refunds} &= (1 - e^{-1/2} + 1)(100)(\$200) \\ &\quad + (e^{-1/2} - e^{-1})(100)(\$100) \\ &\approx \$10{,}255.90\end{aligned}$$

The correct answer choice is **d**.

39. For a uniform distribution,

$$f(x) = \frac{1}{b-a} \text{ for } x \text{ in } [a,\ b].$$

$$f(x) = \frac{1}{36-20} = \frac{1}{16} \text{ for } x \text{ in } [20,\ 36]$$

(a) $\mu = \frac{1}{2}(20+36) = \frac{1}{2}(56)$

$$= 28 \text{ days}$$

(b) $P(30 < X \leq 36)$

$$= \int_{30}^{36} \frac{1}{16}\,dx = \frac{1}{16}x\Big|_{30}^{36}$$

$$= \frac{1}{16}(36-30)$$

$$= 0.375$$

41. We have an exponential distribution, with a = 1. $f(t) = e^{-t}$, $[0,\ \infty)$

(a) $\mu = \frac{1}{1} = 1$ hr

(b) $P(T < 30 \text{ min})$

$$= \int_0^{0.5} e^{-t}\,dt$$

$$= -e^{-t}\Big|_0^{0.5}$$

$$1 - e^{-0.5} \approx 0.3935$$

43. $f(x) = ae^{-ax}$ for $[0,\ \infty)$

Since $\mu = 25$ and $\mu = \frac{1}{a}$,

$$a = \frac{1}{25} = 0.04.$$

This, $f(x) = 0.04e^{-0.04x}$.

(a) We must find t such that $P(X \leq t) = 0.90$.

$$\begin{aligned}\int_0^t 0.04e^{-0.04x}\,dx &= 0.90 \\ -e^{-0.04x}\Big|_0^t &= 0.90 \\ -e^{-0.04t} + 1 &= 0.90 \\ 0.10 &= -e^{-0.04t} \\ -0.04t &= \ln 0.10 \\ t &= \frac{\ln 0.10}{-0.04} \\ t &\approx 57.56\end{aligned}$$

The longest time within which the predator will be 90% certain of finding a prey is approximately 58 min.

(b) $P(X \geq 60)$

$$\begin{aligned}&= \int_{60}^{\infty} 0.04e^{-0.04x}\,dx \\ &= \lim_{b\to\infty} \int_{60}^{b} 0.04e^{-0.04x}\,dx \\ &= \lim_{b\to\infty} (-e^{-0.04x})\Big|_{60}^{b} \\ &= \lim_{b\to\infty} [-e^{-0.04b} + e^{-0.04(60)}] \\ &= 0 + e^{-2.4} \\ &\approx 0.0907\end{aligned}$$

The probability that the predator will have to spend more than one hour looking for a prey is approximately 0.09.

45. For an exponential distribution, $f(x) = ae^{-ax}$ for $[0, \infty)$.

Since $\mu = \frac{1}{a} = 12.1$, $a = \frac{1}{12.1}$

(a) $P(X \geq 20) = \int_{20}^{\infty} \frac{1}{12.1}e^{-x/12.1}dx$

$$\begin{aligned}&= \lim_{b\to\infty} \int_{20}^{b} \frac{1}{12.1}e^{-x/12.1}dx \\ &= \lim_{b\to\infty} \left(-e^{-x/12.1}\Big|_{20}^{b}\right) \\ &= \lim_{b\to\infty} (-e^{-b/12.1} + e^{-20/12.1}) \\ &= e^{-20/12.1} \\ &\approx 0.19\end{aligned}$$

(b) $P(10 \le X \le 20) = \int_{10}^{20} \frac{1}{12.1} e^{-x/12.1} dx$

$$= -e^{-x/12.1}\Big|_{10}^{20}$$
$$= -e^{-20/12.1} + e^{-10/12.1}$$
$$\approx 0.25$$

47. We have an exponential distribution, with $a = 0.229$.
So $f(t) = 0.229e^{-0.229t}$, for $[0, \infty)$.

(a) The life expectancy is

$$\mu = \frac{1}{a} = \frac{1}{0.229} \approx 4.36 \text{ millennia.}$$

The standard deviation is

$$\sigma = \frac{1}{a} = \frac{1}{0.229} \approx 4.36 \text{ millennia.}$$

(b) $P(T \ge 2) = \int_2^{\infty} 0.229e^{-0.229t} dt$

$$= 1 - \int_0^2 0.229e^{-0.229t} dt$$
$$= 1 + \left(e^{-0.229t}\Big|_0^2\right)$$
$$= 1 + [e^{-0.229(2)} - 1]$$
$$= e^{-0.458}$$
$$\approx 0.63$$

49. For an exponential distribution, $f(x) = ae^{-ax}$ for $[0, \infty)$.

Since $\mu = \frac{1}{a} = 8, a = \frac{1}{8}$.

(a) $P(X \ge 10) = \int_{10}^{\infty} \frac{1}{8} e^{-x/8} dx$

$$= 1 - \int_0^{10} \frac{1}{8} e^{-x/8} dx$$
$$= 1 + \left(e^{-x/8}\Big|_0^{10}\right)$$
$$= 1 + [e^{-10/8} - 1]$$
$$= e^{-10/8}$$
$$\approx 0.29$$

(b) $P(X < 2) = \int_0^2 \frac{1}{8} e^{-x/8} dx$

$$= -e^{-x/8}\Big|_0^2$$
$$= -e^{-2/8} + 1$$
$$\approx 0.22$$

51. We have an exponential distribution $f(x) = ae^{-ax}$ for $x \ge 0$. Since $a = \frac{1}{90}$, $f(x) = \frac{1}{90}e^{-x/90}$ for $x \ge 0$.

(a) The probability that the time for a goal is no more than 71 minutes is

$$P(0 < X < 71) = \int_0^{71} \frac{1}{90} e^{-x/90} dx$$
$$= -e^{-x/90}\Big|_0^{71}$$
$$= -e^{-71/90} + 1$$
$$\approx 0.5457.$$

(b) The probability that the time for a goal is 499 minutes or more is

$$P(X \ge 499) = \int_{499}^{\infty} \frac{1}{90} e^{-x/90} dx$$
$$= -e^{-x/90}\Big|_{499}^{\infty}$$
$$= 0 + e^{-499/90}$$
$$\approx 0.003909.$$

Chapter 11 Review Exercises

1. In a probability function, the y-values (or function values) represent probabilities.

3. A probability density function f for $[a, b]$ must satisfy the following two conditions:

(1) $f(x) \ge 0$ for all x in the interval $[a, b]$;

(2) $\int_a^b f(x)\, dx = 1$.

5. $f(x) = \sqrt{x}$; $[4, 9]$

$$\int_4^9 x^{1/2}\, dx = \frac{2}{3}x^{3/2}\Big|_4^9$$
$$= \frac{2}{3}(27 - 8)$$
$$= \frac{38}{3} \ne 1$$

$f(x)$ is not a probability density function.

7. $f(x) = 0.7e^{-0.7x}$; $[0, \infty)$

$$\int_0^{\infty} 0.7e^{-0.7x}\, dx = -e^{-0.7x}\Big|_0^{\infty}$$
$$= \lim_{b\to\infty} (-e^{-0.7b}) + e^0$$
$$= \lim_{b\to\infty} \left(-\frac{1}{e^{0.7b}}\right) + 1$$
$$= 0 + 1 = 1$$

$f(x) \geq 0$ for all x in $[0, \infty)$.
Therefore, $f(x)$ is a probability density function.

9. $f(x) = kx^2;\ [1,\ 4]$

$$\int_1^4 kx^2\,dx = \left.\frac{kx^3}{3}\right|_1^4 = 21k$$

Since $f(x)$ is a probability density function,

$$21k = 1$$
$$k = \frac{1}{21}.$$

11. $f(x) = \frac{1}{10}$ for $[10,\ 20]$

(a) $P(10 \leq X \leq 12)$

$$= \int_{10}^{12} \frac{1}{10}\,dx = \left.\frac{x}{10}\right|_{10}^{12} = \frac{1}{5} = 0.2$$

(b) $P\left(\frac{31}{2} \leq X \leq 20\right)$

$$= \int_{31/2}^{20} \frac{1}{10}\,dx = \left.\frac{x}{10}\right|_{31/2}^{20} = 2 - \frac{31}{20} = \frac{9}{20} = 0.45$$

(c) $P(10.8 \leq X \leq 16.2)$

$$= \int_{10.8}^{16.2} \frac{1}{10}\,dx = \left.\frac{x}{10}\right|_{10.8}^{16.2} = 0.54$$

13. The distribution that is tallest or most peaked has the smallest standard deviation. This is the distribution pictured in graph (b).

15. $f(x) = \frac{2}{9}(x-2);\ [2,\ 5]$

(a)
$$\mu = \int_2^5 \frac{2x}{9}(x-2)dx = \int_2^5 \frac{2}{9}(x^2 - 2x)dx = \left.\frac{2}{9}\left(\frac{x^3}{3} - x^2\right)\right|_2^5 = \frac{2}{9}\left(\frac{125}{3} - 25 - \frac{8}{3} + 4\right) = 4$$

(b)
$$\text{Var}(X) = \int_2^5 \frac{2x^2}{9}(x-2)dx - (4)^2 = \int_2^5 \frac{2}{9}(x^3 - 2x^2)dx - 16 = \left.\frac{2}{9}\left(\frac{x^4}{4} - \frac{2x^3}{3}\right)\right|_2^5 - 16 = \frac{2}{9}\left(\frac{625}{4} - \frac{250}{3} - 4 + \frac{16}{3}\right) - 16 = 0.5$$

(c) $\sigma = \sqrt{0.5} \approx 0.7071$

(d)
$$\int_2^m \frac{2}{9}(x-2)\,dx = \frac{1}{2}$$
$$\left.\frac{1}{9}(m-2)^2\right|_2^m = \frac{1}{2}$$
$$\frac{1}{9}\left[(m-2)^2 - 0\right] = \frac{1}{2}$$
$$m^2 - 4m + 4 = \frac{9}{2}$$
$$m^2 - 4m - \frac{1}{2} = 0$$
$$m = \frac{4 \pm 3\sqrt{2}}{2} \approx -0.121, 4.121$$

We reject -0.121 since it is not in $[2, 5]$. So, $m = 4.121$

(e)
$$\int_2^x \frac{2}{9}(t-2)\,dt = \left.\frac{1}{9}(t-2)^2\right|_2^x = \frac{1}{9}[(x-2)^2 - 0] = \frac{(x-2)^2}{9}, 2 \leq x \leq 5$$

17. $f(x) = 5x^{-6}$; $[1, \infty)$

(a) $$\mu = \int_1^\infty x \cdot 5x^{-6}\,dx = \int_1^\infty 5x^{-5}\,dx$$
$$= \lim_{b\to\infty} \int_1^b 5x^{-5}\,dx = \lim_{b\to\infty} \left.\frac{5x^{-4}}{-4}\right|_1^b$$
$$= \lim_{b\to\infty} \frac{5}{4}\left(1 - \frac{1}{b^4}\right) = \frac{5}{4}$$

(b) $$\text{Var}(X) = \int_1^\infty x^2 \cdot 5x^{-6}\,dx - \left(\frac{5}{4}\right)^2$$
$$= \lim_{b\to\infty} \int_1^b 5x^{-4}\,dx - \frac{25}{16}$$
$$= \lim_{b\to\infty} \left.\frac{5x^{-3}}{-3}\right|_1^b - \frac{25}{16}$$
$$= \lim_{b\to\infty} \frac{5}{3}\left(1 - \frac{1}{b^3}\right) - \frac{25}{16}$$
$$= \frac{5}{3} - \frac{25}{16} = \frac{5}{48} \approx 0.1042$$

(c) $\sigma \approx \sqrt{\text{Var}(X)} \approx 0.32$

(d) $$\int_1^m 5x^{-6}dx = \frac{1}{2}$$
$$-x^{-5}\big|_1^m = \frac{1}{2}$$
$$-m^{-5} + 1 = \frac{1}{2}$$
$$m^{-5} = \frac{1}{2}$$
$$m^5 = 2$$
$$m = \sqrt[5]{2} \approx 1.149$$

(e) $$\int_1^x 5t^{-6}dt = -t^{-5}\big|_1^x$$
$$= -x^{-5} + 1$$
$$= 1 - \frac{1}{x^5}, x \ge 1$$

19. $f(x) = 4x - 3x^2$; $[0, 1]$
$$\mu = \int_0^1 x(4x - 3x^2)dx$$
$$= \int_0^1 (4x^2 - 3x^3)dx$$
$$= \left.\left(\frac{4x^3}{3} - \frac{3x^4}{4}\right)\right|_0^1$$
$$= \frac{4}{3} - \frac{3}{4} = \frac{7}{12} \approx 0.5833$$

Find m such that
$$\int_0^m (4x - 3x^2)dx = \frac{1}{2}.$$
$$\int_0^m (4x - 3x^2)dx = (2x^2 - x^3)\big|_0^m$$
$$= 2m^2 - m^3 = \frac{1}{2}$$

Therefore,
$$2m^3 - 4m^2 + 1 = 0.$$

This equation has no rational roots, but trial and error used with synthetic division reveals that $m \approx -0.4516$, 0.5970, and 1.855. The only one of these in $[0, 1]$ is 0.5970.

$$P\left(\tfrac{7}{12} < X < 0.5970\right)$$
$$= \int_{7/12}^{0.5970} (4x - 3x^2)dx$$
$$= 2x^2 - x^3\big|_{7/12}^{0.5970}$$
$$= 2(0.5970)^2 - (0.5970)^3 - 2\left(\frac{7}{12}\right)^2$$
$$+ \left(\frac{7}{12}\right)^3 \approx 0.0180$$

21. $f(x) = 0.01e^{-0.01x}$ for $[0, \infty)$ is an exponential distribution.

(a) $\mu = \dfrac{1}{0.01} = 100$

(b) $\sigma = \dfrac{1}{0.01} = 100$

(c) $$P(100 - 100 < X < 100 + 100)$$
$$= P(0 < X < 200)$$
$$= \int_0^{200} 0.01e^{-0.01x}\,dx$$
$$= -e^{-0.01x}\Big|_0^{200}$$
$$= 1 - e^{-2} \approx 0.8647$$

For Exercises 23–29, use the table in the Appendix for the areas under the normal curve.

23. Area to the left of $z = -0.43$ is 0.3336.
Percent of area is 33.36%.

25. Area between $z = -1.17$ and $z = -0.09$ is

$$0.4641 - 0.1210 = 0.3431.$$

Percent of area is 34.31%.

27. The region up to 1.2 standard deviations below the mean is the region to the left of $z = -1.2$. The area is 0.1151, so the percent of area is 11.51%.

29. 52% of area is to the right implies that 48% is to the left.

$$P(z < a) = 0.48 \text{ for } a = -0.05$$

Thus, 52% of the area lies to the right of $z = -0.05$.

31. $f(x) = e^{-x}$ for $[0, \infty)$
$f(x) = 1e^{-1x}$

(a) This is an exponential distribution with $a = 1$.

(b) The domain of f is $[0, \infty)$.
The range of f is $(0, 1]$.

(c)

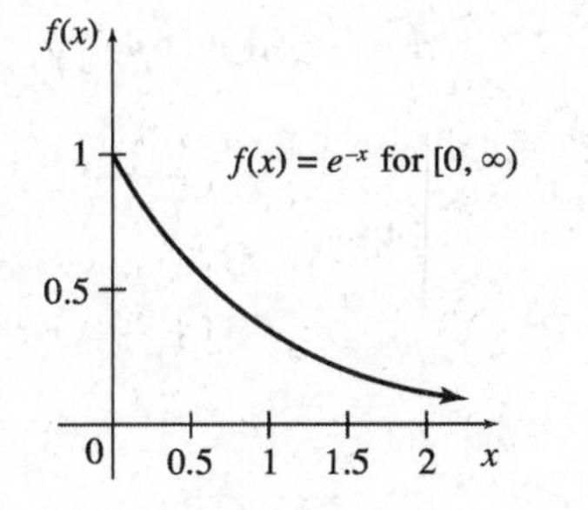

(d) For an exponential distribution, $\mu = \frac{1}{a}$ and $\sigma = \frac{1}{a}$.
Thus

$$\mu = \frac{1}{1} = 1 \quad \text{and} \quad \sigma = \frac{1}{1} = 1.$$

(e)
$$\begin{aligned} P(\mu - \sigma \le X \le \mu + \sigma) &= P(1 - 1 \le X \le 1 + 1) \\ &= P(0 \le X \le 2) \\ &= \int_0^2 e^{-x}\, dx \\ &= -e^{-x}\Big|_0^2 \\ &= -e^{-2} + 1 \\ &\approx 0.8647 \end{aligned}$$

33. $f(x) = \dfrac{xe^{-x/2}}{4}$ for x in $[0, \infty)$

(a) $P(0 \le X \le \infty) = \displaystyle\int_0^\infty \frac{xe^{-x/2}}{4}\, dx$

For all $x \ge 0, e^{-x/2} > 0$ so that $f(x) \ge 0$ for x in $[0, \infty)$.

Evaluate $\displaystyle\int xe^{-x/2} dx$ using integration by parts.

Let $u = x$ and $dv = e^{-x/2} dx$
Then $du = dx$ and $v = -2e^{-x/2}$.

$$\begin{aligned} \frac{1}{4}\int xe^{-x/2} dx &= \frac{1}{4}\left(-2xe^{-x/2} + \int 2e^{-x/2} dx\right) \\ &= \frac{1}{4}\left(-2xe^{-x/2} - 4e^{-x/2}\right) \\ &= -\frac{1}{2}xe^{-x/2} - e^{-x/2} \end{aligned}$$

Therefore,

$$\begin{aligned} \int_0^\infty \frac{xe^{-x/2}}{4}\, dx &= \left(-\frac{1}{2}xe^{-x/2} - e^{-x/2}\right)\Big|_0^\infty \\ &= \lim_{b\to\infty}\left(-\frac{1}{2}xe^{-x/2} - e^{-x/2}\right)\Big|_0^b \\ &= \lim_{b\to\infty}\left(-\frac{1}{2}be^{-b/2} - e^{-b/2}\right) - (0 - 1) \end{aligned}$$

Recall from the section on l'Hospital's rule that $\lim\limits_{b\to\infty} be^{-b/2} = 0$. Therefore,

$$\int_0^\infty \frac{xe^{-x/2}}{4}\, dx = 0 - (-1) = 1.$$

(b)
$$\begin{aligned} P(0 \le X \le 3) &= \int_0^3 \frac{xe^{-x/2}}{4}\, dx \\ &= \left(-\frac{1}{2}xe^{-x/2} - e^{-x/2}\right)\Big|_0^3 \\ &= \left(-\frac{3}{2}e^{-3/2} - e^{-3/2}\right) - (0 - 1) \\ &= 1 - \frac{5}{2}e^{-3/2} \\ &\approx 0.4422 \end{aligned}$$

35. $f(x) = \frac{3}{4}(x^2 - 16x + 65)$ for $[8, 9]$

$$P(8 \le X < 8.50)$$
$$= \int_8^{8.5} \frac{3}{4}(x^2 - 16x + 65)dx$$
$$= \frac{3}{4}\left(\frac{x^3}{3} - 8x^2 + 65x\right)\Bigg|_8^{8.5}$$
$$= \frac{3}{4}[\frac{8.5^3}{3} - 8(8.5)^2 + 65(8.5) - \frac{8^3}{3}$$
$$+ 8(8)^2 - 65(8)]$$
$$\approx 0.406$$

37. **(a)** $\mu = 8$

$$\frac{1}{a} = 8$$
$$a = \frac{1}{8}$$
$$f(x) = \frac{1}{8}e^{-x/8} \text{ for } [0, \infty)$$

(b) Expected number $= \mu = 8$

(c) $\sigma = \mu = 8$

(d) $P(5 < X < 10) = \int_5^{10} \frac{1}{8}e^{-x/8}\, dx$

$$= -e^{-x/8}\Big|_5^{10}$$
$$= -e^{-10/8} + e^{-5/8}$$
$$\approx 0.2488$$

39. $\mu = 46.2,\ \sigma = 15.8,\ x = 60$

$$z = \frac{x - \mu}{\sigma} = \frac{60 - 46.2}{15.8} \approx 0.8734$$

0.8734 is the z-score for the area of about 0.8078 (from the table).

$$P(X \ge 60) \approx P(z \ge 0.8734) \approx 1 - 0.8078 = 0.1922$$

41. $f(x) = 0.01e^{-0.01x}$ for $[0, \infty)$ is an exponential distribution.

$$P(0 \le X \le 100)$$
$$= \int_0^{100} 0.01e^{-0.01x}\, dx$$
$$= -e^{-0.01x}\Big|_0^{100}$$
$$= 1 - \frac{1}{e}$$
$$\approx 0.6321$$

43. $f(x) = \frac{6}{15{,}925}(x^2 + x)$ for $[20, 25]$

(a) $\mu = \int_{20}^{25} \frac{6}{15{,}925}(x^2 + x)dx$

$$= \frac{6}{15{,}925}\int_{20}^{25}(x^3 + x^2)dx$$
$$= \frac{6}{15{,}925}\left[\frac{x^4}{4} + \frac{x^3}{3}\right]\Bigg|_{20}^{25}$$
$$= \frac{6}{15{,}925}$$
$$\cdot\left[\frac{(25)^4}{4} + \frac{(25)^3}{3} - \frac{(20)^4}{4} - \frac{(20)^3}{3}\right]$$
$$\approx 22.68°\text{C}$$

(b) $P(X < \mu)$

$$= \int_{20}^{22.68} \frac{6}{15{,}925}(x^2 + x)dx$$
$$= \frac{6}{15{,}925}\left[\frac{x^3}{3} + \frac{x^2}{2}\right]\Bigg|_{20}^{22.68}$$
$$= \frac{6}{15{,}925}$$
$$\cdot\left[\frac{(22.68)^3}{3} + \frac{(22.68)^2}{2} - \frac{(20)^3}{3} - \frac{(20)^2}{2}\right]$$
$$\approx 0.4819$$

45. Normal distribution, $\mu = 2.2$ g, $\sigma = 0.4$ g, $X =$ tension

$$P(X < 1.9) = P\left(\frac{x - 2.2}{0.4} < \frac{1.9 - 2.2}{0.4}\right)$$
$$= P(z < -0.75)$$
$$\approx 0.2266.$$

47. **(a)**

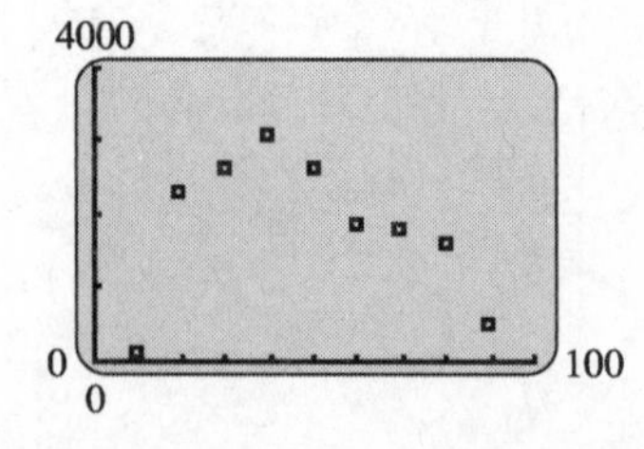

Of the types of functions available using the regression feature of a graphing utility, a polynomial function best matches the data.

(b) The function

$$\begin{aligned} N(x) &= -0.000853613x^4 + 0.196608x^3 \\ &\quad - 16.6309x^2 + 577.248x - 4040.47 \end{aligned}$$

provided by the calculator models this data well, as illustrated by the following graph.

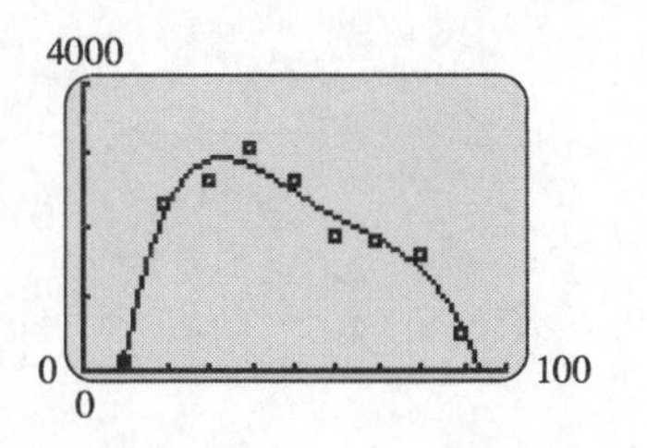

(c) Using the integration feature on our calculator, we find that

$$\int_{9.2}^{93.5} N(x)\,dx \approx 167{,}355.$$

So, $k = \frac{1}{167{,}355}$ and

$$S(x) = \frac{1}{167{,}355} N(x).$$

(d) Using the integration feature on our calculator,

$$P(25 \le X \le 35) = \int_{25}^{35} S(x)dx \approx 0.1731$$

$$P(45 \le X \le 65) = \int_{45}^{65} S(x)dx \approx 0.2758$$

$$P(X \ge 55) = \int_{55}^{93.5} S(x)dx \approx 0.348.$$

From the table, the actual probabilities are

$$P(25 \le X \le 35) = \frac{2521}{16{,}594} \approx 0.1579$$

$$P(45 \le X \le 65) = \frac{2677 + 1860}{16{,}594} \approx 0.2734$$

$$P(X \ge 55) = \frac{1860 + 1791 + 1606 + 524}{16{,}594}$$

$$\approx 0.3484.$$

(e) Using the integration feature on our calculator,

$$E(X) = \int_{9.2}^{93.5} x \cdot S(x)\,dx \approx 46.84 \text{ years.}$$

(f) Using the integration feature on our calculator,

$$\begin{aligned} \text{Var}(X) &= \int_{9.2}^{93.5} x^2 \cdot S(x)\,dx - 46.84^2 \\ &\approx 409.5654 \\ \sigma &= \sqrt{\text{Var}(X)} = \sqrt{409.5654} \\ &\approx 20.24 \text{ years.} \end{aligned}$$

(g) $\displaystyle\int_{9.2}^{m} S(x)dx = \frac{1}{2}$

One method for approximating m using our calculator is to enter $\int_{9.2}^{m} S(x)\,dx$ into Y1, fnInt (Y1,X,9.2,X) (the left side of the equation) into Y2, and $\frac{1}{2}$ into Y3. Set the window to $0 \le X \le 100$ and $0 \le \text{Y} \le 1$. Turn off the graph of Y1 and graph Y2 and Y3 in the same window. (The process is slow.) Use the intersect feature to find where the graphs cross. This occurs at $m \approx 44.8$ years.

49. Normal distribution, $\mu = 40$, $\sigma = 13$, $x =$ "take"

$$\begin{aligned} P(X > 50) &= P\left(\frac{X - 40}{13} > \frac{50 - 40}{13}\right) \\ &= P(Z > 0.77) \\ &= 1 - P(Z \le 0.77) \\ &= 1 - 0.7794 \\ &= 0.2206 \end{aligned}$$

Chapter 11 Test

[11.1]

Decide whether the functions defined as follows are probability density functions on the given intervals. If not, tell why.

1. $f(x) = 4$; $[3, 7]$

2. $f(x) = \frac{1}{5}$; $[5, 10]$

3. $f(x) = \frac{1}{2}(x - 4)$; $[4, 6]$

4. $f(x) = e^{-3x}$; $[0, \infty)$

5. Find the value of k that will make $f(x) = kx^{1/3}$ a probability density function on the interval $[1, 8]$.

6. The probability density function for a random variable x is defined by

$$f(x) = 0.2 \text{ for } x \text{ in } [12, 17].$$

Find the following probabilities.

(a) $P(X \leq 14)$ **(b)** $P(13 < X < 16)$ **(c)** $P(X < 15)$

7. The probability density function of a random variable x is defined by

$$f(x) = 3x^2 \text{ for } x \text{ in } [0, 1].$$

Find the following probabilities.

(a) $P(X \leq 1)$ **(b)** $P(0.2 \leq X \leq 0.7)$ **(c)** $P(X \geq 0.3)$

8. The time in years until a particular radioactive particle decays is a random variable with probability density function

$$f(t) = 0.04e^{-0.04t} \text{ for } t \text{ in } [0, \infty).$$

Find the probability that a certain such particle decays in less than 60 years.

[11.2]

Find the expected value and the standard deviation for each probability density function defined as follows.

9. $f(x) = \frac{1}{4}$; $[1, 5]$

10. $f(x) = 2x$; $[0, 1]$

11. $f(x) = \frac{3}{2}(x - 1)^2$; $[0, 2]$

12. $f(x) = \frac{1}{4\sqrt{x}}$; $[1, 9]$

13. Suppose that the time spent waiting in the waiting room of Dr. Jones' office is a random variable with probability density function defined by

$$f(x) = \frac{x\,(30-x)}{4500} \text{ for } x \text{ in } (0,30)\,.$$

Find the average waiting time. Find the standard deviation.

For the probability density functions defined as follows, find (a) the probability that the value of the random variable will be less than the mean and (b) the probability that the value of the random variable will be within one standard deviation of the mean.

14. $f(x) = 2\,(1-x)$ for x in $[0,1]$

15. $f(x) = 1 - x^{-1/2}$ for x in $[1,4]$

16. For the probability density function defined by

$$f(x) = \frac{1}{8}x \text{ for } x \text{ in } [0,4]\,,$$

find the probability that the value of the random variable will be between the median and the mean.

[11.3]

17. The maximum daily temperature in degrees Celsius in a certain city is uniformly distributed over the interval $[10, 40]$.

(a) What is the expected maximum temperature?

(b) What is the probability that the temperature will be less than 25°C?

18. The number of repairs required by a new product each month is exponentially distributed with an average of 6.

(a) What is the probability density function for this distribution?

(b) What is the probability that the number of repairs is between 3 and 5?

(c) What is the standard deviation?

Find the percent of area under a normal curve for each of the following.

19. The region to the left of $z = -0.57$

20. The region to the right of $z = 1.49$

21. The region between $z = 1.78$ and $z = -1.30$

Find a z-score satisfying the given condition.

22. 19% of the area to the left of z.

23. 91% of the area is to the right of z.

24. The heights of the students in a calculus class are normally distributed with mean 65 inches and standard deviation 10 inches.

(a) Find the probability that a student's height is between 60 inches and 70 inches.

(b) Find the probability that a student's height is more than 72 inches.

25. On a certain day, the amount spent for lunch in a particular restaurant was $3.75 with a standard deviation of $0.25. What percent of the customers spent between $3.60 and $4.10? Assume that the amount spent is normally distributed.

Chapter 11 Test Answers

1. No; $\int_3^7 4\,dx \neq 1$

2. Yes

3. Yes

4. No; $\int_0^\infty e^{-3x}\,dx \neq 1$

5. $\frac{4}{45}$

6. **(a)** 0.4 **(b)** 0.6 **(c)** 0.6

7. **(a)** 1 **(b)** 0.335 **(c)** 0.973

8. 0.9093

9. 03; 1.155

10. 0.6667;0.2357

11. 1; 0.7746

12. 4.333; 2.329

13. 15; 6.708

14. **(a)** 0.5556 **(b)** 0.6285

15. **(a)** 0.4668 **(b)** 0.6059

16. 0.0556

17. **(a)** 25° **(b)** 0.5

18. **(a)** $f(x) = 0.1667e^{-0.1667x}$ **(b)** 0.1719 **(c)** 6

19. 28.43%

20. 6.81%

21. 86.57%

22. −0.88

23. −1.34

24. **(a)** 0.383 **(b)** 0.242

25. 64.49%

Chapter 12

SEQUENCES AND SERIES

12.1 Geometric Sequences

1. $a_1 = 2,\ r = 3,\ n = 4$

Since $n = 4$, we must find a_1, a_2, a_3, and a_4 with $a = a_1 = 2$.

$$\begin{aligned} a_1 &= 2 \\ a_2 &= 2(3)^{2-1} = 2(3)^1 = 2(3) = 6 \\ a_3 &= 2(3)^{3-1} = 2(3)^2 = 2(9) = 18 \\ a_4 &= 2(3)^{4-1} = 2(3)^3 = 2(27) = 54 \end{aligned}$$

The first four terms of this geometric sequence are 2, 6, 18, and 54.

3. $a_1 = \frac{1}{2},\ r = 4,\ n = 4$

Since $n = 4$, we must find a_1, a_2, a_3, and a_4 with $a = a_1 = \frac{1}{2}$.

$$\begin{aligned} a_1 &= \frac{1}{2} \\ a_2 &= \frac{1}{2}(4)^{2-1} = \frac{1}{2}(4)^1 = \frac{1}{2}(4) = 2 \\ a_3 &= \frac{1}{2}(4)^{3-1} = \frac{1}{2}(4)^2 = \frac{1}{2}(16) = 8 \\ a_4 &= \frac{1}{2}(4)^{4-1} = \frac{1}{2}(4)^3 = \frac{1}{2}(64) = 32 \end{aligned}$$

The first four terms of this geometric sequence are $\frac{1}{2}$, 2, 8, and 32.

5. $a_3 = 6,\ a_4 = 12,\ n = 5$

Since $n = 5$, we must find a_1, a_2, a_3, a_4, and a_5 with $r = \frac{a_4}{a_3} = \frac{12}{6} = 2$. To find a, use $a_3 = 6$, $r = 2$, and $n = 3$ in the formula.

$$\begin{aligned} 6 &= a(2)^{3-1} \\ 6 &= a(2)^2 \\ 6 &= 4a \\ \frac{3}{2} &= \frac{3}{2} = a \end{aligned}$$

$$\begin{aligned} a_1 &= \frac{3}{2} \\ a_2 &= \frac{3}{2}(2)^{2-1} = \frac{3}{2}(2)^1 = \frac{3}{2}(2) = 3 \\ a_3 &= \frac{3}{2}(2)^{3-1} = \frac{3}{2}(2)^2 = \frac{3}{2}(4) = 6 \\ a_4 &= \frac{3}{2}(2)^{4-1} = \frac{3}{2}(2)^3 = \frac{3}{2}(8) = 12 \\ a_5 &= \frac{3}{2}(2)^{5-1} = \frac{3}{2}(2)^4 = \frac{3}{2}(16) = 24 \end{aligned}$$

The first five terms of this geometric sequence are $\frac{3}{2}$, 3, 6, 12, and 24.

7. $a_1 = 4,\ r = 3$

Since we want a_5, use $n = 5$ in the formula with $a = a_1 = 4$ and $r = 3$.

$$\begin{aligned} a_5 &= 4(3)^{5-1} = 4(3)^4 = 4(81) = 324 \\ a_n &= 4(3)^{n-1} \end{aligned}$$

9. $a_1 = -3,\ r = -5$

Since we want a_5, use $n = 5$ in the formula with $a = a_1 = -3$ and $r = -5$.

$$\begin{aligned} a_5 &= -3(-5)^{5-1} = -3(-5)^4 \\ &= -3(625) = -1875 \\ a_n &= -3(-5)^{n-1} \end{aligned}$$

11. $a_2 = 12, r = \dfrac{1}{2}$

To find a, use $a_2 = 12$, $r = \frac{1}{2}$, and $n = 2$ in the formula.

$$12 = a\left(\frac{1}{2}\right)^{2-1} = \frac{a}{2}$$

$$a = 24$$

Since we want a_5, use $n = 5$ in the formula with $a = 24$ and $r = \frac{1}{2}$.

$$a_5 = 24\left(\frac{1}{2}\right)^{5-1} = 24\left(\frac{1}{2}\right)^4 = 24\left(\frac{1}{16}\right) = \frac{3}{2}$$

$$a_n = 24\left(\frac{1}{2}\right)^{n-1} \text{ or } \frac{24}{2^{n-1}}$$

13. $a_4 = 64$, $r = -4$
To find a, use $a_4 = 64$, $r = -4$, and $n = 4$ in the formula.

$$\begin{aligned} 64 &= a(-4)^{4-1} \\ 64 &= a(-4)^3 \\ 64 &= a(-64) \\ -1 &= a \end{aligned}$$

Since we want a_5, use $n = 5$ in the formula with $a = -1$ and $r = -4$.

$$\begin{aligned} a_5 &= -1(-4)^{5-1} = -1(-4)^4 \\ &= -1(256) = -256 \\ a_n &= -(-4)^{n-1} \end{aligned}$$

15. 6, 12, 24, 48, ...

$$r = \frac{12}{6} = \frac{24}{12} = \frac{48}{24} = 2$$

Since $r = 2$ and $a = a_1 = 6$, $a_n = 6(2)^{n-1}$.

17. $\frac{3}{4}, \frac{3}{2}, 3, 6, 12, \ldots$

$$r = \frac{\frac{3}{2}}{\frac{3}{4}} = \frac{3}{\frac{3}{2}} = \frac{6}{3} = \frac{12}{6} = 2$$

Since $r = 2$ and $a = a_1 = \frac{3}{4}$, $a_n = \frac{3}{4}(2)^{n-1}$.

19. 4, 8, −16, 32, 64, −128, ...
Since $\frac{8}{4} = 2$ and $\frac{-16}{8} = -2$, the ratio is not constant, so the sequence is not geometric.

21. $-\frac{5}{8}, \frac{5}{12} - \frac{5}{18}, \frac{5}{27}, \ldots$

$$r = \frac{\frac{5}{12}}{-\frac{5}{8}} = \frac{-\frac{5}{18}}{\frac{5}{12}} = \frac{\frac{5}{27}}{-\frac{5}{18}} = -\frac{2}{3}$$

Since $r = -\frac{2}{3}$ and $a = a_1 = -\frac{5}{8}$, $a_n = -\frac{5}{8}\left(-\frac{2}{3}\right)^{n-1}$.

23. 3, 6, 12, 24, ...

Since $a = a_1 = 3$ and $r = \frac{6}{3} = 2$,

$$\begin{aligned} S_5 &= \frac{3(2^5 - 1)}{2 - 1} \\ &= \frac{3(32 - 1)}{1} \\ &= 93. \end{aligned}$$

The sum of the first five terms of this geometric sequence is 93.

25. $12, -6, 3, -\frac{3}{2}, \ldots$

Since $a = a_1 = 12$ and $r = \frac{-6}{12} = -\frac{1}{2}$,

$$\begin{aligned} S_5 &= \frac{12\left[\left(-\frac{1}{2}\right)^5 - 1\right]}{\left(-\frac{1}{2}\right) - 1} \\ &= \frac{12\left[-\frac{1}{32} - 1\right]}{-\frac{3}{2}} \\ &= \frac{33}{4}. \end{aligned}$$

The sum of the first five terms of this geometric sequence is $\frac{33}{4}$.

27. $a_1 = 3$, $r = -2$

Since $a = a_1 = 3$,

$$S_5 = \frac{3\left[(-2)^5 - 1\right]}{-2 - 1} = \frac{3(-32 - 1)}{-3}.$$

The sum of the first five terms of this geometric sequence is 33.

29. $a_1 = 6.324, r = 2.598$
Since $a = a_1 = 6.324$,

$$\begin{aligned} S_5 &= \frac{6.324(2.598^5 - 1)}{2.598 - 1} \\ &\approx \frac{6.324(118.3575 - 1)}{1.598} \\ &\approx 464.4. \end{aligned}$$

The sum of the first five terms of this sequence is about 464.4.

31. For $\sum_{i=0}^{7} 8(2)^i$, use the formula with $a = 8, r = 2$, and $n = 8$.

$$\begin{aligned} S_8 &= \frac{8(2^8 - 1)}{2 - 1} \\ &= \frac{8(256 - 1)}{1} \\ &= 2040 \end{aligned}$$

33. For $\sum_{i=0}^{8} \frac{3}{2}(4)^i$, use the formula with $a = \frac{3}{2}$, $r = 4$, and $n = 9$.

$$\begin{aligned} S_9 &= \frac{\frac{3}{2}(4^9 - 1)}{4 - 1} \\ &= \frac{\frac{3}{2}(262{,}144 - 1)}{3} \\ &= \frac{262{,}143}{2} \end{aligned}$$

35. For $\sum_{i=0}^{4} \frac{3}{4}(-3)^i$, use the formula with $a = \frac{3}{4}$, $r = -3$, and $n = 5$.

$$\begin{aligned} S_5 &= \frac{\frac{3}{4}[(-3)^5 - 1]}{-3 - 1} \\ &= \frac{\frac{3}{4}(-243 - 1)}{-4} \\ &= \frac{183}{4} \end{aligned}$$

37. For $\sum_{i=0}^{8} 64\left(\frac{1}{2}\right)^i$, use the formula with $n = 9$, $r = \frac{1}{2}$, and $a = 64$.

$$\begin{aligned} S_9 &= \frac{64\left[\left(\frac{1}{2}\right)^9 - 1\right]}{\frac{1}{2} - 1} \\ &= \frac{64\left(\frac{1}{512} - 1\right)}{-\frac{1}{2}} \\ &= \frac{511}{4} \end{aligned}$$

Therefore, $\sum_{i=0}^{8} 64\left(\frac{1}{2}\right)^i = \frac{511}{4}$.

39. (a) A machine that loses 20% of its value maintains 80% of its value. If the initial value of the machine is \$12,000, then its value at the end of the first year will be 80% of \$12,000, and its value at the end of each subsequent year will be 80% of its value at the end of the previous year. This means that the end-of-year values form a geometric sequence with $r = 0.80$. If we let $a_1 = 12{,}000$, then a_2 will represent the value at the end of the first year, a_3 will represent the value at the end of the second year, and so on. Thus, to find the value of the machine at the end of the fifth year, we are looking for a_6 in the geometric sequence with $n = 6$, $r = 0.80$ and $a = a_1 = 12{,}000$.

$$\begin{aligned} a_6 &= 12{,}000\,(0.80)^{6-1} = 12{,}000\,(0.80)^5 \\ &= 12{,}000\,(0.32768) = 3932.16 \end{aligned}$$

Therefore, the value of the machine at the end of the fifth year is about \$3932.

(b) To find the value of the machine at the end of the eighth year, we are looking for a_9 in the geometric sequence.

$$\begin{aligned} a_9 &= 12{,}000\,(0.80)^{9-1} = 12{,}000\,(0.80)^8 \\ &= 12{,}000\,(0.16777216) \approx 2013.27 \end{aligned}$$

Therefore, the value of the machine at the end of the eighth year is about \$2013.

41. The amounts saved form a geometric sequence with $a_1 = 1$ and $r = 2$. To determine the amount saved on January 31, we want to use the formula to find a_n with $n = 31$, $r = 2$, and $a = a_1 = 1$.

$$a_{31} = 1\,(2)^{31-1} = 2^{30} = 1{,}073{,}741{,}824$$

To determine the total amount saved during January, we want to use the formula to find S_n with $n = 31$, $r = 2$, and $a = a_1 = 1$.

$$\begin{aligned} S_{31} &= \frac{1\,(2^{31} - 1)}{2 - 1} = \frac{2{,}147{,}483{,}648 - 1}{1} \\ &= 2{,}147{,}483{,}647 \end{aligned}$$

The savings on January 31 would be $\$2^{30}$ or \$1,073,741,824 and for the month of January would be $\$2^{31} - \$1$ or \$2,147,483,647.

43. If we let a_1 represent the initial bacteria population, then the population at the end of the first hour will be $a_1 + 5\%\,(a_1)$, or $105\%\,(a_1)$. The populations form a geometric sequence with $r = 105\% = 1.05$. To determine the percent increase in the population after 7 hours, we want to use the formula to find a_n with $n = 8$, $r = 1.05$, and $a = a_1$.

$$a_6 = a\,(1.05)^{8-1} \approx 1.40710a$$

Since $a = a_1$, $a_8 \approx 1.40710a_1$. This means that the population at the end of 7 hours is about 141% of the initial population.

After five hours, the population has increased by about 41%.

45. The number of rotations form a geometric sequence with $r = \frac{3}{4}$ and $a_1 = 400$. If we let $a_1 = 400$, then a_2 represents the number of rotations at the end of the first minute, a_3 is the number of rotations at the end of the second minute, and so on. To determine the number of rotations in the fifth minute, we want a_6. Use the formula to find a_n with $n = 6$, $r = \frac{3}{4}$, and $a = a_1 = 400$.

$$a_6 = 400\left(\frac{3}{4}\right)^{6-1} = 400\left(\frac{3}{4}\right)^5$$
$$= 400\left(\frac{243}{1024}\right) = 94.921875$$

In the fifth minute after the rider's feet are removed from the pedals, the wheel will rotate about 95 times.

47. **(a)** Initially, there are 64 teams that play $a_1 = 32$ games in round one. After each round, the number of teams (and the number of games) decreases by $\frac{1}{2}$. At the end of the first round, there are 32 teams left to play $a_2 = 16$ games in round two. At the end of the second round, there will be 16 teams left to play $a_3 = 8$ games in round 3, and so on. Note that $a_1 = 2^5$, $a_2 = 2^4$, $a_3 = 2^3$, so that by the sixth round there will be 2 teams left playing $a_6 = 2^0 = 1$ game to decide the champion. Therefore, to determine the total number of games, we need to sum the terms of the geometric sequence $a_1 = 2^5$, $a_2 = 2^4$, $a_3 = 2^3$, $a_4 = 2^2$, $a_5 = 2^1$, $a_6 = 1$ or $1 + 2 + 2^2 + 2^3 + 2^4 + 2^5$.

(b) To find the sum from part (a), use the formula to find S_n with $n = 6$, $r = \frac{1}{2}$, and $a = a_1 = 2^5$.

$$S_6 = \frac{2^5\left[\left(\frac{1}{2}\right)^6 - 1\right]}{\frac{1}{2} - 1} = \frac{\left(\frac{1}{2} - 32\right)}{-\frac{1}{2}} = 63,$$

so 63 games must be played to produce the champion.

(c) To generalize the methods of (a) and (b) for 2^n teams rather than $64 = 2^6$ teams, first note that 2^n teams play 2^{n-1} games in the initial round. In part (a), note that it required 6 rounds to determine a champion when 2^6 teams are initially involved. Similarly, for 2^n teams it will require n rounds. So, the total number of games is given by

$$S_n = \frac{a\left(r^n - 1\right)}{r - 1}$$

with $a = a_1 = 2^{n-1}$, and $r = \frac{1}{2}$, which is the desired sum

$$2^{n-1} + 2^{n-2} + \cdots + 2^2 + 2^1 + 2^0.$$

$$S_n = \frac{2^{n-1}\left[\left(\frac{1}{2}\right)^n - 1\right]}{\frac{1}{2} - 1} = \frac{\frac{1}{2} - 2^{n-1}}{-\frac{1}{2}}$$
$$= -2\left(\frac{1}{2} - 2^{n-1}\right)$$
$$= -1 + 2^n$$
$$= 2^n - 1.$$

12.2 Annuities: An Application of Sequences

1. $R = 120, i = 0.05, n = 10$

$$S = R \cdot s_{\overline{n}|i}$$
$$= 120 \cdot s_{\overline{10}|0.05}$$
$$= 120\left[\frac{(1.05)^{10} - 1}{0.05}\right]$$
$$\approx 120(12.57789254)$$
$$\approx 1509.35$$

The amount is \$1509.35.

3. $R = 9000, i = 0.06, n = 18$

$$S = 9000 \cdot s_{\overline{18}|0.06}$$
$$= 9000\left[\frac{(1.06)^{18} - 1}{0.06}\right]$$
$$\approx 9000(30.90565255)$$
$$\approx 278{,}150.87$$

The amount is \$278,150.87.

5. $R = 11{,}500$, $i = 0.055, n = 30$

$$S = 11{,}500 \cdot s_{\overline{30}|0.055}$$
$$= 11{,}500\left[\frac{(1.055)^{30} - 1}{0.055}\right]$$
$$\approx 11{,}500(72.43547797)$$
$$\approx 833{,}008.00$$

The amount is \$833,008.00.

7. $R = 10{,}500$

Interest is earned semiannually for 7 years, so $i = \frac{0.10}{2} = 0.05$ and $n = 2 \cdot 7 = 14$ periods.

$$\begin{aligned} S &= 10{,}500\left[\frac{(1.05)^{14}-1}{0.05}\right] \\ &\approx 10{,}500(19.59863199) \\ &\approx 205{,}785.6359 \end{aligned}$$

The amount is \$205,785.64.

9. $R = 1800$

Interest is earned quarterly for 12 years, so $i = \frac{0.08}{4} = 0.02$ and $n = 4 \cdot 12 = 48$ periods.

$$\begin{aligned} S &= 1800\left[\frac{(1.02)^{48}-1}{0.02}\right] \\ &\approx 1800(79.35351927) \\ &\approx 142{,}836.33 \end{aligned}$$

The amount is \$142,836.33.

11. This describes an ordinary annuity with $S = 10{,}000$, $i = 0.08$, and $n = 12$ periods.

$$\begin{aligned} 10{,}000 &= R \cdot s_{\overline{12}|0.08} \\ 10{,}000 &= R\,(18.97712646) \\ R &\approx 526.95 \end{aligned}$$

The periodic payment should be \$526.95.

13. This describes an ordinary annuity with $S = 50{,}000$, $i = 0.03\left(= \frac{12\%}{4}\right)$, and $n = 8 \cdot 4 = 32$ periods.

$$\begin{aligned} 50{,}000 &= R \cdot s_{\overline{32}|0.03} \\ 50{,}000 &\approx R\,(52.50275852) \\ R &\approx 952.33 \end{aligned}$$

The periodic payment should be \$952.33.

15. $R = 5000$, $i = 0.06$, and $n = 11$ payments.

$$a_{\overline{11}|0.06} = \left[\frac{1-(1.06)^{-11}}{0.06}\right] = 7.886874577,$$

so

$$P \approx 5000\,(7.886874577) \approx 39{,}434.37,$$

or \$39,434.37.

17. $R = 1400, i = \frac{0.06}{2} = 0.03$, and $n = 2 \cdot 8 = 16$ periods

$$\begin{aligned} P &= 1400\left[\frac{1-(1.03)^{-16}}{0.03}\right] \\ &\approx 1400(12.56110203) \\ &\approx 17{,}585.54 \end{aligned}$$

The present value is \$17,585.54.

19. $R = 50{,}000$, $i = \frac{0.08}{4} = 0.02$, and $n = 10 \cdot 4 = 40$ payments.

$$a_{\overline{40}|0.02} = \left[\frac{1-(1.02)^{-40}}{0.02}\right] \approx 27.35547924,$$

so

$$P \approx 50{,}000\,(27.35547924) \approx 1{,}367{,}773.96,$$

or \$1,367,773.96.

21. $a_{\overline{15}|0.005} = \left[\frac{1-(1.05)^{-15}}{0.05}\right] \approx 10.37965804$, so $P \approx 10{,}000\,(10.37965804) \approx 103{,}796.58$, or \$103,796.58.
A lump sum deposit of \$103,796.58 today at 5% compounded annually will yield the same total after 15 years as deposits of \$10,000 at the end of each year for 15 years at 5% compounded annually.

23. $a_{\overline{15}|0.08} = \left[\frac{1-(1.08)^{-15}}{0.08}\right] = 8.559478688$, so $P \approx 10{,}000(8.559478688) \approx 85{,}594.79$, or \$85,594.79.
A lump sum deposit of \$85,594.79 today at 8% compounded annually will yield the same total after 15 years as deposits of \$10,000 at the end of each year for 15 years at 8% compounded annually.

25. \$2500 is the present value of this annuity of R dollars, with 6 periods, and $i = \frac{16\%}{4} = 4\% = 0.04$ per period.

$$\begin{aligned} P &= R \cdot a_{\overline{n}|i} \\ 2500 &= R \cdot a_{\overline{6}|0.04} \\ R &= \frac{2500}{a_{\overline{6}|0.04}} \\ &\approx \frac{2500}{5.242136857} \\ &\approx 476.90 \end{aligned}$$

Each payment is \$476.90.

27. \$90,000 is the present value of this annuity of R dollars, with 12 periods, and $i = 8\% = 0.08$ per period.

$$\begin{aligned} P &= R \cdot a_{\overline{n}|i} \\ 90{,}000 &= R \cdot a_{\overline{12}|0.08} \\ R &= \frac{90{,}000}{a_{\overline{12}|0.08}} \\ &\approx \frac{90{,}000}{7.536078017} \\ &\approx 11{,}942.55 \end{aligned}$$

Each payment is \$11,942.55.

29. \$55,000 is the present value of this annuity of R dollars, with $i = \frac{0.06}{12} = 0.005$, and $n = 36$ periods.

$$\begin{aligned} P &= R \cdot a_{\overline{n}|i} \\ R &= \frac{55{,}000}{a_{\overline{36}|0.005}} \\ &\approx \frac{55{,}000}{32.87101624} \\ &\approx 1673.21 \end{aligned}$$

Each payment is \$1673.21.

31. Pat's payments form an ordinary annuity with $R = 12{,}000$, $n = 9$, and $i = 0.06$. The amount of this annuity is

$$S = 12{,}000\left[\frac{(1.06)^9 - 1}{0.06}\right].$$

The number in brackets, $s_{\overline{9}|0.06}$, is 11.49131598, so that

$$S \approx 12{,}000\,(11.49131598) \approx 137{,}895.79,$$

or \$137,895.79.

33. This ordinary annuity will amount to \$10,000 in 8 years at 16% compounded quarterly. Thus, $S = 10{,}000$, $n = 8 \cdot 4 = 32$, and $i = \frac{16\%}{4} = 4\% = 0.04$, so

$$\begin{aligned} 10{,}000 &= R \cdot s_{\overline{32}|0.04} \\ R &= \frac{10{,}000}{s_{\overline{32}|0.04}} \\ &\approx \frac{10{,}000}{62.70146867} \\ &\approx 159.49, \end{aligned}$$

or \$159.49.

35. This ordinary annuity will amount to \$24,000 in 5 years at 12% compounded quarterly. Thus, $S = 24{,}000$, $i = \frac{0.12}{4} = 0.03$, and $n = 4 \cdot 5 = 20$ periods.

$$\begin{aligned} S &= R \cdot s_{\overline{n}|i} \\ R &= \frac{24{,}000}{s_{\overline{20}|0.03}} \\ &\approx \frac{24{,}000}{26.87037449} \\ &\approx 893.18 \end{aligned}$$

Each payment is \$893.18.

37. Interest of $\frac{6\%}{2} = 3\%$ is earned semiannually. In $65 - 40 = 25$ years, there are $25 \cdot 2 = 50$ semiannual periods. Since

$$\begin{aligned} s_{\overline{50}|0.03} &= \left[\frac{(1.03)^{50} - 1}{0.03}\right] \\ &\approx 112.7968673, \end{aligned}$$

the \$1000 semiannual deposits will produce a total of

$$\begin{aligned} S &= 1000\,(112.7968673) \\ &\approx 112{,}796.87, \end{aligned}$$

or \$112,796.87.

39. Interest of $\frac{10\%}{2} = 5\%$ is earned semiannually. In $65 - 40 = 25$ years, there are $25 \cdot 2 = 50$ semiannual periods. Since

$$s_{\overline{50}|0.05} = \left[\frac{(1.05)^{50} - 1}{0.05}\right] \approx 209.3479957,$$

the \$1000 semiannual deposits will produce a total of

$$S \approx 1000\,(209.3479957) \approx 209{,}348.00,$$

or \$209,348.00.

41. This ordinary annuity will amount to \$40,000 in 7 years at 6% compounded annually. Thus, $S = 40{,}000$, $n = 7$, and $i = 0.06$, so

$$\begin{aligned} 40{,}000 &= R \cdot s_{\overline{7}|0.06} \\ R &= \frac{40{,}000}{s_{\overline{7}|0.06}} \approx \frac{40{,}000}{8.39383765} \approx 4765.40, \end{aligned}$$

or \$4765.40.

43. **(a)** The total amount of interest paid is $(60{,}000)\,(0.08)\,(7) = 33{,}600$, or \$33,600. This total is divided into $7 \cdot 4 = 28$ equal quarterly interest payments. Since $\frac{33{,}600}{28} = 1200$, each quarterly interest payment will be \$1200.

(b) This ordinary annuity will amount to \$60,000 in 7 years at 6% compounded semiannually. Thus, $S = 60{,}000$, $n = 7 \cdot 2 = 14$, and $i = \frac{6\%}{2} = 3\% = 0.03$, so

$$60{,}000 = R \cdot s_{\overline{14}|0.03}$$

$$R = \frac{60{,}000}{s_{\overline{14}|0.03}} \approx \frac{60{,}000}{17.08632416} \approx 3511.58,$$

or \$3511.58.

(c)

Payment Number	Amount of Deposit	Interest Earned	Total in Account
1	\$3511.58	\$0	\$3511.58
2	\$3511.58	$(3511.58)\,(0.03) =$ \$105.35	$3511.58 + 3511.58 + 105.35 =$ \$7128.51
3	\$3511.58	$(7128.51)\,(0.03) =$ \$213.86	$7128.51 + 3511.58 + 213.86 =$ \$10,853.95
4	\$3511.58	$(10{,}853.95)\,(0.03) =$ \$325.62	$10{,}853.95 + 3511.58 + 325.62 =$ \$14.691.15
5	\$3511.58	$(14{,}691.15)\,(0.03) =$ \$440.73	$14{,}691.15 + 3511.58 + 440.73 =$ \$18,643.46
6	\$3511.58	$(18{,}643.46)\,(0.03) =$ \$559.30	$18{,}643.46 + 3511.58 + 559.30 =$ \$22,714.34
7	\$3511.58	$(22{,}714.34)\,(0.03) =$ \$681.43	$22{,}714.34 + 3511.58 + 681.43 =$ \$26,907.35
8	\$3511.58	$(26{,}907.35)\,(0.03) =$ \$807.22	$26{,}907.35 + 3511.58 + 807.22 =$ \$31,226.15
9	\$3511.58	$(31{,}226.15)\,(0.03) =$ \$936.78	$31{,}226.15 + 3511.58 + 936.78 =$ \$35,674.51
10	\$3511.58	$(35{,}674.51)\,(0.03) =$ \$1070.24	$35{,}674.51 + 3511.58 + 1070.24 =$ \$40,256.33
11	\$3511.58	$(40{,}256.33)\,(0.03) =$ \$1207.69	$40{,}256.33 + 3511.58 + 1207.69 =$ \$44,975.60
12	\$3511.58	$(44{,}975.60)\,(0.03) =$ \$1349.27	$44{,}975.60 + 3511.58 + 1349.27 =$ \$49,836.45
13	\$3511.58	$(49{,}836.45)\,(0.03) =$ \$1495.09	$49{,}836.45 + 3511.58 + 1495.09 =$ \$54,843.12
14	\$3511.58	$(54{,}843.12)\,(0.03) =$ \$1645.29	$54{,}843.12 + 3511.58 + 1645.29 =$ \$59,999.99

So that the final total in the account is \$60,000, add \$0.01 to the last amount of deposit.
Thus, line 14 of the table will be:

14	\$3511.59	$(54{,}843.12)\,(0.03) =$ \$1645.29	$54{,}843.12 + 3511.59 + 1645.29 =$ \$60,000.00

45. We want to find the present value of an annuity of \$2000 per year for 9 years at 8% compounded annually.

$$a_{\overline{9}|0.08} = \left[\frac{1-(1.08)^{-9}}{0.08}\right] \approx 6.246887911, \text{ so}$$

$$P \approx 2000\,(6.246887911) \approx 12{,}493.78,$$

or \$12,493.78. A lump sum deposit of \$12,493.78 today at 8% compounded annually will yield the same total after 9 years as deposits of \$2000 at the end of each year for 9 years at 8% compounded annually.

47. \$22,000 is the present value of this annuity with interest $i = \frac{0.09}{12} = 0.0075$ per month and of $n = 4 \cdot 12 = 48$ monthly payments.

$$S = R \cdot a_{\overline{n}|i}$$

$$R = \frac{22{,}000}{a_{\overline{48}|0.075}} \approx \frac{22{,}000}{40.18478189} \approx 547.47$$

The amount of each payment is \$547.47.

Since 48 payments of \$547.47 each are made, $48(\$547.47) = \$26,278.56$ will be paid in total. This means that \$26,278.56 – \$22,000 = \$4278.56 is the total amount of interest Veronica will pay.

49. \$150,000 is the future value of an annuity over 79 yr compounded quarterly. So, there are $79(4) = 316$ payment periods.

(a) The interest per quarter is $\frac{5.25\%}{4} = 1.3125\%$. Thus, $S = 150,000, n = 316, i = 0.013125$, and we must find the quarterly payment R in the formula

$$S = R\left[\frac{(1+i)^n - 1}{i}\right]$$
$$150{,}000 = R\left[\frac{(1.013125)^{316} - 1}{0.013125}\right]$$
$$R \approx 32.4923796$$

She would have to put \$32.49 into her savings at the end of every three months.

(b) For a 2% interest rate, the interest per quarter is $\frac{2\%}{4} = 0.5\%$. Thus, $S = 150{,}000$, $n = 316, i = 0.005$, and we must find the quarterly payment R in the formula

$$S = R\left[\frac{(1+i)^n - 1}{i}\right]$$
$$150{,}000 = R\left[\frac{(1.005)^{316} - 1}{0.005}\right]$$
$$R \approx 195.5222794$$

She would have to put \$195.52 into her savings at the end of every three months.

For a 7% interest rate, the interest per quarter is $\frac{7\%}{4} = 1.75\%$. Thus, $S = 150{,}000$, $n = 316, i = 0.0175$, and we must find the quarterly payment R in the formula

$$S = R\left[\frac{(1+i)^n - 1}{i}\right]$$
$$150{,}000 = R\left[\frac{(1.0175)^{316} - 1}{0.0175}\right]$$
$$R \approx 10.9663932$$

She would have to put \$10.97 into her savings at the end of every three months.

51. The present value, P, is 249,560, $i = \frac{0.0775}{12} \approx 0.0064583333$, and $n = 12 \cdot 25 = 300$.

$$249{,}560 = R \cdot a_{\overline{300}|0.0064583333} = R\left[\frac{1-(1+0.0064583333)^{-300}}{0.0064583333}\right] \approx R\left[\frac{1-0.1449639356}{0.0064583333}\right] \approx R\left[\frac{0.8550360644}{0.0064583333}\right]$$

or $\quad R \approx 1885.00.$

Monthly payments of \$1885.00 will be required to amortize the loan.

Use the formula for the unpaid balance of a loan with $R = 1885, i \approx 0.0064583333, n = 12 \cdot 15 = 300$, and $x = 12 \cdot 5 = 60$.

$$y = R\left[\frac{1-(1+i)^{-(n-x)}}{i}\right] \approx 1885\left[\frac{1-(1+0.0064583333)^{-240}}{0.0064583333}\right] \approx 229{,}612.44$$

The unpaid balance after 5 years is \$229,612.44.

53. The present value, P, is 353,700, $i = \frac{0.0795}{12} = 0.006625$, and $n = 12 \cdot 30 = 360$ periods.

$$P = R \cdot a_{\overline{n}|i}$$

$$R = \frac{353{,}700}{a_{\overline{360}|0.006625}} = 353{,}700\left[\frac{1-(1.006625)^{-360}}{0.006625}\right] \approx \frac{353{,}700}{136.9334083} \approx 2583.01$$

Monthly payments of \$2583.01 will be required to amortize the loan.

To find the unpaid balance, use the formula for the unpaid balance of a loan with $R = 2583.01, i = 0.006625, n = 360$, and $x = 12 \cdot 5 = 60$.

$$y = R \cdot \left[\frac{1-(1+i)^{-(n-x)}}{i}\right] = 2583.01\left[\frac{1-(1+0.006625)^{-300}}{0.006625}\right] \approx 2583.01(130.1224488) \approx 336{,}107.59$$

The unpaid balance after 5 years is \$336,107.59.

55. (a) \$25,000 is the present value of this annuity of 8 years with interest of 6% per year.

$$a_{\overline{8}|0.06} = 6.209793811, \text{ so}$$
$$25{,}000 = R(6.209793811)$$
$$R \approx 4025.90$$

Annual withdrawals of \$4025.90 each will be needed.

(b) \$25,000 is the present value of this annuity of 12 years with interest of 6% per year.

$$a_{\overline{12}|0.06} = 8.38384394, \text{ so}$$
$$25{,}000 = R(8.38384394)$$
$$R \approx 2981.93$$

Annual withdrawals of \$2981.93 each will be needed.

57. First, find the amount of each payment. \$4000 is the present value of an annuity of R dollars, with 4 periods, and $i = 0.08$ per period.

$$P = R \cdot a_{\overline{n}|i}$$
$$4000 = R \cdot a_{\overline{4}|0.08}$$
$$R = \frac{4000}{a_{\overline{4}|0.08}} \approx \frac{4000}{3.31212684} \approx 1207.68$$

Each payment is \$1207.68.

Payment Number	Amount of Payment	Interest for Period	Portion to Principal	Principal at End of Period
0	—	—	—	\$4000.00
1	\$1207.68	(4000) (0.08) (1) = \$320.00	1207.68 − 320.00 = \$887.68	4000.00 − 887.68 = \$3112.32
2	\$1207.68	(3112.32) (0.08) (1) = \$248.99	1207.68 − 248.99 = \$958.69	3112.32 − 958.69 = \$2153.63
3	\$1207.68	(2153.63) (0.08) (1) = \$172.29	1207.68 − 172.29 = \$1035.39	2153.63 − 1035.39 = \$1118.24
4	\$1207.68	(1118.24) (0.08) (1) = \$89.46	1207.68 − 89.46 = \$1118.22	1118.24 − 1118.22 = \$0.02

So that the final principal is zero, add \$0.02 to the last payment. Thus, line 4 of the table will be:

4	\$1207.70	(1118.24) (0.08) (1) = \$89.46	1207.70 − 89.46 = \$1118.24	1118.24 − 1118.24 = \$0

59. The firm's total cost is $8\,(1048) = \$8384.00$. After making a down payment of \$1200, a balance of $8384 - 1200 =$ \$7184 is owed. To amortize this balance, first find the amount of each payment. \$7184 is the present value of an annuity of R dollars, with $4 \cdot 12 = 48$ periods, and $i = \frac{10.5\%}{12} = .875\% = 0.00875$ per period.

$$\begin{aligned} P &= R \cdot a_{\overline{n}|i} \\ 7184 &= R \cdot a_{\overline{48}|0.00875} \\ R &= \frac{7184}{a_{\overline{48}|0.00875}} \approx \frac{7184}{39.05734359} \approx 183.93 \end{aligned}$$

Each payment is \$183.93.

Payment Number	Amount of Payment	Interest for Period	Portion to Principal	Principal at End of Period
0	—	—	—	\$7184.00
1	\$183.93	$(7184.00)(0.105)\left(\frac{1}{12}\right) = \$62.86$	$183.93 - 62.86 = \$121.07$	$7184.00 - 121.07 = \$7062.93$
2	\$183.93	$(7062.93)(0.105)\left(\frac{1}{12}\right) = \$61.80$	$183.93 - 61.80 = \$122.13$	$7062.93 - 122.13 = \$6940.80$
3	\$183.93	$(6940.80)(0.105)\left(\frac{1}{12}\right) = \$60.73$	$183.93 - 60.73 = \$123.20$	$6940.80 - 123.20 = \$6817.60$
4	\$183.93	$(6817.60)(0.105)\left(\frac{1}{12}\right) = \$59.65$	$183.93 - 59.65 = \$124.28$	$6817.60 - 124.28 = \$6693.32$
5	\$183.93	$(6693.32)(0.105)\left(\frac{1}{12}\right) = \$58.57$	$183.93 - 58.57 = \$125.36$	$6693.32 - 125.36 = \$6567.96$
6	\$183.93	$(6567.96)(0.105)\left(\frac{1}{12}\right) = \$57.47$	$183.93 - 57.47 = \$126.46$	$6567.96 - 126.46 = \$6441.50$

12.3 Taylor Polynomials at 0

1.

Derivative	Value at 0
$f(x) = e^{-2x}$	$f(0) = 1$
$f^{(1)}(x) = -2e^{-2x}$	$f^{(1)}(0) = -2$
$f^{(2)}(x) = 4e^{-2x}$	$f^{(2)}(0) = 4$
$f^{(3)}(x) = -8e^{-2x}$	$f^{(3)}(0) = -8$
$f^{(4)}(x) = 16e^{-2x}$	$f^{(4)}(0) = 16$

$$\begin{aligned} P_4(x) &= f(0) + \frac{f^{(1)}(0)}{1!}x + \frac{f^{(2)}(0)}{2!}x^2 + \frac{f^{(3)}(0)}{3!}x^3 + \frac{f^{(4)}(0)}{4!}x^4 = 1 + \frac{-2}{1!}x + \frac{4}{2!}x^2 + \frac{-8}{3!}x^3 + \frac{16}{4!}x^4 \\ &= 1 - 2x + 2x^2 - \frac{4}{3}x^3 + \frac{2}{3}x^4 \end{aligned}$$

3.

Derivative	Value at 0
$f(x) = e^{x+1}$	$f(0) = e$
$f^{(1)}(x) = e^{x+1}$	$f^{(1)}(0) = e$
$f^{(2)}(x) = e^{x+1}$	$f^{(2)}(0) = e$
$f^{(3)}(x) = e^{x+1}$	$f^{(3)}(0) = e$
$f^{(4)}(x) = e^{x+1}$	$f^{(4)}(0) = e$

$$\begin{aligned} P_4(x) &= f(0) + \frac{f^{(1)}(0)}{1!}x + \frac{f^{(2)}(0)}{2!}x^2 + \frac{f^{(3)}(0)}{3!}x^3 + \frac{f^{(4)}(0)}{4!}x^4 \\ &= e + \frac{e}{1!}x + \frac{e}{2!}x^2 + \frac{e}{3!}x^3 + \frac{e}{4!}x^4 \\ &= e + ex + \frac{e}{2}x^2 + \frac{e}{6}x^3 + \frac{e}{24}x^4 \end{aligned}$$

5.

Derivative	Value at 0
$f(x) = \sqrt{x+9} = (x+9)^{1/2}$	$f(0) = 3$
$f^{(1)}(x) = \frac{1}{2}(x+9)^{-1/2} = \frac{1}{2(x+9)^{1/2}}$	$f^{(1)}(0) = \frac{1}{6}$
$f^{(2)}(x) = -\frac{1}{4}(x+9)^{-3/2} = -\frac{1}{4(x+9)^{3/2}}$	$f^{(2)}(0) = -\frac{1}{108}$
$f^{(3)}(x) = \frac{3}{8}(x+9)^{-5/2} = \frac{3}{8(x+9)^{5/2}}$	$f^{(3)}(0) = \frac{1}{648}$
$f^{(4)}(x) = -\frac{15}{16}(x+9)^{-7/2} = -\frac{15}{16(x+9)^{7/2}}$	$f^{(4)}(0) = -\frac{5}{11{,}664}$

$$\begin{aligned} P_4(x) &= f(0) + \frac{f^{(1)}(0)}{1!}x + \frac{f^{(2)}(0)}{2!}x^2 + \frac{f^{(3)}(0)}{3!}x^3 + \frac{f^{(4)}(0)}{4!}x^4 \\ &= 3 + \frac{\frac{1}{6}}{1!}x + \frac{-\frac{1}{108}}{2!}x^2 + \frac{\frac{1}{648}}{3!}x^3 + \frac{-\frac{5}{11{,}664}}{4!}x^4 \\ &= 3 + \frac{1}{6}x - \frac{1}{216}x^2 + \frac{1}{3888}x^3 - \frac{5}{279{,}936}x^4 \end{aligned}$$

7.

Derivative	Value at 0
$f(x) = \sqrt[3]{x-1} = (x-1)^{1/3}$	$f(0) = -1$
$f^{(1)}(x) = \frac{1}{3}(x-1)^{-2/3} = \frac{1}{3(x-1)^{2/3}}$	$f^{(1)}(0) = \frac{1}{3}$
$f^{(2)}(x) = -\frac{2}{9}(x-1)^{-5/3} = -\frac{2}{9(x-1)^{5/3}}$	$f^{(2)}(0) = \frac{2}{9}$
$f^{(3)}(x) = \frac{10}{27}(x-1)^{-8/3} = \frac{10}{27(x-1)^{8/3}}$	$f^{(3)}(0) = \frac{10}{27}$
$f^{(4)}(x) = -\frac{80}{81}(x-1)^{-11/3} = -\frac{80}{81(x-1)^{11/3}}$	$f^{(4)}(0) = \frac{80}{81}$

$$\begin{aligned} P_4(x) &= f(0) + \frac{f^{(1)}(0)}{1!}x + \frac{f^{(2)}(0)}{2!}x^2 + \frac{f^{(3)}(0)}{3!}x^3 + \frac{f^{(4)}(0)}{4!}x^4 \\ &= -1 + \frac{\frac{1}{3}}{1!}x + \frac{\frac{2}{9}}{2!}x^2 + \frac{\frac{10}{27}}{3!}x^3 + \frac{\frac{80}{81}}{4!}x^4 \\ &= -1 + \frac{1}{3}x + \frac{1}{9}x^2 + \frac{5}{81}x^3 + \frac{10}{243}x^4 \end{aligned}$$

9.

Derivative	Value at 0
$f(x) = \sqrt[4]{x+1} = (x+1)^{1/4}$	$f(0) = 1$
$f^{(1)}(x) = \frac{1}{4}(x+1)^{-3/4} = \frac{1}{4(x+1)^{3/4}}$	$f^{(1)}(0) = \frac{1}{4}$
$f^{(2)}(x) = -\frac{3}{16}(x+1)^{-7/4} = -\frac{3}{16(x+1)^{7/4}}$	$f^{(2)}(0) = -\frac{3}{16}$
$f^{(3)}(x) = \frac{21}{64}(x+1)^{-11/4} = \frac{21}{64(x+1)^{11/4}}$	$f^{(3)}(0) = \frac{21}{64}$
$f^{(4)}(x) = -\frac{231}{256}(x+1)^{-15/4} = -\frac{231}{256(x+1)^{15/4}}$	$f^{(4)}(0) = -\frac{231}{256}$

$$\begin{aligned}
P_4(x) &= f(0) + \frac{f^{(1)}(0)}{1!}x + \frac{f^{(2)}(0)}{2!}x^2 + \frac{f^{(3)}(0)}{3!}x^3 + \frac{f^{(4)}(0)}{4!}x^4 \\
&= 1 + \frac{\frac{1}{4}}{1!}x + \frac{-\frac{3}{16}}{2!}x^2 + \frac{\frac{21}{64}}{3!}x^3 + \frac{-\frac{231}{256}}{4!}x^4 \\
&= 1 + \frac{1}{4}x - \frac{3}{32}x^2 + \frac{7}{128}x^3 - \frac{77}{2048}x^4
\end{aligned}$$

11.

Derivative	Value at 0
$f(x) = \ln(1-x)$	$f(0) = 0$
$f^{(1)}(x) = -\frac{1}{1-x} = \frac{1}{x-1} = (x-1)^{-1}$	$f^{(1)}(0) = -1$
$f^{(2)}(x) = -(x-1)^{-2} = -\frac{1}{(x-1)^2}$	$f^{(2)}(0) = -1$
$f^{(3)}(x) = 2(x-1)^{-3} = \frac{2}{(x-1)^3}$	$f^{(3)}(0) = -2$
$f^{(4)}(x) = -6(x-1)^{-4} = -\frac{6}{(x-1)^4}$	$f^{(4)}(0) = -6$

$$\begin{aligned}
P_4(x) &= f(0) + \frac{f^{(1)}(0)}{1!}x + \frac{f^{(2)}(0)}{2!}x^2 + \frac{f^{(3)}(0)}{3!}x^3 + \frac{f^{(4)}(0)}{4!}x^4 \\
&= 0 + \frac{-1}{1!}x + \frac{-1}{2!}x^2 + \frac{-2}{3!}x^3 + \frac{-6}{4!}x^4 \\
&= -x - \frac{1}{2}x^2 - \frac{1}{3}x^3 - \frac{1}{4}x^4
\end{aligned}$$

13.

Derivative	Value at 0
$f(x) = \ln\left(1+2x^2\right)$	$f(0) = 0$
$f^{(1)}(x) = \dfrac{4x}{1+2x^2}$	$f^{(1)}(0) = 0$
$f^{(2)}(x) = \dfrac{4-8x^2}{(1+2x^2)^2}$	$f^{(2)}(0) = 4$
$f^{(3)}(x) = \dfrac{32x^3-48x}{(1+2x^2)^3}$	$f^{(3)}(0) = 0$
$f^{(4)}(x) = \dfrac{-192x^4+576x^2-48}{(1+2x^2)^4}$	$f^{(4)}(0) = -48$

$$\begin{aligned} P_4(x) &= f(0) + \frac{f^{(1)}(0)}{1!}x + \frac{f^{(2)}(0)}{2!}x^2 + \frac{f^{(3)}(0)}{3!}x^3 + \frac{f^{(4)}(0)}{4!}x^4 \\ &= 0 + \frac{0}{1!}x + \frac{4}{2!}x^2 + \frac{0}{3!}x^3 + \frac{-48}{4!}x^4 \\ &= 2x^2 - 2x^4 \end{aligned}$$

15.

Derivative	Value at 0
$f(x) = xe^{-x}$	$f(0) = 0$
$f^{(1)}(x) = -xe^{-x} + e^{-x}$	$f^{(1)}(0) = 1$
$f^{(2)}(x) = xe^{-x} - 2e^{-x}$	$f^{(2)}(0) = -2$
$f^{(3)}(x) = -xe^{-x} + 3e^{-x}$	$f^{(3)}(0) = 3$
$f^{(4)}(x) = xe^{-x} - 4e^{-x}$	$f^{(4)}(0) = -4$

$$\begin{aligned} P_4(x) &= f(0) + \frac{f^{(1)}(0)}{1!}x + \frac{f^{(2)}(0)}{2!}x^2 + \frac{f^{(3)}(0)}{3!}x^3 + \frac{f^{(4)}(0)}{4!}x^4 \\ &= 0 + \frac{1}{1!}x + \frac{-2}{2!}x^2 + \frac{3}{3!}x^3 + \frac{-4}{4!}x^4 \\ &= x - x^2 + \frac{1}{2}x^3 - \frac{1}{6}x^4 \end{aligned}$$

17.

Derivative	Value at 0
$f(x) = (9-x)^{3/2}$	$f(0) = 27$
$f^{(1)}(x) = -\dfrac{3}{2}(9-x)^{1/2}$	$f^{(1)}(0) = -\dfrac{9}{2}$
$f^{(2)}(x) = \dfrac{3}{4}(9-x)^{-1/2} = \dfrac{3}{4(9-x)^{1/2}}$	$f^{(2)}(0) = \dfrac{1}{4}$
$f^{(3)}(x) = \dfrac{3}{8}(9-x)^{-3/2} = \dfrac{3}{8(9-x)^{3/2}}$	$f^{(3)}(0) = \dfrac{1}{72}$
$f^{(4)}(x) = \dfrac{9}{16}(9-x)^{-5/2} = \dfrac{9}{16(9-x)^{5/2}}$	$f^{(4)}(0) = \dfrac{1}{432}$

$$\begin{aligned} P_4(x) &= f(0) + \frac{f^{(1)}(0)}{1!}x + \frac{f^{(2)}(0)}{2!}x^2 + \frac{f^{(3)}(0)}{3!}x^3 + \frac{f^{(4)}(0)}{4!}x^4 \\ &= 27 + \frac{-\frac{9}{2}}{1!}x + \frac{\frac{1}{4}}{2!}x^2 + \frac{\frac{1}{72}}{3!}x^3 + \frac{\frac{1}{432}}{4!}x^4 \\ &= 27 - \frac{9}{2}x + \frac{1}{8}x^2 + \frac{1}{432}x^3 + \frac{1}{10{,}368}x^4 \end{aligned}$$

19.

Derivative	Value at 0
$f(x) = \frac{1}{1+x} = (1+x)^{-1}$	$f(0) = 1$
$f^{(1)}(x) = -(1+x)^{-2} = -\frac{1}{(1+x)^2}$	$f^{(1)}(0) = -1$
$f^{(2)}(x) = 2(1+x)^{-3} = \frac{2}{(1+x)^3}$	$f^{(2)}(0) = 2$
$f^{(3)}(x) = -6(1+x)^{-4} = -\frac{6}{(1+x)^4}$	$f^{(3)}(0) = -6$
$f^{(4)}(x) = 24(1+x)^{-5} = \frac{24}{(1+x)^5}$	$f^{(4)}(0) = 24$

$$\begin{aligned} P_4(x) &= f(0) + \frac{f^{(1)}(0)}{1!}x + \frac{f^{(2)}(0)}{2!}x^2 + \frac{f^{(3)}(0)}{3!}x^3 + \frac{f^{(4)}(0)}{4!}x^4 \\ &= 1 + \frac{-1}{1!}x + \frac{2}{2!}x^2 + \frac{-6}{3!}x^3 + \frac{24}{4!}x^4 \\ &= 1 - x + x^2 - x^3 + x^4 \end{aligned}$$

For Exercises **21**–**33** each approximation can be determined using the Taylor polynomials from Exercises 1–13, respectively.

21. Using the result of Exercise 1, with $f(x) = e^{-2x}$ and $P_4(x) = 1 - 2x + 2x^2 - \frac{4}{3}x^3 + \frac{2}{3}x^4$, we can approximate $e^{-0.04}$ by evaluating $f(0.01) = e^{-2(0.02)} = e^{-0.04}$. Using $P_4(x)$ from Exercise 1 with $x = 0.02$ gives

$$P_4(0.02) = 1 - 2(0.02) + 2(0.02)^2 - \frac{4}{3}(0.02)^3 + \frac{2}{3}(0.02)^4 \approx 0.096078944.$$

To four decimal places, $P_4(0.02)$ approximates the value of $e^{-0.04}$ as 0.9608.

23. Using the result of Exercise 3, with $f(x) = e^{x+1}$ and $P_4(x) = e + ex + \frac{e}{2}x^2 + \frac{e}{6}x^3 + \frac{e}{24}x^4$, we can approximate $e^{1.02}$ by evaluating $f(0.02) = e^{0.02+1} = e^{1.02}$. Using $P_4(x)$ from Exercise 3 with $x = 0.02$ gives

$$\begin{aligned} P_4(0.02) &= e + e(0.02) + \frac{e}{2}(0.02)^2 + \frac{e}{6}(0.02)^3 + \frac{e}{24}(0.02)^4 \\ &\approx 2.773194764. \end{aligned}$$

To four decimal places, $P_4(0.02)$ approximates the value of $e^{1.02}$ as 2.7732.

25. Using the result of Exercise 5, with $f(x) = \sqrt{x+9}$ and $P_4(x) = 3 + \frac{1}{6}x - \frac{1}{216}x^2 + \frac{1}{3888}x^3 - \frac{5}{279{,}936}x^4$, we can approximate $\sqrt{8.92}$ by evaluating $f(-0.08) = \sqrt{-0.08+9} = \sqrt{8.92}$. Using $P_4(x)$ from Exercise 5 with $x = -0.08$ gives

$$\begin{aligned} P_4(-0.08) &= 3 + \frac{1}{6}(-0.08) - \frac{1}{216}(-0.08)^2 + \frac{1}{3888}(-0.08)^3 - \frac{5}{279{,}936}(-0.08)^4 \\ &\approx 2.986636905. \end{aligned}$$

To four decimal places, $P_4(-0.08)$ approximates the value of $\sqrt{8.92}$ as 2.9866.

27. Using the result of Exercise 7, with $f(x) = \sqrt[3]{x-1}$ and $P_4(x) = -1 + \frac{1}{3}x + \frac{1}{9}x^2 + \frac{5}{81}x^3 + \frac{10}{243}x^4$, we can approximate $\sqrt[3]{-1.05}$ by evaluating $f(-0.05) = \sqrt[3]{-0.05-1} = \sqrt[3]{-1.05}$. Using $P_4(x)$ from Exercise 7 with $x = -0.05$ gives

$$\begin{aligned} P_4(-0.05) &= -1 + \frac{1}{3}(-0.05) + \frac{1}{9}(-0.05)^2 + \frac{5}{81}(-0.05)^3 + \frac{10}{243}(-0.05)^4 \\ &\approx -1.016396348. \end{aligned}$$

To four decimal places, $P_4(-0.05)$ approximates the value of $\sqrt[3]{-1.05}$ as -1.0164.

29. Using the result of Exercise 9, with $f(x) = \sqrt[4]{x+1}$ and $P_4(x) = 1 + \frac{1}{4}x - \frac{3}{32}x^2 + \frac{7}{128}x^3 - \frac{77}{2048}x^4$, we can approximate $\sqrt[4]{1.06}$ by evaluating $f(0.06) = \sqrt[4]{0.06+1} = \sqrt[4]{1.06}$. Using $P_4(x)$ from Exercise 9 with $x = 0.06$ gives

$$\begin{aligned} P_4(0.06) &= 1 + \frac{1}{4}(0.06) - \frac{3}{32}(0.06)^2 + \frac{7}{128}(0.06)^3 - \frac{77}{2048}(0.06)^4 \\ &\approx 1.014673825. \end{aligned}$$

To four decimal places, $P_4(0.06)$ approximates the value of $\sqrt[4]{1.06}$ as 1.0147.

31. Using the result of Exercise 11, with $f(x) = \ln(1-x)$ and $P_4(x) = -x - \frac{1}{2}x^2 - \frac{1}{3}x^3 - \frac{1}{4}x^4$, we can approximate $\ln 0.97$ by evaluating $f(0.03) = \ln(1-0.03) = \ln 0.97$. Using $P_4(x)$ from Exercise 11 with $x = 0.03$ gives

$$\begin{aligned} P_4(0.03) &= -(0.03) - \frac{1}{2}(0.03)^2 - \frac{1}{3}(0.03)^3 - \frac{1}{4}(0.03)^4 \\ &\approx -0.0304592025. \end{aligned}$$

To four decimal places, $P_4(0.03)$ approximates the value of $\ln 0.97$ as -0.0305.

33. Using the result of Exercise 13, with $f(x) = \ln(1+2x^2)$ and $P_4(x) = 2x^2 - 2x^4$, we can compute $\ln 1.008$ by evaluating $f\left(\frac{\sqrt{10}}{50}\right) = \ln\left[1 + 2\left(\frac{\sqrt{10}}{50}\right)^2\right] = \ln[1 + 2(0.004)] = \ln 1.008$. Using $P_4(x)$ with $x = \frac{\sqrt{10}}{50}$ gives

$$\begin{aligned} P_4\left(\frac{\sqrt{10}}{50}\right) &= 2\left(\frac{\sqrt{10}}{50}\right)^2 - 4\left(\frac{\sqrt{10}}{50}\right)^4 \\ &= 0.007968 \end{aligned}$$

To four decimal places, $P_4\left(\frac{\sqrt{10}}{50}\right)$ approximates $\ln 1.008$ as 0.0080.

35. (a)

Derivative	Value at 0
$f(x) = (a+x)^{\frac{1}{n}}$	$f(0) = a^{\frac{1}{n}}$
$f^{(1)}(x) = \frac{1}{n}(a+x)^{\frac{1}{n}-1}$	$f^{(1)}(0) = \frac{1}{n}a^{\frac{1}{n}-1} = \frac{a^{\frac{1}{n}}}{na}$
$f^{(2)}(x) = \frac{1}{n}\left(\frac{1}{n}-1\right)(a+x)^{\frac{1}{n}-2}$	$f^{(2)}(0) = \frac{1}{n}\left(\frac{1-n}{n}\right)a^{\frac{1}{n}-2} = \frac{(1-n)a^{\frac{1}{n}}}{n^2a^2}$
$f^{(3)}(x) = \frac{1}{n}\left(\frac{1}{n}-1\right)\left(\frac{1}{n}-2\right)(a+x)^{\frac{1}{n}-3}$	$f^{(3)}(0) = \frac{1}{n}\left(\frac{1-n}{n}\right)\left(\frac{1-2n}{n}\right)a^{\frac{1}{n}-3} = \frac{(1-n)(1-2n)a^{\frac{1}{n}}}{n^3a^3}$
$f^{(4)}(x) = \frac{1}{n}\left(\frac{1}{n}-1\right)\left(\frac{1}{n}-2\right)\left(\frac{1}{n}-3\right)(a+x)^{\frac{1}{n}-4}$	$f^{(4)}(0) = \frac{1}{n}\left(\frac{1-n}{n}\right)\left(\frac{1-2n}{n}\right)\left(\frac{1-3n}{n}\right)a^{\frac{1}{n}-4} = \frac{(1-n)(1-2n)(1-3n)a^{\frac{1}{n}}}{n^4a^4}$

$$\begin{aligned} P_4(x) &= a^{1/n} + \frac{\frac{a^{1/n}}{na}}{1!}x + \frac{\frac{(1-n)a^{1/n}}{n^2a^2}}{2!}x^2 + \frac{\frac{(1-n)(1-2n)a^{1/n}}{n^3a^3}}{3!}x^3 + \frac{\frac{(1-n)(1-2n)(1-3n)a^{1/n}}{n^4a^4}}{4!}x^4 \\ &= a^{1/n} + \frac{xa^{1/n}}{na} + \frac{x^2a^{1/n}(1-n)}{2!n^2a^2} + \frac{x^3a^{1/n}(1-n)(1-2n)}{3!n^3a^3} + \frac{x^4a^{1/n}(1-n)(1-2n)(1-3n)}{4!n^4a^4} \\ &= a^{1/n} + \frac{xa^{1/n}}{na}\left[1 + \frac{x(1-n)}{2!na} + \frac{x^2(1-n)(1-2n)}{3!n^2a^2} + \frac{x^3(1-n)(1-2n)(1-3n)}{4!n^3a^3}\right] \end{aligned}$$

For values of x that are small compared with a, the terms after 1 in the brackets get closer and closer to zero. For example, consider the term $\frac{x(1-n)}{2!na}$ where $x = \frac{1}{10}a$. Then $\frac{x(1-n)}{2!na} = \frac{1-n}{20n}$. For even the smallest value of n, $n = 2$, the value of this term is $-\frac{1}{40}$. This term, and the terms which follow will have little impact on our approximation. Thus the value of the expression is approximately 1, so $P_4(x) \approx a^{1/n} + \frac{xa^{1/n}}{na}$. Thus, if x is small compared with a, the Taylor polynomial $P_4(x)$ approximates $(a+x)^{1/n}$ as $a^{1/n} + \frac{xa^{1/n}}{na}$.
(b) Using the result of part (a), we can approximate $\sqrt[3]{66}$ by evaluating $f(2)$ given $f(x) = (64+x)^{1/3}$ since $f(2) = (64+2)^{1/3} = 66^{1/3} = \sqrt[3]{66}$. Using $P_4(x)$ from part (a) with $x = 2$, $a = 64$, and $n = 3$ gives

$$(64+2)^{1/3} = 64^{1/3} + \frac{2 \cdot 64^{1/3}}{3 \cdot 64} = 4 + \frac{8}{192} = 4.041\bar{6}$$

To five decimal places, $P_4(x) \approx (64+x)^{1/3}$ approximates the value of $\sqrt[3]{66}$ as 4.04167. To five decimal places, a calculator gives an approximation of 4.04124.

37.
$$\begin{aligned} P_4(x) &= f(0) + \frac{f^{(1)}(0)}{1!}x + \frac{f^{(2)}(0)}{2!}x^2 + \frac{f^{(3)}(0)}{3!}x^3 + \frac{f^{(4)}(0)}{4!}x^4 \\ &= 1 + \frac{1}{1}x + \frac{2}{2}x^2 + \frac{6}{6}x^3 + \frac{24}{24}x^4 \\ &= 1 + x + x^2 + x^3 + x^4 \end{aligned}$$

For a polynomial of degree n with $f^{(n)}(0) = n!$ and $0! = 1$,

$$\begin{aligned} P_n(x) &= f(0) + \frac{f^{(1)}(0)}{1!}x + \frac{f^{(2)}(0)}{2!}x^2 + \cdots + \frac{f^{(n-1)}(0)}{(n-1)!}x^{n-1} + \frac{f^{(n)}(0)}{n!}x^n \\ &= 1 + \frac{1!}{1!}x + \frac{2!}{2!}x^2 + \cdots + \frac{(n-1)!}{(n-1)!}x^{n-1} + \frac{n!}{n!}x^n \\ &= 1 + x + x^2 + \cdots + x^{n-1} + x^n \end{aligned}$$

39. (a)

Derivative	Value at 0
$f(N) = e^{\lambda N}$	$f(0) = 1$
$f^{(1)}(N) = \lambda e^{\lambda N}$	$f^{(1)}(0) = \lambda$
$f^{(2)}(N) = \lambda^2 e^{\lambda N}$	$f^{(2)}(0) = \lambda^2$

$$P_2(N) = f(0) + \frac{f^{(1)}(0)}{1!}N + \frac{f^{(2)}(0)}{2!}N^2 = 1 + \frac{\lambda}{1!}N + \frac{\lambda^2}{2!}N^2 = 1 + \lambda N + \frac{\lambda^2}{2}N^2$$

The Taylor polynomial of degree 2 at $N = 0$ for $e^{\lambda N}$ is $P_2(N) = 1 + \lambda N + \frac{\lambda^2}{2}N^2$.

(b) Substituting $P_2(N) = 1 + \lambda N + \frac{\lambda^2}{2}N^2$ for $e^{\lambda N}$ in the original equation gives:

$$\begin{aligned} \frac{1 + \lambda N + \frac{\lambda^2}{2}N^2}{\lambda} - N &= \frac{1}{\lambda} + k \\ \left(1 + \lambda N + \frac{\lambda^2}{2}N^2\right) - \lambda N &= 1 + \lambda k \\ \frac{\lambda^2}{2}N^2 &= \lambda k \\ N^2 &= \frac{2k}{\lambda} \\ N &= \pm\sqrt{\frac{2k}{\lambda}} \end{aligned}$$

Since N is a positive quantity, use the positive square root. So, $N = \sqrt{\frac{2k}{\lambda}}$.

41.

Derivative	Value at 0
$P(x) = \ln(100+3x)$	$P(0) = 4.605$
$P^{(1)}(x) = \dfrac{3}{100+3x}$	$P^{(1)}(0) = 0.03$
$P^{(2)}(x) = -\dfrac{9}{(100+3x)^2}$	$P^{(2)}(0) = -0.0009$

$$\begin{aligned} P_2(x) &= P(0) + \frac{P^{(1)}(0)}{1!}x + \frac{P^{(2)}(0)}{2!}x^2 \\ &= 4.605 + \frac{0.03}{1}x + \frac{-0.0009}{2}x^2 \\ &= 4.605 + 0.03x - 0.00045x^2 \\ P_2(0.6) &= 4.605 + 0.03(0.6) - 0.00045(0.6)^2 \\ &= 4.622838 \end{aligned}$$

The Taylor polynomial $P_2(0.6)$ approximates $P(0.6)$ as 4.622838. The approximate profit is 4.623 thousand dollars, or \$4623.
Finding $P(0.6)$ by direct substitution gives:

$$\begin{aligned} P(0.6) &= \ln 100 + 3(0.6) \\ &= \ln 101.8 \\ &= 4.623010104 \end{aligned}$$

Direct substitution also estimates the profit as 4.623 thousand dollars, or \$4623.

43.

Derivative	Value at 0
$R(x) = 500\ln\left(4+\dfrac{x}{50}\right)$	$R(0) = 500(1.386) = 693$
$R^{(1)}(x) = \dfrac{500}{200+x}$	$R^{(1)}(0) = 2.5$
$R^{(2)}(x) = -\dfrac{500}{(200+x)^2}$	$R^{(2)}(0) = -0.0125$

$$P_2(x) = R(0) + \frac{R^{(1)}(0)}{1!}x + \frac{R^{(2)}(0)}{2!}x^2 = 693 + \frac{2.5}{1}x + \frac{-0.0125}{2}x^2 = 693 + 2.5x - 0.00625x^2$$

$$P_2(10) = 693 + 2.5(10) - 0.00625(10)^2 = 717.375$$

The Taylor polynomial $P_2(10)$ approximates $R(10)$ as 717.375. Thus, the approximate revenue is 717.
Finding $R(10)$ by direct substitution gives:

$$R(10) = 500\ln\left(4+\frac{10}{50}\right) \approx 717.5422626$$

Direct substitution estimates the revenue as 718.

45.

Derivative	Value at 0
$P(k) = 1 - e^{-2k}$	$P(0) = 1 - e^0 = 0$
$P^{(1)}(k) = 2e^{-2k}$	$P^{(1)}(0) = 2$

$$P_1(k) = P(0) + \frac{P^{(1)}(0)}{1!}k = 0 + \frac{2}{1!}k = 2k$$

So, if k is small, that is, close to 0, then

$$P(k) \approx P_1(k) = 2k.$$

12.4 Infinite Series

1. $20 + 10 + 5 + \frac{5}{2} + \ldots$ is a geometric series with $a = a_1 = 20$ and $r = \frac{1}{2}$. Since r is in $(-1, 1)$, the series converges and has sum

$$\frac{a}{1-r} = \frac{20}{1-\frac{1}{2}} = \frac{20}{\frac{1}{2}} = 40.$$

3. $2 + 6 + 18 + 54 + \ldots$ is a geometric series with $a = a_1 = 2$ and $r = 3$. Since $r > 1$, the series diverges.

5. $27 + 9 + 3 + 1 + \ldots$ is a geometric series with $a = a_1 = 27$ and $r = \frac{1}{3}$. Since r is in $(-1, 1)$, the series converges and has sum

$$\frac{a}{1-r} = \frac{27}{1-\frac{1}{3}} = \frac{27}{\frac{2}{3}} = \frac{81}{2}.$$

7. $100 + 10 + 1 + \ldots$ is a geometric series with $a = a_1 = 100$ and $r = \frac{1}{10}$. Since r is in $(-1, 1)$, the series converges and has sum

$$\frac{a}{1-r} = \frac{100}{1-\frac{1}{10}} = \frac{100}{\frac{9}{10}} = \frac{1000}{9}.$$

9. $\frac{5}{4} + \frac{5}{8} + \frac{5}{16} + \ldots$ is a geometric series with $a = a_1 = \frac{5}{4}$ and $r = \frac{1}{2}$. Since r is in $(-1, 1)$, the series converges and has sum

$$\frac{a}{1-r} = \frac{\frac{5}{4}}{1-\frac{1}{2}} = \frac{\frac{5}{4}}{\frac{1}{2}} = \frac{5}{2}.$$

11. $\frac{1}{3} - \frac{2}{9} + \frac{4}{27} - \frac{8}{81} + \cdots$ is a geometric series with $a = a_1 = \frac{1}{3}$ and $r = -\frac{2}{3}$. Since r is in $(-1, 1)$, the series converges and has sum

$$\frac{a}{1-r} = \frac{\frac{1}{3}}{1-\left(-\frac{2}{3}\right)} = \frac{\frac{1}{3}}{1+\frac{2}{3}} = \frac{\frac{1}{3}}{\frac{5}{3}} = \frac{1}{5}.$$

13. $e - 1 + \frac{1}{e} - \frac{1}{e^2} + \cdots$ is a geometric series with $a = a_1 = e$ and $r = -\frac{1}{e}$. Since r is in $(-1, 1)$, the series converges and has sum

$$\frac{a}{1-r} = \frac{e}{1-\left(-\frac{1}{e}\right)} = \frac{e}{1+\frac{1}{e}} = \frac{e}{\frac{e+1}{e}} = \frac{e^2}{e+1}.$$

15. $S_1 = a_1 = \frac{1}{1} = 1$

$S_2 = a_1 + a_2 = 1 + \frac{1}{2} = \frac{3}{2}$

$S_3 = a_1 + a_2 + a_3 = 1 + \frac{1}{2} + \frac{1}{3} = \frac{11}{6}$

$S_4 = a_1 + a_2 + a_3 + a_4 = 1 + \frac{1}{2} + \frac{1}{3} + \frac{1}{4} = \frac{25}{12}$

$S_5 = a_1 + a_2 + a_3 + a_4 + a_5 = 1 + \frac{1}{2} + \frac{1}{3} + \frac{1}{4} + \frac{1}{5} = \frac{137}{60}$

17. $S_1 = a_1 = \frac{1}{2(1)+5} = \frac{1}{7}$

$S_2 = a_1 + a_2 = \frac{1}{2(1)+5} + \frac{1}{2(2)+5} = \frac{1}{7} + \frac{1}{9} = \frac{16}{63}$

$S_3 = a_1 + a_2 + a_3 = \frac{1}{2(1)+5} + \frac{1}{2(2)+5} + \frac{1}{2(3)+5} = \frac{1}{7} + \frac{1}{9} + \frac{1}{11} = \frac{239}{693}$

$S_4 = a_1 + a_2 + a_3 + a_4$
$= \frac{1}{2(1)+5} + \frac{1}{2(2)+5} + \frac{1}{2(3)+5} + \frac{1}{2(4)+5} = \frac{1}{7} + \frac{1}{9} + \frac{1}{11} + \frac{1}{13} = \frac{3800}{9009}$

$S_5 = a_1 + a_2 + a_3 + a_4 + a_5$
$= \frac{1}{2(1)+5} + \frac{1}{2(2)+5} + \frac{1}{2(3)+5} + \frac{1}{2(4)+5} + \frac{1}{2(5)+5}$
$= \frac{1}{7} + \frac{1}{9} + \frac{1}{11} + \frac{1}{13} + \frac{1}{15} = \frac{22{,}003}{45{,}045}$

19. $S_1 = a_1 = \frac{1}{(1+1)(1+2)} = \frac{1}{6}$

$S_2 = a_1 + a_2 = \frac{1}{(1+1)(1+2)} + \frac{1}{(2+1)(2+2)} = \frac{1}{6} + \frac{1}{12} = \frac{1}{4}$

$S_3 = a_1 + a_2 + a_3$
$= \frac{1}{(1+1)(1+2)} + \frac{1}{(2+1)(2+2)} + \frac{1}{(3+1)(3+2)} = \frac{1}{6} + \frac{1}{12} + \frac{1}{20} = \frac{3}{10}$

$S_4 = a_1 + a_2 + a_3 + a_4 = \frac{1}{(1+1)(1+2)} + \frac{1}{(2+1)(2+2)} + \frac{1}{(3+1)(3+2)} + \frac{1}{(4+1)(4+2)}$
$= \frac{1}{6} + \frac{1}{12} + \frac{1}{20} + \frac{1}{30} = \frac{1}{3}$

$S_5 = a_1 + a_2 + a_3 + a_4 + a_5$
$= \frac{1}{(1+1)(1+2)} + \frac{1}{(2+1)(2+2)} + \frac{1}{(3+1)(3+2)} + \frac{1}{(4+1)(4+2)} + \frac{1}{(5+1)(5+2)}$
$= \frac{1}{6} + \frac{1}{12} + \frac{1}{20} + \frac{1}{30} + \frac{1}{42} = \frac{5}{14}$

21. **(a)** It follows from Viète's formula that

$$\frac{2}{\pi} \approx \frac{\sqrt{2}}{2} \cdot \frac{\sqrt{2+\sqrt{2}}}{2} \cdot \frac{\sqrt{2+\sqrt{2+\sqrt{2}}}}{2}$$

$$= \frac{\sqrt{2}\left(\sqrt{2+\sqrt{2}}\right)\left(\sqrt{2+\sqrt{2+\sqrt{2}}}\right)}{8}$$

$$\approx 0.641$$

Thus, $\pi \approx \frac{2}{0.641} \approx 3.12$.

It follows from Liebniz's formula that

$$\frac{\pi}{4} \approx 1 - \frac{1}{3} + \frac{1}{5} - \frac{1}{7} \approx 0.724.$$

Thus, $\pi \approx 4\,(0.724) \approx 2.90$.

Therefore, approximating π by multiplying the first three terms of Viète's formula together is more accurate than approximating π by adding the first four terms of Liebniz's formula together.

(b) Using the graphing calculator with $Y_1 = 4{*}\text{sum}\left(\text{seq}\left(\frac{(-1)^{N-1}}{2N-1}, N, 1, X\right)\right)$ and the table function gives the following values for Y_1 for $35 \le X \le 41$.

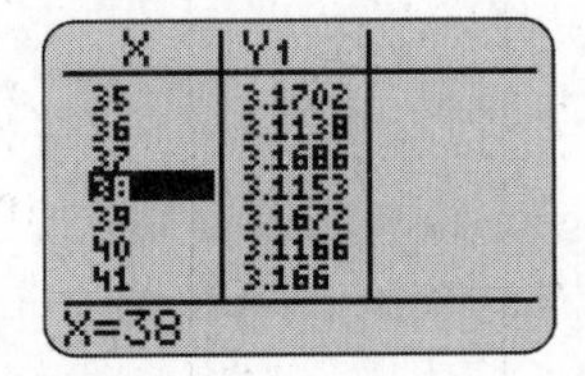

Thus, 38 terms of the second formula must be added together to produce the same accuracy as the product of the first three terms of the first formula.

23. **(a)** With $a = a_1 = 1000$ and $r = 0.1$, the production manager should order

$$\frac{a}{1-r} = \frac{1000}{1-0.1} = \frac{1000}{0.9} = \frac{10{,}000}{9} \approx 1111 \text{ units.}$$

(b) $0.9x = 1000$

$$x = \frac{1000}{0.9} = \frac{10{,}000}{9} \approx 1111 \text{ units}$$

25. Let $p_0 = p$, the probability that the policyholder files no claims during the period. Then, because $p_{n+1} = \left(\frac{1}{5}\right) p_n$, we have that

$$p_0 = p$$

$$p_1 = \frac{1}{5}p_0 = \frac{1}{5}p$$

$$p_2 = \frac{1}{5}p_1 = \frac{1}{5} \cdot \frac{1}{5}p = \left(\frac{1}{5}\right)^2 p$$

$$\vdots$$

$$p_n = \frac{1}{5}p_{n-1} = \frac{1}{5} \cdot \left(\frac{1}{5}\right)^{n-1} p = \left(\frac{1}{5}\right)^n p.$$

The total probability is the sum of the probabilities of the policyholder filing no claim, 1 claim, 2 claims, ..., n claims, ..., and so on.

$$S = p + p_1 + p_2 + \ldots + p_n + \ldots = p + p\left(\frac{1}{5}\right) + p\left(\frac{1}{5}\right)^2 + \ldots + p\left(\frac{1}{5}\right)^n + \ldots$$

This is a geometric series with $a = p$ and $r = \frac{1}{5}$.

$$S = \frac{a}{1-r} = \frac{p}{1-\frac{1}{5}} = \frac{p}{\frac{4}{5}} = \frac{5}{4}p$$

And, since the sum of all probabilities must equal 1,

$$\frac{5}{4}p = 1$$

$$p = \frac{4}{5}.$$

The probability that a policyholder files more than one claim is

$$p_{n>1} = 1 - (p_0 + p_1) = 1 - \left(p + \frac{1}{5}p\right) = 1 - \frac{6}{5}p = 1 - \frac{6}{5}\left(\frac{4}{5}\right) = 1 - \frac{24}{25} = 0.04.$$

The correct answer choice is **a**.

27. The number of times the wheel rotates in the second minute is given by $400\left(\frac{3}{4}\right)$, in the third minute $400\left(\frac{3}{4}\right)^2$, in the fourth minute $400\left(\frac{3}{4}\right)^3$, and so on. Thus, the total number of rotations of the wheel before coming to a complete stop is given by

$$400 + 400\left(\frac{3}{4}\right) + 400\left(\frac{3}{4}\right)^2 + 400\left(\frac{3}{4}\right)^3 + \ldots.$$

This is an infinite geometric series with $a = a_1 = 400$ and $r = \frac{3}{4}$. This series converges to

$$\frac{a}{1-r} = \frac{400}{1-\frac{3}{4}} = \frac{400}{\frac{1}{4}} = 1600.$$

The wheel makes 1600 rotations before coming to a complete stop.

29. The second triangle has sides 1 meter in length, the third triangle has sides $\frac{1}{2}$ meter in length, the fourth triangle has sides $\frac{1}{4}$ meter in length, and so on. Thus, the total perimeter of all the triangles given by

$$3(2) + 3(1) + 3\left(\frac{1}{2}\right) + 3\left(\frac{1}{2}\right)^2 + \ldots$$

is an infinite geometric series with $a = a_1 = 3(2) = 6$ and $r = \frac{1}{2}$. This series converges to

$$\frac{a}{1-r} = \frac{6}{1-\frac{1}{2}} = \frac{6}{\frac{1}{2}} = 12.$$

So, the total perimeter of all the triangles is 12 meters.

31. **(a)** The tortoise's starting point is 10 meters. Since Achilles runs 10 meters per second, it takes him 1 second to reach the tortoise's starting point. The tortoise has traveled 1 meter in this 1 second since the tortoise's rate is 1 meter per second.

(b) After Achilles reaches the tortoise's starting point, the tortoise has traveled 1 meter. Since Achilles travels 10 meters per second, it takes Achilles $\frac{1}{10}$ second to reach this new point. The tortoise has traveled $\frac{1}{10}$ meter in this $\frac{1}{10}$ second since the tortoise's rate is 1 meter per second.

(c) The time that it takes Achilles to reach the tortoise is given by

$$1+\frac{1}{10}+\frac{1}{100}+\cdots.$$

This is an infinite geometric series with $a=a_1=1$ and $r=\frac{1}{10}$. This series converges to

$$\frac{a}{1-r}=\frac{1}{1-\frac{1}{10}}=\frac{1}{\frac{9}{10}}=\frac{10}{9}.$$

It takes Achilles $\frac{10}{9}$ seconds to catch the tortoise.

(d) Using distance = rate × time, Achilles' distance is given by $10t$ meters in t seconds and the tortoise's distance is given by $t+10$ meters in t seconds. The distances are equal when

$$\begin{aligned} 10t &= t+10 \\ 9t &= 10 \\ t &= \frac{10}{9} \end{aligned}$$

It takes Achilles $\frac{10}{9}$ seconds to catch the tortoise.

12.5 Taylor Series

1. This function most nearly matches $\frac{1}{1-x}$. To get 1 in the denominator, instead of 2, divide the numerator and denominator by 2.

$$\frac{5}{2-x}=\frac{\frac{5}{2}}{1-\frac{x}{2}}$$

Thus, we can find the Taylor series for $\frac{\frac{5}{2}}{1-\frac{x}{2}}$ by starting with the Taylor series for $\frac{1}{1-x}$, multiplying each term by $\frac{5}{2}$, and replacing each x with $\frac{x}{2}$.

$$\begin{aligned} \frac{5}{2-x} &= \frac{\frac{5}{2}}{1-\frac{x}{2}} \\ &= \frac{5}{2}\cdot 1+\frac{5}{2}\left(\frac{x}{2}\right)+\frac{5}{2}\left(\frac{x}{2}\right)^2+\frac{5}{2}\left(\frac{x}{2}\right)^3+\cdots+\frac{5}{2}\left(\frac{x}{2}\right)^n+\cdots \\ &= \frac{5}{2}+\frac{5x}{4}+\frac{5x^2}{8}+\frac{5x^3}{16}+\cdots+\frac{5x^n}{2^{n+1}}+\cdots \end{aligned}$$

The Taylor series for $\frac{1}{1-x}$ is valid when $-1<x<1$. Replacing x with $\frac{x}{2}$ gives

$$-1<\frac{x}{2}<1 \text{ or } -2<x<2.$$

The interval of convergence of the new series is $(-2,2)$.

3. $\dfrac{8x}{1+3x} = x \cdot \dfrac{8}{1-(-3x)}$

Use the Taylor series for $\frac{1}{1-x}$, multiply each term by 8, and replace x with $-3x$. Also, use property (3) with $k=1$.

$$\begin{aligned}\frac{8x}{1+3x} &= x \cdot \frac{8}{1-(-3x)} \\ &= x \cdot 8 \cdot 1 + x \cdot 8\,(-3x) + x \cdot 8\,(-3x)^2 + x \cdot 8\,(-3x)^3 + \ldots + x \cdot 8\,(-3x)^n + \ldots \\ &= 8x - 24x^2 + 72x^3 - 216x^4 + \ldots + (-1)^n \cdot 8 \cdot 3^n x^{n+1} + \ldots\end{aligned}$$

The Taylor series for $\frac{1}{1-x}$ is valid when $-1 < x < 1$. Replacing x with $-3x$ gives

$$-1 < -3x < 1 \text{ or } \frac{1}{3} > x > -\frac{1}{3}.$$

The interval of convergence of the new series is $\left(-\frac{1}{3}, \frac{1}{3}\right)$.

5. $\dfrac{x^2}{4-x} = x^2 \cdot \dfrac{\frac{1}{4}}{1-\frac{x}{4}}$

Use the Taylor series for $\frac{1}{1-x}$, multiply each term by $\frac{1}{4}$, and replace x with $\frac{x}{4}$. Also, use property (3) with $k=2$.

$$\begin{aligned}\frac{x^2}{4-x} &= x^2 \cdot \frac{\frac{1}{4}}{1-\frac{x}{4}} \\ &= x^2 \cdot \frac{1}{4} \cdot 1 + x^2 \cdot \frac{1}{4}\left(\frac{x}{4}\right) + x^2 \cdot \frac{1}{4}\left(\frac{x}{4}\right)^2 + x^2 \cdot \frac{1}{4}\left(\frac{x}{4}\right)^3 + \cdots + +x^2 \cdot \frac{1}{4}\left(\frac{x}{4}\right)^n + \cdots \\ &= \frac{x^2}{4} + \frac{x^3}{16} + \frac{x^4}{64} + \frac{x^5}{256} + \cdots + \frac{x^{n+2}}{4^{n+1}} + \cdots\end{aligned}$$

The Taylor series for $\frac{1}{1-x}$ is valid when $-1 < x < 1$. Replacing x with $\frac{x}{4}$ gives

$$-1 < \frac{x}{4} < 1 \text{ or } -4 < x < 4.$$

The interval of convergence of the new series is $(-4, 4)$.

7. We find the Taylor series of $\ln(1+4x)$ by starting with the Taylor series for $\ln(1+x)$ and replacing each x with $4x$.

$$\begin{aligned}\ln(1+4x) &= 4x - \frac{(4x)^2}{2} + \frac{(4x)^3}{3} - \frac{(4x)^4}{4} + \cdots + \frac{(-1)^n (4x)^{n+1}}{n+1} + \cdots \\ &= 4x - 8x^2 + \frac{64}{3}x^3 - 64x^4 + \cdots + \frac{(-1)^n 4^{n+1} x^{n+1}}{n+1} + \cdots\end{aligned}$$

The Taylor series for $\ln(1+x)$ is valid when $-1 < x \le 1$. Replacing x with $4x$ gives

$$-1 < 4x \le 1 \text{ or } -\frac{1}{4} < x \le \frac{1}{4}.$$

The interval of convergence of the new series is $\left(-\frac{1}{4}, \frac{1}{4}\right]$.

9. We find the Taylor series for e^{4x^2} by starting with the Taylor series for e^x and replacing each x with $4x^2$.

$$\begin{aligned} e^{4x^2} &= 1 + (4x^2) + \frac{1}{2!}(4x^2)^2 + \frac{1}{3!}(4x^2)^3 + \cdots + \frac{1}{n!}(4x^2)^n + \cdots \\ &= 1 + 4x^2 + 8x^4 + \frac{32}{3}x^6 + \cdots + \frac{4^n x^{2n}}{n!} + \cdots \end{aligned}$$

The Taylor series for e^{4x^2} has the same interval of convergence, $(-\infty, \infty)$, as the Taylor series for e^x.

11. Use the Taylor series for e^x and replace x with $-x$. Also, use property (3) with $k = 3$.

$$\begin{aligned} x^3e^{-x} &= x^3 \cdot 1 + x^3(-x) + x^3 \cdot \frac{1}{2!}(-x)^2 + x^3 \cdot \frac{1}{3!}(-x)^3 + \cdots + x^3 \cdot \frac{1}{n!}(-x)^n + \cdots \\ &= x^3 - x^4 + \frac{1}{2}x^5 - \frac{1}{6}x^6 + \cdots + \frac{(-1)^n x^{n+3}}{n!} + \cdots \end{aligned}$$

The Taylor series for x^3e^{-x} has the same interval of convergence, $(-\infty, \infty)$, as the Taylor series for e^x.

13. $\dfrac{2}{1+x^2} = \dfrac{2}{1-(-x^2)}$

Use the Taylor series for $\frac{1}{1-x}$, multiply each term by 2, and replace x with $-x^2$.

$$\begin{aligned} \frac{2}{1+x^2} &= \frac{2}{1-(-x^2)} \\ &= 2 \cdot 1 + 2(-x^2) + 2(-x^2)^2 + 2(-x^2)^3 + \cdots + 2(-x^2)^n + \cdots \\ &= 2 - 2x^2 + 2x^4 - 2x^6 + \cdots + (-1)^n \cdot 2 \cdot x^{2n} + \cdots \end{aligned}$$

The Taylor series for $\frac{1}{1-x}$ is valid when $-1 < x < 1$. Replacing x with $-x^2$ gives

$$-1 < -x^2 < 1 \text{ or } 1 > x^2 > -1.$$

The inequality is satisfied by any x in the interval $(-1, 1)$.

15. $\dfrac{e^x + e^{-x}}{2} = \dfrac{1}{2}e^x + \dfrac{1}{2}e^{-x}$

We find the Taylor series for $\frac{1}{2}e^x$ by starting with the Taylor series for e^x and multiplying each term by $\frac{1}{2}$.

$$\begin{aligned} \frac{1}{2}e^x &= \frac{1}{2} \cdot 1 + \frac{1}{2} \cdot x + \frac{1}{2} \cdot \frac{1}{2!}x^2 + \frac{1}{2} \cdot \frac{1}{3!}x^3 + \cdots + \frac{1}{2} \cdot \frac{1}{n!}x^n + \cdots \\ &= \frac{1}{2} + \frac{1}{2}x + \frac{1}{4}x^2 + \frac{1}{12}x^3 + \cdots + \frac{1}{2n!}x^n + \cdots \end{aligned}$$

We find the Taylor series for $\frac{1}{2}e^{-x}$ by starting with the Taylor series for e^x, multiplying each term by $\frac{1}{2}$, and replacing x with $-x$.

$$\begin{aligned} \frac{1}{2}e^{-x} &= \frac{1}{2} \cdot 1 + \frac{1}{2}(-x) + \frac{1}{2} \cdot \frac{1}{2!}(-x)^2 + \frac{1}{2} \cdot \frac{1}{3!}(-x)^3 + \cdots + \frac{1}{2} \cdot \frac{1}{n!}(-x)^n + \cdots \\ &= \frac{1}{2} - \frac{1}{2}x + \frac{1}{4}x^2 - \frac{1}{12}x^3 + \cdots + \frac{(-1)^n}{2n!}x^n + \cdots \end{aligned}$$

Use property (1) with $f(x) = \frac{1}{2}e^x$ and $g(x) = \frac{1}{2}e^{-x}$.

$$\begin{aligned}
&\frac{e^x + e^{-x}}{2} \\
&= \frac{1}{2}e^x + \frac{1}{2}e^{-x} \\
&= \left(\frac{1}{2} + \frac{1}{2}\right) + \left(\frac{1}{2} - \frac{1}{2}\right)x + \left(\frac{1}{4} + \frac{1}{4}\right)x^2 + \left(\frac{1}{12} - \frac{1}{12}\right)x^3 + \cdots + \left(\frac{1}{2n!} + \frac{(-1)^n}{2n!}\right)x^n + \cdots \\
&= 1 + \frac{1}{2}x^2 + \frac{1}{24}x^4 + \frac{1}{720}x^6 + \cdots + \frac{1 + (-1)^n}{2n!}x^n + \cdots \\
&= 1 + \frac{1}{2}x^2 + \frac{1}{24}x^4 + \frac{1}{720}x^6 + \cdots + \frac{1}{(2n)!}x^{2n} + \cdots
\end{aligned}$$

The new series has the same interval of convergence, $(-\infty, \infty)$, as the Taylor series for e^x.

17. Use the Taylor series for $\ln(1 + x)$ and replace x with $2x^4$.

$$\begin{aligned}
\ln\left(1 + 2x^4\right) &= 2x^4 - \frac{\left(2x^4\right)^2}{2} + \frac{\left(2x^4\right)^3}{3} - \frac{\left(2x^4\right)^4}{4} + \cdots + \frac{(-1)^n\left(2x^4\right)^{n+1}}{n+1} + \cdots \\
&= 2x^4 - 2x^8 + \frac{8}{3}x^{12} - 4x^{16} + \cdots + \frac{(-1)^n \cdot 2^{n+1}x^{4n+4}}{n+1} + \cdots
\end{aligned}$$

The Taylor series for $\ln(1 + x)$ is valid when $-1 < x \le 1$. Replacing x with $2x^4$ gives

$$-1 < 2x^4 \le 1 \text{ or } -\frac{1}{2} < x^4 \le \frac{1}{2}.$$

The inequality is satisfied by any x in the interval $\left[-\frac{1}{\sqrt[4]{2}}, \frac{1}{\sqrt[4]{2}}\right]$.

19. $\dfrac{1 + x}{1 - x} = \dfrac{1}{1 - x} + \dfrac{x}{1 - x}$

The Taylor series for $\frac{1}{1-x}$ is given as

$$\frac{1}{1 - x} = 1 + x + x^2 + x^3 + \cdots + x^n + \cdots$$

We find the Taylor series for $\frac{x}{1-x}$ by starting with the Taylor series for $\frac{1}{1-x}$ and using property (3) with $k = 1$.

$$\begin{aligned}
\frac{x}{1 - x} &= x \cdot \frac{1}{1 - x} \\
&= x \cdot 1 + x \cdot x + x \cdot x^2 + x \cdot x^3 + \cdots + x \cdot x^n + \cdots \\
&= x + x^2 + x^3 + x^4 + \cdots + x^{n+1} + \cdots
\end{aligned}$$

Use property (1) with $f(x) = \frac{1}{1-x}$ and $g(x) = \frac{x}{1-x}$.

$$\begin{aligned}
\frac{1 + x}{1 - x} &= \frac{1}{1 - x} + \frac{x}{1 - x} \\
&= 1 + (1 + 1)x + (1 + 1)x^2 + (1 + 1)x^3 + \cdots + (1 + 1)x^n + \cdots \\
&= 1 + 2x + 2x^2 + 2x^3 + \cdots + 2x^n + \cdots
\end{aligned}$$

21. The Taylor series for e^x is given as

$$\begin{aligned} e^x &= 1 + x + \frac{1}{2!}x^2 + \frac{1}{3!}x^3 + \frac{1}{4!}x^4 + \frac{1}{5!}x^5 + \cdots + \frac{1}{n!}x^n + \cdots \\ &= 1 + x + \frac{x^2}{2} + \frac{x^3}{6} + \frac{x^4}{24} + \frac{x^5}{120} + \cdots + \frac{x^n}{n!} + \cdots \\ &= 1 + x + \frac{x^2}{2}\left[1 + \frac{x}{3} + \frac{x^2}{12} + \frac{x^3}{60} + \cdots + \frac{x^{n-2}}{\frac{n!}{2}} + \cdots\right] \end{aligned}$$

Compare the series $\frac{x}{3} + \frac{x^2}{12} + \frac{x^3}{60} + \cdots + \frac{x^{n-2}}{\frac{n!}{2}} + \cdots$, for $n \geq 3$, with the series $\frac{x}{3} + \frac{x^2}{12} + \frac{x^3}{48} + \cdots + \frac{x^{n-2}}{3 \cdot 4^{n-3}} + \cdots$, for $n \geq 3$. The second series is an infinite geometric series that is larger than or equal to the first series term by term. The geometric series has $a = \frac{x}{3}$ and $r = \frac{x}{4}$, and sums to $\frac{4x}{12-3x}$. Thus, for $x > 0$,

$$1 + x + \frac{x^2}{2} < 1 + x + \frac{x^2}{2}\left[1 + \frac{x}{3} + \frac{x^2}{12} + \frac{x^3}{60} + \cdots + \frac{x^{n-2}}{\frac{n!}{2}} + \cdots\right] < 1 + x + \frac{x^2}{2}\left[1 + \frac{4x}{12 - 3x},\right]$$

or

$$1 + x + \frac{x^2}{2} < e^x < 1 + x + \frac{x^2}{2}\left[1 + \frac{4x}{12 - 3x}\right].$$

For values of x sufficiently close to 0, $\frac{4x}{12-3x}$ approaches 0, and $1 + x + \frac{x^2}{2}\left[1 + \frac{4x}{12-3x}\right] \approx 1 + x + \frac{x^2}{2}$. Thus, $e^x \approx 1 + x + \frac{x^2}{2}$.
For $x < 0$, a similar argument can be made.

23. The Taylor series for e^x is given as

$$e^x = 1 + x + \frac{1}{2!}x^2 + \frac{1}{3!}x^3 + \frac{1}{4!}x^4 + \frac{1}{5!}x^5 + \cdots + \frac{1}{n!}x^n + \cdots$$

For $x > 0$, $\frac{1}{n!}x^n > 0$ when $n \geq 2$. Thus,

$$\frac{1}{2!}x^2 + \frac{1}{3!}x^3 + \frac{1}{4!}x^4 + \frac{1}{5!}x^5 + \cdots + \frac{1}{n!}x^n + \cdots > 0$$

since each term on the left-hand side is positive. Adding $1 + x$ to both sides we have

$$1 + x + \frac{1}{2!}x^2 + \frac{1}{3!}x^3 + \frac{1}{4!}x^4 + \frac{1}{5!}x^5 + \cdots + \frac{1}{n!}x^n + \cdots > 1 + x$$

or $e^x > 1 + x$.

For $x = 0$, $e^x = e^0 = 1 = 1 + 0 = 1 + x$.

For $-1 \leq x < 0$, when b is an even positive integer, $\frac{x^b}{b} > 0$. Also, when $a \geq 3$, $\frac{|x|}{a} < 1$, or $1 + \frac{x}{a} > 0$. Thus, $\frac{x^b}{b}\left(1 + \frac{x}{a}\right) > 0$ for b an even positive integer and $a \geq 3$, since both factors are positive. Thus

$$\frac{x^2}{2!}\left(1 + \frac{x}{3}\right) + \frac{x^4}{4!}\left(1 + \frac{x}{5}\right) + \frac{x^6}{6!}\left(1 + \frac{x}{7}\right) + \cdots + \frac{x^{2n}}{(2n)!} + \cdots > 0,$$

since each term has been shown to be positive. Consider the Taylor series for e^x,

$$\begin{aligned} e^x &= 1 + x + \frac{1}{2!}x^2 + \frac{1}{3!}x^3 + \frac{1}{4!}x^4 + \frac{1}{5!}x^5 + \cdots + \frac{1}{n!}x^n + \cdots \\ &= 1 + x + \frac{x^2}{2!}\left(1 + \frac{x}{3}\right) + \frac{x^4}{4!}\left(1 + \frac{x}{5}\right) + \cdots + \frac{x^{2n}}{(2n)!}\left(1 + \frac{x}{2n+1}\right) + \cdots \end{aligned}$$

We have already shown the series to be positive from the third term on, so $e^x > 1 + x$.

For $x < -1$, $1 + x < 0$ and $e^x > 0$. Thus $e^x > 1 + x$.

25. The area is given by

$$\int_0^{\frac{1}{2}} \frac{1}{1-x^3}\,dx.$$

Find the Taylor series for $\frac{1}{1-x^3}$ by using the Taylor series for $\frac{1}{1-x}$ and replacing x with x^3.

$$\frac{1}{1-x^3} = 1 + (x^3) + (x^3)^2 + (x^3)^3 + (x^3)^4 + \cdots + (x^3)^n + \cdots = 1 + x^3 + x^6 + x^9 + x^{12} + \cdots + x^{3n} + \cdots$$

Using the first five terms of this series gives

$$\int_0^{\frac{1}{2}} \frac{1}{1-x^3}\,dx \approx \int_0^{\frac{1}{2}} (1 + x^3 + x^6 + x^9 + x^{12})\,dx = \left(x + \frac{1}{4}x^4 + \frac{1}{7}x^7 + \frac{1}{10}x^{10} + \frac{1}{13}x^{13}\right)\Big|_0^{\frac{1}{2}}$$

$$= \frac{1}{2} + \frac{1}{64} + \frac{1}{896} + \frac{1}{10.240} + \frac{1}{106{,}496} - 0 \approx 0.5168.$$

27. The area is given by

$$\int_0^1 e^{\sqrt{x}}\,dx.$$

Find the Taylor series for $e^{\sqrt{x}}$ by using the Taylor series for e^x and replacing x with $\sqrt{x}$.

$$e^{\sqrt{x}} = 1 + \sqrt{x} + \frac{\sqrt{x}^2}{2!} + \frac{\sqrt{x}^3}{3!} + \frac{\sqrt{x}^4}{4!} + \cdots + \frac{\sqrt{x}^n}{n!} + \cdots$$

Using the first five terms of this series gives

$$\int_0^1 e^{\sqrt{x}}\,dx \approx \int_0^1 \left(1 + \sqrt{x} + \frac{x}{2} + \frac{x^{3/2}}{6} + \frac{x^2}{24}\right) dx = \left(x + \frac{2}{3}x^{3/2} + \frac{x^2}{4} + \frac{x^{5/2}}{15} + \frac{x^3}{72}\right)\Big|_0^1$$

$$= 1 + \frac{2}{3} + \frac{1}{4} + \frac{1}{15} + \frac{1}{72} - 0 \approx 1.9972$$

29. The area is given by

$$\int_0^{0.6} \frac{1}{\sqrt{2\pi}} e^{-x^2/2} dx = \frac{1}{\sqrt{2\pi}} \int_0^{0.6} e^{-x^2/2} dx.$$

Find the Taylor series for $e^{-x^2/2}$ by using the Taylor series for e^x and replacing x with $-\frac{x^2}{2}$.

$$e^{-x^2/2} = 1 - \frac{1}{2}x^2 + \frac{1}{2!2^2}x^4 - \frac{1}{3!2^3}x^6 + \frac{1}{4!2^4}x^8 + \cdots + \frac{(-1)^n}{n!2^n}x^{2n} + \cdots$$

Using the first five terms of this series gives

$$\frac{1}{\sqrt{2\pi}} \int_0^{0.6} e^{-x^2/2} dx \approx \frac{1}{\sqrt{2\pi}} \int_0^{0.6} \left(1 - \frac{1}{2}x^2 + \frac{1}{2!2^2}x^4 - \frac{1}{3!2^3}x^6 + \frac{1}{4!2^4}x^8\right) dx$$

$$= \frac{1}{\sqrt{2\pi}} \int_0^{0.6} \left(1 - \frac{1}{2}x^2 + \frac{1}{8}x^4 - \frac{1}{48}x^6 + \frac{1}{384}x^8\right) dx$$

$$= \frac{1}{\sqrt{2\pi}} \left(x - \frac{1}{6}x^3 + \frac{1}{40}x^5 - \frac{1}{336}x^7 + \frac{1}{3456}x^9\right)\Big|_0^{0.6}$$

$$= \frac{1}{\sqrt{2\pi}} \left(0.6 - \frac{1}{6}(0.6)^3 + \frac{1}{40}(0.6)^5 - \frac{1}{336}(0.6)^7 + \frac{1}{3456}(0.6)^9 - 0\right)$$

$$\approx \frac{1}{\sqrt{2\,(3.1416)}}(0.565864)$$

$$\approx 0.2257.$$

31. The doubling time n for a quantity that increases at an annual rate r is given by

$$n = \frac{\ln 2}{\ln(1+r)} = \frac{\ln 2}{\ln 1.07} \approx 10.245.$$

It will take about 10.245 years.

According to the Rule of 72, the doubling time is given by

$$\text{Doubling time} \approx \frac{72}{100r} = \frac{72}{100(0.07)} \approx 10.286.$$

It will take about 10.286 years, a difference of 0.041 years, or about 2 weeks.

33. (a) $\sum_{x=1}^{\infty} f(x) = \sum_{x=1}^{\infty} (1-p)^{x-1} p = (1-p)^{1-1} p + (1-p)^{2-1} p + (1-p)^{3-1} p + (1-p)^{4-1} p + \cdots + (1-p)^{n-1} p + \cdots$
$= p + p(1-p) + p(1-p)^2 + p(1-p)^3 + \cdots + p(1-p)^{n-1} + \cdots$

This is an infinite geometric series with $a = p$ and $r = (1-p)$. Thus the series sums to

$$\frac{a}{1-r} = \frac{p}{1-(1-p)} = \frac{p}{1-1+p} = \frac{p}{p} = 1.$$

(b) Let $\quad g(z) = p\sum_{x=1}^{\infty} z^x = p\left(z + z^2 + z^3 + z^4 + \cdots + z^n + \cdots\right).$

Then $\quad g'(z) = p\left(1 + 2z + 3z^2 + 4z^3 + \cdots + nz^{n-1} + \cdots\right)$

and $g'(1-p) = p[1 + 2(1-p) + 3(1-p)^2 + 4(1-p)^3 + \cdots + n(1-p)^{n-1} + \cdots]$

$= p[1(1-p)^{1-1} + 2(1-p)^{2-1} + 3(1-p)^{3-1} + 4(1-p)^{4-1} + \cdots + n(1-p)^{n-1} + \cdots]$

$= p\sum_{x=1}^{\infty} x(1-p)^{x-1} = \sum_{x=1}^{\infty} x(1-p)^{x-1} p = \sum_{x=1}^{\infty} xf(x).$

Using the Taylor series for $\frac{1}{1-x}$, write $g(z)$ as

$$g(z) = p\sum_{x=1}^{\infty} z^x = p\left(z + z^2 + z^3 + z^4 + \cdots + z^n + \cdots\right). = pz\left(1 + z + z^2 + z^3 + \cdots + z^{n-1} + \cdots\right)$$
$$= pz \cdot \frac{1}{1-z} = \frac{zp}{1-z}.$$

Then $\quad g'(z) = \dfrac{(1-z)p - zp(-1)}{(1-z)^2} = \dfrac{p}{(1-z)^2}$, and $g'(1-p) = \dfrac{p}{[1-(1-p)]^2} = \dfrac{p}{p^2} = \dfrac{1}{p}.$

Thus, $\sum_{x=1}^{\infty} xf(x) = g'(1-p) = \dfrac{1}{p}.$

(c) The probability, p, or meeting a foreign born player is $p = \frac{175}{750} = \frac{7}{30}$.
The expected value is given by

$$\sum_{x=1}^{\infty} xf(x) = \frac{1}{p} = \frac{1}{\frac{7}{30}} \approx 4.286.$$

Thus you would expect to meet 4.286 players.

(d) For $x \le 3$, find

$$\begin{aligned}\sum_{x=1}^{3} f(x) &= \sum_{x=1}^{3} (1-p)^{x-1} p = \sum_{x=1}^{3} \left(1-\frac{7}{30}\right)^{x-1}\left(\frac{7}{30}\right)\\ &= \left(1-\frac{7}{30}\right)^{1-1}\left(\frac{7}{30}\right)+\left(1-\frac{7}{30}\right)^{2-1}\left(\frac{7}{30}\right)+\left(1-\frac{7}{30}\right)^{3-1}\left(\frac{7}{30}\right)\\ &\approx 0.5494.\end{aligned}$$

12.6 Newton's Method

1. $f(x) = 5x^2 - 3x - 3$
$f'(x) = 10x - 3$

$f(1) = -1 < 0$ and $f(2) = 11 > 0$ so a solution exists in $(1, 2)$.
Let $c_1 = 1$.

$$c_2 = c_1 - \frac{f(c_1)}{f'(c_1)} = 1 - \frac{-1}{7} = 1.1429$$

$$c_3 = c_2 - \frac{f(c_2)}{f'(c_2)} = 1.1429 - \frac{0.1020}{8.4286} = 1.1308$$

$$c_4 = c_3 - \frac{f(c_3)}{f'(c_3)} = 1.1308 - \frac{0.0007}{8.3075} = 1.1307$$

Subsequent approximations will agree with c_3 and c_4 to the nearest hundredth. Thus, $x = 1.13$.

3. $f(x) = 2x^3 - 6x^2 - x + 2$
$f'(x) = 6x^2 - 12x - 1$

$f(3) = -1 < 0$ and $f(4) = 30 > 0$ so a solution exists in $(3, 4)$.
Let $c_1 = 3$.

$$c_2 = c_1 - \frac{f(c_1)}{f'(c_1)} = 3 - \frac{-1}{17} = 3.0588$$

$$c_3 = c_2 - \frac{f(c_2)}{f'(c_2)} = 3.0588 - \frac{0.0419}{18.4325} = 3.0565$$

Subsequent approximations will agree with c_2 and c_3 to the nearest hundredth. Thus, $x = 3.06$.

5. $f(x) = -3x^3 + 5x^2 + 3x + 2$
$f'(x) = -9x^2 + 10x + 3$

$f(2) = 4 > 0$ and $f(3) = -25 < 0$ so a solution exists in $(2, 3)$.
Let $c_1 = 2$.

$$c_2 = c_1 - \frac{f(c_1)}{f'(c_1)} = 2 - \frac{4}{-13} = 2.3077$$

$$c_3 = c_2 - \frac{f(c_2)}{f'(c_2)} = 2.3077 - \frac{-1.3182}{-21.8521} = 2.2474$$

$$c_4 = c_3 - \frac{f(c_3)}{f'(c_3)} = 2.2474 - \frac{-0.0567}{-19.9824} = 2.2445$$

$$c_5 = c_4 - \frac{f(c_4)}{f'(c_4)} = -2.2445 - \frac{-0.0001}{-19.8960} = 2.2445$$

Subsequent approximations will agree with c_4 and c_5 to the nearest hundredth. Thus, $x = 2.24$.

7. $f(x) = 2x^4 - 2x^3 - 3x^2 - 5x - 8$
$f'(x) = 8x^3 - 6x^2 - 6x - 5$

In the interval $[-2, -1]$: $f(-2) = 38 > 0$ and $f(-1) = -2 < 0$ so a solution exists in $(-2, -1)$.
Let $c_1 = -2$.

$$c_2 = c_1 - \frac{f(c_1)}{f'(c_1)} = -2 - \frac{38}{-81} = -1.5309$$
$$c_3 = c_2 - \frac{f(c_2)}{f'(c_2)} = -1.5309 - \frac{10.7834}{-38.5773} = -1.2513$$
$$c_4 = c_3 - \frac{f(c_3)}{f'(c_3)} = -1.2513 - \frac{2.3817}{-22.5622} = -1.1458$$
$$c_5 = c_4 - \frac{f(c_4)}{f'(c_4)} = -1.1458 - \frac{0.2457}{-18.0356} = -1.1322$$
$$c_6 = c_5 - \frac{f(c_5)}{f'(c_5)} = -1.1322 - \frac{0.0036}{-17.5069} = -1.1319$$

Subsequent approximations will agree with c_5 and c_6 to the nearest hundredth. Thus, $x = -1.13$ is a solution in $[-2, -1]$.
In the interval $[2, 3]$: $f(2) = -14 < 0$ and $f(3) = 58 > 0$ so a solution exists in $(2, 3)$.
Let $c_1 = 2$.

$$c_2 = c_1 - \frac{f(c_1)}{f'(c_1)} = 2 - \frac{-14}{23} = 2.6087$$
$$c_3 = c_2 - \frac{f(c_2)}{f'(c_2)} = 2.6087 - \frac{15.6588}{80.5396} = 2.4143$$
$$c_4 = c_3 - \frac{f(c_3)}{f'(c_3)} = 2.4143 - \frac{2.2460}{58.1188} = 2.3756$$
$$c_5 = c_4 - \frac{f(c_4)}{f'(c_4)} = 2.3756 - \frac{0.0774}{54.1413} = 2.3742$$
$$c_6 = c_5 - \frac{f(c_5)}{f'(c_5)} = 2.3742 - \frac{0.0001}{53.9972} = 2.3742$$

Subsequent approximations will agree with c_5 and c_6 to the nearest hundredth. Thus, $x = 2.37$ is a solution in $[2, 3]$.

9. $f(x) = 4x^{1/3} - 2x^2 + 4$
$f'(x) = \frac{4}{3}x^{-2/3} - 4x$

$f(-3) = -19.7690 < 0$ and $f(0) = 4 > 0$ so a solution exists in $(-3, 0)$.
Let $c_1 = -3$.

$$c_2 = c_1 - \frac{f(c_1)}{f'(c_1)} = -3 - \frac{-19.7690}{12.6410} = -1.4361$$
$$c_3 = c_2 - \frac{f(c_2)}{f'(c_2)} = -1.4361 - \frac{-4.6378}{6.7920} = -0.7533$$
$$c_4 = c_3 - \frac{f(c_3)}{f'(c_3)} = -0.7533 - \frac{-0.7744}{4.6236} = -0.5858$$
$$c_5 = c_4 - \frac{f(c_4)}{f'(c_4)} = -0.5858 - \frac{-0.0332}{4.2477} = -0.5780$$
$$c_6 = c_5 - \frac{f(c_5)}{f'(c_5)} = -0.5780 - \frac{-0.0001}{4.2335} = -0.5780$$

Subsequent approximations will agree with c_5 and c_6 to the nearest hundredth. Thus, $x = -0.58$.

11. $f(x) = e^x + x - 2$
$f'(x) = e^x + 1$

$f(0) = -1 < 0$ and $f(3) = 21.085537$ so a solution exists in $(0, 3)$.
Let $c_1 = 0$.

$$c_2 = c_1 - \frac{f(c_1)}{f'(c_1)} = 0 - \frac{-1}{2} = 0.5$$
$$c_3 = c_2 - \frac{f(c_2)}{f'(c_2)} = 0.5 - \frac{0.14872}{2.6487} = 0.4439$$
$$c_4 = c_3 - \frac{f(c_3)}{f'(c_3)} = 0.4439 - \frac{0.00267}{2.5588} = 0.4429$$

Subsequent approximations will agree with c_3 and c_4 to the nearest hundredth. Thus, $x = 0.44$.

13. $f(x) = x^2e^{-x} + x^2 - 2$
$f'(x) = -x^2e^{-x} + 2xe^{-x} + 2x$

$f(0) = -2$ and $f(3) = 7.4480836$ so a solution exists in $(0,3)$.
Let $c_1 = 0$.

$$c_2 = c_1 - \frac{f(c_1)}{f'(c_1)} = 0 - \frac{-2}{0}$$

Since c_2 is undefined, let $c_1 = 1$.

$$c_2 = c_1 - \frac{f(c_1)}{f'(c_1)} = 1 - \frac{-0.6321}{2.3679} = 1.2670$$

$$c_3 = c_2 - \frac{f(c_2)}{f'(c_2)} = 1.2670 - \frac{0.05746}{2.7956} = 1.2464$$

$$c_4 = c_4 - \frac{f(c_3)}{f'(c_3)} = 1.2464 - \frac{2.1 \cdot 10^{-4}}{2.7629} = 1.2463$$

Subsequent approximations will agree with c_3 and c_4 to the nearest hundredth. Thus, $x = 1.25$.

15. $f(x) = \ln x + x - 2$
$f'(x) = \frac{1}{x} + 1$

$f(1) = -1 < 0$ and $f(4) = 3.3862944 > 0$ so a solution exists in $(1,4)$.
Let $c_1 = 1$.

$$c_2 = c_1 - \frac{f(c_1)}{f'(c_1)} = 1 - \frac{-1}{2} = 1.5$$

$$c_3 = c_2 - \frac{f(c_2)}{f'(c_2)} = 1.5 - \frac{-0.0945}{1.6667} = 1.5567$$

$$c_4 = c_3 - \frac{f(c_3)}{f'(c_3)} = 1.5567 - \frac{-7 \cdot 10^{-4}}{1.6424} = 1.5571$$

Subsequent approximations will agree with c_3 and c_4 to the nearest hundredth. Thus, $x = 1.56$.

17. $\sqrt{2}$ is a solution of $x^2 - 2 = 0$.

$f(x) = x^2 - 2$
$f'(x) = 2x$

Since $1 < \sqrt{2} < 2$, let $c_1 = 1$.

$$c_2 = 1 - \frac{-1}{2} = 1.5$$

$$c_3 = 1.5 - \frac{0.25}{3} = 1.417$$

$$c_4 = 1.417 - \frac{0.00789}{2.834} = 1.414$$

$$c_5 = 1.414 - \frac{-6 \cdot 10^{-4}}{2.828} = 1.414$$

Since $c_4 = c_5 = 1.414$, to the nearest thousandth, $\sqrt{2} = 1.414$.

19. $\sqrt{11}$ is a solution of $x^2 - 11 = 0$.

$f(x) = x^2 - 11$
$f'(x) = 2x$

Since $3 < \sqrt{11} < 4$, let $c_1 = 3$.

$$c_2 = 3 - \frac{-2}{6} = 3.333$$

$$c_3 = 3.333 - \frac{0.10889}{6.666} = 3.317$$

$$c_4 = 3.317 - \frac{0.00249}{6.634} = 3.317$$

Since $c_3 = c_4 = 3.317$, to the nearest thousandth, $\sqrt{11} = 3.317$.

21. $\sqrt{250}$ is a solution of $x^2 - 250 = 0$.

$f(x) = x^2 - 250$
$f'(x) = 2x$

Since $15 < \sqrt{250} < 16$, let $c_1 = 15$.

$$c_2 = 15 - \frac{-25}{30} = 15.833$$

$$c_3 = 15.833 - \frac{0.68389}{31.666} = 15.811$$

$$c_4 = 15.811 - \frac{-0.0123}{31.622} = 15.811$$

Since $c_3 = c_4 = 15.811$, to the nearest thousandth, $\sqrt{250} = 15.811$.

23. $\sqrt[3]{9}$ is a solution of $x^3 - 9 = 0$.

$f(x) = x^3 - 9$
$f'(x) = 3x^2$

Since $2 < \sqrt[3]{9} < 3$, let $c_1 = 2$.

$$c_2 = 2 - \frac{-1}{12} = 2.083$$

$$c_3 = 2.083 - \frac{0.03791}{13.017} = 2.080$$

$$c_4 = 2.080 - \frac{-0.0011}{12.979} = 2.080$$

Since $c_3 = c_4 = 2.080$, to the nearest thousandth, $\sqrt[3]{9} = 2.080$.

25. $\sqrt[3]{100}$ is a solution of $x^3 - 100 = 0$.

$f(x) = x^3 - 100$
$f'(x) = 3x^2$

Since $4 < \sqrt[3]{100} < 5$, let $c_1 = 4$.

$$c_2 = 4 - \frac{-36}{48} = 4.75$$
$$c_3 = 4.75 - \frac{7.1719}{67.688} = 4.644$$
$$c_4 = 4.644 - \frac{0.15592}{64.7} = 4.642$$
$$c_5 = 4.642 - \frac{0.02658}{64.644} = 4.642$$

Since $c_4 = c_5 = 4.642$, to the nearest thousandth, $\sqrt[3]{100} = 4.642$.

27. $f(x) = x^3 - 3x^2 - 18x + 4$
$f'(x) = 3x^2 - 6x - 18$

To find critical points, solve $f'(x) = 3x^2 - 6x - 18 = 0$.

$f''(x) = 6x - 6$

$f'(-2) = 6 > 0$ and $f'(-1) = -9 < 0$ so a solution exists in $(-2, -1)$.

Let $c_1 = -2$.

$$c_2 = -2 - \frac{6}{-18} = -1.67$$
$$c_3 = -1.67 - \frac{0.3867}{-16.02} = -1.65$$
$$c_4 = -1.65 - \frac{0.0675}{-15.9} = -1.65$$

Subsequent approximations will agree with c_3 and c_4 to the nearest hundredth. Thus, $x = -1.65$. Since $f''(-1.65) = -15.9 < 0$, the graph has a relative maximum at $x = -1.65$.

$f'(3) = -9 < 0$ and $f'(4) = 6 > 0$ so a solution exists in $(3, 4)$.

Let $c_1 = 3$.

$$c_2 = 3 - \frac{-9}{12} = 3.75$$
$$c_3 = 3.75 - \frac{1.6875}{16.5} = 3.65$$
$$c_4 = 3.65 - \frac{0.0675}{15.9} = 3.65$$

Subsequent approximations will agree with c_3 and c_4 to the nearest hundredth. Thus, $x = 3.65$. Since $f''(3.65) = 15.9 > 0$, the graph has a relative minimum at $x = 3.65$.

29. $f(x) = x^4 - 3x^3 + 6x - 1$
$f'(x) = 4x^3 - 9x^2 + 6$

To find critical points, solve $f'(x) = 4x^3 - 9x^2 + 6 = 0.$

$f''(x) = 12x^2 - 18x$

$f'(-1) = -7 < 0$ and $f'(0) = 6$ so a solution exists in $(-1, 0)$.

Let $c_1 = -1$.

$$c_2 = -1 - \frac{-7}{30} = -0.77$$
$$c_3 = -0.77 - \frac{-1.162}{20.975} = -0.71$$
$$c_4 = -0.71 - \frac{0.03146}{18.829} = -0.71$$

Subsequent approximations will agree with c_3 and c_4 to the nearest hundredth. Thus, $x = -0.71$. Since $f''(-0.71) = 18.829 > 0$, the graph has a relative minimum at $x = -0.71$.

$f'(1) = 1 > 0$ and $f'(1.5) = -0.75 < 0$ so a solution exists in $(1, 1.5)$.

Let $c_1 = 1$.

$$c_2 = 1 - \frac{1}{-6} = 1.17$$
$$c_3 = 1.17 - \frac{0.08635}{-4.633} = 1.19$$
$$c_4 = 1.19 - \frac{-0.00043}{-4.427} = 1.19$$

Subsequent approximations will agree with c_3 and c_4 to the nearest hundredth. Thus, $x = 1.19$. Since $f''(1.19) = -4.427 < 0$, the graph has a relative maximum at $x = 1.19$.

$f'(1.5) = -0.75 < 0$ and $f'(2) = 2 > 0$ so a solution exists in $(1.5, 2)$.

Let $c_1 = 1.5$.

$$c_2 = 1.5 - \frac{-0.75}{0}$$

Since c_2 is undefined, let $c_1 = 1.6$.

$$c_2 = 1.6 - \frac{-0.656}{1.92} = 1.94$$
$$c_3 = 1.94 - \frac{1.3331}{10.243} = 1.81$$
$$c_4 = 1.81 - \frac{0.23406}{6.7332} = 1.78$$
$$c_5 = 1.78 - \frac{0.04341}{5.9808} = 1.77$$
$$c_6 = 1.77 - \frac{-0.0152}{5.7348} = 1.77$$

Subsequent approximations will agree with c_5 and c_6 to the nearest hundredth. Thus, $x = 1.77$. Since $f''(1.77) = 5.7348 > 0$, the graph has a relative minimum at $x = 1.77$.

31. $f(x) = (x+1)^{1/3}$

$f'(x) = \frac{1}{3}(x-1)^{-2/3}$

$f(0) = -1 < 0$ and $f(2) = 1 > 0$ so a solution exists in $(0,2)$.

Let $c_1 = 0$.

$$c_2 = c_1 - \frac{f(c_1)}{f'(c_1)} = 0 - \frac{-1}{0.3333} = 3$$
$$c_3 = c_2 - \frac{f(c_2)}{f'(c_2)} = 3 - \frac{1.2599}{0.2100} = -3$$
$$c_4 = c_3 - \frac{f(c_3)}{f'(c_3)} = -3 - \frac{-1.5874}{0.1323} = 9$$
$$c_5 = c_4 - \frac{f(c_4)}{f'(c_4)} = 9 - \frac{2}{0.0833} = -15$$

Not only are successive approximations alternating in sign but they are moving further and further apart. Thus, the approximations are not approaching any specific value. The method fails in this case because the derivative, $f'(x) = \frac{1}{3}(x-1)^{-2/3}$, is undefined at $x = 1$; the function has a vertical tangent line there.

33. The process should be used at least until the savings produced is equal to the increased costs incurred, or when $S(x) = C(x)$. Therefore, solve $S(x) - C(x)$.

$$f(x) = S(x) - C(x) = \left(x^3 + 5x^2 + 9\right) - \left(x^2 + 40x + 20\right) = x^3 + 4x^2 - 40x - 11$$
$$f'(x) = 3x^2 + 8x - 40$$

$f'(x) = 3x^2 + 8x - 40$

$f(4) = -43 < 0$ and $f(5) = 14 > 0$ so a solution exists in $(4,5)$.

Let $c_1 = 4$.

$$c_2 = 4 - \frac{-43}{40} = 5.08$$
$$c_3 = 5.08 - \frac{20.122}{78.059} = 4.82$$
$$c_4 = 4.82 - \frac{1.1098}{68.257} = 4.80$$
$$c_5 = 4.80 - \frac{-0.248}{67.52} = 4.80$$

Subsequent approximations will agree with c_4 and c_5 to the nearest hundredth. Thus, $x = 4.80$. The process should be used for at least 4.80 years.

35. $$i_2 = 0.02 - \frac{57(0.02) - 57(0.02)(1+0.02)^{-12} - 600(0.02)^2}{57\left[-1 + (12)(0.02)(1+0.02)^{-12-1} + (1+0.02)^{-12}\right]}$$
$$= 0.02 - \frac{0.0011177798}{-1.480804049}$$
$$= 0.02075485$$

$$i_3 = 0.02075485 - \frac{57(0.02075485) - 57(0.02075485)(1+0.02075485)^{-12} - 600(0.02075485)^2}{57\left[-1 + (12)(0.02075485)(1+0.02075485)^{-12-1} + (1+0.02075485)^{-12}\right]}$$
$$= 0.02075485 - \frac{4.0638916 \cdot 10^{-6}}{-1.583924743}$$
$$= 0.02075742$$

12.7 L'Hospital's Rule

1. The limit in the numerator is 0, as is the limit in the denominator, so that l'Hospital's rule applies. Taking derivatives separately in the numerator and denominator gives

$$\lim_{x\to1}\frac{3x^2+2x-1}{2x-1}=\frac{3(1)^2+2(1)-1}{2(1)-1}=4.$$

By l'Hospital's rule,

$$\lim_{x\to1}\frac{x^3+x^2-x-1}{x^2-x}=4.$$

3. The limit in the numerator is 0, as is the limit in the denominator, so that l'Hospital's rule applies. Taking derivatives separately in the numerator and denominator gives

$$\lim_{x\to0}\frac{5x^4-6x^2+8x}{40x^4-4x+5}=\frac{5(0)^4-6(0)^2+8(0)}{40(0)^4-4(0)+5}=0.$$

By l'Hospital's rule,

$$\lim_{x\to0}\frac{x^5-2x^3+4x^2}{8x^5-2x^2+5x}=0.$$

5. The limit in the numerator is 0, as is the limit in the denominator, so that l'Hospital's rule applies. Taking derivatives separately in the numerator and denominator gives

$$\lim_{x\to2}\frac{\frac{1}{x-1}}{1}=\frac{1}{2-1}=1.$$

By l'Hospital's rule,

$$\lim_{x\to2}\frac{\ln(x-1)}{x-2}=1.$$

7. The limit in the numerator is 0, as is the limit in the denominator, so that l'Hospital's rule applies. Taking derivatives separately in the numerator and denominator gives

$$\lim_{x\to0}\frac{e^x}{4x^3}\text{ which does not exist.}$$

By l'Hospital's rule,

$$\lim_{x\to0}\frac{e^x-1}{x^4}\text{ does not exist.}$$

9. The limit in the numerator is 0, as is the limit in the denominator, so that l'Hospital's rule applies. Taking derivatives separately in the numerator and denominator gives

$$\lim_{x\to0}\frac{e^x+xe^x}{e^x}=\frac{e^0+0\cdot e^0}{e^0}=1.$$

By l'Hospital's rule,

$$\lim_{x\to0}\frac{xe^x}{e^x-1}=1.$$

11. $\lim_{x\to0}e^x=1$ and l'Hospital's rule does not apply. However,

$$\lim_{x\to0}\frac{e^x}{2x^3+9x^2-11x}\text{ does not exist.}$$

13. The limit in the numerator is 0, as is the limit in the denominator, so that l'Hospital's rule applies. Taking derivatives separately in the numerator and denominator gives

$$\lim_{x\to 0} \frac{\frac{1}{2(2+x)^{1/2}}}{1} = \frac{1}{2\,(2)^{1/2}} = \frac{1}{2\sqrt{2}} \text{ or } \frac{\sqrt{2}}{4}.$$

By l'Hospital's rule,

$$\lim_{x\to 0} \frac{\sqrt{2+x}-\sqrt{2}}{x} = \frac{1}{2\sqrt{2}} \text{ or } \frac{\sqrt{2}}{4}.$$

15. The limit in the numerator is 0, as is the limit in the denominator, so that l'Hospital's rule applies. Taking derivatives separately in the numerator and denominator gives

$$\lim_{x\to 4} \frac{\frac{1}{2x^{1/2}}}{1} = \frac{1}{2\cdot 4^{1/2}} = \frac{1}{4}.$$

By l'Hospital's rule,

$$\lim_{x\to 4} \frac{\sqrt{x}-2}{x-4} = \frac{1}{4}.$$

17. The limit in the numerator is 0, as is the limit in the denominator, so that l'Hospital's rule applies. Taking derivatives separately in the numerator and denominator gives

$$\lim_{x\to 8} \frac{\frac{1}{3x^{2/3}}}{1} = \frac{1}{3\cdot 8^{2/3}} = \frac{1}{12}.$$

By l'Hospital's rule,

$$\lim_{x\to 8} \frac{\sqrt[3]{x}-2}{x-8} = \frac{1}{12}.$$

19. The limit in the numerator is 0, as is the limit in the denominator, so that l'Hospital's rule applies. Taking derivatives separately in the numerator and denominator gives

$$\lim_{x\to 1} \frac{9x^8+24x^7+20x^4}{1} = \frac{9+24+20}{1} = 53.$$

By l'Hospital's rule,

$$\lim_{x\to 1} \frac{x^9+3x^8+4x^5-8}{x-1} = 53.$$

21. The limit in the numerator is 0, as is the limit in the denominator, so that l'Hospital's rule applies. Taking derivatives separately in the numerator and denominator gives

$$\lim_{x\to 0} \frac{e^x-e^{-x}}{1} = e^0-e^0 = 0.$$

By l'Hospital's rule,

$$\lim_{x\to 0} \frac{e^x+e^{-x}-2}{x} = 0.$$

23. The limit in the numerator is 0, as is the limit in the denominator, so that l'Hospital's rule applies. Taking derivatives separately in the numerator and denominator gives

$$\lim_{x\to 3} \frac{\frac{x}{(x^2+7)^{1/2}}}{2x} = \lim_{x\to 3} \frac{1}{2\,(x^2+7)^{1/2}} = \frac{1}{2\,(3^2+7)^{1/2}} = \frac{1}{8}.$$

By l'Hospital's rule,

$$\lim_{x\to 3} \frac{\sqrt{x^2+7}-4}{x^2-9} = \frac{1}{8}.$$

25. The limit in the numerator is 0, as is the limit in the denominator, so that l'Hospital's rule applies. Taking derivatives separately in the numerator and denominator gives

$$\lim_{x\to 0} \frac{\frac{1}{3}-\frac{1}{3(1+x)^{2/3}}}{2x} = \frac{\frac{1}{3}-\frac{1}{3(1+0)^{2/3}}}{2\cdot 0} = \frac{0}{0}.$$

Using l'Hospital's rule a second time, gives

$$\lim_{x\to 0} \frac{\frac{2}{9(1+x)^{5/3}}}{2} = \lim_{x\to 0} \frac{1}{9\,(1+x)^{5/3}} = \frac{1}{9\,(1+0)^{5/3}} = \frac{1}{9}.$$

By l'Hospital's rule,

$$\lim_{x\to 0} \frac{1+\frac{1}{3}x-(1+x)^{1/3}}{x^2} = \frac{1}{9}.$$

27. The limit in the numerator is 0, as is the limit in the denominator, so that l'Hospital's rule applies. Taking derivatives separately in the numerator and denominator gives

$$\lim_{x\to 0} \frac{\frac{1}{2(1+x)^{1/2}}+\frac{1}{2(1-x)^{1/2}}}{1} = \lim_{x\to 0}\left(\frac{1}{2\,(1+x)^{1/2}}+\frac{1}{2\,(1-x)^{1/2}}\right) = \frac{1}{2\,(1+0)^{1/2}}+\frac{1}{2\,(1-0)^{1/2}} = 1.$$

By l'Hospital's rule,

$$\lim_{x\to 0} \frac{\sqrt{1+x}-\sqrt{1-x}}{x} = 1.$$

29. $\lim_{x\to 0} \sqrt{x^2-5x+4} = 2$ and l'Hospital's rule does not apply. However,

$$\lim_{x\to 0} \frac{\sqrt{x^2-5x+4}}{x} \text{ does not exist.}$$

31. The limit in the numerator is 0, as is the limit in the denominator, so that l'Hospital's rule applies. Taking derivatives separately in the numerator and denominator gives

$$\lim_{x\to 0} \frac{\ln(x+1)+\frac{5+x}{x+1}}{e^x} = \lim_{x\to 0} \frac{(x+1)\ln(x+1)+5+x}{e^x\,(x+1)} = \frac{(0+1)\ln(0+1)+5+0}{e^0\,(0+1)} = 5.$$

By l'Hospital's rule,

$$\lim_{x\to 0} \frac{(5+x)\ln(x+1)}{e^x-1} = 5.$$

33. We have a limit of the form $0\times\infty$. Rewrite the expression.

$$x^2(\ln x)^2 = \frac{(\ln x)^2}{\frac{1}{x^2}}$$

Now both the numerator and denominator become infinite and 1'Hospital's rule applies to the limit of the form ∞/∞. Differentiate the numerator and denominator.

$$\lim_{x\to 0+} \frac{(\ln x)^2}{\frac{1}{x^2}} = \lim_{x\to 0+} \frac{\frac{2(\ln x)}{x}}{\frac{-2}{x^3}} = \lim_{x\to 0+} -x^2\ln x$$

This problem is similar to what we started with so we handle it in the same manner.

$$\lim_{x\to 0+} -x^2\ln x = \lim_{x\to 0+} \frac{-\ln x}{\frac{1}{x^2}} = \lim_{x\to 0+} \frac{\frac{-1}{x}}{\frac{-2}{x^3}} = \lim_{x\to 0+} \frac{x^2}{2} = 0$$

Therefore, by 1'Hospital's rule,

$$\lim_{x\to 0+} x^2(\ln x)^2 = 0.$$

35. We have a limit of the form $0 \times \infty$. Rewrite the expression.

$$x \ln(e^x - 1) = \frac{(\ln e^x - 1)}{\frac{1}{x}}$$

Now both the numerator and denominator become infinite and l'Hospital's rule applies to the limit of the form ∞/∞. Differentiate the numerator and the denominator.

$$\lim_{x \to 0+} \frac{\ln(e^x - 1)}{\frac{1}{x}} = \lim_{x \to 0+} \frac{\frac{e^x}{(e^x - 1)}}{\frac{-1}{x^2}} = \lim_{x \to 0+} \frac{-x^2 e^x}{e^x - 1}$$

Notice that $\frac{e^x}{e^x - 1} = 1 + \frac{1}{e^x - 1}$.

$$\lim_{x \to 0+} \frac{-x^2 e^x}{e^x - 1} = \lim_{x \to 0+} -x^2 \cdot \left(1 + \frac{1}{e^x - 1}\right) = 0 \cdot 1 = 0$$

Therefore, by l'Hospital's rule,

$$\lim_{x \to 0+} x \ln(e^x - 1) = 0.$$

37. The limit in the numerator is ∞, as in the limit in the denominator, so l'Hospital's rule applies. Taking derivatives separately in the numerator and the denominator gives

$$\lim_{x \to \infty} \frac{\sqrt{x}}{\ln(\ln x)} = \lim_{x \to \infty} \frac{\frac{1}{(2\sqrt{x})}}{\frac{1}{(x \ln x)}} = \lim_{x \to \infty} \frac{\sqrt{x} \ln x}{2} = \infty.$$

By l'Hospital's rule,

$$\lim_{x \to \infty} \frac{\sqrt{x}}{\ln(\ln x)} = \infty.$$

39. The limit in the numerator is ∞, as is the limit in the denominator, so l'Hospital's rule applies. Taking derivatives separately in the numerator and the denominator gives

$$\lim_{x \to 0+} \frac{\ln(e^x + 1)}{5x} = \lim_{x \to 0+} \frac{\frac{e^x}{(e^x + 1)}}{5} = \lim_{x \to 0+} \frac{e^x}{5(e^x + 1)}$$

Notice that $\frac{e^x}{e^x + 1} = 1 - \frac{1}{e^x + 1}$.

$$\lim_{x \to 0+} \frac{e^x}{5(e^x + 1)} = \frac{1}{5} \cdot \lim_{x \to 0+} \frac{e^x}{e^x + 1} = \frac{1}{5} \cdot \lim_{x \to 0+} \left(1 - \frac{1}{e^x + 1}\right) = \frac{1}{5} \cdot 1 = \frac{1}{5}$$

Therefore, by l'Hospital's rule,

$$\lim_{x \to 0+} \frac{\ln(e^x + 1)}{5x} = \frac{1}{5}.$$

41. We have a limit of the form $\infty \cdot 0$. Rewrite the expression as a quotient.

$$x^5 e^{-0.001x} = \frac{x^5}{e^{0.001x}}$$

The limit in the numerator is ∞, as in the limit in the denominator, so l'Hospital's rule applies. Taking derivatives separately in the numerator and the denominator gives

$$\lim_{x \to \infty} \frac{x^5}{e^{0.001x}} = \lim_{x \to \infty} \frac{5x^4}{0.001 e^{0.001x}}$$

This problem is similar to what we started with so we handle it in the same manner. In fact, continue the process four more times.

$$\begin{aligned}\lim_{x\to\infty}\frac{5x^4}{0.001e^{0.001x}} &= \lim_{x\to\infty}\frac{20x^3}{(0.001)^2e^{0.001x}} = \lim_{x\to\infty}\frac{60x^2}{(0.001)^3e^{0.001x}}\\ &= \lim_{x\to\infty}\frac{120x}{(0.001)^4e^{0.001x}} = \lim_{x\to\infty}\frac{120}{(0.001)^5e^{0.001x}}\\ &= 0\end{aligned}$$

Therefore, by l'Hospital's rule,

$$\lim_{x\to\infty} x^5e^{-0.001x} = 0.$$

43. $\lim_{x\to 0}\left(\frac{1}{x}-\frac{1}{x^2}\right) = \lim_{x\to 0}\frac{x-1}{x^2}$ which does not exist.

45. $\lim_{x\to 0}\left(\frac{1}{x}+\frac{1}{\sqrt[3]{x}}\right) = \lim_{x\to 0}\frac{1+x^{2/3}}{x}$ which does not exist.

47. $\lim_{x\to 0} x^2+3 \neq 0$, so l'Hospital's rule does not apply.

Chapter 12 Review Exercises

1.

Derivative	Value at 0
$f(x) = e^{2-x}$	$f(0) = e^2$
$f^{(1)}(x) = -e^{2-x}$	$f^{(1)}(0) = -e^2$
$f^{(2)}(x) = e^{2-x}$	$f^{(2)}(0) = e^2$
$f^{(3)}(x) = -e^{2-x}$	$f^{(3)}(0) = -e^2$
$f^{(4)}(x) = e^{2-x}$	$f^{(4)}(0) = e^2$

$$\begin{aligned}P_4(x) &= f(0) + \frac{f^{(1)}(0)}{1!}x + \frac{f^{(2)}(0)}{2!}x^2 + \frac{f^{(3)}(0)}{3!}x^3 + \frac{f^{(4)}(0)}{4!}x^4\\ &= e^2 + \frac{-e^2}{1!}x + \frac{e^2}{2!}x^2 + \frac{-e^2}{3!}x^3 + \frac{e^2}{4!}x^4\\ &= e^2 - e^2x + \frac{e^2}{2}x^2 - \frac{e^2}{6}x^3 + \frac{e^2}{24}x^4\end{aligned}$$

3.

Derivative	Value at 0
$f(x) = \sqrt{x+1} = (x+1)^{1/2}$	$f(0) = 1$
$f^{(1)}(x) = \frac{1}{2}(x+1)^{-1/2} = \frac{1}{2(x+1)^{1/2}}$	$f^{(1)}(0) = \frac{1}{2}$
$f^{(2)}(x) = -\frac{1}{4}(x+1)^{-3/2} = -\frac{1}{4(x+1)^{3/2}}$	$f^{(2)}(0) = -\frac{1}{4}$
$f^{(3)}(x) = \frac{3}{8}(x+1)^{-5/2} = \frac{3}{8(x+1)^{5/2}}$	$f^{(3)}(0) = \frac{3}{8}$
$f^{(4)}(x) = -\frac{15}{16}(x+1)^{-7/2} = -\frac{15}{16(x+1)^{7/2}}$	$f^{(4)}(0) = -\frac{15}{16}$

$$P_4(x) = f(0) + \frac{f^{(1)}(0)}{1!}x + \frac{f^{(2)}(0)}{2!}x^2 + \frac{f^{(3)}(0)}{3!}x^3 + \frac{f^{(4)}(0)}{4!}x^4$$
$$= 1 + \frac{\frac{1}{2}}{1}x + \frac{-\frac{1}{4}}{2}x^2 + \frac{\frac{3}{8}}{6}x^3 + \frac{-\frac{15}{16}}{24}x^4$$
$$= 1 + \frac{1}{2}x - \frac{1}{8}x^2 + \frac{1}{16}x^3 - \frac{5}{128}x^4$$

5.

Derivative	Value at 0
$f(x) = \ln(2-x)$	$f(0) = \ln 2$
$f^{(1)}(x) = -\frac{1}{2-x} = -(2-x)^{-1}$	$f^{(1)}(0) = -\frac{1}{2}$
$f^{(2)}(x) = -(2-x)^{-2} = -\frac{1}{(2-x)^2}$	$f^{(2)}(0) = -\frac{1}{4}$
$f^{(3)}(x) = -2(2-x)^{-3} = -\frac{2}{(2-x)^3}$	$f^{(3)}(0) = -\frac{1}{4}$
$f^{(4)}(x) = -6(2-x)^{-4} = -\frac{6}{(2-x)^4}$	$f^{(4)}(0) = -\frac{3}{8}$

$$P_4(x) = f(0) + \frac{f^{(1)}(0)}{1!}x + \frac{f^{(2)}(0)}{2!}x^2 + \frac{f^{(3)}(0)}{3!}x^3 + \frac{f^{(4)}(0)}{4!}x^4$$
$$= \ln 2 + \frac{-\frac{1}{2}}{1}x + \frac{-\frac{1}{4}}{2}x^2 + \frac{-\frac{1}{4}}{6}x^3 + \frac{-\frac{3}{8}}{24}x^4$$
$$= \ln 2 - \frac{1}{2}x - \frac{1}{8}x^2 - \frac{1}{24}x^3 - \frac{1}{64}x^4$$

7.

Derivative	Value at 0
$f(x) = (1+x)^{2/3}$	$f(0) = 1$
$f^{(1)}(x) = \frac{2}{3}(1+x)^{-1/3} = \frac{2}{3(1+x)^{1/3}}$	$f^{(1)}(0) = \frac{2}{3}$
$f^{(2)}(x) = -\frac{2}{9}(1+x)^{-4/3} = -\frac{2}{9(1+x)^{4/3}}$	$f^{(2)}(0) = -\frac{2}{9}$
$f^{(3)}(x) = \frac{8}{27}(1+x)^{-7/3} = \frac{8}{27(1+x)^{7/3}}$	$f^{(3)}(0) = \frac{8}{27}$
$f^{(4)}(x) = -\frac{56}{81}(1+x)^{-10/3} = -\frac{56}{81(1+x)^{10/3}}$	$f^{(4)}(0) = -\frac{56}{81}$

$$P_4(x) = f(0) + \frac{f^{(1)}(0)}{1!}x + \frac{f^{(2)}(0)}{2!}x^2 + \frac{f^{(3)}(0)}{3!}x^3 + \frac{f^{(4)}(0)}{4!}x^4$$
$$= 1 + \frac{\frac{2}{3}}{1}x + \frac{-\frac{2}{9}}{2}x^2 + \frac{\frac{8}{27}}{6}x^3 + \frac{-\frac{56}{81}}{24}x^4$$
$$= 1 + \frac{2}{3}x - \frac{1}{9}x^2 + \frac{4}{81}x^3 - \frac{7}{243}x^4$$

9. Using the result of Exercise 1, with $f(x) = e^{2-x}$ and $P_4(x) = e^2 - e^2 x + \frac{e^2}{2}x^2 - \frac{e^2}{6}x^3 + \frac{e^2}{24}x^4$, we can approximate $e^{1.93}$ by evaluating $f(0.07) = e^{2-0.07} = e^{2-0.07} = e^{1.93}$. Using $P_4(x)$ from Exercise 1 with $x = 0.07$ gives

$$\begin{aligned} P_4(0.07) &= e^2 - e^2(0.07) + \frac{e^2}{2}(0.07)^2 - \frac{e^2}{6}(0.07)^3 + \frac{e^2}{24}(0.07)^4 \\ &\approx 6.88951034388. \end{aligned}$$

To four decimal places, $P_4(0.07)$ approximates the value of $e^{1.93}$ as 6.8895.

11. Using the result of Exercise 3, with $f(x) = \sqrt{x+1}$ and $P_4(x) = 1 + \frac{1}{2}x - \frac{1}{8}x^2 + \frac{1}{16}x^3 - \frac{5}{128}x^4$, we can approximate $\sqrt{1.03}$ by evaluating $f(0.03) = \sqrt{0.03+1} = \sqrt{1.03}$. Using $P_4(x)$ from Exercise 3 with $x = 0.03$ gives

$$\begin{aligned} P_4(0.03) &= 1 + \frac{1}{2}(0.03) - \frac{1}{8}(0.03)^2 + \frac{1}{16}(0.03)^3 - \frac{5}{128}(0.03)^4 \\ &\approx 1.01488915586. \end{aligned}$$

To four decimal places, $P_4(0.03)$ approximates the value of $\sqrt{1.03}$ as 1.0149.

13. Using the result of Exercise 5, with $f(x) = \ln(2-x)$ and $P_4(x) = \ln 2 - \frac{1}{2}x - \frac{1}{8}x^2 - \frac{1}{24}x^3 - \frac{1}{64}x^4$, we can approximate $\ln 2.05$ by evaluating $f(-0.05) = \ln(2-(-0.05)) = \ln 2.05$. Using $P_4(x)$ from Exercise 5 with $x = -0.05$ gives

$$\begin{aligned} P_4(-0.05) &= \ln 2 - \frac{1}{2}(-0.05) - \frac{1}{8}(-0.05)^2 - \frac{1}{24}(-0.05)^3 - \frac{1}{64}(-0.05)^4 \\ &\approx 0.717842610677 \text{ (using } \ln 2 = 0.69315). \end{aligned}$$

To four decimal places, $P_4(-0.05)$ approximates the value of $\ln 2.05$ as 0.7178.

15. Using the result of Exercise 7, with $f(x) = (1+x)^{2/3}$ and $P_4(x) = 1 + \frac{2}{3}x - \frac{1}{9}x^2 + \frac{4}{81}x^3 - \frac{7}{243}x^4$, we can approximate $(0.92)^{2/3}$ by evaluating $f(-0.08) = (1+(-0.08))^{2/3} = (0.92)^{2/3}$. Using $P_4(x)$ from Exercise 7 with $x = -0.08$ gives

$$\begin{aligned} P_4(-0.08) &= 1 + \frac{2}{3}(-0.08) - \frac{1}{9}(-0.08)^2 + \frac{4}{81}(-0.08)^3 - \frac{7}{243}(-0.08)^4 \\ &\approx 0.945929091687. \end{aligned}$$

To four decimal places, $P_4(-0.08)$ approximates the value of $(0.92)^{2/3}$ as 0.9459.

17. $9 - 6 + 4 - \frac{8}{3} + \ldots$ is a geometric series with $a = a_1 = 9$ and $r = -\frac{2}{3}$. Since r is in $(-1, 1)$, the series converges and has sum

$$\frac{a}{1-r} = \frac{9}{1-\left(-\frac{2}{3}\right)} = \frac{9}{1+\frac{2}{3}} = \frac{9}{\frac{5}{3}} = \frac{27}{5}.$$

19. $3 + 9 + 27 + 81 + \ldots$ is a geometric series with $a = a_1 = 3$ and $r = 3$. Since $r > 1$, the series diverges.

21. $\frac{2}{5} - \frac{2}{25} + \frac{2}{125} - \frac{2}{625} + \ldots$ is a geometric series with $a = a_1 = \frac{2}{5}$ and $r = -\frac{1}{5}$. Since r is in $(-1, 1)$, the series converges and has sum

$$\frac{a}{1-r} = \frac{\frac{2}{5}}{1-\left(-\frac{1}{5}\right)} = \frac{\frac{2}{5}}{1+\frac{1}{5}} = \frac{\frac{2}{5}}{\frac{6}{5}} = \frac{1}{3}.$$

23. $S_1 = a_1 = \frac{1}{2(1)-1} = 1$

$S_2 = a_1 + a_2 = \frac{1}{2(1)-1} + \frac{1}{2(2)-1} = 1 + \frac{1}{3} = \frac{4}{3}$

$S_3 = a_1 + a_2 + a_3 = \frac{1}{2(1)-1} + \frac{1}{2(2)-1} + \frac{1}{2(3)-1} = 1 + \frac{1}{3} + \frac{1}{5} = \frac{23}{15}$

$S_4 = a_1 + a_2 + a_3 + a_4 = \frac{1}{2(1)-1} + \frac{1}{2(2)-1} + \frac{1}{2(3)-1} + \frac{1}{2(4)-1}$

$= 1 + \frac{1}{3} + \frac{1}{5} + \frac{1}{7} = \frac{176}{105}$

$S_5 = a_1 + a_2 + a_3 + a_4 + a_5 = \frac{1}{2(1)-1} + \frac{1}{2(2)-1} + \frac{1}{2(3)-1} + \frac{1}{2(4)-1} + \frac{1}{2(5)-1}$

$1 + \frac{1}{3} + \frac{1}{5} + \frac{1}{7} + \frac{1}{9} = \frac{563}{315}$

25. This function most nearly matches $\frac{1}{1-x}$. To get 1 in the denominator, instead of 3, divide the numerator and denominator by 3.

$$\frac{4}{3-x} = \frac{\frac{4}{3}}{1-\frac{x}{3}}$$

Thus, we can find the Taylor series for $\frac{\frac{4}{3}}{1-\frac{x}{3}}$ by starting with the Taylor series for $\frac{1}{1-x}$, multiplying each term by $\frac{4}{3}$, and replacing x with $\frac{x}{3}$.

$$\begin{aligned}\frac{4}{3-x} &= \frac{\frac{4}{3}}{1-\frac{x}{3}} \\ &= \frac{4}{3} \cdot 1 + \frac{4}{3}\left(\frac{x}{3}\right) + \frac{4}{3}\left(\frac{x}{3}\right)^2 + \frac{4}{3}\left(\frac{x}{3}\right)^3 + \cdots + \frac{4}{3}\left(\frac{x}{3}\right)^n + \cdots \\ &= \frac{4}{3} + \frac{4x}{9} + \frac{4x^2}{27} + \frac{4x^3}{81} + \cdots + \frac{4x^n}{3^{n+1}} + \cdots\end{aligned}$$

The Taylor series for $\frac{1}{1-x}$ is valid when $-1 < x < 1$. Replacing x with $\frac{x}{3}$ gives

$$-1 < \tfrac{x}{3} < 1 \text{ or } -3 < x < 3.$$

The interval of convergence of the new series is $(-3, 3)$.

27. $\frac{x^2}{x+1} = x^2 \cdot \frac{1}{1-(-x)}$

Use the Taylor series for $\frac{1}{1-x}$ and replace x with $-x$. Also, use property (3) with $k = 2$.

$$\begin{aligned}\frac{x^2}{x+1} &= x^2 \cdot \frac{1}{1-(-x)} \\ &= x^2 \cdot 1 + x^2(-x) + x^2(-x)^2 + x^2(-x)^3 + \cdots + x^2(-x)^n + \cdots \\ &= x^2 - x^3 + x^4 - x^5 + \cdots + (-1)^n x^{n+2} + \cdots\end{aligned}$$

The Taylor series for $\frac{1}{1-x}$ is valid when $-1 < x < 1$. Replacing x with $-x$ gives

$$-1 < -x < 1 \text{ or } 1 > x > -1.$$

The interval of convergence of the new series is $(-1, 1)$.

29. We find the Taylor series for $\ln(1-2x)$ by starting with the Taylor series for $\ln(1+x)$ and replacing each x with $-2x$.

$$\begin{aligned}\ln(1-2x) &= -2x - \frac{(-2x)^2}{2} + \frac{(-2x)^3}{3} - \frac{(-2x)^4}{4} + \cdots + \frac{(-1)^n(-2x)^{n+1}}{n+1} + \cdots \\ &= -2x - 2x^2 - \frac{8}{3}x^3 - 4x^4 - \cdots - \frac{2^{n+1}x^{n+1}}{n+1} - \cdots\end{aligned}$$

The Taylor series for $\ln(1+x)$ is valid when $-1 < x \le 1$. Replacing x with $-2x$ gives

$$-1 < -2x \le 1 \text{ or } \frac{1}{2} > x \ge -\frac{1}{2}.$$

The interval of convergence of the new series is $\left[-\frac{1}{2}, \frac{1}{2}\right)$.

31. We find the Taylor series for e^{-2x^2} by starting with the Taylor series for e^x and replacing each x with $-2x^2$.

$$\begin{aligned}e^{-2x^2} &= 1 + (-2x^2) + \frac{1}{2!}(-2x^2)^2 + \frac{1}{3!}(-2x^2)^3 + \cdots + \frac{1}{n!}(-2x^2)^n + \cdots \\ &= 1 - 2x^2 + 2x^4 - \frac{4}{3}x^6 + \cdots + \frac{(-1)^n \cdot 2^n x^{2n}}{n!} + \cdots\end{aligned}$$

The Taylor series for e^{-2x^2} has the same interval of convergence, $(-\infty, \infty)$, as the Taylor series for e^x.

33. Use the Taylor series for e^x, multiply each term by 2, and replace x with $-3x$. Also, use property (3) with $k = 3$.

$$\begin{aligned}2x^3e^{-3x} &= 2x^3 \cdot 1 + 2x^3(-3x) + 2x^3 \cdot \frac{1}{2!}(-3x)^2 + 2x^3 \cdot \frac{1}{3!}(-3x)^3 + \cdots + 2x^3 \cdot \frac{1}{n!}(-3x)^n + \cdots \\ &= 2x^3 - 6x^4 + 9x^5 - 9x^6 + \cdots + \frac{(-1)^n \cdot 2 \cdot 3^n x^{n+3}}{n!} + \cdots\end{aligned}$$

The Taylor series for $2x^3e^{-3x}$ has the same interval of convergence, $(-\infty, \infty)$, as the Taylor series for e^x.

35. The limit in the numerator is 0, as is the limit in the denominator, so that l'Hospital's rule applies. Taking derivatives separately in the numerator and denominator gives,

$$\lim_{x\to 2} \frac{3x^2 - 2x - 1}{2x} = \frac{3(2)^2 - 2(2) - 1}{2(2)} = \frac{7}{4}.$$

By l'Hospital's rule,

$$\lim_{x\to 2} \frac{x^3 - x^2 - x - 2}{x^2 - 4} = \frac{7}{4}.$$

37. $\lim_{x\to -5} x^3 - 3x^2 + 4x - 1 = -221$ and l'Hospital's rule does not apply. However,

$$\lim_{x\to -5} \frac{x^3 - 3x^2 + 4x - 1}{x^2 - 25} \text{ does not exist.}$$

39. The limit in the numerator is 0, as is the limit in the denominator, so that l'Hospital's rule applies. Taking derivatives separately in the numerator and denominator gives,

$$\lim_{x\to 0} \frac{5e^x}{3x^2 - 16x + 7} = \frac{5e^0}{3(0)^2 - 16(0) + 7} = \frac{5}{7}.$$

By l'Hospital's rule,

$$\lim_{x\to 0} \frac{5e^x - 5}{x^3 - 8x^2 + 7x} = \frac{5}{7}.$$

41. The limit in the numerator is 0, as is the limit in the denominator, so that l'Hospital's rule applies. Taking derivatives separately in the numerator and denominator gives,

$$\lim_{x\to 0}\frac{-e^{2x}-2xe^{2x}}{2e^{2x}}=\frac{-e^{2(0)}-2\,(0)\,e^{2(0)}}{2e^{2(0)}}=-\frac{1}{2}.$$

By l'Hospital's rule,

$$\lim_{x\to 0}\frac{-xe^{2x}}{e^{2x}-1}=-\frac{1}{2}.$$

43. The limit in the numerator is 0, as is the limit in the denominator, so that l'Hospital's rule applies. Taking derivatives separately in the numerator and denominator gives,

$$\lim_{x\to 0}\frac{2-\frac{1}{2}\,(1+x)^{-1/2}}{3x^2}=\lim_{x\to 0}\frac{4\,(1+x)^{1/2}-1}{3x^2\left(2\,(1+x)^{1/2}\right)}=\frac{3}{0}\text{ which does not exist.}$$

By l'Hospital's rule,

$$\lim_{x\to 0}\frac{1+2x-(1+x)^{1/2}}{x^3}\text{ does not exist.}$$

45. We have a limit of the form $\infty\cdot 0$. Rewrite the expression.

$$x^2e^{-\sqrt{x}}=\frac{x^2}{e^{\sqrt{x}}}$$

Now both the numerator and denominator become infinite and 1"Hospital's rule applies to the limit of the form ∞/∞. Differentiate the numerator and the denominator.

$$\lim_{x\to\infty}\frac{x^2}{e^{\sqrt{x}}}=\lim_{x\to\infty}\frac{2x}{\frac{e^{\sqrt{x}}}{(2\sqrt{x})}}=\lim_{x\to\infty}\frac{4x^{3/2}}{e^{\sqrt{x}}}$$

This problem is similar to what we started with so we handle it in the same manner. In fact, continue the process three more times.

$$\lim_{x\to\infty}\frac{4x^{3/2}}{e^{\sqrt{x}}}=\lim_{x\to\infty}\frac{6x^{1/2}}{\frac{e^{\sqrt{x}}}{(2\sqrt{x})}}=\lim_{x\to\infty}\frac{12x}{e^{\sqrt{x}}}=\lim_{x\to\infty}\frac{12}{\frac{e^{\sqrt{x}}}{(2\sqrt{x})}}=\lim_{x\to\infty}\frac{24\sqrt{x}}{e^{\sqrt{x}}}=\lim_{x\to\infty}\frac{\frac{12}{\sqrt{x}}}{\frac{e^{\sqrt{x}}}{(2\sqrt{x})}}=\lim_{x\to\infty}\frac{24}{e^{\sqrt{x}}}=0$$

Therefore, by 1'Hospital's rule,

$$\lim_{x\to\infty}x^2e^{-\sqrt{x}}=0.$$

47. $\lim_{x\to 0}\left(\frac{1}{2x}-\frac{1}{3x}\right)=\lim_{x\to 0}\frac{1}{6x}$ which does not exist.

49. $\lim_{x\to 0}\left(\frac{1}{x}-\frac{1}{\sqrt{x}}\right)=\lim_{x\to 0}\frac{1-x^{1/2}}{x}$ which does not exist.

51. $f(x) = x^3 - 8x^2 + 18x - 12$
$f'(x) = 3x^2 - 16x + 18$

$f(4) = -4 < 0$ and $f(5) = 3 > 0$ so a solution exists in $(4, 5)$.
Let $c_1 = 4$.

$$c_2 = c_1 - \frac{f(c_1)}{f'(c_1)} = 4 - \frac{-4}{2} = 6$$
$$c_3 = c_2 - \frac{f(c_2)}{f'(c_2)} = 6 - \frac{24}{30} = 5.2$$
$$c_4 = c_3 - \frac{f(c_3)}{f'(c_3)} = 5.2 - \frac{5.888}{15.92} = 4.8302$$
$$c_5 = c_4 - \frac{f(c_4)}{f'(c_4)} = 4.8302 - \frac{0.98953}{10.709} = 4.7378$$
$$c_6 = c_5 - \frac{f(c_5)}{f'(c_5)} = 4.7378 - \frac{0.05462}{9.5354} = 4.7321$$
$$c_7 = c_6 - \frac{f(c_6)}{f'(c_6)} = 4.7321 - \frac{4.7 \cdot 10^{-4}}{9.4647} = 4.7321$$

Subsequent approximations will agree with c_6 and c_7 to the nearest hundredth. Thus, $x = 4.73$.

53. $f(x) = x^4 + 3x^3 - 4x^2 - 21x - 21$
$f'(x) = 4x^3 + 9x^2 - 8x - 21$

$f(2) = -39 < 0$ and $f(3) = 42 > 0$ so a solution exists in $(2, 3)$.
Let $c_1 = 2$.

$$c_2 = c_1 - \frac{f(c_1)}{f'(c_1)} = 2 - \frac{-39}{31} = 3.2581$$
$$c_3 = c_2 - \frac{f(c_2)}{f'(c_2)} = 3.2581 - \frac{84.558}{186.81} = 2.8055$$
$$c_4 = c_3 - \frac{f(c_3)}{f'(c_3)} = 2.8055 - \frac{16.796}{115.72} = 2.6604$$
$$c_5 = c_4 - \frac{f(c_4)}{f'(c_4)} = 2.6604 - \frac{1.4037}{96.735} = 2.6459$$
$$c_6 = c_5 - \frac{f(c_5)}{f'(c_5)} = 2.6459 - \frac{0.01411}{94.933} = 2.6458$$

Subsequent approximations will agree with c_5 and c_6 to the nearest hundredth. Thus, $x = 2.65$.

55. $\sqrt{37.6}$ is a solution of $x^2 - 37.6 = 0$.

$$f(x) = x^2 - 37.6$$
$$f'(x) = 2x$$

Since $6 < \sqrt{37.6} < 7$, let $c_1 = 6$.

$$c_2 = c_1 - \frac{f(c_1)}{f'(c_1)} = 6 - \frac{-1.6}{12} = 6.1333$$
$$c_3 = c_2 - \frac{f(c_2)}{f'(c_2)} = 6.1333 - \frac{0.01737}{12.2666} = 6.1319$$
$$c_4 = c_3 - \frac{f(c_3)}{f'(c_3)} = 6.1319 - \frac{0.0002}{12.2638} = 6.1319$$

Since $c_4 = c_3 = 6.1319$, to the nearest thousandth, $\sqrt{37.6} = 6.132$.

57. $\sqrt[3]{94.7}$ is a solution of $x^3 - 94.7 = 0$.

$$f(x) = x^3 - 94.7$$
$$f'(x) = 3x^2$$

Since $4 < \sqrt[3]{94.7} < 5$, let $c_1 = 4$.

$$c_2 = c_1 - \frac{f(c_1)}{f'(c_1)} = 4 - \frac{-30.7}{48} = 4.6396$$

$$c_3 = c_2 - \frac{f(c_2)}{f'(c_2)} = 4.6396 - \frac{5.1715}{64.5777} = 4.5595$$

$$c_4 = c_3 - \frac{f(c_3)}{f'(c_3)} = 4.5595 - \frac{0.08763}{62.3671} = 4.5581$$

$$c_5 = c_4 - \frac{f(c_4)}{f'(c_4)} = 4.5581 - \frac{0.0003}{62.3288} = 4.5581$$

Since $c_3 = c_4 = 4.5581$, to the nearest thousandth, $\sqrt[3]{94.7} = 4.558$.

59. The yearly incomes produced by the mine form a geometric sequence with $r = 118\% = 1.18$ and $a_1 = 750{,}000$. To determine the total amount produced in 8 years, use the formula to find S_n with $n = 8$, $r = 1.18$, and $a = a_1 = 750{,}000$.

$$S_8 = \frac{750{,}000\left[(1.18)^8 - 1\right]}{1.18 - 1} = \frac{750{,}000\,(3.75886 - 1)}{0.18} \approx 11{,}495{,}247$$

The total amount of income produced by the mine in five years is \$11,495,247.

61. Michelle's payments form an ordinary annuity with $R = 491$, $n = 9 \cdot 4 = 36$, and $i = \frac{9.4\%}{4} = 2.35\% = 0.0235$. The amount of this annuity is

$$S = 491\left[\frac{(1.0235)^{36} - 1}{0.0235}\right].$$

The number in brackets, $s_{\overline{36}|0.0235}$, is 55.64299673, so that

$$S = 491\,(55.64299673) \approx 27{,}320.71.$$

or \$27,320.71

63. \$20,000 is the present value of an annuity of R dollars, with 9 periods, and $i = 8.9\% = 0.089$ per period.

$$P = R \cdot a_{\overline{n}|i}$$
$$20{,}000 = R \cdot a_{\overline{9}|0.089}$$
$$R = \frac{20{,}000}{a_{\overline{9}|0.089}} = \frac{20{,}000}{6.019696915} \approx 3322.43$$

Each payment is \$3322.43.

65. The present value, P, is 156,890, $i = \frac{0.0774}{12} = 0.00645$, and $n = 12 \cdot 25 = 300$.

$$156{,}890 = R \cdot a_{\overline{300}|0.00645}$$
$$= R\left[\frac{1 - (1 + 0.00645)^{-300}}{0.00645}\right]$$
$$= R\left[\frac{1 - 0.1453244675}{0.00645}\right]$$
$$= R\left[\frac{0.8546755325}{0.00645}\right]$$
$$R \approx 1184.01$$

Monthly payments of \$1184.01 will be required to amortize the loan.

67. The doubling time n for a quantity that increases at an annual rate r is given by

$$n = \frac{\ln 2}{\ln(1+r)} = \frac{\ln 2}{\ln 1.0325} \approx 21.67.$$

It will take about 21.67 years.

According to the Rule of 70, the doubling time is given by

$$\text{Doubling time } \approx \frac{70}{100r} = \frac{70}{100(0.0325)} = 21.54.$$

It will take about 21.54 years, a difference of 0.13 year, or about 7 weeks.

Chapter 12 Test

[12.1]

For each sequence that is geometric, find r, a_5, and a_n.

1. $-\frac{1}{3}, \frac{2}{3}, -\frac{4}{3}, \frac{8}{3}, \cdots$

2. 5, 8, 11, 14

Find the sum of the first five terms of each geometric sequence.

3. $a_1 = -2$, $r = 4$

4. $\sum_{i=1}^{5} \frac{1}{2}\left(3^i\right)$

5. A certain car annually loses 25% of the value that it had at the beginning of that year. If its initial value is $15,525, find its value (to the nearest dollar) at the following times.

(a) The end of the third year

(b) The end of the fifth year

[12.2]

6. Pat, a 45-year-old woman, deposits $3000 per year until age 65 into an Individual Retirement Account earning 7% per year compounded semiannually. Assuming that Pat makes payments of $1500 at the end of each semiannual period, find the total amount in the account at age 65.

7. Find the amount of each annual payment to be made into a sinking fund so that the fund contains $20,000 at the end of 5 years if the money earns 8% compounded annually.

8. Quinn purchased a condo 5 years ago for $75,000. After putting $10,000 down, she agreed to amortize the balance with monthly payments at an annual interest rate of 7.2% for 30 years.

(a) Find the amount of each payment.

(b) If the payment found in part (a) is made monthly for the remaining 25 years at 7.2% compounded monthly, find the present value of the annuity to the nearest dollar.

(c) If today Quinn sells the condo for $88,000, how much money will she have left over after paying off the unpaid loan balance?

[12.3]

9. Let $f(x) = \ln(1 + x)$

(a) Find the Taylor polynomial of degree 4 at $x = 0$.

(b) Use part (a) to approximate $\ln 1.08$. Round the answer to four decimal places.

[12.4]

Identify the infinite geometric series that converge. Give the sum of the convergent series.

10. $3 + 6 + 12 + 24 + \ldots$

11. $\frac{3}{4} + \frac{3}{16} + \frac{3}{64} + \frac{3}{256} + \cdots$

12. Let $a_n = \frac{1}{3n-1}$. Calculate the first five partial sums.

[12.5]

Find the Taylor series for the functions defined as follows. Give the interval of convergence for each series.

13. $f(x) = \frac{1}{1-2x}$

14. $f(x) = x^2 \ln\left(1 + \frac{x}{4}\right)$

15. Approximate the area of the region bounded by $f(x) = e^{-x^2}$, $x = 0$, $x = 1$, and the x-axis using five terms of the Taylor series. Round to four decimal places.

[12.6]

Use Newton's method to find a solution to the nearest hundredth for each equation in the given interval.

16. $3x^3 - 2x^2 - 5x + 4 = 0$; $[-2, -1]$

17. $3 \ln x - x + 1 = 0$; $[6, 7]$

Use Newton's method to approximate each radical to the nearest thousandth.

18. $\sqrt{17}$

19. $\sqrt[3]{35}$

20. What should you do if you have verified that a function $f(x)$ has a solution in $[a, b]$, but Newton's method is not producing a solution?

[12.7]

Use l'Hospital's rule, where applicable, to find each limit.

21. $\lim\limits_{x \to -4} \frac{x^3 + x^2 - 9x + 12}{x^2 - 16}$

22. $\lim\limits_{x \to 0} \frac{\ln(2 + x) - \ln 2}{x}$

23. $\lim\limits_{x \to 0} \frac{xe^{-3x}}{4e^{2x} - 4}$

24. $\lim\limits_{x \to 0} \frac{\sqrt{x^2 + 4x + 3}}{x}$

Chapter 12 Test Answers

1. $r = -2$; $a_5 = -\frac{16}{3}$; $a_n = \left(-\frac{1}{3}\right)(-2)^{n-1}$

2. Not geometric

3. -682

4. $\frac{363}{2}$

5. **(a)** \$6550 **(b)** 3684

6. \$126,825.42

7. \$3409.13

8. **(a)** \$441.21 **(b)** \$61,314 **(c)** \$26,686

9. **(a)** $x - \frac{x^2}{2} + \frac{x^3}{3} - \frac{x^4}{4}$ **(b)** 0.0770

10. Diverges, $r = 2$

11. Converges to 1

12. $S_1 = \frac{1}{2}$; $S_2 = \frac{7}{10}$; $S_3 = \frac{33}{40}$; $S_4 = \frac{403}{440}$; $S_5 = \frac{3041}{3080}$

13. $1 + 2x + 4x^2 + 8x^3 + \cdots + 2^n x^n + \cdots$; $\left(-\frac{1}{2}, \frac{1}{2}\right)$

14. $\frac{x^3}{4} - \frac{x^4}{2 \cdot 4^2} + \frac{x^5}{3 \cdot 4^3} - \frac{x^6}{4 \cdot 4^4} + \cdots + \frac{(-1)^n x^{n+3}}{(n+1) \cdot 4^{n+1}} + \cdots$; $(-4, 4]$

15. 0.747

16. -1.33

17. 6.71

18. 4.123

19. 3.27

20. Try a better initial guess.

21. $-\frac{31}{8}$

22. $\frac{1}{2}$

23. $\frac{1}{8}$

24. Does not exist.

Chapter 13

THE TRIGONOMETRIC FUNCTIONS

13.1 Definitions of the Trigonometric Functions

1. $60° = 60\left(\frac{\pi}{180}\right) = \frac{\pi}{3}$

3. $150° = 150\left(\frac{\pi}{180}\right) = \frac{5\pi}{6}$

5. $270° = 270\left(\frac{\pi}{180}\right) = \frac{3\pi}{2}$

7. $495° = 495\left(\frac{\pi}{180}\right) = \frac{11\pi}{4}$

9. $\frac{5\pi}{4} = \frac{5\pi}{4}\left(\frac{180°}{\pi}\right) = 225°$

11. $-\frac{13\pi}{6} = -\frac{13\pi}{6}\left(\frac{180°}{\pi}\right) = -390°$

13. $\frac{8\pi}{5} = \frac{8\pi}{5}\left(\frac{180°}{\pi}\right) = 288°$

15. $\frac{7\pi}{12} = \frac{7\pi}{12}\left(\frac{180°}{\pi}\right) = 105°$

17. Let $\alpha =$ the angle with terminal side through $(-3, 4)$. Then $x = -3,\ y = 4$, and

$$\begin{aligned} r &= \sqrt{x^2 + y^2} = \sqrt{(-3)^2 + (4)^2} \\ &= \sqrt{25} = 5. \end{aligned}$$

$$\sin\,\alpha = \frac{y}{r} = \frac{4}{5} \qquad \cot\,\alpha = \frac{x}{y} = -\frac{3}{4}$$

$$\cos\,\alpha = \frac{x}{r} = -\frac{3}{5} \qquad \sec\,\alpha = \frac{r}{x} = -\frac{5}{3}$$

$$\tan\,\alpha = \frac{y}{x} = -\frac{4}{3} \qquad \csc\,\alpha = \frac{r}{y} = \frac{5}{4}$$

19. Let $\alpha =$ the angle with terminal side through $(7, -24)$. Then $x = 7,\ y = -24$, and

$$\begin{aligned} r &= \sqrt{x^2 + y^2} = \sqrt{49 + 576} \\ &= \sqrt{625} = 25. \end{aligned}$$

$$\sin\alpha = \frac{y}{r} = -\frac{24}{25} \qquad \cot\alpha = \frac{x}{y} = -\frac{7}{24}$$

$$\cos\alpha = \frac{x}{r} = \frac{7}{25} \qquad \sec\alpha = \frac{r}{x} = \frac{25}{7}$$

$$\tan\alpha = \frac{y}{x} = -\frac{24}{7} \qquad \csc\alpha = \frac{r}{y} = -\frac{25}{24}$$

21. In quadrant I, all six trigonometric functions are positive, so their sign is +.

23. In quadrant III, $x < 0$ and $y < 0$. Furthermore, $r > 0$.

$\sin\,\theta = \frac{y}{r} < 0$, so the sign is $-$.

$\cos\,\theta = \frac{x}{r} < 0$, so the sign is $-$.

$\tan\,\theta = \frac{y}{x} > 0$, so the sign is $+$.

$\cot\,\theta = \frac{x}{y} > 0$, so the sign is $+$.

$\sec\,\theta = \frac{r}{x} < 0$, so the sign is $-$.

$\csc\,\theta = \frac{r}{y} < 0$, so the sign is $-$.

25. When an angle θ of 30° is drawn in standard position, one choice of a point on its terminal side is $(x, y) = (\sqrt{3}, 1)$. Then

$$r = \sqrt{x^2 + y^2} = \sqrt{3 + 1} = 2.$$

$$\tan\,\theta = \frac{y}{x} = \frac{1}{\sqrt{3}} = \frac{\sqrt{3}}{3}$$

$$\cot\,\theta = \frac{x}{y} = \sqrt{3}$$

$$\csc\,\theta = \frac{r}{y} = 2$$

27. When an angle θ of 60° is drawn in standard position, one choice of a point on its terminal side is $(x, y) = (1, \sqrt{3})$. Then

$$r = \sqrt{x^2 + y^2} = \sqrt{1 + 3} = 2.$$

$$\sin\,\theta = \frac{y}{r} = \frac{\sqrt{3}}{2}$$

$$\cot\,\theta = \frac{x}{y} = \frac{1}{\sqrt{3}} = \frac{\sqrt{3}}{3}$$

$$\csc\,\theta = \frac{r}{y} = \frac{2}{\sqrt{3}} = \frac{2\sqrt{3}}{3}$$

29. When an angle θ of 135° is drawn in standard position, one choice of a point on its terminal side is $(x, y) = (-1, 1)$. Then

$$r = \sqrt{x^2 + y^2} = \sqrt{1+1} = \sqrt{2}.$$

$$\tan\,\theta = \frac{y}{x} = -1$$

$$\cot\,\theta = \frac{x}{y} = -1$$

31. When an angle θ of 210° is drawn in standard position, one choice of a point on its terminal side is $(x, y) = (-\sqrt{3}, -1)$. Then

$$r = \sqrt{x^2 + y^2} = \sqrt{3+1} = 2.$$

$$\cos\,\theta = \frac{x}{r} = -\frac{\sqrt{3}}{2}$$

$$\sec\,\theta = \frac{r}{x} = \frac{2}{-\sqrt{3}} = -\frac{2\sqrt{3}}{3}$$

33. When an angle of $\frac{\pi}{3}$ is drawn in standard position, one choice of a point on its terminal side is $(x, y) = (1, \sqrt{3})$. Then

$$r = \sqrt{x^2 + y^2} = \sqrt{1+3} = 2.$$

$$\sin\,\frac{\pi}{3} = \frac{y}{r} = \frac{\sqrt{3}}{2}$$

35. When an angle of $\frac{\pi}{4}$ is drawn in standard position, one choice of a point on its terminal side is $(x, y) = (1, 1)$.

$$\tan\,\frac{\pi}{4} = \frac{y}{x} = 1$$

37. When an angle of $\frac{\pi}{6}$ is drawn in standard position, one choice of a point on its terminal side is $(x, y) = (\sqrt{3}, 1)$ Then

$$r = \sqrt{x^2 + y^2} = \sqrt{3+1} = 2.$$

$$\csc\,\frac{\pi}{6} = \frac{r}{y} = \frac{2}{1} = 2$$

39. When an angle of 3π is drawn in standard position, one choice of a point on its terminal side is $(x, y) = (-1, 0)$. Then

$$r = \sqrt{x^2 + y^2} = \sqrt{1} = 1.$$

$$\cos\,3\pi = \frac{x}{r} = -1$$

41. When an angle of $\frac{7\pi}{4}$ is drawn in standard position, one choice of a point on its terminal side is $(x, y) = (1, -1)$. Then

$$r = \sqrt{x^2 + y^2} = \sqrt{1+1} = \sqrt{2}.$$

$$\sin\,\frac{7\pi}{4} = \frac{y}{r} = \frac{-1}{\sqrt{2}} = -\frac{\sqrt{2}}{2}$$

43. When an angle of $\frac{5\pi}{4}$ is drawn in standard position, one choice of a point on its terminal side is $(x, y) = (-1, -1)$. Then

$$r = \sqrt{x^2 + y^2} = \sqrt{1+1} = \sqrt{2}.$$

$$\sec\,\frac{5\pi}{4} = \frac{r}{x} = \frac{\sqrt{2}}{-1} = -\sqrt{2}$$

45. When an angle of $-\frac{3\pi}{4}$ is drawn in standard position, one choice of a point on its terminal side is $(x, y) = (-1, -1)$. Then

$$\cot -\frac{3\pi}{4} = \frac{y}{x} = \frac{-1}{-1} = 1$$

47. When an angle of $-\frac{7\pi}{6}$ is drawn in standard position, one choice of a point on its terminal side is $(x, y) = (-\sqrt{3}, 1)$. Then

$$r = \sqrt{x^2 + y^2} = \sqrt{3+1} = 2.$$

$$\sin\left(-\frac{7\pi}{6}\right) = \frac{y}{r} = \frac{1}{2}$$

49. $\sin\,39° \approx 0.6293$

51. $\tan\,123° \approx -1.5399$

53. $\sin\,0.3638 \approx 0.3558$

55. $\cos\,1.2353 \approx 0.3292$

57. $f(x) = \cos(3x)$ is of the form $f(x) = a\cos(bx)$ where $a = 1$ and $b = 3$. Thus, $a = 1$ and $T = \frac{2\pi}{b} = \frac{2\pi}{3}$.

59. $s(x) = 3\sin(880\pi t - 7)$ is of the form $s(x) = a\sin(bt + c)$ where $a = 3, b = 880\pi$, and $c = -7$. Thus, $a = 3$ and $T = \frac{2\pi}{b} = \frac{2\pi}{880\pi} = \frac{1}{440}$.

61. The graph of $y = 2\cos x$ is similar to the graph of $y = \cos x$ except that it has twice the amplitude. (That is, its height is twice as great.)

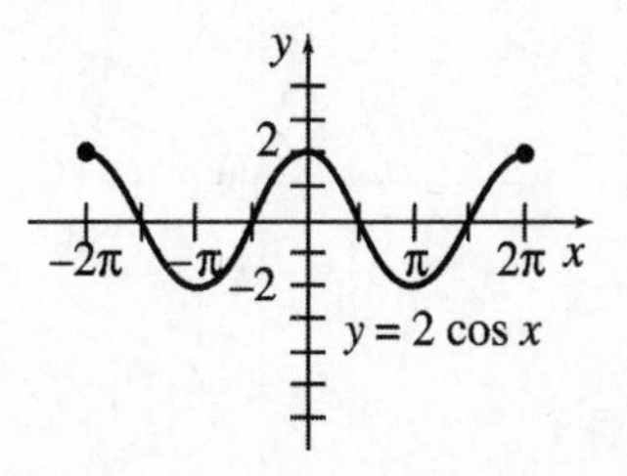

63. The graph of $y = -\frac{1}{2}\cos x$ is similar to the graph of $y = \cos x$ except that it has half the amplitude and is reflected about the x-axis.

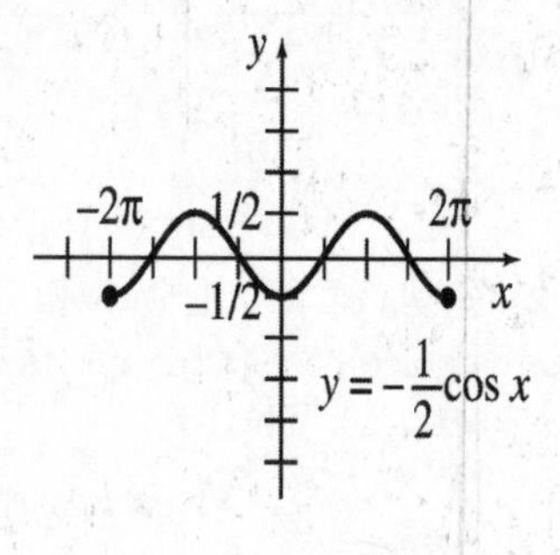

65. $y = 4\sin\left(\frac{1}{2}x + \pi\right) + 2$ has amplitude $a = 4$, period $T = \frac{2\pi}{b} = \frac{2\pi}{\frac{1}{2}} = 4\pi$, place shift $\frac{c}{b} = \frac{\pi}{\frac{1}{2}} = 2\pi$, and vertical shift $d = 2$. Thus, the graph of $y = 4\sin\left(\frac{1}{2}x + \pi\right) + 2$ is similar to the graph of $f(x) = \sin t$ except that it has 4 times the amplitude, twice the period, and is shifted up 2 units vertically. Also, $y = \sin\left(\frac{1}{2}x + \pi\right) + 2$ is shifted 2π units to the left relative to the graph of $g(x) = \sin\left(\frac{1}{2}x\right)$.

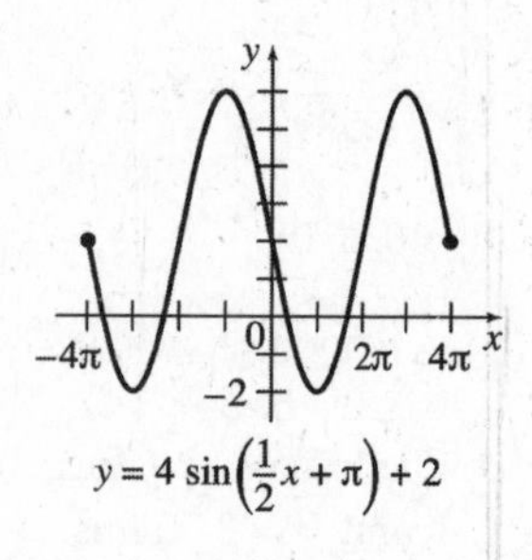

67. The graph of $y = -3\tan x$ is similar to the graph of $y = \tan x$ except that it is reflected about the x-axis and each ordinate value is three times larger in absolute value. Note that the points $\left(-\frac{\pi}{4}, 3\right)$ and $\left(\frac{\pi}{4}, -3\right)$ lie on the graph.

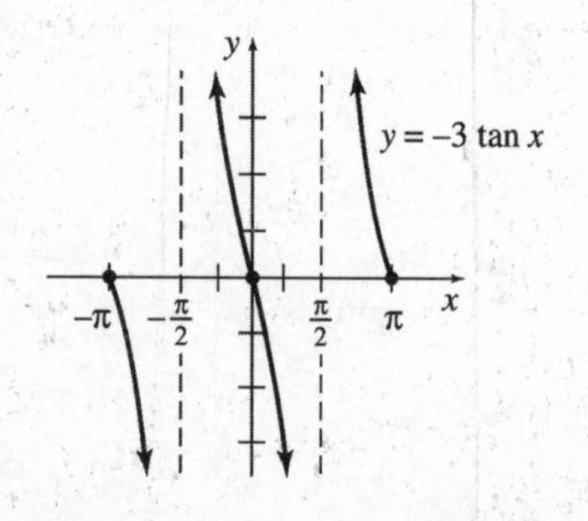

69. **(a)** Enter the data into a graphing calculator and plot.

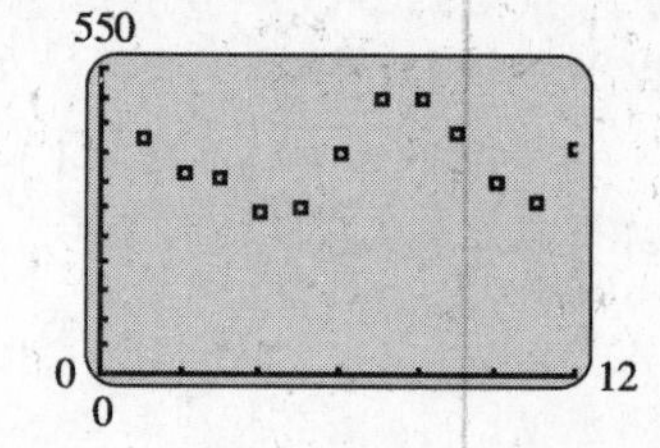

(b) Use the graphing calculator to find a trigonometric function.

```
SinReg
 y = a*sin(bx+c)+d
 a = 82.35275217
 b = .9048814207
 c = 1.138030788
 d = 394.3132805
```

$$c(x) = 82.353\sin(0.905x + 1.138) + 394.313$$

Then graph the function with the data.

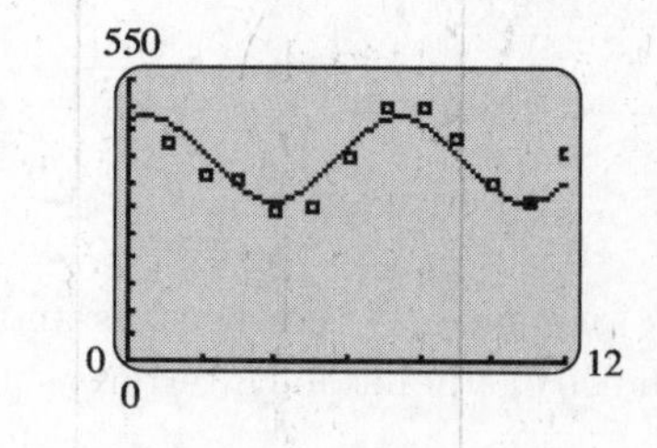

(c) $T = \dfrac{2\pi}{b} \approx 6.9437$

The period is about 6.9 months.

(d) $c(9) \approx 406$

The estimated consumption is about 406 trillion BTUs, which is less than the actual value.

71. **(a)** The period is $\dfrac{2\pi}{\left(\frac{\pi}{14.77}\right)} = 29.54$

There is a lunar cycle every 29.54 days.

(b) $y = 100 + 1.8\cos\left[\frac{(x-6)\pi}{14.77}\right]$ reaches a maximum value when $\cos\left[\frac{(x-6)\pi}{14.77}\right] = 1$ which occurs when

$$x - 6 = 0$$
$$x = 6$$

Six days from April 2, 2007, is April 8, 2007.

$$y = 100 + 1.8\cos\left[\frac{(6-6)\pi}{14.77}\right] = 101.8$$

There is a percent increase of 1.8 percent.

(c) On April 19, $x = 11$.

$$y = 100 + 1.8\cos\left[\frac{(17-6)\pi}{14.77}\right] \approx 98.75$$

The formula predicts that the number of consulations was 98.75% of the daily mean.

73.

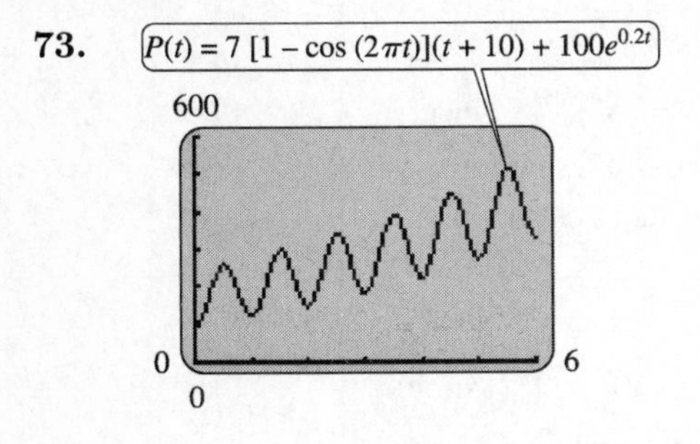

75. Solving $\dfrac{c_1}{c_2} = \dfrac{\sin\theta_1}{\sin\theta_2}$ for c_2 gives

$$c_2 = \frac{c_1 \sin\theta_2}{\sin\theta_1}.$$

$c_1 = 3 \cdot 10^8$, $\theta_1 = 46°$, and $\theta_2 = 31°$ so

$$\begin{aligned} c_2 &= \frac{3 \cdot 10^8(\sin 31°)}{\sin 46°} \\ &= 214{,}796{,}150 \\ &\approx 2.1 \times 10^8 \text{ m/sec.} \end{aligned}$$

77. On the horizontal scale, one whole period clearly spans four squares, so $4 \cdot 30° = 120°$ is the period.

79. $T(x) = 60 - 30\cos\left(\dfrac{x}{2}\right)$

(a) $x = 1$ represents February, so the maximum afternoon temperature in February is

$$T(0) = 60 - 30\cos\frac{1}{2} \approx 34°\text{F}.$$

(b) $x = 3$ represents April, so the maximum afternoon temperature in April is

$$T(3) = 60 - 30\cos\frac{3}{2} \approx 58°\text{F}.$$

(c) $x = 8$ represents September, so the maximum afternoon temperature in September is

$$T(8) = 60 - 30\cos 4 \approx 80°\text{F}.$$

(d) $x = 6$ represents July, so the maximum afternoon temperature in July is

$$T(6) = 60 - 30\cos 3 \approx 90°\text{F}.$$

(e) $x = 11$ represents December, so the maximum afternoon temperature in December is

$$T(11) = 60 - 30\cos\frac{11}{2} \approx 39°\text{F}.$$

81. (a)

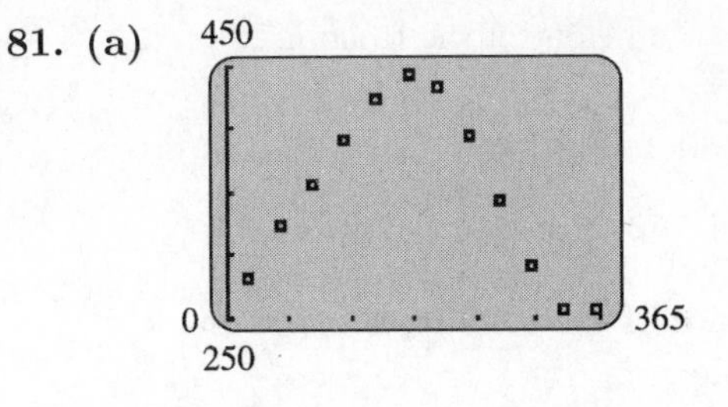

Yes; because of the cyclical nature of the days of the year, it is reasonable to assume that the times of the sunset are periodic.

(b) The function $s(x)$, derived by a $TI - 83$ using the sine regression function under the STAT–CALC menu, is given by

$$s(x) = 94.0872\sin(0.0166x - 1.2213) + 347.4158.$$

(c)

$$\begin{aligned} s(60) &= 94.0872\sin[0.0166(60) - 1.2213] \\ &\quad + 347.4158 \\ &= 326 \text{ minutes} \\ &= \text{5:26 P.M.} \\ s(120) &= 94.0872\sin[0.0166(120) - 1.2213] \\ &\quad + 347.4158 \\ &= 413 \text{ minutes} + 60 \text{ minutes} \\ &\quad \text{(daylight savings)} \\ &= \text{7:53 P.M.} \\ s(240) &= 94.0872\sin[0.0166(240) - 1.2213] \\ &\quad + 347.4158 \\ &= 382 \text{ minutes} + 60 \text{ minutes} \\ &\quad \text{(daylight savings)} \\ &= \text{7:22 P.M.} \end{aligned}$$

(d) The following graph shows $s(x)$ and $y = 360$ (corresponding to a sunset at 6:00 P.M.). These graphs first intersect on day 82. However because of daylight savings time, to find the second value we find where the graphs of $s(x)$ and $y = 360 - 60 = 300$ intersect. These graphs intersect on day 295. Thus, the sun sets at approximately 6:00 P.M. on the 82nd and 295th days of the year.

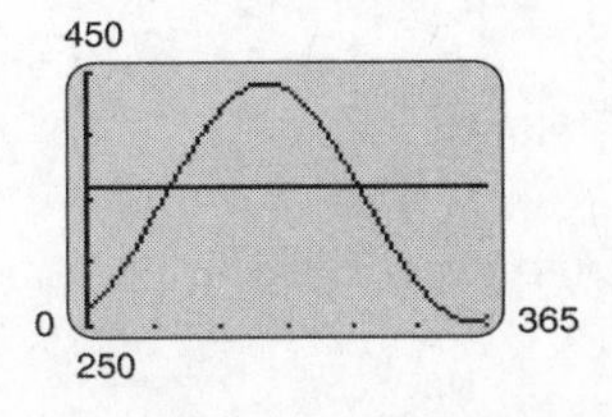

83. Let h = the height of the building.

$$\tan 42.8^\circ = \frac{h}{65}$$
$$h = 65 \tan 42.8^\circ \approx 60.2$$

The height of the building is approximately 60.2 meters.

85. Let θ = the average angle with the horizontal.

$$\tan \theta = \frac{26}{5280}$$

Using the TAN^{-1} key on the calculator,

$$\theta = TAN^{-1}\left(\frac{26}{5280}\right) \approx 0.28^\circ.$$

87. We need to find the values of t for which

$$3.5 \le h(t) \le 4.$$

The following graphs show where $h(t) = \sin\left(\frac{t}{\pi} - 2\right) + 4$ intersects the horizontal lines $y = 3.5$ and $y = 4$.

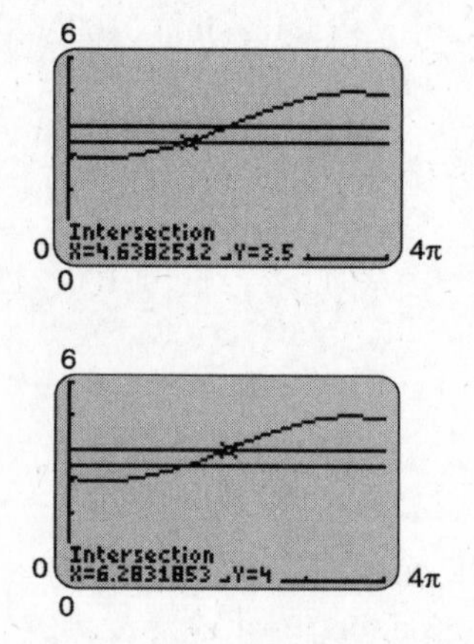

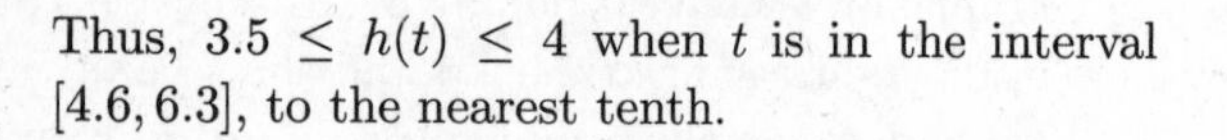

Thus, $3.5 \le h(t) \le 4$ when t is in the interval $[4.6, 6.3]$, to the nearest tenth.

13.2 Derivatives of Trigonometric Functions

1. $y = \frac{1}{2} \sin 8x$

$$\frac{dy}{dx} = \frac{1}{2}(\cos 8x) \cdot D_x(8x)$$
$$= \frac{1}{2}(\cos 8x) \cdot 8$$
$$= 4 \cos 8x$$

3. $y = 12 \tan(9x + 1)$

$$\frac{dy}{dx} = [12 \sec^2(9x+1) \cdot D_x(9x+1)$$
$$= [12 \sec(9x+1)] \cdot 9$$
$$= 108 \sec^2(9x+1)$$

5. $y = \cos^4 x$

$$\frac{dy}{dx} = [4(\cos x)^3] D_x(\cos x)$$
$$= (4 \cos^3 x)(-\sin x)$$
$$= -4 \sin x \cos^3 x$$

7. $y = \tan^8 x$

$$\frac{dy}{dx} = 8(\tan x)^7 \cdot D_x(\tan x)$$
$$= 8 \tan^7 x \sec^2 x$$

9. $y = -6x \cdot \sin 2x$

$$\frac{dy}{dx} = -6x \cdot D_x(\sin 2x) + \sin 2x \cdot D_x(-6x)$$
$$= -6x(\cos 2x) \cdot D_x(2x) + (\sin 2x)(-6)$$
$$= -6x(\cos 2x) \cdot 2 - 6 \sin 2x$$
$$= -6(2x \cos 2x + \sin 2x)$$

11. $y = \frac{\csc x}{x}$

$$\frac{dy}{dx} = \frac{x \cdot D_x(\csc x) - (\csc x) \cdot D_x x}{x^2}$$
$$= \frac{-x \csc x \cot x - \csc x}{x^2}$$
$$= \frac{-(x \csc x \cot x + \csc x)}{x^2}$$

13. $y = \sin e^{4x}$

$$\frac{dy}{dx} = \cos e^{4x} \cdot D_x(e^{4x})$$
$$= (\cos e^{4x}) \cdot e^{4x} \cdot D_x(4x)$$
$$= (\cos e^{4x}) \cdot e^{4x} \cdot 4$$
$$= 4e^{4x} \cos e^{4x}$$

15. $y = e^{\cos x}$

$$\frac{dy}{dx} = e^{\cos x} \cdot D_x(\cos x)$$
$$= e^{\cos x} \cdot (-\sin x)$$
$$= (-\sin x)e^{\cos x}$$

17. $y = \sin(\ln 3x^4)$

$$\frac{dy}{dx} = [\cos(\ln 3x^4)] \cdot D_x(\ln 3x^4) = \cos(\ln 3x^4) \cdot \frac{D_x(3x^4)}{3x^4}$$

$$= \cos(\ln 3x^4)\frac{12x^3}{3x^4} = \cos(\ln 3x^4) \cdot \frac{4}{x} = \frac{4}{x}\cos(\ln 3x^4)$$

19. $y = \ln|\sin x^2|$

$$\frac{dy}{dx} = \frac{D_x(\sin x^2)}{\sin x^2} = \frac{(\cos x^2) \cdot D_x(x^2)}{\sin x^2} = \frac{(\cos x^2) \cdot 2x}{\sin x^2} = \frac{2x \cos x^2}{\sin x^2}$$

or $2x \cot x^2$

21. $y = \dfrac{2 \sin x}{3 - 2 \sin x}$

$$\frac{dy}{dx} = \frac{(3-2\sin x)D_x(2\sin x) - (2\sin x) \cdot D_x(3 - 2\sin x)}{(3 - 2\sin x)^2}$$

$2\cos x \cdot (3 - 2\sin x) - 2\sin x \cdot -2\cos x$

$$= \frac{(3 - 2\sin x) \cdot 2D_x(\sin x) - (2\sin x) \cdot [-2D_x(\sin x)]}{(3 - 2\sin x)^2}$$

$$= \frac{6\cos x - 4\sin x\cos x + 4\sin x\cos x}{(3 - 2\sin x)^2} = \frac{6\cos x}{(3 - 2\sin x)^2}$$

23. $y = \sqrt{\dfrac{\sin x}{\sin 3x}} = \left(\dfrac{\sin x}{\sin 3x}\right)^{1/2}$

$$\frac{dy}{dx} = \frac{1}{2}\left(\frac{\sin x}{\sin 3x}\right)^{1/2} \cdot D_x\left(\frac{\sin x}{\sin 3x}\right) = \frac{1}{2}\left(\frac{\sin 3x}{\sin x}\right)^{-1/2} \cdot \left[\frac{(\sin 3x) \cdot D_x(\sin x) - (\sin x) \cdot D_x(\sin 3x)}{(\sin 3x)^2}\right]$$

$$= \frac{1}{2}\left(\frac{\sin 3x}{\sin x}\right)^{-1/2} \cdot \left[\frac{(\sin 3x)(\cos x) - (\sin x)(\cos 3x) \cdot D_x(3x)}{\sin^2 3x}\right] = \frac{1(\sin 3x)^{1/2}}{2(\sin x)^{1/2}} \cdot \frac{\sin 3x \cos x - 3\sin x\cos 3x}{\sin^2 3x}$$

$$= \frac{(\sin 3x)^{1/2}(\sin 3x\cos x - 3\sin x\cos 3x)}{2(\sin x)^{1/2}(\sin^2 3x)} = \frac{\sqrt{\sin 3x}\,[\sin 3x\cos x - 3\sin x\cos 3x]}{2\sqrt{\sin x}\,(\sin^2 3x)}$$

25. $y = 3\tan\left(\dfrac{1}{4}x\right) + 4\cot 2x - 5\csc x + e^{-2x}$

$$\frac{dy}{dx} = 3\tan\left(\frac{1}{4}x\right) + 4\cot 2x - 5\csc x + e^{-2x}$$

$$= 3\sec^2\left(\frac{1}{4}x\right) \cdot D_x\left(\frac{1}{4}x\right) + 4(-\csc^2 2x) \cdot D_x(2x) - 5(-\csc x\cot x) + e^{-2x} \cdot D_x(-2x)$$

$$= 3\sec^2\left(\frac{1}{4}x\right) \cdot \frac{1}{4} - (4\csc^2 2x) \cdot 2 + 5\csc x\cot x + e^{-2x} \cdot (-2)$$

$$= \frac{3}{4}\sec^2\left(\frac{1}{4}x\right) - 8\csc^2 2x + 5\csc x\cot x - 2e^{-2x}$$

27. $y = \sin x; x = 0$

Let $f(x) = \sin x$.
Then $f'(x) = \cos x$, so

$$f'(0) = \cos 0 = 1.$$

The slope of the tangent line to the graph of $y = \sin x$ at $x = 0$ is 1.

29. $y = \cos x;\, x = -\dfrac{5\pi}{6}$

Let $f(x) = \cos x$.
Then $f'(x) = -\sin x$, so

$$f'\left(-\frac{5\pi}{6}\right) = -\sin\left(-\frac{5\pi}{6}\right) = \frac{1}{2}.$$

The slope of the tangent line to the graph of $y = \cos x$ at $x = -\dfrac{5\pi}{6}$ is $\dfrac{1}{2}$.

31. $y = \tan x;\, x = 0$

Let $f(x) = \tan x$.
Then $f'(x) = \sec^2 x$, so

$$f'(0) = \sec^2 0 = \frac{1}{\cos^2 0} = 1.$$

The slope of the tangent line to the graph of $y = \tan x$ at $x = 0$ is 1.

33. Since $\cot x = \frac{\cos x}{\sin x}$, by using the quotient rule,

$$\begin{aligned}
D_x(\cos x) &= D_x\left(\frac{\cos x}{\sin x}\right) \\
&= \frac{(\sin x)(-\sin x) - (\cos x)(\cos x)}{\sin^2 x} \\
&= \frac{-\sin^2 x - \cos^2 x}{\sin^2 x} \\
&= -\frac{\sin^2 x + \cos^2 x}{\sin^2 x} \\
&= -\frac{1}{\sin^2 x} \\
&= -\csc^2 x.
\end{aligned}$$

35. Since $\csc x = \frac{1}{\sin x} = (\sin x)^{-1}$,

$$\begin{aligned}
D_x(\csc x) &= D_x\left(\frac{1}{\sin x}\right) = D_x(\sin x)^{-1} \\
&= -1(\sin x)^{-2}\cos x \\
&= -\frac{\cos x}{\sin^2 x} \\
&= -\frac{1}{\sin x}\cdot\frac{\cos x}{\sin x} \\
&= -\csc x \cot x.
\end{aligned}$$

37. $R(x) = 120\cos 2\pi x + 150$

(a) $R'(x) = 120 \cdot (-\sin 2\pi x)(2\pi) + 0$
$= -240\pi \sin 2\pi x$

(b) Replace x with $\dfrac{1}{12}$ (for $\dfrac{1}{12}$ of a year).

$$\begin{aligned}
R'\left(\frac{1}{12}\right) &= -240\pi \sin 2\pi\left(\frac{1}{12}\right) \\
&= -240\pi \sin\frac{\pi}{6} \\
&= -240\pi\left(\frac{1}{2}\right) \\
&= -120\pi
\end{aligned}$$

$R'(x)$ for August 1 is $-\$120\pi$ per year.

(c) January 1 is 6 months, or $\frac{6}{12} = \frac{1}{2}$ of a year from July 1. Replace x with $\frac{1}{2}$.

$$\begin{aligned}
R'\left(\frac{1}{2}\right) &= -240\pi \sin 2\pi\left(\frac{1}{2}\right) \\
&= -240\pi \sin \pi \\
&= -240\pi(0) \\
&= 0
\end{aligned}$$

$R'(x)$ for January 1 is \$0 per year.

(d) June 1, is $\frac{11}{12}$ of a year from July 1. Replace x with $\frac{11}{12}$.

$$\begin{aligned}
R'\left(\frac{11}{12}\right) &= -240\pi \sin 2\pi\left(\frac{11}{12}\right) \\
&= -240\pi \sin\frac{11\pi}{6} \\
&= -240\pi\left(-\frac{1}{2}\right) \\
&= 120\pi
\end{aligned}$$

$R'(x)$ for June 1 is $\$120\pi$ per year.

39. **(a)** The following is a table of values for

$$y = \frac{1}{5}\sin \pi(t-1).$$

t	0	0.5	1.0	1.5	2.0	2.5	2.0
y	0	−0.2	0	0.2	0	−0.2	0

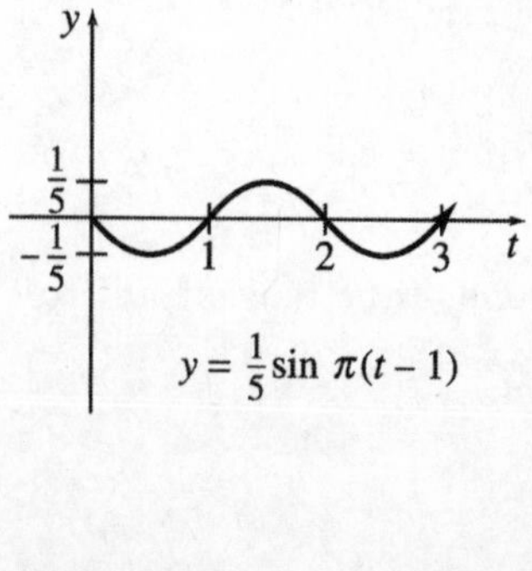

$y = \frac{1}{5}\sin \pi(t-1)$

(b) $v(t) = \frac{dy}{dt} = \left[\frac{1}{5} \cos \pi(t-1)\right] \cdot D_x [\pi(t-1)] = \frac{\pi}{5} \cos \pi(t-1)$

$a(t) = v'(t) = \frac{d^2y}{dt^2} = \left[-\frac{\pi}{5} \sin \pi(t-1)\right] \cdot D_x [\pi(t-1)] = -\frac{\pi^2}{5} \sin \pi(t-1)$

(c) $\frac{d^2y}{dt^2} + \pi^2 y = -\frac{\pi^2}{5} \sin [\pi(t-1)] + (\pi^2)\frac{1}{5} \sin \pi(t-1) = 0$

(d) Since the constant of proportionality is positive, the force and acceleration are in the same direction.

At $t = 1.5$ sec,

$$a(t) = -\frac{\pi^2}{5} \sin \pi(t-1)$$

$$a(1.5) = -\frac{\pi^2}{5} \sin \pi(0.5) = -\frac{\pi^2}{5} \sin \frac{\pi}{2} = -\frac{\pi^2}{5} \cdot 1 = -\frac{\pi^2}{5} < 0,$$

and

$$y = \frac{1}{5} \sin \pi(t-1) = \frac{1}{5} \sin \pi(0.5) = \frac{1}{5} \sin \frac{\pi}{2} = \frac{1}{5} \cdot 1 = \frac{1}{5}.$$

Thus, at $t = 1.5$, acceleration is negative, the arm is moving clockwise, and the arm is at an angle of $\frac{1}{5}$ radian from vertical.

At $t = 2.5$ sec,

$$a(2.5) = -\frac{\pi^2}{5} \sin \pi(1.5) = -\frac{\pi^2}{5} \sin \frac{3\pi}{2} = -\frac{\pi^2}{5}(-1) = \frac{\pi^2}{5} > 0,$$

and

$$y = \frac{1}{5} \sin \frac{3\pi}{2} = \frac{1}{5}(-1) = -\frac{1}{5}.$$

Thus, at $t = 2.5$, acceleration is positive, the arm is moving counterclockwise, and the arm is at angle of $-\frac{1}{5}$ radian from vertical.

At $t = 3.5$ sec,

$$a(3.5) = -\frac{\pi^2}{5} \sin \pi(2.5) = -\frac{\pi^2}{5} \sin \frac{5\pi}{2} = -\frac{\pi^2}{5} \cdot 1 = -\frac{\pi^2}{5} < 0,$$

and

$$y = \frac{1}{5} \sin \frac{5\pi}{2} = \frac{1}{5} \cdot 1 = \frac{1}{5}.$$

Thus, at $t = 3.5$, acceleration is negative, the arm is moving clockwise, and the arm is at an angle of $\frac{1}{5}$ radian from vertical.

41. (a) $f(t) = 1000e^{2\sin(t)}$
$f(0.2) = 1000e^{2\sin(0.2)} \approx 1488$

(b) $f(t) = 1000e^{2\sin(t)}$
$f(1) = 1000e^{2\sin(1)} \approx 5381$

(c) Since $f'(t) = 2000\cos(t)e^{2\sin(t)}$, $f'(0)$ is given by

$$f'(0) = 2000\cos(0)e^{2\sin(0)} = 2000.$$

(d) Since $f'(t) = 2000\cos(t)e^{2\sin(t)}$, $f'(0.2)$ is given by

$$f'(0.2) = 2000\cos(0.2)e^{2\sin(0.2)} \approx 2916.$$

(e)

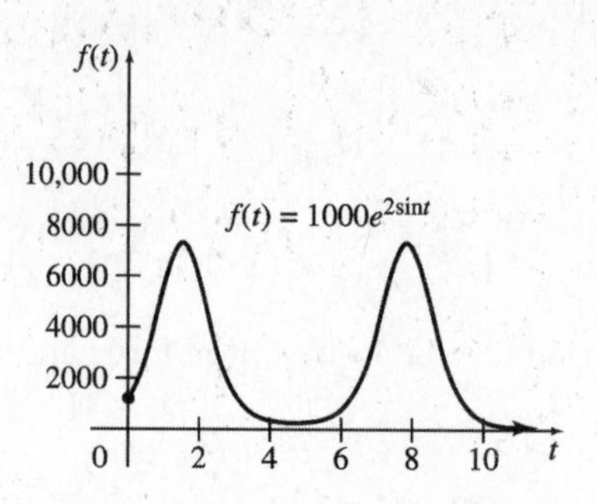

(f) Set $f'(t) = 2000\cos(t)e^{2\sin(t)}$ equal to zero to find critical values.

$$2000\cos(t)e^{2\sin(t)} = 0$$
$$\cos(t)e^{2\sin(t)} = 0$$

$\cos(t) = 0$ or $e^{2\sin(t)} = 0$

$t = \frac{\pi}{2} + n\pi$ $\quad$ $e^{2\sin(t)} \neq 0,$

for n any integer $\quad$ for all t

We will use the second derivative test.

$$f''(t) = -2000\sin(t)e^{2\sin(t)} + 4000\cos^2(t)e^{2\sin(t)}$$
$$= 2000e^{2\sin t}(-\sin t + 2\cos^2 t)$$

Since $2000\mathrm{e}^{2\sin t} > 0$ for all t, $f''(t) > 0$ when $-\sin t + 2\cos^2 t > 0$, when $\sin t < 2\cos^2 t$. Also, $f''(t) < 0$ when $-\sin t + 2\cos^2 t < 0$, when $\sin t > 2\cos^2 t$. Since $\sin(\frac{\pi}{2} + n\pi) < 0 = 2\cos^2(\frac{\pi}{2} + n\pi)$ for n any odd integer, and $\sin(\frac{\pi}{2} + n\pi) > 0 = 2\cos^2(\frac{\pi}{2} + n\pi)$ for n any even integer, f has a maximum at $t = \frac{\pi}{2} + n\pi$ for n any even integer and f has a minimum at $t = \frac{\pi}{2} + n\pi$ for n any odd integer. This is equivalent to f having a maximum for $t = \frac{\pi}{2} + 2\pi n$ for n any integer and f having a minimum for $t = \frac{3\pi}{2} + 2\pi n$ for n any integer. To find the maximum and minimum values we will find $f(\frac{\pi}{2})$ and $f(\frac{3\pi}{2})$ respectively.

$$f\left(\tfrac{\pi}{2}\right) = 1000e^{2\sin(\frac{\pi}{2})} \approx 7390$$
$$\left(\tfrac{3\pi}{2}\right) = 1000e^{2\sin(\frac{3\pi}{2})} \approx 135$$

43. **(a)** Using the graphing calculator to graph $P(t) = 0.003\sin(220\pi t) + \frac{0.003}{3}\sin(660\pi t) + \frac{0.003}{5}\sin(1100\pi t) + \frac{0.003}{7}\sin(1540\pi t)$ in a $[0, 0.01]$ by $[-0.004, 0.004]$ viewing window, gives the following graph.

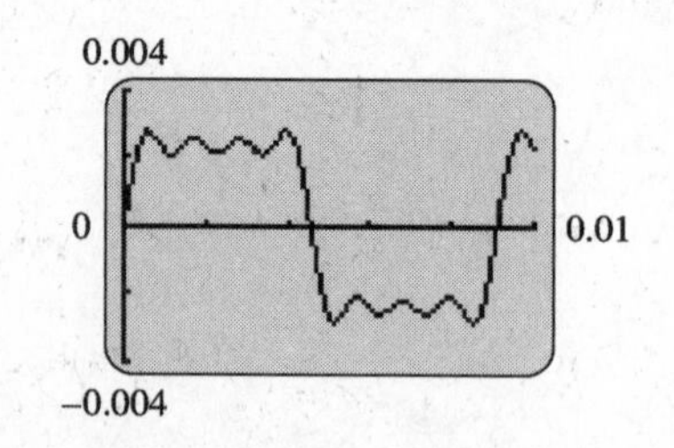

(b) Using the graphing calculator to find $P'(x)$ and evaluate $P'(0.002)$ gives .151. However, the slope of the graph of $P(x)$ at $x = 0.002$ appears to be approximately -1. Thus, we calculate $P'(x)$ by hand and use the calculator to evaluate $P'(0.002)$. Since $P'(t) = 0.66\pi\cos(220\pi t) + 0.66\pi\cos(660\pi t) + 0.66\pi\cos(1100\pi t) + 0.66\pi\cos(1540\pi t)$, $P'(0.002)$ is given by $P'(0.002) = 0.66\pi\cos[220\pi(0.002)] + 0.66\pi\cos[660\pi(0.002)] + 0.66\pi\cos[1100\pi(0.002)] + 0.66\pi\cos[1540\pi(0.002)] \approx -1.05$ pounds per square foot per second.
At 0.002 seconds the pressure is decreasing at a rate of about 1.05 pounds per square foot per second.

45. **(a)** $y = x\tan\alpha - \frac{16x^2}{V^2}\sec^2\alpha = 40\tan\left(\frac{\pi}{4}\right) - \frac{16(40)^2}{44^2}\sec^2\left(\frac{\pi}{4}\right) = 40(1) - \frac{16(40)^2}{44^2}(2) \approx 13.55$ feet

(b)
$$y = x\tan\alpha - \frac{16x^2}{V^2}\sec^2\alpha$$
$$0 = x\tan\alpha - \frac{16x^2}{V^2}\sec^2\alpha$$
$$0 = x\left(\tan\alpha - \frac{16x}{V^2}\sec^2\alpha\right)$$
$$0 = \tan\alpha - \frac{16x}{V^2}\sec^2\alpha \quad (\text{for } x \neq 0)$$
$$\frac{16x}{V^2}\sec^2\alpha = \tan\alpha$$
$$x = \frac{V^2}{16}\cdot\frac{\tan\alpha}{\sec^2\alpha} = \frac{V^2}{16}\cdot\sin\alpha\cos\alpha = \frac{V^2}{32}\cdot 2\sin\alpha\cos\alpha = \frac{V^2}{32}\sin(2\alpha)$$

(c) $x = \frac{V^2}{32}\sin(2\alpha) = \frac{44^2}{32}\sin\left[2\left(\frac{\pi}{3}\right)\right] \approx 52.39$ feet

(d) $\frac{dx}{d\alpha} = \frac{V^2}{32}\cos(2\alpha)\cdot D_\alpha(2\alpha) = \frac{V^2}{16}\cos(2\alpha)$

Find critical values.
$$\frac{V^2}{16}\cos(2\alpha) = 0$$
$$\cos(2\alpha) = 0$$
$$2\alpha = \frac{\pi}{2} + n\pi, \text{ for } n \text{ any integer}$$
$$\alpha = \frac{\pi}{4} + \frac{n\pi}{2}, \text{ for } n \text{ any integer}$$

Since $0 < \alpha < \frac{\pi}{2}$, $\frac{dx}{d\alpha} = 0$ for $\alpha = \frac{\pi}{4}$. Furthermore, $\frac{d^2x}{d\alpha^2} = -\frac{V^2}{8}\sin(2\alpha)$ which is less than zero at $\alpha = \frac{\pi}{4}$.

Therefore, x is maximized when $\alpha = \frac{\pi}{4}$.

(e) Since 60 miles per hour is 88 feet per second, evaluate
$$x = \frac{V^2}{32}\sin(2\alpha) \text{ when } V = 88 \text{ and } \alpha = \frac{\pi}{4}.$$
$$= \frac{88^2}{32}\sin\left[2\left(\frac{\pi}{4}\right)\right] = 242 \text{ feet}$$

47. $s(t) = \sin t + 2\cos t$
$v(t) = s'(t) = \cos t - 2\sin t$

(a) $v(0) = 1 - 2(0) = 1$

(b) $v\left(\frac{\pi}{4}\right) = \frac{\sqrt{2}}{2} - 2\left(\frac{\sqrt{2}}{2}\right) = -\frac{\sqrt{2}}{2} \approx -0.7071$

(c) $v\left(\frac{3\pi}{2}\right) = 0 - 2(-1) = 2$

$a(t) = v'(t) = -\sin t - 2\cos t$

(d) $a(0) = 0 - 2(1) = -2$

(e) $a\left(\frac{\pi}{4}\right) = -\frac{\sqrt{2}}{2} - 2\left(\frac{\sqrt{2}}{2}\right) = -\frac{3\sqrt{2}}{2} \approx -2.1213$

(f) $a\left(\frac{3\pi}{2}\right) = -(-1) - 2(0) = 1$

49. From the figure, we see that

$$\tan\theta = \frac{x}{60}$$
$$60\tan\theta = x.$$

Differentiating with respect to time, t, gives

$$60\sec^2\theta \cdot \frac{d\theta}{dt} = \frac{dx}{dt}$$
$$\frac{d\theta}{dt} = \frac{\frac{dx}{dt}}{60\sec^2\theta}.$$

Since

$$\frac{dx}{dt} = 600,$$
$$\frac{d\theta}{dt} = \frac{600}{60\sec^2\theta} = \frac{10}{\sec^2\theta}.$$

(a) When the car is at the point on the road closest to the camera $\theta = 0$. Thus,

$$\frac{d\theta}{dt} = \frac{10}{(\sec 0)^2} = \frac{10}{1^2} = 10\,\text{radians/min}$$
$$\frac{d\theta}{dt} = \frac{10\text{ radians}}{\text{min}} \cdot \frac{1\text{ rev}}{2\pi\text{ radians}} = \frac{5}{\pi}\text{ rev/min}$$

(b) Six seconds later is $\frac{1}{10}$ of a minute later, and the car has traveled 60 feet. Thus,

$$\tan\theta = \frac{60}{60} \text{ and } \theta = \frac{\pi}{4}.$$
$$\frac{d\theta}{dt} = \frac{10}{\left(\sec\frac{\pi}{4}\right)^2} = \frac{10}{(\sqrt{2})^2} = 5\,\text{radians/min}.$$

This is one-half the previous value, so

$$\frac{d\theta}{dt} = \frac{1}{2}\cdot\frac{5}{\pi}\text{ rev/min} = \frac{5}{2\pi}\text{ rev/min}.$$

51. Let x represent the length of the ladder.

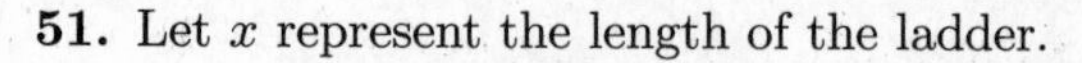

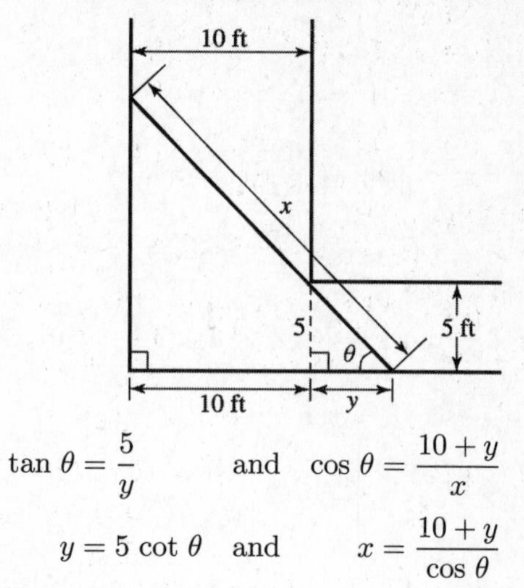

$$\tan\theta = \frac{5}{y} \quad \text{and} \quad \cos\theta = \frac{10+y}{x}$$
$$y = 5\cot\theta \quad \text{and} \quad x = \frac{10+y}{\cos\theta}$$

Thus,

$$x = \frac{10 + 5\cot\theta}{\cos\theta} = 10\sec\theta + 5\csc\theta$$
$$\frac{dx}{d\theta} = 10\sec\theta\tan\theta - 5\csc\theta\cot\theta$$
$$= 10\left(\frac{1}{\cos\theta}\right)\left(\frac{\sin\theta}{\cos\theta}\right) - 5\left(\frac{1}{\sin\theta}\right)\left(\frac{\cos\theta}{\sin\theta}\right)$$
$$= \frac{10\sin\theta}{\cos^2\theta} - \frac{5\cos\theta}{\sin^2\theta}$$

If $\frac{dx}{d\theta} = 0$, then

$$\frac{10\sin\theta}{\cos^2\theta} = \frac{5\cos\theta}{\sin^2\theta}$$
$$10\sin^3\theta = 5\cos^3\theta$$
$$\frac{\sin^3\theta}{\cos^3\theta} = \frac{1}{2}$$
$$\tan^3\theta = \frac{1}{2}$$
$$\tan\theta = \sqrt[3]{\frac{1}{2}}$$
$$\theta \approx 0.6709\text{ radian.}$$

If $\theta < 0.6709$ radian, $\frac{dx}{d\theta} < 0$.

If $\theta > 0.6709$ radian, $\frac{dx}{d\theta} > 0$.

Thus, x will be minimum when $\theta = 0.6709$.
If $\theta = 0.6709$, then

$$x = \frac{10 + 5\cot 0.6709}{\cos 0.6709} \approx 20.81.$$

The length of the longest possible ladder is approximately 20.81 feet.

13.3 Integrals of Trigonometric Functions

1. $\int \cos 3x\,dx$

Let $u = 3x$, so $du = 3\,dx$ or $\frac{1}{3}\,du = dx$.

$$\begin{aligned}\int \cos 3x\,dx &= \int \cos u \cdot \frac{1}{3}\,du\\ &= \frac{1}{3}\int \cos u\,du\\ &= \frac{1}{3}\sin u + C\\ &= \frac{1}{3}\sin 3x + C\end{aligned}$$

3. $\int (3\cos x - 4\sin x)\,dx$

$$\begin{aligned}&= \int 3\cos x\,dx - \int 4\sin x\,dx\\ &= 3\int \cos x\,dx - 4\int \sin x\,dx\\ &= 3\sin x + 4\cos x + C\end{aligned}$$

5. $\int x \sin x^2\,dx$

Let $u = x^2$.

Then $du = 2x\,dx$

$\frac{1}{2}\,du = x\,dx.$

$$\begin{aligned}\int x\sin x^2\,dx &= \int \sin u \cdot \frac{1}{2}\,du\\ &= \frac{1}{2}\int \sin u\,du\\ &= -\frac{1}{2}\cos u + C\\ &= -\frac{1}{2}\cos x^2 + C\end{aligned}$$

7. $-\int 3\sec^2 3x\,dx$

Let $u = 3x$, so $du = 3\,dx$ or $\frac{1}{3}\,du = dx$.

$$\begin{aligned}-\int 3\sec^2 3x\,dx &= -\int 3\sec^2 u \cdot \frac{1}{3}\,du\\ &= -\int \sec^2 u\,du\\ &= -\tan u + C\\ &= -\tan 3x + C\end{aligned}$$

9. $\int \sin^7 x \cos x\,dx$

Let $u = \sin x$.

Then $du = \cos x\,dx$.

$$\begin{aligned}\int \sin^7 x(\cos x)\,dx &= \int u^7\,du\\ &= \frac{1}{8}u^8 + C\\ &= \frac{1}{8}\sin^8 x + C\end{aligned}$$

11. $\int 3\sqrt{\cos x}\,(\sin x)\,dx$

Let $u = \cos x$, so $du = -\sin x\,dx$ or $-du = \sin x\,dx$.

$$\begin{aligned}&\int 3\sqrt{\cos x}\,(\sin x)\,dx\\ &= \int 3\sqrt{u}\,(-du)\\ &= -3\int u^{1/2}\,du\\ &= -3\left(\frac{2}{3}u^{3/2}\right) + C\\ &= -2u^{3/2} + C\\ &= -2(\cos x)^{3/2} + C\end{aligned}$$

13. $\int \frac{\sin x}{1+\cos x}\,dx$

Let $u = 1 + \cos x$.

Then $du = -\sin x$

$-du = \sin x\,dx.$

$$\begin{aligned}\int \frac{\sin x}{1+\cos x}\,dx &= \int \frac{1}{u}(-du) = -\int \frac{1}{u}\,du\\ &= -\ln|u| + C\\ &= -\ln|1+\cos x| + C\end{aligned}$$

15. $\int 2x^7 \cos x^8\,dx$

Let $u = x^8$, so $du = 8x^7\,dx$ or $\frac{1}{8}\,du = x^7\,dx$.

$$\begin{aligned}\int 2x^7\cos x^8\,dx &= \int 2(\cos u)\cdot\frac{1}{8}\,du\\ &= \frac{1}{4}\int \cos u\,du\\ &= \frac{1}{4}\sin u + C\\ &= \frac{1}{4}\sin x^8 + C\end{aligned}$$

17. $\displaystyle\int \tan \frac{1}{3}x\,dx$

Let $u = \dfrac{1}{3}x$, so $du = \frac{1}{3}\,dx$ or $3\,du = dx$.

$$\begin{aligned}\int \tan \frac{1}{3}x\,dx &= \int (\tan u)\cdot 3\,du\\ &= 3\int \tan u\,du\\ &= -3\ln|\cos u| + C\\ &= -3\ln\left|\cos \frac{1}{3}x\right| + C\end{aligned}$$

19. $\displaystyle\int x^5 \cot x^6\,dx$

Let $u = x^6$, so $du = 6x^5\,dx$ or $\frac{1}{6}\,du = x^5\,dx$.

$$\begin{aligned}\int x^5 \cot x^6\,dx &= \int (\cot u)\cdot \frac{1}{6}\,du\\ &= \frac{1}{6}\int \cot u\,du\\ &= \frac{1}{6}\ln|\sin u| + C\\ &= \frac{1}{6}\ln\left|\sin x^6\right| + C\end{aligned}$$

21. $\displaystyle\int e^x \sin e^x\,dx$

Let $u = e^x$.
Then $du = e^x\,dx$.

$$\begin{aligned}\int e^x \sin e^x\,dx &= \int \sin u\,du\\ &= -\cos u + C\\ &= -\cos e^x + C\end{aligned}$$

23. $\displaystyle\int e^x \csc e^x \cot e^x\,dx$

Let $u = e^x$, so that $du = e^x\,dx$.

$$\begin{aligned}\int e^x \csc e^x \cot e^x\,dx &= \int \csc e^x \cot e^x (e^x\,dx)\\ &= \int \csc u \cot u\,du\\ &= -\csc u + C\\ &= -\csc e^x + C\end{aligned}$$

25. $\displaystyle\int -6x\cos 5x\,dx$

Let $u = -6x$ and $dv = \cos 5x\,dx$.

Then $du = -6\,dx$ and $v = \frac{1}{5}\sin 5x$.

$$\begin{aligned}&\int -6x\cos 5x\,dx\\ &= (-6x)\left(\frac{1}{5}\sin 5x\right) - \int\left(\frac{1}{5}\sin 5x\right)(-6\,dx)\\ &= -\frac{6}{5}\sin 5x + \frac{6}{5}\int \sin 5x\,dx\\ &= -\frac{6}{5}x\sin 5x + \frac{6}{5}\cdot\frac{1}{5}(-\cos 5x) + C\\ &= -\frac{6}{5}x\sin 5x - \frac{6}{25}\cos 5x + C\end{aligned}$$

27. $\displaystyle\int 4x\sin x\,dx$

Let $u = 4x$ and $dv = \sin x\,dx$.

Then $du = 4\,dx$ and $v = -\cos x$.

$$\begin{aligned}&\int 4x\sin x\,dx\\ &= 4x(-\cos x) - \int (-\cos x)\cdot 4\,dx\\ &= -4x\cos x + 4\int \cos x\,du\\ &= -4x\cos x + 4\sin x + C\end{aligned}$$

29. $\displaystyle\int -6x^2\cos 8x\,dx$

Let $u = -6x^2$ and $dv = \cos 8x\,dx$.

Then $du = -12x\,dx$ and

$$v = \frac{1}{8}\sin 8x.$$

$$\begin{aligned}&\int -6x^2\cos 8x\,dx\\ &= (-6x^2)\left(\frac{1}{8}\sin 8x\right) - \int\left(\frac{1}{8}\sin 8x\right)(-12x\,dx)\\ &= -\frac{3}{4}x^2\sin 8x + \frac{3}{2}\int x\sin 8x\,dx\end{aligned}$$

In $\displaystyle\int x\sin 8x\,dx$, let

$$u = x \quad\text{and}\quad dv = \sin 8x\,dx.$$

Then $\quad du = dx \quad$ and $\quad v = -\dfrac{1}{8}\cos 8x.$

$$\int -6x^2 \cos 8x\,dx$$
$$= -\frac{3}{4}x^2 \sin 8x + \frac{3}{2}\left[-\frac{1}{8}x \cos 8x - \int \left(-\frac{1}{8}\cos 8x\right) dx\right]$$
$$= -\frac{3}{4}x^2 \sin 8x - \frac{3}{16}x \cos 8x + \frac{3}{16}\int \cos 8x\,dx$$
$$= -\frac{3}{4}x^2 \sin 8x - \frac{3}{16}x \cos 8x + \frac{3}{16}\cdot\frac{1}{8}\sin 8x + C$$
$$= -\frac{3}{4}x^2 \sin 8x - \frac{3}{16}x \cos 8x + \frac{3}{128}\sin 8x + C$$

31. $\int_0^{\pi/4} \sin x\,dx = -\cos x \Big|_0^{\pi/4}$
$$= -\cos\frac{\pi}{4} - (-\cos 0)$$
$$= -\frac{\sqrt{2}}{2} + 1$$
$$= 1 - \frac{\sqrt{2}}{2}$$

33. $\int_0^{\pi/6} \tan x\,dx = -\ln|\cos x| \Big|_0^{\pi/6}$
$$= -\ln\left|\cos\frac{\pi}{6}\right| - (-\ln|\cos 0|)$$
$$= -\ln\frac{\sqrt{3}}{2} + \ln 1$$
$$= -\ln\frac{\sqrt{3}}{2} + 0$$
$$= -\ln\frac{\sqrt{3}}{2}$$

35. $\int_{\pi/2}^{2\pi/3} \cos x\,dx = \sin x \Big|_{\pi/2}^{2\pi/3}$
$$= \sin\frac{2\pi}{3} - \sin\frac{\pi}{2}$$
$$= \frac{\sqrt{3}}{2} - 1$$

37. Use the fnInt function on the graphing calculator to enter fnInt$(e\hat{}(-x)\sin(x), x, 0, b)$, for successively larger values of b, which returns a value of 0.5 for sufficiently large enough b. Thus, an estimate of $\int_0^\infty e^{-x}\sin x\,dx$ is 0.5.

39. $S(t) = 500 + 500\cos\left(\frac{\pi}{6}t\right)$

Total sales over a year's time are approximated by the area under the graph of S during any twelve-month period due to periodicity of S.

Total sales
$$\approx \int_0^{12} S(t)\,dt$$
$$= \int_0^{12}\left[500 + 500\cos\left(\frac{\pi}{6}t\right)\right] dt$$
$$= \int_0^{12} 500\,dt + 500\int_0^{12}\cos\left(\frac{\pi}{6}t\right) dt$$
$$= 500t\Big|_0^{12} + 500\left(\frac{6}{\pi}\sin\left(\frac{\pi}{6}t\right)\Big|_0^{12}\right)$$
$$= 6000 + \frac{3000}{\pi}\left[\sin\left(\frac{\pi}{6}\cdot 12\right) - \sin\left(\frac{\pi}{6}\cdot 0\right)\right]$$
$$= 6000 + \frac{3000}{\pi}(\sin 2\pi - \sin 0)$$
$$= 6000 + \frac{3000}{\pi}(0 - 0) = 6000$$

41. $T(t) = 50 + 50\cos\left(\frac{\pi}{6}t\right)$

Since T is periodic, the number of animals passing the checkpoint is equal to the area under the curve for any 12-month period. Let t vary from 0 to 12.

$$\text{Total} = \int_0^{12}\left[50 + 50\cos\left(\frac{\pi}{6}t\right)\right] dt$$
$$= \int_0^{12} 50\,dt + \frac{6}{\pi}\int_0^{12} 50\cos\left(\frac{\pi}{6}t\right)\left(\frac{\pi}{6}dt\right)$$
$$= 50t\Big|_0^{12} + \frac{300}{\pi}\sin\left(\frac{\pi}{6}t\right)\Big|_0^{12}$$
$$= (600 - 0) + \frac{300}{\pi}(0 - 0)$$
$$= 600 \text{ (in hundreds)}$$

The total number of animals is 60,000.

43. The total amount of daylight is given by
$$\int_0^{365} N(t)dt = \int_0^{365} [183.549\sin(0.0172t - 1.329) + 728.124]dt$$
$$= \left[-\frac{183.549}{0.0172}\cos(0.0172t - 1.329) + 728.124t\right]\Big|_0^{365}$$
$$\approx 265{,}819.0192 \text{ minutes}$$
$$\approx 4430 \text{ hours.}$$

The result is relatively close to the actual value.

45. The amount of water in the tank is
$$\int_0^k \sec^2 t\,dt = \tan t\big|_0^k$$
$$= \tan k - \tan 0$$
$$= \tan k \text{ ("tank").}$$

Chapter 13 Review Exercises

5. $90° = 90\left(\frac{\pi}{180}\right) = \frac{90\pi}{180} = \frac{\pi}{2}$

7. $225° = 225\left(\frac{\pi}{180}\right) = \frac{5\pi}{4}$

9. $360° = 2\pi$

11. $5\pi = 5\pi\left(\frac{180°}{\pi}\right) = 900°$

13. $\frac{9\pi}{20} = \frac{9\pi}{20}\left(\frac{180°}{\pi}\right) = 81°$

15. $\frac{13\pi}{20} = \frac{13\pi}{20}\left(\frac{180°}{\pi}\right) = 117°$

17. When an angle of 60° is drawn in standard position, one choice of a point on its terminal side is $(x, y) = (1, \sqrt{3})$. Then

$$r = \sqrt{x^2 + y^2} = \sqrt{1+3} = 2,$$

so

$$\sin 60° = \frac{y}{r} = \frac{\sqrt{3}}{2}.$$

19. When an angle of $-45°$ is drawn in standard position, one choice of a point on its terminal side is $(x, y) = (1, -1)$. Then

$$r = \sqrt{x^2 + y^2} = \sqrt{1+1} = \sqrt{2},$$

so

$$\cos(-45°) = \frac{x}{r} = \frac{1}{\sqrt{2}} = \frac{\sqrt{2}}{2}.$$

21. When an angle of 120° is drawn in standard position, one choice of a point on its terminal side is $(x, y) = (-1, \sqrt{3})$. Then

$$r = \sqrt{x^2 + y^2} = \sqrt{1+3} = 2,$$

so

$$\csc 120° = \frac{r}{y} = \frac{2}{\sqrt{3}} = \frac{2\sqrt{3}}{3}.$$

23. When an angle of $\frac{\pi}{6}$ is drawn in standard position, one choice of a point on its terminal side is $(x, y) = (\sqrt{3}, 1)$. Then

$$r = \sqrt{x^2 + y^2} = \sqrt{3+1} = 2,$$

so

$$\sin\frac{\pi}{6} = \frac{y}{r} = \frac{1}{2}.$$

25. When an angle of $\frac{5\pi}{3}$ is drawn in standard position, one choice of a point on its terminal side is $(x, y) = (1, -\sqrt{3})$. Then

$$r = \sqrt{x^2 + y^2} = \sqrt{1+3} = 2,$$

so

$$\sec\frac{5\pi}{3} = \frac{r}{x} = \frac{2}{1} = 2.$$

27. $\sin 47° \approx 0.7314$

29. $\tan 115° \approx -2.1445$

31. $\sin 2.3581 \approx 0.7058$

33. $\cos 0.5934 \approx 0.8290$

35. Because the derivative of $y = \sin x$ is $y' = \cos x$, the slope of $y = \sin x$ varies from -1 to 1.

37. The graph of $y = \frac{1}{2}\tan x$ is similar to the graph of $y = \tan x$ except that each ordinate value is multiplied by a factor of $\frac{1}{2}$. Note that the points $\left(\frac{\pi}{4}, \frac{1}{2}\right)$ and $\left(-\frac{\pi}{4}, -\frac{1}{2}\right)$ lie on the graph.

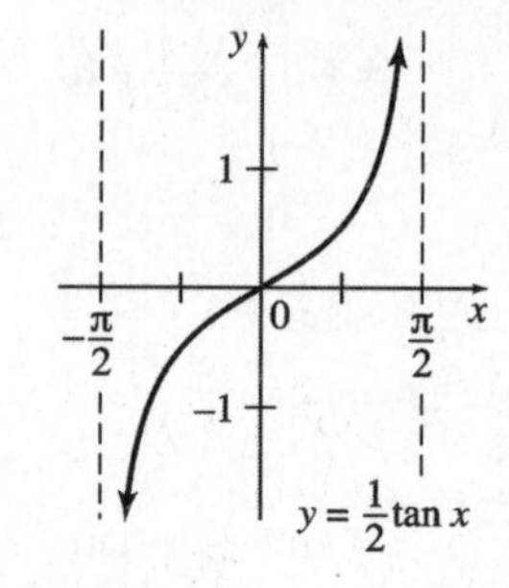

39. The graph of $y = -\frac{2}{3}\sin x$ is similar to the graph of $y = \sin x$ except that it has two-thirds the amplitude and is reflected about the x-axis.

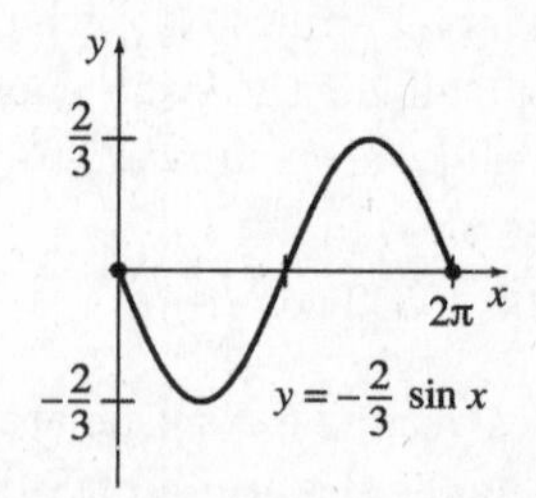

41. $y = 2 \tan 5x$

$$\frac{dy}{dx} = 2 \sec^2 5x \cdot D_x(5x) = 10 \sec^2 5x$$

43. $y = \cot(6 - 3x^2)$

$$\frac{dy}{dx} = [-\csc^2(6 - 3x^2)] \cdot D_x(6 - 3x^2) = 6x \csc^2(6 - 3x^2)$$

45. $y = 2 \sin^4(4x^2)$

$$\frac{dy}{dx} = [8 \sin^3(4x^2)] \cdot D_x[\sin(4x^2)] = 8 \sin^3(4x^2) \cdot \cos(4x^2) \cdot D_x(4x^2) = 64x \sin^3(4x^2) \cos(4x^2)$$

47. $y = \cos(1 + x^2)$

$$\frac{dy}{dx} = [-\sin(1 + x^2)] \cdot D_x(1 + x^2) = -2x \sin(1 + x^2)$$

49. $y = e^{-2x} \sin x$

$$\frac{dy}{dx} = e^{-2x} \cdot D_x(\sin x) + \sin x \cdot D_x(e^{-2x}) = e^{-2x}(\cos x) + (\sin x)(e^{-2x}) \cdot D_x(-2x) = e^{-2x}(\cos x) + (\sin x)(e^{-2x})(-2) = e^{-2x}(\cos x - 2 \sin x)$$

51. $y = \dfrac{\cos^2 x}{1 - \cos x}$

$$\frac{dy}{dx} = \frac{(1 - \cos x)(-2 \cos x \sin x) - (\cos^2 x)(\sin x)}{(1 - \cos x)^2} = \frac{-2 \cos x \sin x + \cos^2 x \sin x}{(1 - \cos x)^2}$$

53. $y = \dfrac{\tan x}{1 + x}$

$$\frac{dy}{dx} = \frac{(1 + x)(\sec^2 x) - (\tan x)(1)}{(1 + x)^2} = \frac{\sec^2 x + x \sec^2 x - \tan x}{(1 + x)^2}$$

55. $y = \ln|5 \sin x|$

$$\frac{dy}{dx} = \frac{1}{5 \sin x} \cdot D_x(5 \sin x) = \frac{\cos x}{\sin x}$$

or $\cot x$

57. $\int \cos 5x \, dx$

Let $u = 5x$, so $du = 5\,dx$ or $\frac{1}{5}\,du = dx$.

$$\int \cos 5x \, dx = \int \cos u \cdot \frac{1}{5} du = \frac{1}{5} \int \cos u \, du = \frac{1}{5} \sin u + C = \frac{1}{5} \sin 5x + C$$

59. $\int \sec^2 5x \, dx$

Let $u = 5x$, so that $du = 5\,dx$.

$$\int \sec^2 5x \, dx = \frac{1}{5} \int (\sec^2 5x)(5\,dx) = \frac{1}{5} \int \sec^2 u \, du = \frac{1}{5} \tan u + C = \frac{1}{5} \tan 5x + C$$

61. $\int 4 \csc^2 x \, dx = -4 \int -\csc^2 x \, dx = -4 \cot x + C$

63. $\int 5x \sec 2x^2 \tan 2x^2 \, dx$

Let $u = 2x^2$, so that $du = 4x\,dx$.

$$\int 5x \sec 2x^2 \tan 2x^2 \, dx = \frac{5}{4} \int \sec 2x^2 \tan 2x^2 (4x\,dx) = \frac{5}{4} \int \sec u \tan u \, du = \frac{5}{4} \sec u + C = \frac{5}{4} \sec 2x^2 + C$$

65. $\int \cos^8 x \sin x \, dx$

Let $u = \cos x$, so $du = -\sin x\,dx$.

$$\int \cos^8 x \sin x \, dx = -\int u^8 \, du = -\frac{1}{9} u^9 + C = -\frac{1}{9} \cos^9 x + C$$

67. $\int x^2 \cot 8x^3\, dx$

Let $u = 8x^3$, so that $du = 24x^2\, dx$.

$$\int x^2 \cot 8x^3\, dx$$
$$= \frac{1}{24}\int (\cot 8x^3)(24x^2)\, dx$$
$$= \frac{1}{24}\int \cot u\, du$$
$$= \frac{1}{24} \ln |\sin u| + C$$
$$= \frac{1}{24} \ln \left|\sin 8x^3\right| + C$$

69. $\int (\cos x)^{-4/3} \sin x\, dx$

Let $u = \cos x$, so that $du = -\sin x\, dx$.

$$\int (\cos x)^{-4/3} \sin x\, dx$$
$$= -\int (\cos x)^{-4/3} (\sin x)\, dx$$
$$= -\int u^{-4/3}\, du$$
$$= 3u^{-1/3} + C$$
$$= 3(\cos x)^{-1/3} + C$$

71. $\int_0^{\pi/2} \cos x\, dx = \sin x \Big|_0^{\pi/2}$

$$= \sin \frac{\pi}{2} - \sin 0$$
$$= 1 - 0 = 1$$

73. $\int_0^{2\pi} (10 + 10 \cos x)\, dx$

$$= \int_0^{2\pi} 10\, dx + \int_0^{2\pi} 10 \cos x\, dx$$
$$= 10(2\pi) + 10 \int_0^{2\pi} \cos x\, dx$$
$$= 20\pi + 10(\sin x) \Big|_0^{2\pi}$$
$$= 20\pi + 10(\sin 2\pi - \sin 0)$$
$$= 20\pi + 10(0 - 0)$$
$$= 20\pi$$

75. **(a)** Enter the data into a graphing calculator and plot.

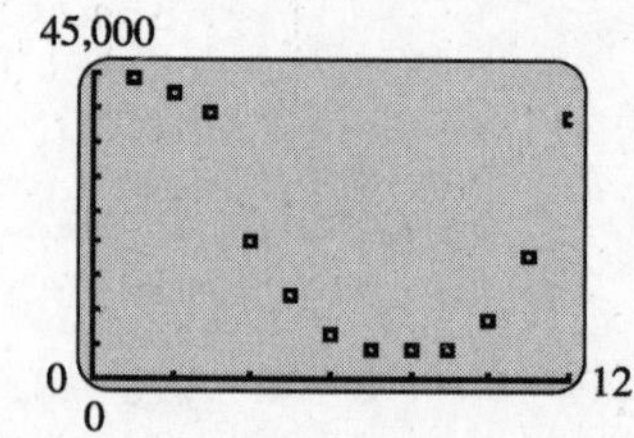

(b) Use the graphing calculator to find a trigonometric function.

```
SinReg
 y = a*sin(bx+c)+d
 a = 22977.42648
 b = .4474476078
 c = 1.308313854
 d = 23709.34648
```

$C(x) = 22{,}977.4 \sin(0.447448x + 1.30831) + 23{,}709.3$

(c)

$$\int_0^{12} C(x)dx$$
$$\approx \int_0^{12} [22{,}977.4 \sin(0.447448x + 1.30831) + 23{,}709.3]dx$$
$$= -\frac{22{,}977.4}{0.447448} \cos(0.447448x + 1.30831) + 23{,}709.3x \Big|_0^{12}$$
$$\approx 250{,}429$$

The estimated total consumption is about 250,429 million cubic feet. The actual total consumption was 244,381 million cubic feet.

(d) $T = \dfrac{2\pi}{b} \approx 14.0422$

The period is about 14.0 months.

77. We draw θ in standard position.

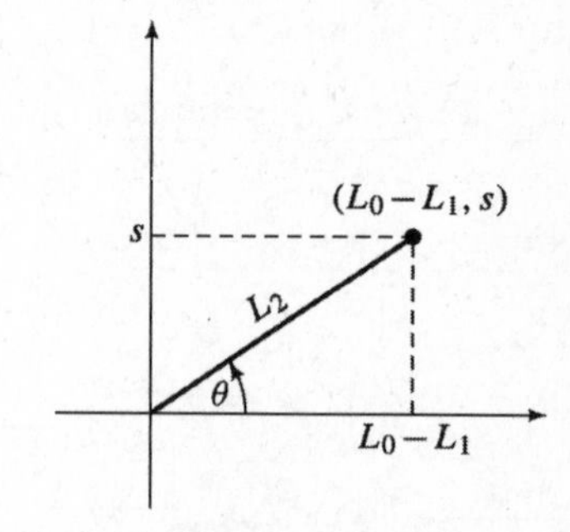

From this diagram, we see that

$$\sin \theta = \frac{s}{L_2}.$$

79. Using the sketch of θ in the solution to Exercise 77 and the definition of cotangent, we see that

$$\cot\theta = \frac{L_0 - L_1}{s}.$$

81. The length of AD is L_1, and the radius of that section of blood vessel is r_1, so the general equation

$$k = \frac{L}{r^4}$$

is similar to

$$R_1 = k\frac{L_1}{{r_1}^4}$$

for that particular segment of the blood vessel.

83.
$$\begin{aligned} R &= R_1 + R_2 \\ &= k\frac{L_1}{{r_1}^4} + k\frac{L_2}{{r_2}^4} \\ &= k\left(\frac{L_1}{{r_1}^4} + \frac{L_2}{{r_2}^4}\right) \end{aligned}$$

85. Since k, L_1, L_0, s, r_1, and r_2 are all constants, the only letter left as a variable is θ, so the differentiation indicated in the symbol R' must be differentiation with respect to θ.

$$\begin{aligned} R' &= D_\theta R \\ &= D_\theta\left(k\frac{L_0}{{r_1}^4} - \frac{sk}{{r_1}^4}\cot\theta + \frac{s}{{r_2}^4}\cdot\frac{k}{\sin\theta}\right) \\ &= kD_\theta\left(\frac{L_0}{{r_1}^4} - \frac{s}{{r_1}^4}\cot\theta + \frac{s}{{r_2}^4}\cdot\frac{1}{\sin\theta}\right) \\ &= k\left[D_\theta\left(\frac{L_0}{{r_1}^4}\right) - D_\theta\left(\frac{s}{{r_1}^4}\cot\theta\right) + D_\theta\left(\frac{s}{{r_2}^4}\cdot\frac{1}{\sin\theta}\right)\right] \\ &= k\left[0 - \frac{s}{{r_1}^4}D_\theta(\cot\theta) + \frac{s}{{r_2}^4}D_\theta\left(\frac{1}{\sin\theta}\right)\right] \\ &= k\left[-\frac{s}{{r_1}^4}(-\csc^2\theta) + \frac{s}{{r_2}^4}\left(\frac{-\cos\theta}{\sin^2\theta}\right)\right] \\ &= k\left(\frac{s}{{r_1}^4}\frac{1}{\sin^2\theta} - \frac{s}{{r_2}^4}\cdot\frac{\cos\theta}{\sin^2\theta}\right) \\ &= \frac{ks}{\sin^2\theta}\left(\frac{1}{{r_1}^4} - \frac{\cos\theta}{{r_2}^4}\right) \\ &= \frac{ks\csc^2\theta}{{r_1}^4} - \frac{ks\cos\theta}{{r_2}^4\sin^2\theta} \end{aligned}$$

87. If the left side of the equation in the solution to Exercise 85 is multiplied by $\frac{\sin^2\theta}{s}$, we get

$$\begin{aligned} &\frac{\sin^2\theta}{s}\cdot\frac{ks}{\sin^2\theta}\left(\frac{1}{{r_1}^4} - \frac{\cos\theta}{{r_2}^4}\right) \\ &= k\left(\frac{1}{{r_1}^4} - \frac{\cos\theta}{{r_2}^4}\right) \end{aligned}$$

This gives the equation

$$\frac{k}{{r_1}^4} - \frac{k\cos\theta}{{r_2}^4} = 0.$$

89. If $r_1 = 1$ and $r_2 = \frac{1}{4}$, then

$$\begin{aligned} \cos\theta &= \left(\frac{\frac{1}{4}}{1}\right)^4 = \left(\frac{1}{4}\right)^4 \\ &= \frac{1}{256} \approx 0.0039, \end{aligned}$$

from which we get

$$\theta \approx 90°.$$

91.
$$\begin{aligned} s(t) &= A\cos(Bt + C) \\ s'(t) &= -A\sin(Bt + C)\cdot D_t(Bt + C) \\ &= -A\sin(Bt + C)\cdot B \\ &= -AB\sin(Bt + C) \\ s''(t) &= -AB\cos(Bt + C)\cdot D_t(Bt + C) \\ &= -AB\cos(Bt + C)\cdot B \\ &= -B^2A\cos(Bt + C) \\ &= -B^2s(t) \end{aligned}$$

93. (a)
$$\begin{aligned} y &= x\tan\alpha - \frac{16x^2}{V^2}\sec^2\alpha + h \\ &= 39\tan\frac{\pi}{24} - \frac{16(39)^2}{73^2}\sec^2\frac{\pi}{24} + 9 \\ &\approx 9.5 \end{aligned}$$

Yes, the ball will make it over the net since the height of the ball is about 9.5 feet when x is 39 feet.

(b) Entering

$$Y_1 = 39\tan x - \frac{16(39)^2}{44^2}\sec^2 x + 9 \text{ and}$$

$$Y_2 = \frac{44^2\sin x\cos x + 44^2\cos^2 x\sqrt{\tan^2 x + \frac{576}{44^2}\sec^2 x}}{32}$$

into the graphing calculator and using the table function, indicates that the tennis ball will clear the net and travel between 39 and 60 feet for $0.18 \le x \le 0.41$ or $0.18 \le \alpha \le 0.41$ in radians. In degrees,

$$10.3 \le \alpha \le 23.5.$$

X	Y1	Y2
.17	2.754	44.141
.18	3.1103	44.827
.22	4.5223	47.571
.34	8.6526	55.486
.38	10.002	57.9
.41	11.006	59.602
.42	11.339	60.146

X=.42

(c) Using Y_2 from part (b) and the graphing calculator, we get

```
nDeriv(Y2,X,π/8)
         57.01184054
```

Note that $57.1\frac{\text{feet}}{\text{radian}} \approx 0.995\frac{\text{feet}}{\text{degree}}$. The distance the tennis ball travels will increase by approximately 1 foot by increasing the angle of the tennis racket by one degree.

95. Refer to the figure below.

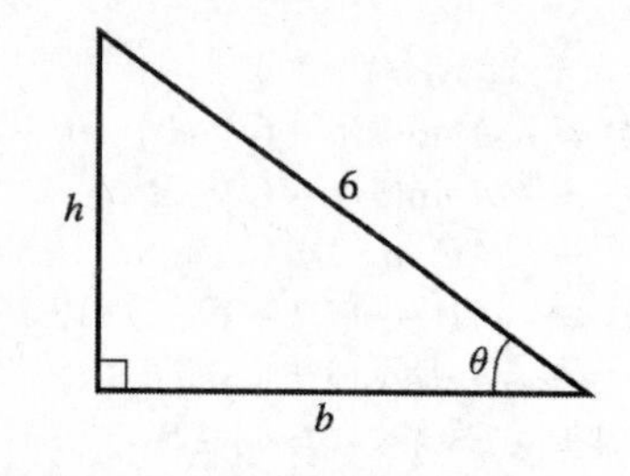

Let A be the area of the triangle.

$$A = \frac{1}{2}bh$$

In this triangle,

$$\sin\theta = \frac{h}{6} \quad \text{and} \quad \cos\theta = \frac{b}{6}$$

$$h = 6\sin\theta \quad \text{and} \quad b = 6\cos\theta.$$

Thus,

$$\begin{aligned} A &= \frac{1}{2}(6\cos\theta)(6\sin\theta) \\ &= 18\sin\theta\cos\theta \\ &= 9(2\sin\theta\cos\theta) \\ A &= 9\sin(2\theta) \\ \frac{dA}{d\theta} &= 9\cos(2\theta)\cdot 2 = 18\cos(2\theta). \end{aligned}$$

If $\frac{dA}{d\theta} = 0$,

$$\begin{aligned} 18\cos(2\theta) &= 0 \\ \cos 2\theta &= 0 \\ 2\theta &= \frac{\pi}{2} \\ \theta &= \frac{\pi}{4}. \end{aligned}$$

If $\theta < \frac{\pi}{4}$, $\frac{dA}{d\theta} > 0$.

If $\theta > \frac{\pi}{4}$, $\frac{dA}{d\theta} < 0$.

Thus, A is maximum when $\theta = \frac{\pi}{4}$ or $45°$.

Chapter 13 Test

[13.1]

1. Convert 72° to radian measure. Express your answer as a multiple of π.

2. Convert $\frac{7\pi}{20}$ to degree measure.

3. Evaluate $\tan \frac{11\pi}{6}$ without using a calculator.

4. Use a calculator to find the value of $\sin 2.986$.

5. Graph $y = -4 \sin x$ over a one-period interval.

6. Graph $y = \frac{1}{2} \tan x$ over a one-period interval.

[13.2]

Find the derivative of each of the following functions.

7. $y = -3 \cos 4x$

8. $y = \cot^2 \dfrac{x}{2}$

9. $y = 3x^2 \sin x$

10. $y = \dfrac{\sin^2 x}{\sin x + 1}$

11. $y = \ln |3 \cos x|$

12. $e^{\cos x}$

13. Find the slope of the tangent to the graph of $y = \tan x$ at $x = \frac{\pi}{3}$.

14. Find the maximum and minimum values of y for the simple harmonic motion given by $y = 3 \sin (2x + 3)$.

15. Explain why the slope of $y = \tan x$ is never zero.

[13.3]

Find each integral.

16. $\displaystyle\int 3 \sin 2x \, dx$

17. $\displaystyle\int 4 \sec^2 x \, dx$

18. $\displaystyle\int x \cos \left(2x^2\right) \, dx$

19. $\displaystyle\int \tan 3x \, dx$

20. $\displaystyle\int (\cos x)^2 \sin x \, dx$

21. $\displaystyle\int_{\pi/6}^{\pi/3} \cos x \, dx$

22. $\displaystyle\int_0^{\pi/4} \sec^2 x \, dx$

23. $\displaystyle\int_0^{\pi/4} x \tan x^2 \, dx$

24. If b is a positive real number, explain why $\displaystyle\int_{-b}^{b} \sin x \, dx$ will always be zero.

Chapter 13 Test Answers

1. $\frac{2\pi}{5}$

2. $63°$

3. $-\frac{\sqrt{3}}{3}$

4. 0.1550

5.

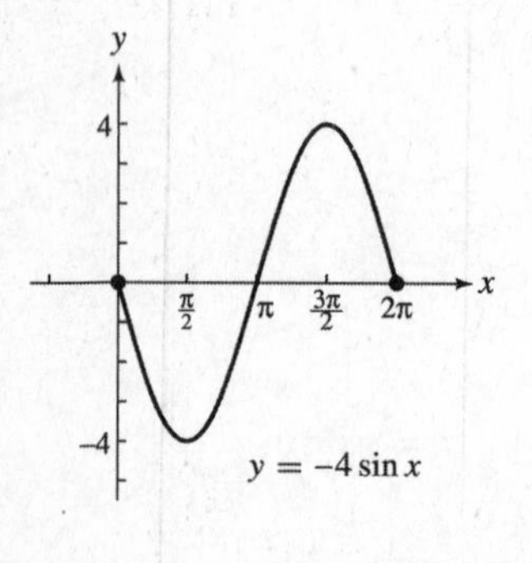

6.

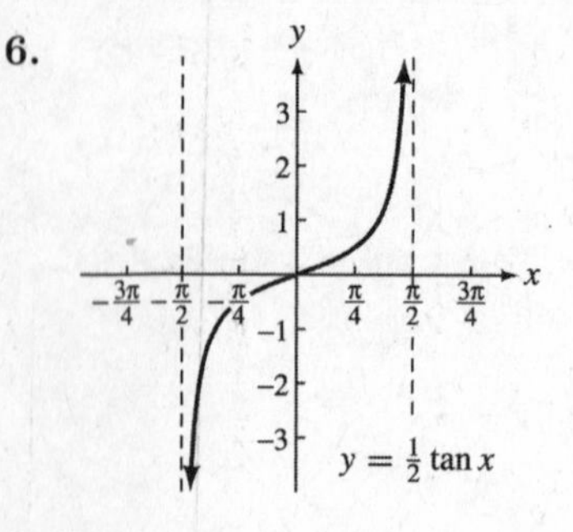

7. $12 \sin 4x$

8. $-\cot \frac{x}{2} \csc^2 \frac{x}{2}$

9. $6x \sin x + 3x^2 \cos x$

10. $\frac{\sin^2 x \cos x + 2 \sin x \cos x}{(\sin x + 1)^2}$

$-\tan x$

12. $-\sin x e^{\cos x}$

13. 4

14. $3, -3$

15. The slope of $y = \tan x$ is given by $y' = \sec^2 x$. Since $\sec x$ is never zero, the slope of $y = \tan x$ is never zero.

16. $-\frac{3}{2} \cos 2x + C$

17. $4 \tan x + C$

18. $\frac{1}{4} \sin (2x^2) + C$

19. $-\frac{1}{3} \ln |\cos 3x| + C$

20. $-\frac{1}{3} \cos^3 x + C$

21. $\frac{\sqrt{3}-1}{2}$ or 0.366

22. 1

23. $-\frac{1}{2} \ln \left(\cos \frac{\pi^2}{16}\right)$ or 0.101851

24.

$$\begin{aligned} \int_{-b}^{b} \sin x \, dx &= -\cos x \Big|_{-b}^{b} \\ &= -\cos b - [-\cos (-b)] \\ &= -\cos b + \cos (-b) \end{aligned}$$

For the cosine function, $\cos(-x) = \cos x$ for all values of x. Thus, $-\cos b + \cos(-b) = 0$ for all values of b, and $\int_{-b}^{b} \sin x \, dx = 0$ for all real numbers b.